Developments in environmental biology of fishes 12

Series Editor
EUGENE K. BALON

The biology of *Latimeria chalumnae* and evolution of coelacanths

The biology of *Latimeria chalumnae* and evolution of coelacanths

Editors:
JOHN A. MUSICK, MICHAEL N. BRUTON & EUGENE K. BALON

Reprinted from *Environmental biology of fishes* 32 (1–4), 1991
with addition of species and subject index

Springer-Science+Business Media, B.V.

Library of Congress Cataloging-in-Publication Data

The biology of Latimeria chalumnae and evolution of coelacanths/
    editors, John A. Musick, Michael N. Bruton & Eugene K. Balon.
        p.   cm. – (Developments in environmental biology of fishes: 12)
    Reprinted from Environmental biology of fishes 32 (1–4), 1991.
    Includes bibliographical references and indexes.
    ISBN 978-0-7923-1289-5   ISBN 978-94-011-3194-0 (eBook)
    DOI 10.1007/978-94-011-3194-0
    1. Coelacanth. 2. Coelacanth–Evolution. I. Musick, John A.
II. Bruton, M. N. (Michael N.) III. Balon, Eugene K. IV. Series.
QL638.L26B56 1991
597'.46–dc20   91-14259

ISBN 978-0-7923-1289-5

Cover design by using the posterior part of a drawing (p. 7)
by Sarah Landry

Contents

6

→

This pencil drawing on rag vellum by Sarah Landry (original 95 × 49 cm) is a composite of the Harvard University's specimen CCC no. 47 (for details), photographs of living individuals by Hans Fricke, and the photograph of a cast by Herbert Axelrod. Sarah Landry of Arlington, Massachusetts, is the author of animal societies illustrations in E.O. Wilson's Sociobiology.

The coelacanth perestroika

Nina Riabova, age 10, painted this *Latimeria* mother- and child-picture the night after she listened to Eugene Balon's lecture and saw Mike Vincent's film 'The Story of the Coelacanth' (Moscow, June 1990). She attended the coelacanth lecture with her father, Igor Riabov, who is himself an accomplished artist and the head of the NPP Ecological Monitoring Team (Chernobyl) of the A.N. Severtsov Institute of Evolutionary Morphology and Ecology of Animals, USSR Academy of Sciences.

Environmental Biology of Fishes **32**: 9–13, 1991.

Prelude: the mystery of a persistent life form

The few surviving bluesmen linger on like coelacanths, waiting for the next academic researcher to come fishing . . .

Phil Patton (1990) in 'Blues for Cousin Charlie', Esquire 114 (4): 54–58.

Nearly one hundred and fifty years ago Agassiz, Thiollière and Huxley distinguished the first fossil coelacanths, and later many more were recognized in rocks between 380 and 66 million years old. Exactly 100 years ago Woodward published the first review on these fishes. Fossils younger than 66 million years were never found, as if all coelacanths had become extinct at that time, very much like the dinosaurs. The bony structures in these fossil crossopterygians, especially their limbs, placed them close to the ancestor of the first amphibians and all other land vertebrates.

It is of no surprise, therefore, that the December 1938 find of a living coelacanth, when announced to the world by J.L.B. Smith in March 1939, caused disbelief and created the greatest biological sensation of this century. Finding a living coelacanth – morphologically so similar to the fossil specimens left in rocks more than 66 million years ago – was as inconceivable as meeting a living dinosaur on a weekend walk.

A special issue of the International Society of Cryptozoology Newsletter devoted to the 50th anniversary of this find opens with the following lines: 'The day was December 22, 1938. In Europe, the clouds of war continued to gather following Hitler's takeover of Austria and Czechoslovakia. In isolationist and complacent America, Hollywood was putting the finishing touches to *Gone With the Wind* and *The Wizard of Oz*. And in East London, South Africa, a 32-year-old naturalist named Marjorie Courtenay-Latimer was putting the finishing touches to the mounting of a fossil reptile in its new display case at the East London Museum. At 10 a.m. the telephone rang. It was a call that was to change Courtenay-Latimer's life – and the history of zoology – forever'. (. . .)

'In a way [continues Richard Greenwell in The Explorers Journal 68: 117–123, 1990], the story of the coelacanth, besides being scientifically important, represents the perfect human drama. The old fish, the sea captain, the young naturalist, the desperate professor, the prime minister, the Air Force crew, and the impossibility of it all becoming possible. I doubt if any novel or movie script could fully capture the personalities of the individuals or the dynamics of the story'.

The living coelacanth is a single advanced life form which has survived with little relatively change for nearly 400 million years. Horace Shipp put this idea into the first two stanzas of his dialogue on evolution (Ichthos, Special edition 2, December 1988):

> There lived a happy coelacanth
> In dim, primordial seas,
> He ate and mated, hunted, slept,
> Completely at his ease.
> Dame Nature urged: 'Evolve!'
> He said: 'Excuse me, Ma'am,
> You get on making Darwin,
> I'm staying as I am.'

The fishes changed their fishy shapes,
The reptiles stormed the land,
The algae turned to trees, the apes
To men, we understand.
The coelacanth remained
A monster and a myth;
he said: 'There's nothing to be gained
By my becoming Smith.'

While some of the coelacanth's relatives became implicated in the ancestry of all terrestrial vertebrates, the aquatic descendants developed structural solutions to life totally absent in other animals (as if to disprove the decline in structural diversity from the Cambrian suggested by S.J. Gould in 'Wonderful Life').

For example, instead of the calcified vertebrae that normally reinforce the axial skeleton, the coelacanths evolved a strong-walled elastic tube which is as far transformed from the notochord as are the segmented vertebrae. Unfortunately, the fluid in this 'chordal hose' gave rise to a dangerous fallacy: the rarity, longevity and survival of the coelacanths for millions of years created a superstition that a life-prolonging elixir existed in the 'notochordal fluid', and thus a black market opened . . .

Instead of a solid braincase, *Latimeria* evolved a two-part neurocranium with an intracranial joint which is operated by a special basicranial muscle (Fig. 1). It is the only animal with that structure living today. This intracranial joint and other unique rotational joints in the head may together with the rostral organ explain the special suction biting and headstanding behavior observed in *Latimeria* by Hans Fricke. For I envisage that this sluggish, drifting predator detects with its electrosensory organs prey which is hiding in cavities; then the beast performs a headstand and, thus positioned most effectively, slurps the prey by opening both jaws. It would, therefore, rarely take a free swimming or dead prey or bait!

Fricke's dives in the research submersibles GEO and JAGO opened a new era in coelacanth research. Past studies of preserved specimens, caught as an incidental bycatch, were supplemented for the first time by studies of free-living coelacanths in their natural habitat. The first footage taken from the submersible revealed the entirely unfishlike movements of this creature. Its mode of locomotion is a combination of flying and gliding, interspersed with headstands and belly-up drifts which appear to defy gravity.

The narrow range of habitats in which the coelacanth has been encountered has led us to realize how vulnerable this ancient relict is. We were therefore hopeful when, during our last visit to the Comoros, our report on the preservation strategy of marine resources (including the coelacanth) was received with attention and interest by the local authorities (Fig. 2).

For each of us the fascination with the coelacanth started at different times and occasions. Mike Bruton was born and bred in East London and since childhood exposed to the most famous exhibit of this town. It began as a student dream for Hans Fricke when he read J.L.B. Smith's 'Old Fourlegs' and decided that one day he would see the live animal. My speculative and mischievous but surprisingly correct prediction of the size of a newborn *Latimeria* presented in a paper at the Ichthyological Congress in Paris in 1976 was motivated by J. Millot's orotund rejection, twenty years earlier, of my plea to obtain at least one scale of *Latimeria*. At that time it was needed to complete an 'Atlas of fish scales' contracted by archaeologists. From its African origin and initial French territorial defence the study of *Latimeria* became increasingly more international. At the inception of this volume a nationally motivated competition to display the first living coelacanth in a public aquarium threatened an ugly escalation. Ultimately even these interests joined ranks not only in their contributions to knowledge on the creature but to its conservation. The international fraternity of coelacanth fanatics consolidated around this volume. Perhaps the most important consequence was that 'the coelacanth was taken out of the marketplace' (Nat. Geogr. Mag., April 1990).

I was privileged to join the area and actors of this celebrated biological drama two years before its golden

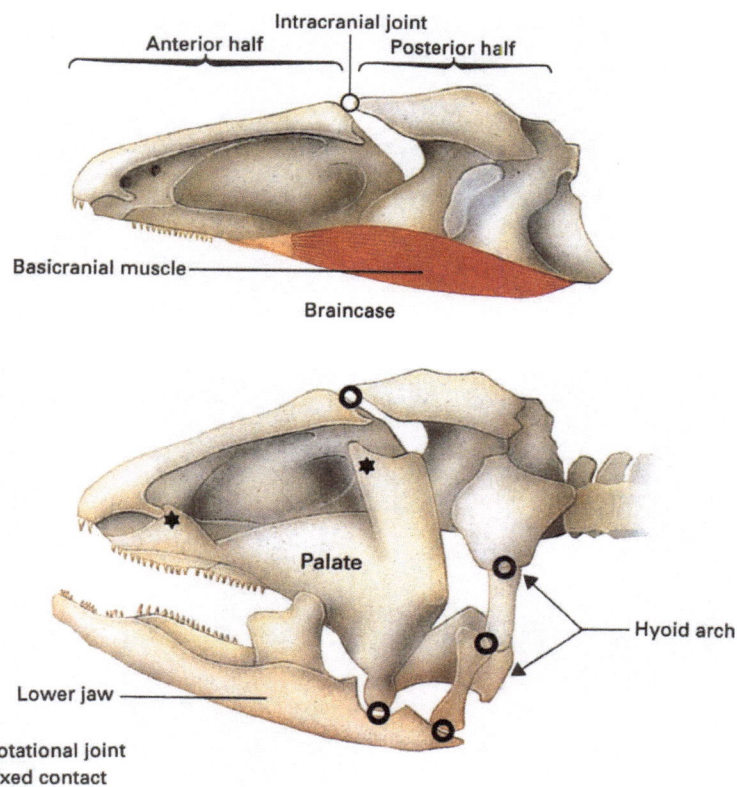

Fig. 1. The intracranial joint is unique to *Latimeria chalumnae* amongst all living vertebrates. It has been interpreted as a four-bar linkage system with rotational joints at the anterior and posterior half of the neurocranium, hyoid arch, lower jaw and palate. These joints probably enable elevation of the upper jaw (where a maxilla is totally absent) along with opening of the lower jaw to form an extended buccal cavity, the closure (bite) of which is facilitated by the powerful basicranial muscle. Reprinted by permission from Forey (1990) and Pergamon Press.

jubilee and to realize the publication of this volume two years after that anniversary. The coelacanth keeps influencing our lives forever – for it was the coelacanth that led to the establishment of the J.L.B. Smith Institute of Ichthyology in Grahamstown where I was invited by Mike Bruton to arrive in December 1986 to begin a sabbatical year. *Latimeria chalumnae* was on our minds when arranging first visits to Margaret Smith, Marjorie Courtenay-Latimer, Hendrik Goosen and, ultimately, to the true home of the creature – the Comoro Islands. Some were visited at the last possible moment (Margaret Smith passed away less than a week after our departure and Hendrik Goosen in January 1990), others at crossroads to significant changes (from domination by mercenaries to dignified self rule). New insights into coelacanth natural history were facilitated by a novel interpretation of earlier data and during expeditions to the Comoro Islands by retracing the route of the second specimen, studying unrecorded specimens, interviewing fishermen and describing their fishing crafts. And all the while we were privileged to meet and befriend all the special people attracted to the area – by the same coelacanth – from around the world.

Nearly every time a specimen of *Latimeria* becomes available to scientists, new structures are discovered. Usually their precise function remains unknown. In time, however, we may develop methods which will

Fig. 2. M. Ali Mroudjae, Minister of the Interior, Damir Ben Ali, Director of the Centre National de Documentation et de la Recherche Scientifique des Comores (both in foreground), and other Comoran officials at the ceremony devoted to the presentation of Bruton's report on the conservation of marine resources (including the coelacanth) in the Federal Islamic Republic of the Comoros. Itsandra Hotel, 16.5.1990; photo by E.K. Balon.

enable us to understand the mechanisms and functions of the unique solutions that *Latimeria* evolved to survive unchanged longer than any other similar organism. For this reason alone, our efforts to protect the creature are of utmost importance.

Despite its size, however, even this volume represents an incomplete understanding of coelacanths. While we were assembling this volume, Saddam Hussein invaded Kuwait and the coelacanth reported from the Science and Natural History Museum of that country (CCC no. c) might have fallen victim to the invaders looting. In time more fossils, even better interpreted, will be found, and individual *Latimeria,* either landed in spite of efforts to avoid capture or observed undisturbed from submersibles, will yield new unknown details. New information is forthcoming even while I am writing these lines:

We still do not know the number and structure of chromosomes in *Latimeria.* Our earlier attempts were futile in the absence of live cells. In May 1990 we left a specially prepared kit for sampling fresh tissues with Michel de San, Director of the European Economic Community Fisheries Development Project in Moroni, to whom all *Latimeria* catches off Grand Comoro are now reported. His first sample from a fresh coelacanth arrived in Guelph dry; an improved sampling kit has since been sent to await the next specimen.

We now know that *Latimeria* is not oviparous but rather a complex case of viviparity. There seem to be

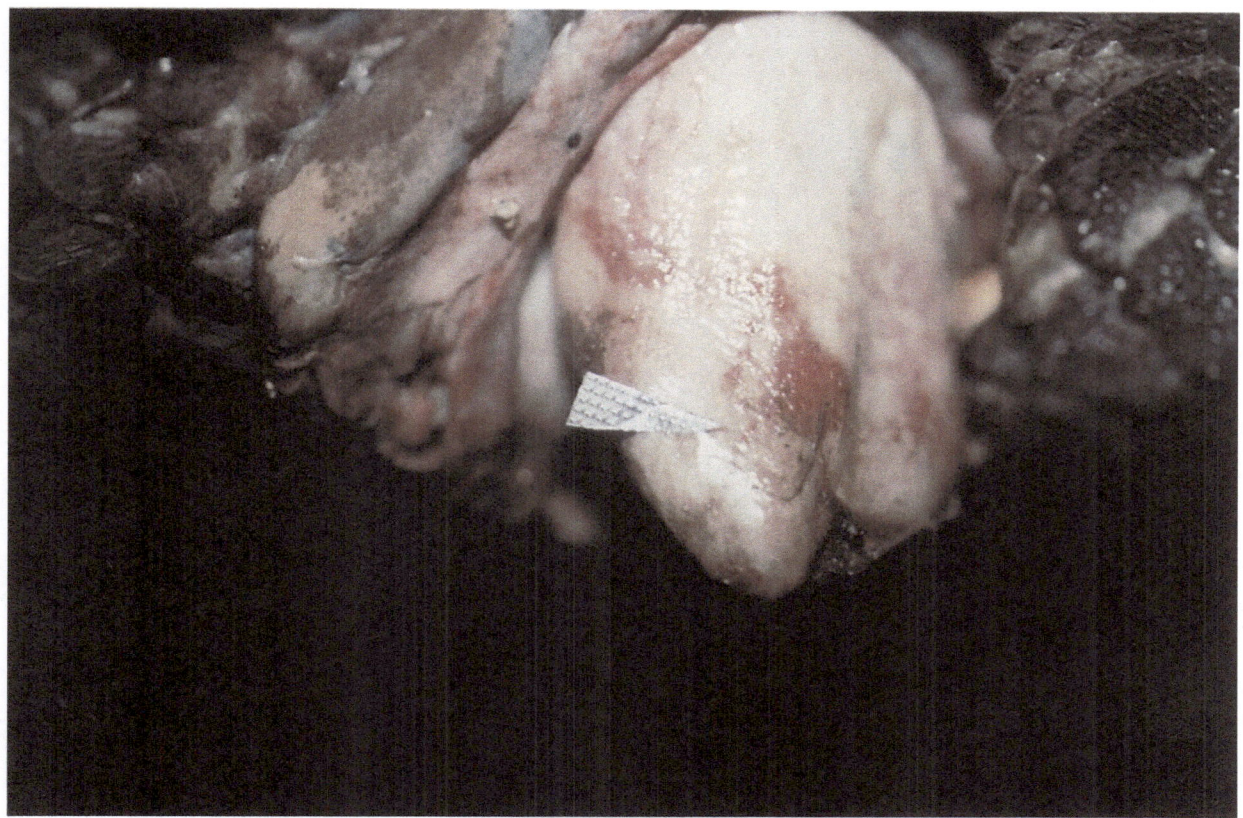

Fig. 3. The extruded (trailing) oviduct of the 168 cm long female (CCC no. 154) during preliminary examination in Moroni 17.5.1990. Photo by E.K. Balon.

four types of nutrient provision for the embryo: supply in the form of yolk and uterine histotrophe, via the placenta and by oophagy. Each occurs in other animals but never have all four been found together. More data, however, are needed to confirm this combination and the tempting conclusion that, given enough evolutionary time, almost anything is possible. Besides, we are still ignorant about the mechanism of internal fertilization in *Latimeria*. The male genitalia lack an intromittent organ. The large female available to us in May 1990 seems to have an externally trailing 'oviduct' (Fig. 3), samples of which are being studied histologically. A dissection of the entire preserved specimen is also planned.

What will we learn next?

Guelph, 28 December 1990   Eugene K. Balon

Latimeria chalumnae spends much of its time sculling slowly using the second dorsal and anal fins. It habitually performs headstands, holding its snout close to the substrate. The precise function of this behavior is not entirely clear (but see p. 10). Reprinted by permission from Forey (Science Progress 74, Fig. 2).

Environmental Biology of Fishes **32**: 15–19, 1991.

Introduction: the recent chronology and contributions

This volume is the product of several groups of scientists working independently who finally came together to share their knowledge of coelacanths and more efficiently utilize *Latimeria chalumnae* tissues, as well as to work for the conservation of *Latimeria*, the last survivor of an entire class of fishes that once dominated the earth's shallow seas.

In April 1987 Eugene Balon, Mike Bruton, Christine Flegler-Balon, Hans Fricke, and Raphael Plante met in Moroni, the capital of the Comoros, to discuss research on *Latimeria* and to try to develop a conservation strategy for the species. Out of this meeting were born the Coelacanth Conservation Council and a plan to produce a volume on *L. chalumnae* in honor of the 50th anniversary of J.L.B. Smith's original description of the species in 1939.

Expeditions from the J.L.B. Smith Institute of Ichthyology in Grahamstown in 1986, 1987, 1988 and 1990 endeavored to establish the conservation status of the coelacanth and the socio-economic and biological factors that threatened its survival (Fig. 1). Fricke's teams, on the other hand, conducted the first ever series of observations from research submersibles on living coelacanths in their natural habitat. Their combined results indicated that, while the traditional Comoran fishery posed little threat to the small population of coelacanths, the increased commercialization of the fishery may threaten its survival. There was clearly a need for internationally coordinated actions to save this remarkable animal.

At about the same time, a small group of members from The Explorers Club (that venerable New York institution once led by Roy Chapman Andrews, Carl Akely and other of their adventurous ilk) joined forces with the New York Aquarium to mount an expedition to the Comoros (Fig. 2). Their goal was to try to determine whether it would be possible to capture a coelacanth and return it alive for study at the Aquarium (a project since suspended until more is known of the animal's living requirements). During an expedition in November 1986, Explorers Club members Jerome Hamlin, Paul Rodzianko and Peter Stevens were able to obtain two fresh specimens captured by local fishermen, as well as two others which had been previously captured and frozen. All of these were returned frozen to North America, where one was deposited in the Royal Ontario Museum in Toronto (CCC no. 138), and the three remaining were offered to the American scientific community. One of these was given to the American Museum of Natural History (CCC no. 139) to be studied by William Bemis. The other two specimens were awarded to the Virginia Institute of Marine Science (VIMS) at the College of William and Mary, where J.A. Musick had assembled a research team from several institutions including Johns Hopkins University, Southern Massachusetts University, University of Kansas, and the University of Washington. William Phoel, a biologist with the National Marine Fisheries Service and a member of the Explorers Club, became the liaison between the club and the research team.

In November 1987 two fish (CCC no. 140, 141) were transported by truck from a freezer at the New York Aquarium to VIMS and stored frozen until 3 January 1988, when the research team assembled. Studies began on 3 January when CCC no. 141 was taken (still frozen) to Riverside Hospital in Newport News, Virginia, for computed tomography (CT scan). On 4 January, while both specimens were allowed to thaw slowly (at ca. 5–10°C), the research team held a planning session to determine research needs by tissue type. Dissection of CCC no. 141 commenced at 0830 h on 5 January. The specimen was partially thawed externally, but its internal organs remained largely frozen. Immediately upon dissection, tissue samples were placed in dry ice for biochemical and physiological studies, or in 10% buffered formalin for gross morphological work. At 1500 h this specimen was returned to the Radiology Unit at Riverside Hospital for Mag-

Fig. 1. Members of the 1990 expedition to the Comoros at the Itsandra Hotel. From left to right Mike Bruton, Roy Reynolds, Eugene Balon (inside the car) and Robin Stobbs. Photographed by Richard Cloutier.

netic Resonance Imaging. This continued with the help of John Daimler (head of Radiology at the Hospital) and Mark Brown of the Siemens Corporation, until 0400 h on 6 January. The specimen was then returned to VIMS where between 0530 and 0630 h the brain and pituitary were removed. These were still partially frozen and in excellent condition. The dissection of CCC no. 140 began 6 January at 1030 h and was completed by about 1230 h.

The primary focus of the biological research initiated at VIMS was to employ methodologies (biochemical, radiological) not generally available to academic scientists over a decade earlier when the last major studies were done on *Latimeria*. In addition to the research of members of the immediate dissection team (Fig. 3), frozen or preserved tissues were subsequently shipped to over 60 scientists around the world, including workers in Australia, Germany and Japan. The results of many of these studies and others were presented in a 'Symposium on the Biology and Evolution of Coelacanths' arranged by J.A. Musick in June 1989 at the 69th Annual Meeting of the American Society of Ich-

Fig. 2. Logo of The Explorers Club coelacanth expeditions.

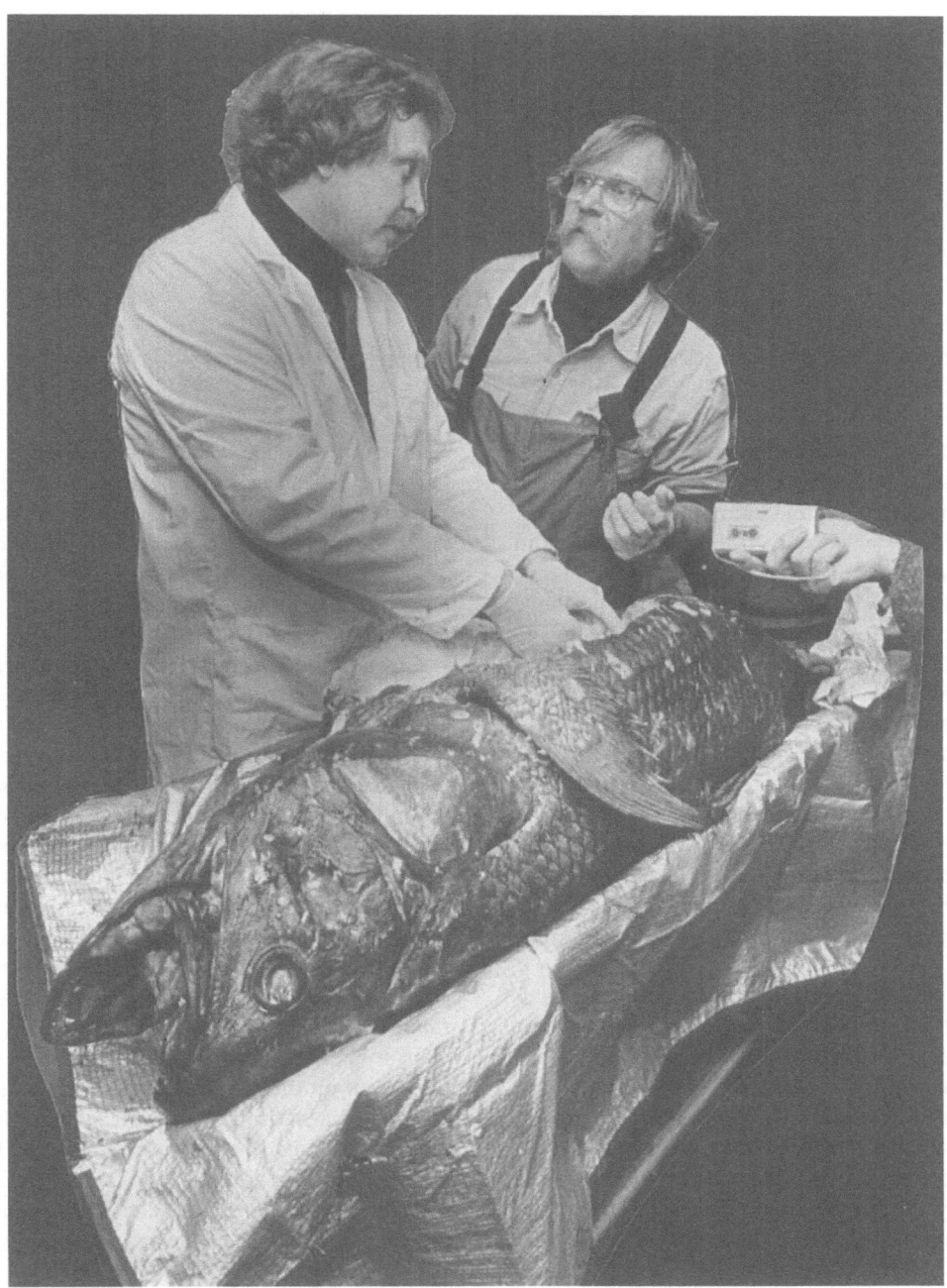

Fig. 3. Craig Sullivan and Jack Musick during dissection of their large female at VIMS. Photo by Kenneth Lyons, courtesy of Daily Press, Inc., Newport News, Virginia.

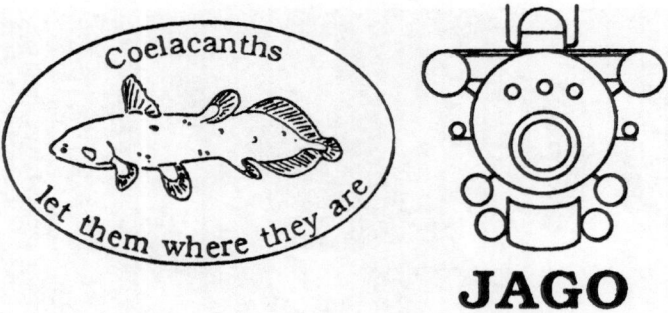

Fig. 4. The message from Fricke's submersible JAGO left in the electronic trap set by a Japanese expedition to capture a live coelacanth in 1989.

thyologists and Herpetologists in San Francisco. Many of these papers, as well as contributions from others unable to attend the symposium (e.g. Fricke, Forey) or conceived later are included in the present volume.

Thus the most recent scientific techniques have been used to study this ancient fish. A decade ago the hypothesis that *Latimeria* was closely related to chondrichthyans was being widely debated, however, even though research in the present volume (Mangum) shows that hemoglobins in the two groups react similarly to high levels of urea, urea synthesis is a primitive adaptation and widespread among vertebrates (Griffith). This information and several recent morphological and biochemical papers have rendered the chondrichthyan/crossopterygian hypothesis most improbable (Griffith, Hillis et al.)

The debate still continues as to whether the Actinistia or the Dipnoi are more closely related to the Rhipidistia, and therefore to the tetrapods. Two works herein provide evidence that actinistians are more closely related to the tetrapods (Waehneldt et al., Schultze) while one disagrees with this proposal (Stock et al.). The very rapid radiation of the osteichthyans into major taxa during the Devonian makes both the biochemical and morphological data hard to interpret (Stock et al., Schultze). Whereas the problem of the interrelationships among the Dipnoi, tetrapods, and actinistians is slowly approaching some resolution, the interrelationships and evolution within the actinistians are far better understood than a decade ago (Cloutier, Cloutier

& Forey, Forey). *Latimeria* is considered to be a fairly specialized coelacanth, yet its basic body plan and probably its basic biology is conservative. Indeed, *Latimeria* may have outlived the intermediate hosts that once were occupied by its parasites because larger, older specimens have very few parasites (Thoney & Hargis).

Through modern technology (manned research submersibles) we have learned more about the ecology and behavior of *Latimeria* during the last four years than we did during the first 48 years of its recognized existence. Fricke et al. have also made preliminary estimates of population size. With this and additional information on the low reproductive potential (Wourms et al., Balon), Bruton & Armstrong have been able to apply some numerical estimates widely used in fisheries biology to assess the impact of man's harvest on the very small population of *Latimeria*. Setter & Brown show that *Latimeria* has very little genetic variability, as with many relict species with small populations. In addition, even on the remote oceanic Comoran archipelago coelacanths carry significant body burdens of toxic chlorinated hydrocarbons like DDT (Hale et al.). These factors all lead to the conclusion that *Latimeria* is in immediate need of strict conservation measures. Such measures are proposed and examined in detail by Bruton & Stobbs, who call for the immediate decommercialization of the coelacanth following the 1989 decision to place the species in the highest category of protection in the Convention on International Trade in Endangered Species (CITES).

At about the same time the Coelacanth Conservation Council was created in the Comoros and the Explorers Club team was active there, the Japanese organized teams to study coelacanths, and even attempted to convey a live specimen to an aquarium in Japan (Fig. 4). Much of their research, virtually unknown to western workers, has not yet found its way into the formal scientific literature, although preliminary results have been published in popular Japanese periodicals (mainly Newton, the Graphic Science Magazine) and inhouse journals. We are fortunate to have in this volume contributions from our Japanese colleagues (Uyeno, Uyeno & Tsutsumi). It is obvious that Japanese and western workers have been treading some of the same ground in ignorance of each other's work.

Finally in this 50th anniversary volume we have attempted an inventory of all known specimens of *Latimeria* (Bruton & Coutouvidis), which reveals that at least 172 specimens have become available for scientific study in the past 52 years. The volume is concluded with a bibliography of 455 references in 23 languages on fossil coelacanths (Forey & Cloutier), and a further bibliography of 822 publications in 16 languages on the living coelacanth (Bruton et al.). This book has become the largest and most comprehensive volume so far published on the coelacanths, with contributions from 44 authors from 6 countries in 27 separate articles. We hope that it will stimulate further research on extinct and living coelacanths as well as contribute to a coordinated effort to protect the only known population of *Latimeria chalumnae* as a world heritage.

Acknowledgements

Thanks are due to J. Hamlin, P. Rodzianko and D. Wilkinson of The Explorers Club for making available specimens procured in the Comoros, to W. Phoel of the National Marine Fisheries Service and the Explorers Club for logistic and moral support, to G. Brown of the University of Washington and President of the Society for the Preservation of Old Fishes (SPOOF) for providing a conduit for communication among scientists in need of material from *Latimeria*, to J. Desfosse and C. Tabit and all the students and staff at VIMS who aided in the endeavor, especially C. Baldwin for invaluable aid in technical editing of some manuscripts, to K. Callis and B. McMillan for organizational and editorial aid, and to all the members of the VIMS dissection team, some of whom authored articles in this volume but also to those who did not, like R. Chapman, C. Sullivan and E. Wiley. Special thanks are offered to F. Perkins, Director of VIMS, for providing support during all phases of this project. We are also most grateful to the officials in the Comoros who made the field work possible, especially the Minister of the Interior M. Ali Mroudjae, the Director of the Cabinet Abderemane Mohamed Sidi, the Director of the Centre National de Documentation et de la Recherche Scientifique des Comones Damir Ben Ali, his staff member Abdu Shakur Aboud, and to the South African consul Marco Boni. We also thank the staff of the J.L.B. Smith Institute of Ichthyology who provided administrative and technical back-up for articles from Grahamstown, and Christine Flegler-Balon, marie Rush and David Noakes (Director of the Institute of Ichthyology in Guelph who discovered the 'Blues for Cousin Charlie') for their continuous support, advice and understanding. Special thanks are due to Herbert R. Axelrod for support and contributions towards the cost of the Guelph specimen as well as color reproductions in this volume and to Roy Reynolds for permission to use his drawings from our 1990 expedition. We also thank Sarah Landry for giving us permission to use her drawing, and Don McAllister for sharing with us the news of its existence.

24 December 1990

John A. Musick
Michael N. Bruton
Eugene K. Balon

20

Systematics and evolution

Environmental Biology of Fishes **32**: 23–58, 1991.
© 1991 *Kluwer Academic Publishers.*

Patterns, trends, and rates of evolution within the Actinistia

Richard Cloutier
Museum of Natural History, Department of Systematics and Ecology, The University of Kansas, Lawrence, KS 66045–2454, U.S.A.
Present address: Department of Palaeontology, British Museum (Natural History), Cromwell Road, London SW7 5BD, England

Received 21.8.1989   Accepted 30.7.1990

Key words: Coelacanth, Rates of morphological evolution, Evolutionary morphology, Cladistics, Phylogeny, Evolution, Paleontology, Tempo, Bradytely, Diversity

Synopsis

The interrelationships of 31 actinistian species (including *Latimeria chalumnae*) are analyzed based on a cladistic analysis of 75 osteological characters. Inference of evolutionary trends (e.g., modification of body shape and skull morphology) from the phylogenetic patterns demonstrates that the morphology of actinistians is less conservative than has been proposed previously. This empirical cladistic approach supports two distinct tempos of evolution during an evolutionary history of 380 million years. Along a phylogenetic pathway originating with a Devonian stem-species and ending with the living *Latimeria chalumnae* (including 101 morphological changes and 18 cladogenetic events), the first tempo occurred during the Devonian – Permian periods as a decreasing rate of morphological changes, which was followed by a stabilizing tempo during the Permian – Recent periods. The decreasing tempo is characterized by a sequence of gradual versus quantum temporal changes and low versus faster rates, whereas the stabilizing tempo primarily is gradual and low. In contrast to a common assumption, no significant correlation was found between the rates of morphological evolution and the temporal diversity of species.

Introduction

More than 150 years ago, Agassiz described the first fossil coelacanth from the Late Permian of Germany. In 1938, *Latimeria chalumnae* was found, the only living representative of a group of sarcopterygian fishes that was thought to be extinct since the Late Cretaceous[1], approximately 66 million years ago. The evolution of actinistians (coelacanths) has been cited as one of the classic examples of conservative (i.e., having no major phenotypic changes since the Devonian) and bradytelic (i.e., having a slow rate of morphological change through time) evolution.

Evidences in the fossil record indicate that actinistians originated as early as the Middle Devonian (specifically the Givetian); thus, they span a period of time of approximately 380 million years. Although the Actinistia is an ancient group, it is thought by many authors to be an evolutionarily conservative group (Eastman 1908, Moy-Thomas 1939, Simpson 1944, Schaeffer 1948, 1952a, b, Echols 1963, Jarvik 1964, 1980, Moy-Thomas & Miles 1971, Lund & Lund 1985, Elliott 1987, Rob-

---

[1] Ørvig (1986) identified a fragment of bone possibly referable to an actinistian from the Paleocene of southern Sweden.

ineau 1987, Schultze 1987, Balon et al. 1988). Simpson (1944) pointed out that the actinistians were a classic example of a taxon exhibiting 'low-rate' of evolution. However, Schaeffer (1952a, b) provided morphological documentation of the rates of actinistian evolution – namely, initial rapid evolution during the Devonian – Carboniferous periods followed by a constant low rate of evolution (i.e., bradytely). Unfortunately, Schaeffer's evolutionary rates cannot be corroborated because he did not detail the characters that he used explicitly, and his phyletic scheme primarily was based on the ages of the fossils. Forey's (1984, 1988) investigations were the first in which all assumptions were stated clearly. Forey (1988) concluded that the rate of morphological evolution of actinistians is relatively low and that, in contrast to earlier studies, there is no distinct pattern of rates.

As defined by Simpson (1944) and Westoll (1949), rates of evolution are dependent on the number of morphological changes and the length of geological time through which they occur. Derstler (1982) pointed out that cladogenesis has to be taken into consideration in the study of evolutionary rates. Finally, Forey (1984, 1988) added a species-diversity factor to the criteria used previously (i.e., time, changes, and cladogenesis).

In this study, I investigate the potential correlations among time, species diversity, cladogenesis, morphological changes, and rates of evolution for the Actinistia. In order to address and interpret these correlations, three objectives had to be achieved: (1) hypothesis of the phylogenetic patterns of the actinistians; (2) summarization of the major evolutionary trends within the group; and (3) quantification of the phylogenetic patterns in terms of rates of morphological evolution, as defined by the number of evolutionary changes per million years. The questions of genomic and taxic (speciation and extinction) rates are not considered.

Materials and methods

Phylogenetic methods

The interrelationships among fossil and living actinistians were studied using the principles of phylogenetic systematics (Hennig 1966). Although the Hennigian principles are accepted, characters and relationships were analyzed more pragmatically by providing detailed character descriptions, polarizing character-states based on out-group comparisons, coding of actual taxa, and determination of trees inferred by computer algorithms.

Operational taxonomic units (OTUs)

The OTU used is the species as recognized by the presence of autapomorphy(ies). The OTUs were selected based primarily on two criteria: (1) completeness of the morphological information, and (2) maximization of the stratigraphic history represented by the fossils. The completeness criterion is such that more than 60% of the characters had to be available for coding. Specimens of most of the species included in the analysis were examined (23 examined of 31 analyzed); those unavailable for examination were coded based on information in the literature.

Characters

Characters were divided into what seemed to be simple, independent characters. All characters were unweighted and divided into plesiomorphic and apomorphic states. Character states were polarized using out-group comparison (Maddison et al. 1984, Watrous & Wheeler 1981) – i.e., two taxonomic out-groups and one functional out-group. Multiple-state characters were analyzed unordered. Characters with ambiguous polarities were analyzed unpolarized.

Out-group selection

The selection of out-groups follows Cloutier (1990). Figure 1 depicts the relationships among the out-groups and the in-group (i.e., Actinistia). The first taxonomic out-group (TOG1) is a clade defined by [[Dipnoi + Youngolepiformes + Porolepiformes] + ['Osteolepiformes' + [Panderichthyiformes + Tetrapoda]]]. The second taxonomic out-group (TOG2) is the Onychodontiformes, and the functional out-group (FOG) is the Late Devonian (Early Frasnian) actinistian *Miguashaia bureaui* Schultze 1973 from the Escuminac Formation of Québec, Canada.

Technical specifications

Phylogenetic relationships were hypothesized using the PAUP program (Phylogenetic Analysis Using Parsimony), Version 2.4 of Swofford (1985). The analysis included 75 characters corresponding to 101 apomorphic states; Characters 9, 45, and 50 were unpolarized, and Characters 5, 7, 8, 12, 20, 21, 23, 25, 30, 44, 70, and 74 were unordered. The following specifications were set to generate the trees: ROOT = ANCESTOR; SWAP = GLOBAL; MULPARS ON; OPTION = FARRIS; MAXTREE = 468 (because of memory limitation of the mainframe IBM-VM); HOLD = 20. The reconstructed, hypothetical ANCESTOR was coded as '9' for unpolarized characters.

Analysis of rates of morphological evolution

Simpson (1949, p. 205) wrote '. . . the expression "rate of evolution" has so many possible meanings as to be almost meaningless without further qualification'. Simpson (1949), Stanley (1975), Raup (1987), and Schoch (1986) recognized and defined three types of rates of evolution – genetic, morphologic[al], and taxonomic. In this paper I will address only the rates of morphological evolution as defined by the number of evolutionary morphological changes per unit of time.

The concepts of mode and tempo in evolution, and the associated notion of evolutionary rates, were first elaborated by Simpson (1944). In order to evaluate rates of morphological evolution, Simpson (1944, 1949) promoted the use of quantitative differences of a single character over a given period of time. Applying the main concepts of Simpson (1944), Westoll (1949) investigated rates of evolution of dipnoans in terms of degree of 'primitiveness'. He reconstructed a dipnoan archetype from which he identified 21 characters ('fac-

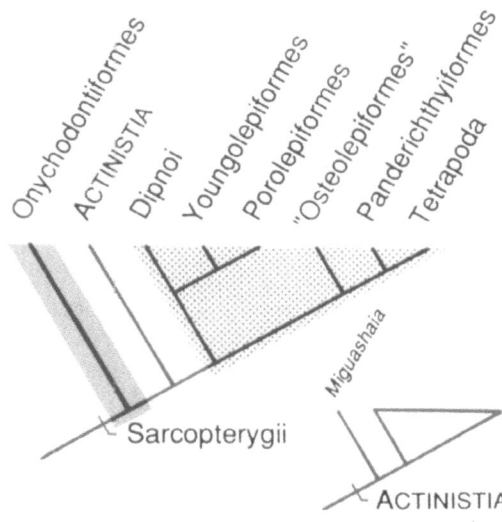

Fig. 1. Phylogenetic position of the Actinistia among the Sarcopterygii. A– Interrelationships among Sarcopterygii showing the relative position of the Actinistia with respect to the taxonomic out-groups. Light-stipled is the first taxonomic out-group (TOG1) and the darker-stipled is the second taxonomic out-group (TOG2). B– Position of the functional out-group (FOG), *Miguashaia bureaui,* with respect to the remaining actinistians.

tors') to analyze. Each character was divided into a 'graded series of steps', and each step was attributed a score (the highest score attributed to the most primitive condition of each character); these steps corresponded to transformation series encompassing the variation observed between the presumed ancestral condition and the living species sequenced in a stratigraphic order. A total score that described the degree of primitiveness (hypothetical ancestor being scored 100) was provided for each of 13 genera based on the summation of the 21 individual scorings. The genera were arranged in stratigraphic order. In summary, Westoll's method consisted of evaluating the degree of divergence from an hypothetical, ancestral type during the evolution of the group.

In his study on evolutionary rates in dipnoans and actinistians, Schaeffer (1952a, b, 1953) converted the score obtained from Westoll's method into a value corresponding to a rate of acquisition of advanced characters. Schaeffer (1952a, b) de-

26

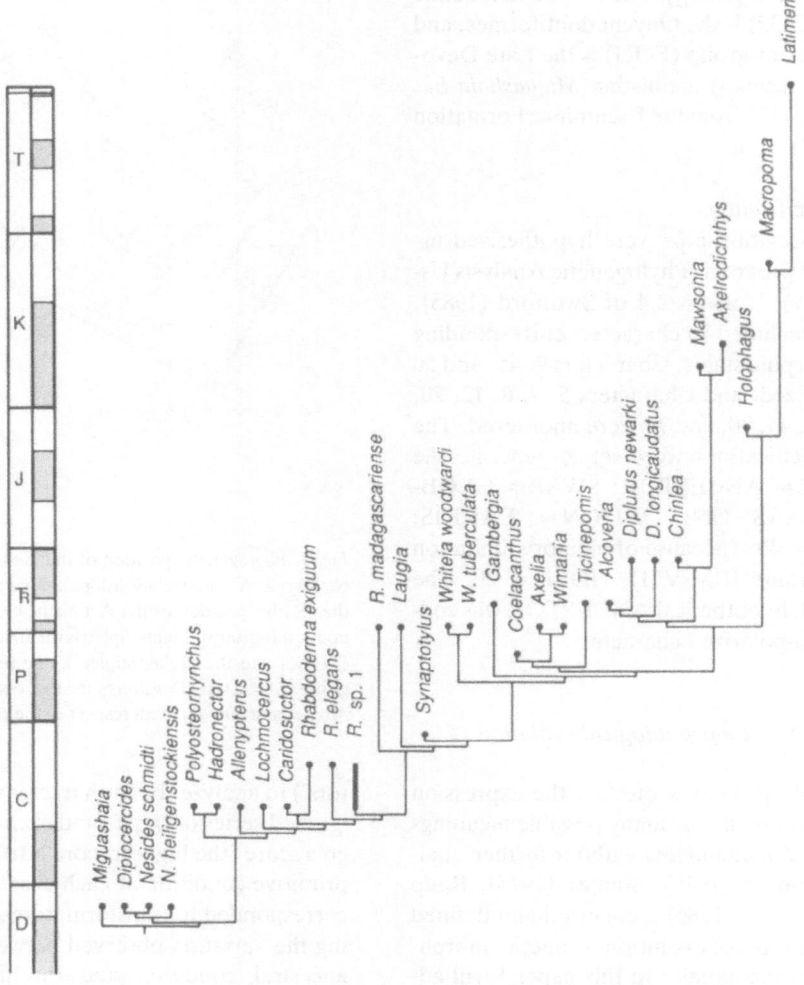

Fig. 2. Temporal cladogram of actinistian species superimposed on a proportional geological time scale. The topology corresponds to that of Figure 3. Species are represented by circle (punctual occurrence) or bar (range occurrence). The geological time scale is from Palmer (1983).

fined rates of morphological evolution in terms of acquisition of advanced characters per million years for the approximated time interval between successive genera in the stratigraphic sequence. The stratigraphic order of the fossil record was used by Schaeffer (1952a, b, 1953) as reflecting a phylogenetic sequence.

Finally, Derstler (1982) estimated rates of morphological evolution by noting the accumulation of derived character-states in a lineage of hy-

pothetical ancestors (stem-species). Derstler used the WAGNER program to reconstruct his phylogeny of stylophoran carpoid echinoderms. Rates were calculated as the difference of the scoring between two consecutive stem-species (i.e., the number of character-states along an internodal branch) divided by the difference of the age of the two consecutive cladogenetic events (i.e., Rate = $[\text{Score}_i - \text{Score}_{i+1}] \star [\text{Age}_{i+1} - \text{Age}_i + 1.0]^{-1}$). Forey (1988) used a similar method in which the

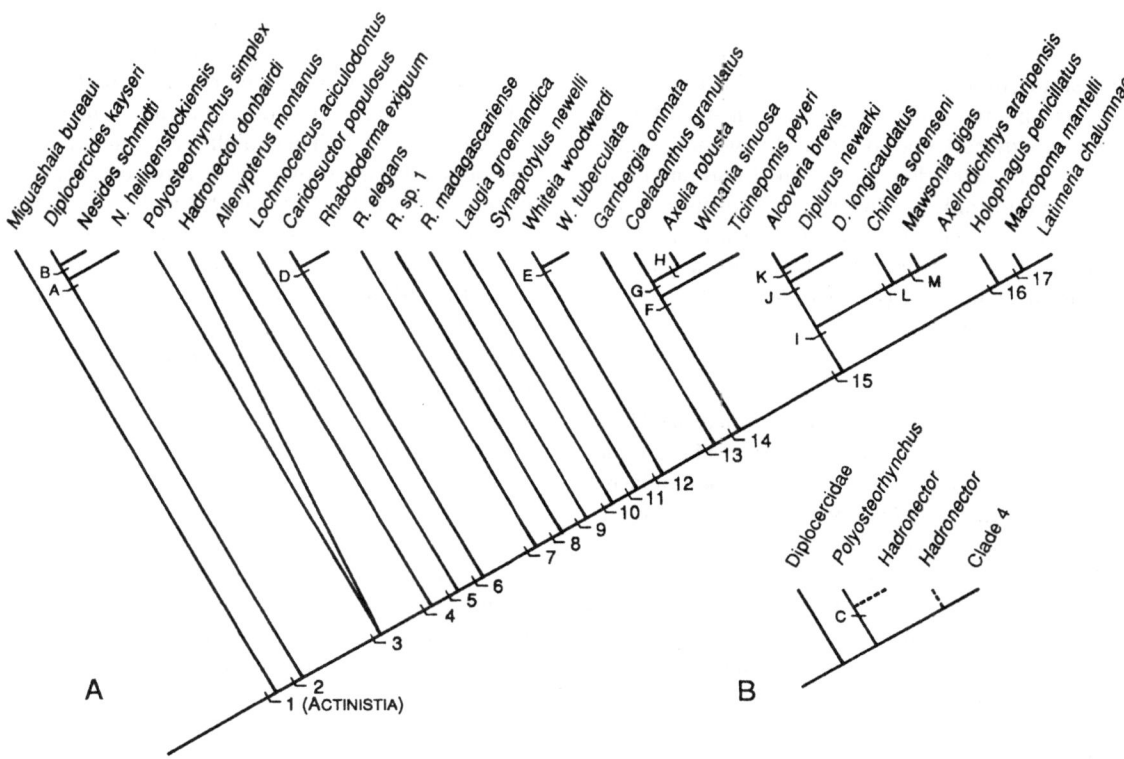

Fig. 3. Interrelationships of 31 actinistian species. A– Adams consensus tree based on 2 equally parsimonious trees at 253 steps [C.I. = 0.391; M.C.I. (Adams) = 0.708]. B– Two equally parsimonious alternative positions of *Hadronector*. The listing of characters for each node is provided in Appendix 2.

number of character-states at a node (generated from the PAUP program) was divided by the estimated time separating two consecutive cladogenetic events.

Specific terminology

An *hierarchical pathway* corresponds to a sequence of hypothetical ancestors (i.e., stem-species) between an origin and a selected terminal taxon. For any given tree, there is a series of *phylogenetic pathways* (subset of hierarchical pathways) in which the selected origin is a hypothetical ancestor to the terminal taxa selected. Phylogenetic pathways are unidirectional whereas hierarchical pathways are bidirectional. The phylogenetic pathway encompasses a series of consecutive cladogenetic events (i.e., branching points or nodes on the tree).

The *cladogenetic rank* corresponds to the relative position of a cladogenetic event on the cladogram in reference to the first rank at the stem-species node (in terms of number of events).

In order to apply the terminology used in this paper, examples are given in reference to Figures 2 and 3. With respect to this cladogram including 31 species and 30 hypothetical ancestors, there are 3660 hierarchical pathways joining any two taxa (e.g., Node 1 to *Latimeria*, *Lochmocercus* to *Axelia*, Node M to *Hadronector*, etc.) of which 636 are phylogenetic pathways (e.g., actinistian stem-species to *Latimeria*, diplocercid stem-species to *Diplocercides*, *Hadronector* hypothetical ancestor to *Axelrodichthys*, etc.). In this study, the actinistian hypothetical ancestor (stem-species) corresponds to the first phylogenetic event as well as the first cladogenetic rank of the actinistian phylogenetic pathway. The consecutive series of hypothetical ancestors (nodes) between the actinistian stem-species and the closest hypothetical ancestor to *Synaptotylus* corresponds to a phylogenetic pathway starting at cladogenetic rank 1 and ending at cladogenetic rank 11. In terms of absolute rank, both *Diplocercides* and *Allenypterus* are at rank 4 because they are separated from the actinistian hypothetical ancestor (i.e., stem-species) by 4 cladogenetic events.

Morphological changes

Each branch on the cladogram is characterized by a series of character-states including (1) unique shared-derived character-states, (2) homoplastic character-states, and (3) reversals. The total number of changes at a cladogenetic event corresponds to the sum of all character-state changes. This corresponds to the definition of changes used by Forey (1988), but differs from that of Forey (1984) who selected only the fully congruent, uniquely shared-derived characters on the tree.

Temporal cladogram

The cladogram(s) generated from the PAUP program was superimposed on a geological time scale (Palmer 1983). The topology hypothesized in the phylogenetic analysis is identical to that used in the temporal cladogram. Each species was mapped accordingly on the time scale. Most species are represented by dots, because the span of time for which the species is known is geologically brief. However, in the case of *Rhabdoderma elegans,* a bar corresponds to the time span between the oldest and the youngest recorded specimens (Fig. 2). The time of cladogenesis between sister-groups is extrapolated from the age of the oldest member of the inclusive clade plus 2 million years (My). This method provides the minimum age of the most recent common ancestors (Paul 1982).

Calculations of rates

The underlying assumptions of using a cladogram to calculate rates of morphological evolution are: (1) the tree reflects the history of the group, and (2) the distribution of the characters on the tree re-

flects the true distribution of characters during the evolution of the group (i.e., unbiased sampling of characters). There are three possible methods by which to calculate rates depending upon the period of time considered. Time might be estimated as (1) the period between two consecutive cladogenetic events (as used by Derstler 1982 and Forey 1988), (2) a fixed period of time in millions of years, and (3) duration of geological period (as used by Schaeffer 1952a, b). Methods 2 ($rate_f$) and 3 ($rate_g$) are independent of the phylogenetic scheme used. The three methods for calculation of rates of morphological evolution are formulated as follows:

Rate between consecutive cladogenetic events:

$$rate_c = change_i \ / \ (t_{i-1} - t_i),$$

where 'i' is the rank of a cladogenetic event, '$change_i$' is the number of character-states at cladogenetic event 'i', 't_i' is the estimated time of cladogenesis, and 't_{i-1}' is the estimated time of previous cladogenesis.

Rate during fixed period of 50 million years:

$$rate_f = \sum_{t_{if}}^{t_{sf}} change \ / \ (t_{if} - t_{sf}),$$

where 't_{sf}' is the superior limit of a fixed period, 't_{if}' is the inferior limit of the same fixed period of time, '$(t_{if} - t_{sf})$' is a fixed period of 50 million years except for the most recent period which is in the range $50 \geq t > 0$.

Rate during a geological period:

$$rate_g = \sum_{t_{ib}}^{t_{sb}} change \ / \ t_{ib} - t_{sb},$$

where 't_{ib}' is the age of the inferior boundary of geological period 'g', and 't_{sb}' is the age of the superior boundary of geological period 'g'.

The geological time scale follows the standard stratigraphic scale of Palmer (1983); the use of this time scale is based on the critique by Menning (1989). The three calculated rates are compared in this paper. For each one of these calculations, the total number of changes along the phylogenetic pathway (stem-species – *Latimeria*) that occurs during the chosen period of time is divided by the duration of the time period itself (in million years).

Analysis of correlation

Gauthier et al. (1988) estimated the coefficient of correlation (*r*) between the age of fossils and clado-genetic rank; ages were arcsine-transformed. However the coefficient of correlation is inappropriate for ordinal variables (Dixon & Massey 1983) such as cladogenetic rank and non-normal distribution (Snedecor & Cochran 1980). None of the variables studied was normally distributed, and the best-known procedure for analyzing such data is to convert variables to rankings (Snedecor & Cochran 1980). The nonparametric rank-order correlation or Spearman correlation (r_s, see Dagnelie 1977) was used because of the ordinal nature of the data. The formula for the Spearman correlation coefficient is as follows:

$$r_s = 1 - \{(6 \star \sum d^2) / [n \star (n^2 - 1)]\},$$

where 'd' is the difference between rank-values for each observation and 'n' is the sample size.

Equal ranks were readjusted according to the procedure described by Dagnelie (1977). For the different rank correlations investigated in this analysis, the first ranking was attributed as follows: (1) the oldest event for the *age rank,* (2) the stem-species of Actinistia for the *cladogenetic rank,* (3) the lowest number of species for the *diversity rank,* (4) the lowest number of changes (per cladogenetic event or period of time) for the *morphological changes,* and (5) the lowest value for the *rate of morphological evolution.*

Phylogeny

Description of characters

Seventy-five characters for a total of 101 apomor-phic character-states were used to analyze relationships among 31 actinistian species. Table 1 provides a synoptic list of the characters used in this analysis (including 54 characters used for the study of the interrelationships of Paleozoic actinistians by Cloutier 1990) with their respective plesiomorphic and apomorphic character-states. For detailed descriptions of Characters 1–54 see Cloutier (1990). Additional character-states have been added to the original description of Characters 8 (5: one prepa-rietal per row), 21 (4: two extrascapulars; 5: four extrascapulars), 25 (3: bar-like squamosal), and 44 (0: more than 15 lepidotrichia in first dorsal fin; 2: 8–9 lepidotrichia; 4: 11–15 lepidotrichia). For the purpose of this study 21 characters are added (Characters 55–75) and described below. The ple-siomorphic and apomorphic character-states are given for each character as well as a brief description and justification of the polarity. The characters described herein are listed in a morphological order from anterior to posterior by structural complexes. Because of ambiguity in the out-group comparison the following characters are unpolarized: 9, 45, and 50. The following characters are unordered: 5, 7, 8, 12, 20, 21, 23, 25, 30, 44, 70, and 74. Appendix 1 is the data matrix for the 31 taxa.

55. Premaxilla
The rostral organ of *Latimeria chalumnae* (Millot & Anthony 1954, 1956, 1965, Bemis & Hethering-ton 1982) consists of a pair of sensory organs located in a median rostral cavity of the ethmosphenoid part of the neurocranium. Three pairs of canals (anterior, posterior superior, and posterior inferior) exit from the rostral cavity to the exterior of the skull. The posterior superior and posterior inferior canals of the rostral organ perforate the tectal (when present; Character 22), whereas the anterior tube perforates the dorsal lamina of the premaxilla (when the dorsal lamina is present; Character 5). Lund & Lund (1985) identified an opening in the premaxilla that possibly corresponds to the anterior tube of the rostral organ in some Carboniferous actinistians. The TOGs lack the rostral organ (and its associated pore in the premaxilla). The dorsal lamina of the premaxilla of the FOG is not perforated by a large pore; this either indicates that

Table 1. List of the 75 characters and respective character-states, used in the PAUP analysis, including the first 54 characters (Cloutier 1990). 0 = plesiomorphic states, 1–5 apomorphic states, * = unpolarized. Multiple states are unordered.

1. Cosmine:
   0 = presence of cosmine;
   1 = absence of cosmine.
2. Scale ornamentation:
   0 = scale with longitudinal ridges and tubercules;
   1 = scale only with longitudinal ridges.
3. Sensory canal pores, size:
   0 = small supraorbital canal pores;
   1 = large supraorbital canal pores.
4. Premaxillae, number:
   0 = paired (and unpaired) premaxillae;
   1 = fragmented premaxillae.
5. Dorsal lamina of premaxilla, completeness of dorsal margin:
   0 = dorsal lamina of premaxilla complete;
   1 = dorsal lamina of premaxilla invaginated medially;
   2 = dorsal lamina of premaxilla reduced.
6. Dorsal lamina of premaxilla, perforation:
   0 = dorsal lamina perforated by rostral commissure;
   1 = dorsal lamina not perforated by rostral commissure.
7. Preparietal, number of rows:
   0 = mosaic of elements (undifferentiated preparietals);
   1 = presence of one row of preparietals across the skull roof;
   2 = presence of two rows of preparietals across the skull roof;
   3 = presence of three rows of preparietals across the skull roof.
8. Preparietals, number:
   0 = more than five preparietals per row;
   1 = two preparietals per row;
   2 = three preparietals per row;
   3 = four preparietals per row;
   4 = five preparietals per row;
   5 = one preparietal per row.
9*. Bifurcating supraorbital pores:
   0 = series of single supraorbital pores between the orbit and the parietal;
   1 = series of bifurcating supraorbital pores between the orbit and the parietal.
10. Parietal size:
   0 = parietal bigger than most posterior preparietal;
   1 = parietal relatively the same size as the most posterior preparietal.
11. Descending process of parietal:
   0 = absence of the parietal process;
   1 = presence of the parietal process.
12. Postparietal shape:
   0 = L-shaped postparietal;
   1 = inverted-L-shaped postparietal;
   2 = rectangular postparietal.
13. Descending process of postparietal:
   0 = absence of descending process of postparietal;
   1 = presence of descending process of postparietal.
14. Postparietal-postorbital articulation:
   0 = postparietal does not contact postorbital;
   1 = postparietal contact postorbital.

Table 1. Continued

15. Postparietal-extrascapular articulation:
    0 = postparietal articulating with the lateral extrascapular;
    1 = postparietal not articulating with the lateral extrascapular.
16. Medial branch of otic sensory canal:
    0 = absence of a medial branch of the otic sensory canal;
    1 = presence of a medial branch of the otic sensory canal.
17. Intertemporal:
    0 = presence of intertemporal;
    1 = absence of an independent intertemporal.
18. Supratemporal:
    0 = presence of supratemporal;
    1 = absence of an independent supratemporal.
19. Descending process of tabular:
    0 = absence of descending process of tabular;
    1 = presence of descending process of tabular.
20. Otic canal:
    0 = otic canal passing through the lateral series (intertemporal and/or supratemporal);
    1 = otic canal passing between postparietal and the lateral series (intertemporal and/or supratemporal);
    2 = otic canal passing through postparietal.
21. Extrascapulars, number:
    0 = three extrascapulars;
    1 = five extrascapulars;
    2 = seven extrascapulars (Forey 1981; his Character 18);
    3 = more than seven extrascapulars;
    4 = two extrascapulars;
    5 = four extrascapulars.
22. Tectal:
    0 = presence of tectal;
    1 = absence of tectal.
23. Cheek bones, articulation:
    0 = complete suture among cheek bones;
    1 = loose articulation of the postspiracular;
    2 = loose articulation of the postspiracular and postorbital.
24. Postspiracular size:
    0 = large postspiracular;
    1 = reduced or absent postspiracular.
25. Squamosal shape:
    0 = triangular squamosal;
    1 = quadrilateral squamosal;
    2 = pentagonal squamosal;
    3 = bar-like squamosal.
26. Quadratojugal:
    0 = presence of the quadratojugal;
    1 = absence of an independent quadratojugal.
27. Jugal canal trajectory:
    0 = jugal canal transverses approximately in the middle of the squamosal;
    1 = jugal canal transverses along ventral margin of squamosal.
28. Operculum shape:
    0 = quadrilateral operculum;
    1 = triangular operculum.
29. Suboperculum:
    0 = presence of suboperculum;
    1 = absence of suboperculum.

32

Table 1. Continued

30. Suboperculum position:
    0 = suture between operculum and suboperculum;
    1 = suture between suboperculum and preopercular;
    2 = suboperculum isolated.
31. Suboperculum size:
    0 = suboperculum longer than deep;
    1 = suboperculum deeper than long.
32. Suboperculum shape:
    0 = quadrilateral suboperculum;
    1 = triangular suboperculum.
33. Pterygoid-quadrate fusion:
    0 = presence of an isolated bony quadrate;
    1 = absence of an isolated bony quadrate.
34. Anterior dorsal lamina of parasphenoid:
    0 = absence of anterior dorsal lamina of parasphenoid;
    1 = presence of anterior dorsal lamina of parasphenoid.
35. Buccohypophysical foramen of parasphenoid:
    0 = presence of buccohypophysial foramen;
    1 = absence of buccohypophysial foramen.
36. Basipterygoid process:
    0 = presence of basipterygoid process;
    1 = absence of basipterygoid process
37. Dentary, orientation of:
    0 = straight dentary;
    1 = diagonal dentary.
38. Coronoid IV shape:
    0 = sub-triangular Coronoid IV;
    1 = sub-quadrilateral Coronoid IV.
39. Cleithral articulation:
    0 = sutures between cleithrum and clavicle, and between extracleithrum and clavicle of equal length;
    1 = suture between cleithrum and clavicle shorter than suture between extracleithrum and clavicle.
40. Extracleithrum:
    0 = absence of extracleithrum;
    1 = presence of extracleithrum.
41. Extracleithrum size:
    0 = extracleithrum shorter than cleithrum;
    1 = extracleithrum equal to, or deeper than, cleithrum.
42. Pectoral fin insertion:
    0 = low insertion of pectoral fin;
    1 = high insertion of pectoral fin.
43. Pelvic fin position:
    0 = posterior insertion of pelvic fin;
    1 = anterior insertion of pelvic fin (abdominal position).
44. First dorsal lepidotrichia, number:
    0 = more than 15 lepidotrichia in first dorsal fin;
    1 = ten lepidotrichia in first dorsal fin;
    2 = eight to nine lepidotrichia in first dorsal fin;
    3 = fewer than eight lepidotrichia in first dorsal fin;
    4 = 11–15 lepidotrichia in first dorsal fin.

Table 1. Continued

45*. Ventral margin of basal plate of first dorsal fin:
    0 = smooth ventral margin;
    1 = emarginate ventral margin.
46. Anterodorsal process of basal plate of second dorsal fin:
    0 = absence of anterodorsal process of basal plate of second dorsal fin;
    1 = presence of anterodorsal process of basal plate of second dorsal fin.
47. Posteroventral process of basal plate of second dorsal fin:
    0 = absence of posteroventral process of basal plate of second dorsal fin;
    1 = presence of posteroventral process of basal plate of second dorsal fin.
48. Posteroventral process of basal plate of second dorsal fin, development:
    0 = posteroventral process poorly developed;
    1 = posteroventral process well developed.
49. Basal plate of anal fin:
    0 = anal basal plate with single process;
    1 = anal basal plate with bifurcated process.
50*. Anterodorsal process of basal plate of anal fin:
    0 = anterodorsal process of basal plate of anal fin short and broad;
    1 = anterodorsal process of basal plate of anal fin long and narrow.
51. Tail:
    0 = heterocercal tail;
    1 = diphycercal tail.
52. Supplementary caudal fin:
    0 = absence of supplementary caudal fin;
    1 = presence of supplementary caudal fin.
53. Caudal lepidotrichia:
    0 = ratio of 3 : 1 or 2 : 1 between the number of caudal lepidotrichia to the number of radials;
    1 = ratio of 1 : 1 between the number of caudal lepidotrichia to the number of radials.
54. Lepidotrichium distal end:
    0 = distal end of lepidotrichium bifurcate;
    1 = distal end of lepidotrichium single.
55. Premaxilla:
    0 = premaxilla without pore for anterior rostral tube;
    1 = premaxilla with pore for anterior rostral tube.
56. Tabular:
    0 = posterior margin of tabular level with posterior margin of postparietal;
    1 = tabular extending beyond posterior margin of the postparietal and medial extrascapulars.
57. Fusion of tabular:
    0 = tabular and postparietal separated;
    1 = tabular and postparietal fused.
58. Lateral process of parietal:
    0 = absence of the parietal process;
    1 = presence of the parietal process.
59. Extrascapular:
    0 = extrascapulars posterior to the postparietal;
    1 = extrascapulars forming part of the skull table.

Table 1. Continued

---

60. Occipital commissure:
    0 = occipital commissure without anterior ramifications;
    1 = anterior branches developed from occipital commissure.
61. Width of cheek:
    0 = normal postorbital distance;
    1 = narrow postorbital distance.
62. Anterior process of postorbital:
    0 = absence of anterior process of postorbital;
    1 = presence of anterior process of postorbital.
63. Postorbital:
    0 = postorbital of normal size;
    1 = postorbital reduced to a small narrow element lying posterodorsally to the orbit.
64. Lacrimojugal:
    0 = lacrimojugal extending only to anterior of orbit;
    1 = lacrimojugal extending to anterior of snout.
65. Lacrimojugal:
    0 = lacrimojugal not expanded anteriorly;
    1 = lacrimojugal expanded anteriorly.
66. Dentary:
    0 = absence of posterodorsal hook on dentary;
    1 = presence of posterodorsal hook on dentary.
67. Ascending process of prootic:
    0 = absence of ascending process of prootic;
    1 = presence of ascending process of prootic.
68. Contact of Coronoid IV:
    0 = abutting with angular;
    1 = suturing with external side of angular and prearticular.
69. Pleural ribs:
    0 = absence of well-developed ribs;
    1 = presence of well-developed ribs.
70. Number of lepidotrichia on pectoral fin:
    0 = fifty or more lepidotrichia;
    1 = between 30 and 49 lepidotrichia;
    2 = between 20 and 29 lepidotrichia;
    3 = between ten and 19 lepidotrichia;
    4 = less than ten lepidotrichia.
71. Position of pelvic fin:
    0 = pelvic fin inserting in an abdominal position or more posteriorly;
    1 = pelvic fin inserting ventrally or anteriorly to the pectoral fins.
72. Spines on lepidotrichia of first dorsal fin:
    0 = absence of denticules on anterior lepidotrichia of first dorsal fin;
    1 = presence of spines on anterior lepidotrichia of first dorsal fin.
73. Basal plate of first dorsal fin:
    0 = hemispherical basal plate of the first dorsal fin;
    1 = triangular basal plate of the first dorsal fin.
74. Number of lepidotrichia on second dorsal fin:
    0 = more than 25 lepidotrichia;
    1 = between 20 and 25 lepidotrichia;
    2 = between 15 and 19 lepidotrichia;
    3 = between ten and 14 lepidotrichia;
    4 = less than ten lepidotrichia.
75. Lepidotrichia:
    0 = narrow lepidotrichia;
    1 = broad lepidotrichia.

---

the rostral organ is absent in the FOG, or that if it is present, the anterior rostral tube does not pass in the premaxilla.

0 = premaxilla without pore for anterior rostral tube;

1 = premaxilla with pore for anterior rostral tube.

56. Tabular

The tabular (= 'supratemporal' of Schaeffer 1952a and Forey 1981) is the lateral-line bone that supports a segment of the otic canal and of the occipital commissure. Primitively, the tabular is located lateral to the postparietal and anterior to the lateral extrascapular. In primitive actinistians and the OGs, the posterior margin of the tabular is at the same level as the posterior margin of the postparietal. Extension of the posterior margin of the tabular beyond the posterior margin of the postparietal to form a shelf lateral to the skull table is considered to be an apomorphic character-state.

0 = posterior margin of tabular level with posterior margin of postparietal;

1 = tabular extending beyond posterior margin of the postparietal and medial extrascapulars.

57. Tabular, fusion of

Forey (1981) considered the fusion of the tabular and postparietal to be the result of late ontogenetic change (i.e., adult fusion). Rieppel (1980) considered the fusion of these elements as a synapomorphy for *Ticinepomis* plus *Holophagus*. The fusion of the tabular with the postparietal is considered to be an apomorphic condition. If there is intraspecific variation among large specimens of a given species (when the sample size allows this comparison) – i.e., some specimens with tabular and postparietal fused and some non-fused – the apomorphic character-state is coded when more than 80% have the two elements fused; a 'U' is coded when 20–80% shows either condition.

0 = tabular and postparietal separated;

1 = tabular and postparietal fused.

58. Lateral process of parietal

The paired parietals are located anterior to the intracranial joint; thus, a small gap exists between the posterior margin of the parietal and the anterior margin of the postparietal. The lateral process of the parietal is formed from the posterolateral margin of the parietal and is directed ventrolaterally toward the dorsal part of the postorbital. The lateral process is present in *Coelacanthus, Axelia, Wimania,* and *Latimeria,* but not in the out-groups; thus, its presence is considered to be derived.

0 = absence of the parietal process;

1 = presence of the parietal process.

59. Extrascapular

The extrascapulars (= 'median postparietals' of Maisey 1986) usually are located posterior to the tabulars and postparietals. Their number varies from three to nine (Character 21). In certain species (e.g., *Chinlea* [Elliott 1987], *Axelrodichthys* and *Mawsonia* [Maisey 1986]), the extrascapulars form an integral part of the posterior skull table between the tabulars, and have interdigitating sutures with the postparietals. The presence of extrascapulars between the tabulars forming an integrated part of the posterior skull table is considered to be derived.

0 = extrascapulars posterior to the postparietal;

1 = extrascapulars forming part of the skull table.

60. Occipital commissure

The occipital commissure (= 'supratemporal commissural canal' of Schaeffer 1952a; 'supratemporal commissure' of Forey 1981) transverses the extrascapulars and the tabulars. Anterior ramifications of the occipital commissure extending to the anterior part of the extrascapulars and/or the posterior part of the postparietals are found in some species. Forey (1981; his Character 17) considered the presence of anterior branches developed from the occipital commissure to be a derived condition. The OGs have a simple occipital commissure.

0 = occipital commissure without anterior ramifications;

1 = anterior branches developed from occipital commissure.

61. Cheek, size of

Stensiö (1921) noted the narrow space between the

orbit and the operculum in some Triassic actinistians. Martin & Wenz (1984) considered the presence of a narrow postorbital distance as found in some Triassic genera (e.g., *Garnbergia, Diplurus, Chinlea, Ticinepomis, Hainbergia,* and *Alcoveria*) to be derived. The cheek is considered to be narrow if it is less than a fourth of the length of the orbit; the 'normal' condition corresponds to a postorbital region broader than a fourth of the length of the orbit.

0 = normal postorbital distance;
1 = narrow postorbital distance.

62. *Postorbital, anterior process of*
The cheek region of actinistians is composed of the postorbital, postspiracular, squamosal, preopercular, and quadratojugal. The postorbital forms the posterior margin of the orbit, and articulates anteroventrally with the lacrimojugal. In some taxa (e.g., *Mawsonia*), the anteroventral corner of the postorbital extends anteriorly beyond the posteroventral margin of the orbit. The process may reach the anterior margin of the lacrimojugal at a level anterior to the orbit. The presence of an anteriorly developed process on the anteroventral margin of the postorbital is considered to be derived.

0 = absence of anterior process of postorbital;
1 = presence of anterior process of postorbital.

63. *Postorbital*
The size of the postorbital varies among actinistians. However, Rieppel (1980) pointed out that the postorbital is greatly reduced in some species (e.g., *Coelacanthus granulatus, Ticinepomis peyeri*). The postorbital in some species (e.g., *C. granulatus*) is reduced to a narrow bony element simply surrounding a segment of the infraorbital canal. The reduced postorbital represents an apomorphic condition.

0 = postorbital of normal size;
1 = postorbital reduced to a small narrow element lying posterodorsal to the orbit.

64. *Lacrimojugal*
Stensiö (1921) noted that extension of the lacrimojugal far anterior to the orbit (in known Triassic species) was characteristic of the actinistians. However, in most Paleozoic actinistians, the lacrimojugal does not extend anterior to the anterior margin of the orbit. The lacrimojugal in all actinistians seems to be elongated in comparison to the condition found in other osteichthyans because of the fusion of two bones. However, in some actinistian species, the lacrimojugal is markedly longer. The anterior end of the lacrimojugal (expanded or not; see Character 65) could either reach the anterior level of the orbit or extend further anteriorly. In the latter condition, the lateral rostral constitutes part of the lateral snout rather than forming an anterior part of the preorbital region. Anterior development of the lacrimojugal may be associated with the absence of the tectal (Character 22).

0 = lacrimojugal extending only to anterior of orbit;
1 = lacrimojugal extending to anterior of snout.

65. *Lacrimojugal*
In some actinistians (e.g., *Mawsonia*), the anterior part of the lacrimojugal expands dorsally to form part of the anterior margin of the orbit and nearly reaches the row of supraorbitals. In the OGs, the lacrimal is not expanded anteriorly.

0 = lacrimojugal not expanded anteriorly;
1 = lacrimojugal expanded anteriorly.

66. *Dentary*
The short dentary lies dorsal to the splenial and anterior to the angular. The posterodorsal margin of the dentary may bear a posteriorly oriented, toothless hook. Martin & Wenz (1984) reported the presence of the dentary hook in *Garnbergia, Holophagus, Macropoma,* and 'more recent coelacanths'. The presence of the posterodorsal hook on the dentary is considered to be derived; the hook is absent in the TOGs and FOG.

0 = absence of posterodorsal hook on dentary;
1 = presence of posterodorsal hook on dentary.

67. *Ascending process of prootic*
Forey (1981) considered the presence of an ascending process of the prootic to be derived.

0 = absence of ascending process of prootic;

1 = presence of ascending process of prootic.

68. Coronoid, contact of

Vertical orientation of Coronoid IV is considered to be an actinistian synapomorphy (Cloutier 1990). In most actinistians (including the FOG – *Miguashaia*), Coronoid IV abuts the medial side of the prearticular. In the plesiomorphic condition (FOG), a relatively flat coronoid abuts the medial surface of the angular, whereas in the apomorphic condition, the coronoid articulates with the prearticular and the lateral side of the angular (e.g., *Axelrodichthys, Mawsonia*).

    0 = Coronoid IV abutting with angular;
    1 = Coronoid IV articulating with external side of angular and prearticular.

69. Pleural ribs

Most actinistians lack pleural ribs; however, elongated, well-developed, ossified pleural ribs are present in some species. Among actinistians, pleural ribs are known in *Diplurus newarki* (Schaeffer 1941, 1952a), *D. longicaudatus* (Schaeffer 1948), *Changxingia aspratilis* (Wang & Liu 1981), *Chinlea sorenseni* (Schaeffer 1967), *Mawsonia gigas* (De-Carvalho 1982), and *Axelrodichthys araripensis* (Maisey personal communication, personal observation). Short, ossified pleural ribs occur in *Coelacanthus* and possibly *Laugia* (Moy-Thomas & Westoll 1935, Stensiö 1932). The presence of well-developed, ossified pleural ribs is considered to be derived.

    0 = absence of well-developed ribs;
    1 = presence of well-developed ribs.

70. Pectoral fin, number of lepidotrichia

The pectoral fins of actinistians are lobed and elongated. The number, size, and segmentation of the lepidotricha vary. Meristic values illustrate the variation: *Allenypterus montanus*: 6–8; *Axelrodichthys araripensis*: 12–15; *Caridosuctor populosus*: 13–15; *Coelacanthus granulatus*: 16; *Diplurus longicaudatus*: 19; *D. newarki*: 13; *Garnbergia ommata*: 18; *Hadronector donbairdi*: 11; *Holophagus penicillatus*: 13; *H. picenus*: 17; *Latimeria chalumnae*: 32–33; *Lochmocercus aciculodontus*: 12; *Po-*

lyosteorhynchus simplex: 8–9; *Rhabdoderma elegans*: 11; *Ticinepomis peyeri*: 17; *Trachymetopon*: 40. When the number of lepidotrichia varies intraspecifically, the coding for the species reflects the condition in the majority of the specimens examined. The range of variation in the number of lepidotrichia has been divided arbitrarily into five character-states according to the relative abundance of each count in the sample analyzed.

    0 = more than 50 lepidotrichia;
    1 = 30–49 lepidotrichia;
    2 = 20–29 lepidotrichia;
    3 = 10–19 lepidotrichia;
    4 = fewer than 10 lepidotrichia.

71. Pelvic fin, position

Andrews et al. (1967) proposed that the family Laugidae was defined by two species *Laugia groenlandica* and *Coccoderma suevicum*, and diagnosed by the position of the pelvic fins below, or anterior to, the pectoral fins. The apomorphic character-state of Character 43 (Cloutier 1990) is clarified with Character 71.

    0 = pelvic fin inserting in an abdominal position or more posteriorly;
    1 = pelvic fin inserting ventrally or anteriorly to the pectoral fins.

72. Lepidotrichia of first dorsal fin, spines

The lepidotrichia of the unpaired fins may bear one or two rows of spiny denticles. Martin & Wenz (1984) considered the absence of spines on the anterior lepidotrichia of the first dorsal fin in *Chinlea sorenseni*, '*Coelacanthus*' *lunzensis*, and *Garnbergia ommata* to be derived.

    0 = absence of denticles on anterior lepidotrichia of first dorsal fin;
    1 = presence of denticles on anterior lepidotrichia of first dorsal fin.

73. First dorsal fin, basal plate

Schaeffer (1941) reported that in most genera the basal plate of the first dorsal fin is triangular with the apex directed anteriorly; *Rhabdoderma* is an exception because the plate is 'kidney-shaped'. A hemispherical-shaped basal plate is known in additional Carboniferous actinistians (e.g., *Hadronector, Caridosuctor*). This character complements the description of Character 45 (see Cloutier 1990).

0 = hemispherical basal plate of the first dorsal fin;
1 = triangular basal plate of the first dorsal fin.

74. Second dorsal fin, number of lepidotrichia

The number of lepidotrichia on the second dorsal fin generally is used as a taxonomic feature to identify 'closely related' species within a genus, but never as a phylogenetic character. Every characteristic of an organism has the potential to carry phylogenetic information; therefore, the number of lepidotrichia in the second dorsal fin is used in this analysis as a phylogenetic character. The coding for intraspecifically variable species reflects the condition in the majority of the individuals observed or described in the literature. In the species analyzed, the number of lepidotrichia varies from six to 28 – e.g., *Allenypterus montanus*: 6–9; *Axelrodichthys araripensis*: 9–10; *Caridosuctor populosus*: 14–19; *Coelacanthus granulatus*: 18–20; *Diplurus longicaudatus*: 14; *D. newarki*: 13; *Hadronector donbairdi*: 9–16; *Holophagus grandis*: 13; *H. picenus*: 12; *H. penicellatus*: 20–21; *Latimeria chalumnae*: 32; *Lochmocercus aciculodontus*: 11; *Polyosteorhynchus simplex*: 7–10; *Ticinepomis peyeri*: 22–23. The FOG (*Miguashaia*) has 28 lepidotrichia in the second dorsal fin.

0 = more than 25 lepidotrichia;
1 = 20–25 lepidotrichia;
2 = 15–19 lepidotrichia;
3 = 10–14 lepidotrichia;
4 = fewer than 10 lepidotrichia.

75. Lepidotrichia

The structure and composition of the lepidotrichia vary among species of actinistians. Characteristics such as number of segments per lepidotrichia, and shape and size of the segments vary but are difficult to define in terms of phylogenetic characters. Character 75 deals with the general shape of the lepidotrichia. Generally, they are narrow (individual segment higher than broad); however, in some forms (e.g., *Holophagus*), the lepidotrichia are ovoid in profile, being relatively broad (individual segments are two to four times broader than high). Stensiö (1921, p. 130) pointed out that lepidotrichia usually become narrower from proximal to distal ends, except in some Mesozoic forms '. . . that expand first in distal direction, decreasing then towards the ends'. Differential preservation of elements could account for a compression of certain elements during fossilization, but one would expect other elements to be deformed or broaden (e.g., neural spines) which is not the case for the species coded with the apomorphic character-state.

0 = narrow lepidotrichia;
1 = broad lepidotrichia.

Phylogenetic patterns among actinistians

The strict and Adams consensus trees based on the two equally parsimonious trees found (253 steps, C.I. = 0.391) are given in Figure 3A. The difference between the two trees (Fig. 3B) is in the relative position of *Hadronector* and *Polyosteorhynchus*; (1) [[*Hadronector* + *Polyosteorhynchus*] + more advanced actinistians], and (2) [*Polyosteorhynchus* + [*Hadronector* + more advanced actinistians]]. The Mickevich's Consensus Information index for the strict and the Adams consensus trees is 0.708.

Because the main objectives of this paper are to determine patterns and trends during the evolution of the actinistians, I will not discuss the detailed character distribution. However, Appendix 2 provides the character-state distribution for one of the two trees; the tree selected is the one in which *Hadronector* is the sister-group of *Polyosteorhynchus*.

Phylogenetic relationships of Latimeria chalumnae

Frequently, the morphology of *L. chalumnae* has been interpreted with respect to the phylogenetic position of *Latimeria* among living sarcopterygians (Forey 1980, 1987, Trueb & Cloutier 1990). However, the morphology of living taxa is dependent on the relationships with other living relatives as well as the extinct, related forms. The phylogenetic position of *L. chalumnae* is addressed herein with respect to its close relatives among extinct actinistians.

Within the family Coelacanthidae, the following generic relationships are supported (Fig. 4D): [[[*Alcoveria* + *Diplurus*] + [*Chinlea* + [*Mawsonia* + *Axelrodichthys*]]] + [*Holophagus* + [*Macropoma* + *Latimeria*]]]. This topology is congruent with the hypotheses proposed by Forey (1981, 1984, 1991; cf. Fig. 4A) and similar to that of Forey (1988; cf. Fig. 4C). Maisey (1986) proposed a different scheme of relationships among these genera (Fig. 4B): [*Holophagus* + [[*Mawsonia* + *Axelrodichthys*] + [*Macropoma* + *Latimeria*]]].

Forey (1981, 1984) considered *Macropoma* to be the sister-group of *Latimeria* based on the presence of an ascending process of the prootic and anterior branches of the occipital commissure (= 'supratemporal commissure' of Forey). The same sister-group relationship is corroborated in this analysis based on two additional characters: quadrilateral subopeculum [32(0)] and presence of 30–49 lepidotrichia in pectoral fin [70(1)].

Forey (1984, p. 168) differentiated *Macropoma* from *Latimeria* based on five minor morphological differences: (1) *Latimeria* has skull roofing bones sunken below the skin surface whereas *Macropoma* has ornamented bones; (2) the cheek bones of *Macropoma* fit more closely together than in *Latimeria;* (3) the snout of *Latimeria* is covered by small ossicles, whereas that of *Macropoma* is covered by a single bone; (4) the 'swim bladder' of *Latimeria* is soft-walled, whereas that of *Macropoma* has a mineralized wall; and (5) minor morphometric and scale ornamentation differences. In addition to these differences, *Latimeria* differs from *Macropoma* in that the supraorbital canal pores are relatively larger (Character 3), and the premaxillae are fragmented rather than paired (Character 4). There are three preparietals per row instead of two (Character 8), and the postparietal is L-shaped rather than inverted-L-shaped (Character 12). The descending process of tabular is absent (Character 19), and there are more than seven extrascapulars rather than seven (Character 21). The articulation of the postspiracular and postorbital is loose (Character 23). The squamosal is

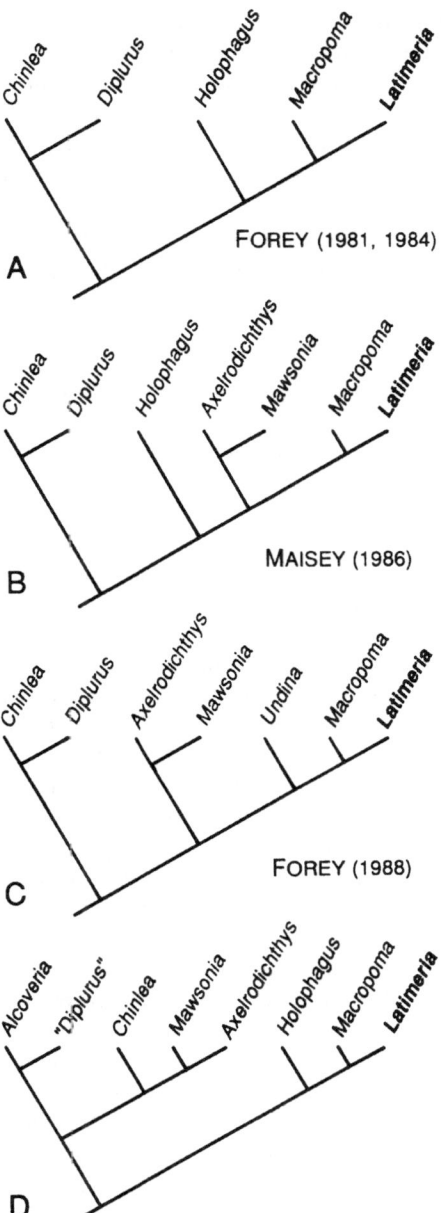

Fig. 4. Four hypotheses of interrelationships of advanced Coelacanthidae including *Latimeria*. A– Hypothesis of Forey (1981, 1984). B– Hypothesis of Maisey (1986). C– Hypothesis of Forey (1988). D– Hypothesis proposed in this analysis (cf. Fig. 3).

pentagonal rather than triangular (Character 25), and the operculum is quadrilateral rather than triangular (Character 28). The posteroventral process of the basal plate of the second dorsal fin is well developed (Character 48). The lateral parietal process is present (Character 58), and the lacrimojugal is not expanded anteriorly (Character 65). The second dorsal fin has more than 25, rather than 10–14 lepidotrichia (Character 74), and the lepidotrichia are not expanded (Character 75).

Evolutionary trends within the Actinistia

Most of the evolutionary studies of actinistians (Schaeffer 1941, 1948, 1952a, Schaeffer & Gregory 1961, Bjerring 1973, Jarvik 1980) do not provide information on the sequences of evolutionary changes nor on the congruence among diverse morphological patterns. Moreover, most of these studies make the following assumptions: (1) the stratigraphic sequence of the fossils reflects the phylogeny of the actinistians; (2) actinistians are conservative because of gross phenetic similarities; and (3) a higher taxon (e.g., Actinistia) evolves as a unit. Thereafter, three assumptions are required in order to summarize the evolutionary trends of a group (Cloutier 1990). First, only monophyletic groups can be studied in terms of phylogenetic trends. Although the monophyly of the actinistians is essential for the study, it should be noted that the *taxon* Actinistia does not evolve as an integrated unit; instead, it is the individual lineages of actinistians that evolve. Second, the most parsimonious distribution of characters and the topology of the taxa reflect the history of a monophyletic group. Third, the nested patterns of characters that determine the historical constraints are the result of evolutionary processes (but patterns have to be identified prior to the identification of trends and processes).

The monophyletic status of the Actinistia has never been questioned; however, the diagnostic features of the actinistians are controversial. Cloutier (1990) compiled a list of 11 actinistian synapomorphies: absence of (1) maxilla, (2) surangular, (3) branchiostegal rays, (4) submandibulars,

and (5) distal radials at the first and second dorsal fins; (6) presence of extracleithrum; (7) fusion of lacrimal with jugal; (8) pterygoid triangular; (9) dentary short, (10) angular highly profile; and (11) Coronoid IV oriented vertically. An additional five characters are potential synapomorphies for the actinistians, but their conditions are unknown or unclear in the basal taxa (Cloutier 1990): presence of (1) rostral organ, (2) ventral process of lateral rostral, (3) numerous supraorbitals, (4) tandem double-jaw articulation, and (5) posteriorly expanded U-shaped urohyal. These characters describe the basic historical constraints on the morphology of the group during the Middle or Late Devonian.

Transformation series used to study the interrelationships of the Actinistia and their evolution are either historically variable or conservative. Some morphological systems, such as the lower jaw (Lund et al. 1985), the appendicular skeleton, and the general body shape (Schaeffer 1948) are relatively conservative throughout the history of the actinistians. Alternatively, the two major trends in modification correspond to the (1) reduction and (2) increase in the number of elements. Jarvik (1980) determined a series of general evolutionary trends common to numerous groups of lower vertebrates; actinistians share four of these: (1) retrogressive development of the skeleton; (2) progressive development of the skeleton (e.g., development of ventral process on skull roof bones); (3) decrease in the number of dermal fin rays (and reduction to a 1 lepidotrichium: 1 radial ratio); and (4) reduction of basipterygoid process. These four trends are but a few of those present within the actinistians. The trends described below are categorized morphologically (cranial and postcranial) and discussed in terms of reduction, increase, and conservatism.

Cranial morphology
The evolutionary morphology of the head of actinistians has received much attention (Schaeffer 1952a, Schaeffer & Gregory 1961, Bjerring 1973, Lund & Lund 1985, Lund et al. 1985). According to the literature, there have been reductions in the numbers of bones of the (1) cheek region (Schaeff-

er 1952a, Jarvik 1964, 1980), (2) opercular and gular series (Jarvik 1963, McAllister 1968), and (3) posterolateral portion of the skull table (Andrews 1973). Because of the predominance of reductive trends, it has been suggested that paedomorphosis could be a primary process in the evolution of actinistians (Lund & Lund 1985, Forey 1988). Alternatively, there has been a trend to increase the number of elements belonging to a series or structural complex – e.g., the premaxillae, preparietals, and extrascapulars.

As mentioned by Jarvik (1964), the cheek region of actinistians has been modified through a reduction in the number of elements (fusion, loss, or both). Also the size of the cheek bones has been reduced so that none articulates with another. In some cases (e.g., *Coelacanthus granulatus*, *Axelia robusta*), the only ossified tissues present on the cheek are small tubes that surround the sensory canals (Westoll 1937). Historically, the postspiracular was reduced in size, followed by the fusion of the quadratojugal with the preopercular, and finally the postspiracular was lost.

Morphology of the skull roof also has changed during the evolution of the actinistians and shows trends toward both reductions and increases. Reduction involves the preparietals and the posterolateral bones of the skull table (i.e., intertemporal, supratemporal, and tabular). The plesiomorphic actinistian conditions of these two series of bones are a mosaic of dermal bones covering the snout region anterior to the paired parietals and a skull table of Type-X (i.e., intertemporal, supratemporal, and tabular bordering the postparietal laterally) rather than a Type-Y in which the postparietal has an inverted-L-shape formed by an anterolateral process (as proposed by Andrews 1973). After the differentiation of the preparietals into two rows, there is a decremental tendency toward reduction of the number of preparietals per row (from four to two). In the skull table, the intertemporal is lost relatively early in the history of the group; then the supratemporal is either lost or fused to the postparietal. Type-Y is achieved independently twice during the evolution of the Actinistia. Based on the trends observed in the dermal bones, one could assume that there is a relative

reduction in the size of the anterior part of the skull (i.e., pre-intracranial joint) in relation to the posterior part; however, Forey (1988) reported that through geological time, the ethmosphenoid part of the skull becomes relatively longer than the otico-occipital part.

The number of extrascapulars increases from three to nine, and the number of premaxillae increases from two to six. Among post-Carboniferous species, two different trends with respect to the extrascapulars are prevalent. (1) The number of extrascapulars increases to nine (in *Latimeria*) and each extrascapular is smaller and does not articulate with other dermal bones (e.g., *Diplurus*, *Latimeria*; Wenz 1975). Alternatively (2), the number of extrascapulars is reduced (to two or three) and the extrascapulars are incorporated in the skull table (e.g., *Chinlea*, *Axelrodichthys*, *Mawsonia*).

The reconstruction of the actinistian plesiomorphic condition and the evolution of the lateral-line system in actinistians have been the subject of various controversies (Stensiö 1947, Chang 1982, Hensel 1986, Schultze 1987, Northcutt 1989). Based on the phylogenetic position of the Actinistia among Sarcopterygii and the interrelationships of the actinistians proposed, the plesiomorphic condition of the sensory canal is as follows. The supraorbital canal connects with the otic canal, and the otic canal is located lateral to the postparietal. A series of three extrascapulars carries the occipital commissure. The anterior, middle, and posterior pit-lines are located on the postparietals (condition known in *Nesides schmidti*) whereas in more advanced species the anterior pit-line is absent. The oral canal is absent. The lateral-line system of the head in actinistians tends to become more complex. There is an incremental addition of sensory canal segments that branch off the main canals: (1) median branch of the otic canal; (2) posterior branch of the mandibular canal; and (3) anterior branches of the occipital commissure. In *Latimeria*, there is an hypertrophy of secondary collaterals (i.e., canaliculi) of the sensory canal (Hensel 1986). Numerous species are characterized by enlargement of the pores of the supraorbital canal; in contrast to the general assumption (Andrews 1973, Panchen & Smithson 1987), the presence of large pores is not a synapomorphy of actinistians (Cloutier 1990). In two Carboniferous genera (i.e., *Hadronector, Polyosteorhynchus*), there are bifurcating pores of the supraorbital canals. These conditions suggest that there is a general trend to increase the area covered by the mechanoreceptors of the lateral line. Complementary to the lateral-line system, there is the putative electroreception system. The presence of the rostral organ has been considered a distinctive feature of actinistians (Andrews 1973, Bjerring 1973, Forey 1981, 1984, Panchen & Smithson 1987); however, it is unclear if the basal taxon *Miguashaia* had a rostral organ. Presence of the rostral canal can be traced in some of the Devonian species (i.e., *Nesides heiligenstockiensis*) by the presence of pores for the superior and inferior posterior rostral tubes in the tectal. In *Hadronector*, there is intraspecific variation in the number of pores in the tectal (one or two).

In contrast to the trends for modification, the cranial morphology of actinistians also exhibits conservative features. Lund & Lund (1985) and Lund et al. (1985) explained the conservatism of the jaw apparatus as a response to a specialization in the feeding mechanism. A reorganization of the maxillary arcade, lower jaw, and suspensorium are the major early historical constraints characterizing the skull of actinistians. The basal historical constraints on the lower jaw involve the modification of Coronoid IV, fusion of the angular and surangular, tandem articulation, and reduction of the dentary. In parallel, the maxillary arcade is also modified; fusion of the lacrimal and jugal structurally replaces the maxilla, which is lost. Lauder (1980) demonstrated that in *Latimeria chalumnae* the primary feeding mechanism is suction. Lund et al. (1985) proposed that the cranial morphology of the actinistians is adapted for suction feeding; according to their study, the morphological characters involved in the suction feeding mechanism are the presence of an intracranial joint, a posteriorly projecting antotic process, thick and fleshy lips, a prominent coronoid flange at the corner of the lips, and long gular plates. Most of the characteristics allowing the special feeding mechanism were established in the Devonian.

Postcranial morphology

Schaeffer (1948) emphasized that the postcranial skeleton was conservative in response to body-locomotion constraints. The mode of locomotion is constrained possibly by the historical morphological changes rather than the converse. Historically, the conservative trend of the postcranial skeleton follows a series of three structural modifications: (1) development of a diphycercal tail associated with the presence of a supplementary caudal lobe; (2) a shift in the position of the pectoral (dorsally) and pelvic (anteriorly) fins; and (3) a change to a 1 : 1 ratio between the number of caudal lepidotrichia and radials (Schaeffer 1952a, Jarvik 1980). Thus, these characters are historical constraints that were stabilized during the early Carboniferous. Modifications of the postcranial skeleton occur with respect to the basal plates and lepidotrichia.

There is a trend towards an incremental increase in the number of processes on the basal plates of the unpaired fins. This historical, gradual increase in the number of processes seems to have provided complex and differentiated bases for individual muscle attachments on the basal plates. These muscle insertions (cf. Millot & Anthony [1958] for comparison with *Latimeria*) ultimately provided more dexterity and control of the unpaired lobed fins; Fricke et al. (1987) reported that the paired and unpaired lobed fins are able to generate forward thrust in the slow-swimming movements of *L. chalumnae*.

The primary modifications of the lepidotrichia during actinistian evolution are: (1) branched to unbranched ending, (2) spine-free lepidotrichia to spiny dorsal lepidotrichia, (3) broadening of lepidotrichia, (4) reduction in the number of lepidotrichia per radial in the caudal fin. Lundberg & Marsh (1976) suggested that *L. chalumnae* spends much of its time resting on (or moving on) its paired fins on the sea floor because the lepidotrichia are unbranched and the segments greatly foreshortened (as in the catostomid cyprinoids). Fricke et al. (1987) did not observe *Latimeria* to contact the substrate with its paired fins. The branching condition of the lepidotrichia in actinistians is an historical constraint (set during the Early Carboniferous), rather than a functional response to the behavior as suggested by Lundberg & Marsh (1976). There is a reduction in the number of lepidotrichia of the first dorsal fin (from more than 10 to 8), which stabilized to eight during the Triassic; the number of lepidotrichia of the second dorsal fin after an initial decrease shows an incremental increase.

Based on the characters used in this analysis, it is inferred that the caudal fin was modified relatively early during the evolution of the Actinistia (Late Devonian), and that during the Carboniferous, the mode of locomotion of early actinistians already was similar to that of *Latimeria chalumnae* (see Fricke et al. 1987).

In summary, actinistians are conservative with respect to some morphological systems (e.g., body shape, lower jaw), but also display numerous variations and trends of modifications (e.g., reduction and increase in the number of integrated elements). Even in historically conservative trends, transformations occur. A basic morphological design was established by the Early Carboniferous, but various alterations modified the basic design throughout the evolution of the group. Our idea of *Latimeria chalumnae* as a 'living-fossil' has to take into account that evolutionary modifications (or sequential historical constraints) have occurred through time (ca. 380 My), and that *L. chalumnae* is not a primitive actinistian (e.g., Balon et al. 1988, Balon 1991).

Rate of evolution

In order to address the question of the trends in rates of morphological evolution within the actinistians, interactions among evolutionary rates, time, cladogenesis, morphological changes, and diversity were investigated. The *time*-factor is considered in terms of both the geological time scale (i.e., *age*) and the estimated period of time separating two consecutive cladogenetic events (i.e., *duration*). The *cladogenesis*-factor refers to the rank of a cladogenetic event from the origin of the actinistian clade. The *change*- and *diversity*-factors relate to the number of character-state changes at a cladogenetic event and the number of nominal species occurring during a given period of time, respectively. The potential interactions among these six factors are quantified in terms of correlation. For the purpose of this study, the phylogenetic pathway is determined from the origin that corresponds to the stem-species of the actinistian clade, and *Latimeria chalumnae* – the selected terminal taxon. The selection of these two taxa allows one to investigate the evolution of the group in relation to *Latimeria*, and at the same time, to maximize the period of time covered by the species analysed in relation to the actual time span of the actinistian history.

Cladogenesis and time

It has been demonstrated that use of a stratigraphic argument in phylogenetics may result in incorrect polarization of characters and incorrect determination of primitive taxa (for review see Schoch 1986). Alternatively, the fossil record provides the only direct means for concluding that a phylogenetically interpreted cladogram may actually represent a statement about phylogenetic history (Hill & Camus 1986a, b).

Figure 2 illustrates the topology of the actinistians (Fig. 3A) combined with the actual time of occurrence of the fossils. In order to evaluate the fit between the phylogeny and the fossil record, cladogenetic and age ranks were compared. The adjusted cladogenetic ranks for the 31 actinistian species included in the analysis were plotted against the rank of the biostratigraphic age of these species (Fig. 5). A positive rank-correlation between the cladogenetic rank and the age rank was found ($r_s = 0.905$; $p < 0.01$). Although, this provides additional corroboration of the phylogenetic patterns, it is not considered to be a test for the accuracy of the topology as was proposed by Paul (1982) and Lazarus & Prothero (1984). A positive correlation is not a test of the topological accuracy because even though a positive correlation is expected from a topology that reflects the true evolutionary sequence (thus fits the stratigraphy), it is possible to

44

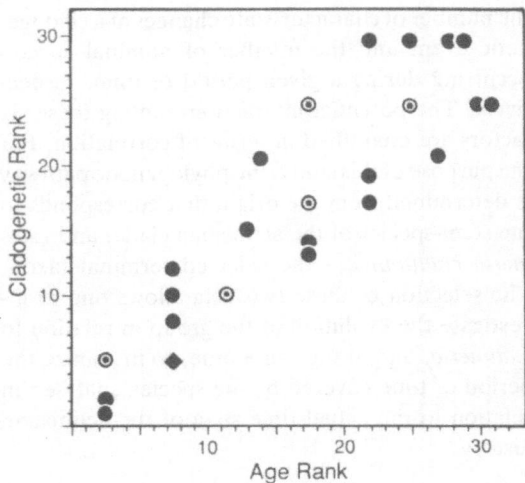

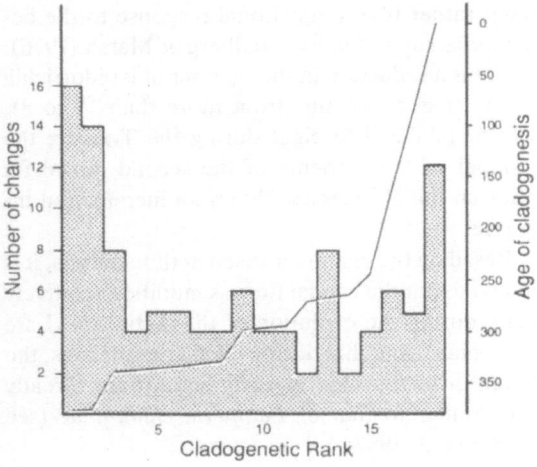

Fig. 5. Scatter diagram showing the relation between age rank and cladogenetic rank for the 30 cladogenetic events hypothesized in the actinistian cladogram (Fig. 3) ($r_s = 0.9278$). Full circles correspond to one observation, open circles with central dot to two observations, and diamond to three observations.

Fig. 6. Histogram of the number of morphological changes (left) and age of cladogenesis (right) per cladogenetic event.

create a branching sequence of unrelated taxa which would have a perfect stratigraphical fit.

The phylogenetic positions of *Rhabdoderma madagascariense* and *Laugia groenlandica* seem to be the least congruent with the biostratigraphic pattern of the tree. The period of time separating consecutive cladogenetic events is the greatest for post-Triassic species; this either reflects a natural process (corresponding to longer period of anagenesis) or is an artifact of the fossil record.

Cladogenesis and number of changes
Schaeffer (1952a) recognized 52 morphological character changes in actinistians during the Devonian, six or seven changes during the Early Carboniferous, three during the Early Permian, and two during the Triassic. Although Schaeffer (1952a) discussed the morphological trends, the characters were not listed clearly. In addition, it is likely that several of his Devonian changes are not historical changes at the level of the actinistians (cf. 'Characterization of the Actinistia' in Cloutier 1990). Furthermore, Westoll's (1949) and Schaeffer's (1952a,

b) approaches do not take into account the phylogeny of the group.

The distribution of the character-states on the cladogram was taken from the apomorphy list provided by PAUP. The distribution of two character-states (8[3] and 48[1]) were readjusted on the tree owing to their logical association with other characters (i.e., Character 8 with 7, and Character 48 with 47). The first cladogenetic event is characterized by the presence of the 11 synapomorphies reported by Cloutier (1990; including Character 40[1]) in addition to five other possible synapomorphies (cf. above 'Evolutionary trends within the Actinistia'). Coding for the latter five characters is unavailable for *Miguashaia;* although these five characters were added to the total number of morphological changes for the first cladogenetic event, discovery of additional specimens of *Miguashaia* may reveal that some (or all) of these five features characterize the second cladogenetic event.

There is a total of 101 character-states or changes (on a total of 253 steps) divided into 17 cladogenetic event (on a total of 30 cladogenetic events) along the phylogenetic pathway investigated in this study (Fig. 3A). Thus, there is an average of 5.941 character-states per cladogenetic event during the evolution of the group; the average increases to 6.278 if the 12 autapomorphies of *Latimeria chalumnae* are included and the *Latimeria*-branch is considered as the 18th event (anagenetic rather than cladogenetic). Figure 6 represents the histogram of the number of character-state changes per cladogenetic event along the phylogenetic pathway. The first cladogenetic event (Node 1) is characterized by the greatest number of changes (16); the second (Node 2) and last events (*Latimeria* branch) also are supported by numerous changes (14 and 12, respectively). None of the events is supported by fewer than two changes. There is no significant correlation between the number of changes and the cladogenetic rank ($r_s = -0.223$; $p > 0.05$), suggesting that the number of morphological changes is independent of the position of the node on the tree.

Cladogenesis, age, and number of changes
In addition to the comparison between the number of changes and the cladogenetic rank, Figure 6 provides a time scale for the age of cladogenesis for the 18 events. The age of cladogenesis is estimated from the position of a cladogenetic event on the temporal cladogram (Fig. 2). Most of the cladogenetic events (1–15) along the main pathway cover the first 130 million years of the history of the Actinistia. The numbers of changes do not fluctuate to match with the age of cladogenesis.

Rates, cladogenesis, and time
Figure 7 provides paired comparisons among the number of morphological changes per event, the duration between consecutive cladogenetic events, and the evolutionary rates. These paired comparisons are given in relation to the main phylogenetic pathway that is represented along the X-axis in terms of the cladogenetic rank of the 18 events. Figure 7A shows that, in general, the number of changes varies synchronously with the duration except for cladogenetic events 3, 12, and 13. As expected (because of mathematical dependence), Figure 7B shows that the number of changes varies synchronously with the rate of morphological evolution; however, event 9 and the last four cladogenetic events are exceptions. The relation between the duration and rates is completely asynchronous

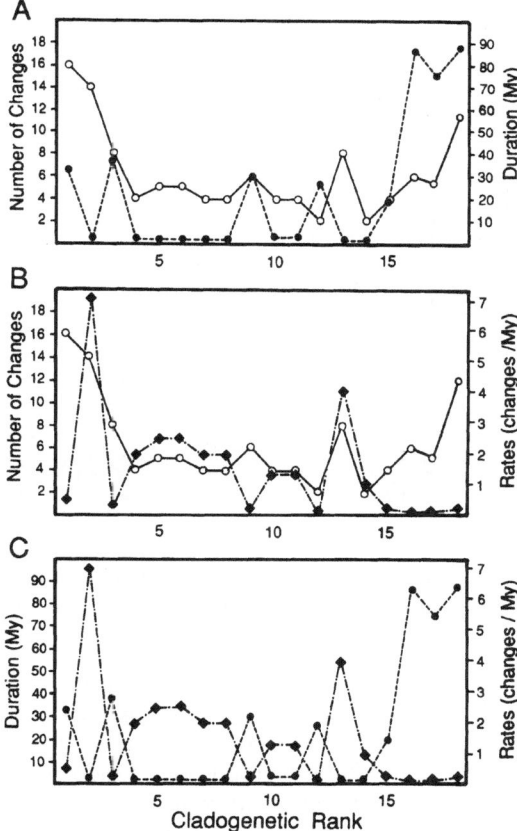

Fig. 7. Relationship among number of changes (open circles), time duration between cladogenetic events (full circles), and rates of morphological evolution (diamonds) with respect to the 18 cladogenetic events along the main phylogenetic pathway in terms of their cladogenetic rank. A– Comparison between number of morphological changes (number of character-states characterizing each event) and the duration (estimated period of time between two consecutive cladogenetic events), showing the partially asynchronous relation of the curves. B– Comparison between number of morphological changes and the rates$_c$, showing the synchroneity between the two curves. C– Comparison between duration and rates$_c$ showing the total asynchroneity of the curves.

(Fig. 7C). This pattern indicates a series of alternations of slow rates of evolution over long periods of cladogenetic time versus high rates over short periods of cladogenetic time. The period covered by cladogenetic events 4–8 is relatively stable for the three paired comparisons.

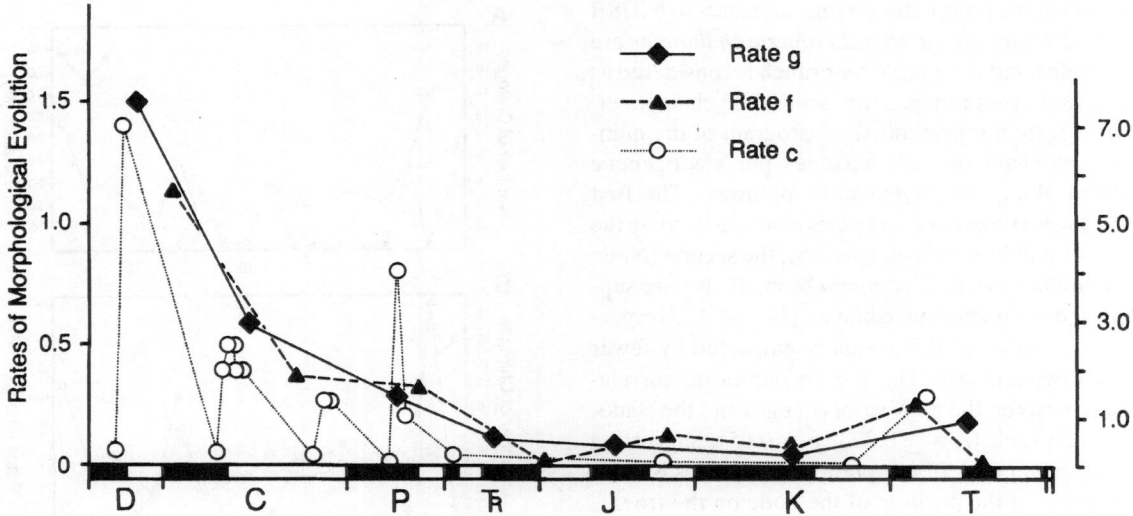

Fig. 8. Rates of morphological evolution of actinistians. Diamond-shaped plots correspond to the rate$_g$ calculated per geological period; the triangles correspond to rate$_f$ evaluated by period of 50 My; and the open circles correspond to rate$_c$ evaluated per cladogenetic events. Left scale for rate$_g$ and rate$_f$ and right scale for rate$_c$. Time scale of Palmer (1983).

The three modes of rate calculation (rate$_c$, rate$_g$, and rate$_f$; see above *'Calculations of rates'*) were calculated for the pathway stem-species – *Latimeria* and plotted along a proportional time axis (Fig. 8). The rates calculated with respect to the cladogenetic events correspond to a sequence of slow versus faster rates, and gradual versus quantum tempos. The greatest intercladogenetic rate$_c$ preceded the branching of the Diplocercidae during the Middle to Late Devonian. The only bradytelic period indicated by rate$_c$ covers the period between the Early Triassic and Late Cretaceous. The increase in rate in the Tertiary is a result of the long period of anagenesis for *Latimeria*. There is a general trend toward diminishment of the rate of evolution which is evidenced by the negative correlation between age and rate ranks ($r_s = -0.586$; p < 0.05).

The main bias of rate$_c$ is that the calculation is partially dependent on the number of species included in the analysis; however, rate$_g$ and rate$_f$ are not. The pattern generated using the rates$_f$ calculated from periods of 50 million years is a decreasing, stepwise stabilizing curve. The rate$_f$ curve indicates two distinct tempos: an initial reduction followed by a stabilizing low rate$_f$. Because of the number of

changes associated with *Latimeria chalumnae*, the rate is increased slightly during the past 70 million years. There is a significant negative correlation between age and rate ranks ($r_s = -0.732$; p < 0.05) indicating this decreasing trend. The curve generated for rate$_g$ is similar to the one for rate$_f$; however, a highly significant negative correlation between age and rate ranks was found ($r_s = -0.857$; p < 0.01).

In Figure 9, the curve of rate$_g$ can be compared with the curve proposed by Schaeffer (1952a, b); the time axis has been readjusted to correspond to that of Palmer (1983). However, the values of the rate provided by Schaeffer (1952a) have not been recalculated in function with the new time scale. The rate$_g$ curve inferred in this analysis differs from that of Schaeffer (1952a, b) in several respects. The initial magnitude of the calculated rates is higher (1.5 changes per My instead of no changes). The initial tachytelic tempo is absent. The greatest Frasnian rate is lower (1.5 instead of 3.7 changes My^{-1}). The Devonian decreasing tempo is not as drastic. A post-Triassic stabilizing bradytelic tempo is present, and finally, the rate increases in the Tertiary.

Rates and cladogenesis

Although there is no significant correlation between the number of changes and the period of time spanned between two cladogenetic events ($r = 0.404$; $p > 0.05$), there is a logarithmic relation between the rate and the duration. The rate$_c$ follows a logarithmic function of the time between two cladogenetic events (rate$_c = 4.3013 \star$ duration$^{-0.8966}$; $R^2 = 0.869$). This logarithmic relation indicates that these two variables are inversely correlated, and is a reflection of the presence of two antagonistic classes of tempos (i.e., faster and quantum versus slower and gradual) in the stabilizing phase of the trend.

Rates and diversity

The term 'diversity' refers to the total number of known actinistian species (i.e., nominal species and undetermined remains of problematic generic assignments) occurring during a given period of time. The data for the apparent specific diversity are taken from Cloutier & Forey (1991). In previous studies, the importance of taxic diversity in relation to evolutionary rates has been emphasized (Forey 1988) or negated (Schaeffer 1952a, b). According to Schaeffer (1952a, b) the greatest rate of evolution (between Late Devonian and Late Carboniferous) is not associated with the highest diversity of species (Triassic). However, Forey (1984) concluded that the '. . . rate of acquisition of characters has remained approximately the same throughout coelacanth history'. In contrast, Forey (1988) subsequently proposed that the times of maximum species diversity and of greatest rate of morphological change coincide in the Triassic. These diametrically opposed views concerning the relation between species diversity and morphological rates reflect different methodological approaches to the solution of the problem.

Of the 121 species of actinistians (Fig. 10; Cloutier & Forey 1991), 31 species (including *Latimeria chalumnae*) were included in the cladistic analysis. The greatest diversity occurred during the Scythian (Early Triassic). However, there is a general trend toward reduction of species diversity (not statistically significant); this is evidenced by a negative correlation between the number of species and

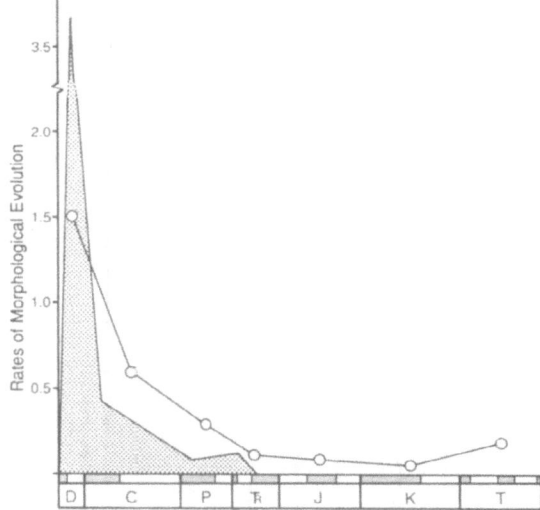

Fig. 9. Comparison of two curves of rate$_g$ of morphological evolution. Schaeffer's (1952a, b) hypothesis is represented as the curve deliniated by the shaded area. The open circles represent the empirical curve evaluated in this study. The time axis for Schaeffer (1952a, b) has been readjusted to the time scale of Palmer (1983), but the values for the rate$_g$ are as read from Schaeffer's graphic.

their geological and biostratigraphic occurrence (Cloutier & Forey 1991). This trend does not take into consideration fluctuations in diversity or gaps in the fossil record.

Before comparing the rates to the specific diversity, it is imperative to verify if the temporal diversity incorporated in the phylogenetic analysis is proportionally representative of the total apparent diversity. The temporal diversity included in the phylogenetic analysis was shown to be representative of the total temporal diversity in terms of proportional representation per geological period; this was tested by comparison of the diversity ranks in the two samples ($r_s = 0.881$; $p < 0.01$). However, there is no significant correlation between the rates of morphological evolution and the total apparent diversity (rate$_g$: $r_s = 0.071$, $p > 0.05$; Fig. 10). Thus, contrary to Forey's (1988) interpretation, the rate of morphological evolution within the actinistians is not correlated with the number of species occurring during a given period of time; despite the fact that both are decreasing trends

48

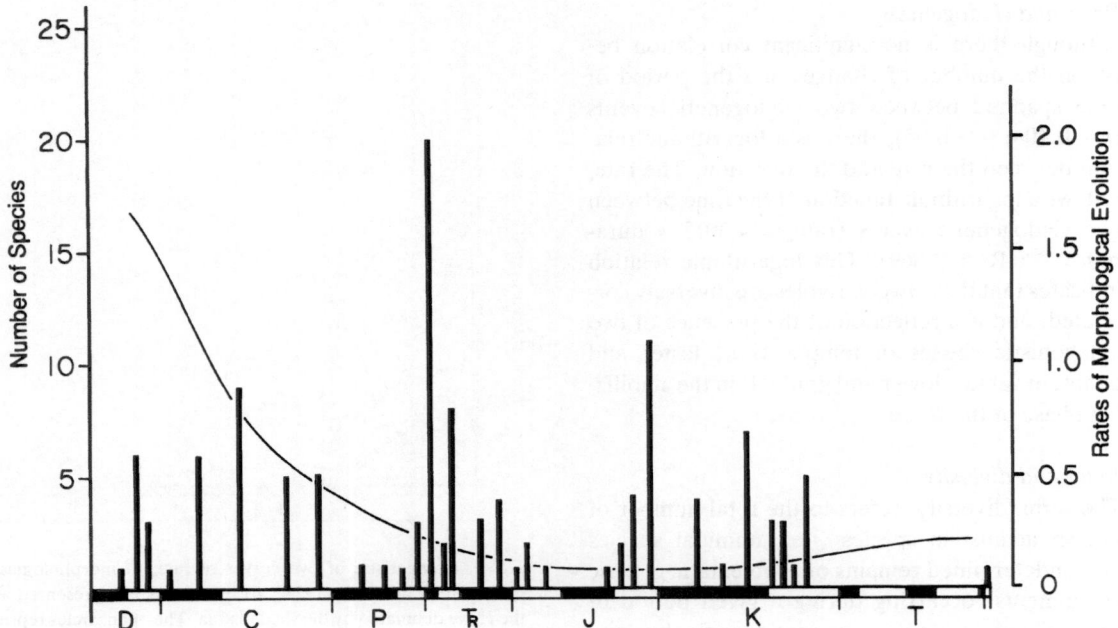

Fig. 10. Histogram of the number of nominal actinistian species through geological time (based on the data collected by Cloutier & Forey 1991) compared to the rates of morphological evolution. Each bar corresponds to a biostratigraphic period.

(diversity: $r_s = -0.452$; rate$_g$: $r_s = -0.857$), they are not dependent. The data on actinistians substantiate the conclusions of Vrba (1980) and Eldredge (1984) that there is no necessary correlation between amount of morphological changes and speciation.

There are a few prominent patterns and trends involving the interrelationships among the five basic components of coelacanth evolution – namely, time, diversity, cladogenesis, number of morphological changes, and rates of morphological evolution. There is a high positive correlation between the age of the fossil and its phylogenetic position, and between species diversity and phylogeny. These two positive correlations indicate the integration between cladogenesis and the pattern of diversity through time. Alternatively, there is a high negative correlation between (1) time and diversity, (2) time and number of changes, and (3) time and rates. These negative correlations correspond to two distinct reductive trends. First, there is a reduction in temporal diversity, and the sec-

ond, a diminishment in the rates of evolution. However, these two reductive trends are not correlated, because species diversity is not correlated with rates of morphological evolution.

Conclusions

Cracraft (1981) reviewed several problems associated with the traditional methods of measuring rates of morphological evolution, as follow: (1) assumption of ancestral – descendant relationships, (2) non-recognition of species as the evolutionary unit, (3) measure of rate between two taxa, and (4) use of quantitative characters. The method used in this study of rates of morphological evolution in actinistians differs from previous studies with respect to four points (including points 1, 2, and 4 of Cracraft 1981). First, sister-group relationships rather than ancestral – descendant relationships are hypothesized. Second, the species rather than the genus (Schaeffer 1952a, b, Forey 1984,

1988) were used as the basic evolutionary unit. Third, all morphological changes characterizing a cladogenetic event were used rather than solely the unique shared-derived character-states (Forey 1984), unpolarized characters (Schaeffer 1952a, b), or quantitative characters (e.g., Simpson 1944, 1949). Finally, the time used corresponds to the radiometric geological time scale of Palmer (1983).

Gould & Eldredge (1977) proposed that clades evolve primarily through punctuated equilibrium. The punctuated equilibrium model holds that very slow evolution is the norm and that rapid spurts of changes are rare but of great consequence (Eldredge & Gould 1972, Stanley 1985). The stabilizing rate$_c$ postulated for the actinistians could represent an alternation of gradualism and punctuated equilibria through geological time. The actinistian case seems to support Gould & Eldredge's (1977) model because most changes occur during the quantum tempos of actinistian evolution. Related to the punctuational view, Stanley (1984) pointed out that Simpson's notion of three tempos of evolution (i.e., tachytelic, horotelic, bradytelic) is not representative of the various patterns of evolution reported. The morphological evolution of the actinistians can be summarized in terms of a minimal number of patterns (i.e., rate$_c$: sequence of slow versus faster rates, and gradual versus quantum tempos; rate$_f$: decreasing, stepwise stabilizing curve; rate$_g$: stabilizing bradytelic tempo), rather than an exclusively bradytelic model.

The 121 actinistians recognized constitute a monophyletic group which is not as morphologically conservative as has been proposed. The phylogenetic patterns proposed herein are highly correlated with the fossil record. Based on the phylogenetic and morphological patterns, evolutionary tempos were inferred. The first tempo, a decreasing rate of morphological changes, occurred during the Devonian – Permian periods. This was followed by a stabilizing tempo during the Permian – Recent periods. In contrast to previous studies, the rate of morphological evolution within actinistians is shown not to be correlated with temporal species diversity.

Acknowledgements

L.S. Ford, R. Kaesler, H.-P. Schultze, L. Trueb, and E.O. Wiley read preliminary drafts of the manuscript and provided valuable comments. Special thanks are extended to P.L. Forey who reviewed the manuscript and with whom I discussed actinistian evolution. I thank the following curators for allowing me access to their collections: M. Arsenault (Musée d'Histoire naturelle de Miguasha, Miguasha, Canada), D.S. Berman (Carnegie Museum of Natural History, Pittsburgh, U.S.A.), H.C. Bjerring (Swedish Museum of Natural History, Stockholm, Sweden), P.L. Forey (British Museum [Natural History], London, UK), D. Goujet (Muséum national d'Histoire naturelle, Paris, France), L. Grande (Field Museum of Natural History, Chicago, U.S.A.), H. Jahnke (Institut und Museum für Geologie und Paläontologie, Göttingen, FRG), R. Laub (Buffalo Museum of Science, Buffalo, U.S.A.), and J.G. Maisey (American Museum of Natural History, New York, U.S.A.). The preparation of this paper has been supported by the Panorama Society Fund, Museum of Natural History, and the Graduate School of The University of Kansas, Lawrence. I thank also the Thomas J. Dee Fellowship Fund at the Field Museum of Natural History, Chicago, funding that allowed me to visit their collections.

References cited

Andrews, S.M. 1973. Interrelationships of crossopterygians. pp. 138–177 In: P.H. Greenwood, R.S. Miles & C. Patterson (ed.) Interrelationships of Fishes, Zool. J. Linn. Soc. Lond. 53, Suppl. 1.

Andrews, S.M., B.G. Gardiner, R.S. Miles & C. Patterson. 1967. Pisces. pp. 637–683. In: W.B. Harland, C.H. Holland, M.R. House, N.F. Hughes, et al. (ed.) The Fossil Record: A Symposium with Documentation, Geol. Soc. Lond.

Balon, E.K. 1991. Probable evolution of the coelacanth's reproductive style: lecithotrophy and orally feeding embryos in cichlid fishes and in Latimeria chalumnae. Env. Biol. Fish. 32: 249–265. (this volume)

Balon, E.K., M.N. Bruton & H. Fricke. 1988. A fiftieth anniversary reflection on the living coelacanth, Latimeria chalumnae: some new interpretation of its natural history and conservation status. Env. Biol. Fish. 23: 241–280.

50

Bemis, W.E. & T.E. Hetherington. 1982. The rostal [sic] organ of *Latimeria chalumnae:* morphological evidence of an electroreceptive function. Copeia 1982: 467–471.

Bjerring, H.C. 1973. Relationships of coelacanthiforms. pp. 179–205. *In:* P.H. Greenwood, R.S. Miles & C. Patterson (ed.) Interrelationships of Fishes, Zool. J. Linn. Soc. Lond. 53, Suppl. 1.

Chang, M.-M. 1982. The braincase of *Youngolepis*, a Lower Devonian crossopterygian from Yunnan, South-Western China. Dept. of Geology, Univ. of Stockholm, Stockholm. 113 pp.

Cloutier, R. 1990. Interrelationships of Palaeozoic actinistians: patterns and trends. *In:* M.M. Chang, G. Zhang & Y. Liu (ed.) Proceedings of the Fifth Symposium on Early Vertebrate Studies and Related Problems in Evolutionary Biology, Beijing, People's Republic of China, Oceanography Publ. House, Beijing. (In press).

Cloutier, R. & P.L. Forey. 1991. Diversity of extinct and living actinistian fishes (Sarcopterygii). Env. Biol. Fish. 32: 59–74. (this volume)

Cracraft, J. 1981. Pattern and process in paleobiology: the role of cladistic analysis in systematic paleontology. Paleobiology 7: 456–468.

Dagnelie, P. 1977. Théorie et méthodes statistiques. Applications agronomiques. Vol. 1, 2nd ed. Presses Agronomiques de Gembloux, Gembloux. 378 pp.

DeCarvalho, M.S.S. 1982. O genero *Mawsonia* na ictiofaunula do Cretaceo do Estado da Bahia. An. Acad. Bras. Cien. 54: 519–539.

Derstler, K. 1982. Estimating the rate of morphological change in fossil groups. pp. 131–136. *In:* B. Mamet & M.J. Copeland (ed.) Proc. Third North Amer. Paleontol. Convent., Vol. 1.

Dixon, W.J. & F.J. Massey. 1983. Introduction to statistical analysis. 4th ed. McGraw-Hill, New York. 678 pp.

Eastman, C.R. 1908. Devonian fishes of Iowa. Iowa Geol. Surv. Annual Report (1907) 18: 29–386.

Echols, J. 1963. A new genus of Pennsylvanian fish (Crossopterygii, Coelacanthiformes) from Kansas. Univ. Kans. Mus. Nat. Hist. Publ. 12: 475–501.

Eldredge, N. 1984. Simpson's inverse: bradytely and the phenomenon of living fossils. pp. 272–277. *In:* N. Etheridge & S.M. Stanley (ed.) Living Fossils, Springer, New York.

Eldredge, N. & S.J. Gould. 1972. Punctuated equilibia: an alternative to phyletic gradualism. pp. 82–115. *In:* T.J.M. Schopf (ed.) Models in Paleobiology, Freeman, Cooper & Co., San Francisco.

Elliott, D.K. 1987. A new specimen of *Chinlea sorenseni* from the Chinle Formation, Dolores River, Colorado. J. Arizona-Nevada Acad. Sci. 22: 47–52.

Forey, P.L. 1980. *Latimeria:* a paradoxical fish. Proc. R. Soc. Lond. 208B: 369–384.

Forey, P.L. 1981. The coelacanth *Rhabdoderma* in the Carboniferous of the British Isles. Palaeontology 24: 203–229.

Forey, P.L. 1984. The coelacanth as a living fossil. pp. 166–169. *In:* N. Etheridge & S.M. Stanley (ed.) Living Fossils, Springer, New York.

Forey, P.L. 1987. Relationships of lungfishes. pp. 75–91. *In:* W.E. Bemis, W.W. Burggren & N.E. Kemp (ed.) The Biology and Evolution of Lungfishes, J. Morphol. Suppl. 1.

Forey, P.L. 1988. Golden jubilee for the coelacanth *Latimeria chalumnae*. Nature 336 (6201): 727–732.

Forey, P.L. 1991. *Latimeria chalumnae* and its pedigree. Env. Biol. Fish. 32: 75–97. (this volume)

Fricke, H., O. Reinicke, H. Hofer & W. Nachtigall. 1987. Locomotion of the coelacanth *Latimeria chalumnae* in its natural environment. Nature 329 (6137): 331–333.

Gauthier, J., A.G. Kluge & T. Rowe. 1988. Amniote phylogeny and the importance of fossils. Cladistics 4: 105–210.

Gould, S.J. & N. Eldredge. 1977. Punctuated equilibria: the tempo and mode of evolution reconsidered. Paleobiology 3: 115–151.

Hennig, W. 1966. Phylogenetic systematics. University of Illinois Press, Urbana. 263 pp.

Hensel, K. 1986. Morphologie et interprétation des canaux et canalicules sensoriels céphaliques de *Latimeria chalumnae* Smith, 1939 (Osteichthyes, Crossopterygii, Coelacanthiformes). Bull. Mus. Natl. Hist. Nat. Paris, 4e sér. 8: 379–407.

Hill, C.R. & J.M. Camus. 1986a. Evolutionary cladistics of marattialean ferns. Bull. Br. Mus. nat. Hist. (Bot.) 14: 219–300.

Hill, C.R. & J.M. Camus. 1986b. Pattern cladistics or evolutionary cladistics? Cladistics 2: 362–375.

Jarvik, E. 1963. The composition of the intermandibular division of the head in fish and tetrapods and the diphyletic origin of the tetrapod tongue. K. Sven. VetenskapsAkad. Handl. 9: 1–74.

Jarvik, E. 1964. Specializations in early vertebrates. Ann. Soc. R. Zool. Belg. 94: 11–95.

Jarvik, E. 1980. Basic structure and evolution of vertebrates, Vol. 1 and 2. Academic Press, New York. 575 and 337 pp.

Lauder, G.V. 1980. On the evolution of the jaw adductor musculature in primitive gnathostome fishes. Breviora 460: 1–10.

Lazarus, D.B. & D.R. Prothero. 1984. The role of stratigraphic and morphologic data in phylogeny. J. Paleontol. 58: 163–172.

Lund, R. & W.L. Lund. 1985. Coelacanths from the Bear Gulch Limestone (Namurian) of Montana and the evolution of the Coelacanthiformes. Bull. Carnegie Mus. Nat. Hist. 25: 1–74.

Lund, W.L., R. Lund & G.A. Klein. 1985. Coelacanth feeding mechanisms and ecology of the Bear Gulch coelacanths. C.R. Neuv. Cong. Internatl. Stratigr. Géol. Carbonifère 5: 492–500.

Lundberg, J.G. & E. Marsh. 1976. Evolution and functional anatomy of the pectoral fin rays in cyprinoid fishes, with emphasis on the suckers (family Catostomidae). Amer. Midl. Nat. 96: 332–349.

Maddison, W.P., M.J. Donoghue & D.R. Maddison. 1984. Outgroup analysis and parsimony. Syst. Zool. 33: 83–103.

Maisey, J.G. 1986. Coelacanths from the Lower Cretaceous of Brazil. Amer. Mus. Novit. 2866: 1–30.

Martin, M. & S. Wenz. 1984. Découverte d'un nouveau Coelacanthidé, *Garnbergia ommata* n.g., n.sp., dans le Muschelkalk superieur du Baden-Württemberg. Stuttg. Beitr. Naturkd., Ser. B, 105: 1–17.

McAllister, D.E. 1968. The evolution of branchiostegals and associated opercular, gular, and hyoid bones and the classification of teleostome fishes, living and fossil. Bull. Can. Nat. Mus. 221: 1–239.

Menning, M. 1989. A synopsis of numerical time scales 1917–1986. Episodes 12: 3–5.

Millot, J. & J. Anthony. 1954. Tubes rostraux et tubes nasaux de *Latimeria* (Coelacanthidae). C.R. Séances Acad. Sci. 239: 1241–1243.

Millot, J. & J. Anthony. 1956. L'organe rostral de *Latimeria* (crossoptérygien coelacanthidé). Ann. Sci. nat. Zool. 18: 381–389.

Millot, J. & J. Anthony. 1958. Anatomie de *Latimeria chalumnae* 1. Squelette, muscles et formations de soutien. C.N.R.S., Paris. 122 pp.

Millot, J. & J. Anthony. 1965. Anatomie de *Latimeria chalumnae* 2. C.N.R.S., Paris. 130 pp.

Moy-Thomas, J.A. 1939. Palaeozoic fishes. Chemical Publishing Co. Inc., New York. 149 pp.

Moy-Thomas, J.A. & R.S. Miles. 1971. Palaeozoic fishes, 2nd ed. W.B. Saunders Company, Philadelphia. 259 pp.

Moy-Thomas, J.A. & T.S. Westoll. 1935. On the Permian coelacanth, *Coelacanthus granulatus* Ag. Geol. Mag. 72: 446–457.

Northcutt, R.G. 1989. The phylogenetic distribution and innervation of craniate mechanoreceptive lateral lines. pp. 17–78. *In:* S. Coombs, P. Görner & H. Munz (ed.) The Mechanosensory Lateral Line: Neurobiology and Evolution, Springer Verlag, New York.

Ørvig, T. 1986. A vertebrate bone from the Swedish Paleocene. Geol. Fören. Stockholm Förhandl. 108: 139–141.

Palmer, A.R. 1983. The decade of North American geology. 1983. Geologic time scale. Geology 11: 503–504.

Panchen, A.L. & T.R. Smithson. 1987. Character diagnosis, fossils and the origin of tetrapods. Biol. Rev. 62: 341–438.

Paul, C.R.C. 1982. The adequacy of the fossil record. pp. 75–117. *In:* K.A. Joysey & A.E. Friday (ed.) Problems of Phylogenetic Reconstruction, Academic Press, London.

Raup, D.M. 1987. Major features of the fossil record and their implications for evolutionary rate studies. pp. 1–14. *In:* K.S.W. Campbell & M.F. Day (ed.) Rates of Evolution, Allen & Enwin, London.

Rieppel, O. 1980. A new coelacanth from the Middle Triassic of Monte San Giorgio, Switzerland. Eclogae Geol. Helv. 73: 921–939.

Robineau, D. 1987. Sur la signification phylogénétique de quelques caractères anatomiques remarquables du coelacanthe *Latimeria chalumnae* Smith, 1939. Ann. Sci. natl., Zool., Paris, ser. 13, 8: 43–60.

Schaeffer, B. 1941. A revision of *Coelacanthus newarki* and notes on the evolution of the girdles and basal plates of the median fins in the Coelacanthini. Amer. Mus. Novit. 1110: 1–17.

Schaeffer, B. 1948. A study of *Diplurus longicaudatus* with notes on the body form and locomotion of the Coelacanthini. Amer. Mus. Novit. 1378: 1–32.

Schaeffer, B. 1952a. The Triassic coelacanth fish *Diplurus*, with observations on the evolution of the Coelacanthini. Bull. Amer. Mus. Nat. Hist. 99(2): 25–78.

Schaeffer, B. 1952b. Rates of evolution in the coelacanth and dipnoan fishes. Evolution 6: 101–111.

Schaeffer, B. 1953. *Latimeria* and the history of coelacanth fishes. Trans. N.Y. Acad. Sci., ser. 2, 15: 170–178.

Schaeffer, B. 1967. Late Triassic fishes from the western United States. Bull. Amer. Mus. Nat. Hist. 135(6): 285–342.

Schaeffer, B. & J.T. Gregory. 1961. Coelacanth fishes from the continental Triassic of the western United States. Amer. Mus. Novit. 2036: 1–18.

Schoch, R.M. 1986. Phylogeny reconstruction in paleontology. Van Nostrand Reinhold, New York. 353 pp.

Schultze, H.-P. 1973. Crossopterygier mit heterozerker Schwanzflosse aus dem Oberdevon Kanadas, nebst einer Beschreibung von Onychodontida-Resten aus dem Mitteldevon Spaniens und aus dem Karbon der USA. Palaeontographica (A) 143: 188–208.

Schultze, H.-P. 1987. Dipnoans as sarcopterygians. pp. 39–74. *In:* W.E. Bemis, W.W. Burggren & N.E. Kemp (ed.) The Biology and Evolution of Lungfishes, J. Morphol. Suppl. 1.

Simpson, G.G. 1944. Tempo and mode in evolution. Columbia University Press, New York. 237 pp.

Simpson, G.G. 1949. Rates of evolution in animals. pp. 205–228. *In:* G.L. Jepsen, E. Mayr & G.G. Simpson (ed.) Genetics, Paleontology and Evolution, Princeton University Press, Princeton.

Snedecor, G.W. & W.G. Cochran. 1980. Statistical methods, 7th ed. Iowa State University Press, Ames. 507 pp.

Stanley, S.M. 1975. A theory of evolution above the species level. Proc. Nat. Acad. Sci. USA 72: 646–650.

Stanley, S.M. 1984. Does bradytely exist? pp. 278–280. *In:* N. Etheridge & S.M. Stanley (ed.) Living Fossils, Springer, New York.

Stanley, S.M. 1985. Rates of evolution Paleobiology 11: 13–26.

Stensiö, E.A. 1921. Triassic fishes from Spitzbergen, Pt. 1. Holzhausen, Vienna. 307 pp.

Stensiö, E.A. 1932. Triassic fishes from East Greenland. Medd. Groenl. 83(3): 1–305.

Stensiö, E.A. 1947. The sensory lines and dermal bones of the cheek in fishes and amphibians. K. Sven. VetenskapsAkad. Handl. 3(24): 1–195.

Swofford, D.L. 1985. PAUP: Phylogenetic analysis using parsimony. User's manual. Illinois Natural History Survey, Champaign.

Trueb, L. & R. Cloutier. 1990. A phylogenetic investigation of the inter- and intrarelationships of the Lissamphibia (Amphibia: Temnospondyli). *In:* H.-P. Schultze & L. Trueb (ed.) Origins of Major Groups of Tetrapods: Controversies and Consensus, Cornell University Press, Ithaca (in press).

Vrba, E.S. 1980. Evolution, species and fossils: how does life evolve? S. Afr. J. Sci. 76: 61–84.

Wang, N. & H. Liu. 1981. Coelacanth fishes from the marine

Permian of Zhejiang, South China. Vertebr. PalAsia. 19(4): 305–312. (In Chinese).

Watrous, L.E. & Q.D. Wheeler. 1981. The out-group comparison method of character analysis. Syst. Zool. 30: 1–11.

Wenz, S. 1975. Un nouveau coelacanthidé du Crétacé inférieur du Niger, remarques sur la fusion des os dermiques. Colloq. Int. C.N.R.S. 218: 175–190.

Westoll, T.S. 1937. On the cheek-bones in teleostome fishes. J. Anat. 71: 362–382.

Westoll, T.S. 1949. On the evolution of the Dipnoi. pp. 121–184. *In:* G.L. Jepsen, E. Mayr & G.G. Simpson (ed.) Genetics, Paleontology and Evolution, Princeton University Press, Princeton.

Appendix 1

Data matrix of 75 osteological characters for 31 actinistian species. 0 = plesiomorphic state; 1–7 = apomorphic states; L = logical impossibility coding; N = non-available coding; U = unclear condition. Unordered characters are: 5, 7, 8, 12, 20, 21, 23, 25, 30, 44, 70, and 74. Unpolarized characters are: 9, 45, and 50. Description of the characters and character states in text. ALCOVBR = *Alcoveria brevis*, ALLENMO = *Allenypterus montanus*, AXELIRO = *Axelia robusta*, AXELRAR = *Axelrodichthys araripensis*, CARIDPO = *Caridosuctor populosus*, CHINLSO = *Chinlea sorenseni*, COELAGR = *Coelacanthus granulatus*, DIPLOKA = *Diplocercides kayseri*, DIPLULO = *Diplurus longicaudatus*, DIPLUNE = *D. newarki*, GARNBOM = *Garnbergia ommata*, HADRODO = *Hadronector donbairdi*, HOLOPPE = *Holophagus penicillata*, LATIMCH = *Latimeria chalumnae*, LAUGIGR = *Laugia groenlandica*, LOCHMAC = *Lochmocercus aciculodontus*, MACROMA = *Macropoma mantelli*, MAWSOGI = *Mawsonia gigas*, MIGUABU = *Miguashaia bureaui*, NESIDSC = *Nesides schmidti*, NESIDHE = *Nesides heiligenstockiensis*, POLYOSI = *Polyosteorhynchus simplex*, RHABDEL = *Rhabdoderma elegans*, RHABDEX = *Rhabdoderma exiguum*, RHABDMA = *Rhabdoderma madagascariense*, RHABSP1 = *Rhabdoderma* sp. 1 of Forey 1981, SYNAPNE = *Synaptotylus newelli*, TICINPE = *Ticinepomis peyeri*, WHITEWO = *Whiteia woodwardi*, WHITETU = *W. tuberculata*, WIMANSI = *Wimania sinuosa*.

| | | | | | | | | | | | Characters | | | | | | | | |
Taxa	1	2	3	4	5	6	7	8	9	10	11	12	13	14	15	16	17	18	19
ALCOVBR	1	N	1	0	0	N	N	N	0	N	N	2	N	N	N	N	1	1	N
ALLENMO	1	1	1	0	2	1	2	3	0	0	N	2	0	0	1	0	1	0	0
AXELIRO	1	1	1	N	N	N	2	5	1	0	1	1	1	0	1	0	1	1	1
AXELRAR	1	1	0	0	2	1	2	1	0	1	1	0	1	1	0	1	1	1	0
CARIDPO	1	1	1	0	1	0	2	2	0	0	N	0	0	1	1	1	1	0	1
CHINLSO	1	1	0	0	2	1	2	0	0	0	1	1	1	1	0	N	1	1	0
COELAGR	1	0	N	1	2	1	2	2	0	0	N	1	N	0	1	1	1	1	1
DIPLOKA	1	0	0	N	N	N	0	L	0	0	0	1	0	1	0	0	1	1	0
DIPLULO	1	1	1	N	N	N	N	N	N	N	N	1	N	N	0	N	1	1	N
DIPLUNE	1	1	1	0	2	1	2	2	0	0	1	1	1	1	0	1	1	1	0
GARNBOM	1	1	0	0	2	N	2	N	0	1	1	1	1	U	N	N	1	1	1
HADRODO	1	1	1	0	0	0	3	3	1	0	N	2	0	0	0	0	1	0	0
HOLOPPE	1	1	0	0	1	1	2	1	0	0	N	0	N	1	1	1	1	1	1
LATIMCH	1	0	1	1	2	1	2	1	0	0	1	0	1	1	1	1	1	1	0
LAUGIGR	1	1	1	0	0	0	2	N	0	0	1	1	0	1	0	1	1	1	1
LOCHMAC	1	N	1	0	2	1	2	N	0	0	N	2	N	1	1	0	1	0	0
MACROMA	1	0	0	0	2	1	2	2	0	0	1	1	1	1	N	1	1	1	1
MAWSOGI	1	N	N	N	N	N	2	2	0	0	1	1	1	1	0	1	1	1	1
MIGUABU	0	0	0	0	0	0	0	L	U	N	0	0	0	0	0	0	0	0	0
NESIDSC	1	N	0	N	N	N	U	U	0	0	0	1	0	1	N	0	1	1	0
NESIDHE	1	N	0	0	2	U	N	N	0	N	N	1	N	1	0	0	1	1	0
POLYOSI	1	1	1	0	2	0	2	0	1	0	N	2	N	0	0	0	1	0	0
RHABDEL	1	1	1	0	1	0	2	2	0	1	1	1	0	0	1	0	1	0	1
RHABDEX	1	1	N	0	N	N	2	N	N	N	N	0	0	N	N	0	N	N	N
RHABDMA	1	0	U	N	N	N	N	N	N	N	N	2	N	1	1	N	1	1	1
RHABSP1	1	1	0	0	1	0	2	2	0	1	1	2	0	1	0	1	1	1	1
SYNAPNE	1	1	N	N	N	N	2	N	0	N	N	N	1	N	N	N	1	1	0
TICINPE	1	1	0	0	0	1	N	N	0	0	N	2	N	0	0	N	1	1	N
WHITEWO	1	1	1	0	1	0	2	2	0	1	1	1	1	1	1	1	1	1	1
WHITETU	1	0	1	N	N	N	2	2	0	0	N	1	N	1	N	1	1	1	1
WIMANSI	1	1	0	N	N	N	2	5	0	0	1	1	1	1	N	0	1	1	1

Appendix 1. Continued.

	Characters																	
Taxa	20	21	22	23	24	25	26	27	28	29	30	31	32	33	34	35	36	37	38
ALCOVBR	N	N	N	1	1	N	N	N	0	1	L	L	L	N	N	N	N	1	1
ALLENMO	1	1	0	0	0	2	0	0	0	0	0	1	1	0	0	1	N	0	1
AXELIRO	2	3	1	2	1	3	1	L	1	1	L	L	L	0	0	1	0	N	N
AXELRAR	2	0	1	2	1	1	1	1	0	N	N	N	N	0	1	1	1	0	U
CARIDPO	0	1	0	0	0	1	0	0	1	0	0	1	1	1	0	0	N	0	1
CHINLSO	1	5	1	1	1	1	1	0	1	1	L	L	L	0	N	N	1	0	1
COELAGR	2	3	1	2	1	U	0	L	1	1	L	L	L	0	0	1	1	0	1
DIPLOKA	2	0	0	1	N	2	N	0	0	N	N	N	N	0	0	0	0	1	1
DIPLULO	N	2	N	N	N	N	N	0	1	N	N	N	N	0	N	N	N	1	1
DIPLUNE	2	2	1	1	1	0	1	U	0	1	L	L	L	0	1	1	1	1	0
GARNBOM	N	N	1	1	1	0	1	N	1	1	L	L	L	0	N	N	N	1	N
HADRODO	0	0	0	0	0	0	0	0	0	0	0	1	1	0	0	N	N	1	1
HOLOPPE	N	2	1	1	1	0	1	1	1	0	0	1	1	0	1	1	1	0	0
LATIMCH	2	3	1	2	1	2	1	1	0	0	2	1	0	0	1	1	1	1	1
LAUGIGR	2	1	0	2	1	0	1	1	1	1	L	L	L	0	N	1	1	0	1
LOCHMAC	0	1	0	0	0	2	0	0	1	0	0	1	0	0	N	N	1	0	1
MACROMA	2	2	1	1	1	0	1	1	1	0	2	1	0	0	1	1	1	1	1
MAWSOGI	2	4	1	2	1	0	1	0	1	1	L	L	L	0	1	N	1	N	1
MIGUABU	0	0	U	0	N	N	0	N	0	0	0	0	0	N	0	0	N	0	0
NESIDSC	2	N	0	1	N	0	0	0	0	0	2	1	1	0	0	0	0	1	1
NESIDHE	N	0	0	1	0	0	0	0	0	0	0	1	1	N	N	N	N	0	0
POLYOSI	1	0	0	0	0	1	0	0	0	0	0	0	0	N	0	1	N	0	1
RHABDEL	0	1	0	0	1	0	0	0	1	N	N	N	N	0	0	1	1	1	1
RHABDEX	N	N	N	N	N	N	N	N	1	N	N	N	N	0	0	0	N	0	1
RHABDMA	N	1	0	1	1	2	1	0	1	0	1	1	1	N	N	N	N	N	N
RHABSP1	2	1	0	0	1	2	1	0	1	0	1	1	1	0	0	1	1	0	0
SYNAPNE	N	N	N	N	N	N	N	0	1	0	N	N	1	0	N	1	0	0	1
TICINPE	2	4	1	2	1	3	1	N	0	1	L	L	L	1	N	N	N	1	1
WHITEWO	2	1	0	1	1	1	1	0	0	0	2	1	1	0	0	1	1	1	1
WHITETU	2	N	1	1	1	1	0	0	1	0	2	1	U	N	N	N	N	1	1
WIMANSI	N	N	1	2	1	1	1	0	1	1	L	L	L	0	0	1	0	1	1

Appendix 1. Continued.

	Characters																	
Taxa	39	40	41	42	43	44	45	46	47	48	49	50	51	52	53	54	55	56	57
ALCOVBR	1	1	0	1	1	2	N	N	N	N	N	N	1	1	1	1	N	N	N
ALLENMO	1	1	0	1	1	0	0	0	0	L	0	1	1	1	0	1	1	0	0
AXELIRO	N	N	N	1	N	N	N	N	N	N	N	N	1	N	1	1	N	1	0
AXELRAR	1	1	0	1	1	1	0	1	1	0	0	1	1	1	1	1	N	1	0
CARIDPO	1	1	1	1	1	0	1	1	1	1	0	U	1	1	1	1	1	0	0
CHINLSO	1	1	0	1	1	2	1	1	1	0	0	1	1	N	1	1	0	1	0
COELAGR	1	1	0	1	1	1	0	1	1	0	0	0	1	1	1	1	0	0	0
DIPLOKA	L	0	L	N	N	N	N	N	N	N	N	N	1	N	0	1	N	0	0
DIPLULO	1	1	0	1	1	4	0	1	1	0	0	1	1	1	1	1	N	1	0
DIPLUNE	1	1	0	1	1	2	0	1	1	0	1	1	1	1	1	1	0	1	0
GARNBOM	N	N	N	1	1	2	0	1	1	1	N	N	1	N	N	1	N	1	0
HADRODO	1	1	0	1	1	0	0	1	0	L	0	1	1	1	1	1	1	0	0
HOLOPPE	1	1	0	1	1	1	0	1	0	L	1	1	1	1	1	1	0	0	0
LATIMCH	1	1	0	1	1	2	0	1	1	1	0	1	1	1	1	1	0	1	0
LAUGIGR	L	0	0	1	1	1	0	1	1	1	1	0	1	1	1	1	1	0	0
LOCHMAC	0	1	1	1	1	3	1	0	0	L	N	N	1	1	0	1	N	0	0
MACROMA	1	1	0	1	1	2	0	1	1	0	0	1	1	1	1	1	0	1	0
MAWSOGI	N	N	N	N	N	N	N	N	N	N	N	N	N	N	N	N	N	1	0
MIGUABU	0	1	0	0	0	0	0	N	N	N	N	N	0	0	0	0	0	0	0
NESIDSC	N	N	N	N	N	N	N	N	N	N	N	N	N	N	N	N	N	0	0
NESIDHE	L	0	L	0	0	0	N	N	N	N	N	N	1	N	0	1	N	0	0
POLYOSI	1	1	1	1	1	3	1	0	0	L	0	1	1	1	1	1	0	0	0
RHABDEL	1	1	0	1	1	1	1	1	0	L	N	N	1	1	1	1	1	0	0
RHABDEX	1	1	0	1	1	0	1	N	0	N	N	N	1	1	1	1	N	0	N
RHABDMA	N	N	N	1	1	1	N	1	0	L	N	N	1	1	1	1	N	0	0
RHABSP1	1	1	0	1	1	1	1	1	0	L	0	1	1	1	1	1	1	0	0
SYNAPNE	N	N	N	N	1	N	0	1	N	N	N	N	1	1	1	1	N	0	0
TICINPE	1	1	0	0	1	2	0	N	N	N	N	N	1	1	1	1	N	1	1
WHITEWO	1	1	0	1	1	2	N	N	N	N	N	N	1	1	1	1	N	1	0
WHITETU	1	1	1	1	1	2	0	N	N	N	N	N	1	1	1	1	N	1	0
WIMANSI	N	N	N	N	N	N	N	N	N	N	N	N	N	N	N	N	N	1	0

Appendix 1. Continued.

	Characters																	
Taxa	58	59	60	61	62	63	64	65	66	67	68	69	70	71	72	73	74	75
ALCOVBR	N	N	N	1	0	0	0	N	N	N	0	0	3	0	1	N	3	0
ALLENMO	0	0	0	0	0	0	0	0	0	N	0	0	4	0	0	0	4	0
AXELIRO	1	0	0	1	0	1	N	N	N	N	0	0	N	N	N	N	N	0
AXELRAR	0	1	0	1	0	0	0	1	1	1	1	1	3	0	1	1	4	0
CARIDPO	0	0	0	0	0	0	0	0	0	N	0	0	3	0	0	0	2	0
CHINLSO	0	1	0	1	0	0	0	1	1	0	0	1	2	0	0	1	3	0
COELAGR	1	0	0	1	0	1	0	0	0	0	0	0	3	0	1	1	2	0
DIPLOKA	0	0	0	0	0	N	N	N	N	N	0	0	N	0	N	N	N	0
DIPLULO	N	0	N	1	0	0	0	0	1	N	0	1	3	0	1	1	3	0
DIPLUNE	0	0	0	1	0	0	0	0	1	0	0	1	3	0	1	1	3	0
GARNBOM	1	0	N	1	0	0	0	1	1	N	0	0	3	0	0	1	2	0
HADRODO	0	0	0	0	0	0	0	0	0	N	0	0	3	0	0	1	3	0
HOLOPPE	0	0	0	0	0	0	0	1	N	0	0	0	3	0	1	1	1	1
LATIMCH	1	0	1	0	0	0	0	0	1	1	0	0	1	0	1	1	0	0
LAUGIGR	0	0	0	0	0	0	0	0	0	N	0	0	3	1	0	1	2	0
LOCHMAC	0	0	0	0	0	0	0	0	0	N	0	0	3	0	0	0	3	0
MACROMA	0	0	1	0	0	0	0	1	1	1	0	0	N	0	1	1	3	1
MAWSOGI	1	1	0	1	1	0	1	1	N	1	1	0	N	N	N	N	N	N
MIGUABU	0	0	0	0	0	0	0	0	0	N	0	0	0	0	0	0	0	0
NESIDSC	0	0	0	0	0	0	0	0	0	0	0	0	N	N	N	N	N	0
NESIDHE	0	0	0	0	0	0	0	U	0	N	0	0	N	0	0	N	3	0
POLYOSI	0	0	0	0	0	0	0	0	0	N	0	0	4	0	0	0	4	0
RHABDEL	0	0	0	0	0	0	0	0	0	N	0	0	3	0	0	N	3	0
RHABDEX	N	N	N	0	N	N	0	0	0	N	0	0	3	0	0	0	2	0
RHABDMA	0	0	0	0	0	0	0	0	N	N	0	0	N	0	N	1	2	0
RHABSP1	0	0	0	0	0	0	0	1	0	0	0	0	3	0	0	0	3	0
SYNAPNE	0	0	0	0	N	0	N	N	N	N	0	0	N	0	N	N	N	0
TICINPE	0	0	N	1	0	1	0	0	1	N	0	0	3	0	1	1	1	1
WHITEWO	0	0	0	0	0	0	0	0	0	N	0	0	3	0	0	1	3	0
WHITETU	0	0	0	0	0	0	0	0	0	N	0	0	N	0	N	1	3	0
WIMANSI	1	0	0	0	0	0	0	0	1	0	0	0	N	N	N	N	N	N

Appendix 2

Diagnostic information for the nodes and terminal taxa illustrated in Figure 2. The taxonomic names are as in Appendix 1 and the characters and character states are as in Table 1. The character states are provided in parentheses following the character number. Character with an asterix indicates uniquely shared derived characters.

Node	Apomorphy list
1 (Actinistia)	40(1) + (10 synapomorphies and 5 potential synapomorphies)
2	1*(1), 5(2), 12(2), 17*(1), 31(1), 32(1), 38(1), 39(1), 51*(1), 52*(1), 54*(1), 55(1), 70(4), 74(3)
3	2(1), 3(1), 7*(2), 11*(1), 35(1), 36(1), 42(1), 43*(1)
4	6(1), 15(1), 21(1), 25(2)
5	8(2), 14(1), 28(1), 45(1), 70(3)
6	5(1), 6(0), 19(1), 46(1), 53(1)
7	10(1), 24*(1), 30(1), 44(1)
8	16(1), 18(1), 20(2), 26(1)
9	10(0), 23(1), 45(0), 50(0), 73(1), 74(2)
10	12(1), 25(0), 30(2), 47(1)
11	13*(1), 22(1), 44(2), 55(0)
12	37(1), 56(1)
13	3(0), 5(2), 6(1), 21(4), 29(1), 61(1), 65(1), 66(1)
14	48(0), 72(1)
15	21(2), 34*(1), 50(1), 74(3)
16	8(1), 12(0), 27(1), 29(0), 61(0), 75(1)
17	2(0), 32(0), 60*(1), 67(1), 70(1)
A (Diplocercidae)	12(1), 14(1), 18(1), 20(2), 23(1), 40(0)
Node B	30(2), 37(1)
C (Hadronectoridae)	9(1), 53(1)
Node D	12(0), 25(1), 35(0), 74(2)
Node E	25(1), 74(3)
Node F	14(0), 23(2), 25(3), 63(1), 65(0)
Node G	4(1), 21(3), 44(1), 58(1)
Node H	8(5), 16(0), 36(0)
Node I	15(0), 19(0), 69(1)
Node J	3(1), 65(0)
Node K	28(0), 49(1)
Node L	21(5), 25(1), 37(0), 59*(1)
Node M	21(4), 23(2), 44(1), 67(1), 68*(1), 74(4)
ALCOBRE	5(0), 12(2), 69(0)
ALLENMO	20(1), 74(4)
AXELRAR	8(1), 10(1), 12(0), 21(0), 27(1), 28(0)
AXELROB	3(1), 9(1)
CARIDPO	16(1), 33(1), 41(1), 47(1)
CHINSOR	8(0), 20(1), 45(1), 70(2), 72(0)
COELAGR	2(0), 26(0), 37(0), 56(0), 66(0)
DIPLOKA	25(2)
DIPLULO	44(4)
DIPLUNE	38(0)
GARNBOM	10(1), 58(1)
HADRODO	5(0), 7*(3), 37(1), 46(1), 70(3), 73(1)
HOLOPPE	5(1), 30(0), 37(0), 38(0), 44(1), 47(0), 49(1), 56(0), 74(1)
LATIMCH	3(1), 4(1), 19(0), 21(3), 23(2), 25(2), 28(0), 48(1), 58(1), 65(0), 74(0), 75(0)
LAUGIGR	5(0), 15(0), 23(2), 27(1), 29(1), 40(0), 49(1), 71*(1)
LOCHMAC	32(0), 39(0), 41(1), 44(3)

58

Appendix 2. Continued.

MACRMAN	8(2), 12(1)
MAWSGIG	19(1), 25(0), 58(1), 62*(1), 64*(1), 69(0)
NESIDHE	38(0)
POLYOSI	8(0), 20(1), 25(1), 31(0), 32(0), 41(1), 44(3), 45(1), 55(0), 74(4)
RHABDEL	12(1), 14(0), 25(0), 37(1)
RHABDMA	2(0)
RHABSP1	3(0), 15(0), 38(0), 65(1)
SYNAPNE	19(0), 36(0)
TICINPE	5(0), 12(2), 15(0), 28(0), 33(1), 42(0), 57*(1), 74(1), 75(1)
WHITEWO	10(1), 22(0), 28(0)
WHITETU	2(0), 26(0), 41(1)
WIMANSI	14(1), 25(1), 61(0), 63(0)

Environmental Biology of Fishes **32**: 59–74, 1991.
© 1991 *Kluwer Academic Publishers.*

Diversity of extinct and living actinistian fishes (Sarcopterygii)

Richard Cloutier[1,2] & Peter L. Forey[2]

[1] *Museum of Natural History, Department of Systematics and Ecology, The University of Kansas, Lawrence, KS 66045-2454, U.S.A.*

[2] *Department of Palaeontology, British Museum (Natural History), Cromwell Road, London SW7 5BD, England*

Received 15.4.1990   Accepted 11.8.1990

Key words: Coelacanth, Fossil record, Species diversity

Synopsis

A total of 121 actinistian species belonging to 47 genera and 17 undetermined actinistians is reported from the literature. There are 69 valid species with fair assessment of their phylogenetic position; 21 valid species with poor assessment of their phylogenetic position; 31 actinistian incertae sedis; and 18 taxa that had been identified incorrectly as actinistians or are nomen nuda. The fossil record of the actinistians covers a history of approximately 380 million years. The greatest diversity occurred during the Scythian (Early Triassic).

Introduction

The Actinistia (= Coelacanthi, Coelacanthia, Coelacanthiformes, Coelacanthii, Coelacanthina), represented today by a single living species, *Latimeria chalumnae* Smith, has a long fossil record beginning with the Middle Devonian *Euporosteus eifelianus* (Givetian) and ending with the Late Cretaceous *Macropoma lewesiensis* (Santonian). Most of us are familiar with the past 65 million year stratigraphic gap (i.e., lack of fossils). Nevertheless the group had been known from fossils prior to the discovery of *L. chalumnae* for more than a century. The present contribution provides a list of actinistian taxa with our assessments of synonyms. In presenting this taxonomic list, we hope to provide some basis for assessment of species diversity, and also to give some idea of the basis of knowledge since it remains true that many species are known only by very fragmentary remains.

Actinistian systematic overview

The systematics of actinistians is poorly resolved. Prior to the discovery of *Latimeria chalumnae* in 1938, actinistians were thought to belong to a single crossopterygian family – the Coelacanthidae. Some authors (Berg 1940, Vorobyeva & Obruchev 1964, Lehman 1966, Romer 1966, Andrews et al. 1967, Thomson 1969, Moy-Thomas & Miles 1971, Andrews 1973, Carroll 1988) have addressed the classification of actinistians, but these attempts have resulted in the recognition of gradal and monotypic groups (Forey 1981, Cloutier 1990). No generally accepted phylogeny or higher taxonomic scheme is used for the Actinistia. However, the most commonly referred classifications are those of Berg (1940), Vorobyeva & Obruchev (1964), Andrews et al. (1967), Lund & Lund (1985), and Carroll (1988). Moy-Thomas (1937), Forey (1981), and Cloutier (1990) have investigated the systematics of Paleozoic actinistians and provided lists of Paleozoic taxa. However, the clarification of interrelationships of post-Paleozoic actinistians, which corresponds to most of the specific diversity, has only recently been attempted (Cloutier 1991, Forey 1991).

The compiled tabulation (Table 1) provides a list of all genera, species, and undetermined actinistians mentioned in the literature. This list is supplemented by geological, stratigraphic, and geographic information for each taxon. The primary author for each taxon is provided and included in the bibliography article (Forey & Cloutier 1991); however, authors of subsequent taxonomic changes are not given in the list but are incorporated in the bibliography article. For each genus, the type species is listed first, followed by other referred species alphabetically and the undetermined species.

Assessments about synonymies of taxa have been followed only when it was justified in the literature or following work in progress by one or other of us. Many actinistian species were described originally under different genera; in order to indicate these nomenclatural changes, we follow the convention adopted in the *Handbook of Paleoichthyology* (see e.g., Denison 1978) by refering to the original generic assignment as the first entry of synonymy. Spelling of species names is in agreement with the *International Code of Zoological Nomenclature* (Ride et al. 1985). Among the species that have been recognized herein, it remains likely that some are synonyms or not valid, but further systematic studies are needed and beyond the scope of this paper. For example, there are three species in the genus *Rhipis*, all of which are Late Cretaceous taxa known solely from scales, and two of them come from the same locality; thus it is possible that these forms represent a single species. For this reason a brief statement of parts of the animal known is given with each entry.

Our tabulation is divided into four categories reflecting our systematic interpretation of these taxa. For each category the genera are listed alphabetically. Category 'A' deals with the genera and species that can be placed with some confidence within at least one of the cladistic classifications given in Forey (1991) and Cloutier (1991). This category includes species that can be positioned in reference to terminal taxa, nodes, restricted clades, or a restricted range of nodes on the cladograms.

Category 'B' lists the genera and species that are considered to be valid species but for which little else may be said about their relationships (i.e., phylogenetic position). This category accounts for the Actinistia incertae sedis. These taxa are important for estimating total diversity but are of little use when estimating rates of clade evolution (morphological and taxonomic).

Category 'C' lists genera and species of those remains that can only be identified as actinistians. Their status as valid species is questioned. Finally, category 'D' includes the taxa that are undefined (nomen nuda), or those that have been cited as coelacanths but are referable to other fish groups.

Based on our tabulation, there are 69 valid species with fair assessment of their phylogenetic position (Category A); 21 valid species with poor assessment of their phylogenetic position (Category B); 31 actinistian incertae sedis (Category C); and 18 taxa that had been recognized in the literature as actinistians but are not (Category D). There are 47 recognized genera, 29 of which are monotypic and 17 undetermined actinistians (most of which are partial remains). Thus, there is a total of 90 valid species and 31 incertae sedis actinistians distributed from the Givetian (Middle Devonian) to the Recent; this total differs from that of Forey (1984) who reported approximately 70 species belonging to 28 actinistian genera.

Fossil actinistians have been found on most continents with the exception of Antarctica and Australia. They are predominant in Europe (47 valid species and 17 incertae sedis), Africa (17 valid species and 6 incertae sedis), and North America (19 valid species and 5 incertae sedis); this probably reflects the relative bias of collecting rather than corresponding to some real biogeographic patterns.

Less than half (46%) of the actinistian taxa (categories A–C) are known from entire skeletons varying in state of preservation. Of the remaining taxa, 36% are known almost exclusively from head or head fragments, 10% from scales solely, and 8% from partial trunk.

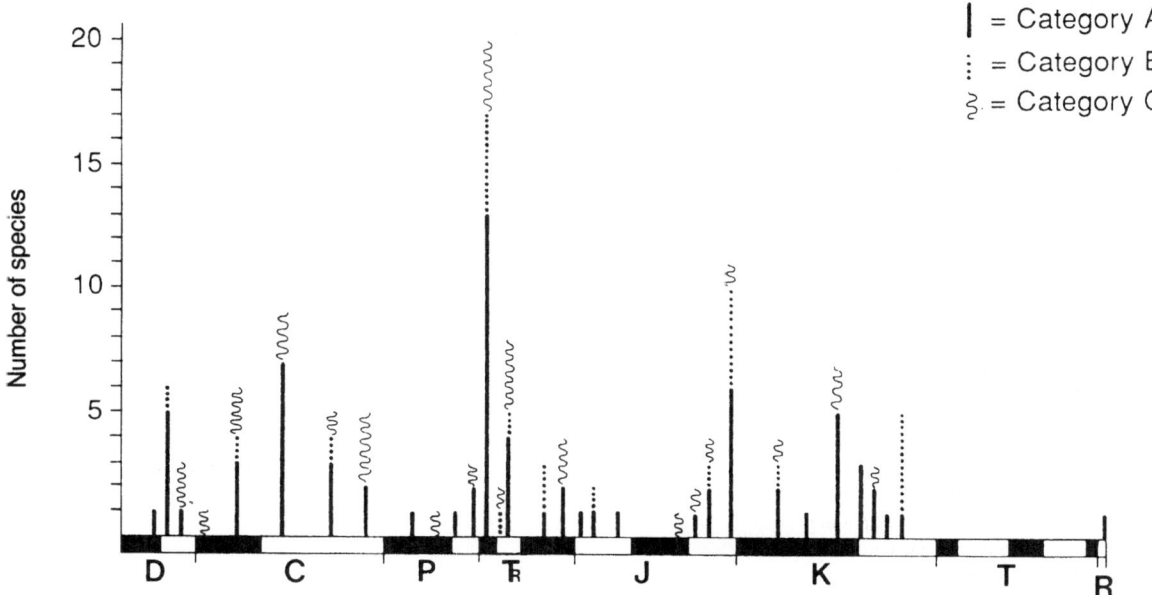

Fig. 1. Diversity of actinistian species through geological time based on the listing of Table 1. Category A is represented by a straight bar, category B by a dotted line, and category C by an undulating line. Each bar corresponds to the number of species for a given biostratigraphic stage.

Patterns of diversity

Histograms of generic diversity through geological time have been provided by Schaeffer (1952a, 1953), Thomson (1977), and Forey (1984, 1988). As pointed out by Smith & Patterson (1988), monotypic and paraphyletic taxa are problematic in generic-level studies because they provide historically uninformative patterns. Among actinistians there is a high proportion of monotypic genera (see above) and others are known to be non-monophyletic (e.g., *Rhabdoderma* and *Coelacanthus*). However, temporal diversity studied at the specific level can be used to address historical questions (e.g., patterns and trends of species diversity).

Here we use Table 1 to provide some details of species diversity. We plot the number of species of categories A–C (above) against time (Fig. 1). The geological time scale of Palmer (1983) is used for the determination of age-boundaries and duration of geological units (i.e., age, period, and stage); in contrast to Palmer (1983), the Tertiary is divided into five epochs (Pliocene, Miocene, Oligocene,

Eocene, Paleocene; Holmes 1959) rather than 15 stages. The geological stage was used as the basic biostratigraphic unit for the count of taxa. In the diversity histograms, the bars were plotted according to the total diversity (i.e., number of taxa) occuring during a given stage. The bars were set on a proportional time axis (X-axis) at the mid-time between consecutive boundaries of each stage (i.e., $[t_0–t_1]/2$).

The proportional distribution of valid species (categories A and B) and known taxa (categories A–C; provided in parentheses) through geological era is as follows: 28.9% (33.05%) Paleozoic, 70% (66.12%) Mesozoic, 0% (0%) Cenozoic, and 1.1% (0.83%) Quaternary. There are four primary modes in the species diversity: (1) Frasnian with 6 species, (2) Namurian with 7 species (plus 2 incertae sedis), (3) Scythian with 17 species (plus 4 incertae sedis), and (4) Tithonian with 10 species (plus 1 incerta sedis). There are two major geological gaps in the fossil record of the actinistians: (1) Middle Jurassic (gap of ca. 29 My) and (2) Tertiary (gap of ca. 66 My).

Table 1. List of actinistian species with their respective geological, stratigraphic, geographical, and morphological information. Fm. = Formation, Gr. = Group, Mbr. = Member, Subgr. = Subgroup, Supergr. = Supergroup

'Category A'

Alcoveria Beltan 1972
A. brevis Beltan 1972 (type species)
    Middle Triassic; Ladinian.
    Muschelkalk.
    Spain.   entire skeleton

Allenypterus Melton 1969
A. montanus Melton 1969 (type species)
    Lower Carboniferous; Namurian A (E_2B).
    Bear Gulch Limestone Mbr., Heath Fm., Big Snowy Gr.
    Montana, USA.   entire skeleton

Axelia Stensiö 1921
A. robusta Stensiö 1921 (type species)
    Lower Triassic; Smitho-Spathian.
    Sticky Keep Fm., Sassendalen Gr.
    West Spitsbergen.   entire skeleton
A. elegans Stensiö 1921
    Lower Triassic; Smitho-Spathian.
    Sticky Keep Fm., Sassendalen Gr.
    West Spitsbergen.   entire skeleton

Axelrodichthys Maisey 1986
A. araripensis Maisey 1986 (type species)
    Lower Cretaceous; Apto-Albian.
    Lower Romualdo Mbr., Santana Fm.
    Ceara, Brazil.   entire skeleton

Caridosuctor Lund & Lund 1984
C. populosus Lund & Lund 1984 (type species)
    Lower Carboniferous: Namurian A (E_2B).
    Bear Gulch Limestone Mbr., Heath Fm., Big Snowy Gr.
    Montana, USA.   entire skeleton

Chagrinia Schaeffer 1962
C. enodis Schaeffer 1962 (type species)
    Upper Devonian; Famennian.
    Chagrin Shale.
    Ohio, USA.   entire skeleton

Changxingia Wang & Liu 1980
C. aspratilis Wang & Liu 1980 (type species)
    Upper Permian; Tatarian.
    Changxing Fm.
    Zhejiang Province, China.   entire skeleton

Chinlea Schaeffer 1967
C. sorenseni Schaeffer 1967 (type species)
    Upper Triassic; Carnian – Norian.
    (1) Chinle Fm. and (2) Tecovas Fm., Dockum Gr.
    (1) Colorado, New Mexico, Utah, (2) Texas, USA.   entire skeleton
C. sp. Ash 1978
    Upper Triassic.
    Ciniza Lake Beds; Chinle Fm.
    New Mexico, USA.   scales

Table 1. Continued.

Coccoderma Quenstedt 1858	
C. suevicum Quenstedt 1858 (type species) [*C. nudum* Reis 1888]	
Upper Jurassic; Tithonian.	
Solnhofen Lithographic Limestone.	
Germany.	entire skeleton
C. harlemensis (Winkler 1871) [*Coelacanthus, Undina*]	
Upper Jurassic; Tithonian.	
Solnhofen Lithographic Limestone.	
Germany.	entire skeleton
C. sp. Fabre et al. 1982	
Lower Cretaceous; Barriasian.	
France.	entire skeleton
Coelacanthus Agassiz 1839	
C. granulatus Agassiz 1839 (type species) [*C. hassiae* Münster 1842, *C. gracilis* Agassiz 1844,	
C. caudalis Egerton 1850, *C. macrocephalus* Willemoes-Suhm 1869, *C. granulosus* Agassiz]	
Upper Permian; Ufimian.	
Kupferschiefer.	
Germany and England.	entire skeleton
'*C.*' *madagascariense* Woodward 1910 [*Rhabdoderma*]	
Lower Triassic; Dienerian.	
Middle Sakamena Gr.	
Madagascar.	entire skeleton
Diplocercides Stensiö 1922	
D. kayseri (von Koenen 1895) (type species) [*Holoptychius, Coelacanthus, Nesides schmidti*	
Stensiö 1937]	
Upper Devonian; Frasnian.	
Germany [also Poland according to Gorizdro-Kulczycka (1950)].	entire skeleton
D. davisi (Moy-Thomas 1937) [*Rhabdoderma, R.* (?) *abdenense* Moy-Thomas 1937]	
Lower Carboniferous; Visean P_1.	
Ireland and Scotland.	head fragments, pectoral girdle
D. heiligenstockiensis (Jessen 1966) [*Nesides*]	
Upper Devonian; Frasnian.	
Upper Plattenkalk.	
Germany.	entire skeleton
D. jaekeli Stensiö 1922	
Upper Devonian; Frasnian.	
Germany.	partial head
D. sp. Janvier 1974	
Upper Devonian; Frasnian.	
Iran.	lower jaws
Diplurus Newberry 1878	
D. longicaudatus Newberry 1878 (type species) [*Rhabdiolepis, R. gwyneddensis* Bock 1959,	
R. striata Bock 1959]	
Lower Jurassic; Hettangian.	
Newark Supergr.	
Connecticut, New Jersey, Pennsylvania, Virginia, USA.	entire skeleton
D. newarki (Bryant 1934), Schaeffer 1954 [*Coelacanthus, Osteopleurus, O. milleri* Shainin 1943,	
O. milleri grantonensis Shainin 1943, *Pariostegus myops* Cope 1868]	
Upper Triassic; Carnian.	
Newark Supergr.	
New Jersey, Pennsylvania, Virginia, USA.	entire skeleton

64

Table 1. Continued.

Euporosteus Jaekel 1927
E. eifelianus Jaekel 1927 (type species)
    Middle Devonian; Givetian.
    Germany.   partial head

Garnbergia Martin & Wenz 1984
G. ommata Martin & Wenz 1984 (type species)
    Middle Triassic; Ladinian.
    Upper Muschelkalk.
    Germany.   head, partial trunk

Hadronector Lund & Lund 1984
H. donbairdi Lund & Lund 1984 (type species)
    Lower Carboniferous; Namurian A (E_2B).
    Bear Gulch Limestone Mbr., Heath Fm., Big Snowy Gr.
    Montana, USA.   entire skeleton

Heptanema Bellotti 1857
H. paradoxum Bellotti 1857 (type species)
    Middle Triassic; Ladinian.
    Italy.   entire skeleton

Holophagus Egerton 1861
H. gulo Egerton 1861 (type species) [*Trachymetopon liassicum* Hennig 1951]
    Lower Jurassic; Sinemurian.
    England.   entire skeleton

Indocoelacanthus Jain 1974
I. robustus Jain 1974 (type species)
    Lower Jurassic; Toarcian.
    Kota Fm.
    India.   entire skeleton

Latimeria Smith 1939
L. chalumnae Smith 1939 (type species) [*Malania anjouanae* Smith 1953,
  Latimeria anjouanae (Smith) Lenoble & Le Grand 1954]
    Recent.
    Comoro Islands.   entire anatomy

Laugia Stensiö 1932
L. groenlandica Stensiö 1932 (type species) [*L.* ? sp. Stensiö 1932]
    Lower Triassic; Scythian.
    Wordie Creek Fm., Nordenskiöld Subgr.
    East Greenland.   entire skeleton

Libys Münster 1842
L. polypterus Münster 1842 (type species)
    Upper Jurassic; Tithonian.
    Solnhofen Lithographic Limestone.
    Germany.   head, trunk

L. superbus Zittel 1887
    Upper Jurassic; Tithonian.
    Solnhofen Lithographic Limestone.
    Germany   entire skeleton

Lochmocercus Lund & Lund 1984
L. aciculodontus Lund & Lund 1984 (type species)
    Lower Carboniferous; Namurian A (E_2B).
    Bear Gulch Limestone Mbr., Heath Fm., Big Snowy Gr.
    Montana, USA.   entire skeleton

Table 1. Continued.

Lualabaea de Saint-Seine 1955
L. lerichei de Saint-Seine 1955 (type species)
    Upper Jurassic; Kimmeridgian.
    Lualabaea Series.
    Zaire.   entire skeleton
L. henryi de Saint-Seine 1955
    Upper Jurassic; Kimmeridgian.
    Lualabaea Series.
    Zaire.   scales, fin rays
Macropoma Agassiz 1835
M. lewesiensis (Mantell 1822) (type species) [*Amia*?; *Macropoma mantelli* Agassiz 1835]
    Upper Cretaceous; Cenomanian – Santonian
    Chalk.
    England.   entire skeleton
M. praecursor Woodward 1909
    Lower–Upper Cretaceous; Albian – Cenomanian.
    Gault and Chalk.
    England.   head, partial trunk
M. speciosum Reuss 1857 [*M. forte* Fritsch 1878]
    Upper Cretaceous; Turonian.
    Czechoslovakia.   entire skeleton
Macropomoides Woodward 1942
M. orientalis Woodward 1942 (type species) [Coelacanthe 'B' and Coelacanthe 'C' of Gaudant 1975]
    Upper Cretaceous; Cenomanian.
    Lebanon.   entire skeleton
Mawsonia Woodward 1907
M. gigas Woodward 1907 (type species) [*Mawsonia minor* Woodward 1908]
    Lower Cretaceous; Apto-Albian.
    Romualdo Mbr., Santana Fm.
    Ceara, Brazil.   entire skeleton
M. lavocati Tabaste 1963
    Lower Cretaceous; Albian.
    Morocco.   head fragments
M. libyca Weiler 1935
    Lower Cretaceous; Albian.
    Egypt.   head fragments
M. tegamensis Wenz 1973
    Lower Cretaceous; Aptian.
    Niger.   head
M. ubangiana Casier 1961
    Lower Cretaceous; Neocomian.
    Zaire.   head fragments
Miguashaia Schultze 1973
M. bureaui Schultze 1973 (type species)
    Upper Devonian; Frasnian.
    Escuminac Fm.
    Québec, Canada.   entire skeleton
Mylacanthus Stensiö 1921
M. lobatus Stensiö 1921 (type species)
    Lower Triassic; Smitho-Spathian.
    Sticky Keep Fm., Sassendalen Gr.
    West Spitsbergen.   head fragments, scales

Table 1. Continued.

M. spinosus Stensiö 1921
    Lower Triassic; Smitho-Spathian.
    Sticky Keep Fm., Sassendalen Gr.
    West Spitsbergen.   head fragments

Piveteauia Lehman 1952
P. madagascariensis Lehman 1952 (type species)
    Lower Triassic; Dienerian.
    Middle Sakamena Gr.
    Madagascar.   entire skeleton

Polyosteorhynchus Lund & Lund 1984
P. simplex Lund & Lund 1984 (type species)
    Lower Carboniferous; Namurian A (E_2B).
    Bear Gulch Limestone Mbr., Heath Fm., Big Snowy Gr.
    Montana, USA.   entire skeleton

Rhabdoderma Reis 1888
R. elegans (Newberry 1856) (type species) [*Coelacanthus; C. lepturus* Agassiz 1844,
    C. robustus Newberry 1856, *C. ornatus* Newberry 1856, *C. elongatus* Huxley 1866,
    C. summiti Wellburn 1903, *C. watsoni* Aldinger 1931, *Conchiopsis filiferus* Cope 1873,
    C. anguliferus Cope 1873, *C. corrugatus* Moy-Thomas 1935, *Hoplopygus binneyi* Agassiz 1844]
    Upper Carboniferous; Namurian – Stephanian.
    Allegheny Gr.
    Ohio, USA; England, Scotland, Wales, Ireland, France, Belgium, Holland, Germany,
      Ukraine.   entire skeleton
R. aldingeri Moy-Thomas 1937
    Upper Carboniferous; Namurian A (E_2).
    Wales.   head, partial trunk
R. ardrossense Moy-Thomas 1937
    Lower Carboniferous; Visean P_1.
    Calciferous Sandstone Series.
    Scotland.   entire skeleton
R. exiguum (Eastman 1902) [*Coelacanthus*]
    Upper Carboniferous; Westphalian D.
    Carbondale Fm.
    Illinois, USA.   entire skeleton
R. huxleyi (Traquair 1881) [*Coelacanthus, Dumfregia*]
    Lower Carboniferous; Visean C_2S_1.
    Upper Border Gr.
    Scotland.   entire skeleton
R. tingleyense (Davis 1884) [*Coelacanthus; C. mucronatus* Pruvost 1913, *C. granulostriatus*
    Moy-Thomas 1935]
    Upper Carboniferous; Westphalian A–C.
    England, Wales, France, Belgium, Holland, Germany.   entire skeleton

Sassenia Stensiö 1921
S. tuberculata Stensiö 1921 (type species)
    Lower Triassic; Smitho-Spathian.
    Sticky Keep Fm., Sassendalen Gr.
    West Spitsbergen.   partial head

Scleracanthus Stensiö 1921
S. asper Stensiö 1921 (type species)
    Lower Triassic; Smitho-Spathian.
    Sticky Keep Fm., Sassendalen Gr.
    West Spitsbergen.   head fragments, scales

Table 1. Continued.

Spermatodus Cope 1894
S. pustulosus Cope 1894 (type species)
    Lower Permian.
    Admiral Fm.
    Texas, USA.   head
Synaptotylus Echols 1963
S. newelli (Hibbard 1933) (type species) [*Coelacanthus, C. arcuatus* Hibbard 1933]
    Upper Carboniferous; Stephanian.
    Lansing Gr.
    Kansas, USA.   entire skeleton
Ticinepomis Rieppel 1980
T. peyeri Rieppel 1980 (type species)
    Middle Triassic; Ladinian.
    Grenzbitumen horizon.
    Switzerland.   entire skeleton
Undina Münster 1834
U. penicillata Münster 1834 (type species) [*Holophagus, Coelacanthus, Coelacanthus striolaris*
  Muñster 1842, *C. koehleri* Münster 1842, *C. major* Wagner 1863, *Undina acutidens* Reis 1888]
    Upper Jurassic; Tithonian.
    Solnhofen Lithographic Limestone.
    Germany.   entire skeleton
U. cirinensis Thiollière 1854 [*Undina minuta* Wagner 1863, *Coelacanthus minutus*
  Willemoes-Suhm 1869]
    Upper Jurassic; Tithonian.
    Solnhofen Lithographic Limestone.
    France and Germany.   entire skeleton
U. purbeckensis Woodward 1916
    Upper Jurassic; Purbeckian.
    England.   partial head, trunk
Whiteia Moy-Thomas 1935
W. woodwardi Moy-Thomas 1935 (type species) [*Coelacanthus evolutus* Beltan 1980]
    Lower Triassic; Dienerian.
    Middle Sakamena Gr.
    Madagascar.   entire skeleton
W. africanus (Broom 1905) [*Coelacanthus*]
    Lower Triassic; Scythian.
    Upper Beaufort Beds; Beaufort Series.
    Republic of South Africa.   head, partial trunk
W. tuberculata Moy-Thomas 1935
    Lower Triassic; Dienerian.
    Middle Sakamena Gr.
    Madagascar.   entire skeleton
W. sp. Gardiner 1966
    Lower Triassic; Smithian.
    Toad, Grayling, and Sulphur Mountain Fms.
    Alberta, British Columbia, Canada.   entire skeleton
Wimania Stensiö 1921
W. sinuosa Stensiö 1921 (type species) [*Leioderma sinuata* Stensiö 1918]
    Lower Triassic; Smitho-Spathian.
    Sticky Keep Fm., Sassendalen Gr.
    West Spitsbergen.   head fragments, scales

Table 1. Continued.

Youngichthys Wang & Liu 1981
Y. xinhuainsis Wang & Liu 1981 (type species)
    Upper Permian; Tatarian.
    Changxing Fm.
    Zhejiang Province, China   entire skeleton

'Category B'

Bogdanovia Obrucheva 1955
B. orientalis Obrucheva 1955 (type species)
    Upper Devonian; Frasnian.
    Central Kazakhstan, USSR.   head fragments
Coccoderma Quenstedt 1858
C. bavaricum Reis 1888
    Upper Jurassic; Tithonian.
    Solnhofen Lithographic Limestone.
    Germany.   head and trunk fragments
C. gigas Reis 1888
    Upper Jurassic; Tithonian.
    Solnhofen Lithographic Limestone.
    Germany   lower jaw
C. substriolatum (Huxley 1866), Reis 1888
    Upper Jurassic; Kimmeridgian.
    Kimmeridge Clay.
    England.   partial head
Coelacanthus Agassiz 1839
Coelacanthus banffensis Lambe 1916
    Lower Triassic; Smithian.
    Spray River Fm.
    Alberta, Canada   partial trunk
C. gracilis Agassiz 1844
    ? Middle Triassic.
    ? Germany.   caudal skeleton
'C'. lunzensis Reis 1900
    Upper Triassic; Carnian.
    Lunz Sandstone.
    Austria.   entire skeleton
C. welleri Eastman 1908
    Lower Carboniferous; Tournaisian or Lower Visean.
    Kinderhook Limestone.
    Iowa, USA.   trunk
Graphiurichthys White & Moy-Thomas 1937
G. callopterus (Kner 1866) (type species)
    Upper Triassic; Carnian.
    Raibl Beds.
    Austria.   entire skeleton
Hainbergia Schweizer 1966
H. granulata Schweizer 1966 (type species)
    Middle Triassic; Ladinian.
    Muschelkalk.
    Germany.   entire skeleton
Mawsonia Woodward 1907
M. sp. Wenz 1981
    Lower Cretaceous; Neocomian.
    Niger.   lower jaw

Table 1. Continued.

Moenkopia Schaeffer & Gregory 1961
M. wellesi Schaeffer & Gregory 1961 (type species)
    Middle Triassic; Anisian.
    Moenkopi Fm.
    Arizona, USA.   head fragments
Rhipis de Saint-Seine 1950
R. moorseli de Saint-Seine 1950 (type species) [*R. moorseli* forma *undulatus* Casier 1965]
    Upper Cretaceous.
    Kinko beds; Kwango Series.
    Zaire.   scales
R. tuberculatus Casier 1965
    Upper Cretaceous.
    Kinko beds; Kwango Series.
    Zaire.   scales
R. sp. Casier 1965 [*R.* sp. indét. Casier 1965]
    Upper Cretaceous.
    Kwango Series.
    Congo.   scales
Sassenia Stensiö 1921
S. (?) *guttata* (Woodward 1912) [*Coelacanthus*]
    Lower Triassic; Smitho-Spathian.
    Sticky Keep Fm., Sassendalen Gr.
    West Spitsbergen.   partial head, trunk
Sinocoelacanthus Liu 1964
S. fengshanensis Liu 1964 (type species)
    Lower Triassic.
    Lolou Series.
    Kwangsi Province, China.   caudal skeleton
Undina Münster 1834
U. (?) *barroviensis* Woodward 1890
    Lower Jurassic; Sinemurian.
    England.   entire skeleton
U. grandis Eastman 1914
    Upper Jurassic; Tithonian.
    Solnhofen Lithographic Limestone.
    France.   trunk
U. willemoesi (Vetter 1881) [*Macropoma, Heptanema*]
    Upper Jurassic; Tithonian.
    Solnhofen Lithographic Limestone.
    Germany.   entire skeleton
Wimania Stensiö 1921
W. (?) *multistriata* Stensiö 1921
    Lower Triassic; Smitho-Spathian.
    Sticky Keep Fm., Sassendalen Gr.
    West Spitsbergen.   head fragments, scales

'Category C'

Coelacanthopsis curta Traquair 1901 (type species)
    Lower Carboniferous; Visean.
    Scotland.   entire skeleton
Coelacanthus dendrites Gardiner
    Permian.
    Republic of South Africa.   scales

Table 1. Continued.

C. phillipsii Agassiz 1844
    Upper Carboniferous; Westphalian A.
    Halifax Hard Bed.
    England.   caudal skeleton
C. stensioei Aldinger 1931
    Upper Carboniferous; Namurian E_1.
    Germany.   head fragments
C. sp. Aldinger 1931
    Upper Carboniferous; Namurian E.
    Germany.   head fragments
C. sp. indet. of Chabakov 1927
    Upper Carboniferous; Stephanian B.
    Ukraine, USSR.   scale
C. sp. of Fletcher 1884
    Upper Carboniferous.
    Coal Measures.
    Nova Scotia, Canada.   gular plate
Mawsonia sp. Campos & Wenz 1982
    Lower Cretaceous; Apto-Albian.
    Romualdo Mbr., Santana Fm.
    Ceara, Brazil.   head
Mylacanthus? sp. or *Scleracanthus?* sp. Stensiö 1921
    Lower Triassic; Smitho-Spathian.
    Sticky Keep Fm., Sassendalen Gr.
    West Spitsbergen.   partial head
cf. *Rhabdoderma* Forey & Young 1985
    Carboniferous; Dinantian.
    Scotland.   partial head
Undina picena Costa 1862 [*Urocomus*]
    Upper Triassic; Norian.
    Italy.   entire skeleton
Undina sp. Andersson 1916
    Middle Triassic; Ladinian.
    Italy.   head and trunk fragments
Wimania? sp. Stensiö 1921
    Lower Triassic; Smitho-Spathian.
    Sticky Keep Fm., Sassendalen Gr.
    West Spitsbergen.   partial trunk
Actinistia gen. et sp. indet. of Lelièvre & Janvier 1988
    Upper Devonian; Famennian.
    Morocco.   head fragments, scale
? Actinistia gen. et sp. indet. of Janvier, Lethiers, Monod & Balkas 1984
    Lower Carboniferous; Tournaisian.
    Köprülü Shales.
    Turkey.   scale
Actinistia indet. of Schultze & Möller 1986
    Middle Triassic; Anisian.
    Muschelkalk.
    Germany.   scales
Coelacanth of Patton & Tailleur 1964
    Lower Triassic; Scythian.
    Shublick Fm.
    Alaska, USA.   partial trunk

Table 1. Continued.

[Coelacanth] identical scales of Gardiner 1973	
? Upper Permian – Lower Triassic.	
Madumabisa Shales.	
Zimbabwe.	scales
[Coelacanthe] Basisphénoide **1** of Wenz 1979	
Upper Jurassic; Oxfordian.	
France.	basisphenoid
[Coelacanthe] Basisphénoide **2** of Wenz 1979	
Middle Jurassic; Callovian.	
France.	basisphenoid
[Coelacanthe] Carré-entoptérygoide **1** of Wenz 1979	
Upper Jurassic; Kimmeridgian.	
France.	quadrate, pterygoid fragment
Coelacanthe indéterminé of Martin & Wenz 1984	
Middle Triassic; Ladinian.	
Upper Muschelkalk.	
Germany.	scales
Coelacanthidae cf. **Undina** of Forey, Monod & Patterson 1985	
Upper Jurassic; Tithonian.	
Akkuyu Fm.	
Turkey.	partial head
Coelacanthidae indet. of Schultze & Chorn 1988	
Upper Carboniferous; Stephanian.	
Bern Limestone Fm.	
Kansas, USA.	articular, scales
Coelacanthidae gen. et sp. **indet.** of Martin 1981	
Upper Triassic.	
Morocco.	angular
Coelacanthidae gen. et sp. indet. of Zidek 1975	
Upper Carboniferous; Stephanian.	
Wild Cow Fm.	
New Mexico, USA.	scales, trunk fragments
Coelacanthidae nov. gen. of Dehm 1956	
Middle Triassic.	
France.	partial trunk
Coelacanthoidea of Gall, Grauvogel & Lehman 1974	
Lower Triassic; Scythian.	
Buntsandstein.	
France.	entire skeleton
Forme 'B' of Campos & Wenz 1982	
Lower Cretaceous; Apto-Albian.	
Romualdo Mbr., Santana Fm.	
Ceara, Brazil.	head
Undescribed actinistian of Cloutier (1990)	
Upper Devonian; Famennian.	
New York, USA.	entire skeleton
Undet. coelacanth of Berger 1832	
Upper Triassic.	
Keuper.	
Germany.	partial head

Table 1. Continued.

'Category D'

Bunoderma baini de Saez 1940
    Upper Jurassic.
    Argentina.
	(not positively determinable)

Coelacanthus abdenesis Traquair 1903
    Lower Carboniferous; Visean.
    Scotland.
	(nomen nudum)

C. distans Wellburn 1902
    Upper Carboniferous; Namurian.
    England.
	(nomen nudum)

C. giganteus Winkler 1880
    Triassic.
    Germany.
	(nomen nudum)

C. hindi Wellburn 1902
    Upper Carboniferous; Namurian.
    England.
	(nomen nudum)

C. minor Agassiz 1844
    Middle Triassic; ? Ladinian.
    Lunéville, France.
	(nomen nudum)

C. munsteri Agassiz 1844 [*Undina*]
    Carboniferous.
    Coal Lebach.
	(nomen nudum, dipnoan
	Conchopoma gadiforme
	Kner 1868)

C. spinatus Wellburn 1902
    Upper Carboniferous; Namurian.
    England.
	(nomen nudum)

C. tuberculatus Wellburn 1920
    Upper Carboniferous; Namurian.
    England.
	(nomen nudum)

C. woodwardi Wellburn 1902
    Upper Carboniferous; Namurian.
    England.
	(nomen nudum)

Dictyonosteus arcticus Stensiö 1918
    Middle Devonian.
    West Spitsbergen.
	(not positively determinable)

Rhabdoderma (?) *aegyptiaca* Heide 1955
    Lower Carboniferous.
    Egypt.
	(rhizodont)

Celacantideo of Richter 1985
    Lower Permian.
    Irati Fm.
    Brazil.
	(not positively determinable)

Coelacanth bone of Ørvig 1986
    Palaeocene.
    Sweden.
	(not positively determinable)

Coelacanth remains of Gardiner 1966
    Upper Devonian
    Alberta, Canada.
	(dipnoan)

Coelacanthidae gen. et sp. indet. of Dziewa 1980
    Early Triassic.
    Knocklofty Fm.
    Tasmania, Australia.
	(not positively determinable)

Coelacanthidae genus non det. of Woodward 1895 ['Coelacanth' of Schaeffer 1941]
    Upper Jurassic.
    Talbragar Beds.
    New South Wales, Australia.
	(actinopterygian)

Coelacanthinien genre indét. of Casier 1961
    Lower Cretaceous.
    Congo.
	(not positively determinable)

Trends of diversity

Schaeffer (1952b, p. 109) maintained that the origination (speciation) rate of actinistians '... remained consistently low throughout their long history'. Raup et al. (1973) have suggested that the pattern of actinistian species diversity is one which shows an initial burst of speciation followed by a drastic decrease in diversity ending in a single surviving species.

The variation of the diversity through the duration of the group can be evaluated using non-parametric Spearman's rank-correlation coefficient. This coefficient is used to quantify the correlation between the number of taxa and the geological age. This is done in order to describe the trends of diversity through time. Rank-correlations were calculated for the two different units of time: geological period ($r_{s \, period}$) and biostatigraphic stage ($r_{s \, stage}$). The rankings were attributed as follows: rank 1 equals Middle Devonian, rank 2 equals Late Devonian, rank 3 equals Early Carboniferous, ..., and rank 20 equals Quaternary for the geological period; rank 1 equals Givetian, rank 2 equals Frasnian, rank 3 equals Famennian, ..., and rank 48 equals Holocene for the biostratigraphic stage. The number of species occurring in any one of the geological periods and biostatigraphic stages are ranked numerically; for instance, the Early Triassic is the geological period and the Scythian is the biostratigraphic stage containing the most species and for the purpose of computation this is given the highest rank. For equal rankings of species numbers, the computation procedure described by Dagnelie (1977) is invoked.

These rankings were then applied to the non-parametric Spearman's rank-correlation coefficient algorithm [see Cloutier (1991) for discussion of the method].

The history of the actinistians ranged over 20 (14 with actinistians) geological periods (from Middle Devonian to Quaternary) including 48 (31 with actinistians) biostratigraphic stages (from Givetian to Holocene) for a corresponding duration of 380 million years. The coefficients of rank-correlation calculated for the geological periods and the biostratigraphic stages show a low negative correlation (for categories A and B: $r_{s \, period} = -0.4459$, $r_{s \, stage} = -0.2455$; for categories A–C: $r_{s \, period} = -0.4944$, $r_{s \, stage} = -0.3551$) between the diversity-rank and time-rank. None of these correlations are significant. Therefore, these results support Schaeffer's (1952b) conclusion that there has been no significant decreasing or increasing trends in species diversity through time.

Some of the peaks of actinistian diversity are coincident to major geological events. These events are not interpreted as causal effects on the diversity. They might have an influence on actual diversity because of vicariance events or habitat diversification, or an effect on apparent diversity by increasing the likelihood of fossilisation owing to an increase of sedimentation, for example. The Frasnian peak succeeds the Caledonian orogeny which affected Europe and precedes the Acadian orogeny which affected North America; most of the Frasnian species are found in north-east America and Europe. A second peak of species diversity, chiefly reflecting North American actinistian fauna, occurred during the Namurian, stage associated with the beginning of the Sudetian orogenic phase. The greatest diversity occurred during the Scythian which corresponds to a worldwide marine transgression (Schaeffer & Mangus 1976). The greatest actinistian diversity occurring during the Tithonian is documented mostly in Europe, and may be associated with the Late Kimmerian orogenic phase. Finally, the Albian peak may be broadly contemporaneous with the Subhercynian/Austrian orogeny which influenced western Africa; most of the Cretaceous actinistians are found in northern Africa and South America.

References cited

See 'References cited' in Forey & Cloutier (1991) for all taxonomic references.

Cloutier, R. 1991. Patterns, trends, and rates of evolution within the Actinistia. Env. Biol. Fish. 32: 23–58. (this volume)
Dagnelie, P. 1977. Théorie et méthodes statistiques. Applications agronomiques. Vol. 1, 2nd ed. Presses Agronomiques de Gembloux, Gembloux. 378 pp.
Denison, R.H. 1978. Placodermi. pp. 1–62. In: H.-P. Schultze

(ed.) Handbook of Paleoichthyology, Vol. 2., Gustav. Fischer Verlag, Stuttgart.

Forey, P.L. 1991. Latimeria chalumnae and its pedigree. Env. Biol. Fish. 32: 75–97. (this volume)

Forey, P.L. & R. Cloutier. 1991. Literature relating to fossil coelacanths. Env. Biol. Fish. 32: 391–401. (this volume)

Holmes, A. 1959. A revised geological time scale. Trans. Edinburgh Geol. Soc. 17: 183–216.

Lenoble, J. & Y. Le Grand. 1954. Le tapis de l'oeil du coelacanthe (*Latimeria anjouanae* [Smith]). Bull. Mus. Hist. nat., Paris 26: 460–463.

Palmer, A.R. 1983. The decade of North American geology. 1983. Geologic time scale. Geology 11: 503–504.

Raup, D.M., S.J. Gould, T.J.M. Schoff & D.S. Simberloff. 1973. Stochastic models of phylogeny and the evolution of diversity. J. Geol. 81: 525–542.

Ride, W.D.L., C.W. Sabrosky, G. Bernardi & R.V. Melville (ed.). 1985. International Code of Zoological Nomenclature.

XX General Assembly of the International Union of Biological Sciences. International Trust for Zoological Nomenclature. H. Charlesworth & Co. Ltd., Huddersfield. 338 pp.

Schaeffer, B. & M. Mangus. 1976. An early Triassic fish assemblage from British Columbia. Bull. Amer. Mus. Nat. Hist. 156: 517–563.

Smith, A.B. & C. Patterson. 1988. The influence of taxonomic method on the perception of patterns of evolution. Evol. Biol. 23: 127–216.

Thomson, K.S. 1977. The pattern of diversification among fishes. pp. 377–404. *In:* A. Hallam (ed.) Patterns of Evolution, as Illustrated by the Fossil Record, Development in Palaeontology and Stratigraphy, Vol. 5, Elsevier, Amsterdam.

Vorobyeva, E.I. & D.V. Obruchev. 1967. Subclass Sarcopterygii. pp. 420–509. *In:* Y.A. Orlov (ed.) Fundamentals of Paleontology, Vol. XI Agnatha, Pisces, Israel Program for Scientific Translations, Jerusalem. (Russian original 1964).

Environmental Biology of Fishes **32**: 75–97, 1991.
© 1991 *Kluwer Academic Publishers.*

Latimeria chalumnae and its pedigree

Peter L. Forey

Department of Palaeontology, British Museum (Natural History), Cromwell Road, London SW7 5BD, England

Received 10.4.1990   Accepted 4.10.1990

Key words: Coelacanth, Evolution, Phylogeny, Cladistics, Classification, Intracranial joint, Fins, Vertebral column

Synopsis

Latimeria is the product of a long coelacanth lineage, usually viewed as having changed very little. In this paper a classification of better known coelacanth genera is proposed based on a cladistic computer analysis of 56 morphological characters. Biometrical data are then matched with the classification to explore the possibility of identifying subtle change. It is concluded that throughout coelacanth history there have been changes in the structure of the vertebral column involving an overall increase in the number of vertebral elements, and a consequent crowding of these elements within the abdominal region. These changes may be associated with increasing lobation of the second dorsal and anal fins. In the skull, parameters involving the intracranial joint have also changed in such a way that the anterior part of the skull has lengthened in relation to the posterior part and this may be associated with an increase in length of the basicranial muscle.

Abbreviations used in text figures: Ang – angular, a.o.r – anterior opening of the rostral organ, Art – articular, ba.cr.m – basicranial muscle, Basi – basisphenoid, Boc – basioccipital, bpt.pr – basipterygoid process, c.p.l – cheek pit line, De – dentary, Esc – extrascapular, eth.sp – ethmosphenoid, f.e – frontoethmoid, Fr – frontal, Fr.d – descending process of frontal, intr.j – intracranial joint, io.s – interorbital septum, j.sc – jugal sensory canal, L.e – lateral ethmoid, m.Cor – modified coronoid, Mm – mentomeckelian, m.ot.sc – medial branch of otic canal, Op – operculum, o.p.l – oral pit line, ot.occ – otico-occipital, Pa – parietal, Pa.d – descending process of parietal, Par – parasphenoid, Part – prearticular, pa.s – parietal shield, p.Cor – principal coronoid, Po – postorbital, Pop – preoperculum, p.o.r – posterior openings of the rostral organ, Pmx – premaxilla, Pre – preorbital, Pro – prootic, Pro.p – posterior process of prootic, Rart – retroarticular, Sc.o – sclerotic ossicle, So – supraorbital, Soc – supraoccipital, Sop – suboperculum, Sp – spiracular, Spl – splenial, Sq – squamosal, Stt – supratemporal, Stt.com – supratemporal commissure, Stt.d – descending process of supratemporal, Par.a.w – ascending wing of parasphenoid, Te – tectal, X – level of vagus exit.

Introduction

This volume celebrates the coelacanth, *Latimeria chalumnae* Smith. Included are assessments of the evolutionary significance, biology and conservation of this unique species; and as part of that assessment there are several accounts of anatomy (Cloutier 1991, Schultze 1991, Schultze & Cloutier 1991) and biology (Balon 1991, Fricke et al. 1991, Wourms et al. 1991) which together stress the individuality of this remarkable fish amongst the Recent fish fauna. For instance, *Latimeria* is the only living animal with an intracranial joint, and potentially will yield information about the mechanics and function of this highly unusual structure which was such a prominent feature of a wide variety of Paleozoic sarcopterygian fishes. *Latimeria* moves in a rather unusual way maintaining a stiff, inflexible body while sculling along using a lobed second dorsal fin and a lobed anal fin. The paired fins are also lobed and move with a co-ordination similar to that in the limbs of tetrapods (Fricke et al. 1987). *Latimeria* also behaves in a curious way by performing headstands, positioning its snout a few centimeters above the substrate while keeping station against a strong sea current.

These characteristics set *L. chalumnae* apart from other living fishes. But it is pertinent to ask how *Latimeria* came by some of the more obvious characters, since *Latimeria* is the product of a long history of coelacanth fishes beginning in the Palaeozoic. This paper attempts to detail some of that history by using a classification as a source for explanation.

Latimeria chalumnae is the sole living representative of a lineage extending back to at least the Upper Devonian, some 360 million years ago. In all probability the coelacanth lineage is somewhat older since the probable sister-groups, Dipnoi or Dipnoi + Tetrapoda (Cloutier 1991, Forey 1989, Forey et al. 1990) are known in the Lower Devonian. Coelacanths have a reasonably continuous fossil record from the Upper Devonian to the Upper Cretaceous, although they are never numerous and many of the coelacanth fossils are very poor. To date there are approximately 120 species known, with the peak of diversity occurring in the Triassic. The majority were marine fishes, like *Latimeria,* but there is no evidence that any occupied the steep marine slopes in deep water which are the home of *Latimeria* today (Fricke & Plante 1988). Species numbers and habitats are indicated in Figure 1 with some idea of the geographical distribution of the coelacanth record included in the Appendix.

Coelacanths have remained relatively conservative throughout their history and this fact in itself poses some interesting problems for explanations of the rates of morphological evolution (Cloutier

1991, Forey 1984, 1989, Forey et al. 1990). There are nevertheless some interesting differences between various fossil coelacanth species and *Latimeria*. For example, compared with many Palaeozoic coelacanths, *Latimeria* has a weakly ossified neurocranium, and to compensate for this several supporting struts projecting ventrally from the dermal skull roof have developed to form braces. *Latimeria* has a rather differently proportioned skull when compared with most Palaeozoic forms. The lobation of the second dorsal and the anal fins appears to be more extreme in *Latimeria*. There is an increased number of vertebrae compared with that in many fossil coelacanths, and this increase appears to be confined to the abdominal region (see below). Some of these features have already been noted by Schaeffer (1952) who attempted to document skeletal changes in chronological sequence with the intention of discovering evolutionary trends. That attempt was undertaken without a clear idea of the interrelationships between species and therefore it was impossible to distinguish trends illustrated by their distribution amongst taxa from 'trends' selected by citing particular examples. However, it is possible that the characters we see in *Latimeria* are the result of separate morphological transformation series and to test this thesis I have attempted to produce a classification on which several parameters may be plotted.

In this paper I choose to examine skeletal features associated with the intracranial joint, the vertebral column and the median fins, as these appear to be particular features of the skeleton of *Latimeria* which set it apart from contemporary fishes.

Character analysis and coelacanth classification

This section attempts to produce a classification which can then be used to analyse certain biometric variables such as skull dimensions, proportions within the vertebral column and lobation of the median fins.

The monophyly of coelacanths is not questioned: it has been justified on several occasions (Forey 1981, 1984, Rosen et al. 1981, Cloutier 1990). Coelacanths show several unique features shared by all

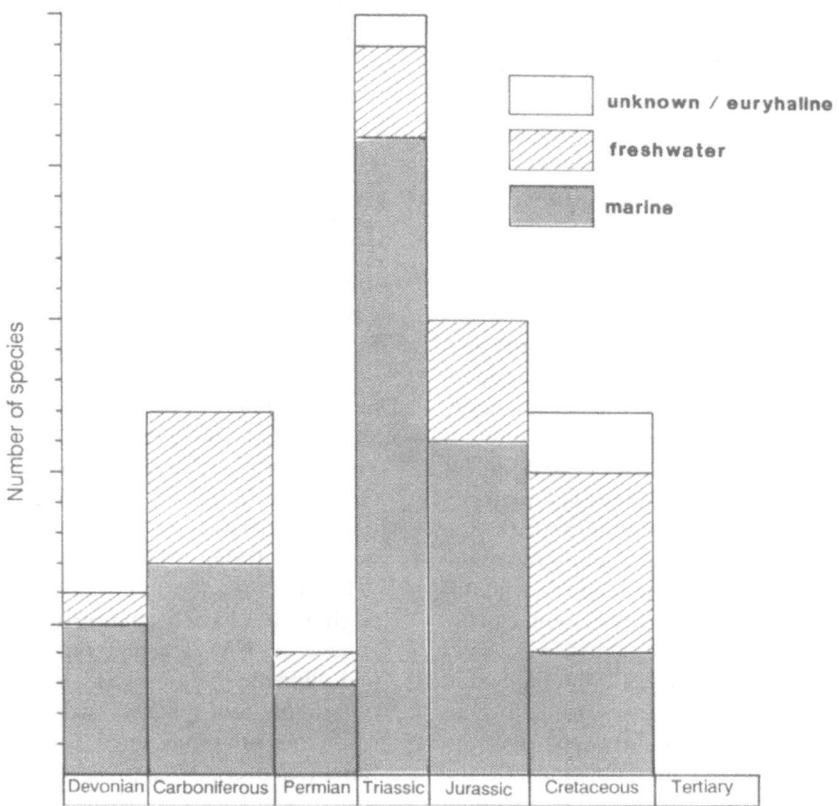

Fig. 1. Species numbers and habitat of coelacanths through time. In this diagram only named species are included: those records simply noted as coelacanth indet. or sp. are omitted. M = marine record, FW = freshwater, ? = unknown environment or at present debated. The records are as follows (full details in Appendix):

Devonian – *Chagrinia enodis* [M], *Diplocercides kayseri* [M], *Diplocercides jaekeli* [M], *Diplocercides heiligenstockiensis* [M], *Euporosteus eifeliensis* [M], *Miguashaia bureaui* [FW].

Carboniferous – *Allenypterus montanus* [M], *Diplocercides davisi* [M], *Hadronector donbairdi* [M], *Lochmocercus aciculodontus* [M], *Polyosteorhynchus simplex* [M], *Rhabdoderma elegans* [FW], *Rhabdoderma tinglyense* [FW], *Rhabdoderma ardrossense* [M], *Rhabdoderma* (?) *alderingi* [FW], *Rhabdoderma huxleyi* [M], *Rhabdoderma exiguum* [FW], *Synaptotylus newelli* [FW].

Permian – *Spermatodus pustulosus* [FW], *Coelacanthus granulatus* [M], *Changxingia aspratilis* [M], *Youngichthys xinhuanisis* [M].

Triassic – *Alcoveria brevis* [M], *Axelia robusta* [M], *Axelia elegans* [M], *Chinlea sorenzeni* [FW], *Coelacanthus lunzensis* [FW], *Diplurus newarki* [FW], *Garnbergia ommata* [M], *Graphiurichthys callopterus* [M], *Hainbergia granulata* [M], *Heptanema paradoxum* [M], *Laugia groenlandica* [M], *Moenkopia wellesi* [FW], *Mylacanthus lobatus* [M], *Mylacanthus spinosus* [M], *Piveteauia madagascariensis* [M], *Sassenia tuberculata* [M], *Sassenia groenlandica* [M], *Scleracanthus asper* [M], *Sinocoelacanthus fengshanensis* [?], *Ticinepomis peyeri* [M], *Wimania sinuosa* [M], *Whiteia africanus* [M], *Whiteia groenlandica* [M], *Whiteia tuberculata* [M], *Whiteia woodwardi [M]*, *Wimania* [?] *multistriata* [M].

Jurassic – *Bunoderma baini* [M], *Coccoderma bavaricum* [M], *Coccoderma gigas* [M], *Coccoderma suevicum* [M], *Diplurus longicaudatus* [FW], *Holophagus gulo* [M], *Indocoelacanthus robustus* [FW], *Libys polypterus* [M], *Libys superbus* [M], *Lualubaea henryi* [FW], *Lualubaea lerichei* [FW], *Undina* [?] *barroviensis* [M], *Undina cirinensis* [M], *Undina purbeckensis* [M], *Undina penicillata* [M].

Cretaceous – *Axelrodichthys araripensis* [?], *Macropoma lewesiensis* [M], *Macropoma praecursor* [M], *Macropoma speciosum* [M], *Macropomoides orientalis* [M], *Mawsonia gigas* [?], *Mawsonia lavocati* [FW], *Mawsonia libyca* [FW], *Mawsonia tegamensis* [FW], *Mawsonia ubangiensis* [FW], *Rhipis moorseli* [FW], *Rhipis tuberculata* [FW].

members and a suite of attributes which are common to the vast majority and make coelacanths one of the more easily recognisable groups of fishes.

Coelacanth fishes may be described as sarcopterygians in which: (1) the dorsal fins are each supported by a single basal element; the anterior dorsal fin is sail-like without a basal lobe, the second dorsal and the anal fins are lobed fins containing a skeleton of several segments which resemble the endoskeleton of the paired fins; (2) the paired fins have short lobes in which the endoskeleton consists of four or five axial segments supporting reduced radials; (3) the body is often plump and the caudal fin is usually diphycercal, consisting of outwardly equal main lobes on either side of the vertebral column and a small supplementary caudal fin; (4) the fin rays are segmented, but in the vast majority of coelacanths these remain unbranched; (5) the head is always divided by a prominent intracranial joint in which the rear half of the braincase extends anteriorly to form a track-and-groove joint with the basisphenoid; (6) the extrascapular series shows elements with a laterad overlap; (7) the snout contains a large rostral organ which is probably concerned with electroreception and which opens to the surface through three paired openings; (8) the palate is rigidly attached to the ethmosphenoid portion of the braincase such that these two move as a unit; (9) the palate is characterised by a large triangular pterygoid; (10) the hyomandibular is short and stout, free from the palatoquadrate and only indirectly involved with jaw suspension; (11) the premaxilla is small and a maxilla is absent; (12) in the lower jaw the dentary is very short, there are only two infradentaries of which the angular is the larger component and the posteriormost coronoid is modified; (13) there is a tandem jaw articulation in which the usual quadrate/articular joint is accompanied by an articulation between the retroarticular and an elongate symplectic; (14) the gular series is represented as a single pair of large plates; a median gular is absent; (15) branchiostegal rays are absent; and (16) the shoulder girdle shows a high fin insertion and there is an extracleithrum sutured to the cleithrum and clavicle. Of these features those numbered 1, 5, 7, 9, 11, 12, 13, 15, 16

may be regarded as autapomorphies of coelacanths.

Classification

The characters used in this analysis reflect those most obviously visible in fossil coelacanths. Throughout coelacanth history there have been subtle changes in braincase morphology, shoulder girdle anatomy and several features associated with the branchial arches. However, some of these details are known from so few taxa that they are not included here. These characters include the varying development of the lateral ethmoid region and the supraoccipital region of the braincase; the differing pattern of basibranchial and gill arch dentition; and the differing shapes and relationships of the anocleithrum, cleithrum and extracleithrum.

In order to conduct a cladistic computer analysis, 56 characters have been recorded for 36 coelacanth genera, and for actinopterygians, porolepiforms and dipnoans representing three monophyletic outgroups. Some genera were not included because they are represented by extremely poor material (*Alcoveria, Chagrinia, Changxingia, Heptanema, Youngichthys, Graphiurichthys, Hainbergia, Sinocoelacanthus, Undina picena*). Other coelacanths are represented by very incomplete material (*Rhipis* – scales; *Euporosteus* – fronto-ethmoid portion of neurocranium; *Moenkopia* – basisphenoid; *Mylacanthus* and *Scleracanthus* – fragments of opercular apparatus, cheek and skull roof).

In this analysis, genera rather than species are chosen as terminal taxa. This is because, within coelacanths, most genera are either monospecific or known only by one well-known species accompanied by far more imperfectly-known species. Exceptionally, *Macropoma* is represented by three reasonably well-known species, and *Undina* by at least two well-known species.

The data matrix used here indicates some extent of our ignorance of many taxa. The taxa have been arranged to show those where over 75% of the chosen characters are known, 50–74% are known and 25–49% are known. Some of the taxa included in previous attempts at coelacanth classification (*Mylacanthus, Scleracanthus, Heptanema* and *Euporosteus* – see Schaeffer 1941) are omitted from the computer analysis because they have less than 25% data.

The following represents a list of the characters used together with their coding: where relevant, explanatory notes are given. Where a structure is present I have used a coding of '1', irrespective of any implication of plesiomorphic or apomorphic state. Figures 2–4 illustrate many of the features used.

Characters analysed

1. Interorbital septum present (1).
2. Basipterygoid process developed (1).
3. Parasphenoid with ascending wings (1). The wings are developed to brace the lateral ethmoids and participate in the articulation with the autopalatine.
4. Buccohypophysial canal perforating the parasphenoid (1).
5. Otico-occipital portion of the neurocranium formed as distinct ossifications (1).
6. Prootic with posterior processes developed as a complex suture with the basisphenoid (1).
7. Ascending process on the prootic developed to meet the descending process of the parietal (1) (postparietal of some authors).
8. Vagus foramen bone-enclosed (1).
9. Premaxilla splint-like (1). In most coelacanths, as in primitive actinopterygians, the premaxilla has a 'pars dentalis' and dorsal flange that may or may not carry the rostral commissure. In some coelacanths the premaxilla is reduced to the tooth-bearing part only and this is regarded as derived.
10. Several internasals (postrostrals of other authors). (1). This is a median series of bones wedged between the nasals and reaching back to the frontals in some coelacanths. In most coelacanths at least one internasal is retained.
11. Number of frontals. The frontals are here re-garded as large paired elements, lying immediately anterior to the intracranial joint, and roofing the orbit. It is assumed that two pairs of frontals is the plesiomorphic coelacanth condition (0), one pair (1).
12. Frontals similar or dissimilar in size. This character relates to the relative sizes of the two pairs of frontals above the eye. For those few coelacanths which have only one frontal then this character is scored as not applicable. The coding is: similar = 0, dissimilar = 1.
13. Frontal descending process. This is a dermal process developed from the ventral surface of the frontal and serving to brace the skull roof against the basisphenoid. Process present (1).
14. Number of supraorbitals/tectals. This character is run unordered and all possible outgroups have fewer than 6. six – eight = 0, 10 – 13 = 1.
15. Intertemporal. The presence of an intertemporal (supratemporal) is regarded as the plesiomorph condition. Absence is derived, but it is not possible to be certain whether absence has been arrived at by loss or fusion. In any event it results in a type of skull table designated as 'type Y' by Andrews (1973). Porolepiforms also have a 'type Y' skull table. Scoring is: intertemporal present = 1, absent = 0.
16. Parietal descending process. This is a dermal process developed from the ventral surface of the parietal (postparietal) serving to brace the skull roof against the prootic region of the otico-occipital portion of the neurocranium. Process present (1).
17. Supratemporal descending process. This is a dermal process from the ventral surface of the supratemporal (tabular) serving to brace the skull roof against the otico-occipital portion of the neurocranium. It is usually a strap-like process which is directed anteroventrally to lie in front of the hyomandibular facet. Process present (1).
18. Number of extrascapulars. The plesiomorph condition assumed here shows three extrascapulars, a median plus a lateral; in coelacanths, as in porolepiforms, the overlap is laterad. Within coelacanths there may be more than three ex-

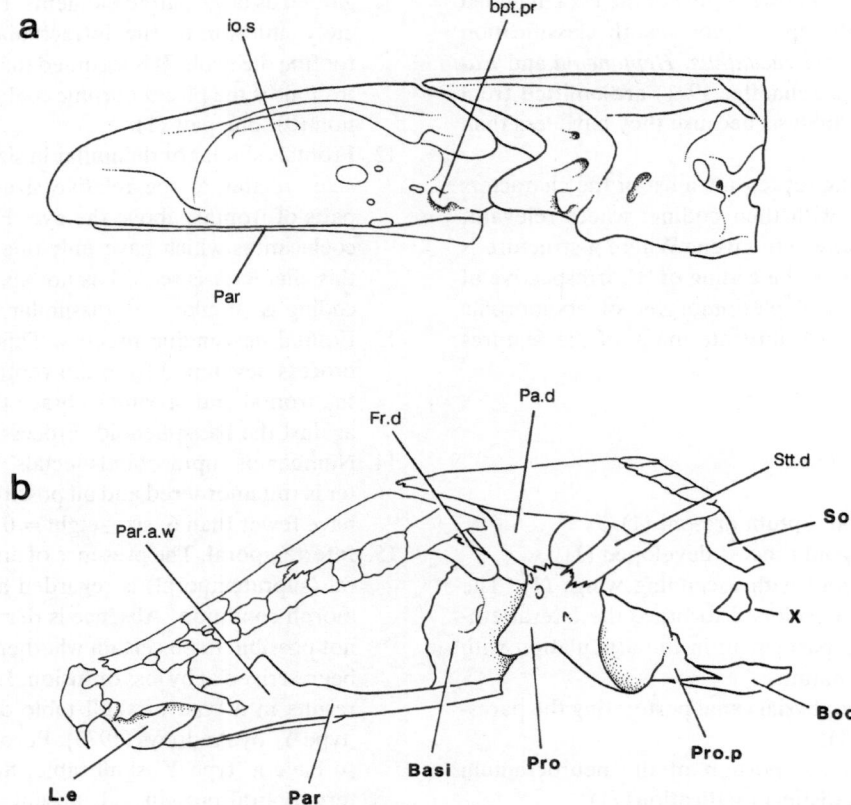

Fig. 2. Details of the coelacanth neurocranium used in the character analysis: a – *Diplocercides* showing a solidly ossified braincase, b – *Macropoma lewesiensis* showing a largely cartilaginous braincase.

trascapulars. The coding is as follows: three extrascapulars = 0, five extrascapulars = 1, more than seven extrascapulars = 2. In most coelacanths it is assumed that the lateralmost extrascapular has 'fused' with the supratemporal (see character 36). In those cases the supposedly fused bone is counted as an extrascapular.

19. Extrascapular contact. An extrascapular may be sutured (0) to the parietal shield or may lie free (1): the former condition is regarded as plesiomorphic.

20. Preorbital. This is a bone which is apparently without equivalent in rhipidistians but which may be represented in fossil lungfishes by bone '2'. It is wedged between the tectal and the lateral rostral and forms the anterior margin of the orbit. When present it is always perforated by the posterior openings of the rostral organ. When it is absent the lachrymal meets the tectal series and is usually notched by the posterior openings of the rostral organ. Preorbital present (1).

21. Spiracular bone. This is a small bone (sometimes called postspiracular) wedged between the postorbital and the operculum. There are some taxa (e.g. *Whiteia*) in which the bone may be absent as an individual variation. If the bone has been recorded in any specimen it is coded as being present in that taxon; spiracular present (1).

22. Non-overlapping cheek plates. The cheek of

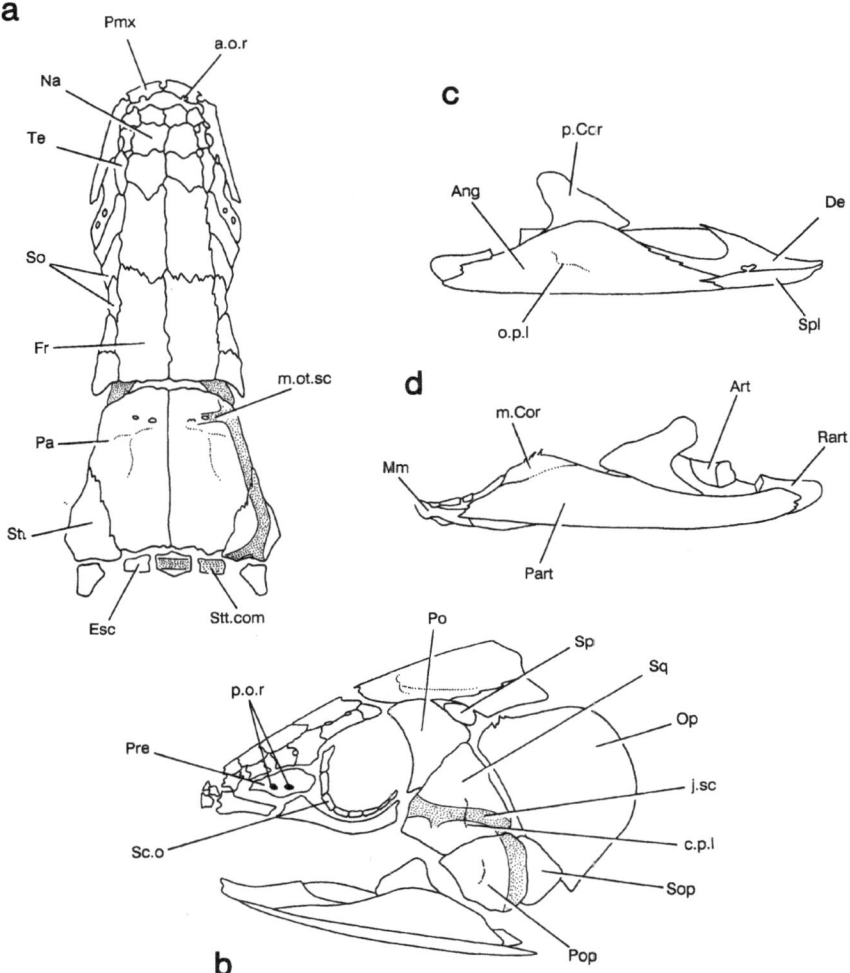

Fig. 3. Dermal bones, sensory lines and pit lines used in the character analysis: a – dorsal view of the braincase of *Whiteia woodwardi,* b – lateral view of the skull of *Rhabdoderma elegans,* c – lateral view of the lower jaw of *Macropoma lewesiensis,* d – mesial view of the lower jaw of *Macropoma lewesiensis.*

primitive bony fishes consists of a variable number of bones which are closely fitting and show various degrees of overlap. Separation of the cheek bones so that they lie free from one another is regarded as the derived condition (1).

23. Preoperculum present (1).
24. Suboperculum present (1).
25. Retroarticular and articular ossifications. The posterior end of Meckel's cartilage primitively

ossifies from two centres in bony fishes, articular and retroarticular (Nelson 1973). In *Latimeria* and many fossil coelacanths, these ossifications remain separate, while in others they fuse to form a compound articular/retroarticular. It is assumed here that a single ossification is the plesiomorphic adult condition, even though it may develop from two centres. Persistence of two centres into the adult is regarded as derived (1). This can provide some diffi-

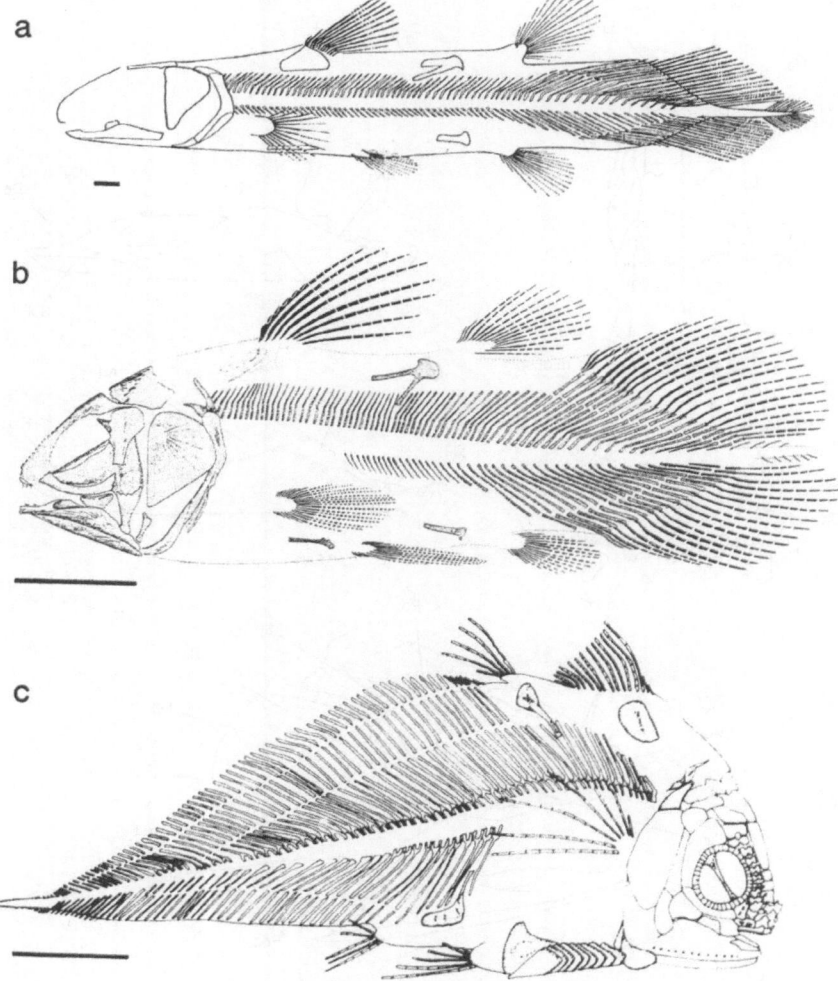

Fig. 4. Diagram illustrating three differently-shaped coelacanths: a – *Coelacanthus granulatus* (from Moy-Thomas & Westoll 1935), b – *Macropomoides orientalis,* c – *Allenypterus montanus* (from Lund & Lund 1985).

culty in coding for certain taxa, particularly those known only from few specimens because there is no guarantee that they are adults.

26. Dentary teeth separate from dentary. Teeth firmly attached to the supporting dentary are here regarded as a plesiomorphic feature of adult bony fishes since this is the common condition in outgroups. An edentulous dentary or a dentary bearing separate tooth plates as the adult condition is regarded as derived (1). In most fossil coelacanths it is impossible to distinguish between an edentulous dentary and one having separate tooth plates because the tooth plates drift away during fossilisation.

27. Crushing dentition. The dentition in many coelacanths consists of small villiform teeth with occasional larger and stouter conical teeth developed on the palatine, ectopterygoids and that coronoid which lies opposite the posterior end of the dentary (character 28). This type of

dentition probably served for holding prey before ingestion. Some coelacanths as adults have small sphaeroidal teeth, particularly developed on the parasphenoid, pterygoid and coronoids and these appear to be used for crushing (1).

28. Modified coronoid opposite dentary. The coronoid series is divisible into an anterior series of three or four elements lying medial to the dentary and separated by a diastema from the highly distinctive principal coronoid located immediately anterior to the jaw articulation. Wihin the anterior series some coelacanths show an enlarged coronoid closely associated with the dentary and bearing enlarged teeth. This is here regarded as a derived condition (1) and, in all probability, is associated with the development of enlarged teeth on the opposing dermopalatine and ectopterygoid.

29. Dentary with prominent lateral swelling (1).

30. Hook-shaped dentary. The dentary of all coelacanths is a relatively small element, reaching only a small distance along the jaw margin. As in other bony fishes, it is regarded as being primitively shallow. The hook-shape posterior end of the dentary, as seen in many coelacanths, is regarded as a derived condition (1) and is no doubt associated with the elaboration of the muscular lip fold as seen in *Latimeria.*

31. Principal coronoid sutured to the angular (1). The usual condition shows no sutural union between the coronoid and the angular.

32. Ornament on the lower jaw restricted. The plesiomorph condition is assumed to be a uniform covering of ornament on the dentary, splenial and angular. In coelacanths, other than *Miguashaia,* there is an area devoid of ornament on the posterolateral aspect of the dentary which provides an insertion area for the labial fold. A derived condition within coelacanths is considered to be the restriction of ornament to the angular (1).

33. Sclerotic ossicles. As in all sarcopterygians the sclerotic ring consists of many (c. 20) ossicles. A sclerotic ring is recorded for actinopterygians but here only four ossicles are present. Sclerotics present (1).

34. Premaxilla containing the anterior opening of the rostral organ (1).

35. Posterior openings of the rostral organ confluent (1). The identification of this character in fossil material is only possible where the preorbital is present and is therefore linked to character 20. In those taxa where the preorbital is absent the character is scored as not applicable.

36. Otic canal and supratemporal commissure join within the supratemporal. This is regarded as the derived condition and is assumed to reflect a phylogenetic fusion of the lateralmost extrascapular and the supratemporal. This results in an embayed posterior margin of the skull roof (1).

37. Medial branch of the otic canal developed and opening within the parietal (1). In some coelacanths, such as *Caridosuctor* and *Rhabdoderma,* there is variation. If a medial branch occurs in any individual, this character is coded '1'.

38. Anterior sensory pores on head substantially larger than the posteriorly located pores (1).

39. Jugal sensory canal running along ventral margin of the squamosal (1). This character reflects the fact that in some coelacanths the ventral half of the squamosal appears to have been reduced or lost.

40. Supratemporal commissure with anterior branches penetrating the parietals (1).

41. Subopercular branch of the mandibular sensory canal (1). *Latimeria* is unusual amongst living vertebrates in the development of a prominent posterior branch of the mandibular sensory canal which loops posteriorly and ramifies within a large subopercular flap of skin (Hensel 1986). The presence of this subopercular branch can be inferred by the pattern of pores within the posterior end of the angular.

42. Dentary sensory pore present. This pore, which is regarded as a derived feature (1), is located near the dentary/splenial suture. It is located above the trajectory of the mandibular sensory canal but is connected with the main canal and usually is substantially larger than the pores leading from the main line of the mandibular sensory canal. Some porolepiforms and

84

dipnoans have a series of three large pores within the lower jaw. Presumed primitive members of the porolepiforms (e.g. *Porolepis*) do not show these enlarged pores.

43. Oral pit line confined to the angular. A long oral pit line (infradentary pit line) extending forward from the angular onto the dentary and splenial is regarded as a primitive condition; abbreviation and restriction of this line to the angular is derived (1). No distinction is made here as to whether the line is continuous or broken into short segments.

44. Pit lines on cheek bones. The maximum development consists of a pit line on the squamosal and a short vertical line on the preoperculum. Pit lines present (1).

45. Diphycercal tail with supplementary lobe (1).

46. 'Asymmetrical' diphycercal tail. This character refers to a condition where one of the principal caudal lobes contains more radials and is substantially longer than the other (1).

47. Caudal fin rays equal in number to endoskeletal supports. In the majority of bony fishes the caudal lepidotrichia far outnumber the endoskeletal supports, and this is interpreted as the plesiomorph condition. In most coelacanths there is a one-to-one correspondence between the fin rays and the radial supports (1).

48. Unbranched fin rays. This is regarded as the derived condition (1). The condition is elsewhere known in some actinopterygians.

49. Fin rays with denticulate ornamentation. Smooth, unsculptured fin rays are regarded as plesiomorphic. Denticulate covering is considered derived (1), despite the fact that the form of the covering may vary.

50. Expanded fin rays. This derived condition usually involves the inner rays of the pelvic fins and is perhaps best exemplified by *Laugia* and *Coccoderma* spp., where it may be a feature of mature individuals. In some other coelacanths the fin rays of the median fins may also be expanded. This character has been coded as: fin rays not expanded = 0, paired fin rays expanded = 1, paired and median fin rays expanded = 2.

51. Anterior position of the pelvics (1). This character is scored as present even if it is only shown in some individuals of a species.

52. Long ossified ribs (1).

53. Neural arches expanded. In most coelacanths the neural arches are very slender elements and lie between the dorsal and ventral roots of the spinal nerves such that there are no notches or foramina in the arch. In some coelacanths the neural arches immediately behind the head are low, expanded and confluent with one another so that the spinal nerves issue through foramina. The expanded condition is regarded as derived (1).

54. Scale ornament differentiated. The plesiomorph condition is assumed to be a scale with a dense covering of uniform tubercles and/or ridges. In some coelacanths the central tubercles are very much enlarged, hollow and surrounded by much smaller tubercles and this is regarded as a derived condition (1).

55. Lateral line openings per scale. The plesiomorph condition is assumed to be a single opening (0) per lateral line scale. In many coelacanths there are multiple openings (1) per scale and these are often associated with a complex system of secondary tubules.

56. Rugose ornamentation of the dermal bones (1). In most coelacanths the ornamentation consists of ridges and tubercles of dentine and enamel. Rugose ornamentation describes the condition where dentine and enamel are absent and the underlying bone shows irregular vermiform patterning. Porolepiforms have been scored '0' on the basis that presumed primitive members (*Porolepis, Youngolepis*) show a superficial cosmine layer.

Results

The taxa with characters as listed in Table 1 were analysed with the PAUP computer program (Swofford 1985) and the Hennig86 computer program (Farris 1988) in an attempt to find the most parsimonious tree(s). Several runs with both programs were performed under different branch-swapping routines in order to construct a tree of coelacanth genera. It is not intended to give all the results of all the runs here since this is not the primary aim of the paper. But it is worth noting that the scope of the tree topologies remained the same, irrespective of which taxon (actinopterygians, porolepiforms or lungfishes) was used as the outgroup. The actual tree lengths did vary considerably and this probably reflected the fact that for several characters the outgroup/ancestor could only be coded as missing/non-applicable. The solution presented here used porolepiforms as the ancestor and gave the shortest trees.

Runs using all taxa coded in Table 1 resulted in too many trees being useful; this happened when running data using either PAUP (global swap branch-swapping option) or Hennig86 ('bb' branch

swapping option). This was because taxa with less than 50% real data created too many alternatives.

Therefore the solution given here and accepted for the purposes of deriving a useful classification utilises only those taxa having greater than 50% coded data. The strict consensus tree shown here results from combining 12 trees (PAUP) or 8 trees (Hennig86), containing alternative solutions at two nodes; the results of the PAUP analysis will be used. The consensus tree is given in Figure 5 and the character optimisation is given in the figure legend. One node specifies a trichotomy between *Macropoma*, *Latimeria* and *Macropomoides* and represents three alternative solutions. This uncertainty was caused by the program seeking alternative states for 26/56 of the 'missing' states in *Macro-*

Table 1. Data matrix used in PAUP analysis. The taxa have been arranged so that the first 14 have more than 75 percent real data, the next 12 have 50–74 percent and the remaining six less than 49 percent real data. Only taxa within the first two sets were used in the computer analysis. ? = details unclear, N = not applicable, U = unknown since that part of anatomy not yet found. Amongst taxa used in this analysis the data for *Hadronector* and *Polyosteorhynchus* were taken directly from Lund & Lund (1985).

Rhabdoderma elegans	1 0 0 1 0 0 0 1 0 0 0 0 1 0 0 0 1 1 0 1 1 0 1 1 0 1 0 1 0 1 0 0 0 0 1 1 0 0 1 0 0 0 0 1 1 1 1 0 1 1 0 0 0 0 0 0 0 0
Latimeria	0 0 1 0 1 1 1 0 1 0 0 1 1 1 0 1 1 2 1 0 1 1 1 1 1 1 0 1 0 1 0 N 0 0 N 1 1 0 1 1 1 1 1 0 1 1 1 1 0 0 1 1 1 0
Laugia	1 0 0 1 0 0 0 1 1 0 1 N 1 0 0 0 1 1 0 1 0 1 0 0 0 1 0 0 0 1 0 0 0 0 1 1 0 0 0 1 1 0 1 1 1 1 0 1 1 0 1 0 1 0 ? 0
Whiteia	0 0 0 ? 1 1 0 1 0 0 0 0 1 0 0 1 1 1 1 1 1 0 1 1 1 1 0 1 0 1 0 1 1 1 0 1 1 0 0 0 ? 1 1 1 1 0 1 1 1 0 0 0 0 0 1 0
Diplocercides	1 1 0 1 0 0 0 1 0 1 0 1 0 0 0 0 0 0 1 1 0 1 1 0 0 0 0 0 0 0 0 0 1 ? 0 0 0 0 0 0 0 0 0 1 ? 0 0 1 0 0 ? 0 0 0 0 0
Macropoma	0 0 1 0 1 1 1 1 0 0 0 1 1 1 0 1 1 2 1 0 0 1 1 0 1 1 0 ? 0 1 0 1 0 0 N 1 1 0 1 1 1 1 1 0 ? 0 1 1 1 0 0 0 1 1 1 0
Axelrodichthys	0 0 0 0 1 1 0 ? 1 0 0 0 1 0 0 1 0 1 0 0 0 1 1 0 1 1 0 1 1 1 1 1 0 0 N 1 1 0 1 0 ? 1 1 ? ? 0 1 1 1 0 0 ? 0 0 ? 1
Undina	0 0 1 ? 1 1 ? ? 1 ? 0 0 1 1 0 1 1 2 1 0 0 1 ? 1 1 0 1 0 0 1 1 0 N 1 1 0 1 0 ? 1 1 ? 1 0 1 1 0 1 0
Diplurus	0 0 0 0 1 1 ? ? 1 0 0 1 1 1 0 1 0 2 1 0 1 1 1 0 1 1 ? ? 0 0 0 N 0 0 N 1 1 0 ? 0 0 1 N 0 1 0 1 1 1 0 0 1 1 0 0 0
Coelacanthus	1 0 0 1 ? 1 ? ? 1 ? 0 1 1 0 0 0 1 1 0 0 0 1 0 0 0 1 0 1 0 0 0 N 1 0 N 1 1 ? 1 0 0 1 N ? 1 1 1 1 0 0 0 0 0 0 ? 0
Holophagus	0 0 ? ? 1 1 ? 1 0 ? ? 1 1 0 1 1 ? ? 0 ? 1 1 1 1 1 0 1 0 1 0 0 1 0 N 1 1 0 1 0 1 1 1 ? 1 0 1 1 1 1 0 0 1 0 ? 0
Mawsonia	0 0 0 0 1 1 0 ? ? 0 0 0 1 0 0 1 0 1 0 0 0 1 ? ? 1 1 0 1 1 1 1 1 0 ? N 1 1 0 1 0 ? 1 ? 0 ? 0 1 1 ? 0 0 ? 0 0 ? ? 1
Allenypterus	1 ? 0 ? ? ? ? ? 0 0 1 ? 1 0 ? ? 1 0 1 1 0 1 1 1 0 1 0 ? 0 0 0 0 1 ? 0 0 0 0 0 2 0 0 0 1 ? 1· 1 0 1 0 0 0 0 0 0 0 0
Coccoderma	1 ? ? ? ? ? ? ? ? ? 0 1 1 0 0 0 1 1 0 ? 0 1 1 0 ? 1 0 1 0 0 0 0 1 1 ? 0 1 1 0 0 0 1 1 0 1 0 1 1 0 1 1 0 ? 0 1 0
Polyosteorhynchus	1 ? ? ? ? ? ? 0 0 0 1 ? 0 ? ? 1 0 0 1 1 0 1 1 1 ? 0 0 0 0 0 ? 1 1 ? 0 0 0 0 0 0 ? 1 1 1 1 1 0 0 0 0 1 0 ? 0
Chinlea	? ? ? ? ? ? ? ? 1 0 0 0 1 1 0 ? ? 1 0 0 0 1 1 0 ? 1 0 1 0 1 0 1 0 0 N 1 1 0 1 0 ? 1 1 ? ? ? ? 1 ? 0 0 1 ? 0 ? 0
Caridosuctor	? ? ? ? ? ? ? ? 0 0 0 0 ? 0 0 ? ? 1 0 1 ? 0 1 1 0 1 0 1 0 0 0 0 1 1 0 0 1 0 ? 0 0 1 1 ? 1 1 1 1 0 0 0 0 1 0 ? 0
Hadronector	? ? ? ? ? ? ? ? ? 1 0 1 ? 0 1 ? ? 0 0 1 1 0 1 1 0 0 0 0 0 0 0 0 1 0 0 0 0 0 0 1 0 0 0 ? 0 0 1 1 1 0 1 1 0 0 0 0 ? ? 0
Libys	0 0 1 ? 1 1 ? ? 1 ? ? ? 1 ? 0 1 1 1 1 ? 1 0 1 0 1 0 N 1 0 ? 1 1 0 ? 0 1 1 1 0 ? 1 1 1 1 1 0 0 0 0 1 0
Sassenia	1 0 0 1 0 1 0 1 ? 0 0 1 1 0 0 0 1 ? ? 1 1 0 1 1 ? 1 ? 1 0 1 0 1 1 ? 1 1 1 ? ? 0 0 0 1 1 1 ? ? ? ? ? ? 0 ? ? 0 ? 0
Spermatodus	1 0 0 1 0 ? 0 1 0 0 0 1 1 0 0 0 1 ? 1 1 1 ? 0 1 ? 0 1 1 1 ? ? 0 ? 1 1 1 1 1 0 0 0 0 ? 1 1 U U U U U U ? U U 0 ? 0
Miguashia	1 ? 0 0 ? ? ? ? 0 ? 0 ? 0 0 1 0 0 0 0 ? ? 0 1 1 ? 0 0 0 0 ? 0 ? 1 ? ? 0 0 0 0 0 0 0 0 ? 1 0 ? 0 0 0 0 0 0 ? ? 0 0 ?
Piveteauia	0 0 0 ? 1 1 ? 1 ? ? 0 ? 1 0 0 1 1 ? ? ? 1 0 1 1 ? ? 0 ? ? ? ? ? 1 ? ? 1 1 0 0 0 ? ? ? 1 1 1 1 1 1 0 0 0 ? 0 1 0
Axelia	0 0 0 0 1 1 0 0 ? 0 1 ? 1 0 0 1 1 2 1 1 ? 0 1 1 ? ? 1 ? 1 ? ? ? ? 1 1 ? 0 1 2 0 0 0 0 ? ? 0 ? ? 1 1 ? 0 0 ? 0 ? ?
Rhabdoderma huxleyi	? ? ? ? ? ? ? 0 ? ? ? 1 ? 0 ? ? ? 0 ? ? ? ? ? ? 1 0 ? 0 0 0 ? 1 1 ? 0 1 0 ? 0 0 ? 1 ? 1 1 0 1 1 0 0 0 0 0 0
Macropomoides	0 ? ? ? ? ? ? ? 0 ? ? ? 1 ? 0 ? ? ? ? 0 ? 1 ? ? ? 1 0 1 0 1 0 N 0 ? N 1 ? ? ? ? 1 1 1 ? 1 1 1 1 1 0 0 0 ? 1 1 1
Ticinepomis	? ? ? ? ? ? ? ? 0 0 ? ? ? ? 0 ? ? ? ? ? ? ? 1 1 0 ? 1 ? 1 0 1 0 1 0 ? ? 1 ? 0 ? ? ? ? 1 ? ? 1 0 1 1 1 0 0 0 ? 0 ? 0
Wimania	0 0 0 0 1 1 0 0 ? 0 1 N 1 ? 0 1 1 ? ? ? ? 1 1 0 ? ? 0 ? ? 0 ? ? ? ? ? 1 ? ? ? ? 0 0 ? ? ? ? ? ? ? ? ? ? ? 0 ? ? 0 ? ?
Lualubaea	? ? ? ? ? ? ? ? ? ? ? 0 0 ? ? 0 ? ? 1 0 ? 0 ? ? ? ? ? ? ? 1 1 0 N 0 ? ? ? 1 ? ? 0 ? 1 1 ? ? 0 1 1 1 1 0 ? 0 ? 1 ? 1
Garnbergia	? ? ? ? ? ? ? ? 1 ? ? ? 1 ? 0 ? ? ? 1 ? 0 ? ? 0 ? 1 1 0 ? 1 0 ? 1 0 1 0 0 1 0 0 ? ? ? ? ? ? 1 ? U U U 1 0 0 U ? ? 0 ? 0
Lochmocercus	? ? ? ? ? ? ? ? ? ? ? ? ? ? ? ? ? ? ? 1 ? 0 1 1 ? 0 ? ? ? 0 0 ? 1 ? 0 ? 0 ? ? ? ? ? 0 ? ? 1 0 0 1 0 0 0 0 ? ? 0
Indocoelacanthus	? U U U U U U U ? 0 ? ? ? 0 ? ? ? ? ? 1 1 1 ? ? 1 ? ? 0 1 0 1 0 U ? ? 0 ? ? ? ? ? ? 0 U U U U U U U U U 0 ? ?
Porolepiforms	1 1 0 1 0 0 0 1 0 1 ? ? 0 0 0 0 0 0 0 0 ? 0 1 1 0 0 0 0 0 0 0 0 1 1 ? 0 0 0 0 0 0 0 0 1 0 ? 0 0 0 0 0 0 0 0 0 0 0

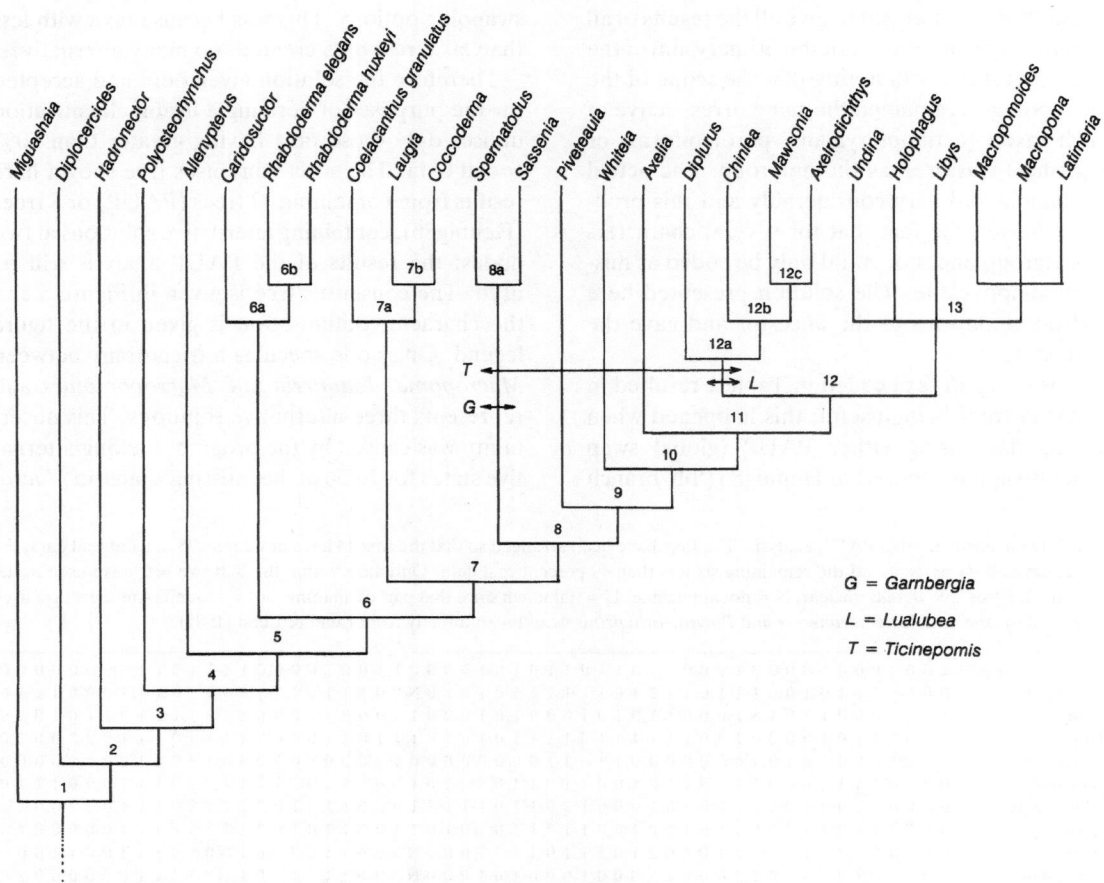

Fig. 5. Classification of coelacanths. Consensus tree produced from six trees obtained by a PAUP analysis of coelacanth genera with more than 50 percent data in Table 1 (Swap = global, multistate characters unordered). Statistics for the six equally parsimonious trees: Consistency index = 0.533; length = 107. Some taxa with less than 50 percent data may be included on the classification within limits. These are shown as horizontal lines. See text for discussion of these taxa. The character transformations are as follows, with those occurring once asterisked: node 1; 20(0–1)*, 34(0–1)*: node 2; 11(1–0)*, 45(0–1)*, 48(0–1)*: node 3; 2(1–0)*, 13(0–1)*, 17(0–1)*, 43(0–1)*, 47(0–1)*: node 4: 10(1–0)*, 26(0–1)*, 46(0–1): node 5; 18(0–1)*, 28(0–1)*: node 6; 37(0–1)*, 42(0–1)*: node 6a; 12(1–0): node 6b; 46(1–0): node 7; 6(0–1)*, 32(0–1)*, 55(0–1)*: node 7a; 9(0–1), 21(1–0), 22(0–1), 24(1–0), 34(1–0): node 7b; 38(0–1)*, 44(1–0), 50(0–1), 51(0–1)*: node 8; 19(0–1)*, 30(0–1)*, 36(0–1): node 8a; 35(0–1)*: node 9; 1(1–0)*, 5(0–1)*, 16(0–1)*, 49(0–1)*: node 10; 25(0–1), 46(1–0): node 11; 4(1–0), 18(1–2)*, 44(1–0): node 12; 9(0–1), 14(0–1), 20(1–0), 22(0–1), 34(1–0), 39(0–1): node 12a; 17(1–0)*, 24(1–0), 33(1–0), 52(0–1)*: node 12b; 12(1–0), 18(2–1), 19(1–0), 21(1–0): node 12c; 14(1–0), 19(1–0), 29(0–1)*, 31(0–1)*, 56(0–1): node 13; 3(0–1)*, 41(0–1)*: node 14; 7(0–1)*, 33(1–0)*, 40(0–1)*, 53(0–1)*, 54(0–1)*. Character transformations leading to terminal taxa: node 1-*Miguashaia;* 15(0–1): node 3-*Hadronector;* 15(0–1): node 4-*Polyosteorhynchus;* 4(1–0), 25(0–1), 53(0–1): node 5-*Allenypterus;* 47(1–0), 14(0–1): node 6a-*Caridosuctor;* 53(0–1): node 7a-*Coelacanthus;* 36(0–1), 20(0–1), 23(1–0), 39(0–1): node 7b-*Laugia;* 6(1–0), 11(0–1), 23(1–0), 53(0–1): node 7b-*Coccoderma;* 46(1–0): node 8a-*Spermatodus;* 27(0–1): node 10-*Whiteia;* 12(1–0): node 11-*Axelia;* 8(1–0), 11(0–1), 27(0–1): node 12a-*Diplurus;* 30(1–0), 53(0–1), 55(1–0); node 13-*Undina;* 12(1–0), 21(1–0): node 13-*Holophagus;* 9(1–0), 32(1–0): node 13-*Libys;* 18(2–1): node 14-*Macropomoides;* 46(0–1), 56(0–1): node 14-*Macropoma;* 21(1–0), 24(1–0): node 14-*Latimeria;* 8(1–0).

pomoides. Three of the four characters placed at the node leading to the trichotomy are unknown in *Macropomoides*.

The cause of the tetrachotomy between *Libys, Undina, Holophagus* and the taxon *Macropoma + Macropomoides + Latimeria* is more complicated. Some of the alternatives included unsupported nodes (e.g. all those alternatives which placed *Libys* as plesiomorphic to the other taxa at this hierarchical level; and that alternative which treated *Libys, Undina* and *Holophagus* as a monophyletic group). Other alternatives were suggested as a result of differing interpretations of missing data. Resolution at this level requires more data.

A few other aspects of the tree require comment. The node leading to *Laugia, Coccoderma* and *Coelacanthus*, and specifying them as a monophyletic group, is supported entirely by homoplasies. Furthermore, two of these characters (loss of spiracular – 21, and loss of suboperculum – 24) refer to character transformations which appear to have taken place many times in coelacanth history. One character supporting this node is that the cheek plates do not overlap (22). This is also a character of most Mesozoic coelacanths and the Recent *Latimeria*. However, the cheek plates of *Laugia, Coccoderma* and *Coelacanthus* are also very much reduced and it may be possible to regard this latter feature as additional support for such a grouping.

The node leading to many Mesozoic coelacanths and *Latimeria* (node 12) is also supported exclusively by homoplasies. One of these concerns the non-overlapping cheek plates referred to above at a different node. Another character is loss of preorbital (character 20), also seen in *Coelacanthus*. It is clear that this node requires additional justification.

Some of the other taxa excluded from this analysis because of lack of data may be included tentatively within this classification on the basis of particular characters. For example, *Lualubea* shows a skull roof pattern which is very similar to that seen in the *Diplurus + Chinlea + Mawsonia + Axelrodichthys* clade. In these coelacanths the extrascapulars are closely associated with the parietals, the parietal shield is parallel-sided posteriorly and tapers rapidly to a relatively narrow intracranial joint margin, and the skull bones are ornamented with coarse rugae. On these features, and in the absence of contradictory evidence, *Lualubea* may be placed within this clade.

Based on currently available information, the lower and upper limits of other taxa may be fixed within the context of this tree. *Ticinepomis* may be placed within the range of *Spermatodus + Sassenia* to the clade leading to *Diplurus/Chinlea* since it has the derived condition of a hook-shaped dentary (character 30) but is known to lack a splint-like maxilla (character 9). *Garnbergia* may be interchangeable with *Spermatodus + Sassenia* since, like *Ticinepomis*, it shows a hook-shaped dentary but it lacks denticulate ornament on the fin rays (character 49). From rather limited information about *Lochmocercus*, it appears as though it is interchangeable with *Hadronector*. *Lochmocercus* has a diphycercal tail (character 45) but according to Lund & Lund (1985, p. 45) it retains teeth which are attached directly to the dentary.

These taxa are included in Figure 5 against horizontal lines. *Indocoelacanthus* shows a hook-shaped dentary and therefore belongs in the cladistically derived half of the cladogram, but there is little evidence to set the upper limit and it is not shown in Figure 5.

Many of the character changes reflected in the tree are self-explanatory but it is worth emphasising here some of the changes in particular areas of anatomy which may be correlated and which have influenced coelacanth evolution. As a deduction from the tree there are several important changes in braincase morphology when we compare the fully ossified neurocranium of *Diplocercides* with the largely cartilaginous condition in *Latimeria*.

Several important changes appear to have taken place at node 9 (Fig. 5) where reduction in ossification is marked by the loss of an ossified interorbital septum (character 1) and the presence of discrete ossifications within the otico-occipital portion (character 5). At this hierarchical level a decrease in ossification may be associated with the development of a descending process of the parietal (character 16). Other processes from the skull roof (characters 13 and 17) specify a more inclusive group and, in terms of an evolutionary tree, are interpreted as having developed earlier (node 3). One final modification associated with the reduction of neurocranial ossification is the development of an additional dorsal process on the prootic, but this is only seen in the cladistically most derived coelacanths, including *Latimeria* (node 14).

A different area of coelacanth anatomy which appears to have undergone subtle changes is the lower jaw. There is considerable interspecific variation between species in the shapes of the principal coronoid, the angular, and the dentary and splenial. The significance of these changes remains unclear. There are, however, two features which may have functional importance. The acquisition of a modified coronoid (character 28) at node 5 and the appearance of a hook-shaped dentary (character 30) at node 8 reflect a more powerful dentition and possibly a more muscular lip fold.

A third area in which there appears to have been sequential changes concerns the elaboration of the sensory canal system. At node 6 there is the development of a large and specialised sensory pore opening within the dentary and the appearance of a medial branch of the otic sensory canal within the parietal. At a less inclusive node (13) two further elaborations are seen: the supratemporal commissure extends forward as a series of branches within the parietals, and the mandibular canal sends off branches within a large subopercular flap of skin such as is seen in *Latimeria*.

This classification illustrates that, at least in these three areas of anatomy, there have been transformations of morphologically and functionally related characters. Despite the obvious shortcomings of this classification (homoplasy and the inability to include all taxa) I wish to explore the hypothesis that there have been subtle changes in proportions and meristic parameters throughout coelacanth history.

Deductions from the classification

A. The vertebral column. One of the remarkable features of the biology of *Latimeria* is the mode of slow movement in which the body is held rigid with the thrust being created principally by sculling movements of the second dorsal and the anal fins (Fricke et al. 1987). The ability to keep a body rigid is, in part, dependent on the solidity or inflexibility of the vertebral column. At first sight this may seem curious because *Latimeria* is a notochordal fish, without the development of centra and strongly interlocking vertebrae. The neural arches are slender and thinly ossified. As Andrews (1978) has documented, the lack of ossification is compensated for by the large size of the notochord which is surrounded by several layers of tough connective tissue and is filled with a viscous jelly-like fluid (Griffith 1980, Locket 1980). This results in a thick and relatively inflexible rod which is of constant cross section throughout much of the body but which tapers rapidly at the level of those neural arches which support the caudal fin. In addition, the abdominal neural arches are closely spaced with many narrow abdominal segments, and this probably also contributes to the rigidity. Although specimens of *Latimeria* do vary, it appears as though the abdominal neural arches are more closely packed than those in the caudal region. Averaging the measurements from three specimens, the mean value for the spacing of abdominal neural arches is approximately 65% of that in the caudal region.

There is some evidence to suggest that, with exceptions, there has been an increase in the total number of vertebrae, as well as a relative crowding in the abdominal region. Thus, if we compare *Latimeria* (91–93 neural arches) with *Rhabdoderma* (45–50) there is a marked difference in their number and relative spacing between the abdominal and the caudal regions.

Some parameters of the vertebral column have been plotted against the classification in Figure 6. Admittedly, it is difficult to gather counts for most coelacanth taxa and many are unknown in this respect. Even when information is available it is gathered from very few specimens (see legend to Fig. 6). The distinction between the abdominal and caudal regions has been taken as the first neural spine to which the anterior caudal radial is attached. This rather unusual marker has been adopted because it is usually very difficult to decide

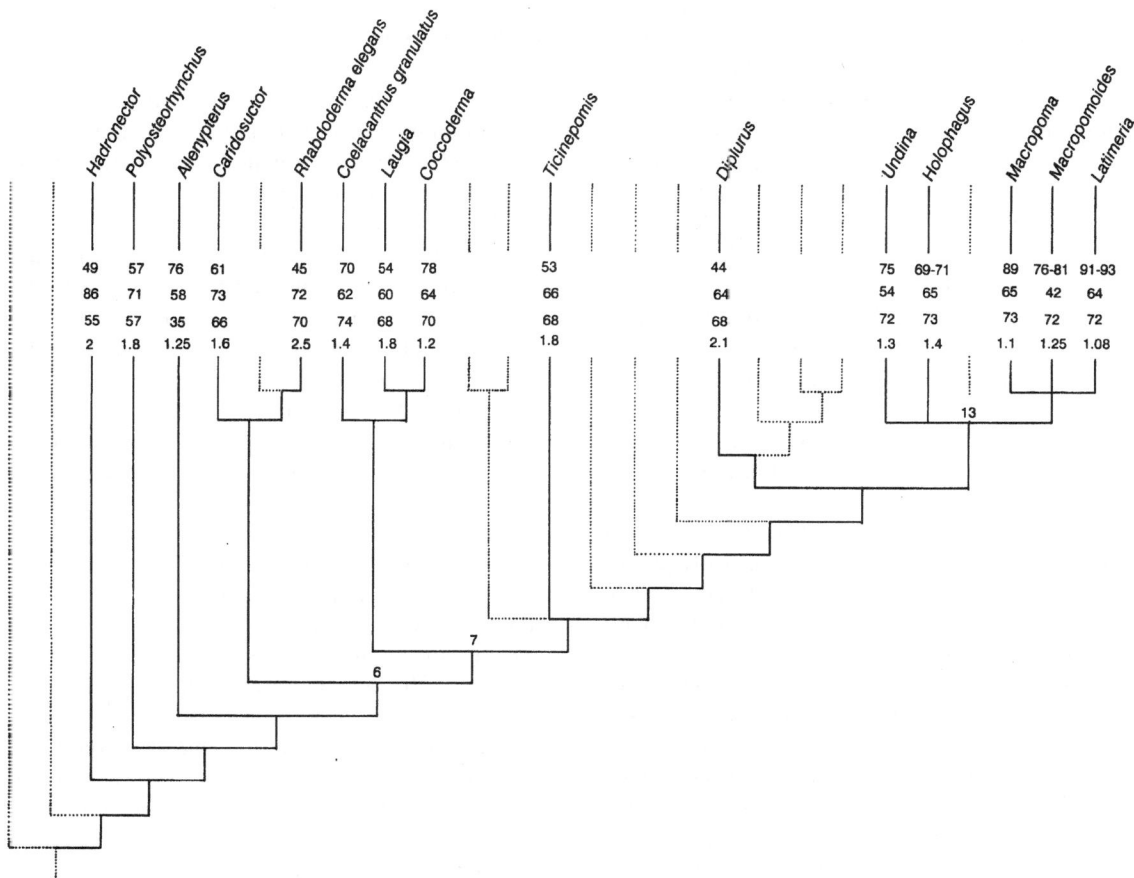

Fig. 6. Biometrics of the coelacanth vertebral column plotted against a classification. Original data on file in Department of Palaeontology, British Museum (Natural History).

The total number of vertebrae is given as the top line of figures and is counted as the number of neural arches from the occiput to the base of the supplementary lobe. The second line of figures indicates the relative spacing of the abdominal vertebrae and is calculated as the percent spacing (abdominal length/number of vertebrae) against the spacing of the caudal vertebrae, taken as 100 percent. Variation in spacing within the regions is ignored. The third line of figures gives the percent number of abdominal vertebrae. The fourth line gives an indication of the average spacing throughout the column and is calculated as length/number of vertebrae as percentage of length. Data for *Polyosteorhynchus* and *Hadronector* taken from the figures of Lund & Lund (1985); for *Ticinepomis* taken from Rieppel (1980); for *Diplurus* taken from Schaeffer (1952) and American Museum of Natural History specimen AMNH 14944. Other data: *Allenypterus* – Field Museum PF.10939 (cast); *Caridosuctor* – Field Museum PF.12920 (cast); *Rhabdoderma elegans* – from Forey (1981); *Coelacanthus granulatus* – British Museum (Natural History) BMNH P.3335, Hancock Museum G50.66; *Undina penicillata* – cast of holotype; *Coccoderma nudum* – Bayerische Staatssammlung für Palaontologie und historische Geologie 1870.XIV.23 (holotype); *Holophagus gulo* – British Museum (Natural History) BMNH P.7795, P.3344; *Macropoma speciosum* – BMNH P.9007 and from Tima (1986); *Latimeria* – from figures in Anthony (1956), Millot & Anthony (1958). Andrews (1978).

where the first complete haemal spine occurs (the usual marker for the boundary between abdominal and caudal regions). As justification for this decision it is noteworthy that in many coelacanths this is the level at which the notochord begins to taper rapidly posteriorly (see Andrews 1978, pl. 1B).

Granted that there are problems with sampling, some general observations might be made. There does appear to be a general increase in the total number of vertebrae with *Latimeria* and cladistically derived taxa at node 13 showing a high vertebral count (69–93). This compares with much lower counts for cladistically more plesiomorphic taxa. Two exceptions amongst this latter category are *Coelacanthus* (Fig. 4a) and *Coccoderma,* both with higher numbers of vertebrae (70 and 78 respectively). Significantly, these coelacanths are more elongate than most.

The percentage of abdominal vertebrae increases from less than to more than 60% at node 6 and thereafter remains relatively constant at 66–74% of the total number.

Perhaps the most significant parameter is the relative spacing of the vertebrae between the abdominal and the caudal regions. Using the spacing of the caudal vertebrae as 100%, the abdominal spacing drops from 86–72% in cladistically plesiomorphic forms, with one important exception in *Allenypterus* (see below) to 42–66% at node 7, with the majority of coelacanths showing figures of 60–65%. Thus, there does appear to be a crowding of vertebrae within the abdominal region and this is not related to overall length. *Macropomoides* (abdominal vertebrae spaced at 42% of caudal spacing) shows extremely crowded abdominal elements and is exceptional in this respect (Fig. 4b).

Crowding of the abdominal vertebrae may be a device for increasing the rigidity of the vertebral column anteriorly; and may be compensation for the lack of centra. *Latimeria* also shows further evidence for this since the ventral arcual elements of the abdominal region are in the form of irregular cartilages, fused to one another over several segments (Andrews 1978, Fig. 2b), an arrangement which must also help to increase rigidity. This feature has not been observed in fossil coelacanths but it is rare to find the ventral arcual elements well preserved, except in those few forms which have ossified ribs (those specified by node 12a in Fig. 5).

From these figures I deduce that, historically, the vertebral column of coelacanths has become more rigid over the anterior (abdominal) half with the implication that flexibility has decreased. This may be correlated with the observations of *Latime-ria* movement made by Fricke et al. (1987), that during slow movement the body is not flexed.

B. Second dorsal and anal fins. Another feature of slow swimming in *Latimeria* is the sculling action produced by the coordinated movements of the second dorsal and the anal fins, which appears to produce most of the forward thrust. Both these fins, which are of equal length to one another, are borne on prominent muscular peduncles with the fin rays arranged in a near symmetrical fashion around the end of the lobe. During movement both the second dorsal and the anal fin peduncles move from one side to the other in unison, the anal having the greater amplitude. The fin rays are capable of creating a thrust by a coordinated rotating movement around the long axis (Fricke et al. 1987).

The effectiveness of the sculling motion depends upon the flexibility and the degree of development of the lobes. There is some evidence to suggest that amongst fossil taxa the lobation of the second dorsal fin and the anal fin has increased throughout coelacanth history. Unfortunately, this feature is not quantifiable in the same way as are the vertebral characters. Indeed, the endoskeleton of the second dorsal and the anal fins is rarely preserved, having been formed by cartilage in life as in *Latimeria. Laugia* is one of the rare exceptions, showing a completely ossified endoskeleton. Usually, just a deeply embedded basal is preserved. However, the degree of lobation may be inferred from the degree of symmetry expressed by the fin rays and this is known for a number of taxa. Some of these are shown in Figure 7.

The correlation of increased lobation of the second dorsal and the anal fins, and increase in vertebral stiffening within the abdominal region, may be significant. These two phenomena may indicate that coelacanth fishes changed their mode of slow locomotion from one produced primarily by trunk flexibility to one of sculling with the second dorsal and the anal fins. It is probably also significant that the second dorsal and the anal fins are opposite, and thus the thrust can be coordinated. Fast movement and accelerated thrust in *Latimeria* appear to be produced by lateral movement of the very broad, fan-shaped tail.

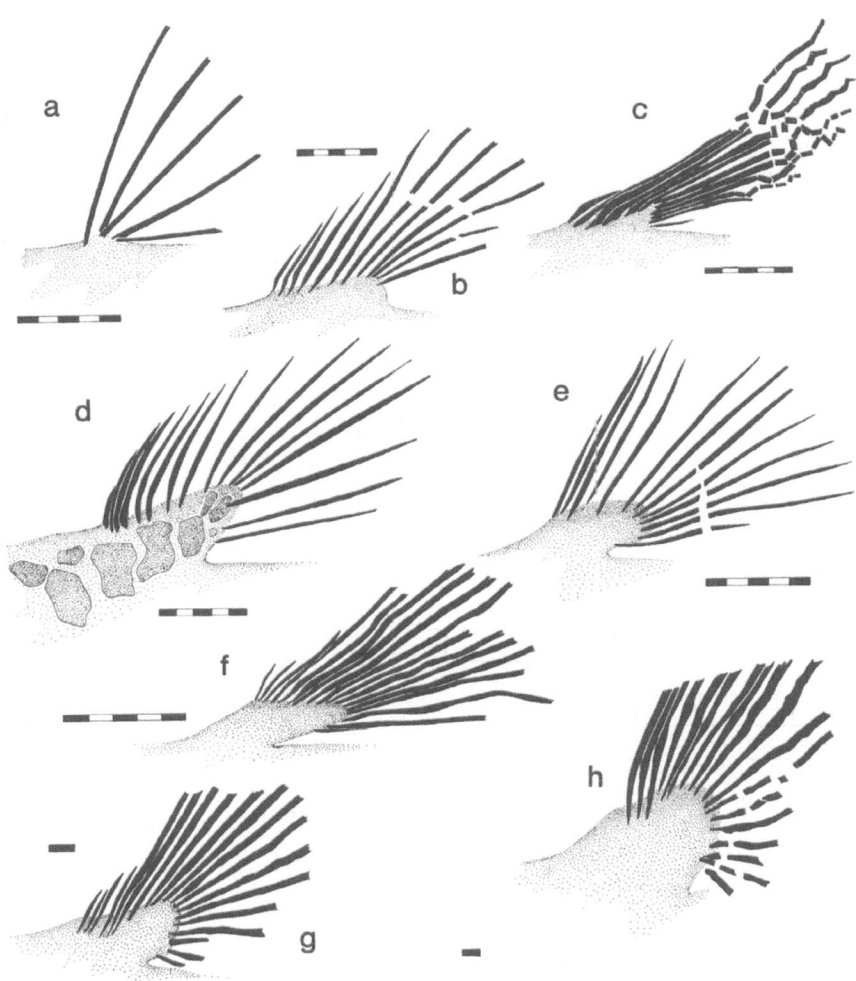

Fig. 7. Camera lucida drawings of the second dorsal fin in certain coelacanths. Some drawings reversed. Segmentation of fin rays omitted. a – *Allenypterus montanus* Field Museum PF. 10939, b – *Caridosuctor populosus* Field Museum PF. 12920, c – *Rhabdoderma elegans* Carnegie Museum CM 43965, d – *Laugia groenlandica* Mineralogical Museum Copenhagen MG VP 3170b (holotype), e – *Whiteia tuberculata* British Museum (Natural History) BMNH P.17214 (holotype), f – *Macropomoides orientalis* Museum National d'histoire Naturelle, Paris MNHN HDJ.73.21, g – *Undina penicillata* British Museum (Natural History) BMNH P.5543, h – *Holophagus gulo* British Museum (Natural History) BMNH P.3344. Scale bar 5 mm.

The exceptional taxon with regard to the biometrical ratios of the vertebral column discussed above is *Allenypterus* (Fig. 4c). Amongst coelacanths, this had a very unusual, dacryform body, with a short abdominal region and a long tapering tail which is outwardly dissimilar to the broad fanlike tail of other coelacanths. The long tail (measuring 80% of the postcranial length) accounts for the high number of vertebrae; there is also crowding of the abdominal neural arches, as in cladistically more derived forms.

The second dorsal and the anal fins in *Allenypterus* are very small when compared with those in other coelacanths, and neither is lobed. The second dorsal is also situated considerably in advance of the anal fin. Similarly, the long tapering caudal fin

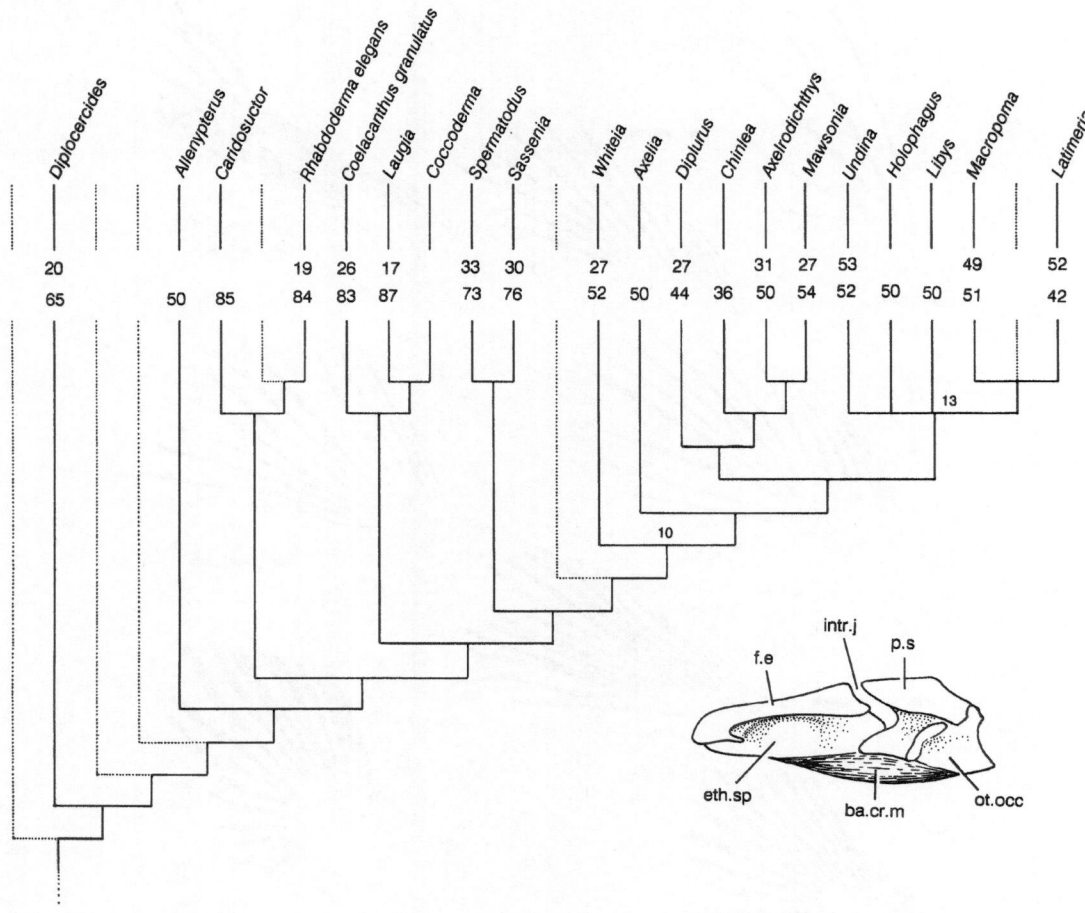

Fig. 8. Figure illustrating two parameters related to the change in skull proportions relevant to the intracranial jòint. The upper line of figures represents the percentage of the parasphenoid which is untoothed and inferred to have been covered by basicranial muscle. The lower line represents the length of the parietal shield as a percentage of the frontoethmoid shield. Original data on file in Department of Palaeontology, British Museum (Natural History). See text for discussion.

is highly asymmetrical with most of the fin rays forming a long dorsal fringe. Thus, although *Allenypterus* is precocious in the strengthening of the abdominal region of the vertebral column, the remainder of the axial skeleton suggests that fast and slow movements were produced by undulation of the long tail. Perhaps during slow movement the long upper fringe of caudal fin rays could produce longitudinal waves of movement, similar to those produced by the elongate anal fin of gymnotid teleosts (knife-fishes). Thus, *Allenypterus* probably stands apart from other coelacanths in its mode of locomotion.

C. Features related to the intracranial joint. The intracranial joint is particularly well developed in coelacanths although there is debate about its precise mechanics as well as its function (Thomson 1970, Alexander 1973, Robineau & Anthony 1973, Adamicka & Ahnelt 1976, Lauder 1980). The detailed mechanics of the joint and the parts played by the relevant muscles remain to be worked out. Here, I point out some changes in skull parameters which have taken place throughout coelacanth history, and which may have a bearing on the history of this unusual structure.

The intracranial joint divides the neurocranium into orbitosphenoid and otico-occipital moieties. These are roofed by the fronto-ethmoid and parietal shields respectively. Measurements of the neurocranial components are difficult to obtain, chiefly because most coelacanths included by node 9 (Fig. 5) have largely cartilaginous neurocrania. However, measurements of the overlying dermal roof are easier to obtain.

In Figure 8 the midline length of the parietal shield is plotted on the classification as a percentage of the midline length of the fronto-ethmoid shield. This shows that either there is a marked decrease in the relative length of the parietal shield or, conversely, an increase in the length of the fronto-ethmoid shield. A particularly marked change appears to have taken place at node 10 when the length of the parietal shield drops to half or less that of the fronto-ethmoid shield.

The relative length of the fronto-ethmoid and orbitosphenoid may have implications for the mechanics of the joint. The anterior half of the joint rotates in the vertical plane relative to the otico-occipital, taking the palate with it, at the same time bringing about a forward shift of the jaw joint. For *Latimeria,* Thomson (1970, Fig. 6) estimated that the front part of the skull moves through about 15°. Assuming that the joint of other coelacanths was equally flexible, this means that the absolute distance through which the snout will move is related to the length of the anterior moiety. Potentially therefore the movement is greater in *Latimeria* than in, for instance, *Rhabdoderma* or *Coelacanthus.*

Another factor in the operation of the intracranial joint is the development of the large paired basicranial muscle which spans the joint ventrally and (presumably) serves to close it. In *Latimeria* this muscle is attached to the basioccipital and prootic behind the joint, and reaches forward by tendons to insert on the anterior half of the parasphenoid. As Bjerring (1967) has noted, a basicranial muscle was probably present in the early coelacanth *Diplocercides* and the rhipidistian *Eu-*

sthenopteron (subcranial muscle, Jarvik 1980, Fig. 94C), but here it probably inserted on the posterior edge of the parasphenoid.

The parasphenoid is completely toothed in *Diplocercides,* whereas in *Latimeria* the forward extent of the muscle and tendon restricts the teeth to the anterior half of the bone. Thus, the extent of the muscle is probably inversely correlated with the tooth cover on the parasphenoid. Muscle length can be plotted as the percentage of the untoothed parasphenoid (Fig. 8). If the percentage untoothed parasphenoid is correlated with the length of the basicranial muscle then it can be inferred that this muscle increased considerably at node 13, extending along at least 50% of the parasphenoid.

It is tempting to correlate this increase with the changing proportions of the ethmosphenoid and otico-occipital moieties. That is, it is possible that the increasing length of the basicranial muscle is a response to the increasing length of the anterior moiety of the skull. The correlation cannot be a simple 'cause and effect' because the significant change in skull ratio takes place at a more universal level than the significant change in inferred muscle length. Since we understand so little about the mechanics of the joint in *Latimeria,* further speculation is fraught with too many uncertainties. The closure of the joint may well be effected by muscles other than the basicranial muscle (e.g. adductor mandibulae) and for a proper understanding the history of other muscles will have to be considered. For now, the inferred correlations of joint parameters with a classification are presented here to direct attention to the fact that some changes have occurred in an orderly, non-random fashion, a conclusion that seeks explanation.

Acknowledgements

I wish to thank Lance Grande of the Field Museum, Chicago and acknowledge the J. Dee Fellowship for funds to study coelacanths. The following people have kindly allowed me to study coelacanths in their care and I acknowledge their interest and hospitality while in their respective institutions: S.E. Bendix-Algreen, Geologisk Museum, Copenhagen University; P. Wellnhofer, Bayerische Staatssammlung für Paläontologie und historische Geologie, Munich; M. Gayet, Museum National d'Histoire Naturelle, Paris; J. Maisey, American Museum of Natural History, New York. I would also like to thank D. Hamilla, Sandusky, Ohio, for sending many peels of specimens of *Rhabdoderma* now in the Carnegie Museum of Natural History. I am also grateful to Richard Cloutier for discussions about coelacanths, to him and to Colin Patterson and Humphry Greenwood, all of the British Museum (Natural History) for reading and commenting on the manuscript, and to Jack Musick and Mike Bruton for the invitation to contribute to this volume.

References cited

Adamicka, P. & H. Ahnelt. 1976. Beitrage zur funktionellen Analyse und zur Morphologie des Kopfes von *Latimeria chalumnae* Smith. Annalen des (K.K.) Naturhistorischen (Hof) Museums 80: 251–272.

Alexander, R. Mc. 1973. Jaw mechanisms of the coelacanth *Latimeria chalumnae*. Copeia 1973: 156–158.

Andrews, S.M. 1973. Interrelationships of crossopterygians. pp. 137–177. *In*: P.H. Greenwood, R.S. Miles & C. Patterson. (ed.) Interrelationships of Fishes, Academic Press, London.

Andrews, S.M. 1978. The axial skeleton of the coelacanth, *Latimeria*. pp. 271–288. *In*: S.M. Andrews, R.S. Miles & A.D. Walker (ed.) Problems in Vertebrate Evolution, Academic Press, London.

Anthony, J. 1956. Opération coelacanthe. Arthaud, Paris. 197 pp.

Balon, E.K. 1991. Probable evolution of the coelacanth's reproductive style: lecithotrophy and orally feeding embryos in cichlid fishes and in *Latimeria chalumnae*. Env. Biol. Fish. 32: 249–265. (this volume)

Bjerring, H.C. 1967. Does a homology exist between the basicranial muscle and the polar cartilage? Colloques internationaux du Centre National de la Recherche Scientifique 163: 223–268.

Cloutier, R. 1990. Interrelationships of Paleozoic actinistians: patterns and trends. *In*: M.M. Chang, Y. Lui & G. Zhang (ed.) Early Vertebrate Studies and Related Problems of Evolutionary Biology, Science Press, Beijing. (in press).

Cloutier, R. 1991. Patterns, trends, and rates of evolution within the Actinistia. Env. Biol. Fish. 32: 23–58. (this volume)

Farris, J.S. 1988. Hennig86 reference. Users Manual. Stony Brook, New York.

Forey, P.L. 1981. The coelacanth *Rhabdoderma* in the Carboniferous of the British Isles. Palaeontology 24: 203–229.

Forey, P.L. 1984. The coelacanth as a living fossil. pp. 166–169. *In*: N. Eldredge & S.M. Stanley (ed.) Living Fossils, Springer Verlag, New York.

Forey, P.L. 1989. Le Coelacanthe. La Recherche 215: 1318–1326.

Forey, P.L., B.G. Gardiner & C. Patterson. 1990. The coelacanth, lungfish and cow revisited. *In*: H.-P. Schultze & L. Treub (ed.) Origins of Major Groups of Tetrapods: Contoversies and Consensus, Cornell University Press. Ithaca (in press).

Fricke, H., K. Hissmann, J. Schauer, O. Reinicke, L. Kasang & R. Plante. 1991. Habitat and population size of the coelacanth *Latimeria chalumnae* at Grand Comoro. Env. Biol. Fish. 32: 287–300. (this volume)

Fricke, H. & P. Plante. 1988. Habitat requirements of the living coelacanth *Latimeria chalumnae* at Grande Comore, Indian Ocean. Naturwissenschaften 75: 149–151.

Fricke, H., O. Reinicke, H. Hofer & W. Nachtigall. 1987. Locomotion of the coelacanth, *Latimeria chalumnae*. in its natural environment. Nature 329: 331–333.

Griffith, R.W. 1980. Chemistry of the body fluids of the coelacanth, *Latimeria chalumnae*. Proceedings of the Royal Society of London 208: 329–347.

Hensel, K. 1986. Morphologie et interprétation des canaux et canalicules sensonels céphaliques de *Latimeria chalumnae* Smith, 1939 (Osteichthyes, Crossopterygii, Coelacanthiformes). Bull. Mus. d'Hist. nat., Paris 8: 379–407.

Jarvik, E. 1980. Basic structure and evolution of vertebrates. Volume I. Academic Press, London. 575 pp.

Lauder, G.V. 1980. The role of the hyoid apparatus in the feeding mechanism of the coelacanth *Latimeria chalumnae*. Copeia 1980: 1–9.

Locket, A. 1980. Some advances in coelacanth biology. Proc. Roy. Soc. London, B. 208: 265–307.

Lund, R. & W.L. Lund. 1985. Coelacanths from the Bear Gulch Limestone (Namurian) of Montana and the evolution of the coelacanthiformes. Bulletin of the Carnegie Museum of Natural History 25: 1–74.

Millot, J. & J. Anthony. 1958. Anatomie de *Latimeria chalumnae*, 1, squelette, muscles et formations de soutien. C.N.R.S., Paris. 122 pp.

Moy-Thomas, J.A. & T.S. Westoll, 1935. On the Permian Coelacanth, *Coelacanthus granulatus* Ag. Geological Magazine 72: 446–457.

Nelson, G.J. 1973. Relationships of clupeomorphs, with remarks on the structure of the lower jaw in fishes. pp. 333–349. *In*: P.H. Greenwood, R.S. Miles & C. Patterson. (ed.) Interrelationships in Fishes, Academic Press, London.

Rieppel, O. 1980. A new coelacanth from the Middle Triassic of Monte San Giorgio, Switzerland. Eclogae Geologicae Helvetiae, Basel 73: 921–939.

Robineau, D. & J. Anthony. 1973. Biomechanique du crane de *Latimeria chalumnae*. Compte Rendu Hebdomadaire des Seances de l'Academie des Sciences, Paris D 276: 1305–1308.

Rosen, D.E., P.L. Forey, B.G. Gardiner & C. Patterson. 1981. Lungfishes, tetrapods, paleontology and plesiomorphy. Bull. Amer. Mus. Nat. Hist. 167: 159–276.

Schaeffer, B. 1941. A revision of *Coelacanthus newarki* and notes on the evolution of the girdles and basal plates of the median fins in the coelacanthini. American Museum Novitates 1110: 1–17.

Schaeffer, B. 1952. The Triassic coelacanth *Diplurus,* with observations on the evolution of the coelacanthini. Bull. Amer. Mus. Nat. Hist. 99: 25–78.

Schultze, H.-P. 1991. CT scan reconstruction of the palate region of *Latimeria chalumnae.* Env. Biol. Fish. 32: 183–192. (this volume)

Schultze, H.-P. & R. Cloutier. 1991. Computed Tomography and Magnetic Resonance Imaging studies of *Latimeria chalumnae.* Env. Biol. Fish. 32: 159–182. (this volume)

Swofford, D.L. 1985. Phylogenetic analysis using parsimony. Version 2.4. User Manual. Illinois Natural History Survey, Champaign.

Thomson, K.S. 1970. Intracranial movement in the coelacanth *Latimeria chalumnae* Smith (Osteichthyes, Crossopterygii). Postilla 149: 1–12.

Tima, V. 1986. Revision of *Macropoma speciosum* Reuss, 1857. (Crossopterygii, Coelacanthiformes). Věstník Ustředního Ústavu Geologickeho 61: 209–216.

Wourms, J.P., J.W. Atz & M.D. Stribling. 1991. Viviparity and the maternal-embryonic relationship in the coelacanth *Latimeria chalumnae.* Env. Biol. Fish. 32: 225–248. (this volume)

Appendix

Alphabetical listing of taxa, with stratigraphic and geographical information, used to generate histogram in Figure 1. Full references to authors following specific names in Forey & Cloutier (1991).

Alcoveria brevis Beltan 1972, Middle Triassic (Ladinian), Taragana, Spain.

Allenypterus montanus Melton 1969, Lower Carboniferous (Namurian A), Heath Formation, Bear Gulch Limestone Member, Montana, U.S.A.

Axelia elegans Stensio 1921, Lower Triassic (Scythian), Sassendalen Group, Sticky Keep Formation, West Spitzbergen.

Axelia robusta Stensio 1921, Lower Triassic (Scythian), Sassendalen Group, Sticky Keep Formation, West Spitzbergen.

Axelrodichthys araripensis Maisey 1986, Lower Cretaceous (Aptian), Santana Formation, Ceara, Brazil.

Caridosuctor populosum Lund & Lund 1984, Lower Carboniferous (Namurian A), Heath Formation, Bear Gulch Limestone Member, Montana, U.S.A.

Chagrinia enodis Schaeffer 1962, Upper Devonian (Fammenian), Chagrin Shale, Ohio, U.S.A.

Changxingia aspratilis Wang & Liu 1981, Upper Permian, Changxing Formation, Zhejiang Province, China.

Chinlea sorenseni Schaeffer 1967, Upper Triassic (Carnian), Chinle Formation of Utah, Colorado and New Mexico and the Tecovas Formation of Texas.

Coccoderma bavaricum Reis 1888. Upper Jurassic (Kimmeridgian), Bavaria, Germany.

Coccoderma gigas Reis 1888, Upper Jurassic (Kimmeridgian), Bavaria, Germany.

Coccoderma suevicum Quenstedt 1858, Upper Jurassic (Kimmeridigian), Bavaria, Germany.

Coelacanthus granulatus Agassiz 1839 [*C. granulosus* Agassiz 1844; *C. hassiae* Munster 1842; *C. caudalis* Egerton 1850; *C. macrophalus* Willemoes-Suhn 1869], Upper Permian (Gaudaloupian), Marl Slate, Durham and Northumberland, England and Kuperschiefer of Germany.

Coelacanthus lunzensis Teller, Upper Triassic (Carnian), Lunz Sandstone, Austria.

Diplocercides davisi (Moy-Thomas) 1937, [*Rhabdoderma* (?) *abdenense* Moy-Thomas 1937]. Lower Carboniferous (Visean), Carboniferous Limestone, Armagh, Ireland and Calciferous Sandstone, Fife, Scotland.

Diplocercides heiligenstockiensis (Jessen 1966), Upper Devonian (Frasnian), Upper Plattenkalk, Bergisch-Gladbach, Germany.

Diplocercides jaekeli Stensio 1922, Upper Devonian (Frasnian), Wildungen, Germany.

Diplocercides kayseri (v. Koenen) 1895, Upper Devonian (Frasnian), Gerolstein and Wildungen, Germany.

Diplurus longicaudatus (Newberry) 1878, Lower Jurassic (Hettangian – Sinemurian), various formations of Connecticut, New Jersey, Virginia, U.S.A.

Diplurus newarki (Bryant) 1934, Upper Triassic (Carnian), various formations of Connecticut, New Jersey, Virginia, U.S.A.

Euporosteus eifeliensis Jaekel 1927, Upper Devonian (Frasnian), Gerolstein, Germany.

Garnbergia ommata Martin & Wenz 1984, Middle Triassic (Ladinian), Baden-Württemberg, Germany.

Graphiurichthys callopterus (Kner) 1866, Upper Triassic (Norian), Raibl, Austria.

Hadronector donbairdi Lund & Lund 1984. Upper Carboniferous (Namurian A), Heath Formation, Bear Gulch Limestone Member, Montana, U.S.A.

Hainbergia granulata Schweitzer 1966, Middle Triassic (Ladinian), Göttingen, Germany.

Heptanema paradoxum Belloti 1857, Middle Triassic (Ladinian), Perledo, Italy.

Holophagus gulo Egerton 1861, [*Trachymetopon liassicum* HENNIG 1951], Lower Jurassic, (Sinemurian), Dorset, England, and Germany.

Indocoelacanthus robustus Jain 1974, Lower Jurassic, Kota Formation, India.

Laugia groenlandica Stensiö 1932, Lower Triasic (Scythian), Wordie Creek Formation, Cape Stosch, East Greenland.

Libys polypterus Münster 1842, Upper Jurassic (Kimmeridgian), Bavaria, Germany.

Libys superbus Zittel 1887, Upper Jurassic (Kimmeridgian), Bavaria, Germany.

Lochmocercus aciculodontus Lund & Lund 1984, Lower Carboniferous (Namurian A), Heath Formation, Bear Gulch Limestone Member, Montana, U.S.A.

Lualubaea henryi Saint-Seine 1955, Upper Jurassic (? Kimmeridgian), Lualubaea Series, Maosaosa, Zaire.

Lualubaea lerichei Saint-Seine 1955, Upper Jurassic (? Kimmeridgian), Lualubaea, Series, Maosaosa, Zaire.

Macropoma lewesiensis (Mantell) 1822, Upper Cretaceous (Cenomanian – Santonian), S.E. England.

Macropoma praecursor Woodward 1909, Upper Cretaceous (Cenomanian), S.E. England.

Macropoma speciosum Reuss 1857, Upper Cretaceous (Turonian), Czechoslovakia.

Macropomoides orientalis Woodward 1942, Upper Cretaceous (Middle Cenomanian), Lebanon.

Mawsonia gigas Woodward 1907 [*M. minor* Woodward 1908], Lower Cretaceous, Ilhas Formation, Bahia, Brazil.

Mawsonia lavocati Tabaste 1963, Lower Cretaceous (Albian), southern Morocco, Algeria.

Mawsonia libyca Weiler 1935, Lower Cretaceous (Albian), Baharia Formation, Baharije, Egypt.

Mawsonia tegamensis Wenz 1975, Lower Cretaceous (Aptian), Tegame Formation, Gadoufaoua, Niger.

Mawsonia ubangiensis Casier 1961, Lower Cretaceous (Neocomian), Bokungu Series, Ubangi, Zaire.

Miguashaia bureaui Schultze 1973, Upper Devonian (Frasnian), Escuminac Formation, Quebec, Canada.

Moenkopia wellesi Schaeffer & Gregory 1961, Lower Triassic (Anisian), Moenkopi Formation, Arizona.

Mylacanthus lobatus Stensiö, Lower Triassic (Scythian), Sassendalen Group, Sticky Keep Formation, West Spitzbergen.

Mylacanthus spinosus Stensiö, Lower Triassic (Scythian), Sassendalen Group, Sticky Keep Formation, West Spitzbergen.

Piveteauia madagascariensis Lehman 1952, Lower Triassic (Scythian), Middle Sakemena Group, northern Madagascar.

Polyosteorhynchus simplex Lund & Lund 1984, Lower Carboniferous (Namurian A), Heath Formation, Bear Gulch Limestone Member, Montana, U.S.A.

Rhabdoderma (?) *alderingi* Moy-Thomas 1937. Upper Carboniferous (Namurian A), Clwyd, Wales.

Rhabdoderma ardrossense Moy-Thomas 1937, Lower Carboniferous (Visean P1), Calciferous Sandstone Series, Fifeshire, Scotland.

Rhabdoderma elegans (Newberry) 1856. [*Coelacanthus lepturus* Agassiz 1844, *C.robustus* Newberry 1856, *C. ornatus* Newberry 1856, *C. elongatus* Huxley 1866, *C. summiti* Wellburn 1903, *C. watsoni* Aldinger 1931, *Conchiopsis filiferus* Cope 1873, *C. anguliferus* Cope 1873, *Hoplopygus binneyi* Agassiz

1844, *Rhabdoderma corrugatum* Moy-Thomas 1935]. Upper Carboniferous (Westphalian D) of Linton, Ohio, U.S.A. This species is also known from Cannelton, Pennsylvania and from many Upper Carboniferous (Namurian – Westphalian) localities in the British Isles, northern France, Belgium, Holland, Germany and the Stephanian of the Ukraine.

Rhabdoderma exiguum (Eastman) 1902, Upper Carboniferous (Westphalian D), Francis Creek Shale, Carbondale Formation, Illinois, U.S.A.

Rhabdoderma huxleyi (Traquair) 1881. Lower Carboniferous (Visean B), Calciferous Sandstone, Glencartholm Volcanic Group, Dumfriesshire, Scotland.

Rhabdoderma madagascariensis (Woodward) 1910. Lower Triassic (Scythian), Middle Sakamena Group, northern Madagascar.

Rhabdoderma tinglyense (Davis) 1884. [*Coelacanthus mucronatus* Pruvost 1914, *C. granulostriatus* Moy-Thomas 1935], Upper Carboniferous (Westphalian) of the British Isles, northern France, Belgium, Holland and Germany.

Rhipis moorseli Saint-Seine 1950, Lower Cretaceous, Kimbau, Zaire.

Rhipis tuberculata Casier 1965. Lower Cretaceous, Kinko, Zaire.

Sassenia groenlandica Forey et al. 1990. Lower Triassic (Scythian), Wordie Creek Formation, Cape Stosch, East Greenland.

Sassenia (?) *guttata* (Woodward) 1912, Lower Triassic (Scythian), Sassendalen Group, Sticky Keep Formation, West Spitzbergen.

Sassenia tuberculata Stensiö 1921. Lower Triassic (Scythian), Sassendalen Group, Sticky Keep Formation, West Spitzbergen.

Scleracanthus asper Stensiö, Lower Triassic (Scythian), Sassendalen Group, Sticky Keep Formation, West Spitzbergen.

Sinocoelacanthus fengshanensis Liu 1964, Lower Triassic, Fengshan District, Kwangsi, China.

Spermatodus pustulosus Cope 1894, Lower Permian (Admiral Formation, Wichita Group) of Texas, U.S.A.

Synaptotylus newelli (Hibbard) 1933. [*Coelacanthus arcuatus* Hibbard 1933]. Upper Carboniferous (Stephanian), Kansas, U.S.A.

Ticinepomis peyeri Rieppel 1980, Middle Triassic (Ladinian), Grenzbitumen Horizon, Monte San Giorgo, Switzerland.

Undina (?) *barroviensis* Woodward 1890, Lower Jurassic (Lower Lias), Leicestershire and Warwickshire, England.

Undina cirinensis Thiolliere 1854 [*U. minuta* Wagner 1863], Upper Jurassic (Kimmeridgian), Bavaria, Germany; Cerin, France.

Undina penicillata Münster 1834 [*Undina acutidens* Reis 1888; *Coelacanthus striolaris* Münster 1842; *Coelacanthus kohleri* Münster 1842], Upper Jurassic (Kimmeridgian), Bavaria, Wurttemburg, Germany.

Undina picena (Costa 1862), Upper Triassic (Norian), Italy.

Undina purbeckensis Woodward 1916, Upper Jurassic (Purbeckian), Dorset, England.

Whiteia africanus (Broom) 1905, Lower Triassic, Orange Free State, Republic of South Africa.

Whiteia groenlandica Forey et al. 1990, Lower Triassic (Scythian), Wordie Creek Formation, Cape Stosch, East Greenland.

Whiteia tuberculata Moy-Thomas 1935, Lower Triassic (Scythian), Middle Sakemena Group, northern Madagascar.

Whiteia woodwardi Moy-Thomas 1935. [*Coelacanthus evolutus* Beltan 1980], Lower Triassic (Scythian), Middle Sakemena Group, northern Madagascar.

Wimania (?) *multistriata* Stensiö 1921, Lower Triassic (Scythian), Sassendalen Group, Sticky Keep Formation, West Spitzbergen

Wimania sinuosa Stensiö 1921, [*Leioderma sinuata* Stensiö 1918]. Lower Triassic (Scythian), Sassendalen Group, Sticky Keep Formation, West Spitzbergen.

Youngichthys xinhuanisis Wang & Liu 1981, Upper Permian, Changxing Formation, Zhejiang Province, China.

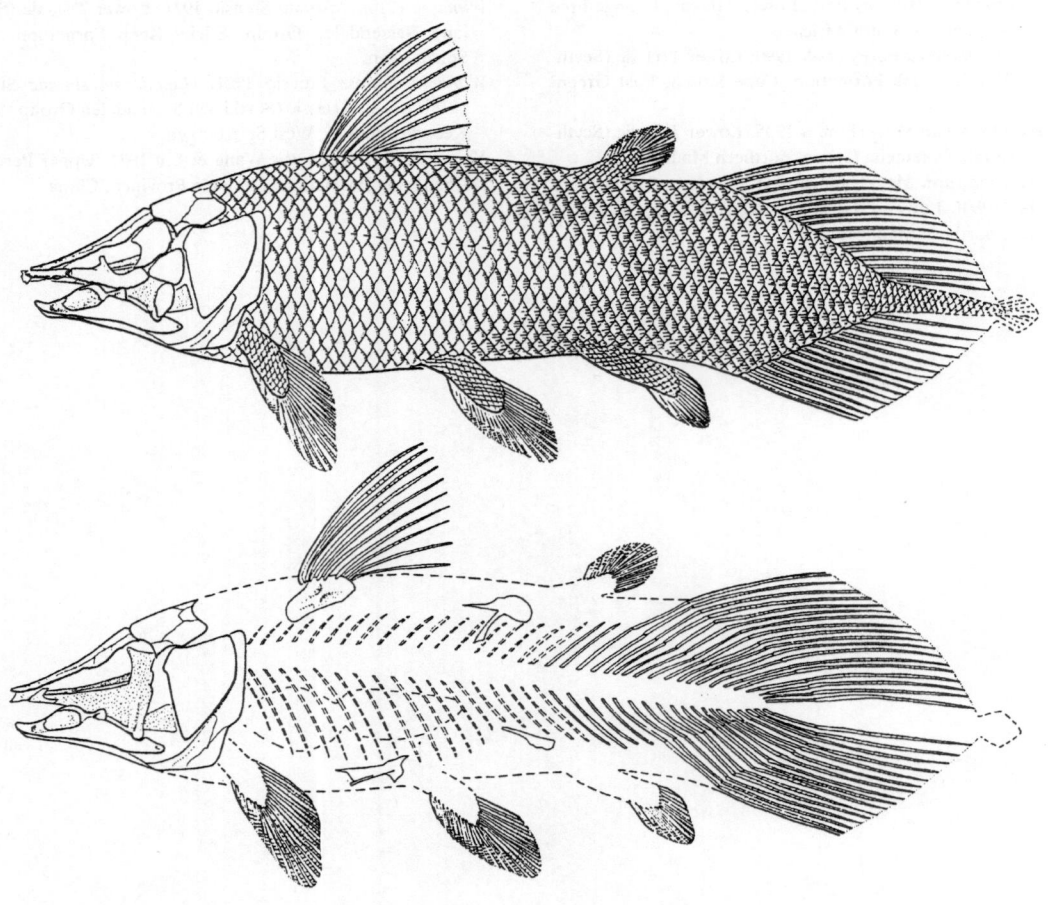

Reconstruction of the 110 million old *Axelrodichthys araripensis* from the just published book on **Santana Fossils** by John G. Maisey (by permission of T.F.H. Publications).

Environmental Biology of Fishes **32**: 99–117, 1991.
© 1991 *Kluwer Academic Publishers.*

A phylogenetic analysis of the 18S ribosomal RNA sequence of the coelacanth *Latimeria chalumnae*

David W. Stock, Kirk D. Moberg, Linda R. Maxson[1] & Gregory S. Whitt
Department of Ecology, Ethology & Evolution, University of Illinois, Urbana, IL 61801, U.S.A.
[1] *Present address: Department of Biology, Pennsylvania State University, University Park, PA 16802, U.S.A.*

Received 21.5.1989   Accepted 30.7.1990

Key words: Molecular phylogeny, Maximum parsimony, Chondrichthyes, Dipnoi, Actinopterygii, Tetrapoda, Sarcopterygii

Synopsis

Approximately 98% of the sequence of the 18S ribosomal RNA (rRNA) of the coelacanth *Latimeria chalumnae* was determined by a combination of direct RNA sequencing and sequencing of rRNA genes amplified by the polymerase chain reaction. This sequence was compared with 18S rRNA sequences of similar length from seven other vertebrate species, representing the taxa Petromyzontiformes, Holocephali, Elasmobranchii, Actinopterygii, Dipnoi, Amphibia, and Amniota, in order to determine the most likely sister group of the coelacanth. Maximum parsimony analysis of these sequences resulted in a single most parsimonious tree containing a number of anomalous relationships among these groups. A bootstrap analysis showed that none of the relationships in this tree was significantly supported at the 95% level, however. Addition of data from 15 other vertebrates (providing multiple representatives of most of the higher taxa) resulted in similar ambiguous groupings, as did a number of methods of editing the sites compared (designed to eliminate rapidly evolving positions). These results may be due to a relatively rapid radiation of the major lineages of osteichthyans, the resolution of which will require molecular information from a larger portion of the coelacanth genome.

Introduction

The phylogenetic relationships of *Latimeria chalumnae* Smith have been among the most controversial issues in vertebrate systematics. Coelacanths are most commonly united with lungfishes and tetrapods in the taxon Sarcopterygii, which forms the sister group to the Actinopterygii within the Osteichthyes (Romer 1955). Within the Sarcopterygii, however, relationships among the three taxa are unclear. The 'traditional' view has been that coelacanths are the sister group of tetrapods among extant vertebrates (Westoll 1961, Romer 1966, Holmes 1985, Schultze 1986, Fritzsch 1987),

although a number of authors have suggested that lungfishes are the sister group of tetrapods, with coelacanths forming the sister group of this pair (Miles 1977, Forey 1980, 1986, Gardiner 1980, 1984, Rosen et al. 1981, Maisey 1986, Mommsen & Walsh 1989). The final possibility, a sister group relationship between coelacanths and lungfishes, has only been proposed recently (Northcutt 1986, Forey 1988). This heterogeneity of interpretations reflects a level of uncertainty that has led some authors to represent relationships within the Sarcopterygii as an unresolved trifurcation (Nelson 1969, Bjerring 1973, Schultze 1981, Panchen & Smithson 1987).

Even the generally accepted sarcopterygian affinities of coelacanths have been challenged by several studies. Coelacanths have variously been considered to be the sister group of the Chondrichthyes (Løvtrup 1977, Lagios 1979, 1982), the sister group of all other osteichthyans (Wiley 1979), or the sister group of the Actinopterygii plus Tetrapoda (von Wahlert 1968). In summary, since virtually all extant gnathostome higher taxa have been suggested by one or more authors to be the sister group of coelacanths, the most conservative conclusion is that *Latimeria* is a gnathostome of uncertain affinity.

Most of the studies listed above have utilized morphological characters, although a few have used physiological (Løvtrup 1977, Lagios 1979, Mommsen & Walsh 1989) or karyological (Dingerkus 1979) data. One source of phylogenetic information that has yet to be adequately exploited for the determination of coelacanth relationships is the comparison of macromolecular sequences. The only macromolecules from the coelacanth that have been sequenced to date are triosephosphate isomerase (Kolb et al. 1974), α and β parvalbumins (Jauregui-Adell & Pechere 1978, Pechere et al. 1978), and a portion of the 28S ribosomal RNA (Hillis & Dixon 1989, Hillis et al. 1991). None of these molecules has been sequenced in a sample of gnathostome taxa that is adequate to test all of the above phylogenetic hypotheses. The sequence of triosephosphate isomerase, for example, is not known in any other fishes, and the last three sequences have not been determined in lungfishes.

If molecular sequence comparisons are to prove useful in determining the phylogenetic position of the coelacanth, sequences must not only be obtained from a sufficient representation of taxa to test all alternative hypotheses, but the molecule being examined must have sites that have evolved at rates appropriate for reconstructing branching events that occurred over 400 million years ago. For example, Maeda et al. (1984) concluded that parvalbumins evolved too rapidly for resolving such ancient events. On the other hand, triosephosphate isomerase (see the comparison of *Latimeria* and tetrapod sequences by Lu et al. 1984) and 28S rRNA (Hillis & Dixon 1989, Hillis et al.

1991) may prove to be appropriate for such comparisons once data are gathered from a greater number of taxa. One molecule with a highly conserved primary structure that previously has not been examined in the coelacanth or an extensive sample of other gnathostome taxa is the 18S ribosomal RNA (rRNA). The conservative nature of this molecule is demonstrated by its use in phylogenetic studies of the relationship of eukaryotes to the main prokaryotic lineages (Woese 1987, Olsen 1987, 1988, Cedergren et al. 1988, Lake 1988), the interrelationships of lower eukaryotes (McCarroll et al. 1983, Elwood et al. 1985, Herzog & Maroteaux 1986, Sogin & Elwood 1986, Sogin et al. 1986, 1989, Gunderson et al. 1987, Vossbrinck et al. 1987, Clark & Cross 1988), the interrelationships of the higher eukaryotic kingdoms (Hasegawa et al. 1985, Gouy & Li 1989), the interrelationships of higher level taxa of plants (Nairn & Ferl 1988, Zimmer et al. 1989), the interrelationships of animal phyla (Field et al. 1988), the interrelationships of higher level taxa of echinoderms (Raff et al. 1987) and the interrelationships of insect orders (Wheeler 1989). Despite this extreme conservation of certain regions of the 18S rRNA, some regions exhibit sequence variation within the Vertebrata, as pointed out by Raynal et al. (1984) and McCallum & Maden (1985), and as illustrated by the present study.

In vertebrates, the 18S rRNA gene is a moderately repetitive sequence, existing in several hundred to several thousand copies (Long & Dawid 1980). The transcripts of these genes are also highly abundant. This latter characteristic allows direct sequencing of 18S rRNA transcripts from a total cellular RNA preparation (Qu et al. 1983, Lane et al. 1985). This approach avoids the danger of sequencing aberrant copies of the gene that are not expressed, and is extremely rapid (Qu et al. 1983). However, RNA sequencing has a slightly higher error rate than DNA sequencing (Lane et al. 1985), and with the advent of the polymerase chain reaction (PCR), sequencing of rRNA genes is essentially as rapid as direct RNA sequencing. We have used a combination of these techniques to determine the nucleotide sequence of approximately 98% of the 18S rRNA from the coelacanth

Latimeria chalumnae. In this paper, we report the results of a phylogenetic comparison of this sequence with 18S rRNA sequences from seven other species, representing the taxa Petromyzontiformes, Holocephali, Elasmobranchii, Dipnoi, Tetrapoda, and Actinopterygii. The use of the first of these taxa as an outgroup allows the evaluation of all of the previously proposed hypotheses of the phylogenetic position of the coelacanth within the Gnathostomata.

Materials and methods

Source of sequences

The 18S rRNA sequence of *Latimeria chalumnae* (VIMS # 8118) was determined as described in the following sections. The sequence presented here contains several additions to and modifications of the sequence reported by Moberg (1989). Sequences from *Homo sapiens* and *Xenopus laevis* were obtained from Torczynski et al. (1985) and Salim & Maden (1981), respectively. An additional sequence from *Homo sapiens* was reported by McCallum & Maden (1985), and differs from that of Torczynski et al. (1985) by one substitution, one insertion, and one apparent inversion of two nucleotides. Both sequences, however, yield identical results in the phylogenetic analyses reported below. Sequences from representatives of Petromyzontiformes (*Petromyzon marinus*), Holocephali (*Hydrolagus colliei*), Elasmobranchii (*Rhinobatos lentiginosus*), Dipnoi (*Neoceratodus forsteri*), and Actinopterygii (*Atractosteus spatula*) were determined by direct RNA sequencing only, and will be described in future publications addressing a different set of molecular phylogenetic issues (Stock & Whitt unpublished).

Ribosomal RNA extraction

Ribosomal RNA was extracted from *Latimeria chalumnae* using a protocol modified from Loening (1967). Approximately 1 g of frozen skeletal muscle was ground to a fine powder in liquid nitrogen,

using a mortar and pestle. This powder was suspended in 15 ml of homogenization buffer (10 mM Tris-HCl pH 8.0, 1 mM EDTA, 2% (w/v) SDS, 5% (w/v) sodium tri-isopropylnaphthalene sulfonate) and homogenized with a Brinkmann Polytron homogenizer. The homogenate was then extracted twice with an equal volume of phenol, followed by addition of NaCl to 0.5 M. Additional extractions were performed with phenol and phenol/chloroform/isoamyl alcohol (25 : 24 : 1) until no protein was visible at the interface. A final extraction with chloroform/isoamyl alcohol (24 : 1) was followed by ethanol precipitation and resuspension of the nucleic acid pellet in H_2O. DNA and low molecular weight RNA were removed by two cycles of sodium acetate precipitation as described by MacDonald et al. (1987), followed by a final ethanol precipitation. The resulting high molecular weight RNA was resuspended in H_2O to a final concentration of approximately 0.5 μg per μl. All solutions used for RNA extraction and sequencing were treated with diethylpyrocarbonate as described in Maniatis et al. (1982) in order to inhibit ribonuclease activity.

18S rRNA sequencing

Direct sequencing of 18S rRNA using reverse transcriptase and specific oligonucleotide primers was carried out essentially as described by Lane et al. (1985), except that the chase reaction was omitted. The primers used, their nucleotide sequence, and their locations relative to the 18S rRNA sequences of *Latimeria chalumnae* (Appendix 1) and *Homo sapiens* (Torczynski et al. 1985) are shown in Table 1. The products of each chain termination reaction were subjected to electrophoresis on two separate 50 cm 8% polyacrylamide, 8 M urea gels (Lane et al. 1985). The first was a buffer gradient gel (Biggin et al. 1983) run at 1800 V for approximately 3.5 h, and the second gel (with a uniform buffer concentration) was run at 1600 V for 8–10 h. Band compressions were clarified by carrying out additional sequencing reactions with 7-deaza-dGTP in place of dGTP (Barr et al. 1986). Sequencing reactions employing a terminal deoxynucleotidyl transferase chase step, as described by DeBorde et al. (1986), were used to clarify further sequence ambiguities.

PCR amplification of 18S rRNA genes

The polymerase chain reaction (Saiki et al. 1988) was used to amplify the 18S rRNA genes of *L. chalumnae* in order to verify the direct RNA sequence on the opposite (antisense) strand. Genomic DNA was extracted according to the procedure of Maniatis et al. (1982) from the same tissue sample as used for the RNA extraction. Approximately 100 ng of DNA was used as the substrate for a 100 μl PCR containing 10 mM Tris-HCl pH 8.3, 50 mM KCl, 2 mM $MgCl_2$, 0.01% (w/v) gelatin, 200 mM each dNTP, 2.5 units of *Taq* DNA polymerase (Perkin-Elmer Cetus), and 1 μM each amplification primer. The primers used are indicated in

Table 1 and permit the amplification of a 1758 base pair stretch of the 18S rRNA genes of *L. chalumnae*. Amplification was performed in a Perkin-Elmer Cetus thermal cycler. Samples were heated at 94° C for 2 min followed by 35 cycles of the following temperature profile: 1 min at 94° C, 1 min at 60° C, and 2 min at 72° C. After the 35th cycle, a 5 min extension at 72° C was performed.

Single stranded DNA for sequencing was prepared from the double stranded product of the reaction described above by asymmetric PCR (Gyllensten & Erlich 1988). A 10 μl aliquot of the double stranded PCR product was subjected to electrophoresis on a 1% low melting temperature agarose gel (SeaPlaque – FMC). The DNA was visualized by ethidium bromide staining and long wavelength UV illumination. A single band of the predicted molecular weight was excised, placed in 100 μl of H_2O, and melted at 65° C for 5 min. 2 μl of

Table 1. Oligonucleotide primers used for sequencing and PCR amplification of the 18S ribosomal RNA of *Latimeria chalumnae*. The first thirteen primers were used for direct RNA sequencing, while the remaining seven were used for sequencing of PCR-amplified DNA. The two primers used for PCR amplification are indicated in bold. The numbering of the positions to which the primers anneal follows that of Appendix 1 (*Latimeria*) and Torczynski et al. (1985) (*Homo sapiens*). K = G and T, M = C and A, R = A and G, Y = C and T.

Primer sequence	*Latimeria* positions	*Homo* positions
CTAGAATTRCCACAGTTATCC	161–141	165–145
CCATCGAAAGTTGATAGGGCAG	332–311	374–353
TTTCTCAGGCTCCCTCTCCGG	407–387	449–429
ACCGCGGCKGCTGGC[1]	585–571	627–613
TCCAACTACGAGCTT[2]	643–629	685–671
GTCCTATTCCATTATTCC	830–813	876–859
CCGRTCCAAGAATTTCACCTCT	930–909	976–955
GCCCTTCCGTCAATTCCTTTAAGTTTCAGC[3]	1167–1138	1213–1184
GTCAAATTAAGCCGC	1201–1187	1247–1233
AAGAACGGCCATGCACCACC	1297–1278	1343–1324
TCTAAGGGCATCACAGAC	1458–1441	1508–1491
ACGGGCGGTGTGTRC[1]	1657–1643	1707–1693
CACCTACGGAAACCTTGTT	NA	1848–1830
CTGGTTGATCCTGCCAG	4–20	4–20
GTCTGCCCTATCAACT	309–324	351–366
GTGCCAGDMGCCGCGG[1]	569–584	611–626
YAGAGGTGAAATTCTTGG	908–925	954–971
GAAACTTAAAKGAATTG[1]	1141–1157	1187–1203
ARCTTCTTAGAGGGAC	1388–1403	1438–1453
TGYACACACCGCCCGT[1]	1642–1657	1692–1707

[1] Obtained from C.R. Woese, University of Illinois, Urbana, IL.

[2] Obtained from D.L. Nickrent, University of Illinois, Urbana, IL.

[3] Obtained from E.A. Zimmer, Louisiana State University, Baton Rouge, LA.

the melted gel slice was used as the substrate for a 200 μl asymmetric PCR containing the same components as the reaction described above, with the exception that primer 1830 (numbered according to the *H. sapiens* sequence – see Table 1) was present at 1 μM and primer 20 was present at 0.01 μM. The cycling parameters were the same as described above. This procedure resulted in the production of single stranded DNA corresponding to the antisense strand (the strand opposite to that sequenced by direct RNA analysis). The single stranded DNA was then purified by ultrafiltration using a Centricon-100 microconcentrator (Amicon).

Sequencing of asymmetric PCR products

Direct sequencing of asymmetric PCR products was performed using a modified T7 DNA polymerase (Sequenase – US Biochemicals) according to the manufacturer's instructions. Electrophoresis followed the procedure described above for RNA sequencing.

Alignment of sequences

Alignment of 18S rRNA sequences was performed manually, as described by McCarroll et al. (1983), Woese et al. (1983) and Gutell et al. (1985). Due to the extremely high degree of sequence conservation observed within the vertebrates, refinements based on secondary structure were largely unnecessary. Any areas of ambiguous alignment due to length variation were deleted from the phylogenetic analyses for the reasons discussed by Olsen (1988).

Phylogenetic analyses

Maximum parsimony analyses were performed with version 3.0 of the PAUP program written by D. Swofford (Illinois Natural History Survey, Champaign, Illinois). The most parsimonious trees were determined using the EXHAUSTIVE SEARCH or BRANCH AND BOUND options, which are guaranteed to find all most parsimonious trees. The degree of confidence that could be assigned to the various groupings in the most parsimonious trees was determined using the BOOTSTRAP option of PAUP. In this process, an artificial data set of the same size as the original data set is constructed by sampling characters from the original data set with replacement. The most parsimonious tree or trees are then determined from the artificial data set. This process is then repeated (generally 100 or more times) and the frequency of occurrence of each grouping of taxa in the most parsimonious trees is recorded. A grouping is generally considered to be significantly supported if it appears in 95% or more of the most parsimonious trees (Felsenstein 1985).

Results

The sequence of 1779 nucleotides of the 18S rRNA of *Latimeria chalumnae* that we have determined is shown in Appendix 1. Approximately 99% of this sequence was obtained from both strands (i.e. the rRNA transcript itself and the antisense strand of the rRNA genes). The only region of the molecule not sequenced corresponds to the last 39 nucleotides at the 3′ end of the human sequence published by Torczynski et al. (1985). Since this region is highly conserved in both length and sequence among most eukaryotes (see the alignment of Dams et al. 1988), we estimate the total length of the 18S rRNA of *L. chalumnae* to be 1818 nucleotides.

The *L. chalumnae* sequence was aligned with sequences of similar length from representatives of the taxa Petromyzontiformes (*Petromyzon marinus*), Holocephali (*Hydrolagus colliei*), Elasmobranchii (*Rhinobatos lentiginosus*), Actinopterygii (*Atractosteus spatula*), Dipnoi (*Neoceratodus forsteri*) and Tetrapoda (*Xenopus laevis* – Salim & Maden 1981, McCallum & Maden 1985; *Homo sapiens* – Torczynski et al. 1985). Sequences from the first five of these taxa were obtained by direct RNA sequencing only and therefore contained sequence ambiguities comprising up to 4% of their total length. Many of these ambiguities consisted of bands in all lanes at a particular level of the sequencing gel, and occurred at positions known to be occupied by modified nucleotides (i.e. 2'-O-methyl riboses or base modifications) in the sequences of *Xenopus laevis* and *Homo sapiens* (McCallum & Maden 1985). These positions were coded as missing data, but probably did not greatly affect the analysis, since they presumably were identical in the different species. The other type of sequencing ambiguity was band compression caused by undenatured secondary structure of the cDNAs produced in the sequencing reactions (Kramer & Mills 1978). Regions of band compression in one or more of the species were deleted from the phylogenetic analyses.

In addition to regions of ambiguous alignment caused by band compression, some regions of the 18S rRNA could not be unambiguously aligned because of actual length variation among taxa (i.e. due to insertions or deletions) and were excluded from the phylogenetic analyses. These analyses were performed on the remaining 1668 aligned positions, 202 of which were variable among the eight taxa examined. Of these 202 positions, 68 were phylogenetically informative (i.e. were occupied by at least two different nucleotides, each of which occurred in at least two different taxa).

The most parsimonious tree found by analysis of this data set with the EXHAUSTIVE SEARCH option of PAUP required 298 steps and is shown in Figure 1. Inspection of this tree reveals a number of relationships that differ radically from any tree yet proposed for the Gnathostomata. Particularly surprising are the placement of the elasmobranch (*Rhinobatos*) well within the Osteichthyes and apart from the holocephalan (*Hydrolagus*), as well as the paraphyly of the Tetrapoda. Maisey (1986) summarized characters supporting the monophyly of both the Osteichthyes and the Chondrichthyes, whereas tetrapod monophyly is defended by Panchen & Smithson (1987). Even the diphyletic origin of tetrapods from porolepiforms and osteolepiforms advocated by Jarvik (1980) would result in a monophyletic Tetrapoda relative to the species included in our analysis (at least in his classification scheme). The placement of *Atractosteus spatula* and *Neoceratodus forsteri* is also unorthodox, al-though White (1965) and von Wahlert (1968) have proposed that the Dipnoi (represented by *Neoceratodus*) are the sister group of the remainder of the Osteichthyes. We have taken two approaches to assess the significance of these results. The first is to examine previously proposed hypotheses of the position of coelacanths within the Gnathostomata, and to compare the lengths of the corresponding trees calculated from our data set with the distribution of all possible tree lengths. The second approach is a bootstrap analysis, as described by Felsentein (1985).

Figure 2A–G illustrates phylogenetic arrangements of the taxa in this study according to seven of the most commonly cited hypotheses for the relationship of *Latimeria chalumnae* to other gnathostomes. Although not all of the authors listed specified the composition of the Chondrichthyes, we have considered it to comprise the Elasmobranchii and the Holocephali. It should be noted that in the case of the hypothesis of Løvtrup (1977), *L. chalumnae* was specifically postulated to be the sister group of these two taxa. PAUP was used to determine the lengths of each of these seven trees with the 18S rRNA data set, and these lengths are shown in Figure 2. As can be seen from this figure, even the shortest of the trees (307 steps, corresponding to the hypothesis of von Wahlert 1968) is considerably longer than the most parsimonious tree (298 steps). Figure 3 illustrates the distribution of the lengths of all possible trees relating these eight taxa, which range from 298 to 329 steps. The trees representing the hypotheses tested range from being equal in length to or longer than 7.3% of all possible trees (in the case of the hypothesis of von Wahlert 1968) to 60.8% of all possible trees (in the case of the hypothesis of Løvtrup 1977).

Since Lagios (1979) and Jarvik (1980) proposed that *L. chalumnae* was a chondrichthyan without specifying the relationships of taxa within the Chondrichthyes or the Osteichthyes, we have tested the length of the most parsimonious tree consistent with this proposal using the CONSTRAINTS feature of PAUP. The BRANCH AND BOUND option was used to determine the most parsimonious tree meeting the topological constraints that (1) the Chondrichthyes is a monophyletic group consisting of *Hydrolagus, Rhinobatos,* and *Latimeria,* (2) the Osteichthyes is a monophyletic group consisting of the remaining four gnathostome taxa, and (3) *Homo* and *Xenopus* form a monophyletic group. The most parsimonious tree which meets these constraints requires 311 steps and is illustrated in Figure 2H. This tree is also considerably longer than the most parsimonious tree found in the absence of constraints.

Because the most parsimonious tree contains many unorthodox relationships and is separated by a relatively large number of steps from any hypothesis of coelacanth relationship yet proposed, we have investigated the significance of the support for the nodes on this tree using the BOOTSTRAP option of PAUP. One thousand bootstrap replications were performed using the BRANCH AND BOUND option, so that all of the most parsimonious trees were found at each step. The resulting bootstrap consensus tree listing the percentage of trees in which a given branch was found is shown in Figure 4. The topology of this tree differs slightly from that of the most parsimonious tree (Fig. 1), but still contains many unusual relationships. None of the nodes on the consensus tree was found in more than 50% of the trees resulting from the bootstrap procedure, far short of the 95% value suggested by Felsenstein (1985) as a confidence limit for significance.

Felsenstein (1978, 1983) has described models of character evolution under which parsimony methods converge on an incorrect tree as more data are added to an analysis. In particular, unequal rates of evolution can cause taxa with a rapid rate of character evolution to be grouped with each other incorrectly, or to be placed too close to the root of the tree. Examination of the trees in Figure 2, in which branch lengths have been drawn proportional to the number of changes assigned to each branch, reveals that the branches leading to *Homo* and *Hydrolagus* are consistently longer than those leading to their generally acknowledged sister taxa (*Xenopus* and *Rhinobatos,* respectively). The branch lengths in Figure 2 were assigned using the ACCELERATED TRANSFORMATION method of character state optimization in PAUP, but the length differences are also seen with other methods

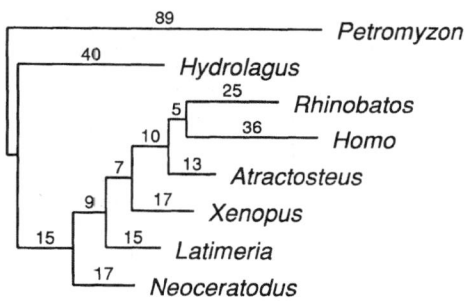

Fig. 1. Most parsimonious tree found by EXHAUSTIVE SEARCH with 1668 aligned positions (202 variable, 68 phylogenetically informative). Length = 289 steps, consistency index (excluding uninformative characters) = 0.570. Numbers above each branch indicate the number of steps assigned to that branch using the ACCELERATED TRANSFORMATION method of character state optimization.

of character state optimization. In an attempt to ameliorate the effects of unequal rates of sequence evolution among taxa, we repeated the EXHAUSTIVE SEARCH and BOOTSTRAP analyses described previously after excluding *Homo* and *Hydrolagus* from consideration (analyses not shown). Although exclusion of these taxa removes the opportunity to use the assumed monophyly of the Tetrapoda and of the Chondrichthyes as a check on the reliability of the trees found, the results of these analyses paralleled the previous ones in that the three equally most parsimonious trees were unorthodox (e.g. did not support the monophyly of the Osteichthyes) and the bootstrap analysis did not show significant support for any of the nodes on the consensus tree.

Hendy & Penny (1989) have suggested that one method of overcoming the tendency of long branches to 'attract' each other in parsimony analyses is to subdivide the long branches by the inclusion of additional taxa. We have attempted this subdivision of branches by including in our analysis sequences from 15 additional gnathostomes. These taxa include representatives of the Elasmobranchii (*Heterodontus francisci, Narcine brasiliensis*), Actinopterygii (*Erpetoichthys calabaricus, Polyodon spathula, Scaphirhynchus platorynchus, Amia calva, Lepomis cyanella*), Dipnoi (*Protopterus* sp., *Lepidosiren paradoxa*), Amphibia (*Hyla cinerea,*

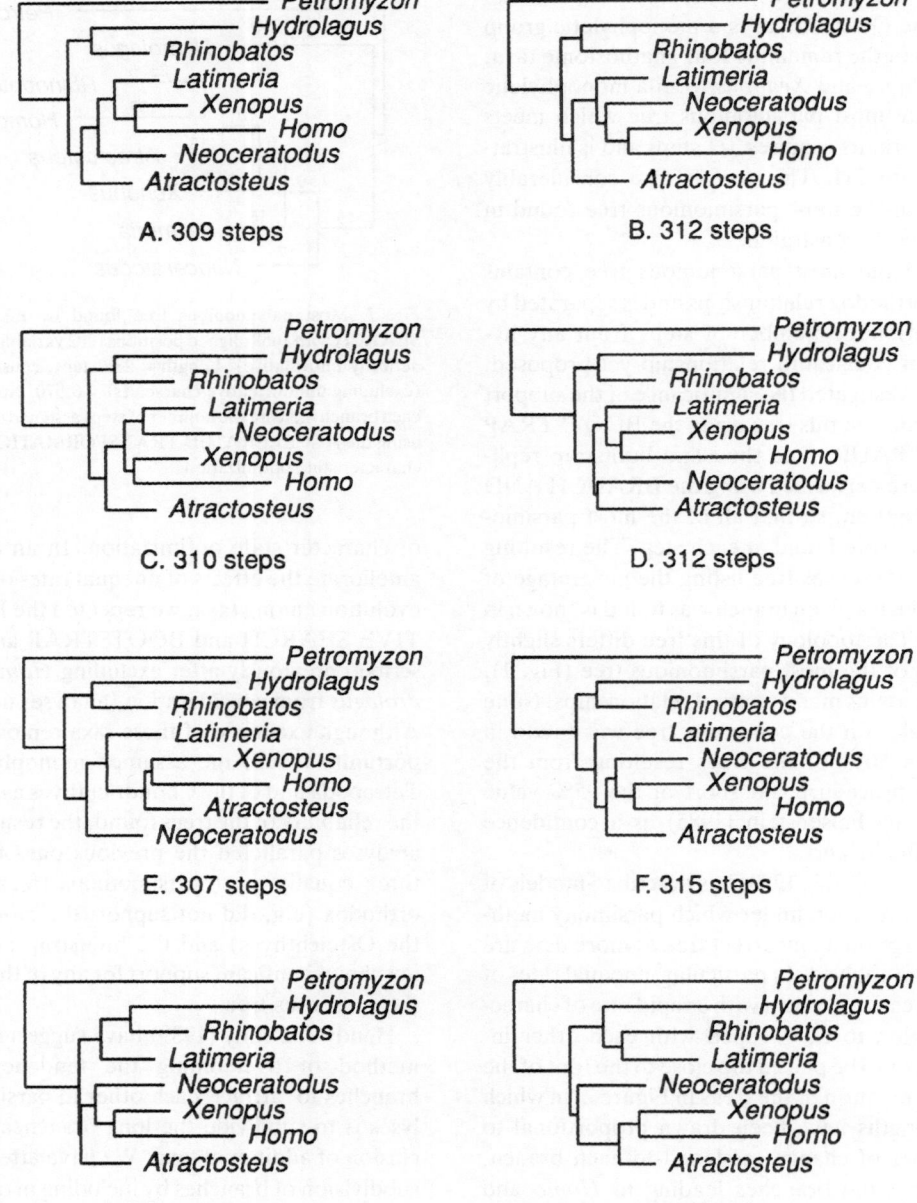

A. 309 steps

B. 312 steps

C. 310 steps

D. 312 steps

E. 307 steps

F. 315 steps

G. 317 steps

H. 311 steps

Fig. 2. Comparison of previously published hypotheses of gnathostome relationship. In A–G, the length of the tree corresponding exactly to the topology illustrated was determined. In H, the shortest tree was determined using BRANCH AND BOUND with the constraints that (1) the Chondrichthyes is a monophyletic group consisting of *Hydrolagus, Rhinobatos,* and *Latimeria,* (2) the Osteichthyes is a monophyletic group consisting of the remaining four gnathostome taxa, and (3) *Homo* and *Xenopus* form a monophyletic group. In all cases, the data set was the same as that described in Figure 1. The length of each branch is drawn in proportion to the number of steps assigned to it using the ACCELERATED TRANSFORMATION method of character state optimization. Consistency indices were calculated after excluding uninformative characters. A– Hypothesis of Schultze (1986). Length = 309 steps, consistency index = 0.531. B– Hypothesis of Forey (1986). Length = 312 steps, consistency index = 0.521. C– Hypothesis of Northcutt (1986). Length = 310 steps, consistency index = 0.528. D– Hypothesis of Bertmar (1968). Length = 312 steps, consistency index = 0.521. E– Hypothesis of von Wahlert (1968). Length = 307 steps, consistency index = 0.538. F– Hypothesis of Wiley (1979). Length = 315 steps, consistency index = 0.512. G– Hypothesis of Løvtrup (1977). Length = 317 steps, consistency index = 0.506. H– Most parsimonious tree consistent with the hypotheses of Lagios (1979) and Jarvik (1980). Length = 311 steps, consistency index = 0.524.

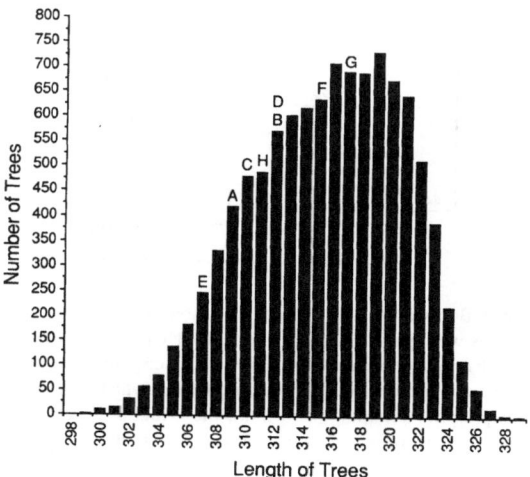

Fig. 3. Distribution of lengths of all possible trees relating the eight taxa, given the data set described in Figure 1. The lengths of previously published hypotheses of gnathostome relationship are indicated by letters corresponding to the tree topologies illustrated in Figure 2 A–H. A– Hypothesis of Schultze (1986). B– Hypothesis of Forey (1986). C– Hypothesis of Northcutt (1986). D– Hypothesis of Bertmar (1968). E– Hypothesis of von Wahlert (1968). F– Hypothesis of Wiley (1979). G– Hypotheses of Løvtrup (1977). H– Most parsimonious tree consistent with the hypothesis of Lagios (1979) and Jarvik (1980).

Ambystoma mexicanum), and Amniota (*Pseudemys scripta, Heterodon platyrhinos, Alligator mississippiensis, Turdus migratorius*). The first nine of these taxa are sequences to be presented elsewhere (Stock & Whitt unpublished), while the remaining six are taken from Moberg (1989). Because of the large number of taxa (23) it was not possible to use the EXHAUSTIVE SEARCH or BRANCH AND BOUND options, which will always find all shortest trees. Instead, a heuristic search procedure with a SIMPLE addition sequence and TREE-BISECTION-RECONNECTION branch swapping was used. Figure 5 shows the strict consensus of the three equally most parsimonious trees found by this method. This tree was drawn by collapsing individual taxa into larger monophyletic groups, all of which were present in all three trees. Comparison of this consensus tree with that for eight taxa shown in Figure 1 reveals the same relationship of the Holocephali and Dipnoi to the other gnathostome taxa. Although the other five taxa are arranged differently in the two trees, in both cases, none of the taxa Tetrapoda, Chondrichthyes, or

Osteichthyes are monophyletic. A bootstrap analysis of the twenty three taxa did not provide significant support for any of the relationships shown in Figure 5, although it did provide significant support for the monophyly of some of the higher taxa listed in the tree.

The analyses performed above, in which all unambiguously alignable sequence positions are used, suggest that a considerable amount of noise exists in the data set. One possible way of overcoming this problem would be to eliminate or give less weight to rapidly evolving positions, since the branching events of interest occurred over 400 million years ago. Parsimony analyses would be expected to be more effective when confined to slowly evolving sites (Felsenstein 1983, although see Hendy & Penny 1989). A number of methods for giving greater weight to slowly evolving positions have been suggested, including (1) weighting based on considerations of secondary structure in ribosomal RNA molecules (Olsen 1988, Wheeler & Hon-

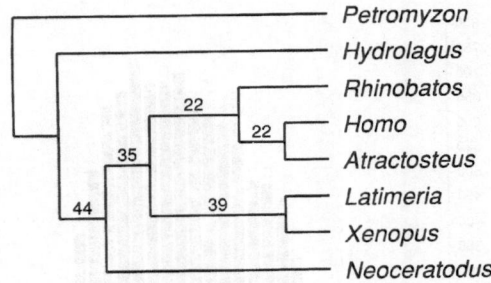

Fig. 4. Bootstrap consensus tree. One thousand bootstrap replications were performed using the BRANCH AND BOUND search option at each step. Numbers above each branch refer to the percentage of most parsimonious trees in which that branch was present. Groupings shown are those that occurred at the highest frequencies and were consistent with all groupings of higher frequency. The original data set used was the same as that described in Figure 1.

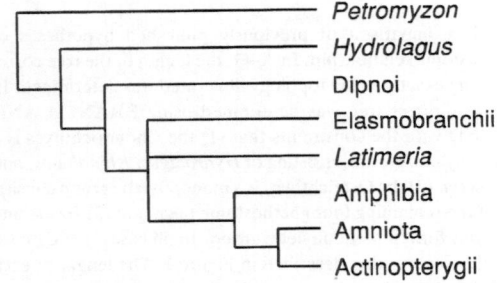

Fig. 5. Strict consensus of three most parsimonious trees determined using a SIMPLE addition sequence and TREE-BISECTION-RECONNECTION branch swapping with twenty three vertebrate species. Illustrated higher taxa represent individual species as follows: Dipnoi = *Neoceratodus forsteri, Protopterus* sp., *Lepidosiren paradoxa;* Elasmobranchii = *Heterodontus francisci, Narcine brasiliensis, Rhinobatos lentiginosus;* Amphibia = *Xenopus laevis, Hyla cinerea, Ambystoma mexicanum;* Amniota = *Pseudemys scripta, Heterodon platyrhinos, Alligator mississippiensis, Turdus migratorius, Homo sapiens;* Actinopterygii = *Erpetoichthys calabaricus, Polyodon spathula, Scaphirhynchus platorhynchus, Atractosteus spatula, Amia calva, Lepomis cyanella.* All illustrated higher taxa were monophyletic in all three trees and relationships within each of these taxa were identical in the three trees. 1668 aligned positions were analysed. Length of trees = 585 steps, consistency index (excluding uninformative positions) = 0.499.

eycutt 1988, Patterson 1989, Smith 1989), (2) differential weighting of transitions and transversions (Lake 1987, Li et al. 1987), and (3) using sites with the least variation within recognized monophyletic groups (Olsen 1987). The results of preliminary applications of these methods to our data set are described below.

Wheeler & Honeycutt (1988), in their analysis of insect 5.8S rRNA sequences, suggested that positions that do not form base pairs in secondary structure models (loop regions) were more reliable for inferring phylogenies than those that are base paired (stem regions). This result was contradicted by Patterson (1989) and Smith (1989), who found sites in paired regions of the 18S rRNA to be more reliable for determining the phylogenetic relationships of metazoan phyla and various echinoderm taxa, respectively. We have used the secondary structure models proposed by Dams et al. (1988) for *Xenopus* and human 18S rRNA to partition the sites in our data set into paired and unpaired subsets. Differences between the two models were generally in regions that were deleted from our initial analyses due to ambiguous alignment, and therefore did not interfere with our assignment of sites to the two classes. Exactly 970 sites, 100 of which were variable and 28 of which were phylogenetically informative, occurred in paired regions, while 698 sites, 102 of which were variable and 40 of

which were phylogenetically informative, occurred in unpaired regions. The 50% majority rule consensus of the seven equally most parsimonious trees found by the EXHAUSTIVE SEARCH option of PAUP using only the paired sites is shown in Figure 6A. This tree is unusual in the non-monophyly of the taxa Osteichthyes, Chondrichthyes, and Tetrapoda, as well as in the placement of the Amniota (represented by the human sequence) as the sister group of all other gnathostome taxa. None of the nodes in this tree were found to be significantly supported by bootstrapping. Figure 6B illustrates the 50% majority rule consensus of the three equally most parsimonious trees found by a similar analysis of only the unpaired sites. This tree agrees with most conventional views of gnathostome phylogeny in the monophyly of the Chondrichthyes and the Sarcopterygii (excluding the Amniota). However, it is unconventional in the placement of the Actinopterygii and the Amniota as successive sister groups to the rest of the gnathostome taxa. As in the analysis of paired sites, none of the nodes in the tree were found to be significantly supported by bootstrapping.

A number of authors (e.g. Lake 1987, Li et al. 1987) have recommended giving greater weight to transversion differences than to transition differences in the construction of phylogenetic trees for distantly related organisms, as the former generally represent less common events. We have used the STEP MATRIX feature of PAUP to assign different weights to transitions and transversions. This feature allows the specification of the number of steps to be added to the tree length for each type of nucleotide transformation. Figure 7 illustrates the results of setting the cost of all transitions at one step and that of all transversions at two steps. The tree is the consensus of two equally parsimonious trees found by the BRANCH AND BOUND option. This tree is identical to that found by the analysis of unpaired sites (Fig. 6B) except for the trifurcation of *Latimeria, Neoceratodus,* and *Xenopus.* Increasing the weight of transversions to tenfold that of transitions resolves the trifurcation in a fashion identical to that of Figure 6B. Use of transversional changes only (i.e. giving transitions a weight of zero) resulted in three equally parsimonious trees, the strict consensus of which amounts to collapsing the branch uniting *Latimeria, Neoceratodus,* and *Xenopus* in Figure 7. Bootstrapping with these various weighting schemes showed no significant support for any of the nodes on the trees. This lack of resolution may be due to the scarcity of sites at which phylogenetically informative transversions have occurred. Only 15 of the 85 variable sites at which transversional differences exist are phylogenetically informative.

The final method of editing sites that we employed, with the goal of removing rapidly evolving ones, was to remove those that are variable within groups that are accepted to be monophyletic, a procedure described by Olsen (1987). Of the higher taxa examined in the present study, the Petromyzontiformes, Holocephali, and Actinistia were represented by a single species and are thus uninformative for editing the positions. In order to remove intrataxon variability, we examined 18S rRNA sequences from three elasmobranch species

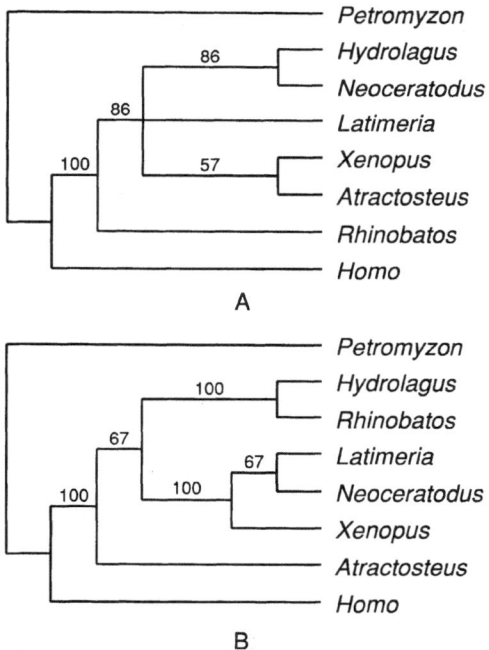

Fig. 6. A– 50% majority rule consensus of the seven equally most parsimonious trees found by EXHAUSTIVE SEARCH with the analysis restricted to positions in paired regions of the 18S rRNA secondary structure models of Dams et al. (1988). Of these 970 positions, 100 were variable and 28 were phylogenetically informative. Numbers above branches represent the percentage of the seven trees in which they were present. Length of trees = 136 steps, consistency index (excluding uninformative positions) = 0.567. B– 50% majority rule consensus of the three equally most parsimonious trees found by EXHAUSTIVE SEARCH with the analysis restricted to positions in unpaired regions of the 18S rRNA secondary structure models of Dams et al. (1988). Of these 698 positions, 102 were variable and 40 were phylogenetically informative. Numbers above branches represent the percentage of the three trees in which they were present. Length of trees = 159 steps, consistency index = 0.591.

(representing three orders), three species of dipnoans (representing all extant genera), three amphibian species (two anurans and a urodele), five amniote species (a turtle, a snake, an alligator, a bird, and a mammal), and six actinopterygian species (representing the taxa Cladistia, Chondrostei, Ginglymodi, Halecomorphi, and Teleostei). The species examined are those mentioned above in the analysis of twenty three taxa, and provide a substantial representation of the diversity in most of the higher taxa. When all positions that are varia-

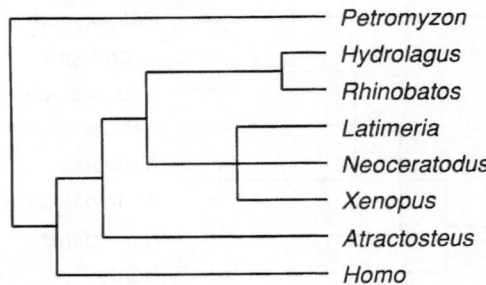

Petromyzon
Hydrolagus
Rhinobatos
Latimeria
Neoceratodus
Xenopus
Atractosteus
Homo

Fig. 7. Strict consensus of the three equally most parsimonious trees found with transversions given a weight of two steps and transitions a weight of one step. A BRANCH AND BOUND search was performed with the data set described in Figure 1. Length of trees = 400 steps.

ble within the Elasmobranchii, the Dipnoi, the Amphibia, the Amniota, or the Actinopterygii are removed, sixty variable positions remain, only eight of which are phylogenetically informative for the eight higher taxa. Phylogenetic analysis of these eight positions does not provide much resolution (not surprisingly) and most of the relationships that are supported are unconventional. One position (657 in the sequence of *Homo sapiens,* and 615 in the *Latimeria chalumnae* sequence) is worthy of note. This position is the only one in the entire data set that is occupied by the same base in all four chondrichthyan taxa and a different base in all of the osteichthyan taxa. Since *Petromyzon* (the outgroup) and the chondrichthyans exhibit a C while *Latimeria* and the osteichthyans have a U, the position provides a potential synapomorphy uniting *Latimeria* with the Osteichthyes.

The extreme decrease in the number of informative positions brought about by this method of editing the sites is due largely to a few taxa which differ extensively from other members of their groups. Most notable are *Narcine brasiliensis, Homo sapiens* and *Erpetoichthys calabaricus.* It might be possible to edit the sites without removing so much information by applying a less stringent criterion than absolute within-group conservation. However, the example mentioned above serves to illustrate the pattern of variation that is likely to be causing the ambiguous results in all of the phylogenetic analyses attempted.

Discussion

The results of our comparison of the 18S rRNA sequence of *Latimeria chalumnae* with those of other vertebrates do not allow firm conclusions to be made as to the phylogenetic position of the coelacanth. The most parsimonious tree contains enough improbable groupings at several different taxonomic levels that it would be imprudent to interpret the position of *L. chalumnae* within this tree literally. Previously proposed hypotheses of coelacanth relationship require a considerably larger number of steps with our data set than does the most parsimonious tree, and none of the nodes in the bootstrap consensus tree are significantly supported. Our attempts to compensate for unequal rates of evolution among taxa and among sites by analyzing different subsets of the data also failed to provide convincing support for any particular hypothesis of coelacanth relationship.

Although all of the above analyses were conducted using the principal of maximum parsimony, preliminary investigations of the data set with the evolutionary parsimony method of Lake (1987), and the least squares additive distance method of DeSoete (1983) (analyses not shown) produced similar ambiguous results. None of the topologies tested by evolutionary parsimony was significantly favored (p < 0.05) over its alternatives. The best tree found by the distance method of DeSoete (1983) was not the same as that found in the parsimony analysis with the complete data set, but shared the same features, i.e. some extremely unorthodox relationships among the major lineages examined, but no strong support for any of these relationships, as judged by short branches uniting the taxa.

A number of aspects of the tempo and mode of 18S rRNA evolution may be contributing to the ambiguous phylogenetic results described above. Obvious possibilities include unequal rates of substitution in different lineages, the presence of rapidly evolving sites within otherwise conserved regions, and structural constraints (e.g. for maintaining base-pairing in regions of secondary structure) leading to an increased probability of convergence. Because of the importance of determining the phylogenetic relationships of *L. chalumnae*, it is reasonable to ask whether other molecules show more promise for resolving this issue. Only three other macromolecules, of which we are aware, have been used to investigate the phylogenetic relationships of the coelacanth. Maeda et al. (1984) compared α parvalbumin sequences from an elasmobranch, seven teleosts, the living coelacanth, two amphibians, a turtle, and a snake, as well as β parvalbumin sequences from the coelacanth, a teleost, two amphibians, and a mammal. Because these two classes of molecules are presumably related by duplication of a common ancestral gene, these authors included both in the same analysis and placed the root of the resulting tree between the two classes of molecules. The most parsimonious tree relating these sequences had several radical departures from generally accepted views on vertebrate phylogeny. The β parvalbumin portion of the tree violated a number of generally accepted ordinal assignments of teleosts, placed the snake within the Teleostei, and placed the turtle within the Amphibia. The coelacanth formed the sister group to the assemblage consisting of all of the previously mentioned taxa, and the elasmobranch was the sister group of all osteichthyan plus tetrapod taxa. The α parvalbumin portion of the tree showed the coelacanth as the sister group of teleosts plus tetrapods and placed the mammal within the Amphibia. Maeda et al. (1984) argued that their data may suggest that actinopterygians are more closely related to tetrapods than is the coelacanth, based on the congruence between the two classes of molecules and the large number of additional steps required in the α parvalbumin tree to place the coelacanth as the sister group of tetrapods. However, they pointed out that the occurrence of drastically different rates of evolution of parvalbumins, large amounts of homoplasy, and frequent gene duplications make the molecules unsuitable for higher level vertebrate taxonomy.

More promising results were obtained by Hillis & Dixon (1989) and Hillis et al. (1991) in comparisons of partial 28S rRNA sequences. The first study included sequences from *L. chalumnae*, a teleost, and four tetrapods, with *Drosophila* as an outgroup, while the second study used a different, but overlapping, set of five tetrapods, and used a composite outgroup with portions of the *Drosophila* sequence replaced by homologous sequences from a lamprey. Both studies reported significant support for the sister group relationship of *Latimeria* and tetrapods, as judged by, respectively, a modification of the Wilcoxon matched-pairs signed rank test described by Templeton (1983) and the bootstrapping technique described by Felsenstein (1985). These results argue against the hypotheses of von Wahlert (1968), Wiley (1979), Lagios (1979, 1982), and Løvtrup (1977). However, the absence of a lungfish sequence prevents the discrimination among more generally accepted hypotheses that place *Latimeria* within the Sarcopterygii. In addition, rejection of hypotheses that place *Latimeria* within the Chondrichthyes would be better supported with the inclusion of elasmobranch or holocephalan sequences. Finally, additional outgroup data are required, since the lamprey sequence is available for only a small portion of the aligned positions, and the *Drosophila* sequence is too divergent to provide an ideal outgroup. Nevertheless, as Hillis et al. (1991) point out, *Latimeria chalumnae* and the tetrapods are united by a long common branch, implying that once additional sequences are determined, it may be possible to provide significant support for a more complete hypothesis of the phylogenetic position of *L. chalumnae* using 28S rRNA sequences.

An alternative possibility, however, is that the 18S rRNA sequences are actually providing an indication of a rapid radiation of osteichthyan (and perhaps gnathostome) higher taxa that may be difficult to resolve with any data set. Maeda & Fitch (1981) described the problems that a rapid radiation of lineages separated by a long period of time from the present would cause for molecular phylogenetic studies. These authors were studying the interrelationships of the major amniote lineages and suggested that the radiation of these lineages may have taken place within a 10 million year span 300 million years ago. As they pointed out, three factors make the resolution of such events extremely difficult. First, during such a short period of time, very few substitutional events will occur along a branch uniting two taxa to the exclusion of others. Even if such events do occur, they are susceptible to being obscured by further substitutions or by parallel substitutions in other lineages. Finally, even if the evidence for relative branching order at a particular site is preserved until the present, its influence in a phylogenetic reconstruction algorithm may be overwhelmed by misleading changes occurring at other sites. Such a radiation would explain the results found in the analysis of sequences of 23 taxa, in which most generally accepted monophyletic lineages (such as the Dipnoi and the Actinopterygii) were recognized, but no resolution of the relationships among these major lineages was possible.

The fossil record of osteichthyans also appears to be consistent with this hypothesis, i.e. that the branching events associated with coelacanth origins occurred during a short time span separated by a much longer period of time from the present. Unambiguous actinopterygian fossils first appear in the fossil record in the Lower Devonian, although isolated scales from the Upper Silurian have been tentatively considered actinopterygian remains (Moy-Thomas & Miles 1971). Dipnoans and porolepiforms also appear in the Lower Devonian, followed in the Middle Devonian by osteolepiforms and actinistians (Dineley 1984, Campbell & Barwick 1986). Even tetrapods had appeared by at least the Upper Devonian (Rackoff 1980, Westoll 1980). The appearances of these lineages thus occurred over a 40–50 million year period 400 million years ago, based on the time scale of Palmer (1983).

The difficulty in distinguishing between various hypotheses of osteichthyan relationships encountered in our analyses may be a reflection of this radiation. More surprising, however, are the tendencies of our analyses to group the elasmobranchs within the Osteichthyes and to violate the monophyly of the Tetrapoda. With respect to the first result, it is worth noting that elasmobranch remains appear no earlier in the fossil record than do several lineages of osteichthyans (Moy-Thomas & Miles 1971). If the fossil record is interpreted literally, it is possible that the divergence between the Chondrichthyes and the Osteichthyes may have occurred only a short time prior to the divergence of the major osteichthyan lineages from one another,

and that the distinctive morphologies of chondrichthyans and osteichthyans may have arisen in a relatively short period of time, after which they became relatively stable. The tetrapod results are interesting with respect to recent proposals that the Devonian *Ichthyostega* is the sister group of the Amphibia (Panchen & Smithson 1988, Benton 1990), rather than of the Tetrapoda, as suggested earlier by Gaffney (1979) and Gardiner (1982). This would push the divergence between the Amphibia and the Amniota back into the Upper Devonian, and make the monophyly of the Tetrapoda more difficult to recover by molecular techniques.

The obvious next step toward a better understanding of coelacanth relationships is to gather sequences from additional molecules, in order to determine if they have been affected by the probable rapid radiation of osteichthyan taxa in a manner similar to the 18S rRNA. It may turn out, as Raff et al. (1989) suggested in their analysis of the metazoan radiation, that the resolution of rapidly branching taxa is not possible through a simple analysis of the sequences of slowly evolving molecules. Such a resolution, as they pointed out, may require the use of shared character states resulting from rare events, such as duplication of particular genomic regions, to order the lineages. Analysis of either type of molecular data set will require a better understanding of the probability of and constraints on character change. Fortunately, there is no shortage of molecular information awaiting future exploitation, as the combined length of the macromolecular sequences gathered from *Latimeria chalumnae* to date represents on the order of one ten thousandth of one percent of the genome of this most intriguing fish.

Acknowledgements

We are grateful to J.A. Musick, R.W. Chapman, and the Explorers Club, as well as to R.W. Murphy (Royal Ontario Museum) for supplying tissues from *Latimeria chalumnae*. In addition, we wish to thank the following individuals for assisting in various aspects of the project: D.L. Swofford for providing us with the PAUP package and for valuable suggestions as to its use; C.R. Woese for assistance with distance analyses and for helpful advice on phylogenetic reconstruction; C.T. Amemiya, W.E. Bemis, J.K. Gibbons, and W. Gobin for providing some of the specimens used in the analysis; and C.R. Woese, E.A. Zimmer, and D.L. Nickrent for the gifts of sequencing primers. Laboratory facilities for sequencing were provided by the Center for Molecular Phylogeny (supported in part by the University of Illinois Research Board and NSF BBS 87–14603, and sponsored by the Natural History Museum of the University of Illinois, T. Uzzell, Director). This study was supported by NSF BSR 87–17417 (to G.S.W.), NSF BSR 88–15362 (to G.S.W. and D.W.S.), NSF BSR 88–96269 (to L.R.M.), and an NSF Predoctoral Fellowship (Genetics) to D.W.S.

References cited

Barr, P.J., R.M. Thayer, P. Laybourn, R.C. Najarian, F. Seela & D.R. Tolan. 1986. 7-deaza-2'-deoxyguanosine-5'-triphosphate: enhanced resolution in M13 dideoxy sequencing. Biotechniques 4: 428–432.

Benton, M.J. 1990. Phylogeny of the major tetrapod groups: morphological data and divergence dates. J. Mol. Evol. 30: 409–424.

Bertmar, G. 1968. Lungfish phylogeny. pp. 259–283. In: T. Orvig (ed.) Current Problems of Lower Vertebrate Phylogeny, Interscience Publishers, New York.

Biggin, M.D., T.J. Gibson & G.F. Hong. 1983. Buffer gradient gels and ^{35}S label as an aid to rapid DNA sequence determination. Proc. Nat. Acad. Sci. USA 80: 3963–3965.

Bjerring, H.C. 1973. Relationships of coelacanthiforms. pp. 179–205. In: P.H. Greenwood, R.S. Miles & C. Patterson (ed.) Interrelationships of Fishes, Academic Press, London.

Campbell, K.S.W. & R.E. Barwick. 1986. Paleozoic lungfishes – a review. J. Morph. Suppl. 1: 93–131.

Cedergren, R., M.W. Gray, Y. Abel & D. Sankoff. 1988. The evolutionary relationships among known life forms. J. Mol. Evol. 28: 98–112.

Clark, C.G. & G.A.M. Cross. 1988. Small-subunit ribosomal RNA sequence from Naegleria gruberi supports the polyphyletic origin of amoebas. Mol. Biol. Evol. 5: 512–518.

Dams, E., L. Hendriks, Y. Van de Peer, J.-M. Neefs, G. Smits, I. Vandenbempt & R. De Wachter. 1988. Compilation of small ribosomal subunit RNA sequences. Nucleic Acids Res. 16: r87–r173.

DeBorde, D.C., C.W. Naeve, M.L. Herlocher & H.F. Maassab. 1986. Resolution of a common RNA sequencing ambiguity by terminal deoxynucleotidyl transferase. Analyt. Biochem. 157: 275–282.

DeSoete, G. 1983. A least squares algorithm for fitting additive trees to proximity data. Psychometrika 48: 621–626.

Dineley, D.L. 1984. Devonian vertebrates in biostratigraphy. Proc. Linn. Soc. N.S.W. 107: 185–196.

Dingerkus, G. 1979. Chordate cytogenetic studies: an analysis of their phylogenetic implications with particular reference to fishes and the living coelacanth. pp. 117–127. In: J.E. McCosker & M.D. Lagios (ed.) The Biology and Physiology of the Living Coelacanth, Occ. Pap. Calif. Acad. Sci. 134, San Francisco.

Elwood, H.J., G.J. Olsen & M.L. Sogin. 1985. The small-subunit ribosomal RNA gene sequences from the hypotrichous ciliates Oxytricha nova and Stylonychia pustulata. Mol. Biol. Evol. 2: 399–410.

Felsenstein, J. 1978. Cases in which parsimony or compatibility methods will be positively misleading. Syst. Zool. 27: 401–410.

Felsenstein, J. 1983. Parsimony in systematics: biological and statistical issues. Ann. Rev. Ecol. Syst. 14: 313–333.

Felsenstein, J. 1985. Confidence limits on phylogenies: an approach using the bootstrap. Evolution 39: 783–791.

Field, K.G., G.J. Olsen, D.J. Lane, S.J. Giovannoni, M.T. Ghiselin, E.C. Raff., N.R. Pace & R.A. Raff. 1988. Molecular phylogeny of the animal kingdom. Science 239: 748–752.

Forey, P.L. 1980. Latimeria: a paradoxical fish. Proc. R. Soc. Lond. B208: 369–384.

Forey, P.L. 1986. Relationships of lungfishes. J. Morph. Suppl. 1: 75–91.

Forey, P.L. 1988. Golden jubilee for the coelacanth Latimeria chalumnae. Nature 336: 727–732.

Fritzsch, B. 1987. Inner ear of coelacanth fish Latimeria has tetrapod affinities. Nature 327: 153–154.

Gaffney, E.S. 1979. Tetrapod monophyly: a phylogenetic analysis. Bull. Carnegie Mus. Nat. Hist. 13: 92–105.

Gardiner, B.G. 1980. Tetrapod ancestry: a reappraisal. pp. 177–185. In: A.L. Panchen (ed.) The Terrestrial Environment and the Origin of Land Vertebrates, Academic Press, New York.

Gardiner, B.G. 1982. Tetrapod classification. Zool. J. Linn. Soc. 74: 207–232.

Gardiner, B.G. 1984. The relationships of the paleoniscid fishes, a review based on new specimens of Mimia and Moythomasia from the upper Devonian of western Australia. Bull. Brit. Mus. Nat. Hist., Geol. 37: 173–427.

Gouy, M. & W.-H. Li. 1989. Molecular phylogeny of the kingdoms Animalia, Plantae, and Fungi. Mol. Biol. Evol. 6: 109–122.

Gunderson, J.H., H. Elwood, A. Ingold, K. Kindle & M.L. Sogin. 1987. Phylogenetic relationships between chlorophytes, chrysophytes, and oomycetes. Proc. Nat. Acad. Sci. USA 84: 5823–5827.

Gutell, R.R., B. Weiser, C.R. Woese & H.F. Noller. 1985. Comparative anatomy of 16S-like ribosomal RNA. Prog. Nucl. Acid Res. and Mol. Biol. 32: 155–216.

114

Gyllensten, U.B. & H.A. Erlich. 1988. Generation of single-stranded DNA by the polymerase chain reaction and its application to direct sequencing of the *HLA-DQA* locus. Proc. Nat. Acad. Sci. USA 85: 7652–7656.

Hasegawa, M., Y. Iida, T. Yano, F. Takaiwa & M. Iwabuchi. 1985. Phylogenetic relationships among eukaryotic kingdoms inferred from ribosomal RNA sequences. J. Mol. Evol. 22: 32–38.

Hendy, M.D. & D. Penny. 1989. A framework for the quantitative study of evolutionary trees. Syst. Zool. 38: 297–309.

Herzog, M. & L. Maroteaux. 1986. Dinoflagellate 17S rRNA sequence inferred from the gene sequence: evolutionary implications. Proc. Nat. Acad. Sci. USA 83: 8644–8648.

Hillis, D.M. & M.T. Dixon. 1989. Vertebrate phylogeny: evidence from 28S ribosomal DNA sequences. pp. 355–367. *In:* B. Fernholm, K. Bremer & H. Jörnvall (ed.) The Hierarchy of Life, Elsevier Science Publishers, Amsterdam.

Hillis, D.M., M.T. Dixon & L.K. Ammerman. 1991. The relationship of the coelacanth *Latimeria chalumnae:* evidence from sequences of vertebrate 28S ribosomal RNA genes. Env. Biol. Fish. 32: 119–130. (this volume)

Holmes, E.B. 1985. Are lungfishes the sister group of tetrapods? Biol. J. Linn. Soc. 25: 379–397.

Jarvik, E. 1980. Basic structure and evolution of vertebrates, Vol 2. Academic Press, London. 337 pp.

Jauregui-Adell, J. & J.-F. Pechere. 1978. Parvalbumins from coelacanth muscle. III. Amino acid sequence of the major component. Biochim. Biophys. Acta 536: 275–282.

Kolb, E., J.I. Harris & J. Bridgen. 1974. Triose phosphate isomerase EC-5.3.1.1 from the coelacanth: an approach to the rapid determination of an amino-acid sequence with small amounts of material. Biochem. J. 137: 185–197.

Kramer, F.R. & D.R. Mills. 1978. RNA sequencing with radioactive chain-terminating ribonucleotides. Proc. Nat. Acad. Sci. USA 75: 5334–5338.

Lagios, M.D. 1979. The coelacanth and the Chondrichthyes as sister groups: a review of shared apomorph characters and a cladistic analysis and reinterpretation. pp. 25–44. *In:* J.E. McCosker & M.D. Lagios (ed.) The Biology and Physiology of the Living Coelacanth, Occ. Pap. Calif. Acad. Sci. 134, San Francisco.

Lagios, M.D. 1982. *Latimeria* and the Chondrichthyes as sister taxa: a rebuttal to recent attempts at refutation. Copeia 1982: 942–948.

Lake, J.A. 1987. A rate independent technique for analysis of nucleic acid sequences: evolutionary parsimony. Mol. Biol. Evol. 4: 167–191.

Lake, J.A. 1988. Origin of the eukaryotic nucleus determined by rate-invariant analysis of rRNA sequences. Nature 331: 184–186.

Lane, D.J., B. Pace, G.J. Olsen, D.A. Stahl, M.L. Sogin & N.R. Pace. 1985. Rapid determination of 16S ribosomal RNA sequences for phylogenetic analyses. Proc. Nat. Acad. Sci. USA 82: 6955–6959.

Li, W.-H., K.H. Wolfe, J. Sourdis & P.M. Sharp. 1987. Reconstruction of phylogenetic trees and estimation of divergence times under nonconstant rates of evolution. Cold Spring Harbor Symp. Quant. Biol. LII: 847–856.

Loening, U.E. 1967. The fractionation of high-molecular-weight ribonucleic acid by polyacrylamide-gel electrophoresis. Biochem. J. 102: 251–257.

Long, E.O. & I.B. Dawid. 1980. Repeated genes in eukaryotes. Ann. Rev. Biochem. 49: 727–764.

Løvtrup, S. 1977. The phylogeny of Vertebrata. John Wiley & Sons, London. 330 pp.

Lu, H.S., P.M. Yuan & R.W. Gracy. 1984. Primary structure of human triose phosphate isomerase EC-5.3.1.1. J. Biol. Chem. 259: 11958–11968.

MacDonald, R.J., G.H. Swift, A.E. Przybyla & J.M. Chirgwin. 1987. Isolation of RNA using guanidinium salts. Methods Enzymol. 152: 219–223.

Maeda, N. & W.M. Fitch. 1981. Amino acid sequence of a myoglobin from lace monitor lizard, *Varanus varius,* and its evolutionary implications. J. Biol. Chem. 256: 4301–4309.

Maeda, N., D. Zhu & W.M. Fitch. 1984. Amino acid sequences of lower vertebrate parvalbumins and their evolution: parvalbumins of boa, turtle, and salamander. Mol. Biol. Evol. 1: 473–488.

Maisey, J.G. 1986. Heads and tails: a chordate phylogeny. Cladistics 2: 201–256.

Maniatis, T., E.F. Fritsch & J. Sambrook. 1982. Molecular cloning: a laboratory manual. Cold Spring Harbor Laboratory, Cold Spring Harbor, New York. 545 pp.

McCallum, F.S. & B.E.H. Maden. 1985. Human 18S ribosomal RNA sequence inferred from DNA sequence: variations in 18S sequences and secondary modification patterns between vertebrates. Biochem. J. 232: 725–733.

McCarroll, R., G.J. Olsen, Y.D. Stahl, C.R. Woese & M.L. Sogin. 1983. Nucleotide sequence of the *Dictyostelium discoideum* small-subunit ribosomal ribonucleic acid inferred from the gene sequence: evolutionary implications. Biochemistry 22: 5858–5868.

Miles, R.S. 1977. Dipnoan (lungfish) skulls and the relationships of the group: a study based on a new species from the Devonian of Australia. Zool. J. Linn. Soc. 61: 1–328.

Moberg, K.D. 1989. Tetrapod phylogeny as inferred from 18S and 28S ribosomal RNA sequence comparisons. PhD Thesis, University of Illinois at Urbana-Champaign. 198 pp.

Mommsen, T.P. & P.J. Walsh. 1989. Evolution of urea synthesis in vertebrates: the piscine connection. Science 243: 72–75.

Moy-Thomas, J. & R.S. Miles. 1971. Paleozoic fishes, 2nd ed. W.B. Saunders Company, Philadelphia. 259 pp.

Nairn, C.J. & R.J. Ferl. 1988. The complete nucleotide sequence of the small-subunit ribosomal RNA coding region for the cycad *Zamia pumila:* phylogenetic implications. J. Mol. Evol. 27: 133–141.

Nelson, G. 1969. Gill arches and the phylogeny of fishes, with notes on the classification of vertebrates. Bull. Amer. Mus. Nat. Hist. 141: 475–552.

Northcutt, R.G. 1986. Lungfish neural characters and their bearing on sarcopterygian phylogeny. J. Morph. Suppl. 1: 277–297.

Olsen, G.J. 1987. Earliest phylogenetic branchings: comparing rRNA-based evolutionary trees inferred with various techniques. Cold Spring Harbor Symp. Quant. Biol. LII: 825–839.

Olsen, G.J. 1988. Phylogenetic analysis using ribosomal RNA. Methods Enzymol. 164: 793–812.

Palmer, A.R. 1983. The decade of North American geology 1983 geologic time scale. Geology 1: 503–504.

Panchen, A.L. & T.R. Smithson. 1987. Character diagnosis, fossils and the origin of tetrapods. Biol. Rev. 62: 341–438.

Panchen, A.L. & T.R. Smithson. 1988. The relationships of the earliest tetrapods. pp. 1–32. In: M.J. Benton (ed.) The Phylogeny and Classification of the Tetrapods, vol. 1: Amphibians, Reptiles, Birds, Clarendon Press, Oxford.

Patterson, C. 1989. Phylogenetic relations of major groups: conclusions and prospects. pp. 471–488. In: B. Fernholm, K. Bremer & H. Jörnvall (ed.) The Hierarchy of Life, Elsevier Science Publishers, Amsterdam.

Pechere, J.-F., H. Rochat & C. Ferraz. 1978. Parvalbumins from coelacanth muscle. II. Amino acid sequence of the two less acidic components. Biochim. Biophys. Acta 536: 269–274.

Qu, L.H., B. Michot & J.-P. Bachellerie. 1983. Improved methods for structure probing in large RNA's: a rapid 'heterologous' sequencing approach is coupled to the direct mapping of nuclease accessible sites. Application to the 5′ terminal domain of eukaryotic 28S rRNA. Nucleic Acids Res. 11: 5903–5920.

Rackoff, J.S. 1980. The origin of the tetrapod limb and the origin of tetrapods. pp. 255–292. In: A.L. Panchen (ed.) The Terrestrial Environment and the Origin of Land Vertebrates, Academic Press, London.

Raff, R.A., J.A. Anstrom, J.E. Chin, K.G. Field, M.T. Ghiselin, D.J. Lane, G.J. Olsen, N.R. Pace, A.L. Parks & E.C. Raff. 1987. Molecular and developmental correlates of macroevolution. pp. 109–138. In: R.A. Raff & E.C. Raff (ed.) Development as an Evolutionary Process, Alan R. Liss, New York.

Raff, R.A., K.G. Field, G.J. Olsen, S.J. Giovannoni, D.J. Lane, M.T. Ghiselin, N.R. Pace & E.C. Raff. 1989. Metazoan phylogeny based on analysis of 18S ribosomal RNA. pp. 247–260. In: B. Fernholm, K. Bremer & H. Jörnvall (ed.) The Hierarchy of Life, Elsevier Science Publishers, Amsterdam.

Raynal, F., B. Michot & J.P. Bachellerie. 1984. Complete nucleotide sequence of mouse 18S rRNA gene: comparison with other available homologs. FEBS Lett. 167: 263–268.

Romer, A.S. 1955. Herpetichthys, Amphibioidea, Choanichthys or Sarcopterygii? Nature 176: 126.

Romer, A.S. 1966. Vertebrate paleontology, 3rd ed. University of Chicago Press, Chicago. 468 pp.

Rosen, D.E., P.L. Forey, B.G. Gardiner & C. Patterson. 1981. Lungfishes, tetrapods, paleontology, and plesiomorphy. Bull. Amer. Mus. Nat. Hist. 167: 163–275.

Salim, M. & B.E.H. Maden. 1981. Nucleotide sequence of Xenopus laevis 18S ribosomal RNA inferred from gene sequence. Nature 291: 205–208.

Saiki, R.K., D.H. Gelfand, S. Stoffel, S.J. Scharf, R. Higuchi, G.T. Horn, K.B. Mullis & H.A. Erlich. 1988. Primer-directed enzymatic amplification of DNA with a thermostable DNA polymerase. Science 239: 487–491.

Schultze, H.-P. 1981. Hennig und der Ursprung der Tetrapoda. Paläont. Z. 55: 71–86.

Schultze, H.-P. 1986. Dipnoans as sarcopterygians. J. Morph. Suppl. 1: 39–74.

Smith, A.B. 1989. RNA sequence data in phylogenetic reconstruction: testing the limits of its resolution. Cladistics 5: 321–344.

Sogin, M.L. & H.J. Elwood. 1986. Primary structure of the Paramecium tetraurelia small-subunit rRNA coding region: phylogenetic relationships within the Ciliophora. J. Mol. Evol. 23: 53–60.

Sogin, M.L., H.J. Elwood & J.H. Gunderson. 1986. Evolutionary diversity of eukaryotic small-subunit rRNA genes. Proc. Nat. Acad. Sci. USA 83: 1383–1387.

Sogin, M.L., J.H. Gunderson, H.J. Elwood, R.A. Alonso & D.A. Peattie. 1989. Phylogenetic meaning of the kingdom concept: an unusual ribosomal RNA from Giardia lamblia. Science 243: 75–77.

Templeton, A.R. 1983. Phylogenetic inference from restriction endonuclease cleavage site maps with particular reference to the evolution of humans and the apes. Evolution 37: 221–244.

Torczynski, R.M., M. Fuke & A.P. Bollon. 1985. Cloning and sequencing of a human 18S ribosomal RNA gene. DNA 4: 283–291.

von Wahlert, G. 1968. Latimeria und die Geschichte der Wirbeltiere. Eine Evolutionsbiologische Untersuchung. Gustav Fischer Verlag, Stuttgart. 125 pp.

Vossbrink, C.R., J.V. Maddox, S. Friedman, B.A. Debrunner-Vossbrinck & C.R. Woese. 1987. Ribosomal RNA sequence suggests microsporidia are extremely ancient eukaryotes. Nature 326: 411–414.

Westoll, T.S. 1961. A crucial stage in vertebrate evolution: fish to land animal. Proc. R. Inst. Gr. Britain 38: 600–618.

Westoll, T.S. 1980. Prologue: problems of tetrapod origin. pp. 1–10. In: A.L. Panchen (ed.) The Terrestrial Environment and the Origin of Land Vertebrates, Academic Press, London.

Wheeler, W.C. 1989. The systematics of insect ribosomal DNA. pp. 307–321. In: B. Fernholm, K. Bremer & H. Jörnvall (ed.) The Hierarchy of Life, Elsevier Science Publishers, Amsterdam.

Wheeler, W.C. & R.L. Honeycutt. 1988. Paired sequence difference in ribosomal RNAs: evolutionary and phylogenetic implications. Mol. Biol. Evol. 5: 90–96.

White, E.I. 1965. The head of Dipterus valenciennesi Sedgwick & Murchison. Bull. Br. Mus. Nat. Hist. (Geol.) 11: 1–45.

Wiley, E.O. 1979. Ventral gill arch muscles and the interrelationships of gnathostomes, with a new classification of the Vertebrata. Zool. J. Linn. Soc. 67: 149–180.

116

Woese, C.R. 1987. Bacterial evolution. Microbiol. Rev. 51: 221–271.

Woese, C.R., R. Gutell & H.F. Noller. 1983. Detailed analysis of the higher-order structure of 16S-like ribosomal ribonucleic acids. Microbiol. Rev. 47: 621–669.

Zimmer, E.A., R.K. Hamby, M.L. Arnold, D.A. Leblanc & E.C. Theriot. 1989. Ribosomal RNA phylogenies and flowering plant evolution. pp. 205–214. In: B. Fernholm, K. Bremer & H. Jörnvall (ed.) The Hierarchy of Life, Elsevier Science Publishers, Amsterdam.

Appendix 1

18S ribosomal RNA sequence determined for *Latimeria chalumnae*. Lower case u at position 1 indicates uncertainty as to the identity of this nucleotide. Ellipses at the end of the sequence indicate the unsequenced 3′ portion of the molecule (estimated to be 39 nucleotides based on comparisons with other eukaryotic sequences). This sequence will be deposited in GenBank.

uACCUGGUUG	AUCCUGCCAG	UAGCAUAUGC	UUGUCUCAAA	GAUUAAGCCA	UGCAUGUCUA	60
AGUACAAACG	GUGCGUACAG	UGAAACUGCG	AAUGGCUCAU	UAAAUCAGUU	AUGGUUCCUU	120
UGAUCGCUCC	AACGUUACUC	GGAUAACUGU	GGUAAUUCUA	GAGCUAAUAC	AUGCCGACGA	180
GCGCUGACCU	UCGGGGAUGC	GUGCAUUUAU	CAGACCAAAA	CCAAUCCGGG	UCCGCCCGGC	240
CGCUUUGGUG	ACUCUAGAUA	ACCUCGGGCC	GAUCGCACGU	CCUCGUGGCG	GCGACGAUUC	300
CUUCGAAUGU	CUGCCCUAUC	AACUUUCGAU	GGUACUUUCU	GUGCCUACCA	UGGUGACCAC	360
GGGUAACGGG	GAAUCAGGGU	UCGAUUCCGG	AGAGGGAGCC	UGAGAAACGG	CUACCACAUC	420
CAAGGAAGGC	AGCAGGCGCG	CAAAUUACCC	ACUCCCGACG	CGGGGAGGUA	GUGACGAAAA	480
AUAACAAUAC	AGGACUCUUU	CGAGGCCCUG	UAAUUGGAAU	GAGUACACUU	UAAAUCCUUU	540
AACGAGGAUC	UAUUGGAGGG	CAAGUCUGGU	GCCAGCAGCC	GCGGUAAUUC	CAGCUCCAGU	600
AGCGUAUAUU	AAAGUUGCUG	CAGUUAAAAA	GCUCGUAGUU	GGAUCUUGGG	AUCGAGCUGG	660
CGGUCCGCCG	CGAGGCGAGC	UACCGCCUGU	CCCAGCCCCU	GCCUCUUGGC	GCUCCCUUGA	720
UGCUCUUAAC	UGAGUGUCCU	GGGGGUCCGA	AGCGUUUACU	UUGAAAAAAU	UAGAGUGUUC	780
AAAGCAGGCC	GGUCGCUUGG	AUACUUCAGC	UAGGAAUAAU	GGAAUAGGAC	UCCGGUUCUA	840
UUUUGUUGGU	UUUCGGAACU	GGGGCCAUGA	UUAAGAGGGA	CGGCCGGGGG	CAUUCGUAUU	900
GUGCCGCUAG	AGGUGAAAUU	CUUGGACCGG	CGCAAGACGG	ACAAAAGCGA	AAGCAUUUGC	960
CAAGAAUGUU	UUCAUUAAUC	AAGAACGAAA	GUCGGAGGUU	CGAAGACGAU	CAGAUACCGU	1020
CGUAGUUCCG	ACCAUAAACG	AUGCCAACUA	CCGAUCCGGC	GGCGUUAUUU	CCAUGACCCG	1080
CCGGGCAGGU	UCCGGGAAAC	CAAAGUCUUU	GGGUUCCGGG	GGGAGUAUGG	UUGCAAAGCU	1140
GAAACUUAAA	GGAAUUGACG	GAAGGGCACC	ACCAGGAGUG	GAGCCUGCGG	CUUAAUUUGA	1200
CUCAACACGG	GAAACCUCAC	CCGGCCCGGA	CACGGAAAGG	AUUGACAGAU	UGAUAGCUCU	1260
UUCUCGAUUC	UGUGGGUGGU	GGUGCAUGGC	CGUUCUUAGU	UGGUGGAGCG	AUUUGUCUGG	1320
UUAAUUCCGA	UAACGAACGA	GACUCCUCCA	UGCUAAAUAG	UUACGCGACC	CCGAGUGGUC	1380
GGCGUCCAAC	UUCUUAGAGG	GACAAGUGAC	GUUUAGCCAC	ACGAGAUUGA	GCAAUAACAG	1440
GUCUGUGAUG	CCCUUAGAUG	UCCGGGGCUG	CACGCGCGCU	ACACUGAAUG	GAUCAGCGUG	1500
UGUCUACCCU	ACACCGACAG	GUGCGGGUAA	CCCGUUGAAC	CCCAUUCGUG	AUAGGGAUCG	1560
GGGAUUGCAA	UUAUUUCCCG	UGAACGAGGA	AUUCCCAGUA	AGUGCGGGUC	AUAAGCUCGC	1620
GUUGAUUAAG	UCCCUGCCCU	UUGUACACAC	CGCCCGUCGC	UACUACCGAU	UGGAUGGUUU	1680
AGUGAGGUCC	UCGGAUCGGA	CCCGCCGGGG	UCGUCCGCGG	CCCUGGCGGA	GCGCUGAGAA	1740
GACGAUCAAA	CUUGACUAUC	UAGAGGAAGU	AAAAGUCGU...			1779

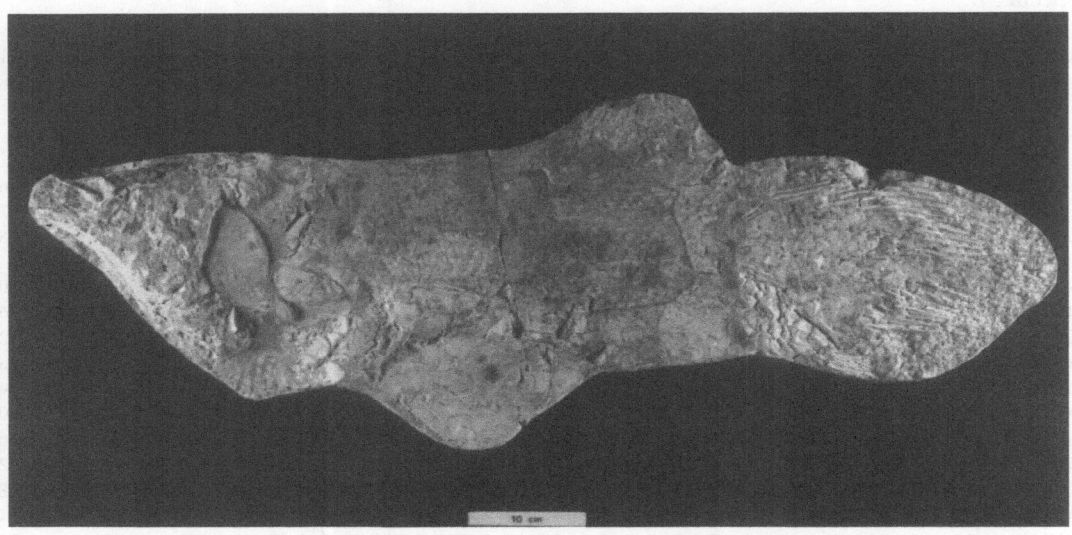

Axelrodichthys araripensis from the Cretaceous Santana formation in Brazil now in the collection of the Institute of Ichthyology, University of Guelph. Photo by J. Eckert.

Environmental Biology of Fishes **32**: 119–130, 1991.
© 1991 *Kluwer Academic Publishers.*

The relationships of the coelacanth *Latimeria chalumnae*: evidence from sequences of vertebrate 28S ribosomal RNA genes

David M. Hillis, Michael T. Dixon & Loren K. Ammerman
Department of Zoology, The University of Texas, Austin, TX 78712, U.S.A.

Received 1.8.1989   Accepted 30.7.1990

Key words: Sarcopterygians, Phylogeny, Outgroup comparison

Synopsis

A subgenomic library created from genomic DNA of *Latimeria chalumnae* was screened for 28S ribosomal RNA (rRNA) clones. The resulting clone was subcloned into a plasmid vector, and over 2 kb of the 28S rRNA region was sequenced. Sequences of 28S rRNA genes were also obtained for *Rhineura floridana* (Squamata), *Cyprinella lutrensis* (Actinopterygii), and *Lampetra aepyptera* (Petromyzontiformes) by cloning and/or amplification by the polymerase chain reaction. The 28S rDNA sequences were aligned for all the above species as well as for the previously published 28S rDNA sequences of the genera *Mus, Rattus,* and *Homo* (Mammalia), *Xenopus* (Amphibia), and *Drosophila* (Insecta). Phylogenetic analysis of these species (using both the insect and lamprey sequences for outgroup comparison, or using only the lamprey sequence in the outgroup) produced a single optimal solution: (Outgroup(*Cyprinella*(*Latimeria*(*Xenopus*(*Rhineura*-(*Homo*(*Rattus*(Mus))))))))). Bootstrap analysis indicated that the placement of *L. chalumnae* on this tree was significant at p < 0.01. Previously published alternative hypotheses of relationships of *Latimeria* require at least 19 additional steps compared to the optimal solution; the rDNA data are sufficient to reject the hypotheses that place *Latimeria* in groups other than the sarcopterygians.

Introduction

There have been more diverse hypotheses about the relationships of coelacanths (Actinistia) than virtually any other major group of vertebrates. Interest in actinistians has been great not only because of the key position they hold in the problem of vertebrate transition from water to land, but also because of the persistence to the present of but a single living member of this group, *Latimeria chalumnae.* The existence of a living species (in addition to the fairly extensive fossil record of coelacanths) permits the collection of a diversity of biological data that can be used to infer relationships. However, past studies on coelacanth phylogeny have generated virtually every conceivable hypothesis for the relationship of this group to lungfishes, tetrapods, actinopterygians, and chondrichthyans (see Forey 1988).

All methods of phylogenetic inference depend on two primary assumptions: (1) the characters examined are heritable; and (2) the characters examined are historically independent. The largest possible set of independent, heritable characters for any organism consists of a subset of its DNA sequences; only a subset of DNA sequences are independent because many repeated sequences evolve in concert (see Hillis 1987). However, molecular biology has tapped a miniscule portion of this reservoir of phylogenetic information to date, and virtually nothing is known about the DNA sequences of *Latimeria chalumnae.* We describe in this paper the creation of a gene library for *Latimeria;* we also describe the first gene that we have screened and partially sequenced from this library. We chose to examine the 28S ribosomal RNA gene because of the conserved nature of this gene among vertebrates (Hillis & Davis 1987) and its potential for elucidating the relationships of coelacanths. Preliminary analysis of a portion of the 28S rRNA gene (Hillis & Dixon 1989) suggested that this sequence could be used to resolve relationships among major groups of vertebrates. We here report the sequence of over 2 kb of the 28S rRNA gene of *L. chalumnae* (along with the homologous genes of other major vertebrate groups) and examine the relationships suggested by these sequences.

Materials and methods

Frozen muscle tissue from *Latimeria chalumnae* was obtained from a specimen procured by the Explorer's Club from Grand Comoro island in November 1986 (VIMS 8118). High molecular weight DNA was isolated from this tissue following the protocol of Hillis & Davis (1986). *L. chalumnae* DNA was cleaved with the restriction enzyme *Eco-RI* and ligated into the lambda vector Lambda Zap II (Stratagene) to produce a subgenomic library (see Hillis et al. 1990). This library was screened (via filter-lift hybridization) with a cloned 28S rDNA gene of *Rana catesbeiana* (pE2528, Hillis & Davis 1987). Positive plaques were selected and the inserts were subcloned in the vector pBluescript (Stratagene). Subclones were verified by restriction digestion, Southern blotting, and sequencing. A verified 28S clone was designated pDH8804.

In addition to the 28S rRNA gene of *L. chalumnae* we also examined the 28S rRNA genes of *Cyprinella* (*Notropis* sensu lato) *lutrensis* (Actinopterygii) and *Rhineura floridana* (Squamata) (see Hillis & Dixon 1989). Furthermore, we compared these sequences to the published 28S DNA sequences of *Mus musculus, Rattus norvegicus,* and *Homo sapiens* (Mammalia; Hassouna et al. 1984, Hadjiolov et al. 1984, Gonzalez et al. 1985), *Xenopus laevis* (Amphibia; Ware et al. 1983), and *Drosophila melanogaster* (Insecta; Tautz et al. 1988).

We also amplified (via the polymerase chain reaction, or PCR; Mullis & Faloona 1987) and sequenced a portion of the 28S rRNA gene of *Lampetra aepyptera* (Petromyzontiformes). PCR amplification was also used to verify pDH8804 as a 28S rDNA clone of *L. chalumnae*. A 700 bp fragment of 28S rDNA was amplified from genomic DNA with modified *Taq* polymerase (AmpliTaq, Cetus) using primers 28v and 28x (Table 1, Fig. 1.). Thirty cycles of melting/annealing/polymerization were performed to produce double-stranded DNA. This product was used as the starting template for asymmetrical amplification to generate single-stranded DNA which was used for direct sequencing (Gyllensten & Erlich 1988). Both an amplification primer (28v) and an internal primer (28w) were used to sequence the amplified DNA.

Plasmid DNA was purified as described by Hillis et al. (1990), denatured in alkali, and sequenced by the base-specific dideoxynucleotide chain termination method (Sanger et al. 1977) using modified T7 DNA polymerase (Tabor & Richardson 1987). DNA amplified via PCR was sequenced in the same manner, except that the single-stranded DNA did not require the treatment with alkali. Primers used are shown in Table 1 and Figure 1. Reaction products were separated on 55 cm 4–6% polyacrylamide field gradient gels (Ansorge & Labeit 1984) and visualized by autoradiography. DNA sequences were aligned against the *Mus* 28S rDNA sequence with the alignment subroutines described by Pustell & Kafatos (1982, 1984, 1986), with adjustments made manually to increase similarity.

All possible tree topologies were examined using the exhaustive search routine of the Phylogenetic Analysis Using Parsimony (PAUP) software package (Swofford 1985 et seq.), with all characters coded as unordered. The insect and lamprey sequences were used for outgroup comparison. The distribution of lengths of all possible trees provides a means of evaluating the relative information to noise ratio in a data set (Fitch 1984, Hillis & Dixon 1989). Confidence limits of nodes on the most parsimonious tree were estimated using bootstrap analysis (Felsenstein 1985). Briefly, this involves sampling characters with replacement from the

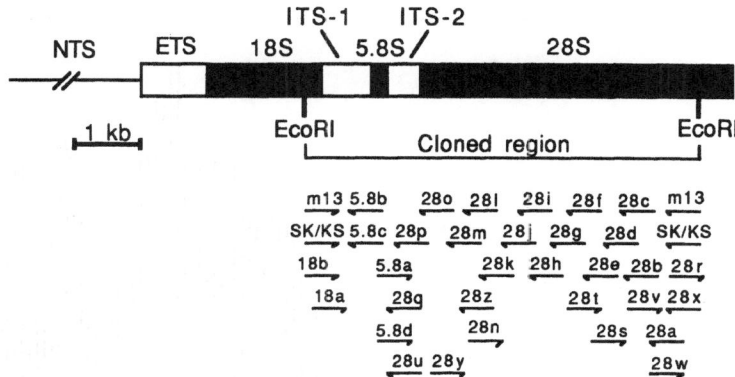

Fig. 1. Cloning and sequencing strategy of ribosomal DNA. Abbreviations: ETS = external transcribed spacer; ITS = internal transcribed spacer; NTS = non-transcribed spacer.

original data matrix to produce new data matrices of equal size to the original, each of which is then analyzed to find the most parsimonious solution. The branch and bound algorithm of PAUP was used to find optimal solutions during bootstrapping. Bootstrap sampling was replicated 1000 times. Another means of assessing relative stability of nodes on a most parsimonious cladogram is con-

Table 1. Primers used to obtain the sequences in Appendix 1. Positions of the primers in the *Mus* sequence correspond to the positions of the RNA nucleotides in the mature transcripts. 'S' stands for the strand synonymous to RNA; 'C' stands for the complementary strand.

Primer	*Mus* position	Strand	Sequence
28a	3913–3927	S	5'-CCTTCTGCTCCACGG-3'
28b	3695–3709	S	5'-AGAGTAGTGGTATTT-3'
28c	3481–3495	S	5'-ACAGTGGGAATCTCG-3'
28d	3284–3298	S	5'-TTAAACAGTCGGATT-3'
28e	2976–2990	S	5'-GTCCAGAGTCGCCGC-3'
28f	2617–2631	S	5'-TCCCGAAGTTACGGA-3'
28g	2386–2400	S	5'-CTGCCCTTCACAAAG-3'
28h	2101–2115	S	5'-CTACCACCAAGATCT-3'
28i	1840–1854	S	5'-GCGCCATCCATTTTC-3'
28j	1665–1679	S	5'-CCAGTTCTGCTTACC-3'
28k	1402–1416	S	5'-CGATTTGCACGTCAG-3'
28l	1131–1145	S	5'-GGTCCGTGTTTCAAG-3'
28r	3828–3842	C	5'-CAGGTGGGGAGTTTG-3'
28s	2559–2573	C	5'-AGGTGAACAGCCTCT-3'
28t	2338–2352	C	5'-ACCGATCCCGGAGAA-3'
28v	3429–3452	C	5'-AAGGTAGCCAAATG(T,C)CTCGTCATC-3'
28w	3565–3588	C	5'-CCTGTTGAGCTTGACTCTAGTCTG-3'
28x	4106–4137	S	5'-GTGAATTCTGCTTCACAATGATAGGAAGAGCC-3'
m13F	NA	S/C	5'-GTTTTCCCAGTCACGAC-3'
m13R	NA	S/C	5'-CAGGAAACAGCTATGAC-3'
SK	NA	S/C	5'-TCTAGAACTAGTGGATC-3'
KS	NA	S/C	5'-CGAGGTCGACGGTATCG-3'

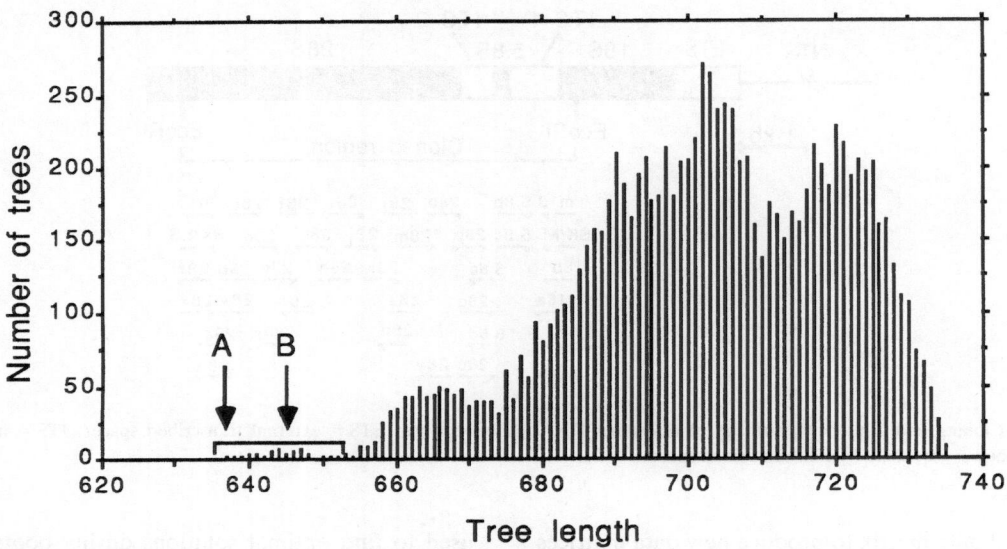

Fig. 2. Distribution of lengths of all possible tree topologies fit to the 28S rDNA sequence data. A = Shortest tree (637 steps); see Figure 3A. B = Distribution of 45 shortest trees used to construct consensus tree (see Figure 3B).

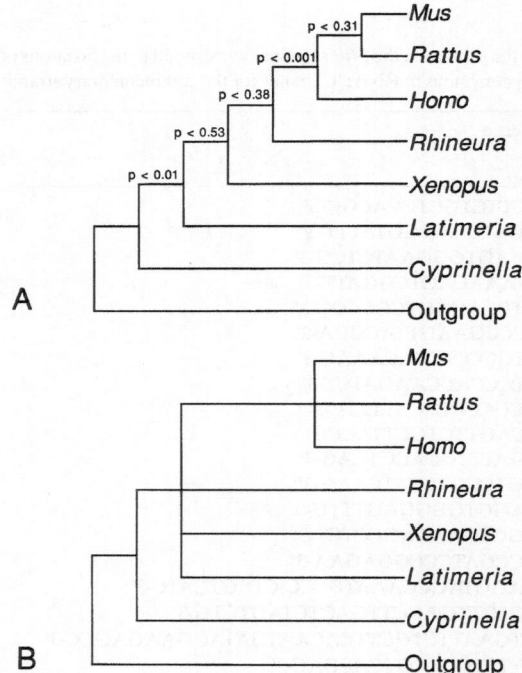

Fig. 3. A = Shortest tree, showing bootstrap probabilities (based on the number of bootstrap trees that do not contain the corresponding node). B = Strict consensus tree of the 45 shortest trees.

sensus analysis of near-parsimonious trees. Strict consensus trees were calculated for all solutions near the most parsimonious tree in order to determine the minimum number of steps needed to break the branch that united *L. chalumnae* to other vertebrates.

Results and discussion

The sequences from over 2 kb of the 28S rRNA genes of the study species are given in Appendix 1. We aligned 1989 nucleotide positions among the species; unambiguous alignment among all the species was not possible in six divergent domains; these regions were not used in the phylogenetic analysis. These divergent domains are characterized by high G/C content and major length variation among the vertebrates, which makes unambiguous alignment between distantly related species difficult or impossible (Appendix 1). Sequence similarities within divergent domains do exist among the more closely related species (e.g. among the mammals), but the similarity among

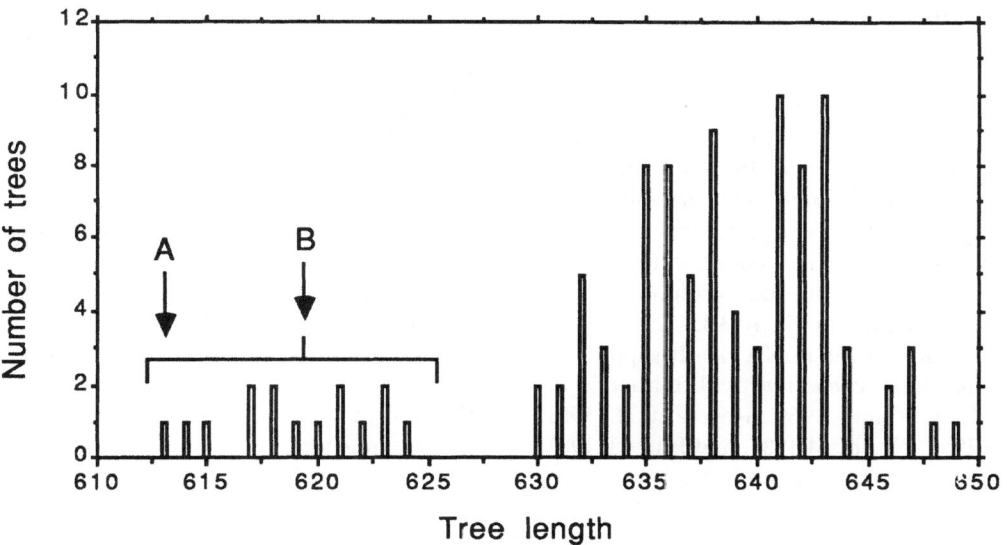

Fig. 4. Distribution of lengths of all possible tree topologies fit to the 28S rDNA sequence data with only one mammal (*Homo*) included. A = shortest tree (613 steps); this tree is the same as Figure 3A with the two rodent lineages removed. B = Distribution of the 15 shortest trees, which are all the possible groupings of *Latimeria* with the tetrapods to the exclusion of the actinopterygian.

distantly related taxa in these regions is not significantly better than random.

The distribution of all possible trees in the phylogenetic analysis was strongly skewed with a long left tail (Fig. 2). Such a distribution indicates that the data matrix has a high information content; there are very few trees that explain the data well and many that are poor solutions to the problem. There is a single most parsimonious tree which places *L. chalumnae* as the sister group to the tetrapods (Fig. 2, 3A). All other nodes of the most parsimonious tree are consistent with traditional views of vertebrate phylogeny (e.g. the two rodents, the three mammals, the four amniotes, and the five tetrapods are united as monophyletic groups).

The distribution of all possible trees (Fig. 2) is skewed partly as a result of the inclusion of three species of mammals in the analysis. However, if only one species of mammal is included the distribution is still strongly skewed, and all solutions that place *L. chalumnae* with the tetrapods are considerably shorter than the remaining possibilities (Fig. 4). As with the full data set, the shortest tree in this reduced data set includes the amniotes, the tetra-

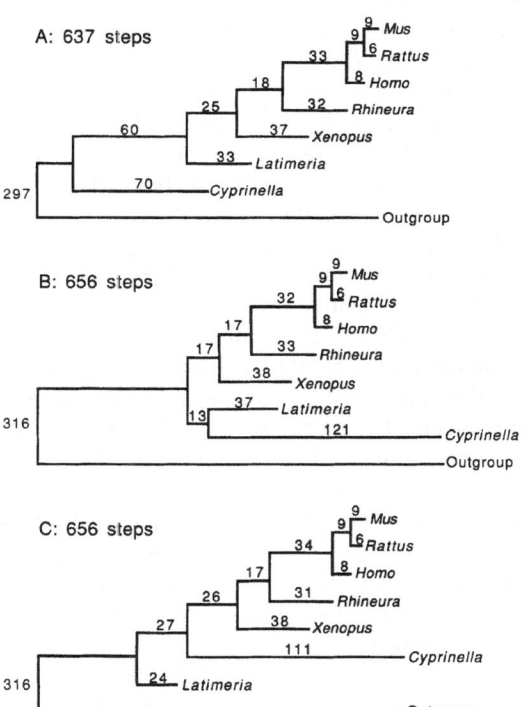

Fig. 5. Branch lengths of the most parsimonious tree (A) and alternative hypotheses of *Latimeria* relationships (B and C).

pods, and the sarcopterygians as monophyletic groups.

Bootstrap analysis of the full data set demonstrates that the branch that connects *Latimeria* with the tetrapods has a 99% confidence interval (Fig. 3A). All of the trees within 15 steps of the most parsimonious solution unite *Latimeria* and the tetrapods; all of these trees also unite the three species of mammals (Fig. 3B). The 45 shortest trees (637 to 652 steps) are the 45 possible resolutions of the consensus tree shown in Figure 3B.

The sequences of 28S rDNA clearly support the relationship of *L. chalumnae* with tetrapods (as suggested by Romer 1966, Rosen et al. 1981, Schultze 1987, Northcutt 1987, and Fritzsch 1987, as well as in the preliminary analysis of a portion of these data by Hillis & Dixon 1989) rather than outside of actinopterygians plus tetrapods (as suggested by von Wahlert 1968, Lagios 1979, 1982, Lævtrup 1977, and Wiley 1979). The shortest trees that place *Latimeria* with actinopterygians or outside of actinopterygians plus tetrapods require significantly more steps than the most parsimonious tree (Fig. 5). These alternatives require a large number of reversals in the lineage leading to the actinopterygians. Based on the rDNA sequences, one can reject hypotheses that place *L. chalumnae* apart from tetrapods with a probability of making a Type I error of $p < 0.01$ (Fig. 3A).

One could argue that the rooting of the tree might not be accurate because of the inclusion of the distantly related *Drosophila* sequence in the outgroup. However, if only the partial lamprey sequence is used for the outgroup, then the results are the same. There is a single most parsimonious tree identical to Figure 3A, and all 37 trees within 12 steps of the most parsimonious tree place *Latimeria* with the tetrapods. A bootstrap analysis again indicates a 99% confidence interval for the *Latimeria*-tetrapod clade. The more complete *Drosophila* sequence lends further support to the position of the root, so it is informative as an additional outgroup.

Among the hypotheses that link coelacanths to other sacropterygians, there is no consensus of opinion concerning the relationships among coelacanths, lungfishes, and tetrapods. Each of the three

possible resolutions has been suggested: coelacanths as the living sister group to lungfishes plus tetrapods (e.g. Rosen et al. 1981); lungfishes as the sister group to coelacanths plus tetrapods (e.g. Schultze 1987, Fritsch 1987); and tetrapods as the sister group to lungfishes plus coelacanths (Northcutt 1987). Because we have not yet sequenced the 28S gene of a lungfish, we cannot address this problem at present. However, the long branch that unites tetrapods and *L. chalumnae* suggests that an abundance of informative sites exist within the 28S gene that should be relevant to this question. In any case, hypotheses that place *Latimeria* outside of the Sarcopterygii are strongly rejected by the 28S rDNA sequences.

Acknowledgements

This work was supported by National Science Foundation grants BSR 8614622 and BSR 8796293 to D.M.H. We thank the Explorer's Club, the Virginia Institute of Marine Science Coelacanth Program (J.A. Musick, P.I.), and E.O. Wiley for providing the *Latimeria* tissue. John Gold and Scott Davis provided the clone of *Cyprinella lutrensis,* and William Ramos and Rafael de Sá provided laboratory assistance.

References cited

Ansorge, W. & S. Labeit. 1984. Field gradients improve resolution on DNA sequencing gels. J. Biochem. Biophys. Meth. 10: 237–243.

Felsentein, J. 1985. Confidence limits on phylogenies: an approach using the bootstrap. Evolution 39: 783–791.

Fitch, W.M. 1984. Cladistic and other methods: problems, pitfalls, and potentials, pp. 221–252. *In*: T. Duncan & T.F. Stuessy (ed.) Cladistics: Perspectives on the Reconstruction of Evolutionary History, Columbia University Press, New York.

Forey, P.L. 1988. Golden jubilee for the coelacanth *Latimeria chalumnae*. Nature 336: 727–732.

Fritzsch, B. 1987. Inner ear of the coelacanth fish *Latimeria* has tetrapod affinities. Nature 327: 153–154.

Gonzalez I.L., J.L. Gorski, T.J. Campen, D.J. Dorney, J.M. Erickson, J.E. Sylvester & R.D. Schmickel. 1985. Variation among human 28S ribosomal RNA genes. Proc. Natl. Acad. Sci. USA 82: 7666–7670.

Gyllensten, V. & H. Erlich. 1988. Generation of single-stranded DNA by the polymerase chain reaction and its applications to direct sequencing of the HLA-DQA locus. Proc. Natl. Acad. Sci. USA 85: 7652–7656.

Hadjiolov, A.A., O.I. Georgiev, V.V. Nosikov & L.P. Yavachev. 1984. Primary and secondary structure of rat 28S ribosomal RNA. Nucl. Acids Res. 12: 3677–3693.

Hassouna, N., B. Michot & J.-P. Bachellerie. 1984. The complete nucleotide sequence of mouse 28S rRNA gene: implications for the process of size increase of the large subunit rRNA in higher eukaryotes. Nucl. Acids Res. 12: 3563–3583.

Hillis, D.M. 1987. Molecular versus morphological approaches to systematics. Ann. Rev. Ecol. Syst. 18: 23–42.

Hillis, D.M. & S.K. Davis. 1986. Evolution of ribosomal DNA: fifty million years of recorded history in the frog genus *Rana*. Evolution 40: 1275–1288.

Hillis, D.M. & S.K. Davis. 1987. Evolution of the 28S ribosomal RNA gene in anurans: regions of variability and their phylogenetic implications. Mol. Biol. Evol. 4: 117–125.

Hillis, D.M. & M.T. Dixon. 1989. Vertebrate phylogeny: evidence from 28S ribosomal DNA sequences, pp. 355–367. *In:* B. Fernholm, K. Bremer & H. Jornvall (ed.) The Hierarchy of Life, Elsevier Science Publishers, Amsterdam.

Hillis, D.M., A. Larson, S.K. Davis & E.A. Zimmer. 1990. Nucleic acids III: sequencing. pp. 318–370. *In:* D.M. Hillis & C. Mortiz (ed.) Molecular Systematics, Sinauer Associates, Sunderland.

Lagios, M.D. 1979. The coelacanth and the Chondrichthyes as sister groups: a review of shared apomorph characters and a cladistic analysis and reinterpretation. Occ. Pap. California Acad. Sci. 134: 25–44.

Lagios, M.D. 1982. *Latimeria* and the Chondrichthyes as sister taxa: a rebuttal to recent attempts at refutation. Copeia 1982: 942–948.

Løvtrup, S. 1977. The phylogeny of Vertebrata. John Wiley and Sons, London. 330 pp.

Mullis, K.B. & F.A. Faloona. 1987. Specific synthesis of DNA *in vitro* via a polymerase catalyzed chain reaction. Meth. Enzymol. 155: 335–350.

Northcutt, R.G. 1987. Lungfish neural characters and their bearing on sarcopterygian phylogeny. J. Morph. Suppl. 1: 277–297.

Pustell J. & F.C. Kafatos. 1982. A convenient and adaptable package of DNA sequence analysis prorams. Nucl. Acids Res. 10: 51–59.

Pustell, J. & F.C. Kafatos. 1984. A convenient and adaptable package of computer programs for DNA and protein sequence management, analysis, and homology determination. Nucl. Acids Res. 12: 643–655.

Pustell, J. & F.C. Kafatos. 1986. A convenient and adaptable microcomputer environment for DNA and protein manipulation and analysis. Nucl. Acids Res. 14: 479–488.

Romer, A.S. 1966. Vertebrate paleontology, 3rd ed. University of Chicago Press, Chicago. 468 pp.

Rosen, D.E., P.L. Forey. B.G. Gardiner & C. Patterson. 1981. Lungfishes, tetrapods, paleontology, and plesiomorphy. Bull. Amer. Mus. Nat. Hist. 167: 159–276.

Sanger, F., S. Nicklen & A.R. Coulson. 1977. DNA sequencing with chain-terminating inhibitors. Proc. Natl. Acad. Sci. USA 74: 5463–5467.

Schultze, H.-P. 1987. Dipnoans as sarcopterygians. J. Morph. Suppl. 1: 39–74.

Swofford, D.L. 1985 et seq. Phylogenetic analysis using parsimony. University of Illinois, Urbana. (computer program).

Tabor, S. & C.C. Richardson. 1987. DNA sequence analysis with a modified bacteriophage T7 DNA polymerase. Proc. Natl. Acad. Sci. USA 84: 4767–4771.

Tautz, D., J.M. Hancock, D.A. Webb, C. Tautz & G.A. Dover. 1988. Complete sequences of the rRNA genes of *Drosophila melanogaster*. Mol. Biol. Evol. 5: 366–376.

von Wahlert, G. 1968. *Latimeria* und die Geschichte der Wirbeltiere: eine evolutionsbiologische Untersuchung. Gustav Fischer Verlag, Stuttgart. 125 pp.

Ware, V.C., B.W. Tague, C.G. Clark, R.L. Gourse, R.C. Brand & S.A. Gerbi. 1983. Sequence analysis of 28S ribosomal DNA from the amphibian *Xenopus laevis*. Nucl. Acids Res. 11: 7795–7817.

Wiley, E.O. 1979. Ventral gill arch muscles and the interrelationships of gnathostomes, with a new classification of the Vertebrata. Zool. J. Linnean Soc. 67: 149–179.

Appendix 1

Aligned sequences of 28S rDNA. The reference numbers correspond to the positions in the *Mus* sequence. Outgroups are designated by OD for outgroup, *Drosophila* and OL for outgroup, *Lampetra*. Positions that vary among the species are marked with an asterisk below the site. Sequences within brackets are not aligned. In the outgroup, positions that could not be aligned with the ingroup taxa are represented by 'N' (five next pages).

126

```
             11|00                11|20                    11|40              11|60
Mus        CACGGGGTCGGCGGCGATGTCGGCTACCCACCCGACCCGTCTTGAAACACGGACCAAGGAGTCTAACGCGTG
Rattus     CACGGGGTCGGCGGCGATGTCGGCTACCCACCCGACCCGTCTTGAAACACGGACCAAGGAGTCTAACGCGTG
Homo       CACGGGGTCGGCGGCGACGTCGGCTACCCACCCGACCCGTCTTGAAACACGGACCAAGGAGTCTAACACGTG
Rhineura   CCCGGGGTCCGCGGCGATGTCGGCCGCCCACCCGACCCGTCTTGAAACACGGACCAAGGAGTCTAACACGCG
Xenopus    CCNAGGGTCCGCGGCGATGTCGGTGTCCCACCCGACCCGTCTTGAAACACGGACCAAGGAGTCTAACGCGCG
Latimeria  CA-GGGGTCTGCGGCGATGTCGGTTTCCCACCCGACCCGTCTTGAAACACGGACCAAGGAGTCTAACGCGTG
Cyprinella CTCGAGGTCCGCGGCGATGTCGGCCACCCACCCGACCCGTCTTGAAACACGGACCAAGGAGTCTAAGCCACG
Outgroup(D)ATTAACAATGCGAAAGATTCAGGATACCTTCGGGACCCGTCTTGAAACACGGACCAAGGAGTCTAACATATG
           ***************  ****   ***   ** **                                *****

             11|80                12|00                   12|20
Mu  CGCGAGTCAGGGGCTCG-TC--CGAAAGCCGCCGTGGCGCAATGAAGGTGAAGGGC [CCCGCCCGGGGGCCC   ] GA
Ra  CGCGAGTCAGGGGCTCG-TC--CGAAAGCCGCCGTGGCGCAATGAAGGTGAAGGGC [CCCGTTCCCGGGGGCCCC] GA
Ho  CGCGAGTCGGGGGCTCGCA---CGAAAGCCGCCGTGGCGCAATGAAGGTGAAGGCC [GGCGCGCTCGCCGGCC  ] GA
Rh  CGCGAGTCAGAGGCTCG-ACC-CGAAAGCC-CCGTGGCGCAATGAAGGTGACGGCG [CGCGCCGGCC        ] GA
Xe  CGCGAGTCGGAGGGACTCTGCGCGAAACC--CTGTGGCGCAATGAAGGTGAGGGCC [GGGGCGCCCCGGCT    ] GA
La  CGCGAGTCAGAGGGCAG-AG--CGAAAGCC-CCATGGCGCAATGAAAGTGAGGCGC [GCGGGCCGGCT       ] GA
Cy  CGCGAGTCAGAGGGT-G-TC--CACGAGCCCCCACGGCGCAATGAAGGTGAGCGGC [GGCGCGCCGGCCC     ] G-
OD  TGCAAGTTATTGGGATA-T-----AAACCTAATA--GCGTAATTAACTTGANNNNN                       NN
      * *  ****  ************** *  ******   *   **  *****                          *

           12|40                12|60                12|80                    13|00
Mu  GGTGGGATCCCGAGGC [CTCTCCAGTCC] GCC-GAGGGCGCACCACCGGCCCGTCTCGCCCGCCGCGCCGGGGAGGTG
Ra  GGTGGGATCCCGAGGC [CTCTCCAGTCC] GCC-GAGGGCGCACCACCGGCCCGTCTCGCCCGCCGCGCCGGGGAGGTG
Ho  GGTGGGATCCCGAGGC [CTCTCCAGTCC] GCC-GAGGG-GCACCACCGGCCCGTCTCGCCCGCCGCGCCGGGGAGGTG
Rh  GGTGGGATCCCGAGGC [GCCGA      ] GCG-GAGGGCGCACCACCGGCCCGTCTCGCCCGCC-CGTCGGGGAGGTG
Xe  GGTGGGATCCCGCCGC [CCCTCCCTCC1] GCCGGCGGGCNCACCACCGGCCCGTCTCCCGCCCCCGTCGGGGNGGTG
La  GGTGGGATCCCCACGC [CTCGT      ] GCG-GGGGGCGCACCACCAGCCCGTCTCACCCGCAGCGTCGGGGAGGTG
Cy  GGTGGGATCCCCTCCG                GAG-GGGGGCGCACCACCGGCCCGTCTCACCCGGTCCGCCGGGGAGGTG
OD  NNNNNNNNNNNNNNNNN               NNNNNNNNNNNNNNNNNNNNNNNNNNNNNNNNNNNNNNNNNNNNNNNNNN
     *****           *** *   *            *          *       *    *** *
    [1: GCCCCCCCGGGGGCGGGGGGGGGGC]
           13|20                13|40                13|60                    13|80
Mu  GAGCACGAGCGTACGCG-TTAGGACCCGAAAGATGGTGAACTATGCTTGGGCAGGGCGAAGCCAGAGGAAACTCTGGTGG
Ra  GAGCACGAGCGTACGCG-TTAGGACCCGAAAGATGGTGAACTATGCTTGGGCAGGGCGAAGC-AGAGGAAACTCTGGTGG
Ho  GAGCACGAGCGCACGTG-TTAGGACCCGAAAGATGGTGAACTATGCCTGGGCAGGGCGAAGCCAGAGGAAACTCTGGTGG
Rh  GAGCGTGAGCGCGCGTGC-TAGGACCCGAAAGATGGTGAACTATGCCTGGGCAGGGCGAAGCCAGAGGAAACTCTGGTGG
Xe  GNGCGTGAGCGCGCGCGATTAGGACCCGAAAGATGGTGAACTATGCCTGGGCAGNGCGAAGCCAGAGGAAACTCTGGTGG
La  GAGCATGAGCGTGCGTGA-TAGGACCCGAAAGATGGTGAACTATGCCTGGGCAGGGCGAAGCCAGAGGAAACTCTGGTGG
Cy  GAGC--GAGAGCGCGCGA-TGGTACCCGAAAGATGGTGAACTATGCCTGGGCAGGGCGAAGCCAGAGGAAACTCTGGTGG
OD  NNNNNNNNNNNNNNNNNNNTGTGACCCGAAAGATGGTGAACTATCTTGATCAGGTTGAAGTCAGGGGAAACCCTGATGG
      **    *  **    *  ** ***        *   *    ** ***    **  ** *   * *

           14|00                14|20                14|40                    14|60
Mu  AGGTCCGTAGCGGTCCTGACGTGCAAATCGGTCGTCCGA-CCTGGGTATAGGGGCGAAAGACTAATCGAACCATCTAGTA
Ra  AGGTCCGTAGCGGTCCTGACGTGCAAATCGGTCGTCCGA-CTTGGGTATAGGGGCGAAAGACTAATCGAACCATCTAGTA
Ho  AGGTCCGTAGCGGTCCTGACGTGCAAATCGGTCGTCCGA-CTTGGGTATAGGGGCGAAAGACTAATCGAACCATCTAGTA
Rh  AGGTCCGTAGCGGTCCTGACGTGCAAATCG-TCGTCCGA-CCTGGGTATAGGGGCGAAAGACTAATCGAACCATCTAGTA
Xe  AGGTCCGTAGCGGTCCTGACGTGCAAATCGGTCGTCCGA-CCTNGGTATAGGGGCGAAAGACGAATCGAACCATCTAGTA
La  AGGTNCGTAGCGGTCCTGACGTGCAAATCGGTCGTCCGAACCTGGGTATAGGGGCGAAAGACTAATCGAACCATCTAGTA
Cy  AGG-CCGCAGCGGTCCTGACGTGCAAATCGGTCGTCCGA-CCTGGGTATAGGGGCGAAAGACTAATCGAACCATCTAGTA
OD  AAGACCGAAACAGTTCTGACGTGCAAATCGATTGTCAGAA-TTGAGTATAGGGGCGAAAGACCAATCGAACCATCTAGTA
      * *   * * *  *              ** *  * *** *                        *

           14|80                15 |00                15|20
Mu  GCTGGTTCCCTCCGAAGTTTCCCTCAGGATAGCTGGCGCTCTCGC [TCCCGACGTA           ] CGCAGTTTTA
Ra  GCTGGTTCCCTCCGAAGTTTCCCTCAGGATAGCTGGCGCTCTCGC [AACGCGTTCGCTCGACAACC ] CGCAGTTTTA
Ho  GCTGGTTCCCTCCGAAGTTTCCCTCAGGATAGCTGGCGCTCTCGC [AGACCCGACGCACCCCCGCCA] CGCAGTTTTA
Rh  GCTGGTTCCCTCCGAAGTTTCCCTCAGGATAGCTGGCGCTCGTCA [CGCGAACCC           ] CGCAGTTTTA
Xe  GCTGGTTCCCTCCGAAGTTTCCCTCAGGATAGCTGGCGCTNGTCC [GT                  ] CGCAGTTTTA
La  GCTGGTTCCCTCCGAAGTTTCCCTCAGGATAGCTGGTGCTCGAGC [GAA                 ] CGCAGTTTTA
Cy  GCTGGTTCCCTCCGAAGTTTCCCTCAGGATAGCTGGCGCTGCCA                          CGCAGTTTTA
OD  GCTGGTTCCTTCCGAAGTTTCCCTCAGGATAGCTGGTGCATTTTA [ATATTATATAA         ] AATAATCTTA
        *                               *   ******                       *** * *
```

```
            15|40                 15|60                15|80                16|00
Mu TCCGGTAAAGCGAATGATTAGAGGTCTTGGGGCCGAAA-CGATCTCAACCTATTCTCAAACTTTAAATGGGTAAGAAGCC
Ra TCCGGTAAAGCGAATGATTAGAGGTCTTGGGGCCGAAA-CGATCTCAACCTATTCTCAAACTTTAAATGGGTAAGAAGCC
Ho TCCGGTAAAGCGAATGATTAGAGGTCTTGGGGCCGAAA-CGATCTCAACCTATTCTCAAACTTTAAATGGGTAAGAAGCC
Rh TCTGGTAAAGCGAATGATTAGAGGTCTTGGGGCCGAAA-CGATCTCAACCTATTCTCAAACTTTAAATGGGTAAGAACCG
Xe TCCGGTAAAGCGAATGATTAGAGGTCTTGGGGCCGAAATCGATCTCAACCTATTCTCAAACTTTAAATGGGTAAGAAGCC
La TCTGGTAAAGCGAATGACTAGAGGTCTTGGGGCCGAAA-CGATCTCAACCTATTCTCAAACTTTAAATGGGTAAGAAGCC
Cy TCCGGTAAAGCGAATGACTAGAGGCCTTGGGGCCGAAA-CGATCTCAACCTATTCTCAAACTTTAAATGGGTAAGAAGCC
OD TCTGGTAAAGCGAATGATTAGAGGCCTTAGGGTCGAAA-CGATCTTAACCTATTCTCAAACTTTAAATGGGTAAGAACCT
   *             *     *   *    *    *      *                                  * *
            16|20                 16|40                16|60                16|80
Mu CGGCTCGCTGGCGTGGAGCCGGG-CGTGGAATGCGA-GT--GCC-TAGTGGGCCACTTTTGGTAAGCAGAACTGGCGCTG
Ra CGGCTCGCTGGCGTGGAGCCGGG-CGTGGA-TGCGA-GT--GCC-TAGTGGGCCACTTTTGGTAAGCAGAACTGGCGCTG
Ho CGGCTCGCTGGCGTGGAGCCGGG--GTGGAATGCGA-GT--GCC-TAGTGGGCCACTTTTGGTAAGCAGAACTGGCGCTG
Rh CGGCTCGCTGGCGTGGAGCCGGG-CGTGGAATGCGA-GC-CGCC-TAGTGGGCCACTTTTGGTAAGCAGAACTGGCGCTG
Xe CGGCTCGCTGGCTTGGAGCCGGGGCGTGGAATGCGNNGCACGCCATAGTGGGCCACTTTTGGTAAGCAGAACTGGCGCTG
La CGGCTCGCTGGCTTGGAGCCGGG-CGTGGAATGCGA-GT--GCC-TAGTGGGCCACTTTTGGTAAGCAGAACTGGCGCTG
Cy CGGCTCGCTGGCTTGGAGCCGGG-CGTGGAATGCGA-GA-GCCC-CAGTGGGCCGCTTTTGGTAAGCAGAACTGGCGCTG
OD TAACTTTCTTGATATGA-CCNNNNNNNNNNNNNNNNNNNNNNNNNNCC-CAGTGGGCCACTTTTGGTAAGCAGAACTGGCGCTG
   ***  **  * ****  **      **     *    *  ****  **     *
            17|00                 17|20                17|40                17|60
Mu CGGGATGAACCGAACGCCGGGTTAAGGCGCCCGATGCCGACGCTCAT-CAGACCCCAGAAAAGGTGTTGGTTGATATAGA
Ra CGGGATGAACCGAACGCCGGGTTAAGGCGCCCGATGCCGACGCTCAT-CAGACCCCAGAAAAGGTGTTGGTTGATATAGA
Ho CGGGATGAACCGAACGCCGGGTTAAGGCGCCCGATGCCGACGCTCAT-CAGACCCCAGAAAAGGTGTTGGTTGATATAGA
Rh CGGGATGAACCGAACGCCGGGTTAAGGCGCCCGATGCCGACGCTCAT-CAGACCCCAGAAAAGGTGTTGGTTGATATAGA
Xe CGGGATGAACCGAACGCCGGGTTAAGGCGCCCGATGCCGACGCTCAT-CAGACCCCAGAAAAGGTGTTGGTTGATATAGA
La CGGGATGAACCGAACGCCGGGTTGTGGCGCCCGATGCCGACGCTCAT-CAGACCCCAGAAAAGGTGTTGGTTGATATAGA
Cy CGGGATGAACCGAACGCCGGGTTAAGGCGCCCGATGCCGACGCTCAT-CAGACCCCAGAAAAGGTGTTGGTTGATATAGA
OD TGGGAGTAACCAAACGTAATGTTACGTGCCCAAATTAACAA-CTCATGCAGATACCATGAAAGGCGTTGGTTGCTTAAAA
   *   **     *        ** ***  **  ****  **     *    **    **  *        *   * ** *
            17|80                 18|00                18|20                18|40
Mu CAGCAGGACGGTGGCCATGGAAGTCGGAATCCGCTAAGGAGTGTGTAACAACTCACCTGCCGAATCAACTAGCCCTGAAA
Ra CAGCAGGACGGTGGCCATGGAAGTCGGAATCCGCTA-GGAGTGTGTAACAACTCACCTGCCGAATCAACTAGCCCTGAAA
Ho CAGCAGGACGGTGGCCATGGAAGTCGGAATCCGCTAAGGAGTGTGTAACAACTCACCTGCCGAATCAACTAGCCCTGAAA
Rh CAGCAGGACGGTGGCCATGGAAGTCGGAATCCGCTAAGGAGTGTGTAACAACTCACCTGCCGAATCAACTAGCCCTGAAA
Xe CAGCAGGACGGTGGCCATGGAAGTTGGAATCCGCTAAGGAGTGTGTAACAACTCACCTGCCGAATCAACTAGCCCTGAAA
La CAGCAGGACGGTGGCCATGGAAGTCGGAATCCGCTAAGGAGTGTGTAACAACTCACCTGCCGAATCAACTAGCCCTGAAA
Cy CAGCAGGACGGTGGCCATGGAAGTCGGAATCCGCTAAGGAGTGTGTAACAACTCACCTGCCGAATCAACTAGCCCTGAAA
OD CAGCAGGACGGTGATCAAGGAAGTCGAAATCCGCTAAGGAGTGTGTAACAACTCACCTGCCGAAGCAACTACGGGTTAAA
              **  *     ** *    *                              *        **** *
            18|60                 18|80                21|00
Mu ATGGATGGCGCTGGAGC-GTCGGGCCCATACCCGGCCGTCGCCG     AGGTGCA-GATCTTGGT-GGTAGTAGC
Ra ATGGATGGCGCTGGAGC-GTCGGGCCCATACCCGGCCGTCGCCG     AGGTGCA-GATCTTGGT-GGTAGTAGC
Ho ATGGATGGCGCTGGAGC-GTCGGGCCCATACCCGGCCGTCGCCG     AGGTGCA-GATCTTGGT-GGTAGTAGC
Rh ATGGATGGCGCTGGAGC-GTCGGGCCCATAGCCGGCCGTCGCCG     AGGTGCA-TATCTTGGT-GGTAGTAGC
Xe ATGGATGGCGCTGGAGC-GTCGGGCCCATACCCGGCCGTCGCCG     GGGTGCA-GATCTTGGT-GGTAGTAGC
La ATGGATGGCGCTGGAGC-GTCGGGCCCATACCCGGCCGTCGCGG     GGGTGCA-GATCTTGGT-GGTAGTAGC
Cy ATGGATGGCGCTGGAGC-GTCGGGCCCATACCCGGCCGTCGAAG     GGGTGCA-GATCTTGGT-GGTAGTAGC
OD ATGGATGGCGCTTAAGTTGTATA-CCTATACATTACCGCTAAAG     TTGATCACGA-GTTAGTCGGTCCTAA-
              **  **  ****  *   *****   *****      ** ** ** **   *   **  **
21|20                 21|40                21|60                21|80
Mu AAATATTCAAA-CG  AGAAC--TTTGAAGGCCGAAGTGGAGAAGGGTTCCATGTGAACAGCAGTTGAACATGGGTCAG
Ra AAATATTCAAA-CG  AGAAC--TTTGAAGGCCGAAGTGGAGAAGGGTTCCATGTGAACAGCAGTTGAACATGGGTCAG
Ho AAATATTCAAA-CG  AGAAC--TTTGAAGGCCGAAGTGGAGAAGGGTTCCATGTGAACAGCAGTTGAACATGGGTCAG
Rh AAATATTCAAA-CG  AGAAC--TTTGAAGGCCGAAGTGGAGAAGGGTTCCATGTGAACAGCAGTTGAACATGGGTCAG
Xe AAATATTCAAA-CG  AGAAC--TTTGAAGGCCGAAGTGGAGAAGGGTTCCATGTGAACAGCAGTTGAACATGGGTCAG
La AAATATTCAAA-CG  AGAAC--TTTGAAGGCCGAAGTGGAGAAGGGTTCCATGTGAACAGCAGTTGAACATGGGTCAG
Cy AAATATTCAAAGCG  AGAGCG-TTTGAAGGCCGAAGTGGAGAAGGGTTCCATGTGAACAGCAGTTGAACATGGGTCAG
OD ---AGTTCAAGGCG[2]AGAAGGGTTTNNNNNNNNNNNNNNNNNNNNNNNNNNNNNNNNNNNNNNNNNNNNNNNNNNNNNNN
   *****       **       ****

[2: CACTGAAGTGG]
```

```
         22|00                22|20                22|40                22|60
Mu TC-GGTCCTGAGAGATGGGCGAGTGCC-GTTCCGAAGGGACGGGCGATGGCCTCCGTTGCCCTCGGCCGATCGAAAGGGA
Ra TC-GGTCCTGAGAGATGGGCGAGTGCC-GTTCCGAAGGGACGGGCGATGGCCTCCGTTGCCCTCAGCCGATCGAAAGGGA
Ho TC-GGTCCTGAGAGATGGGCGAGCGCC-GTTCCGAAGGGACGGGCGATGGCCTCCGTTGCCCTCAGCCGATCGAAAGGGA
Rh TC-GGTCCTAAGAGATAGGCGAGCGCC-GTTCCGAAGGGACGGGCGATGGCCTCCGTTGCCCTCGGCCGATCGAAAGGGA
Xe TC-GGTCCTAAGAGATGGGCGAGCGCC-GTTCGGAAGGGACGGGCGATGGCCTCCGTCGCCCTCGGCCGATCGAAAGGGA
La TC-GGTCCTAAGAGATGGGCGAACGCC-GTTCCGAAGGGACGGGCGATGGCCTCCGTCGCCCTCAGCCGATCGAAAGGGA
Cy TCAGGTCCTAAG-GATGGGCGACCGCCAGTTCGGAAGGGA-GGCCGATGGCCTCCGTCGCCCCCGGCCGATCGAAAGGGA
OD NNNNNNNNNNNNNNNNNNNNNNNNNNNNNNNNNNNNNNNNNNNNNNNNNNNNNNNNNNNNNNNNNNNNNNNNNNNNNNNNN
      *     *     *   *   ** *     *    *       * *               *    * *
         22|80                23|00                          23|20
Mu GTCGGGTTCAGATCCCCGAATCCGGAGTGGCGGAGATGGGCGCC         GCGA   GG---CCAGTG-CGGTAACGCG
Ra GTCGGGTTCAGATCCCCGAATCCGGAGTGGCGGAGATGGGCGCC         GCGA   GGCGTCCAGTGCCGGTAACGCG
Ho GTCGGGTTCAGATCCCCGAATCCGGAGTGGCGGAGATGGGCGCC         GCGA   GGCGTCCAGTG-CGGTAACGCG
Rh GTCGGGTTCAGATCCCCGAATCCGGAGTGGCGGAGACGGGCGCG         -CGA   GGCGTCCAGTG-CGGTAACGCG
Xe GTCGGGTTCAGATCCCCGAACCCGGAGTGGCGGAGACGGGCGCC  [3]    CCGA [4] GGCGTCCAGTG-CGGCGACGCG
La GTTGGGTTCAGATCCCCGAATCCGGAATGGCGGAGAGGGGCGCC         GCTG [5] GGCGTCCAGTG-CGGCAACGCA
Cy GTCGGGTTCAGATCCNCGAACCCGNAGTGGCGNAGACGCCGCG-         GCGA   GGCGCCCAGTG-CGGTAACGCA
OD NNNNNNNNNNNNNNNNNNNNNNNNNNNNNNNNNNNNNNNNNNNNN         NNNN   NNNNNNCATCC-TGGCAACAGG
      *               *        *              * *****      * **       ***  ***** **  ***
[3: CGCGGCCCCCC]
[4: CGCCTCGCGGCGGCGGGGGGGCGGG]
[5: TT]

         23|40                23|60                29|80                30|00
Mu ACCGATCCCGGAGAAGCCG-GCGGGAG         GCGGCGGCGGCGACTCTGGACGCGAGCCGGGCCCTTCCCG
Ra ACCGATCCCGGAGAAGCCG-GCGGGAG         GCGGCGGCGGCGACTCTGGACGCGAGCCGGGCCCTTCCCG
Ho ACCGATCCCGGAGAAGCCG-GCGGGAG         GCGGCGGCGGCGACTCTGGACGCGAGCCGGGCCCTTCCCG
Rh ACCGATCCCGGAGAAGCCGGCGGGCGGGAG      GCGGCGGCGGCGACTCTGGACGCGAGCCGGGCCCTTCCTG
Xe ACCGATCCCGGAGAAGCCGNGNGGGAG         CCGGCGGCGGCGACTCTGGACGCGCGCGCCGGGCCCTTCCTG
La AC-GATCCCGGAGAAGCCG-GCGGGAG         CGGGCGGCGGCGACTCTGGACGCGAGCCGGACCCTTCCTG
Cy AACGAACCTGGAGAAGCTG-GCGAGAG         TGCGCGGCGGCGACTCTGGACGTGAGCCGGGCCCTTCTCG
OD AACGACCATAAGAAGCCG-TCGAGAG          NNNNNNNNNNNNNNNNNNNNNNNNNNNNNNNNNNNNNNNNNN
      **    *   ****      * **   *       ***    **        * *      *         **
         30|20                          32|40                32|60
Mu TGGATCGCCTCAGCTGCGGCGGGCGTCGC  [ 6]  CGCCTCGGCCGGCGCCTAGCAGCCGACTT-AGAACTGGT
Ra TGGATCGCCCCAGCTGCGGCGGGCGTCGC  [ 7]  CGCCTCGGCCGGCGCCTAGCAGCCGACTT-AGAACTGGT
Ho TGGATCGCCCCAGCTGCGGCGGGCGTCGC  [ 8]  CGCCTCGGCCGGCGCCTAGCAGCCGACTT-AGAACTGGT
Rh TGGATCGCCCCAGCTGCGGCCGTCGCCCG  [ 9]  GCCCTCGGCCGGCGCCTAGCAGCTGACTT-AGAACTGGT
Xe TGGATCGCCCCAGCTGCGGCGCGCGCCTC  [10]  GGCCTCGGCCGGCGCCTAGCAGCTGACTT-AGAACTGGT
La TGGATCGCCTCAGCTGCGGCGCGCGCGGG  [11]  CGCCTCGGCCGGCGCCTAGCAGCTGACTT-AGAACTGGT
Cy CGGATCTCCGCACGGCTACGGCTCGCGTCGG [12] CTGCCTCGCCGGGGAGTAGCAGCCGGCTT-AGAACTGGT
OD NNNNNNNNNNNNNNNNNNNNNNNNNNNNNNN       NNNNNNNNNNNNNNNNNNNNNNNNNNNTTCAGAACTGGC
      *     *    *       ***  ****        *** ***   * **       ** *   *         *
[ 6: 195 bases]
[ 7: 166 bases]
[ 8: 283 bases]
[ 9: GCCCTCCTCTCGCGGGGACGGGCGGGCGGTT]
[10: 79 bases]
[11: CGTGGCAGGTCNNT]
[12: ACCTCCGTCCGGGCGTCTCTCCTCGCGCGGGGGGGGGCGTCTGGGCGGGGGCCACCCGGCCGG]

         32|80                33|00                33|20                33|40
Mu GCGGACCAGGGGAATCCGACTGTTTAATTAAAACAAAGCATCGCGAAGGCCCGCGGCGGGTGTTGACGCGATGTGATTTC
Ra GCGGACTAGGGGAATCCGACTGTTTAATTAAAACAAAGCATCGCGAAGGCCCGCGGCGGGTGTTGACGCGATGTGATTTC
Ho GCGGACTAGGGGAATCCGACTGTTTAATTAAAACAAAGCATCGCGAAGGCCCGCGGCGGGTGTTGACGCGATGTGATTTC
Rh GCGGACCAGGGGAATCCGACTGTTTAATTAAAACAAAGCATCGCGAAGGCCCGGTGGGTTGTTGACGCGATGTGATTTC
Xe GCGGACNANGGGAATCCGACTGTTTAATTAAAACAAAGCATCGCGAAGGCCCGAGGCGGGTGTTGACGCGATGTGATTTC
La GCGGACCAGGGGAATCCGACTGTTTAATTAAAACAAAGCATCGCGAAGGCCCGCGGCGGGTGTTGACGCGATGTGATTTC
Cy GCGGACCAGGGGAATCCGACTGTTTAATTAAAACAAAGCATCGCGAAGGCCCGCGGCGGGTGTTGACGCGATGTGATTTC
OD ACGGACTTGGGGAATCCGACTGTCTAATTAAAACAAAGCATTGTGATGGGCCTA-GCGGGTGTTGACACAATGTGATTTC
      *      **                 *                  * *  *  *  *  *** *         * *
```

```
          33|60               33|80               34|00               34|20
Mu TGCCCAGTGCTCTGAATGTCAAAGTGAAGAAATTCAATGAAGCGCGGGTAAACGGCGGGAGTAACTATGACTCTCTTAAG
Ra TGCCCAGTGCTCTGAATGTCAAAGTGAAGAAATTCAATGAAGCGCGGGTAAACGGCGGGAGTAACTATGACTCTCTTAAG
Ho TGCCCAGTGCTCTGAATGTCAAAGTGAAGAAATTCAATGAAGCGCGGGTAAACGGCGGGAGTAACTATGACTCTCTTAAG
Rh TGCCCAGTGCTCTGAATGTCAAAGTGAAGAAATTCAATGAAGCGCGGGTAAACGGCGGGAGTAACTATGACTCTCTTAAG
Xe TGCCCAGTGCTCTGAATGTCAAAGTGAAGAAATTCAATGAAGCGCGGGTAAACGGCGGGAGTAACTATGACTCTCTTAAG
La TGCCCAGTGCTCTGAATGTCAAAGTGAAGAAATTCAATGAAGCGCGGGTAAACGGCGGGAGTAACTATGACTCTCTTAAG
Cy TGCCCAGTGCTCTGAATGTCAAAGTGAAGAAATTCAATGAAGCGCGGGTAAACGGCGGGAGTAACTATGACTCTCTTAAG
OD TGCCCAGTGCTCTGAATGTCAAAGTGAAGAAATTCAAGTAAGCGCGGGTCAACGGCGGGAGTAACTATGACTCTCTTAAG
                                     **       *
          34|40               34|60               34|80                          35|00
Mu GTAGCCAAATGCCTCGTCATCTAATTAGTGACGCGCATGAATGGATGAACGAGATT     Mu CCCACTGTCCCTACCT
Ra GTAGCCAAATGCCTCGTCATCTAATTAGTGACGCGCATGAATGGATGAACGAGATT     Ra CCCACTGTCCCTACCT
Ho GTAGCCAAATGCCTCGTCATCTAATTAGTGACGCGCATGAATGGATGAACGAGATT     Ho CCCACTGTCCCTACCT
Rh GTAGCCAAATGCCTCGTCATCTAATTAGTGACGCGCATGAATGGATGAACGAGATT     Rh CCCACTGTCCCTACCT
Xe GTAGCCAAATGCCTCGTCATCTAATTAGTGACGCGCATGAATGGATGAACGAGATT     Xe CCCACTGTCCCTACCT
La GTAGCCAAATGCCTCGTCATCTAATTAGTGACNCGCATGAATGGATGAACGAGATT     La CCCACTGTCCCTACCT
Cy GTAGCCAAATGCCTCGTCATCTAATTAGTGACGCGCATGAATGGATGAACGAGATT     Cy CCCACTGTCCCTACCT
OD GTAGCCAAATGCCTCGTCATCTAATTAGTGACGCGCATGAATGGATTAACGAGATT     OL CCCACTGTCCCTATNT
                                                    *                         *
          35|20               35|40               35|60               35|80
Mu ACTATCCAGCGAAACCACAGCCAAGGGAACGGGCTTGGCGGAATCAGCGGGGAAAG-AAGACCCTGTTGAGCTTGACTCT
Ra ACTATCCAGCGAAACCACAGCCAAGGGAACGGGCTTGGCGGAATCAGCGGGGAAAG-AAGACCCTGTTGAGCTTGACTCT
Ho ACTATCCAGCGAAACCACAGCCAAGGGAACGGGCTTGGCGGAATCAGCGGGGAAAG-AAGACCCTGTTGAGCTTGACTCT
Rh ACTATCTAGCGAAACCACAGCCAAGGGAACGGGCTTGGCAGAATCAGCGGGGAAAGGAAGACCCTGTTGAGCTTGACTCT
Xe ACTATCTAGCGAAACCACAGCCAAGGGAACGGGCTTGGCGGAATCAGCGGGGAAAG-AAGACCCTGTTGAGCTTGACTCT
La ACTATCTAGCGAAACCACAGCCAAGGGAACGGGCTTGGCAGAATCAGCGGGGAAAG-AAGACCCTGTTGAGCTTGACTCT
Cy GCTATCTAGCGAAACCACAGCCAAGGGAACGGGCTTGGCAGAATCAGCGGGGAAAG-AAGACCCTGTTGAGCTTGACTCT
OL ACTATCCAGCGAAACCACAGCCAAGGGAACGGGCTTNGCCGAATCAGCGGGNAAAG-AAGACCCTGTTGAGCTTGACTCT
      *      *                              *                           *
                    36|00               36|20                               37|00
Mu AGTCTGGCACGGTGAAGAGACATGAGAGGTGTAGAATAAGTGGGAGGCCC [13] GCCGCCGGTGAAATACCACTACT
Ra AGTCTGGCACGGTGAAGAGACATGAGAGGTGTAGAATAAGTGGGAGGCCC [14] GCCGCCGGTGAAATACCACTACT
Ho AGTCTGGCACGGTGAAGAGACATGAGAGGTGTAGAATAAGTGGGAGGCCC [15] GCCGCCGGTGAAATACCACTACT
Rh AGTCTGGCCCTGTGAAGAGACATGAGAGGTGTAGAATAAGTGGGAG-CCC [16] GCCGCCGGTGAAATACCACTACT
Xe AGTCTGCAACTGTGAAGAGACATGAGAGGTGTAGGATAAGTGGGAGGCCC [17] GCCGCCGGTGAAATACCACTACT
La AGTCTGGCACTGTGAAGAGACATGAGAGGTGTAGGATAAGTGGGAGGCCT [CG] GCCGCCGGTGAAATACCACTACT
Cy AGTCTGGCACTGTGAAGAGACATGAGGGGTGTAGAATAAGTGGGAGGCCC [18] GCCGCCGGTGAAATACCACTACT
OL AGCCTGGCACGGTGAAAAGACATGAGAGGTGTAGGATAAGTGGGAGGTGC [19] GTGACCNGTGAAATACCACTACT
      *    *** *       *        *        *          ****        ***
[13: 52 bases]
[14: 55 bases]
[15: 57 bases]
[16: ACGCGG]
[17: 21 bases]
[18: GGGTACCTGGGTCCACGGCGC]
[19: GTCGCTCCCTCGCTAGCTCTNATGATGCAC]

          37|20               37|40               37|60
Mu CTCATCGTTTTTTTCACTGACCCGGTGAGGCGGG--GGGGCGAGCCCC----GAGGGG--CTC---TCGCTTCTGGCGCC-
Ra CTCATCGTTTTTTTCACTGACCCGGTGAGGCGGG--GGGGCGAGCCCC----GAGGGG--CTC---TCGCTTCTGGCTCCG
Ho CTGATCGTTTTTTTCACTGACCCGGTGAGGCGGG--GGGGCGAGCCC-----GAGGGG--CTC---TCGCTTCTGGCTCC-
Rh CTGATCGTTTTTTTCACTGACCCGGTGAGGCGGG--GGGGCGAGCCCC----GAGTGG--CTC---TCGCTTCTGGCTCC-
Xe CTTATCGTTTTTTTCACTTACCCGGTGAGGCGNN--GGGGCGAGCCCC----GAGGGG--CTC---TCGCTTCTGGACCC-
La CTTATCGTTTTTTTCACTTACCCGGTGAGGCGGG--GGGGCGAGTCCC----GAGGGG--CTC---TCGATTCTGGTGAC-
Cy CTTATCGTTTCCTCACTTACCCGGTGAGGGGGG--AGGCC-AGCCCCCG--GGCGGG--CTA---GCGCTTCTGGTGTC-
OL CTGATCGTTTGTTCACTTACCCGGTGAGGTCGGGGAGGCTGAGCTCTCGCTGGCGGGTGCTACATGCGCTTNTGGCTCC-
      *       **    *           **  ***  ***  **  **** ***  **  *****  *      *** *
```

```
          37|80
Mu AA--GCG [TCCGTCCCGCGCGTGCG         ]  GGCGGGCGCGACCCGCTCCGGGGACA-GTGCCAGGTGGGG
Ra AA---CG [CGTCCGCGCGCGC            ]  GGCGGGCGCGACCCGCTCCGGGGACA-GTGCCAGGTGGGG
Ho AA--GCG [CCCGCCCG                 ]  GCCGGGCGCGACCCGCTCCGGGGACA-GTGCCAGGTGGGG
Rh AA--GCG [TCGGCGCGG                ]  GCCGGGCGCGACCCGCTCCGGGGACA-GCGTCAGGTGGGG
Xe AA--GCG [CNCGGCCCCCGC             ]  GCCGGGCGCGACCCGCTCCGAGGACA-GTGGCAGGTGGGG
La AA--GCG [CCGGCTCG                 ]  GCCGGGTGCGACCCGCTCCGGGGACA-GTGGCAGGTGGGG
Cy AA--GCC [GGGGGGGTCCCTCTCCGGAGGTTCCC] CCCCCCGGCGAC--G-TCCC-GGACA-GTGGCAGGTGAGG
OL AAGTGCC [TCTGGCGCCCGAACACTGTTTGACGGT] GCAGAGCACGATCCGCTCCGGGAACATGTGTATAGTGGGG
           *** *                         ******** *** *  ** *  * * **** *
          38|40                       38|60          38|80          39|00
Mu AGTTTGACT  Mu GGGGCGGTACACCTGTCAAACGGTAACGCAGGTGTCCTAAGGCGAGCTCAG-G-GAGG-ACAGA
Ra AGTTTGACT  Ra GGGGCGGTACACCTGTCAAACGGTAACGCAGGTGTCCTAAGGCGAGCTCAG-G-GAGG-ACAGA
Ho AGTTTGACT  Ho GGGGCGGTACACCTGTCAAACGGTAACGCAGGTGTCCTAAGGCGAGCTCAG-G-GAGG-ACAGA
Rh AGTTTGACT  Rh GGGGCGGTACACCTGTCAAACCGTAACGCAGGTGTCCTAAGGCGAGCTCAG-G-GAGG-ACAGA
Xe AGTTTGACT  Xe GGGGCGGTACACCTGTCAAACCGTAACGCAGGTGTCCTAAGGCGAGCTCAG-GCGAGCTACAGA
La AGTTTGACT  La GGGGCGGTACACCTGTCAAACCGTAACGCAGGTGTCCTAAGGCGAGCTCAG-G-GAGG-ACAGA
Cy AGTTTGACT  Cy GGGGCGGTACACCTGTCAAACTGTAACGCAGGTGTCCTAAGGCGAGCTCAG-G-GAG---CAGA
OL AGTTTGACT  OD GGGGCGGTACATCTCTCAAATAATAACGGAGGTGTCCCAAGGCAGCTCAGTGCG-G--ACAGA
                             *  *   ***   *      *  *     *     * * **** *
          39|20          39|40          39|60          39|80
Mu AACCTCCCGTGGAGCAGAAGGGCAAAAGCTCGCTTGATCTTGATTTTCAGTACGAATACAGACCGTGAAA--GCGGG-GC
Ra AACCTCCCGTGGAGCAGAAGGGCAAAAGCTCGCTTGATCTTGATTTTCAGTACGAATACAGACCGAGAAA--GCGGG-GC
Ho AACCTCCCGTGGAGCAGAAGGGCAAAAGCTCGCTTGATCTTGATTTTCAGTACGAATACAGACCGTGAAA--GCGGG-GC
Rh AACCTCCCGTGGAGCAGAAGGGCAAAAGCTCGCTTGATCTTGATTTTCAGTATGAATACAGACCGTGAAA--GCGGG-GC
Xe AACCTCCCGTGGAGCAGAAGGGCAAAAGCTCGCTTGATCTTGATTTTCAGTATGAATACAGACCGTGAAA-CGCGGGNGC
La AACCTCCCGTGGAGCAGAAGGGCAAAAGCTCGCTTGATCTTGATTTTCAGTATGAGTACAGACCGTGAAAGCGCGGG-GC
Cy AACCTCCCGAGGAGCAGAAGGGCAAAAGCTCGCTTGATCTTGATTTTCAGTATGAGTACGGACCGTGAAA--GCGGG-GC
OD AACCACACATAGAGCAAAAGGGCAAATGCTGACTTGATCTCGGTGTTCAGTACACACAGGGACAGCAAAAGC-CG---GC
   * * ***        *      * **      * *      * *     ***** **  * **   *** ***
          40|00          40|20          40|40
Mu CTCA-CGA-TCCTTCTGACCTTTTGGGTTTT-AAGCAGGAGG-T-GTCAGAAAAGTTACCACAGGGATAACTGGCTTGTG
Ra CTCA-CGA-TCCTTCTGACCTTTTGGGTTTT-AAGCAGGAGG-T-GTCAGAAAAGTTACCACAGGGATAACTGGCTTGTG
Ho CTCA-CGA-TCCTTCTGACCTTTTGGGTTTT-AAGCAGGAGG-T-GTCAGAAAAGTTACCACAGGGATAACTGGCTTGTG
Rh CTCA-CGA-TCCTTCTGACTTTTTGGGTTTT-AAGCAGGAGG-T-GTCAGAAAAGTTACCACAGGGATAACTGGCTTGTG
Xe CTCA-CGA-TCCTTCTGACTTTTTGGGTTTT-AAGCAGGAGG-T-GTCAGAAAAGTTACCACAGGGATAACTGGCTTGTG
La CTCA-CGA-TCCTTCTGACTTTTTGGGTTTT-AAGCAGGAGG-T-GTCAGAAAAGTTACCACAGGGATAACTGGCTTGTG
Cy CTCA-CGA-TCCTTCTGGCTTTTTGGGTTTT-AAGCAGGAGGATCGTCAGAAAAGTTACCACAGGGATAACTGGCTTGTG
OD CT-ATCGAATCCTTTTGGTTTAAAGAGTTTTTAA-CAAGAGG-T-GTCAGAAAAGTTACCATAGGGATAACTGGCTTGTG
   * *    *      *  * *** *** *    *   *   *   * *        *
          40|60          40|80          41|00          41|20
Mu GC-GGCCAAGCGTTCATAG--CGA--CGTCGCTTTTTGATCCTTCGATGTCGGCTCTTCCTATCATTGTGAAGCAGAATTC
Ra GC-GGCCAAGCGTTCATAG--CGA--CGTCGCTTTTTGATCCTTCGATGTCGGCTCTTCCTATCATTGTGAAGCAGAATTC
Ho GC-GGCCAAGCGTTCATAG--CGA--CGTCGCTTTTTGATCCTTCGATGTCGGCTCTTCCTATCATTGTGAAGCAGAATTC
Rh GC-GGCCAAGCGTTCATAG--CGA--CGTCGCTTTTTGATCCTTCGATGTCGGCTCTTCCTATCATTGTGAAGCAGAATTC
Xe GCCGGCCAAGCGTTCATAG--CGA--CGTCGCTTTTTGATCCTTCGATGTCGGCTCTTCCTATCATTGTGAAGCAGAATTC
La GC-GGCCAAGCGTTCATAG--CGA--CGTCGCTTTTTGATCCTTCGATGTCGGCTCTTCCTATCATTGTGAAGCAGAATTC
Cy GC-GGCCAAGCGTTCATAGATCGAATCGTCGCTTTTTGATCCTTCGATGTCGGCTCTTCCTATCATTGTGAAGCAGAATTC
OD GC-GGCCAAGCGTTCATAG--CGA--CGTCGCTTTTTGATCCTTCGATGTCGGCTCTTCCTATCATTGTGAAGCAAAATTC
   *          **   **
```

Environmental Biology of Fishes **32**: 131–143, 1991.

Central nervous system myelin proteins of the coelacanth *Latimeria chalumnae:* phylogenetic implications

Thomas V. Waehneldt[1], Joachim Malotka[1], Gunnar Jeserich[2] & Jean-Marie Matthieu[3]
[1] *Max-Planck-Institut für experimentelle Medizin, Forschungsstelle Neurochemie, D-3400 Göttingen, Germany*
[2] *Zoophysiologie/Zellphysiologie, Universität Osnabrück, D-4500 Osnabrück, Germany*
[3] *Laboratoire de Neurochimie, Service de Pédiatrie, CHUV, CH-1011 Lausanne, Switzerland*

Received 1.3.1989   Accepted 12.9.1990

Key words: Proteolipid protein, Immunoblot analysis, *Latimeria chalumnae,* Lungfishes, Tetrapods, Actinistia

Synopsis

Myelin was isolated from the brain of a coelacanth. Its protein components were separated by polyacrylamide gel electrophoresis in the presence of sodium dodecylsulfate (SDS-PAGE). A protein component of 25000 Dalton was predominant; it was not glycosylated but reacted moderately with anti-mammalian CNS myelin proteolipid protein (PLP) antibodies and weakly with anti-lungfish CNS myelin glycosylated proteolipid protein (gPLP) antibodies. A component equivalent to mammalian DM-20 was not detectable. Presumably due to autolysis myelin basic protein (MBP) was not discernible by protein staining but showed up as a single band of 17000 Dalton with anti-mammalian MBP antibodies. Wolfgram protein (WP) was not present upon immunoblotting and the values for the myelin-specific 2′, 3′-cyclic nucleotide 3′-phosphodiesterase (CNPase) were extremely low. These results question a chondrichthyan association of the coelacanth but are strongly in favor of an Actinistia-Tetrapoda sister group relationship, with Dipnoi being most closely related to that combined group.

Introduction

Myelin serves as an insulator to increase the velocity of nerve impulse propagation (Ritchie 1984). It is synthesized by oligodendrocytes in the central nervous system (CNS) and by Schwann cells in the peripheral nervous system (PNS) and consists of spirally fused, axon-enwrapping arrangements of glial plasma membrane extensions (Raine 1984) (Fig. 1). The density of the myelin membrane is exceptionally low due to large amounts of lipids relative to small proportions of a limited number of characteristic myelin proteins (Norton & Cammer 1984).

Mammalian myelin has been the subject of extensive studies. The major myelin protein components in mammals are of two types: one is the myelin basic protein (MBP), present both in CNS and PNS myelin; it is a hydrophilic protein located at the major dense line of compact myelin (Fig. 1), presumably serving as a 'glue' to form the tight cytoplasmic apposition (Kirschner & Ganser 1980, Braun 1984, Lees & Brostoff 1984). The other type is either the unglycosylated proteolipid protein (PLP), found exclusively in CNS myelin, or the P_0 glycoprotein (P_0), detectable only in PNS myelin. Both PLP and P_0 represent extremely hydrophobic proteins with one (P_0) or several (PLP) transmem-

132

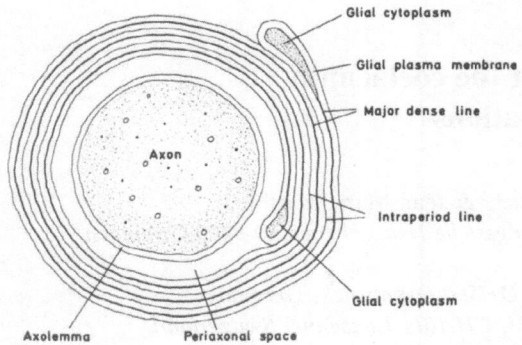

Fig. 1. Schematic representation of mammalian myelination demonstrating the glial plasma membrane enwrapping the axon in a spiral fashion. Compaction produces the thick major dense line (intracellular or cytoplasmic apposition) and the thinner intraperiod line (extracellular apposition). The periaxonal (extracellular) space separates neuronal from glial elements.

brane domains (Laursen et al. 1984, Stoffel et al. 1984, Lemke & Axel 1985) which reflect their exon-intron arrangements (Diehl et al. 1986, Lemke et al. 1988). The nucleotide sequences of PLP and P_0 are entirely dissimilar (Laursen et al. 1984, Stoffel et al. 1984, Lemke & Axel 1985) as are their amino acid sequences (Stoffel et al. 1983, Sakamoto et al. 1987). Nevertheless, both PLP and P_0 are instrumental in maintaining the tight adhesion of the myelin compaction via homophilic interactions at the extracellular apposition (intraperiod line) (Fig. 1). In other words: PLP and P_0 serve an analogous function in myelin compaction despite the absence of sequence homology.

While P_0 is detectable in PNS myelin of all gnathostome classes (Waehneldt et al. 1986a), the Agnatha do not form myelin although they possess glia (Bullock 1974, Bullock et al. 1984). PLP is only found in CNS myelin of tetrapods along with the minor Wolfgram protein (WP). The latter consists of a closely spaced doublet of 45000–50000 Dalton (Waehneldt & Malotka 1980) and is identical with CNPase that displays 2', 3'-cyclic nycleotide 3'-phosphodiesterase activity (Drummond & Dean 1980, Sprinkle et al. 1980, Vogel & Thompson 1988). PLP is not detectable in CNS myelin of teleostean and chondrichthyan fishes, nor is WP/CNPase; PLP is here replaced by P_0-like glycosylat-

ed IP components that crossreact strongly with antibodies against mammalian P_0 (Waehneldt et al. 1985, 1986a). Hence, as regards the major hydrophobic CNS myelin proteins of vertebrates, there exists a discontinuity at the transition from Osteichthyes (glycosylated P_0) to Tetrapoda (unglycosylated PLP [plus WP/CNPase]). Therefore, these proteins are useful molecular phylogenetic markers for the two groups.

In search of possible clues to the change from P_0 to PLP in putative pretetrapodal members the analysis of CNS tissue from the African lungfish revealed the presence of a PLP-reactive major component among the myelin proteins (Waehneldt et al. 1986b); its immunoreactivity with anti-mammalian PLP antibodies was relatively weak but unequivocal, while P_0 immunoreactivity was absent. This was taken as a molecular indication of a dipnoan-amphibian relationship (Waehneldt et al. 1986a) notwithstanding the fact that PLP of the African lungfish as well as that of the South American and Australian lungfishes were glycosylated (gPLP) (Waehneldt et al. 1986b, 1987) and also despite vanishingly low CNPase activities (Waehneldt et al. 1986b) which were comparable to those of other actinopterygian and chondrichthyan fishes (Franz et al. 1981). This situation is somewhat reminiscent of the ambivalent fish-tetrapod considerations with which the lungfish was rated briefly after its discovery more than 150 years ago (for historical background, see Rosen et al. 1981, Forey 1986, 1988).

Today, largely based on morphological characters, the living coelacanth, like the lungfish, is regarded as having close tetrapod affinities, although the relative position of these two groups appears to be a matter of substantial controversy (Romer 1966, Wahlert 1968, Løvtrup 1977, Compagno 1979, Dingerkus 1979, Lagios 1979, 1982, Wiley 1979, Jarvik 1980, Rosen et al. 1981, Holmes 1985, Forey 1986, Schultze 1986, Goodman et al. 1987). We have extended our biochemical analysis of CNS myelin proteins of vertebrates to include those of the coelacanth. We report here that *Latimeria chalumnae* CNS myelin possesses a major hydrophobic component which is characterized by the absence of glycosylation and which shows an immunoreactivity with anti-mammalian PLP antibodies substantially stronger than that of lungfish.

Materials and methods

The female coelacanth (VIMS 8118, 1452 mm in length and 53.75 kg in weight) was caught on native handline near Grand Comoro island in Autumn 1986 and frozen by the Explorer's Club personnel. It was air-shipped and maintained frozen at the New York Aquarium and the Virginia Institute of Marine Science until dissection of the brain and other tissues on 5 Jan 1988. The pituitary gland was excised and preserved; the remaining brain fragments were kept at $-80°C$ until air shipment to Göttingen (FRG).

Preparation of myelin

All steps were carried out at coldroom temperature $(2–4°C)$ unless stated otherwise. Brain fragments (4.19 g) were brought from $-70°C$ to coldroom temperature, immediately minced with a scalpel on a glass plate, transferred to a beaker containing 32 ml 0.3 M sucrose and homogenized with an Ultra-turrax homogenizer equipped with an 8 mm rotating tip (four bursts each of 5 seconds with intermittent cooling of 30 seconds). Additional homogenization was carried out in a teflon-glass homogenizer with 10 up-and-down strokes. Two ml of the total homogenate were set aside for protein determination and assay of CNPase (see below). Approximately 8 ml of the whitish homogenate was layered over 5 ml 0.80 M sucrose in each of four tubes and centrifuged in a swing–out rotor at 32000 rev min^{-1} (120,000 g_{av}) for 2 h. After lifting a thick greyish layer from the top of the gradient the myelin – containing band at the 0.30/0.80 M sucrose interphase was carefully removed with a Pasteur pipette, diluted to 40 ml with water and sedimented to a dense pellet in a fixed angle rotor at 36000 rev min $^{-1}$ (90,000 g_{av}) for 2 h. The washing step was repeated once. After rehomogenization in water and reserving small aliquots for analysis of protein content and CNPase activity the myelin fraction was lyophilized and partially delipidated by twofold extraction with ice-cold ether-ethanol (4 : 3).

Polyacrylamide gel electrophoresis in sodium dodecylsulfate (SDS-PAGE)

The partially delipidated myelin fraction was dissolved at 2 mg ml^{-1} in SDS solution (1% (w.v) SDS, 3% (v/v) 2-mercaptoethanol, 10% (v/v) glycerol, 0.01% (w/v) bromophenolblue) and subjected to slab gel electrophoresis in 1.5 mm thick gels of 14% (w/v) acrylamide in the Tris-glycine system and the presence of 0.1% (w/v) SDS (Laemmli 1970). Protein staining was done overnight with 0.1% (w/v) Coomassie blue R 250 in 50% (v/v) methanol, 7.5% (v/v) acetic acid for 3 h and destaining in 5% (v/v) methanol, 7.5% (v/v) acetic acid and 5% (v/v) glycerol.

Glycoprotein staining

Lectin-binding glycoproteins were visualized by the procedure of Franz et al. (1981) and Schott et al. (1984). The gels were gently shaken in 25% (v/v) isopropanol, 10% (v/v) acetic acid for 30 min, washed with two changes of phosphate-buffered saline (PBS, pH 7.4) for 30 min, and then incubated with Concanavalin A (200 mg 200 ml^{-1} PBS). The remainder of the lectin was removed during a 1 h period with several changes of PBS. This was followed by a 30 min exposure to horseradish peroxidase (HRP; 50 mg 200 ml^{-1} PBS). Excess HRP was eliminated by several washes of PBS during 2 h. After equilibrating the gel for 10 min in 0.05 M sodium phosphate buffer, pH 6.0, the glycoprotein bands were developed for 3–5 min with 10 mg diaminobenzidine (DAB) and 25 μl H$_2$O$_2$ (30%) in 100 ml phosphate buffer, pH 6.0. The reaction was stopped with dilute acetic acid. For equilibration the gels were stored overnight in the protein destaining solution (see previous paragraph).

134

Immunoblotting

Lungfish gPLP antigen was obtained by preparative SDS-PAGE of *Neoceratodus* myelin proteins (Waehneldt et al. 1987) following the method described earlier (Waehneldt et al. 1984). Anti-lungfish gPLP antiserum was raised in rabbits as were all the other antisera used. Anti-human MBP antiserum, anti-rat PLP antiserum, anti-bovine P_0 antiserum, anti-trout 36K antiserum, and anti-porcine WP antiserum were prepared in our laboratories and are identical with those of previous publications (Matthieu et al. 1983, Matthieu & Bürgisser 1983, Waehneldt & Jeserich 1984, Waehneldt et al. 1984, 1985, Karin & Waehneldt 1985, Jeserich & Waehneldt 1986).

Electro-immunoblotting (Western blotting) followed the procedure of Towbin et al. (1979) and Newcombe et al. (1982). After electrophoretic separation the myelin proteins from *L. chalumnae* and those from the other vertebrates were electrotransferred at 500 mA for 3 h onto nitrocellulose sheets in 25 mM Tris, 192 mM glycine, 20% (v/v) methanol and 0.04% (w/v) SDS. The sheets were gently shaken overnight in the coldroom in 3% (w/v) bovine serum albumin in PBS (BSA-PBS), and then incubated for 2 h at room temperature with an appropriate dilution of a corresponding rabbit antimyelin protein antiserum in BSA-PBS. Excess antisera were removed with several changes of PBS during 1 h, followed by an 1 h incubation with horseradish peroxidase-conjugated goat anti-rabbit IgG (1 : 500 in BSA-PBS). Excess second antibody was removed with several changes of PBS during 1 h. Immunoreactive protein bands were stained for 1–2 min with 10 mg diaminobenzidine and 25 μl H_2O_2 (30%) in 100 ml 17 mM ammonium chloride, 3 mM citric acid, pH 5.0. The reaction was stopped by immersing the nitrocellulose sheets in large volumes of water.

Other procedures

Proteins were estimated using the method of Lowry et al. (1951) with bovine serum albumin as the standard. The reaction mixtures of coelacanth total homogenates were turbid owing to excessively high lipid contents and were therefore subjected to ether extraction prior to photometric reading. The activity of 2', 3'-cyclic nucleotide 3'-phosphodiesterase (CNPase; EC 3.1.4.37) was photometrically determined (Prohaska et al. 1973) after activation with 1% (v/v) Triton X-100 or 1% (w/v) sodium deoxycholate.

Results

Pilot experiments had shown that it was impossible to homogenize coelacanth brain tissue in a motor-driven teflon-glass homogenizer, which is conventionally used for homogenization (e.g. mammalian brain). Thorough homogenization was feasible only after pretreatment with a high speed Ultraturrax homogenizer. When coelacanth brain total homogenates were subjected to density centrifugation on small gradient steps (0.2/0.4/0.6/0.8/1.0 M sucrose) most of the myelin fraction floated on 0.8 M sucrose, with traces on 0.6 M sucrose. For simplicity, a two step gradient was therefore used (0.3/0.8 M sucrose), ensuring that the entire myelin fraction was collected on 0.8 M sucrose and that excessive amounts of lipids and oily droplets were clustered at the top of the gradient. The presence of oily droplets has never been observed by these authors in myelin preparations of other vertebrates. A beige-brownish pellet was separated from the myelin layer by a clear 0.8 M sucrose phase. The 0.3 M sucrose phase was also clear.

The yield of coelacanth brain myelin protein (1.98 mg g^{-1} wet tissue) was lower than that of rat (11.3 mg g^{-1} wet tissue) (Franz et al. 1981). This reduction is also reflected in the amount of total brain protein (26.3 mg g^{-1} wet tissue in coelacanth vs. 102 mg g^{-1} wet tissue in rat). Therefore, our limited data indicate that the coelacanth brain is extremely rich in lipids and rather low in protein. However, the relative proportion of myelin protein and total brain protein remains fairly equal in both cases (7.6% and 10.8% of total brain protein is myelin protein in coelacanth and rat, respectively). At this time we cannot say whether these findings are true of the entire brain since the anatomical relationship of the coelacanth brain was not preserved after removal of the pituitary gland.

The electrophoretic profiles of CNS myelin proteins from different vertebrates were characterized by the presence of typical components (Fig. 2A). Rat had four hydrophilic myelin basic proteins. (14, 17, 18.5, 21.5) in addition to the hydrophobic proteolipid protein (PLP), DM-20 and Wolfgram protein (WP). Trout displayed two basic proteins (BP1 and BP2), two glycosylated hydrophobic P_0-immunoreactive proteins (IP1 and IP2) and the teleostean CNS myelin-specific 36K of 36000 Dalton. The lungfishes each showed fairly weak basic proteins (open triangles in Fig. 2A) but a dominant band of glycosylated proteolipid protein (gPLP) of molecular size greater than that of rat PLP. The CNS myelin proteins of these species serve as references for comparison with the pattern of coelacanth CNS myelin proteins (Fig. 2A, lane C). The profile of *L. chalumnae* was characterized by a predominant band of 25000 Dalton migrating like rat PLP, by two major components of approximately 10000 and 12000 Dalton (arrowheads in Fig. 2A), by lack of bands in the 'normal' molecular weight range of myelin basic proteins (approximately 14000–18000 Dalton) and, similar to the other species, by a large number of higher molecular weight minor components the nature of which with few exceptions is not known.

Concanavalin A-binding glycoproteins of the same species are revealed in panel B of Figure 2. Aside from the glycosylation of IP1 and IP2 the reaction of the major hydrophobic components of the lungfishes (gPLP) was very intense. Enzymatic deglycosylation of these components (27500 Dalton in *Neoceratodus forsteri* and 29000 Dalton in *Lepidosiren paradoxa*) led to loss of lectin binding as well as reduction of molecular weight by about 3000 Dalton (cf. Waehneldt et al. 1986b). In contrast to lungfishes and trout, coelacanth and rat were devoid of such prominent glycoprotein patter.. In rat, trace amounts of the myelin/oligodendroglial glycoprotein (MOG; Linington et al. 1984) gave rise to very weak staining at the position of the unglycosylated PLP; whether a similar weak staining at the position of the predominant coelacanth protein may also be due to the presence of a

MOG-homologous glycoprotein remains to be established. At any rate, the disparity in the predominant coelacanth protein of massive Coomassie blue staining on the one hand and the virtual absence of glycoprotein staining on the other hand was overwhelming evidence for lack of glycosylation. This was supported by unaltered electrophoretic mobility after enzymatic deglycosylation (not shown).

Applying anti-rat PLP antibodies for immunodetection the reaction of lungfish gPLP was only weak (Fig. 3A, arrows), in agreement with earlier results (Waehneldt et al. 1986b, 1987). Dimeric forms were not discernible. By contrast, the predominant coelacanth component of 25000 Dalton was distinctly stained, as was a dimeric form. The staining of rat PLP and its homologous DM-20 component was even more intense, and also that of the dimeric form (and of higher oligomers); the rat dimer migrated at virtually the same position as the *L. chalumnae* dimer. The coelacanth, however, did not show immunoreaction in a region similar to DM-20, although a fuzzy Coomassie blue-staining band around 21000 Dalton (Fig. 2A) might invite such speculation.

Conversely, when applying anti-lungfish gPLP antibodies for immunodetection (Fig. 3B) the staining of the predominant coelacanth component was rather weak; its dimer as well as rat PLP could not be clearly discerned. The lungfish gPLP and its aggregated forms strongly reacted in either dipnoan species. Weak bands migrating slightly ahead of lungfish gPLP most probably were degradation and/or deglycosylation products and as such not related to a distinct polypeptide form equivalent to DM-20 (Simons et al. 1987, Nave et al. 1987). It must be mentioned that the immunoreactivity of deglycosylated gPLP with anti-lungfish gPLP antibodies was virtually as intense as that of the native component. Thus, the antigenicity resided largely, if not totally, in the protein backbone. These results show that the predominant 25000 Dalton component of coelacanth is recognized by antibodies both to mammalian PLP and dipnoan gPLP, thereby taking an intermediate position.

Trout CNS myelin IP1 and IP2 proteins reacted intensely with anti-bovine P_0 antibodies (Fig. 4A).

136

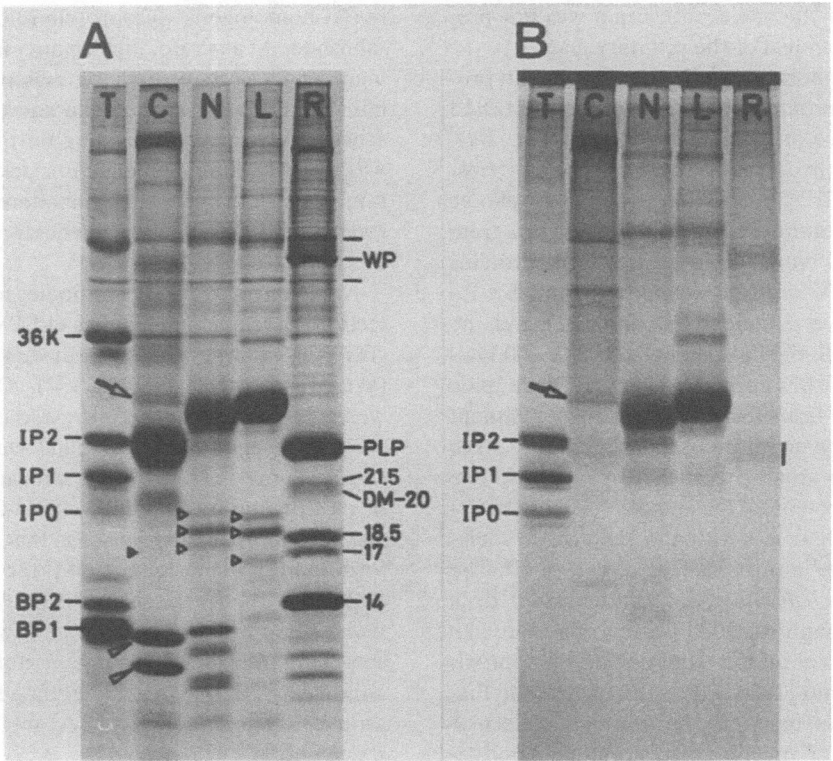

Fig. 2. Electrophoretic separation of vertebrate CNS myelin proteins on 14% SDS gels (SDS-PAGE). Migration is towards the anode (bottom). Protein loads were approximately 20 µg per lane. Gels were stained for protein with Coomassie blue (panel A) and for glycoprotein with Concanavalin A-horseradish peroxidase (panel B). T, trout; C, coelacanth; N, *Neoceratodus;* L, *Lepidosiren;* R, rat. Left margin of panel A: 36K, teleost CNS myelin specific unglycosylated protein of 36000 Dalton; IP1 and IP2, glycosylated hydrophobic trout CNS myelin proteins of 23000 and 26200 Dalton, respectively; IP0, glycosylated autolysis prodnct of IP2 and IP1; BP1 and BP2, hydrophilic trout CNS myelin basic proteins. Right margin of panel A: WP, closely spaced Wolfgram protein doublet of approximately 48000 Dalton (lines above and below WP point at tubulin and actin, respectively (Waehneldt & Malotka 1980); PLP and DM-20, rat CNS myelin unglycosylated proteolipid and intermediate proteins; 14, 17, 18.5, 21.5, hydrophilic rat myelin basic proteins giving their molecular weights in kiloDalton. Arrow, minor glycosylated protein doublet of about 30000 Dalton; arrowheads, major coelacanth protein bands of unknown specificity of molecular weight below 12000 Dalton; open triangles, lungfish myelin basic proteins; closed triangle, position of coelacanth myelin basic protein inferred from immunoblotting (Fig. 4B). Short vertical line on the right margin of panel B shows the position of rat PLP and also of underlying trace amounts of the myelin/oligodendroglial glycoprotein (MOG; Linington et al. 1984).

Neither lungfish nor rat showed any reaction apart from some unspecific binding of trace components of higher molecular weight. A *L. chalumnae* protein doublet of approximately 30000 Dalton stained weakly with Coomassie blue and with lectin (arrow in Fig. 2A, B) but distinctly bound antibodies against mammalian P_0 (Fig. 4A, arrow). It is interesting to note that another P_0-staining doublet

was underlying the predominant 25000 Dalton PLP-reactive protein (Fig. 4A, double arrow). Its P_0-reactivity was weak compared to that of the 30000 Dalton doublet. Taking the enormous difference in Coomassie blue staining of 30000 Dalton doublet vs. the predominant 25000 Dalton component into consideration (Fig. 2A), it must be concluded that we deal here with trace amounts of

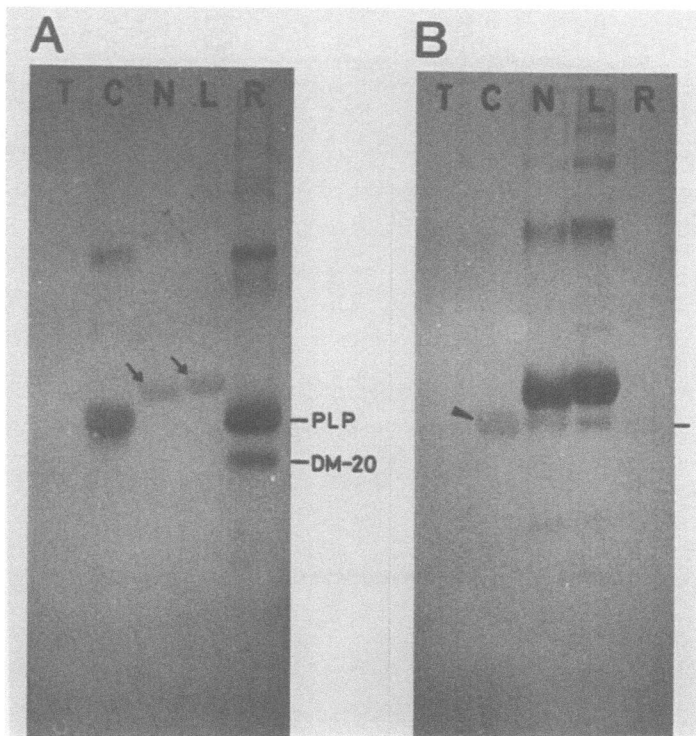

Fig. 3. Electrophoretic separation of vertebrate CNS myelin proteins on 14% SDS gels (SDS-PAGE), followed by electrotransfer to nitrocellulose sheets and immunodetection with polyclonal rabbit antisera. Protein loads were approximately 20 μg per lane. Panel A, anti-rat PLP antiserum (1 : 150); panel B, anti-ceratodidian gPLP antiserum (1 : 1250). Small arrows in panel A point to weak reaction of anti-rat PLP antiserum with ceratodidian (N) and lepidosirenidian (L) gPLP. Conversely, arrowhead in panel B shows the weak reaction of anti-lungfish gPLP antibodies with coelacanthidian PLP; horizontal line at the right margin marks the position of rat PLP which barely reacts with anti-lungfish gPLP antiserum upon visual inspection and is lost in the photographic process. For further abbreviations, see Fig. 2.

other glycosylated protein components (Fig. 2B) with electrophoretic characteristics similar to those of the predominant 25000 Dalton component.

Immunoblotting with anti-myelin basic protein (MBP) antibodies showed typical banding patterns for the vertebrates analyzed herein (Fig. 4B). Coelacanth had a single MBP-immunoreactive component migrating with the 17K protein of rat. Although no Coomassie blue-staining band was clearly visible in that region of the coelacanth gel lane (Fig. 2A, filled triangle) this result attests to the presence of at least low amounts of MBP.

The specific activities of CNPase in total homog-enate and myelin of coelacanth and other vertebrates are shown in Table 1. The values for coelacanth were very low; they were comparable to those for lungfish and other fishes. This was also underlined by the absence of immunoreactivity with antibodies against porcine Wolfgram protein (WP) in coelacanth and lungfish (not shown).

Discussion

Since the discovery of the living crossopterygian coelacanth *L. chalumnae* in 1938 the number of

138

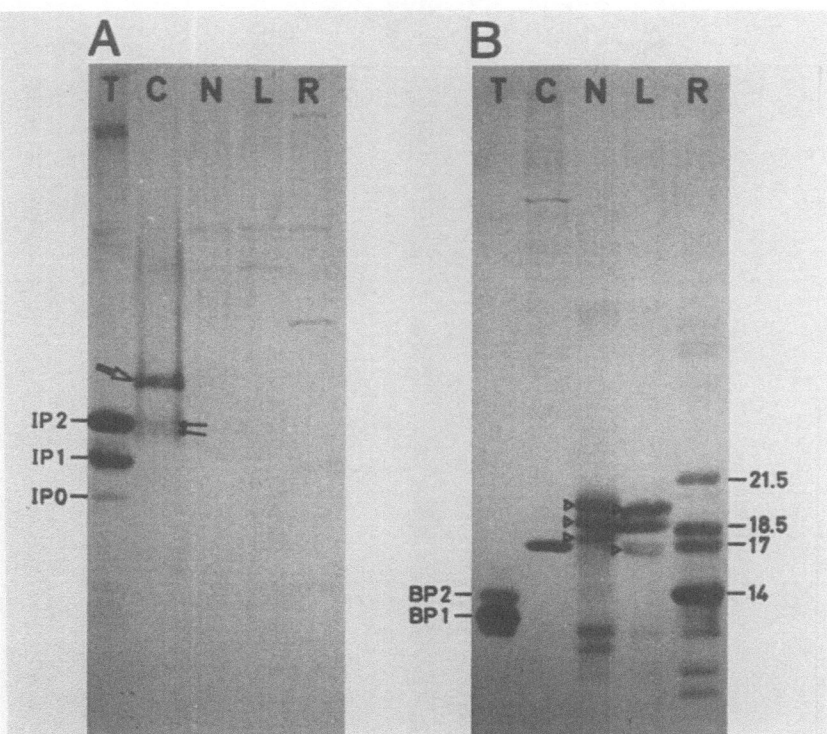

Fig. 4. Electrophoretic separation of vertebrate CNS myelin proteins on 14% SDS gels (SDS-PAGE), followed by immunoblotting. Protein loads were approximately 20 μg per lane, except for coelacanth in panel B (60 μg). Panel A, anti-bovine P_0 antiserum (1 : 500); panel B, anti-human MBP antiserum (1 : 500). Open arrow in panel A points to the weak P_0-immunoreactive 30000 Dalton doublet in coelacanth (cf. Fig. 2A, B). Small arrows show the 25000 Dalton doublet underlying the predominant PLP-reactive coelacanth component. Open triangles in panel B show the positions of the lungfish myelin basic proteins (see also Fig. 2A). For further abbreviations, see Fig. 2.

specimens available for morphological, physiological and biochemical studies has been limited (Forey 1988). Presumably this is the main reason why very little is known so far of the biochemical composition of its nervous system. Tamai et al. (1986) were the first to characterize the myelin proteins of coelacanth brain by SDS electrophoresis and protein staining with Coomassie blue. They showed the presence of a predominant component which comigrated with rat PLP, the presence of another main band with apparent molecular size of 28000 Dalton, and of a slightly and broadly stained double band migrating to the position of rat large myelin basic protein (18500 Dalton). While this pattern was superficially similar to that of the electric ray (Waehneldt et al. 1984) and other chondrichthyan members (Franz et al. 1981), no statements were made concerning the immunological relatedness of these proteins (Waehneldt et al. 1986a).

Our report is the first to demonstrate that the predominant 25000 Dalton component of coelacanth brain myelin is immunologically related both to mammalian PLP and to dipnoan gPLP. A double approach has been taken herein: using anti-mammalian PLP antibodies more epitopes were recognized in coelacanth PLP than in lungfish gPLP (Fig. 3A). This was reconfirmed by an inverse analysis with anti-lungfish gPLP antibodies where coelacanth PLP still gave a distinct signal while rat PLP was barely detectable (Fig. 3B). In each case the strongest reactivity was shown by the 'parent' antigen. These results are a clear indication of a closer relationship between coelacanth and tetrapod PLP than between lungfish and tetrapod PLP. In other words: the predominant 25000 Dalton component of the actinistian coelacanth takes an immunological position between lungfish gPLP and mammalian PLP. This is also strengthened by the lack of PLP glycosylation in coelacanth and tetrapods. We therefore label the coelacanth 25000 Dalton component 'PLP'. The glycosylation of lungfish gPLP thus appears to be a dipnoan specialization.

As reported by Tamai et al. (1986) their 28000 Dalton major Coomassie blue-positive component was also stained with the rather insensitive periodic acid-Schiff (PAS) glycoprotein stain. Although in our myelin preparation we do not see a major component in that gel region it is entirely possible that our minor Concanavalin A- and anti-PNS-P_0-positive 30000 Dalton doublet may be identical with the massively Coomassie blue staining band of the Japanese workers; particularly since their molecular size of 28000 Dalton appears to be slightly underestimated. It must be considered that varying proportions of myelinated PNS nerves emerging from, and thereby contaminating, the starting CNS tissue may account for varying amounts of the component of Tamai et al. (1986) and of our doublet. This could be of special relevance in the case of the anterior brainstem where several large calibre PNS nerves emanate (see Fig. 2 in Northcutt 1986). In this context mention must also be made of the weak anti-P_0-reactive doublet that comigrates with coelacanth PLP (Fig. 4A, double arrow). The shape of its anti-P_0 stain was however entirely different from the broad anti-PLP stain of coelacanth PLP (Fig. 3A). Therefore, we deal here possibly with minute amounts of degradation products of the 30000 Dalton doublet, adventitiously migrating like the predominant coelacanth PLP. Proteolysis products have also been observed in the case of mammalian P_0 (Lees & Brostoff 1984).

In CNS myelin of all gnathostomates myelin basic protein (MBP) is regarded as an indispensable

Table 1. CNPase specific activity in coelacanth brain total homogenate and isolated myelin. Comparison with lungfish and other vertebrates.

	Total homogenate		Myelin	
	DOC	Trit	DOC	Trit
Coelacanth (*Latimeria chalumnae*)	2.1 ± 0.4	5.5 ± 1.0	1.2 ± 0.3	4.8 ± 0.3
Lungfish (*Protopterus dolloi*)	0.9 ± 0.2	2.6 ± 0.2	2.0 ± 0.6	6.0 ± 0.8
Rat (*Rattus rattus*)	216	281	954	1345
Pigeon (*Columbia livia*)	84	223	455	891
Lizard (*Lacerta galoti*)	141	210	331	693
Xenopus (*Xenopus laevis*)	130	664	287	1953
Frog (*Rana temporaria*)	99	257	345	1439
Salamander (*Salamandra salamandra*)	129	280	413	1660
Trout (*Oncorhynchus mykiss*)	17	52	11	35
Sterlet (*Acipenser ruthenus*)	n.d.	17	n.d.	15
Dogfish (*Scyliorhinus caniculus*)	n.d.	n.d.	n.d.	n.d.
Ray (*Raja clavata*)	n.d.	22	7	33
Electric ray (*Torpedo marmorata*)	n.d.	n.d.	n.d.	n.d.

Specific activities are expressed as μmol 2', 3'-cyclic nucleotide AMP hydrolyzed mg^{-1} protein h^{-1}. The coelacanth figures represent the means of six and three determinations ± SEM for total homogenate and isolated myelin, respectively. Enzyme activation was carried out in the presence of sodium deoxycholate (DOC) or Triton X-100 (Trit). The other data are from Franz et al. (1981) except for lungfish which are from Waehneldt et al. (1986b). n.d. = not detectable. For clarity the standard deviations are not shown in tetrapods, actinopterygians and chondrichthyans; in all cases they were below 10%.

140

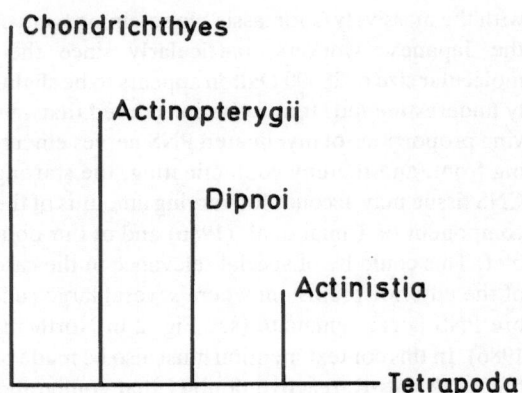

Fig. 5. Phylogeny of the major groups of extant gnathostomes based on the occurrence of CNS myelin proteins. Apart from myelin basic protein (MBP) and myelin-associated glycoprotein (MAG) which are common to all groups (Matthieu et al. 1986b, Waehneldt et al. 1986a), CNS myelin carries the following components: Chondrichthyes = P_0 glycoprotein; Actinopterygii = P_0 glycoprotein and (in Holostei and Teleostei) 36K; Dipnoi = glycosylated PLP; Actinistia = unglycosylated PLP; Tetrapoda = unglycosylated PLP and WP/CNPase. Matthieu et al. (1986a) reported the presence of trace amounts of unglycosylated PLP in brain myelin of the cladistian *Polypterus*. In general, *Polypterus* is regarded as an early actinopterygian candidate (Patterson 1982); this is strengthened by the presence of P_0 as the predominant hydrophobic component, besides low 36K immunoreactivity. However, the simultaneous occurrence in *Polypterus* CNS myelin of P_0 and PLP was, and remains to be, puzzling.

component to hold the cytoplasmic sides of the myelin membrane together (Kirschner & Ganser 1980, Waehneldt et al. 1986a). The lack in coelacanth of Coomassie blue staining band(s) in the typical MBP gel region (about 14000–18000 Dalton) was therefore a surprise. Only when the protein load was substantially increased a single MBP-positive band of approximately 17000 Dalton was detectable upon immunoblotting (Fig. 4B). Since mammalian MBP is prone to postmortem autolysis (Berlet & Volk 1980) and since the frozen coelacanth was brought to room temperature overnight and then subjected to 6–8 h NMR scanning and additional dissection (C.V. Sullivan personal communication) it is very likely that most of the coelacanth's MBP was autolysed prior to the myelin purification step. Future experiments on a rapidly

frozen brain will hopefully establish the in vivo level of MBP and will resolve the question of whether there is only one MBP component (this paper) or possibly two (Tamai et al. 1986).

The presence of CNPase activity and of the Wolfgram protein doublet appears to be characteristic of tetrapods while fishes are essentially devoid of it (Franz et al. 1981, Waehneldt et al. 1985, 1986a). Although the role of CNPase in myelination remains enigmatic – aside from being a structural protein – (Vogel & Thompson 1988), the dramatic increase at the transition from fishes to amphibians can be regarded as an important tetrapodal feature. We observe a several hundred-fold increase of CNPase activity in amphibians when considering the low values in myelin of coelacanth and lungfish in either detergent used (Table 1). Therefore, as regards WP/CNPase, both coelacanth and lungfish clearly range among the fishes.

In conclusion, the biochemical results on myelin proteins form the basis for the branching pattern shown in Figure 5. They are summarized as follows. First, a chondrichthyan association of the coelacanth is extremely unlikely because CNS myelin of rays and sharks is only endowed with P_0-reactive hydrophobic components, in total absence of PLP (Waehneldt et al. 1984, Tai & Smith 1984). This is in contrast to the views of Løvtrup (1977) and Lagios (1979, 1982) who favor a sister group relationship of Actinistia and Chondrichthyes. Second, PLP appears to be a novel acquisition of sarcopterygian vertebrates and is not shared by actinopterygians (Waehneldt et al. 1986a. See, however, legend to Fig. 5 and Matthieu et al. 1986a). In sarcopterygians PLP is a predominant structural CNS myelin component and as such common to lungfishes and the coelacanth as well as tetrapods, while P_0 is absent from their CNS myelin. Third, the closer immunological similarity between coelacanthidian and tetrapodal PLP speaks for a sister group relationship of Actinistia and Tetrapoda, with the lungfishes being adjacent to that combined group, and the actinopterygian fishes joining next. Our scheme is in accord with the views of Romer (1966) and with the studies of Schultze (1986, 1988) and Fritzsch (1987) but disagrees with the analysis of Rosen et al. (1981) who concluded that lungfishes are the sister group of tetrapods. Finally, high levels of WP/CNPase in association with CNS myelin is characteristic of tetrapods. Extremely low levels of CNPase in fishes are of uncertain analytical value.

Acknowledgements

The authors wish to thank V. Neuhoff for generous support and C. Linington and A. Linington-White for critically reading the manuscript. We express our gratitude to the Explorer's Club and the Virginia Institute of Marine Science Coelacanth Program (J.A. Musick) for access to study material and to C.V. Sullivan for excision and shipment of brain tissue.

References cited

Berlet, H.H. & B. Volk. 1980. Studies of human myelin proteins during old age. Mechanisms of Ageing and Development 14: 211–222.

Braun, P.E. 1984. Molecular organization of myelin. pp. 97–116. In: P. Morell (ed.) Myelin, Plenum Press, New York.

Bullock, T.H. 1974. Comparisons between vertebrates and invertebrates in nervous organization. pp. 343–346. In: F.O. Schmitt & F.G. Worden (ed.) The Neurosciences, Third Study Program, MIT Press, Cambridge.

Bullock, T.H., J.K. Moore & R.D. Fields. 1984, Evolution of myelin sheaths: both lamprey and hagfish lack myelin. Neurosci. Lett. 48: 145–148.

Compagno, L.J.V. 1979. Coelacanths: shark relatives or bony fishes? Occ. Pap. Calif. Acad. Sci. 134: 45–52.

Diehl, H.-J., M. Schaich, R.-M. Budzinski & W. Stoffel. 1986. Individual exons encode the integral membran domains of human myelin proteolipid protein. Proc. Nat. Acad. Sci. USA 83: 9807–9811.

Dingerkus, G. 1979. Chordate cytogenetic studies: an analysis of their phylogenetic implications with particular references to fishes and the living coelacanth. Occ. Pap. Calif. Acad. Sci. 134: 111–126.

Drummond, R.J. & G. Dean. 1980. Comparison of 2', 3'-cyclic nucleotide 3'-phosphodiesterase and the major component of Wolfgram protein W1. J. Neurochem. 35: 1155–1165.

Forey, P.L. 1986. Relationships of lungfishes. J. Morph. Supp. 1: 75–91.

Forey, P.L. 1988. Golden jubilee for the coelacanth Latimeria chalumnae. Nature 336: 727–732.

Franz, T., T.V. Waehneldt, V. Neuhoff & K. Wächtler. 1981. Central nervous system myelin proteins and glycoproteins in vertebrates: a phylogenetic study. Brain Res. 226: 245–258.

Fritzsch, B. 1987. Inner ear of the coelacanth fish Latimeria has tetrapod affinities. Nature 327: 153–154.

Goodman, M., M.M. Miyamoto & J. Czelusniak. 1987. Pattern and process in vertebrate phylogeny revealed by coevolution of molecules and morphologies. pp. 141–176. In: C. Patterson (ed.) Molecules and Morphology in Evolution: Conflict or Comprise? Cambridge University Press, Cambridge, England.

Holmes, E.B. 1985. Are lungfishes the sister group of tetrapods? Biol. J. Linn. Soc. 25: 379–397.

Jarvik, E. 1980. Basic structure and evolution of vertebrates. 2 vol., Academic Press, London. 575 pp.

Jeserich, G. & T.V. Waehneldt. 1986. Bony fish myelin: evidence for common major structural glycoproteins in central and peripheral myelin of trout. J. Neurochem. 46: 525–533.

Karin, N.J. & T.V. Waehneldt. 1985. Biosynthesis and insertion of Wolfgram protein into optic nerve membranes. Neurochem. Res. 10: 897–907.

Kirschner, D.A. & A.L. Ganser. 1980. Compact myelin exists in the absence of basic protein in the shiverer mutant mouse. Nature 283: 207–210.

142

Laemmli, U.K. 1970. Cleavage of structural proteins during the assembly of the head of bacteriophage T4. Nature 227: 680–685.

Lagios, M.D. 1979. The coelacanth and the chondrichthyes as sister groups: a review of shared apomorph characters and a cladistic analysis and reinterpretation. Occ. Pap. Calif. Acad. Sci. 134: 25–44.

Lagios, M.D. 1982. *Latimeria* and the chondrichthyes as sister taxa: a rebuttal to recent attempts at refutation. Copeia 1982: 942–948.

Laursen, R.A., M. Samiullah & M.B. Lees. 1984. The structure of bovine brain myelin proteolipid and its organization in myelin. Proc. Nat. Acad. Sci. USA 81: 2912–2916.

Lees, M.B. & S.W. Brostoff. 1984. Proteins of myelin. pp. 197–224. *In:* P. Morell (ed.) Myelin, Plenum Press, New York.

Lemke, G. & R. Axel. 1985. Isolation and sequence of a cDNA encoding the major structural protein of peripheral myelin. Cell 40: 501–508.

Lemke, G., E. Lamar & J. Patterson. 1988. Isolation and analysis of the gene encoding myelin protein zero. Neuron 1: 73–83.

Linington, C., M. Webb & P.L. Woodhams. 1984. A novel myelin-associated glycoprotein defined by a mouse monoclonal antibody. J. Neuroimmunol. 6: 387–396.

Løvtrup, S. 1977. The phylogeny of the vertebrata. John Wiley and Sons, London. 330 pp.

Lowry, O.H., N.J. Rosebrough, A.L. Farr & R.J. Randall. 1951. Protein measurement with the Folin phenol reagent. J. Biol. Chem. 193: 265–275.

Matthieu, J.-M. & P. Bürgisser. 1983. Radioimmunological determination of myelin basic protein in the CSF of neurological patients. pp. 223–226. *In:* H. Peeters (ed.) Protides of the Biological Fluids, Vol. 30, Pergamon Press, Oxford.

Matthieu, H.-M., G. Almazan & T.V. Waehneldt. 1983. Intrinsic myelin proteins are normally synthesized in vitro in the myelin deficient (mld) mutant mouse. Dev. Neurosci. 6: 246–250.

Matthieu, J.-M., M. Eschmann, P. Bürgisser, J. Malotka & T.V. Waehneldt. 1986a. Expression of myelin proteins characteristic of fish and tetrapods by *Polypterus* revitalizes long discredited phylogenetic links. Brain Res. 379: 137–142.

Matthieu, J.-M., T.V. Waehneldt & N. Eschmann. 1986b. Meylin-associated glycoprotein and myelin basic protein are present in central and peripheral nerve myelin throughout phylogeny. Neurochem. Int. 8: 521–526.

Nave, K.-A., C. Lai, F.E. Bloom & R.J. Milner. 1987. Splice site selection in the proteolipid protein (PLP) gene transcript and primary structure of the DM-20 protein of central nervous system myelin. Proc. Nat. Acad. Sci. 84: 5665–5669.

Newcombe, J., P. Glynn & M.L. Cuzner. 1982. The immunological identification of brain proteins on cellulose nitrate in human demyelinating disease. J. Neurochem. 38. 267–274.

Northcutt, R.G. 1986. Lungfish neural characters and their bearing on sarcopterygian phylogeny. J. Morph. Supp. 1: 277–297.

Norton, W.T. & W. Cammer. 1984. Isolation and character-

ization of myelin. pp. 147–195. *In:* P. Morell (ed.) Myelin, Plenum Press, New York.

Patterson, C. 1982. Morphologies and interrelationships of primitive actinopterygian fishes. Amer. Zool. 22: 241–259.

Prohaska, J.R., D.A. Clark & W.W. Wells. 1973. Improved rapidity and precision in the determination of brain 2', 3'-cyclic nucleotide 3'-phosphohydrolase. Analyt. Biochem. 56: 275–282.

Raine, C.S. 1984. Morphology of myelin and myelination. pp. 1–50. *In:* P. Morell (ed.) Myelin, Plenum Press, New York.

Ritchie, J.M. 1984. Physiological basis of conduction in myelinated nerve fibres. pp. 117–145. *In:* P. Morell (ed.) Myelin, Plenum Press, New York.

Romer, A.S. 1966. Vertebrate paleontology, 3rd ed. University of Chicago Press, Chicago. 468 pp.

Rosen, D.E., P.L. Forey, B.G. Gardiner & C. Patterson. 1981. Lungfishes, tetrapods, paleontology, and plesiomorphy. Bull. Amer. Mus. Nat. Hist. 167: 159–276.

Sakamoto, Y., K. Kitamura, K. Yoshimura, T. Nishijima & K. Uyemura. 1987. Complete amino acid sequence of P₀ protein in bovine peripheral nerve myelin. J. Biol. Chem. 262: 4298–4214.

Schott, K.-J., V. Neuhoff, B. Nessel, U. Pötter & J. Schröter. 1984. Staining of concanavalin A-reactive glycoproteins on polyacrylamide gels with horseradish peroxidase – a critical evaluation. Electrophoresis 5: 77–83.

Schultze, H.-P. 1986. Dipnoans as sarcopterygians. J. Morph. Supp. 1: 39–74.

Schultze, H.-P. 1988. Notes on the structure and phylogeny of vertebrate otoliths. Copeia 1988: 257–259.

Simons, R., N. Alon & J.R. Riordan. 1987. Human myelin DM-20 proteolipid protein deletion defined by cDNA sequence. Biochem. Biophys. Res. Comm. 146: 666–671.

Sprinkle, T.J., M.R. Wells, F.A. Garver & D.B. Smith. 1980. Studies on the Wolfgram high molecular weight CNS myelin proteins: relationship to 2', 3'-cyclic nucleotide 3'-phosphodiesterase. J. Neurochem. 35: 1200–1208.

Stoffel, W., H. Hillen, W. Schröder & R. Deutzmann. 1983. The primary structure of bovine brain myelin lipophilin (proteolipid apoprotein). Hoppe-Seyler's Z. Physiol. Chem. 364: 1455–1466.

Stoffel, W., H. Hillen & H. Giersiefen. 1984. Structure and molecular arrangement of proteolipid protein of central nervous system myelin. Proc. Nat. Acad. Sci. USA 81: 5012–5016.

Tai, F.L. & R. Smith. 1984. Comparison of the major proteins in shark myelin with the proteins of higher vertebrates. J. Neurochem. 42: 426–433.

Tamai, Y., H. Kojima & K. Abe. 1986. Chemical characterization of the brain of a coelacanth, *Latimeria chalumnae*. Comp. Biochem. Physiol. 83: 295–299.

Towbin, H., T. Staehelin & J. Gordon. 1979. Electrophoretic transfer of proteins from polyacrylamide gels to nitrocellulose sheets: procedure and some applications. Proc. Nat. Acad. Sci. USA 76: 4350–4354.

Vogel, U.S. & R.J. Thompson. 1988. Molecular structure, localization, and possible functions of the myelin-associated enzyme 2', 3'-cyclic nucleotide 3'-phosphodiesterase. J. Neurochem. 50: 1667–1677.

Waehneldt, T.V. & J. Malotka. 1980. Comparative electrophoretic study of the Wolfgram proteins in myelin from several mammalia. Brain Res. 189: 582–587.

Waehneldt, T.V. & G. Jeserich. 1984. Biochemical characterization of the central nervous system myelin proteins of the rainbow trout, *Salmo gairdneri.* Brain Res. 309: 127–134.

Waehneldt, T.V., M.-L. Kiene, J. Malotka, C. Kiecke & V. Neuhoff. 1984. Nervous system myelin in the electric ray, *Torpedo marmorata:* morphological characterization of the membrane and biochemical analysis of its protein components. Neurochem. Int. 6: 223–235.

Waehneldt, T.V., J. Malotka, N.J. Karin & J.-M. Matthieu. 1985. Phylogenetic examination of vertebrate central nervous system myelin proteins by electro-immunoblotting. Neurosci. Lett. 57: 97–102.

Waehneldt, T.V., J.-M. Matthieu & G. Jeserich. 1986a. Appearance of myelin proteins during vertebrate evolution. Neurochem Int. 9: 463–474.

Waehneldt, T.V., J.-M. Matthieu & G. Jeserich. 1986b. Major central nervous system myelin glycoprotein of the African lungfish (*Protopterus dolloi*) cross-reacts with myelin proteolipid protein antibodies, indicating a close phylogenetic relationship with amphibians. J. Neurochem. 46: 1387–1391.

Waehneldt, T.V., J.-M. Matthieu, J. Malotka & J. Joss. 1987. A glycosylated proteolipid protein is common to CNS myelin of recent lungfish (Ceratodidae, Lepidosirenidae). Comp. Biochem. Physiol. 88: 1209–1212.

Wahlert, G. von 1968. *Latimeria* und die Geschichte der Wirbeltiere. Eine evolutionsbiologische Untersuchung. Gustav Fischer Verlag, Stuttgart. 125 pp.

Wiley, E.O. 1979. Ventral gill arch muscles and the phylogenetic relationship of *Latimeria.* Occ. Pap. Calif. Acad. Sci. 134: 56–67.

Morphology and physiology

Morphology and physiology

Environmental Biology of Fishes **32**: 147–158, 1991.
© 1991 *Kluwer Academic Publishers.*

Innervation of the basicranial muscle of *Latimeria chalumnae*

William E. Bemis[1] & R. Glenn Northcutt[2]
[1] *Department of Zoology, University of Massachusetts, Amherst, MA 01003, U.S.A. &*
Department of Ichthyology, American Museum of Natural History, New York, NY 10024, U.S.A.
[2] *Neurobiology Unit, Scripps Institution of Oceanography & Department of Neurosciences, A-001,*
University of California, San Diego, La Jolla, CA 92093, U.S.A.

Received 26.11.1990   Accepted 5.1.1991

Key words: Nervus abducens, Cranial segmentation, Extrinsic eye muscles, Pisces, Actinistia

Synopsis

New observations based on a serial reconstruction of histological sections of a prenatal coelacanth demonstrate that, in contrast to previous reports, the basicranial muscle is innervated by the abducent nerve rather than the vagal nerve. A detailed account of the course of the abducent nerve and its terminal arborizations in the basicranial muscle and lateral rectus muscle is provided. This finding bears on the phylogenetic derivation of the basicranial muscle in sarcopterygians, its possible homologues in other vertebrates, and patterns of head segmentation in craniates.

Introduction

Latimeria chalumnae is the only living vertebrate with paired basicranial muscles spanning an intracranial joint that divides the neurocranium into anterior (ethmosphenoid) and posterior (oticooccipital) portions (Nelson 1970). The basicranial muscles (subcephalic muscles of some authors) are large, and the joint has well-developed synovial surfaces. There are also specializations that allow for flexure of the organ systems which cross the region, such as the supraorbital sensory canal. This remarkable case of intracranial kinesis has been the subject of a number of functional anatomical interpretations (reviewed by Lauder 1980). Although a complete understanding must await studies of feeding and respiration in living specimens, it is generally agreed that the basicranial muscles act to depress the anterior portion of the neurocranium to produce strong bite-forces at the tooth row. There is less general agreement about the phylo-

genetic origin of the basicranial muscle. Although fossil coelacanths have been shown to possess a comparable system (Millot & Anthony 1958), it is far from clear that the intracranial joint is homologous in those sarcopterygian taxa in which a joint is known (Bjerring 1973). In a particularly straightforward analysis, Nelson (1970) identified four theories regarding the possible origin of the muscle: (1) it is part of the body (trunk) musculature; (2) it is derived from the fifth somite (a post-otic somite); (3) it is derived from the second somite (a pre-otic somite); (4) it is of branchiomeric (visceral) origin.

One might expect that the nerve supply to the muscle would provide unequivocal support for one of these four theories, but this has not been the case. Millot & Anthony (1965, p. 66) reported that the basicranial muscle was innervated by the vagal nerve, and regarded the muscle as a derivative of a post-otic somite, primitively associated with the vagal nerve. Other authors have based their interpretations of the basicranial muscle on the idea that its motor fibers must have secondarily become associated with the vagal nerve. Nelson (1970) made comparisons with the anterior trunk muscles in *Polypterus* and he suggested that in *Latimeria* there has been a secondary association of somatic motor components of the occipital nerves with the vagal nerve. Bjerring (1972, 1973) advanced theories that the basicranial muscle originated from a pre-otic somite. A critical piece of evidence for Bjerring's ideas was the positional relationship of the basicranial muscle and the abducent nerve. Bjerring (1972, Fig. 5) reconstructed the basicranial muscle of the Devonian coelacanth *Nesides* with innervation by the so-called nervus rarus, a previously unknown cranial nerve which he considered had become associated with the vagal nerve, as in modern *Latimeria*. Only Thomson (1967, 1969) followed the strict interpretation that if the muscle is innervated by the vagal nerve, it must have a branchiomeric origin. As pointed out by Nelson (1970), however, the branchial musculature of *Latimeria* is both complete and normal (i.e., primitive) and provides no indication of missing components that could have contributed to the basicranial muscle.

From this background it is clear that any new information about the innervation of the basicranial muscle would be of interest from the standpoint of assessing its phylogenetic origin and possible homologues in other fishes and in tetrapods. New information about the basicranial muscle also has implications about patterns of segmentation in the head, particularly the fate of somitomeres. The new observations reported here are abstracted from a larger work (Northcutt & Bemis unpublished) dealing with the cranial nerves of a prenatal coelacanth, based on our reconstruction of the head of the serially sectioned prenatal pup from the collections of the American Museum of Natural History which was prepared by Michael D. Lagios.

Materials and methods

A few workers have examined slides from this series (e.g., Fritzsch 1987), and information on the specimen and its preparation may be found in Wourms et al. (1991). According to Smith et al. (1975), the mother fish (AMNH 32949; CCC no. 29) was collected off Mutsamudu on Anjouan Island on 8 January 1962 by native fishermen. One of the largest coelacanths ever taken, it was preserved in formalin by G.W. Garrouste. The preservation of histological detail in the sectioned pup is remarkable considering these circumstances.

Five pups were discovered in the mother (Smith et al. 1975). Pup number 1 was sectioned and is designated AMNH 32949h, 303 mm TL (CCC no. 29.1). The famous lateral photograph of one of the prenatal coelacanths is pup number 2, but a color photograph of pup 1 taken prior to sectioning is presented by Myking (1977). The head of pup 1 was prepared as two blocks, A and B, that were separately embedded in nitrocellulose and sectioned at $50 \mu m$ under the direction of M.D. Lagios (not $15 \mu m$ as reported by Fritzsch 1987). Block A included the tip of the snout to just posterior to the eye; block B continued caudally about halfway along the opercular flap, so that most but not all of the gill arches were completely sectioned. The sections through blocks A and B were not made in a perfectly transverse plane, so that the right side of the specimen extends slightly more caudal than the left. The sections were stained in Weigert's hematoxylin and Van Gieson's picro-fuchsin, which produced excellent differentiation of muscles, bones and cartilages, and adequate differentiation for nerves. Approximately every 10th section was stained in an initial run and mounted on lantern slides. Later, the remaining sections were stained and mounted on 5 × 7.5 cm slides.

In the period between the sectioning of the prenatal pup in the 1970's and the start of our study in 1989, detailed records were lost, and a complete resorting and renumbering of the sections was needed before we could begin our analysis. What follows is based on careful study and conjecture based on the original numbers (here termed the packet numbers) recorded on the slides. It appears that during sectioning, block A was cut from caudal to rostral into 550 sections, which were in turn put into packets. Packets A1, A2, A3, A4 and A5 originally included 100 sections each, and packet A6 – including the most anterior part of the prenatal pup – had 50 sections. The packet numbers appear to have originally been stamped either on the sections themselves or more likely on the paper interleaves between the sections; in any event, the packet numbers were separated from the sections prior to mounting them and transcribed to typed paper labels affixed to the slides. There were mistakes in transcription, but in general, the packet numbers of the A series are a good guide to the actual section order, and based on these numbers only about 50 of the sections from block A are missing from the final set.

Block B was also sectioned from the caudal to rostral direction, and again based on the packet numbers, we believe it resulted in about 550 sections. These sections were divided into packets B1, B2, B3, B4, and B5 (each with 100 sections) and a final packet B6 (with 54 sections). The original packet numbers for the B series are less reliable than are those of the A series, and the ordering implied by the packet numbers is wrong in many cases. Also, an estimated 90 sections were lost from packet B3 in the time between preparation of the lantern slide series and the final preparation of the 5 × 7.5 cm slides.

After all of the sections had been stained and mounted, we infer that a second series of numbers was assigned to the slides in an apparent attempt to provide a consistent rostral to caudal numbering scheme. Unfortunately, the slides were so badly out of order at that point that this attempt was a failure.

To rectify these problems, each of the 915 slides made available to us (from what we believe was an original total of 1104 sections) was studied and compared to determine its actual position in the series. We then assigned a new set of 'rostral to caudal' numbers, starting with RC 1 at the tip of the snout. We left 98 blanks to cover specific losses of sections where we could identify gaps of greater than 2 sections both on the basis of the original packet numbers and careful inspection of the sections. In practice, all of these gaps turned out to be in the otic region in the area of packet B3 noted above. Fortunately, the series of lantern slides prepared from every tenth section was available for the otic region, so that the gaps did not prevent a complete reconstruction. We assumed that the remaining small gaps and inconsistencies would work out in the reconstruction itself, and began our work with 1013 sections and blanks accounted for. Throughout the detailed description of the structures, we make reference to our RC numbers; a table correlating our RC numbers with the two previous numbering attempts is available with the sections.

Our reconstruction was based on a 10 × magnification of the sections. The total length of blocks A and B was estimated to be 58 mm by comparing the remaining, unsectioned part of the body of pup 1 with the intact pup 2. We made our principal reconstruction from the dorsal view because this allowed us to use the midline as the line of registration for every section. We reconstructed only the left side of the specimen, but made comparisons between right and left sides and relied on the slight offset of structures (owing to the imperfectly transverse plane) to help in reconstructing the area of the gap in the otic region. Initially, every 25th section was traced at 10 ×, the nerves, muscles, and other structures of interest were plotted, and the outlines were connected. To reconstruct many structures, additional tracings of intervening sections, measurements, and notes were needed, all of which were transcribed to master reconstructions. This method of hand reconstruction – as opposed to computer assisted reconstruction – poses many advantages for material of this type, in which there is an extremely large number of sections subject to varying distortion and for which written notes were needed to fully trace a large number of structures.

Results

The results are divided into sections dealing briefly with the anatomy of the skull and intracranial joint, followed by descriptions of the anatomy of muscles innervated by the abducent nerve, and finally a detailed report on the course of the abducent nerve in the prenatal pup. Throughout this description, it is important to know that all of the cranial structures of interest in the prenatal pup are fully differentiated, although endochondral bones are only

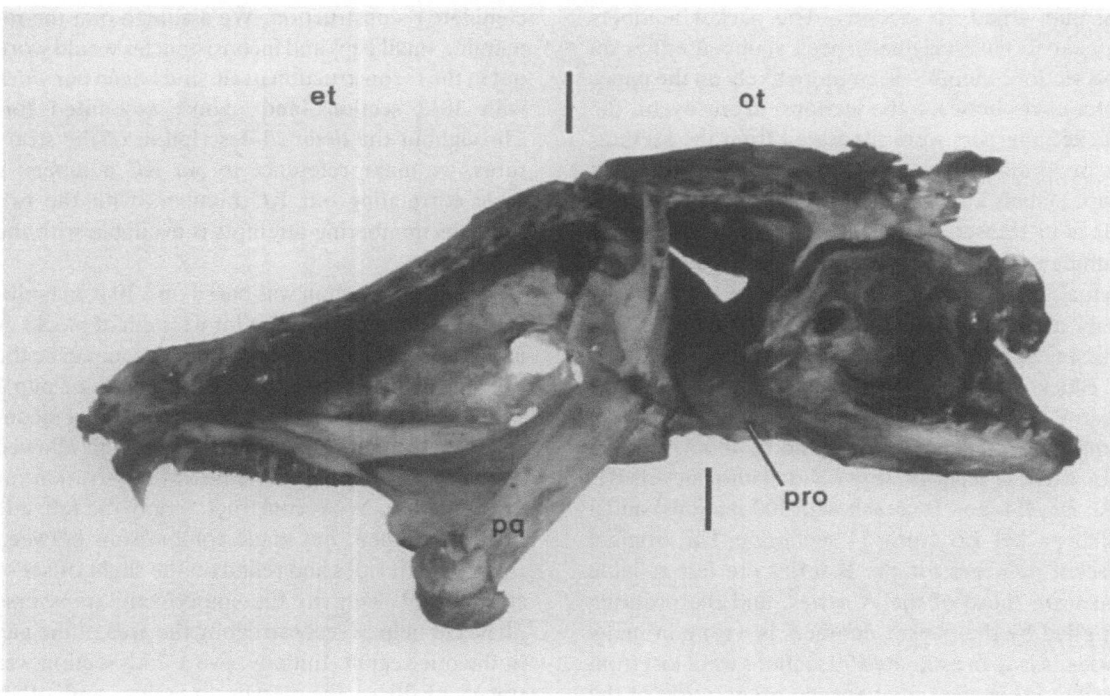

Fig. 1. Lateral view of the skull of a small adult specimen, CCC no. 150 (TL 1230 mm). The ethmosphenoid (et) and oticooccipital (ot) regions of the skull are indicated as is the joint between the two halves of the braincase (vertical lines). The palatoquadrate complex (pq) and prootic bone (pro) are also labeled.

partially ossified. No dramatic qualitative changes between the condition in the prenatal pup and the condition of an adult are to be expected.

Ethmosphenoid and oticooccipital portions of the skull

Lateral and ventral views of a dry skull from an adult specimen (AMNH 56150, TL = 1230 mm, CCC no. 150) are presented in Figures 1 and 2. The ethmosphenoid (et) and oticooccipital (ot) portions of the skull are indicated in Figure 1, with vertical lines demarcating the course of the intracranial joint. The left palatoquadrate complex (pq) is still attached to the neurocranium of this specimen. The principal element of the oticooccipital portion that is important for understanding the following description is the prootic bone (pro) visible in both lateral and ventral views. The portion of the prootic bone labeled in the figure is referred

to as the otical shelf by Millot & Anthony (1958, Fig. 7) and others. The otical shelf of the prootic extends anterior to the portions of the ethmosphenoid that form the intracranial joint. In ventral view, the relationships of the prootic, notochord (not) and parasphenoid (psp) can be seen.

These relations are further demonstrated by section RC 600, midway through the intracranial joint of the prenatal pup (Fig. 3). At this level of the prenatal pup, the anterior tip of the notochord lies in a pit at the posterior end of the ethmosphenoid (not). On the right and left sides, the connecting processes of the ethmosphenoid (pc) can be seen articulating with the prootic bones (pro), which are only partially ossified in the prenatal pup. Dorsal and lateral to the connecting process is the antotic process of the ethmosphenoid (ao) and its articulation with the palatoquadrate complex (pq). This section is entirely posterior to the parasphenoid.

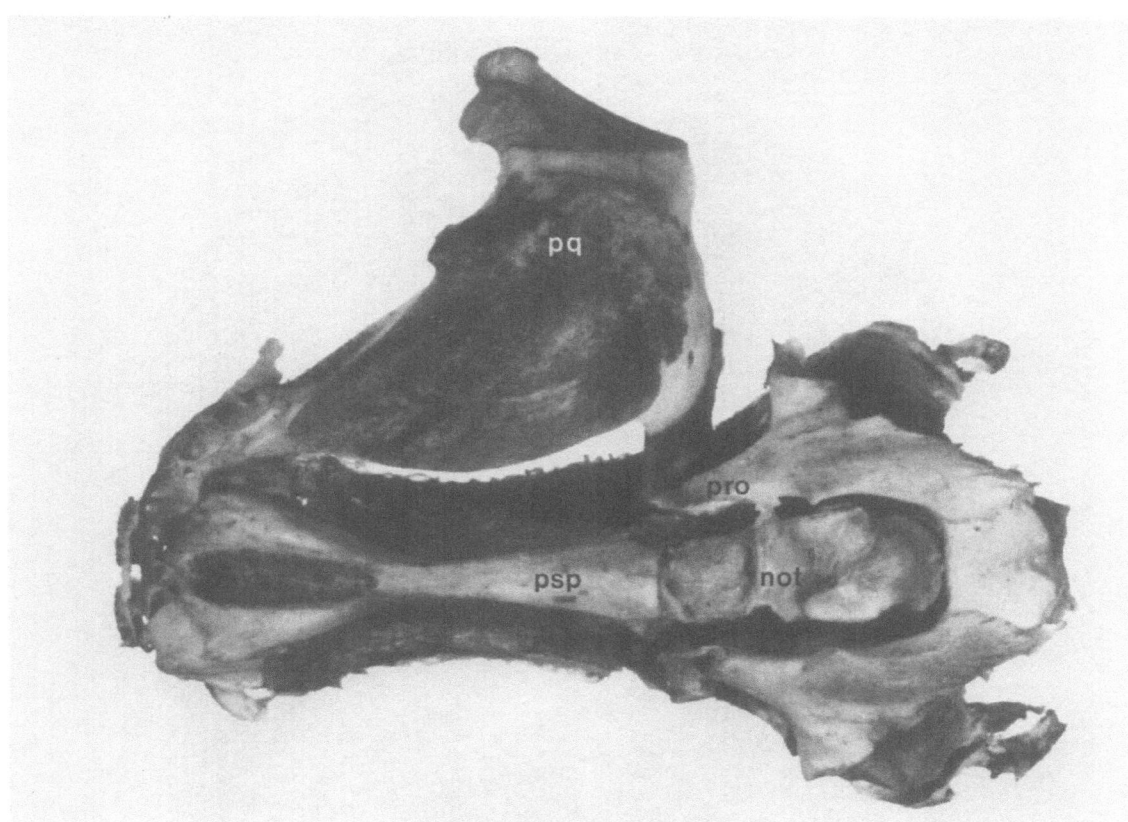

Fig. 2. Ventral view of the skull shown in Figure 1. This figure shows how the notochord (not) ends at the posterior tip of the ethmosphenoid. The parasphenoid bone (psp) is also indicated. Other abbreviations as in Figure 1.

Basicranial muscle

Cross sections of the right and left basicranical muscles are also shown in section RC 600 (Fig. 3, bcm). The left basicranial muscle was traced anteriorly from its origin on the prootic at approximately section RC 887 to its anterior limit at RC 422, nearly half the length of the head as diagrammed in a dorsal reconstruction (Fig. 4). The muscle's origin from the ventral surface of the oticooccipital portion is restricted to the ossified portions of the prootic, ranging from RC 875 to RC 551. In this part of the head, the belly of the muscle is closely associated with the ventral surface of the prootic. At RC 525, the basicranial muscle crosses the ventral portion of the intracranial joint, and becomes closely associated with the ethmosphenoid portion

of the braincase, as traced in sections RC 531 to RC 501. Insertion of the muscle on the parasphenoid bone occurs anterior to about section RC 485 (particularly well shown in RC 452). The muscle finally disappears at level RC 422.

Lateral rectus muscle

All of the eye muscles and the eye itself are distorted slightly due to shrinkage in the orbit. Our reconstruction shows an approximate outline of the eye in which the wrinkling of the sclera has been smoothed out (dashed line), but we have not attempted to compensate for the shrinkage of the extrinsic eye muscles. The lateral rectus muscle (lrm; external rectus of some authors and posterior rectus of Fritzsch et al. 1990) originates far ventral-

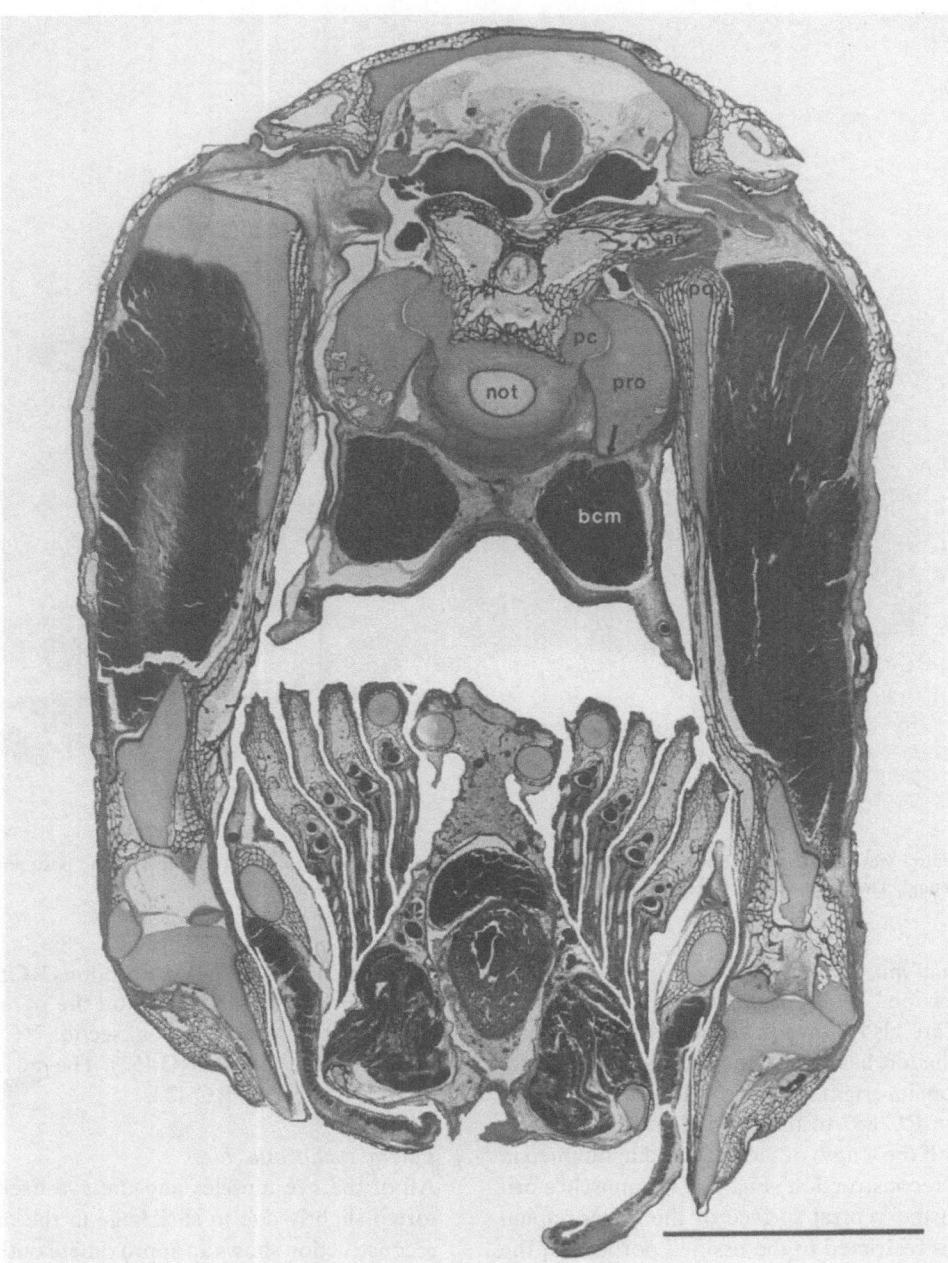

Fig. 3. Transverse section through the intracranial joint of a serially sectioned head of a prenatal coelacanth, CCC no. 29.1, 303 mm TL. This section (RC 600) is from approximately the middle of the head of the specimen. The left side of the specimen is to the right (i.e., the figure is oriented as if the specimen was being viewed head on). The connecting processes of the ethmosphenoid (pc) articulate with the prootic (pro). The antotic process (ao) articulates with the palatoquadrate complex. The arrow indicates the position of the abducent nerve lying between the prootic bone and the basicranial muscle (bcm). Sections through the palatine ramus of the facial nerve are also visible (not labelled) lying adjacent to the dorsolateral corner of the basicranial muscle. Scale bar = 10 mm.

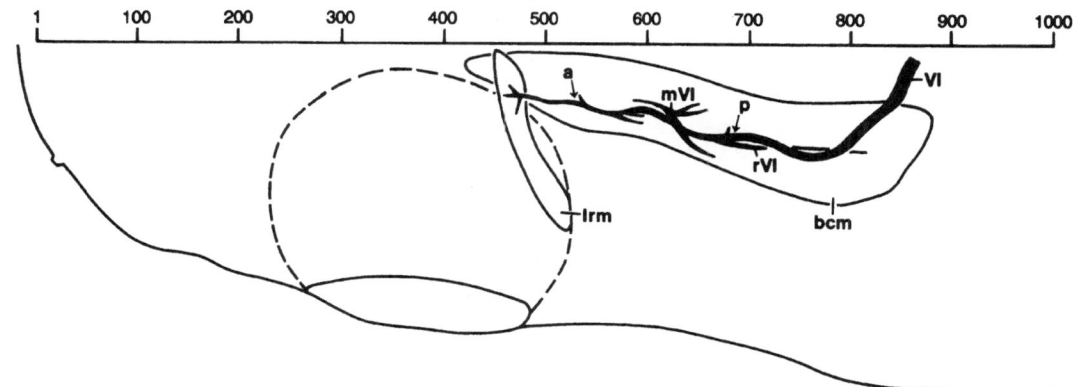

Fig. 4. Serial reconstruction of the abducent nerve and its muscles on the left side of the head of the prenatal pup in dorsal view. The lateral rectus muscle (lrm) is dorsal to the basicranial muscle (bcm) as indicated by the breaks in the outline of the basicranial muscle. Other abbreviations: a, most anterior contact of the main trunk of the abducent nerve with the basicranial muscle; mVI, medial ramus of the abducent nerve; p, most posterior contact of the main trunk of the abducent nerve with the basicranial muscle; rVI, recurrent motor ramus of the abducent nerve; VI, abducent nerve.

ly on the basisphenoid portion of the chondrocranium (e.g., RC 461) just ventral to the origin of the superior rectus muscle. It is a thin sheet, about 5 mm in width (measured in RC 475). It passes posterolaterally and its tendon inserts on the sclera by level RC 525. Description of the eye muscles innervated by oculomotor and trochlear nerves is deferred to our subsequent paper (Northcutt & Bemis unpublished).

Abducent nerve
The abducent nerve of *Latimeria* follows a highly unusual course. The nerve arises from the posterior part of the medulla (Millot & Anthony 1965) as in other craniates. From this point, the route of the abducent nerve can be traced ventrally through the otic region, then forward along the dorsal surface of the basicranial muscle, and finally up into the orbit to innervate the lateral rectus muscle. Its position on the dorsal surface of the basicranial muscle is indicated by an arrow in Figure 3. There are many important and previously unknown details about its path and terminal branches that warrant further description.

The abducent nerve exits the ventral surface of the medulla between levels RC 855 to RC 860. The number of originating rootlets in the prenatal pup

could not be counted as they are separated from the surface of the brain. Millot & Anthony (1965, Fig. 17) show the abducent nerve originating from a variable number of rootlets, four on the right and five on the left, but this should be rechecked when additional well-fixed material becomes available. Tracing forward, the nerve enters a foramen defined by the medial wall of the otic capsule and connective tissue associated with the notochord. At level RC 851, the nerve is enclosed by cartilage on three sides, with the medial wall of its exit foramen still composed of connective tissue. A small blood vessel travels together with the nerve.

At level RC 844, a branch of the octaval nerve that supplies the basilar papilla and papilla neglecta enters the lateral corner of the foramen for the abducent nerve. At RC 820, the abducent nerve has crossed the medial wall of the otic capsule and now lies in the inner ear cavity, medial to the saccule. In RC 810, it lies medial to branches of the octaval nerve, closely pressed against the medial wall of the inner ear chamber. In RC 800, it lies in its own small depression in the cartilage of the wall. Its route continues in this depression, moving ventrally until by RC 771 it become a well-defined foramen in the wall. The nerve lies completely within this foramen by RC 761. Traces of per-

154

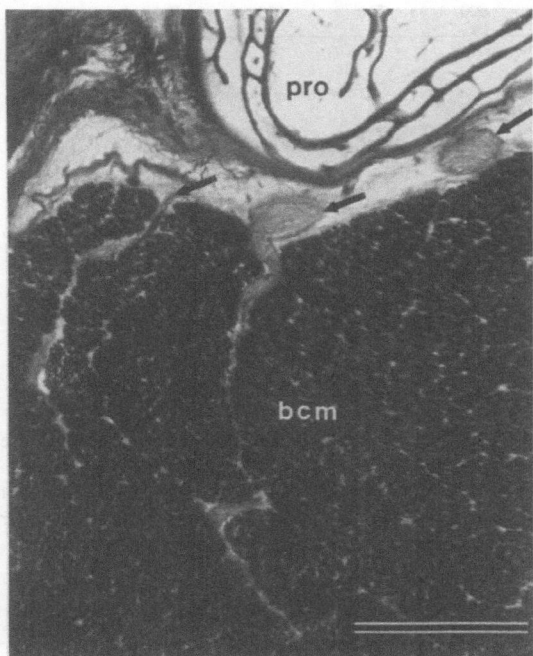

Fig. 5. Photomicrograph of basicranial muscle in section RC 635 showing cross sections of the main trunk of the abducent nerve and motor branches (arrows) entering the belly of the muscle. bcm, basicranial muscle; pro, prootic bone. Scale bar = 1 mm.

ichondral ossification are apparent around the foramen in RC 751. By RC 728, the abducent nerve lies in the midst of trabecular ossifications of the prootic bone (see RC 732). Still moving ventrally, the path of the abducent nerve completes it passage through the prootic bone by about RC 680, where it starts to emerge on the surface of the basicranial muscle. This point is indicated by 'p' on the reconstruction (Fig. 4).

The innervation of the basicranial muscle by the abducent nerve is massive and occurs along the length of the muscle. At RC 680, a small medial ramule leaves the main trunk of the nerve. There is also a larger ramule, the first recurrent motor branch of the abducent nerve (rVI), which can be traced posteriorly along the dorsal surface of the basicranial muscle to about level RC 810. Along its course, the first recurrent throws off fibers which innervate the muscle.

At RC 671, the main trunk of the abducent nerve

finally exits from its foramen through the prootic bone, so that the nerve is now in very close association with the dorsal surface of the muscle. The next major branching, however, does not occur until level RC 630, where there is a branch of a large medial ramus (mVI). This ramus in turn gives rise to two large recurrent branches that enter the main mass of the basicranial muscle and branch repeatedly before they disappear as discrete bundles (at about level RC 655). Also, in the area of RC 630, a small lateral recurrent ramus branches off the main trunk and continues along the dorsolateral border of the muscle as far caudally as level RC 663.

In the region between RC 635 and RC 620, the medial ramus (mVI) angles anteromedially on the dorsal surface of the muscle, giving rise to side branches so that several cross sections are present (e.g., RC 635, Fig. 5). This point on the muscle (almost exactly midway along its rostral to caudal extent) is the major site of innervation for the whole muscle. The medial ramus continues to branch and subdivide until its subdivisions are indistinguishable (approximately RC 588).

The main portion of the abducent nerve continues forward as a well-defined cross section. More small motor branches leave the main nerve (e.g., at RC 580, RC 534), and the nerve takes a slightly more lateral course. Also noteworthy in this region is the palatine ramus of the facial nerve, lying dorsolateral to the basicranial muscle in section RC 600 (Fig. 3). No fibers of the facial nerve were observed to enter the basicranial muscle.

By level RC 525 (where there is a final small medial branch that leaves the main part of the abducent nerve), the path of the abducent nerve begins to swing dorsally, away from the surface of the basicranial muscle at point 'a' in Figure 4 (also see Millot & Anthony 1965, Fig. 23). The route of the nerve also bends laterally, along the anterolateral surface of the prootic (which is entirely cartilaginous at this point; see for example section RC 500). By section RC 488, the route of the abducent nerve has completed its lateral passage of the prootic, and the nerve lies poised to enter the back of the orbit.

In RC 486, the abducent nerve lies between two

major sinuses of the orbital venous sinus system, dorsal to the lateral rectus muscle. By RC 475, the nerve begins its terminal subdivisions as it approaches the posterodorsal edge of the lateral rectus muscle. Three branches can be distinguished tracing forward from RC 472 to RC 460. The medial branch courses along the dorsal surface of the muscle. The lateral branch enters the belly of the muscle at RC 470. The central branch begins to subdivide and disappears into the muscle itself.

Complicating matters at this point is the descent of the profundal nerve through the orbit. The trunk of the profundus lies directly dorsal to the lateral rectus muscle, near the terminal arborizations of the abducent nerve. This is well shown in RC 475. Eventually, the route of the profundus hooks sharply around the anterior edge of the lateral rectus muscle and on to a position deep in the orbit.

Discussion

In this discussion, we first take up the question of the differences between our observations and those of previous workers and then turn briefly to the implications of our findings.

Comparison with previous work
This is the first report of original observations bearing on the path of the abducent nerve and innervation of the basicranial muscle since the elegant dissections by Millot & Anthony (1958, 1965). The anatomical relationships of the basicranial and lateral rectus muscles of adult coelacanths were described by Millot & Anthony (1958, pp. 43–47, Fig. 13). In all essential features of these muscles, the prenatal pup we studied closely resembles the adult. It appears that the basicranial muscles must grow disproportionately to reach their relatively larger size in adults, a change no doubt related to the positive allometric growth of the snout.

We differ from previous interpretations regarding the innervation of the basicranial muscle. This is probably because the abducent nerve is relatively small in the adult specimens in which it was previously traced by dissection; Millot & Anthony (1965, p. 55) have commented on this difficulty. It

has proved much easier to trace the abducent nerve and the innervation of the basicranial muscle in histological sections of the prenatal pup. Our observations regarding the general course of the abducent nerve agree with figures provided by Millot & Anthony (1965, Fig. 23, 25). However, Millot & Anthony did not observe the motor branches of the abducent nerve that supply the basicranial muscle; instead, they described and figured a branch of the vagus (nerf du muscle sous-crânien, Fig. 31) applied to the lateral surface of the muscle as its motor supply. We have carefully examined this region in sections of the prenatal pup, but the only fibers that can be traced here are the pharyngeal branches of the glossopharyngeal and vagal nerves which run in the oral mucosa (Northcutt & Bemis unpublished). We believe that these branches were mistakenly traced to the basicranial muscle by Millot & Anthony (1965).

The previous interpretation that the vagal nerve innervates the basicranial muscle has resulted in several papers, in particular a series by Bjerring (1967, 1968, 1971, 1972, 1973). In early papers in this series, Bjerring regarded the basicranial muscle as a derivative of the second somite. Although there are many aspects of Bjerring's analyses that are difficult to credit, our demonstration that the basicranial muscle is innervated by the abducent nerve does support his basic idea that the basicranial muscle develops from pre-otic paraxial mesoderm and that it is neither a trunk muscle nor a branchiomeric muscle, as has been suggested by others. However, we doubt that the muscle is related to the second of these pre-otic somites as originally interpreted by Bjerring (e.g., 1968; but see Bjerring 1977 for a revised view). Our reasons are explained in the following brief discussion of head segmentation.

Segmentation of the head
In the last ten years increased attention has been given to the comparative embryology of the vertebrate head (reviewed by Northcutt 1990). In particular, studies of avian embryos have contributed much new information about the derivation of cranial myoblasts and connective tissue cells that challenges conventional notions about cranial organi-

zation (Noden 1983, 1988). These studies have shown that the myoblasts of all cranial muscles are derived from paraxial mesoderm, so that there is not a strictly separate developmental origin of the branchiomeric musculature. Moreover, it is clear that some earlier ideas about the number and phylogenetic consistency of segments in the head have been mistaken. In the anterior part of the head, typical somites comparable to those of the trunk generally do not form; instead, there is a series of somitomeres anteriorly, which can be counted from the anterior tip of the paraxial mesoderm to the first somite that is organized as a typical trunk somite, usually located just caudal to the developing otic capsule. There are differences in the number of somitomeres in amniotes compared with anamniotes: 7 in the chick (Noden 1983) and the snapping turtle (Meier & Packard 1984) versus 4 in newts and sharks (Jacobson & Meier 1984, Fig. 11). It is possible, however, to resolve this information into a model that incorporates both recent and classical data and makes clear predictions about the fate of specific cranial segments (Northcutt 1990, Fig. 5).

Several classical studies have traced the early development of the extrinsic eye muscles in cartilaginous fishes from pre-otic condensations present in their embryos (reviewed by Neal 1918). As is well known, these muscles are exceedingly conservative in their structure and development in gnathostomes. Thus, it is reasonable to make predictions based upon the taxa for which developmental studies are available. Accordingly, we predict that the basicranial muscle is derived from the so-called hyoid 'somite' (see Northcutt 1990 for further discussion of terminology).

In amniotes, Noden (1983) attributed the origin of the eye muscles of the chick to specific somitomeres. Those muscles innervated by the oculomotor nerve (medial rectus muscle, superior rectus muscle, inferior rectus muscle, and inferior oblique muscle) are said to be derived from somitomeres 1 and 2. The superior oblique muscle, innervated by the trochlear nerve, is derived from somitomere 3. The lateral rectus and palpebral muscles of the chick are derived from somitomere 5, and both of these muscles are innervated by the abducent

nerve. Although a strictly comparable pattern of somitomeres does not occur in either actinopterygians or elasmobranchs, it should be possible to trace the lateral rectus muscle to a discrete portion of the hyoid 'somite' in the head of various embryonic fishes.

A final topic for discussion is whether we can identify any homologues of the basicranial muscle in other taxa. Lampreys show some interesting and significant deviations from the basic gnathostome pattern of eye muscles and nerves (Fritzsch et al. 1990, Table 1). In particular, the abducent nerve innervates two of the six extrinsic eye muscles of *Lampetra*: the ventral rectus muscle (= lateral rectus muscle of gnathostomes) and the caudal rectus muscle. These muscles have separate pools of motoneurons in the nucleus of the abducent nerve in the brainstem (Fritzsch et al. 1990, Fig. 6). Fritzsch et al. (1990) developed a revised nomenclature of the eye muscles of vertebrates in which they homologized the caudal rectus muscle of lampreys to the retractor bulbi muscle of tetrapods. In addition to the palpebral muscles, there are also many other interesting derivatives of the retractor bulbi system in tetrapods that retain an innervation pattern associated with the abducent nerve including, for example, the tentacle retractor muscle of caecilians (Wake 1985). In view of these observations, it is tempting to speculate that the basicranial muscle of *Latimeria* is homologous with the caudal rectus of lampreys and the retractor bulbi and its derivatives in tetrapods.

Future work

Our results are essentially gross anatomy done with a microscope. Although it seems unlikely that living specimens of *Latimeria* will ever become available for experimental neuroanatomical studies, it may be possible to apply newly developed tracing methods to properly fixed specimens in order to make further studies of the abducent nerve and its muscles. Based on studies by Fritzsch et al. (1990) and Baker (1986), we predict that separate pools of motor neurons should be present in the nucleus of the abducent nerve to supply the basicranial and lateral rectus muscles. There are some pertinent references to this in the literature of *Latimeria*.

Lemire (1971, Plate 2A) reported that there were actually two nuclei of the abducens nerve in the brainstem; however, Kremers & Nieuwenhuys (1979, p. 616) could not substantiate this. It will also be of particular interest to study the early development of the basicranial muscle and the abducent nerve when and if early embryos of *Latimeria* become available.

Conclusions

The basicranial muscle is innervated by the abducent nerve and not by branches of the vagal nerve as previously thought. There is no longer any reason to predict the existence of the so-called nervus rarus in fossil coelacanths and rhipidistians as a separate motor supply to the basicranial muscle. The basicranial muscle may be homologous with the caudal rectus of lampreys and with the retractor bulbi (and its derivatives) in tetrapods. In the absence of any information from early development, the cranial nerves still provide the best guide to homologies of cranial muscles, but it is important to continue efforts to broaden the data base concerning patterns of cranial innervation in fishes.

As a final comment, we note that while, after fifty years, we are at last gaining some exciting new information on the biology of free-living coelacanths (Fricke et al. 1987, 1991), there is still much to be learned about the anatomy of *Latimeria,* particularly its nervous system. In view of the uncertain population status of coelacanths, we urge that all specimens which continue to be taken by local fishermen as an accidental bycatch be preserved in a manner suitable for the greatest array of modern anatomical methods. At a minimum this should involve establishing an appropriate land-based facility dedicated to the handling, dissection, and effective fixation of specimens in Moroni. The usual practice of freezing specimens before fixation is incompatible with the hodological (tracing) and immunohistochemical methods needed for contemporary neuroanatomical studies. A comparatively small effort could have an enormous potential for increasing our knowledge of the biology of coelacanths.

Acknowledgements

This information was not presented at the original symposium but resulted from a separate invitation to contribute a paper to the present anthology in November 1990. Throughout our several years of study of the head of prenatal and adult coelacanths, Gareth Nelson has been extremely supportive. It is a special pleasure to acknowledge support for studies of coelacanths provided by Jane H. Bemis, the Donn Rosen Fund, and members of the Explorers Club expeditions to Grand Comoro in 1986–1988, which resulted in specimens used for this and other studies. In particular, we would like to mention Jerry Hamlin, David Wilkinson, Arnold Neiss, and Paul Rodzianko. We have discussed our findings with Ed Gilland and David Wake. Jim Atz, Hans-Peter Schultze, and Eugene Balon commented on the manuscript. Additional support was received from grants from the Whitehall Foundation, NSF (BSR 8806539), and NIH (NS24669 and NS24689).

References cited

Baker, R. 1985. Brainstem neurons are peculiar for oculomotor organization. pp. 257–271. *In:* H.J. Freund, U. Büttner, B. Cohen & J. Noth (ed.) Progress in Brain Research, Volume 64, Amsterdam.

Bjerring, H.C. 1967. Does a homology exist between the basicranial muscle and the polar cartilage? pp. 223–267. *In:* Problèmes Actuels de Paléontologie (Évolution des Vertébrés), Colloques Internationaux du Centre National de la Recherche Scientifique 163, Paris.

Bjerring, H.C. 1968. The second somite with special reference to the evolution of its myotomic derivatives. pp. 341–357. *In:* T. Ørvig (ed.) Current Problems of Lower Vertebrate Phylogeny, Nobel Symposium 4, Almqvist & Wiksell, Stockholm.

Bjerring, H.C. 1971. The nerve supply to the second metamere basicranial muscle in osteolepiform vertebrates, with some remarks on the basic composition of the endocranium. Acta Zool. 52: 189–225.

Bjerring, H.C. 1972. The *nervus rarus* in coelacanthiform phylogeny. Zool. Scripta 1: 57–68.

Bjerring, H.C. 1973. Relationships of coelacanthiforms. pp. 179–205. *In:* P.H. Greenwood, R.S. Miles & C. Patterson (ed.) Interrelationships of Fishes, Linnean Society of London, Academic Press, London.

Bjerring, H.C. 1977. A contribution to structural analysis of the head of craniate animals. The orbit and its contents in 20–22 mm embryos of the North American actinopterygian *Amia calva* L., with particular reference to the evolutionary significance of an aberrant, nonocular, orbital muscle innervated by the oculomotor nerve and notes on the metameric character of the head in craniates. Zool. Scripta 6: 127–183.

Fricke, H., O. Reinicke, H. Hofer & W. Nachtigall. 1987. Locomotion of the coelacanth *Latimeria chalumnae* in its natural environment. Nature 329: 331–333.

Fricke, H., K. Hissmann, J. Schauer, O. Reinicke, L. Kasang & R. Plante. 1991. Habitat and population size of the coelacanth *Latimeria chalumnae* at Grand Comoro. Env. Biol. Fish. 32: 287–300. (this volume)

Fritzsch, B. 1987. Inner ear of the coelacanth fish *Latimeria* has tetrapod affinities. Nature 327: 153–154.

Fritzsch, B., R. Sonntag, R. Dubuc, Y. Ohta & S. Grillner. 1990. Organization of the six motor nuclei innervating the ocular muscles in lamprey. J. Comp. Neurol. 294: 491–506.

Jacobson, A.G. & S. Meier. 1984. Morphogenesis of the head of a newt: mesodermal segments, neuromeres, and distribution of neural crest. Develop. Biol. 106: 181–193.

Kremers, J.-W., P.M. & R. Nieuwenhuys. 1979. Topological analysis of the brain stem of the crossopterygian fish *Latimeria chalumnae*. J. Comp. Neurol. 187: 613–638.

Lauder, G.V. 1980. The role of the hyoid apparatus in the feeding mechanism of the coelacanth *Latimeria chalumnae*. Copeia 1980: 1–9.

Lemire, M. 1971. Étude architectonique du rhombencéphale de *Latimeria chalumnae* Smith. Bull. Mus. Natn. Hist. Nat., Zoologie 3(2): 41–95.

Meier, S. & D.S. Packard, Jr. 1984. Morphogenesis of the cranial segments and distribution of neural crest in the embryos of the snapping turtle, *Chelydra sepentina*. Develop. Biol. 102: 309–323.

Millot, J. & J. Anthony. 1958. Anatomie de *Latimeria chalumnae*. I. Squelette, muscles et formations de soutien. Centre National de la Recherche Scientifique, Paris. 122 pp.

Millot, J. & J. Anthony. 1965. Anatomie de *Latimeria chalumnae*. II. Système nerveux et organes des sens. Centre National de la Recherche Scientifique, Paris. 130 pp.

Myking, L.M. 1977. Old four legs: the living fossil. Sea Frontiers 23: 334–341.

Neal, H.V. 1918. The history of the eye muscles. J. Morphol. 30: 433–453.

Nelson, G.J. 1970. Subcephalic muscles and intracranial joints of sarcopterygian and other fishes. Copeia 1970: 468–471.

Noden, D.M. 1983. The embryonic origin of avian cephalic and cervical muscles and associated connective tissues. Amer. J. Anat. 168: 257–276.

Noden, D.M. 1988. Interactions and fates of avian craniofacial mesenchyme. Development 103 (supplement): 121–140.

Northcutt, R.G. 1990. Ontogeny and phylogeny: a re-evaluation of conceptual relationships and some applications. Brain Behav. Evol. 36: 116–140.

Smith, C.L., C.S. Rand, B. Schaeffer & J.W. Atz. 1975. *Latimeria*, the living coelacanth, is ovoviviparous. Science 190: 1105–1106.

Thomson, K.S. 1967. Mechanisms of intracranial kinetics in fossil rhipidistian fishes (Crossopterygii) and their relatives. J. Linn. Soc. London (Zoology) 46: 223–253.

Thomson, K.S. 1969. The biology of the lobe-finned fishes. Biol. Rev. 44: 91–154.

Wake, M.H. 1985. The comparative morphology and evolution of the eyes of caecilians (Amphibia, Gymnophiona). Zoomorphologie 105: 277–295.

Wourms, J.P., J.W. Atz & M.D. Stribling. 1991. Viviparity and the maternal-embryonic relationship in the coelacanth, *Latimeria chalumnae*. Env. Biol. Fish. 32: 225–248. (this volume)

Environmental Biology of Fishes **32**: 159–181, 1991.
© 1991 *Kluwer Academic Publishers.*

Computed Tomography and Magnetic Resonance Imaging studies of *Latimeria chalumnae*

Hans-Peter Schultze & Richard Cloutier
Museum of Natural History and Department of Systematics and Ecology, The University of Kansas, Lawrence, KS 66045-2454, U.S.A.

Received 21.8.1989   Accepted 29.7.1990

Key words: Sarcopterygii, Actinistia, Coelacanth, Radiologic techniques, Radiology, Catscan, Morphology

Synopsis

Recent radiologic imaging techniques (CT[Computed Tomography] and MRI[Magnetic Resonance Imaging]) were used to investigate the cranial anatomy of the coelacanth *Latimeria chalumnae*. The non-invasive CT and MRI techniques were performed successfully on a 1.45 m female specimen. This specimen had been frozen a year earlier for future research; the CT was conducted on the frozen animal, whereas the MRI method was performed immediately after thawing. The CT technique provides information about differential density of the organism (especially informative with respect to hard tissues, bone and cartilage), whereas three different types of MRI (proton resonance T_1, T_2 and 'flash') distinguish cartilage, muscles, and different connective tissues. A total of 381 CT cross sections (2 mm thick with 1 mm of overlap) through the head region were used in a computerized three-dimensional reconstruction program to address questions concerning cranial morphology. The results obtained from these radiologic imaging techniques confirmed most of the basic anatomy known from traditional dissections. However, the morphology of complex structures, such as the cartilaginous processes of the neurocranium, and the integration of the branchial arches and palate can only now be described more accurately.

Introduction

Very few living organisms have received as much scientific consideration as the living coelacanth, *Latimeria chalumnae*. Since its discovery in 1938, approximately 200 specimens have been caught (Balon et al. 1988) and of these only a small number have been carefully investigated morphologically. Because of its phylogenetic position with respect to the origin of tetrapods (cf. Forey 1988) and its proposed status as a 'living fossil' (Forey 1984), the anatomy of *Latimeria* has been studied with a broad spectrum of techniques and approaches.

Smith (1940) provided a description of the osteology of *Latimeria chalumnae* based on a dissection of the left side of the holotype. In 1952 the second specimen of *L. chalumnae* was discovered, which allowed Smith (1953a, b) to describe briefly the soft anatomy. Subsequent specimens were studied by a group of French scientists (cf. Millot et al. 1972, Anthony 1980, Balon et al. 1988). These studies resulted in the three volumes of 'Anatomie de *Latimeria chalumnae*' (Millot & Anthony 1958a, 1965, Millot et al. 1978) that are a monumental memoir on the morphology of this fish. These volumes cover the gross anatomy of *L. chalumnae* based on a series of traditional dissections, sections of preserved specimens, and dried skeletons. Although these volumes provide a detailed overview of the anatomy, a collection of more than 170 publications present additional specialized morphological data on *L. chalumnae* (Appendix 1). Histological information concerning *L. chalumnae* has been obtained by transmission and scanning electron microscopy (TEM and SEM) (Appendix 1), and crystallographic study (Carlström 1963, Lange 1983).

Functional morphology of the feeding mechanism and cranial kinesis (Millot & Anthony 1955b, 1955c, Thomson 1966a, b, 1967, 1970, 1973, Cracraft 1968, Alexander 1973, Robineau 1973, 1987, Robineau & Anthony 1973, Anthony & Robineau 1976b, Adamicka & Ahnelt 1976, Millot et al. 1978, Lauder 1980a, Lund & Lund 1985, Lund et al. 1985, Forey 1988) and locomotion (Schaeffer 1948, Wahlert & Wahlert 1962, 1967, Thomson 1966a, Wahlert 1968a, b, Fricke et al. 1987) have complemented the morphological studies. Morphometric studies on *L. chalumnae* have addressed changes in body size and proportions (Anthony & Robineau 1976a, Hureau & Ozouf 1977, McAllister & Smith 1978, Suzuki et al. 1985), gill efficiency (Hughes 1976), and brain proportion (Northcutt et al. 1978).

Most of the aforementioned studies employed destructive and invasive techniques to collect data. Only a few non-invasive investigations have been conducted on *L. chalumnae*; Hobdell & Miller (1969) and Miller (1979) studied the histology of some mineralized tissues using microradiographic techniques, Millot & Anthony (1958a, pl. 1, 2) investigated the general anatomy of the head, Millot (1955) discussed the variation in caudal fin morphology, and Millot & Anthony (1956a, pl. 6b; 1958a) investigated the morphology of the axial skeleton using X-ray pictures. Because of the rarity of *L. chalumnae*, researchers, museums and collectors do not wish to see their specimens destroyed in the process of collecting anatomical data. Recent advances in radiologic technology (cf. Jones & MacFall 1988, Lancaster & Fullerton 1988, White 1988) permit the use of non-invasive techniques on large specimens to obtain such data. Outside the medical field these techniques have been demonstrated to be worthwhile in paleontology (Conroy & Vannier 1984, Zangerl & Schultze 1989) and archeology (e.g. Marx & D'Auria 1986, 1988). Computed tomography (CT) (Suzuki & Hamada

1990), Magnetic Resonance Imaging (MRI) techniques, and traditional dissection were used to study the gross anatomy of a specimen of *L. chalumnae* (Cloutier et al. 1988). This contribution is primarily designed to explain the techniques used and the types of results obtained. Some of our sections (Fig. 1) are however, compared to data gathered through traditional dissections.

Material

The specimen of *Latimeria chalumnae* under investigation is a female (standard length = 1.452 m; weight = 53.75 kg) collected by Explorer's Club members from a native fisherman, November 1986 on the west coast of Grand Comoro island. This specimen (VIMS 8118, CCC no. 141) is now in the collection of the Virginia Institute of Marine Science, School of Marine Science (VIMS), Gloucester Point, U.S.A. The specimen was frozen at $-30°$ C in a freezer in Moroni, Grand Comoro, and maintained frozen at $-30°$ C at the New York Aquarium and the Virginia Institute of Marine Science until dissection on 5 January 1988. While travelling, the specimen was stored on dried ice.

Computed tomographies were taken on the frozen specimen on January 3 over approximately 8 hours at the Radiological Unit of the Riverside Hospital in Newport News, Virginia, U.S.A. The postcranial region of the specimen was dissected the following day at the VIMS. Magnetic resonance imaging (MRI) was performed on the thawed specimen between January 5–6, at the Radiological Unit of the Riverside Hospital in Newport News.

Radiologic techniques

Before discussing the results, we would like to introduce the general principles related to the two radiologic techniques used in this study – Computed Tomography (CT) and Nuclear Magnetic Resonance Imaging (NMRI or simply MRI). Traditional X-ray images are the result of a single planar (two-dimensional) X-ray exposure for which over-

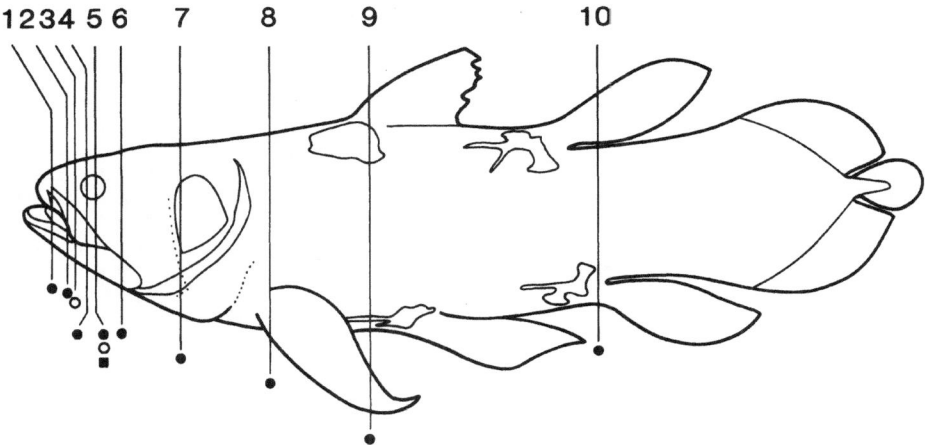

Fig. 1. Position of the 12 scans (cross-sections) presented in this paper on a lateral view of *Latimeria chalumnae*. Section 1 is Figure 5A (CT), section 2 is Figure 2B (CT), section 3 is Figure 3F (MRI-T$_1$), section 4 is Figure 4A and B (CT), section 5 is Figure 7A (CT), 7B (MRI-T$_1$), and 7C (MRI-flash), section 6 is Figure 5B (CT), section 7 is Figure 5C (CT), section 8 is Figure 5D (CT), section 9 is Figure 5E (CT), and section 10 is Figure 5F (CT). Black circles correspond to CT scans, empty circles correspond to MRI-T$_1$ scans, and black square is a MRI-flash scan.

lying and underlying hard tissues are superimposed. Occasionally, the use of stereoscopic X-ray pictures facilitates the interpretation of planar superimposition (Zangerl & Schultze 1989). To produce CT or MRI scans, scanners gather data from sections to be computed in a three-dimensional manner eliminating the poor planar resolution. Images collected with a scanner are therefore 'slices' or 'sections' of the individual at pre-set intervals. To maximize the anatomical resolution, the number of sections is increased, the thickness of each section is reduced (minimum set at 1.5 mm), and each section overlaps adjacent slices.

In a CT scanner, the X-ray tube rotates around the specimen and the bidimensional projection from each direction is obtained and collected in peripheral detectors (Fig. 2A). These projections are then automatically reconstructed into images by computer algorithms (Fig. 2B). Although this is a simplistic explanation of CT, we refer the reader to Brooker (1986) for additional technical information on the scanner, and Kak & Slaney (1988) for information concerning the principles of CT and reconstruction algorithms. Each CT section is composed of pixels (i.e., a digital array of picture elements) that correspond to a voxel unit (i.e., volume

image element to be converted to pixel on the image) in the specimen (Fig. 2C); the planar representation is a matrix of 256 horizontal by 256 vertical pixels (Fig. 2D). The pixels are then combined in a matrix to provide a cross-sectional representation of the anatomy, where digital values correspond to particular characteristics of the tissues they represent. The CT images provide characterization of X-ray attenuation (or absorption) which is dependent primarily on the electron density of most tissues (Valk et al. 1985, Lancaster & Fullerton 1988); this roughly corresponds to tissue density (White 1988). For each two-dimensional coordinate, the digital value of a pixel can vary from 1 to 1250; the 1250-unit scale provides an accurate quantification of X-ray attenuation. Different tissues (e.g. bone, cartilage) potentially are characterized by different values in the original scan data. In addition to all of the digital information contained in each section, the relational position among sections is recorded as well.

Although both CT and MRI are scanning techniques, the fundamental principles and basic data collected by MRI (Valk et al. 1985) are different from that of the CT (Kak & Slaney 1988). MRI scans provide complementary information to CT

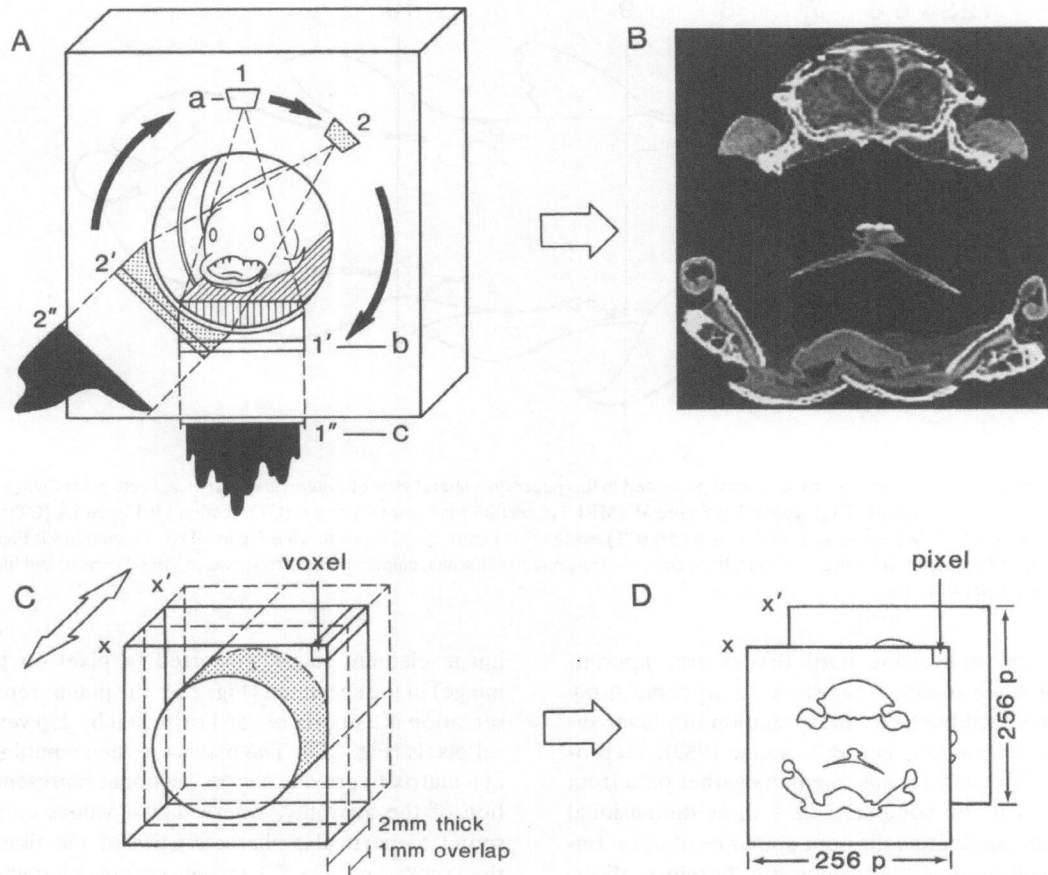

Fig. 2. Schematic representation of data collecting and image processing in Computed Tomography: A – Position of the specimen within the scanner illustrating the rotational collecting of bidimensional projections (1″ and 2″); the X-ray tube (1 and 2) rotates simultaneously with the detector (1′ and 2′). B – Example of a CT section through the anterior part of the ethmosphenoidal region of the head (section 2; C20, scan 89/19). C – Successive overlapping three-dimensional sections (x and x′) as taken in the scanner. D – Schematic representation of the sections (x and x′ shown in C) illustrating the dimension (in pixels) and the relative position of serial sections.

scans. In medical research, CT scanning is used primarily for morphological examination, whereas MRI scanning provides more information on the cellular condition of the tissues (Valk et al. 1985).

The MRI images are dependent on the magnetic properties of the protons in the water molecules of the specimen (Jones & MacFall 1988). The MRI scanner combines the physical properties of protons with respect to their response to magnetic field and radio frequency (Fig. 3A–D). As the first step, protons align themselves parallel or antiparallel to the magnetic field (Fig. 3B). Then radio frequency pulses (at the same frequency as the proton) are used to strike the proton out of alignment (Fig. 3C). Finally, when the radio pulse is stopped, the protons realign themselves back to their normal position in the magnetic field; emitting a weak radio signal proportional to the total number of nuclei present (Fig. 3D). This process of restoring the magnetic vector in line with the static magnetic field, after the radio frequency pulse has ceased, is called *relaxation* (Valk et al. 1985). The relaxation

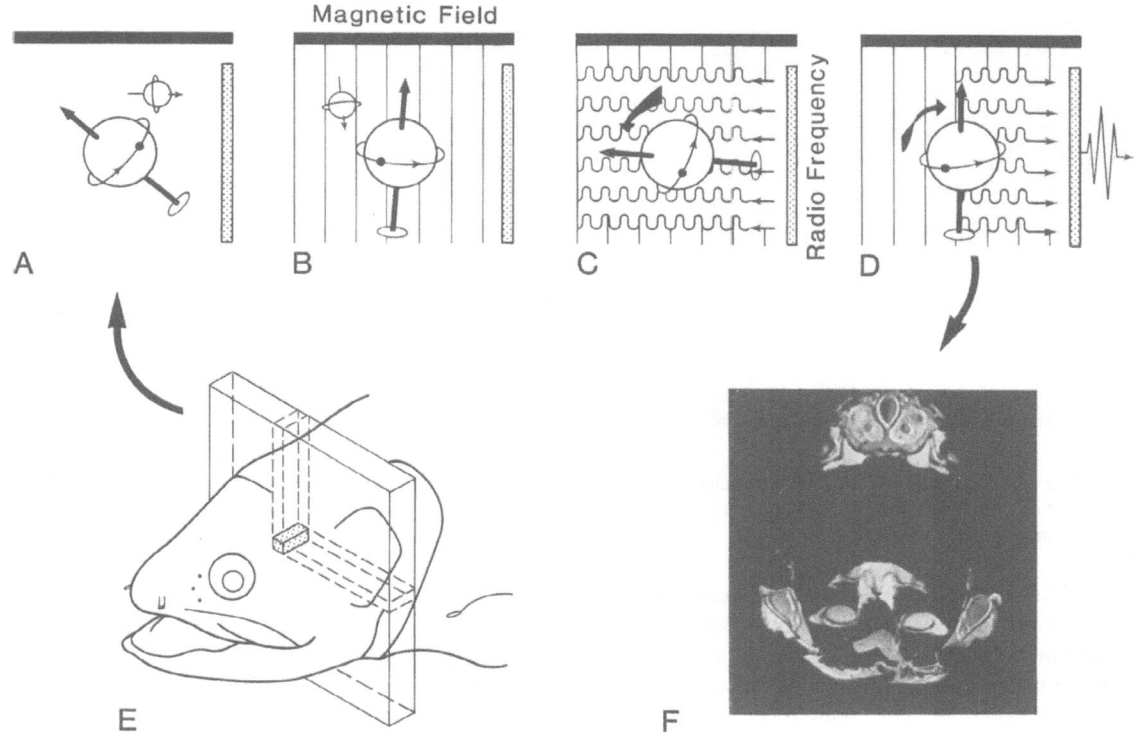

Fig. 3. Specifications of the Magnetic Resonance Imaging technique. A–D. Effects of MRI technique on protons: A – normal resting position of protons (and electron), B – parallel and antiparallel orientation of protons under a magnetic field, C – tilting of proton under the influence of radio waves, and D – return of proton to a normal position without the influence of the radio wave accompanied by emission of a given frequency from which the MRI (T_1 and T_2) is computed. E – Representation of a three-dimensional section on the specimen (in voxel). F – Example of a MRI-T_1 section through the ethmosphenoidal region of the head (section 3; M1-P-1, scan 133/137).

time is divided into two parts – T_1 (longitudinal relaxation time) and T_2 (transverse relaxation time). The relaxation signals are collected by a computer and converted into images (Fig. 3F). An additional category of MRI is referred to as 'flash' for '*f*ast *l*ow *a*ngle *sh*ooting'. The angle of relaxation for the flash scans are lower than those of the T weighted scans. The pulses are recorded for each voxel (Fig. 3E). The amplitude of the signal from any single voxel depends primarily on the number of protons in that volume element MRI (Lancaster & Fullerton 1988): MRI-T_1 weighted images primarily reflect changes in the motion of water molecules owing to the presence of macromolecular structures; and MRI-T_2 weighted images reflect changes in the motion of large molecules such as proteins that

make up the structural fraction of tissues. We refer the reader to Young (1984) for additional detailed information on the MRI technique.

CT and MRI scans are two-dimensional images. A variety of graphic software has been designed to convert a series of planar images into three-dimensional reconstructions. Techniques of computerized three-dimensional reconstruction of CT and MRI scans have been described thoroughly in the literature (e.g. Vannier et al. 1984, Knapp et al. 1985, Woolson et al. 1986, Lancaster & Fullerton 1988), because of their importance in diagnostic radiology. There are two different procedures to select the structures to be reconstructed. Structure(s) to be reconstructed in three-dimensions are extracted from the serial slices by density threshold-

164

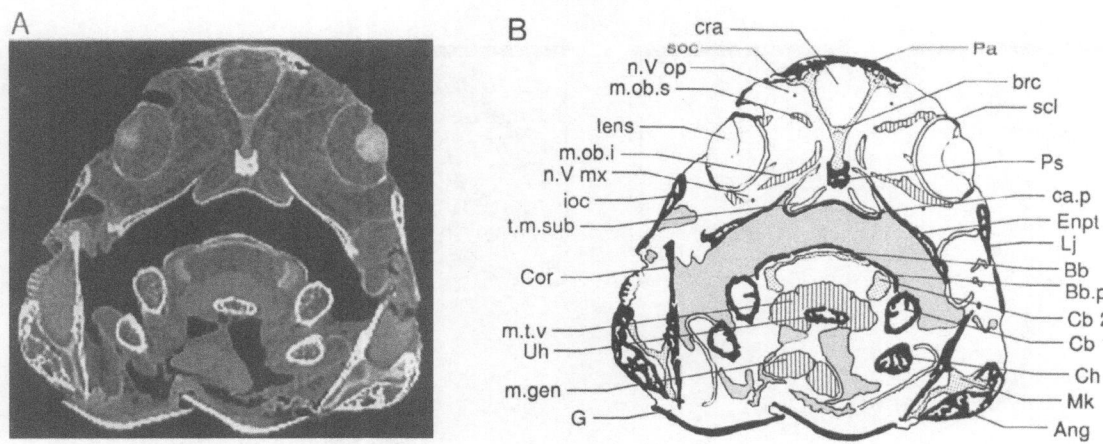

Fig. 4. Cross-section (section 4) in the orbit region of *Latimeria chalumnae*. A – Computed Tomographic scan (C38, scan 161/191). B – Interpretation of the CT scan; black = bone, fine stipple = cartilage, very fine stipple = empty space (e.g. mouth cavity), hatch pattern = muscle. For abbreviations see p. 181.

ing and/or edge extraction. Density thresholding is an automatic process that requires a selection of the digital value of the density, whereas the edge extraction procedure is a manual process that requires the access of every scan in order to digitize the structures to be reconstructed. Selected reconstructed structure(s) can be rotated in any direction.

Techniques applied to the specimen of *Latimeria chalumnae*

Both CT and MRI techniques were used on VIMS 8118. Magnetic tapes and photographic prints are deposited at the Museum of Natural History, The University of Kansas, Lawrence. Sequential numbers referring to the negative plate number and the scan number (e.g. C35, scan 120/191) were assigned to each image for ease of future reference. Figure 1 indicates all of the sections presented in this paper in relation to a schematized lateral view of *L. chalumnae*. Descriptions of the technical specifications for CT, MRI, and three-dimensional reconstruction are provided in the following discussion.

Computed tomography

The CT (Computed Tomography) was performed on a SIEMENS DR-GH scanner at Riverside Hospital, Newport News, Virginia. Scanning parameters were 125 kV, 0.52 AS, and 7 sec scan time.

All the images are cross-sections (i.e. 'transversal', and 'cross horizontal' in medical terminology). Sections in the head region of VIMS 8118 are a series of 2 mm thick sections with 1 mm overlap between each section (Fig. 2C); sections in the pectoral girdle region are 4 mm thick without overlap; and those in the postcranial region are 8 mm thick without overlap (outside of the pectoral girdle). There is a total of 443 sections (381 head and 62 postcranial sections). Sections 4 and 5 (Fig. 4A [C38, scan 161/191] and 7A [C39, scan 165/191], respectively) are an example of the resolution between topographically close sections separated by only 2 mm. Most sections in the head region do not show the dorsal part of the head (corresponding to the dermal skull roof) because this space is used by the computer for the individual header.

CT scans are more informative than MRI in the discrimination between hard tissues (e.g. cartilage and bone) and soft tissues. Nevertheless CT scans are informative with respect to the gross soft anatomy (e.g. branchial axial musculature, tendons, eyes). The human eye is not sensitive enough to allow discrimination between all of the grey-scale differences to distinguish between cartilage and certain tendons in the CT scans; the computer, however, can be programmed for such resolution.

Magnetic resonance imaging

The MRI scans were taken with a SIEMENS MAGNETON IMAGE scanner in Riverside Hospital, Newport News, Virginia.

Three different types of sections are possible with MRI: cross-section (Fig. 7B, C), longitudinal (i.e. 'median', or 'sagittal' in medical terminology) (Fig. 6A), and horizontal (i.e. 'coronal', and 'frontal' in medical terminology).

Multiple techniques (i.e. T_1, T_2, and 'flash') were used on specific regions of the head (e.g. rostral organ, pituitary gland, labyrinth). Figure 7 provides a series of three cross-sections of the same head region contrasting CT, MRI-T_1, and MRI-flash. An approximate incremental grey-scale for the MRI-T_1 scan is related to the water content of the tissue and can be established from black (low proton content) to white (high proton content): bone and eye lens (black), to cartilage-tendon, to muscle, to connective tissues (white). A comparable intensity scale from black to white for the MRI-flash results in: bone (black), to tendon, to connective tissues, to muscles, to cartilage and aqueous vitreum (white). Blood vessels and nerves that are surrounded by connective tissues are therefore highly contrasted.

Computerized three-dimensional reconstruction

Prints of planar CT and MRI images were useful in the predissection evaluation of *L. chalumnae* (e.g. localization of the pituitary gland). Planar prints rather than three-dimensional reconstructions were used during the dissection because of time constraints. Data obtained from the CT scans were transferred from nine-track magnetic tapes to a CEMAX-1000 computer for three-dimensional reconstruction. First an external reconstruction of

the specimen ('ghost' image) was produced. Each CT scan slice was analyzed and only bone density pixels were selected. The three-dimensional bone reconstruction was then imposed over onto the 'ghost' image of the head and rotated. Lateral (Fig. 8) and ventral (Fig. 9) three-dimensional reconstructions of the head region of VIMS 8118 are reproduced herein.

Interpretation and comparison of sections

Although the primary purpose of this paper is to present the radiologic techniques used to study *L. chalumnae*, we would like to compare part of the new information generated through these techniques to that available in the literature. Some of the CT and MRI sections used in this paper correspond to published sections (e.g. section of frozen specimen, reconstruction). We cite references in the literature for visual comparison and briefly discuss some congruences and discrepancies (both technical and morphological) between our observations and the literature. We will focus primarily on the branchial and cranial morphology (e.g. neurocranium, rostral organ) and other hard tissues. Results on the palate of *L. chalumnae* are dealt with in a separate paper (Schultze 1991).

Millot & Anthony (1958a, 1965) illustrated the dissection of *L. chalumnae* by means of photographs of sections and drawings. The photos and drawings provided by Millot & Anthony (1958a) do not include the lower jaw; however, Smith (1940, text-fig. 9) reconstructed a section through the lower jaw. Figure 2B is comparable to pl. 10 of Millot & Anthony (1958a). Figures 4 and 7 can be compared with sections illustrated by Smith (1940, text-fig. 16) and Millot & Anthony (1958a, pl. 14). Our sections are slightly more posterior than the section represented by Millot & Anthony (1958a). The six scans illustrated in Figure 5 correspond to the following plates of Millot & Anthony (1958a): Fig. 5A to pl. 9; Fig. 5B to pl. 14; Fig. 5C is located topographically between the sections figured in pl. 17 and 18; and Fig. 5F is located anterior to the section in pl. 49b. The longitudinal MRI-T_1 scan (Fig. 6A, B) can be compared to Millot & Anthony (1958a,

166

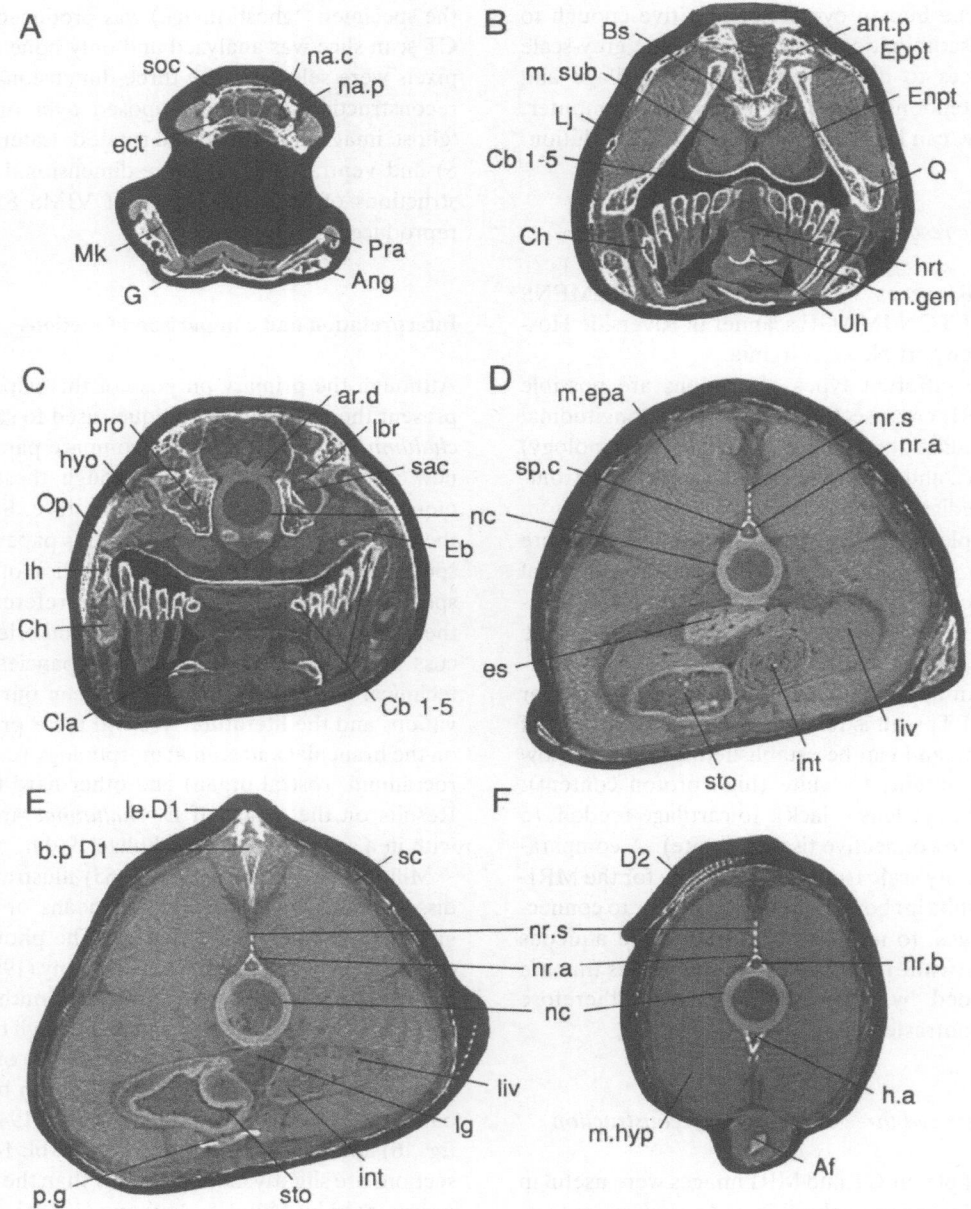

Fig. 5. Series of six computed tomographic cross-sections of *Latimeria chalumnae:* A – Rostral region of the head (section 1; C18, scan 78/19). B – Posterior ethmosphenoidal region of the head (section 6; C55, scan 32/19). C – Posterior region of the oticooccipital region of the head (section 7; C80, scan 133/191). D – Anterior abdominal cavity (section 8; C95, scan 4/42). E – Pelvic fin insertion (section 9; C97, scan 12/42). F – Anal fin level (section 10; C104, scan 40/42). See Figure 1 for position of sections. For abbreviations see p. 181.

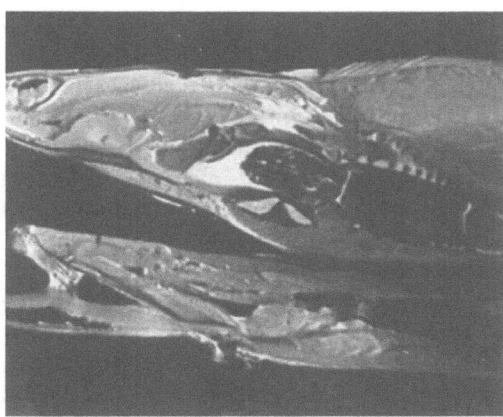
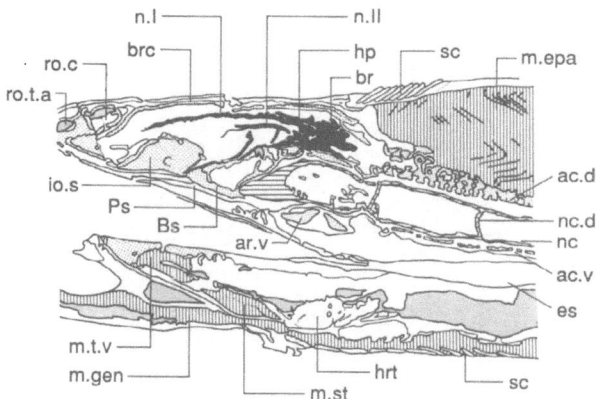

Fig. 6. Longitudinal section of the head of *Latimeria chalumnae.* A – Magnetic Resonance Imaging (T₁) scan (MI-BL-P1, scan 131/140). B – Interpretation of A. For abbreviations see p. 181.

pl. 3, 21; 1958c, fig. 1843–1844), Hughes (1976, fig. 1), Bjerring (1973, fig. 1d), and Anthony (1980, pl. 1). The lateral three-dimensional reconstruction (Fig. 8) can be compared to the planar X-ray of Millot & Anthony (1958a, pl. 1), the lateral reconstructions of Thomson (1967, fig. 14–16), Bjerring (1972, fig. 6c), Alexander (1973, fig. 1), Jarvik (1980, fig. 206a, 219), Lauder (1980a, fig. 1–3), and Forey (1988, fig. 5).

Sections 1–3 are preorbital cross-sections through the anterior ethmosphenoidal region of the head. Of particular interest in this anterior region are the rostral organ and nasal cavity.

On the right side of section 1 (Fig. 5A), detail within the nasal capsule shows the folded olfactory epithelium ('muqueuse olfactive' of Millot & Anthony 1958a, 1965) and the opening for the posterior external narial tube ('canalis buccalis' of Millot & Anthony 1958a) between the lateral rostral (dorsally) and the ectethmoid (ventrally). The posterior part of the anterior rostral tubes meet at the anterior end of the rostral cavity (= 'ethmoidal nasal cavity' of Smith 1940). The trajectory of the three pairs of rostral tubes (seen in MRI-T₁ M1–4 scan 25–36 of 137) is congruent with published descriptions (see Appendix 1). Dorsolaterally, the supraorbital canal is located below, but surrounded by, the supraorbital.

In section 2 (Fig. 2B), the median rostral cavity is seen between the paired nasal cavities. The broad-est region of the parasphenoid (see Schultze 1991) forms most of the buccal floor and attaches dorsolaterally to the ectethmoids. Ventral to the autopalatine (the lateral element of the buccal floor), the teeth of the second dermopalatine protrude. In the anterior part of the lower jaw, Meckel's cartilage is ossified and bordered medially by the prearticular and laterally by the angular. This region is also where the anterior part of the tongue is covered dorsally by two anterior basibranchial tooth plates ('copula' of Smith 1940, Millot & Anthony 1958a).

Sections 3–7 provide cross-sectional views of the branchial arches and musculature. The general morphology of the branchial arches is consistent with that described by Millot & Anthony (1958c, fig. 1837), Nelson (1969, fig. 1, 14, pl. 81.1, 82.2, 83.2), Wiley (1979a, b), and Rosen et al. (1981, fig. 49A).

Section 3 (MRI-T₁; Fig. 3F) is located slightly posterior to section 2 (CT). Ventrolateral to the median rostral cavity, there are two circular structures within the nasal cavity: the posterior external narial tube (dorsolaterally) and cranial nerve I (medioventrally). This section is in the region where ceratobranchial I articulates with the T-shaped basibranchial. The ceratohyals are shown as circular structures ventrolateral to the basibranchial. The geniohyoideus muscles lie ventral to the branchial apparatus and dorsal to the lateral gulars.

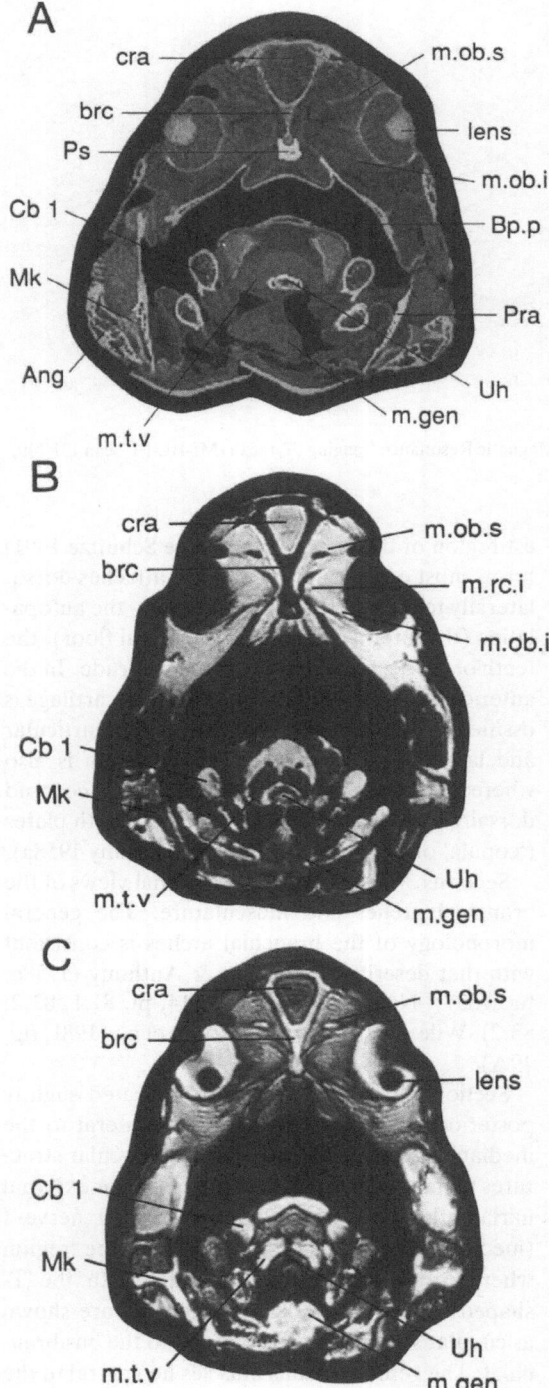

A

cra — — m.ob.s
brc — — lens
Ps —
— m.ob.i
Cb 1 — — Bp.p
Mk —
— Pra
Ang —
— Uh
m.t.v — m.gen

B

cra — — m.ob.s
brc — — m.rc.i
— m.ob.i
Cb 1 —
Mk —
— Uh
m.t.v — m.gen

C

cra — — m.ob.s
brc — — lens
Cb 1 —
Mk —
— Uh
m.t.v — m.gen

Within the lower jaw, the adductor muscle is seen in cross-section above the Meckel's cartilage.

Sections 4–5 are in the orbital region; section 4 is 2 mm anterior to section 5. The eye (in CT) and associated muscles (primarily in MRI) are distinct. The lenses appear as light grey circles (in position) in CT scan and as black circles lower in the eye as a result of thawing of the specimen in the MRI scan. In sections 4–5 (Fig. 4A–B, 7A, respectively), the subcephalic muscles (= 'basicranial m.' of Bjerring 1967; represented in section 6 as circular muscular masses ventral to the neurocranium) are absent, but their anterior tendons are located ventrolateral to the parasphenoid. Coronoid IV (= 'posterior coronoid' of Jarvik 1980) is shown only on the left side because the specimen is at a slight angle in the CT scanner. Meckel's cartilage is Y-shaped in these sections through the middle of the length of the lower jaw (Fig. 4A–B, 7A–C). Meckel's cartilage articulates laterally with the angular, dorsomedially with the prearticular, and ventrally at the suture between the angular and prearticular. Smith (1940) described the basibranchial tooth plates (= 'copula') as a composite ossification made of four fused plates. The basibranchial tooth plates lay on the basibranchial. Millot & Anthony (1958a) and Nelson (1969) reported intraspecific variation with respect to the number of plates that are fused, which is generally five. The CT scan (Fig. 4A–B) shows the buccal surface of the basibranchial tooth plates. They are formed by a pair of lateral plates and a median plate. The basibranchial tooth plates of VIMS 8118 are more similar to those figured by Millot & Anthony (1958a, pl. 45) and Nelson (1969, fig. 14, pl. 81.1) than the one figured by Smith (1940, pl. 23, text-fig. 8). The geniohyoideus muscles are paired oval muscles located ventral to the urohyal, whereas the transversi ventrales 2 muscles are dorsal to the urohyal. The median division of

←

Fig. 7. Comparison of three different kinds of radiologic imaging through the orbital level of *Latimeria chalumnae* (section 5). A – CT scan (C39, scan 165/191). B – MRI-T₁ scan (MI-I-6, scan 7/137). C – MRI-flash scan (MI-F-3, scan 70/137). For abbreviations see p. 181.

the two transversi ventrales is clear in the MR scan (Fig. 7B, C).

Section 6 (Fig. 5B) is located posterior to the orbital region, and anterior to the quadrate-articular articulation (the anterior articulation of the double articulation of the lower jaw), but at the level of the antotic articulation. The entopterygoid sutures dorsally with the epipterygoid (= 'metapterygoid' of Millot & Anthony 1958a) and the quadrate (ventrolaterally). Dorsally the suspensorium articulates with the antotic process of the neurocranium. The articulation between the epipterygoid and the neurocranium is considered fixed in numerous functional investigations (e.g. Thomson 1967, Alexander 1973, Lauder 1980a, Forey 1988). The CT + MRI sections show connective tissue between the epipterygoid and the neurocranium, thus suggesting the possibility of lateral movement. The internal structure of the basisphenoid ossification is bilaterally symmetrical. The large circular subcephalic muscles are located between the ethmosphenoidal region of the neurocranium and the buccal calcareous pavement. Millot & Anthony (1958a) mislabelled the calcareous pavement as the parasphenoid in their pl. 14. Cross-sections of the gill arches (five ceratobranchials and the ceratohyal) are visible. Ventral to the gill arches, the median bony element is the urohyal; the heart (see also Fig. 6) is located dorsal to the urohyal, whereas the paired geniohyoideus muscles are located ventrally.

Section 7 (Fig. 5C) is through the posterior oticooccipital region of the head, as indicated by the presence of the notochord and the labyrinth with its otolith. This section is posterior to the urohyal and gulars but shows the operculum and clavicle. The notochord is bordered dorsally by the posterior extension of the dorsal arcual plates, laterally by the prootics, and ventrally by the paired ventral arcual plates. The notochord shows internal transverse division (seen in MRI-T_2; Fig. 6) that seem to follow the convolutions described by Millot & Anthony (1958a). This section cuts through the anterior part of the saccular otolith on the right side. Laterally the operculum covers the articulation of the hyomandibula with the antotic process of the prootic (seen on the right side). The ventromedial

emargination of the left hyomandibula corresponds to the opening for the hyomandibularis artery and cranial nerve VII. The ossified interhyal (= 'epihyal' of Millot & Anthony 1958a) articulates dorsally with the hyomandibula and ventrally with the ceratohyal. In addition to the ventral gill arch elements, the epibranchial and pharyngobranchial are dorsal to the buccal cavity.

Sections 8–10 are postcranial sections of the trunk and precaudal regions. The epaxial and hypaxial body musculature, the vertical and horizontal septa, the notochord, and the neural spines are visible in Figures 5D–F. The density of the fibrous sheath of the notochord is similar to that of the cartilage. The neural spines and neural arches are ossified; the elements that Andrews (1977) referred to as cartilaginous neural arch bases and illustrated by Millot & Anthony (1958a, pl. 50b) seem to be ossified in VIMS 8118. The anterior pleurocentra are visible only in the longitudinal MRI scan (Fig. 6); this scan can be compared with Andrews' (1977, fig. 2) drawing of the vertebral column of *L. chalumnae*. The spinal cord is identifiable as a distinct light grey spot embedded in lighter density perimeningeal adipose tissue located between the neural arch and the notochord (see Millot & Anthony 1958c, fig. 1838). Our observations of the axial skeleton of *L. chalumnae* confirm the detailed account provided by Andrews (1977). In addition, in these three postcranial scans, a thick fatty layer lies between the scales and the muscles around all of the body.

Section 8 (Fig. 5D) is through the anterior region of the abdominal cavity. Gross soft anatomy of the abdominal cavity can be observed (e.g. esophagus, spiral intestine, two lateral lobes of the liver). The lumen of the esophagus is lined by a dense layer corresponding to the stratified squamous epithelium (see also Fig. 5E). The spiral intestine (Fig. 5D, E) and the stomach contains dense food inclusions. The lobes of the liver are prominent laterally.

Section 9 (Fig. 5E) is through the first dorsal fin insertion. The anterior margin of the first dorsal fin (showing four lepidotrichia on each side in oblique section) and the basal plate can be compared to Millot & Anthony (1958a, pl. 54–56). The dorsal bifurcation of the dorsomedial septum is associated

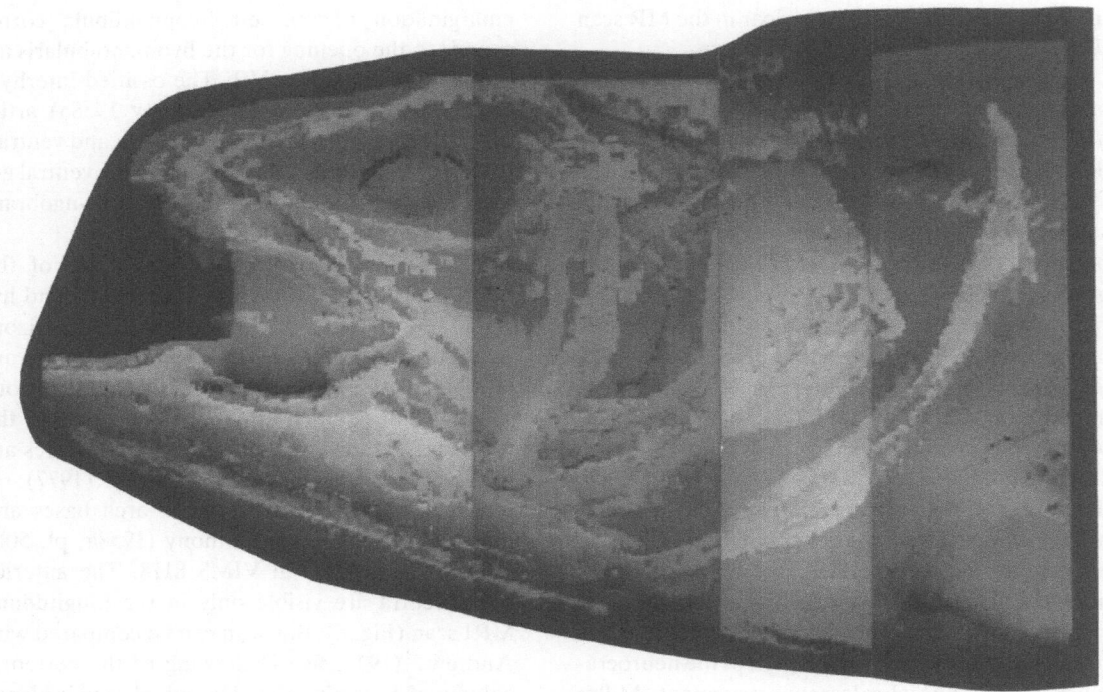

Fig. 8. Computerized three-dimensional reconstruction of the lateral view of the head region of *Latimeria chalumnae.* Bone selected by density thresholding, reconstruction created on a CEMAX-1000 system.

with the presence of the first dorsal basal plate. At this level, the abdominal cavity contains the stomach, lung, spiral intestine, and right lobe of the liver. The single dorsomedial lung (= 'swim bladder', 'air bladder') is below the notochord. The stomach occupies approximately half (the left side) of the abdominal cavity. The paired ossifications ventral to the abdominal cavity correspond to the anterior processes of the pelvic girdle.

Section 10 (Fig. 5F) is through the caudal region posterior to the insertion of the second dorsal and the anal fins. The notochord is bordered dorsally by the neural arches and ventrally by the haemal arches. The apparent segmentation of the neural spines on the CT scan is a result of the posterodorsal inclination of individual spines. There is some distinction between the myoseptum and myomeres similar to the pattern figured by Millot & Anthony (1956a, pl. 4; 1958a, pl. 49b; 1958c, fig.

1838). The internal endoskeletal element of the anal fin is seen in cross-section.

Bones, calcified cartilage, and dense cartilage have been selected for three-dimensional reconstruction and superimposed on the reconstructed external shape in Figures 8 and 9. The reduction of cranial ossification is more evident in the three-dimensional reconstructions than in the individual sections: there is reduction of the cheek bones and opercular series, the neurocranium is poorly ossified, and the pectoral girdle elements are reduced in size.

The lateral view (Fig. 8) provides information on the relative position of the neurocranium, suspensorium, lower jaw, branchial arches, and pectoral girdle. The entopterygoid (fused to the metapterygoid and quadrate) occupies the greatest area of the cranium. The selection of the grey level for the threshold for bone resulted in only the anterior articulation of the lower jaw (involving the quad-

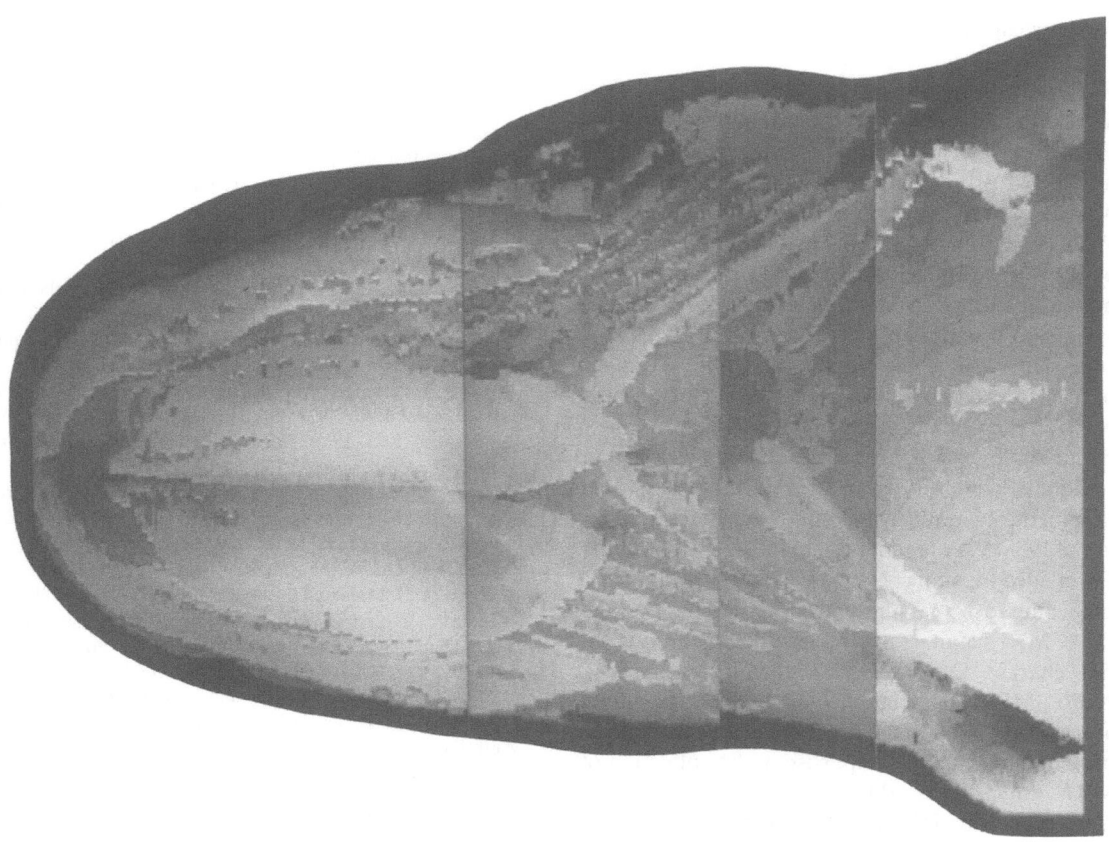

Fig. 9. Computerized three-dimensional reconstruction of the ventral view of the head region of *Latimeria chalumnae*. Bone selected by density thresholding, reconstruction created on a CEMAX-1000 system.

rate and articular) being visible. In contrast to the cross- and longitudinal sections, the intracranial joint is identifiable in the lateral reconstruction. The operculum covers the distal part of the branchial arches. The trajectory of the mandibular canal is indicated by the largest pores on the lateral and ventral projections on the lower jaw.

The ventral view (Fig. 9) shows the relative position of the lower jaw, lateral gulars, branchial arches, and pectoral girdle. The lateral gulars are located between the two rami of the lower jaw, covering the branchial arches and the palate (for a palatal view, see Schultze 1991). The anocleithra are oriented dorsomedially, and the clavicles about ventromedially dorsal to the lateral gular.

Conclusion

New radiologic techniques have been performed successfully on an adult frozen (CT) and thawed (MRI) specimen of *Latimeria chalumnae*. Currently, these non-invasive techniques are limited to large specimens, because these instruments were developed for humans. The accuracy and informativeness of our results on the gross anatomy of the internal structures of *L. chalumnae* are comparable to that used in the diagnostic radiology of humans. Tomographic section analyses and three-dimensional reconstructions surpass what traditional X-ray analysis can offer in terms of facility of interpretation.

Acknowledgements

We first want to thank the Explorers Club, New York and J.A. Musick, Virginia Institute of Marine Sciences (VIMS), Gloucester Point, who accepted our initial project proposal to participate in the dissection of two specimens of *L. chalumnae*. E.O. Wiley, Museum of Natural History, The University of Kansas, proposed the idea of using high-technology to study *L. chalumnae* so that dissection and salvage of organs could be done as quickly as possible. E.O. Wiley participated in the initial stages of the project. J. Musick gave us the opportunity to work on the Coelacanth Project and made contacts with radiology personnel at the Riverside Hospital. Graduate students from VIMS were more than helpful during our stay in Gloucester Point in January of 1988. We are indebted to the personnel of the Riverside Hospital at Newport News, Virginia; and especially to J. Daimler who provided time on the CT + MRI scanners and technical assistance without charge; C. White, CT technician, and V. Neese, MRI technician also gave freely their time and expertise. M. Brown, Siemens representative, provided his valuable expertise in the collection of MRI data. Three-dimensional reconstructions were made possible through the generous invitation of J. Zinreich and the technical assistance of C. Quinn, Department of Radiology, John Hopkins Hospital, Baltimore, Maryland. We thank S.J. Dwyer III, L. Cooke, and R. Laws from the Medical Center of The University of Kansas, Kansas City, who provided negatives from magnetic tape of CT and MRI scans. K. Shaw, Museum of Natural History, The University of Kansas, kindly improved the English; S. Hagen, Division of Biological Sciences, The University of Kansas, drew Figures 1–3, and J. Elder, the same institution, typed the final manuscript. Biomedical Research Fund and the Graduate School of The University of Kansas provided financial support for our trip to Gloucester Point, and P. Humphrey, Museum of Natural History, The University of Kansas, for the trip to Baltimore.

References cited

Adamicka, P. & H. Ahnelt. 1976. Beiträge zur funktionellen Analyse und zur Morphologie des Kopfes von *Latimeria chalumnae* Smith. Ann. Naturhistor. Mus. Wien 80: 251–271.

Ahlberg, P.E. 1989. Paired fin skeletons and relationships of the fossil group Porolepiformes (Osteichthyes: Sarcopterygii). Zool. J. Linn. Soc. 96: 119–166.

Alexander, R.M. 1973. Jaw mechanisms of the coelacanth *Latimeria*. Copeia 1973: 156–158.

Andrews, S.M. 1977. The axial skeleton of the coelacanth, *Latimeria*. pp. 227–288. *In:* S.M. Andrews, R.S. Miles & A.D. Walker (ed.) Problems in Vertebrate Evolution, Linn. Soc. Lond. Symp. Ser. 4, Academic Press, London.

Anthony, J. 1980. Évocation des travaux français sur *Latimeria* notamment depuis 1972. Proc. R. Soc. Lond. B208: 349–367.

Anthony, J. 1984. Le témoignage du coelacanthe *Latimeria*. Rev. Fr. Aquariol. 11: 33–38.

Anthony, J. & J. Millot. 1972. Variations sexuelles de l'appareil excréteur du coelacanthe, *Latimeria chalumnae*. Connexions avec l'appareil génital male. C. R. hedb. Séances Acad. Sci. Paris, Sér. D 276: 2447–2448.

Anthony, J., J. Millot & D. Robineau. 1965. Le coeur et l'aorte ventrale de *Latimeria chalumnae* (poisson coelacanthidé). C.R. hebd. Séances Acad. Sci. Paris, Sér. D 261: 223–226.

Anthony, J. & D. Robineau. 1967. Le cercle céphalique de *Latimeria* (poisson coelacanthidé). C.R. hebd. Séances Acad. Sci. Paris, Sér. D 265: 343–346.

Anthony, J. & D. Robineau. 1968. Branchies et artères branchiales de *Latimeria chalumnae* (poisson coelacanthidé). C.R. hebd. Séances Acad. Sci. Paris, Sér. D 266: 375–378.

Anthony, J. & D. Robineau. 1976a. Sur quelques caractères juvéniles de *Latimeria chalumnae* Smith (Pisces, Crossopterygii, Coelacanthidae). C.R. hebd. Séances Acad. Sci. Paris, Sér. D 283: 1739–1742.

Anthony, J. & D. Robineau. 1976b. Corrélations anatomofonctionnelles au niveau de la tête de coelacanthe; *Latimeria chalumnae*. Bull. Soc. zool. Fr. 101: 181.

Balon, E.K., M.N. Bruton & H. Fricke. 1988. A fiftieth anniversary reflection on the living coelacanth, *Latimeria chalumnae:* some new interpretations of its natural history and conservation status. Env. Biol. Fish. 23: 241–280.

Bemis, W.E. & T.E. Hetherington. 1982. The rost[r]al organ of *Latimeria chalumnae:* morphological evidence of an electroreceptive function. Copeia 1982: 467–471.

Bemis, W.E. & R.G. Northcutt. 1991. Innervation of the basicranial muscle of *Latimeria chalumnae*. Env. Biol. Fish. 32: 147–158. (this volume)

Bernhauser, A. 1961. Zur Knochen-und Zahn-Histologie von *Latimeria chalumnae* (Smith) und einiger Fossil-Formen. Sber. öst. Akad. Wiss. 170: 119–137.

Bjerring, H.C. 1967. Does a homology exist between the basicranial muscle and the polar cartilage? Problèmes Actuels de Paléontologie (Évolution des Vertébrés), Colloq. Internatl. C.N.R.S. 163: 223–267.

Bjerring, H.C. 1968. The second somite with special reference to the evolution of its myotomic derivatives. pp. 341–357. *In:* T. Ørvig (ed.) Current Problems of Lower Vertebrate Phylogeny, Nobel Symposium 4, Almquist & Wiksell, Stockholm.

Bjerring, H.C. 1971. The nerve supply to the second metamere basicranial muscle in osteolepiform vertebrates, with some remarks on the basic composition of the endocranium. Acta Zool. (Stockholm) 52: 189–225.

Bjerring, H.C. 1972. The nervus rarus in coelacanthiform phylogeny. Zool. Scripta 1: 57–68.

Bjerring, H.C. 1973. Relationships of coelacanthiforms. pp. 179–205. *In:* P.H. Greenwood, R.S. Miles & C. Patterson (ed.) Interrelationships of Fishes, Zool. J. Linn. Soc. Lond. 53, Suppl. 1.

Bjerring, H.C. 1977. A contribution to structural analysis of the head of craniate animals. Zool. Scripta 6: 127–183.

Bjerring, H.C. 1978. The 'intracranial joint' versus the 'ventral otic fissure'. Acta Zool. (Stockholm) 59: 203–214.

Bjerring, H.C. 1984. The term 'fossa bridgei' and five endocranial fossae in teleostome fishes. Zool. Scripta 13: 231–238.

Bjerring, H.C. 1985. Facts and thoughts on piscine phylogeny. pp. 31–57. *In:* R.E. Foreman, A. Gorbman, J.M. Dodd & R. Olsson (ed.) Evolutionary Biology of Primitive Fishes, NATO ASI series, Ser. A, life sciences 103, Plenum Press, New York.

Bjerring, H.C. 1986. The rostral organ of *Latimeria chalumnae* (Tofsstjärtfiskarnas elsinnesorgan – 'ett sjätte sinne'). Fauna och Flora (Stockholm) 81: 215–222. (In Swedish.)

Brooker, M.J. 1986. Computed tomography for radiographers. MTP Press, Lancaster. 130 pp.

Carlström, D. 1963. A crystallographic study of vertebrate otoliths. Biol. Bull. 125: 441–463.

Castanet, J., F. Meunier, C. Bergot & Y. François. 1975. Données préliminaires sur les structures histologiques du squelette de *Latimeria chalumnae*. I – Dents, écailles, rayons des nageoires. Problèmes Actuels de Paléontologie (Évolution des Vertébrés), Colloq. Internatl. C.N.R.S. 218: 161–167.

Chavin, W. 1972. Thyroid of the coelacanth, *Latimeria chalumnae* Smith. Nature 239: 340.

Chavin, W. 1976. The thyroid of the sarcopterygian fishes (Dipnoi and Crossopterygii) and the origin of the tetrapod thyroid. Gen. comp. Endocrinol. 30: 142–155.

Cloutier, R. H.-P. Schultze, E.O. Wiley, J.A. Musick, J.C. Daimler, M.A. Brown, S.J. Dwyer III, L.T. Cook & R.L. Laws. 1988. Recent radiologic imaging techniques for morphological studies of *Latimeria chalumnae*. Env. Biol. Fish. 23: 281–282.

Cole, D.F. 1968. The anterior chamber of the coelacanth eye. Br. J. Ophthalmol. 52: 415–418.

Compagno, L.J.V. 1979. Coelacanths: shark relatives or bony fishes? pp. 45–52. *In:* J.E. McCosker & M.D. Lagios (ed.) The Biology and Physiology of the Living Coelacanth, Occ. Pap. Calif. Acad. Sci. 134, San Francisco.

Conroy, G.C. & M.W. Vannier. 1984. Noninvasive three-dimensional computer imaging of matrix-filled fossil skulls by high-resolution computed tomography. Science 226: 456–458.

Cracraft, J. 1968. Functional morphology and adaptation significance of cranial kinesis in *Latimeria chalumnae* (Coelacanthini). Amer. Zool. 8: 354.

Dartnall, H.J.A. 1972. Visual pigment of the coelacanth. Nature 239: 341–342.

Dingerkus, G., H.K. Mok & M.D. Lagios. 1978. The living coelacanth *Latimeria chalumnae* does not have a cloaca. Nature 276: 261–262.

Epple, A. & J.E. Brinn Jr. 1975. Islet histophysiology: evolutionary correlations. Gen. Comp. Endocrinol. 27: 320–349.

Forey, P.L. 1984. The coelacanth as a living fossil. pp. 166–169. *In:* N. Eldredge & S.M. Stanley (ed.) Living Fossils, Springer, New York.

Forey, P.L. 1988. Golden jubilee for the coelacanth *Latimeria chalumnae*. Nature 336: 727–732.

François, Y. 1959. La nageoire dorsale, anatomie comparée et évolution. Année biol. Paris, 35: 81–113.

Francillon, H., F. Meunier, D. Ngo Tuan Phong & A. de Ricqlès. 1975. Données préliminaires sur les structures histologiques du squelette de *Latimeria chalumnae*. II. – Tissus osseux et cartilages. Problèmes Actuels de Paléontologie, Évolution des Vertébrés, Colloq. Internatl. C.N.R.S. 218: 169–174.

Fraschini, A. 1967. Presenza di cellule di Paneth in *Latimeria chalumnae* e loro caracteristiche istochimische. Boll. Soc. Ital. Biol. 43: 1449–1450.

Fricke, H., O. Reinicke, H. Hofer & W. Nachtigall. 1987. Locomotion of the coelacanth *Latimeria chalumnae* in its natural environment. Nature 329: 331–333.

Fritzsch, B. 1987. Inner ear of the coelacanth fish *Latimeria* has tetrapod affinities. Nature 327: 153–154.

Fritzsch, B. & M.H. Wake. 1988. The inner ear of gymnophione amphibians and its nerve supply: a comparative study of regressive events in a complex sensory system (Amphibia, Gymnophiona). Zoomorphol. 1988 (108): 210–217.

Fukuda, Y., N. Kawamoto, I. Obata & Y. Kanie. 1978. Notes of observation by scanning electron microscope on a scale of *Latimeria chalumnae*. Sci. Rep. Yokosuka City Mus. 24: 45–50. (In Japanese.)

Géraudie, J. & F.-J. Meunier. 1980. Elastoidin actinotrichia in coelacanth fins: a comparison with teleosts. Tissue & Cell 12: 637–645.

Giraud, M.M., J. Castanet, F.J. Meunier & Y. Bouligand. 1978a. Organisation spatiale de l'isopédine des écailles du coelacanthe (*Latimeria chalumnae* Smith). C.R. hebd. Séances Acad. Sci. Paris, Sér. D 278: 487–489.

Giraud, M.M., J. Castanet, F.J. Meunier & Y. Bouligand. 1978b. The fibrous structure of the coelacanth scales: A twisted 'plywood'. Tissue & Cell 10: 671–686.

Grady, J.E. 1970. Tooth development in *Latimeria chalumnae* (Smith). J. Morphol. 132: 377–388.

Grassé, P.P. 1976. Précis de Zoologie: Vertébrés. 2. Reproduction, Biologie, Évolution et Systématique. Agnathes, Poissons, Amphibiens et Reptiles. 2nd ed., Masson, Paris. 480 pp.

Grossner, O. 1968. Das Inselorgan des Crossopterygiers *Lati-*

174

meria chalumnae Smith. Z. Zellforsch. mikrosk. Anat. 84: 417–428.

Hafeez, M.A. & M.E. Merhige. 1977. Light and electron microscopic study on pineal complex of coelacanth, *Latimeria chalumnae* Smith. Cell Tissue Res. 178: 249–265.

Hensel, K. 1986. Morphologie et interprétation des canaux et canalicules sensoriels céphaliques de *Latimeria chalumnae* Smith, 1939 (Osteichthyes, Crossopterygii, Coelacanthiformes). Bull. Mus. Natl. Hist. Nat., Paris, 4e sér. 8(2): 379–407.

Hobdell, M.H. & W.A. Miller. 1969. Radiographic anatomy of the teeth and tooth supporting tissues of *Latimeria chalumnae*. Arch. oral Biol. 14: 855–858.

Hughes, G.M. 1972. Gills of a living coelacanth, *Latimeria chalumnae*. Experientia 28: 1301–1302.

Hughes, G.M. 1976. On the respiration of *Latimeria chalumnae*. Zool. J. Linn. Soc. 59: 195–208.

Hughes, G.M. 1980. Ultrastructure and morphometry of the gills of *Latimeria chalumnae* and a comparison with the gills of associated fishes. Proc. R. Soc. Lond. B208: 309–328.

Hureau, J.-C. & C. Ozouf. 1977. Détermination de l'âge et croissance du coelacanthe *Latimeria chalumnae* Smith, 1939 (poisson, crossoptérygien, coelacanthidé). Cybium 2: 129–137.

Imake, H. & W. Chavin. 1973. Ultrastructure of integumental melanophores in the coelacanth. Amer. Zool. 13: 1348.

Imake, H. & W. Chavin. 1984. Ultrastructure of mucous cells in the sarcopterygian integument. Scanning Electron Microsc. 1984: 409–422.

Isokawa, S., Y. Toda & K. Kubota. 1968. A histological observation of a coelacanth (*Latimeria chalumnae*). J. Nihon Univ. Sch. Dent. 14: 102–114.

Janvier, P. 1974. Preliminary report on Late Devonian fishes from central and eastern Iran. Geol. Surv. Iran, Rep. 31: 5–47.

Jarvik, E. 1942. On the structure of the snout of crossopterygians and lower gnathostomes in general. Zool. Bidr. Uppsala 21: 235–675.

Jarvik, E. 1960. Théories de l'évolution des vertébrés reconsidérées à la lumière des récentes découvertes sur les vertébrés inférieurs. Masson & Cie, Paris. 104 pp.

Jarvik, E. 1963. The composition of the intermandibular division of the head in fish and tetrapods and the diphyletic origin of the tetrapod tongue. K. svenska VetenskAkad. Handl., ser. 4, 9(1): 1–74.

Jarvik, E. 1980. Basic structure and evolution of vertebrates. Vol. 1, Academic Press, New York. 575 pp.

Jarvik, E. 1981. Basic structure and evolution of vertebrates. Vol. 2, Academic Press, New York. 337 pp.

Jessen, H. 1966. Die Crossopterygier des Oberen Plattenkalkes (Devon) der Bergisch-Gladbach – Paffrather Mulde (Rheinisches Schiefergebirge) unter Berücksichtigung von amerikanischem und europäischem *Onychodus*-Material. Ark. Zool., ser. 2, 18: 305–389.

Jollie, M. 1972. Chordate morphology. R.E. Krieger, Huntington. 478 pp.

Jones R.D. & J.R. MacFall. 1988. Computers in magnetic resonance imaging. Computers in Physics 1988: 25–30.

Kak, A.C. & M. Slaney. 1988. Principles of computerized tomographic imaging. IEEE Press, New York. 329 pp.

Knapp, R.H., M.W. Vannier & J.L. Marsh. 1985. Generation of three dimensional images from CT scans: technological perspective. Radiol. Technol. 56: 391–398.

Kremers, J.W. 1975. The structure of the brain stem of *Latimeria chalumnae*. Acta Morphol. Neerl.-Scand. 13: 307.

Kremers, J.W.P.M. & R. Nieuwenhuys. 1979. Topological analysis of the brain stem of the crossopterygian fish *Latimeria chalumnae*. J. Comp. Neurol. 187: 613–638.

Laerm, J. 1979. On the origin of rhipidistian vertebrae. J. Paleontol. 53: 175–186.

Lagios, M.D. 1972. Evidence for a hypothalamo-hypophysial portal vascular system in the coelacanth *Latimeria chalumnae* Smith. Gen. Comp. Endocrinol. 18: 73–82.

Lagios, M.D. 1974. Granular epitheloid (juxtaglomerular) cell and renovascular morphology of the coelacanth *Latimeria chalumnae* Smith compared with that of other fishes. Gen. comp. Endocrinol. 22: 296–307.

Lagios, M.D. 1975. The pituitary gland of the coelacanth *Latimeria chalumnae* Smith. Gen. comp. Endocrinol. 25: 126–146.

Lagios, M.D. 1979. The coelacanth and the Chondrichthyes as sister groups: a review of shared apomorph characters and a cladistic analysis and reinterpretation. pp. 25–44. *In:* J.E. McCosker & M.D. Lagios (ed.) The Biology and Physiology of the Living Coelacanth, Occ. Pap. Calif. Acad. Sci. 134, San Francisco.

Lagios, M.D. & S. Staskov-Concannon. 1979. Presumptive interrenal tissue (adrenocortical homolog) of the coelacanth *Latimeria chalumnae*. Gen. Comp. Endocrinol. 37: 404–406.

Lamer, H.I. & W. Chavin. 1975. Ultrastructure of the integumental melanophores of the coelacanth, *Latimeria chalumnae*. Cell Tissue Res. 163: 383–394.

Lancaster, J.L. & G.D. Fullerton. 1988. Computers, physics, and medicin: imaging the body. Computers in Physics 1988: 16–22.

Lange, R.H. 1983. The lipo vitellin phosvitin crystals in the oocyte of *Latimeria chalumnae* coelacanth (Pisces): comparative investigation. C.R. Séances Acad. Sci. Paris, Sér. 3, 297: 393–396.

Lauder, G.V. 1980a. The role of the hyoid apparatus in the feeding mechanism of the coelacanth *Latimeria chalumnae*. Copeia 1980: 1–9.

Lauder, G.V. 1980b. On the relationship of the myotome to the axial skeleton in vertebrate evolution. Paleobiol. 6: 51–56.

Lehman, J.-P. 1966. Crossopterygii. pp. 301–412. *In:* J. Piveteau (ed.) Traité de Paléontologie, Vol. 4, Masson & Cie, Paris.

Lemire, M. 1970. Étude cytoarchitectonique du rhombencéphale de *Latimeria chalumnae*, poisson crossoptérygien, coelacanthidé. C.R. hebd. Séances Acad. Sci. Paris, Sér. D 271: 1994–1997.

Lemire, M. 1971. Étude architectonique du rhombencéphale de

Latimeria chalumnae Smith (poisson, crossoptérygien, coelacanthidé). Bull. Mus. Natl. Hist. Nat., Paris, Sér. 3, 2: 41–96.

Lemire, M. 1976. Caractéristiques ultrastructurales des cellules acineuses de la glande postvanale de coelacanthe (*Latimeria chalumnae* Smith). C.R. hebd. Séances Acad. Sci. Paris, Sér. D 282: 641–644.

Lemire, M. 1977. Étude ultrastructurale du parenchyme sécréteur de la glande postanale du coelacanthe *Latimeria chalumnae* Smith. Annal. Sci. Nat., Zool. biol. animal 19: 227–245.

Lemire, M. & M. Lagios. 1979. Ultrastructure du parenchyme sécréteur de la glande post-anale du coelacanthe *Latimeria chalumnae* Smith. Acta Anat. 104: 1–15.

Lenoble, J. & Y. Le Grand. 1954. Le tapis de l'oeil du coelacanth (*Latimeria anjouanae* (Smith)). Bull. Mus. Natl. Hist. Nat., Paris, Sér. B 26: 460–463.

Locket, N.A. 1973a. Retinal structure in *Latimeria chalumnae*. Phil. Trans. R. Soc. Lond. B266: 493–521.

Locket, N.A. 1973b. Possible discontinuous retinal rod outer segment formation in *Latimeria chalumnae*. Nature 244: 308–309.

Locket, N.A. 1974. The choroidal tapetum lucidum of *Latimeria chalumnae*. Proc. R. Soc. Lond. B186: 281–290.

Locket, N.A. 1980. Some advances in coelacanth biology. Proc. R. Soc. Lond. B208: 265–307.

Lund, R. & W.L. Lund. 1985. Coelacanths from the Bear Gulch Limestone (Namurian) of Montana and the evolution of the Coelacanthiformes. Bull. Carnegie Mus. Nat. Hist. 25: 1–74.

Lund, W.L., R. Lund & G.A. Klein. 1985. Coelacanth feeding mechanisms and ecology of the Bear Gulch coelacanths. C.R. Neuvième Congr. Internatl. Stratigr. Géol. Carbonifère 5: 492–500.

Maisey, J.G. 1987. Notes on the structure and phylogeny of vertebrate otoliths. Copeia 1987: 495–499.

Marx, M. & S.H. D'Auria. 1986. CT examination of eleven Egyptian mummies. RadioGraphics 6: 321–330.

Marx, M. & S.H. D'Auria. 1988. Three-dimensional CT reconstructions of an ancient human Egyptian mummy. Amer. J. Roentgenol. 150: 147–149.

Mathews, M.B. 1975. Connective tissue. Macromolecular structure and evolution. Springer Verlag, Berlin. 318 pp.

McAllister, D.E. 1968. The evolution of branchiostegals and associated opercular, gular, and hyoid bones and the classification of teleostome fishes, living and fossil. Natl. Mus. Can. Bull. 221: 1–239.

McAllister, D.E. 1971. Le vieux quadrupède. Mus. Natl. Sci. Nat. Canada, Coll. Odyssée 1: 1–25.

McAllister, D.E. & C.L. Smith. 1978. Mensurations morphologiques, dénombrements méristiques et taxonomie du coelacanthe, *Latimeria chalumnae*. Naturaliste canad. 105: 63–76.

McAllister, J.A. 1987. Phylogenetic distribution and morphological reassessment of the intestines of fossil and modern fishes. Zool. Jb. Anat. 115: 281–294.

Meinke, D.K. 1982. A light and scanning electron microscope study on the dermal skeleton of *Spermatodus* (Pisces: Coelacanthini) and the evolution of the dermal skeleton in coelacanths. J. Paleontol. 56: 620–630.

Melmed, R.N. & S.J. Holt. 1975. Ultrastructural observations on exocrine and endocrine pancreatic cells of the coelacanth *Latimeria chalumnae* Smith. Israel J. med. Sci. 11: 405.

Meunier, F.J. 1980. Les relations isopédine – tissu osseux dans le post-temporal et les écailles de la ligne latérale de *Latimeria chalumnae* (Smith). Zool. Scripta 9: 307–317.

Meunier, F.-J. & J. Géraudie. 1980. Les structures en contreplaqué du derme et des écailles des vertébrés inférieurs. Ann. Biol. 19: 1–18.

Meunier, F.J., J. Castanet, H. Francillon & Y. François. 1974. Examen microradiographique des écailles de quelques téléostéens. Bull. Ass. Anat. Paris 58: 615–624.

Miller, W.A. 1969. Tooth enamel of *Latimeria chalumnae* Smith. Nature 221: 1244.

Miller, W.A. 1979. Observations on the structure of mineralized tissues of the coelacanth, including the scales and their associated odontodes. pp. 68–78. *In:* J.E. McCosker & M.D. Lagios (ed.) The Biology and Physiology of the Living Coelacanth, Occ. Pap. Calif. Acad. Sci. 134, San Francisco.

Miller, W.A. & M.C. Hobdell. 1968. Preliminary report on the histology of the dental and paradental tissues of *Latimeria chalumnae* Smith with a note on tooth replacement. Arch. oral Biol. 13: 1289–1291.

Millot, J. 1955. Unité spécifique des coelacanthes actuels. La Nature, Rev. Sci. Applic. 3238: 58–59.

Millot, J. 1964. Le diencéphale de *Latimeria chalumnae* Smith (poisson coelacanthidé). C.R. hebd. Séances Acad. Sci. Paris, 258: 5051–5055.

Millot, J. & J. Anthony. 1954. Tubes rostraux et tubes nasaux de *Latimeria* (Coelacanthidae). C.R. hebd. Séances Acad. Sci. Paris, Sér. D 239: 1241–1243.

Millot, J. & J. Anthony. 1955a. Les canaux sensoriels de la tête chez *Latimeria* (Coelacanthidae). C.R. VIe Congr. Internatl. Anat. (Paris): 161–162.

Millot, J. & J. Anthony. 1955b. L'articulation intracrânienne de *Latimeria* (Coelacanthidae). C.R. VIe Congr. Internatl. Anat. (Paris): 161–162.

Millot, J. & J. Anthony. 1955c. Considérations physiomorphologiques sur la tête de *Latimeria* (crossoptérygien coelacanthidé). C.R. Séances Acad. Sci. Paris, Sér. D 241: 114–116.

Millot, J. & J. Anthony. 1956a. Considérations préliminaires sur le squelette axial et le système nerveux centrale de *Latimeria chalumnae* Smith. Mém. Inst. scient. Madagascar A 11: 167–188.

Millot, J. & J. Anthony. 1956b. L'organe rostral de *Latimeria* (crossoptérygien coelacanthidé). Annal. Sci. nat. Zool. 18: 381–389.

Millot, J. & J. Anthony. 1956c. L'organe rostral de *Latimeria* (Coelacanthidae). C.R. hebd. Séances Acad. Sci. Paris, Sér. D 239: 1241–1243.

Millot, J. & J. Anthony. 1956d. Note préliminaire sur le thymus et la glande thyroïde de *Latimeria chalumnae* (crossoptérygien coelacanthidé). C.R. hebd. Séances Acad. Sci. Paris, Sér. D 242: 560–562.

Millot, J. & J. Anthony. 1958a. Anatomie de *Latimeria chalum*-

176

nae. 1. Squelette, muscles et formations de soutien. C.N.R.S., Paris. 122 pp.

Millot, J. & J. Anthony. 1958b. De l'existence chez *Latimeria chalumnae* Smith (Coelacanthidae) d'un organe régulateur du courant sanguin supra-branchial. C.R. hebd. Séances Acad. Sci. Paris, Sér. D 246: 1600–1602.

Millot, J. & J. Anthony. 1958c. Crossoptérygiens actuels, *Latimeria chalumnae,* dernier des crossoptérygiens. pp. 2553–2597. *In:* P.P. Grassé (ed.) Traité de Zoologie, Vol. 13, Masson, Paris.

Millot, J. & J. Anthony. 1959. Les neuromastes du système latéral de *Latimeria chalumnae.* Annal. Sci. Nat. Zool. 1: 317–328.

Millot, J. & J. Anthony. 1960a. Un nouvel aspect du coelacanthe: Le montage complet de son squelette. Sci. nat. Paris 37: 1–3.

Millot, J. & J. Anthony. 1960b. Le cloaque chez les coelacanthes. Bull. Mus. Natl. Hist. Nat. Paris, Sér. 2, 32: 287–289.

Millot, J. & J. Anthony. 1960c. Appareil génital et reproduction des coelacanthes. C.R. hebd. Séances Acad. Sci. Paris, Sér. D 251: 442–443.

Millot, J. & J. Anthony. 1962. Premières précisions sur l'organisation du télencéphale chez *Latimeria chalumnae* (poisson crossoptérygien coelacanthidé). C.R. hebd. Séances Acad. Sci. Paris, Sér. D 254: 2067–2068.

Millot, J. & J. Anthony. 1965. Anatomie de *Latimeria chalumnae.* 2. Système nerveux et organes des sens. C.N.R.S., Paris. 130 pp.

Millot, J. & J. Anthony. 1966. L'organisation générale du prosencéphale de *Latimeria chalumnae* Smith (poisson crossoptérygien coelacanthidé). pp. 50–60. *In:* R. Hassler & H. Stephan (ed.) Evolution of the Forebrain, Plenum, New York.

Millot, J. & J. Anthony. 1972. La glande post-anale de *Latimeria.* Annal. Sci. Nat., Sér. 12, 14: 305–418.

Millot, J. & J. Anthony. 1973a. Le pancréas des crossoptérygiens coelacanthidés. Z. Zellforsch. mikrosk. Anat. 123: 215–223.

Millot, J. & J. Anthony. 1973b. La position ventrale du rein de *Latimeria chalumnae* (poisson coelacanthidé). C.R. hebd. Séances Acad. Sci. Paris, Sér. D 276: 2171–2173.

Millot, J. & J. Anthony. 1973c. Variations sexuelles de l'appareil excréteur du coelacanthe, *Latimeria chalumnae.* Connexions avec l'appareil génital mâle. C.R. hebd. Séances Acad. Sci. Paris, Sér. D 276: 2447–2448.

Millot, J. & J. Anthony. 1973d. L'appareil excréteur de *Latimeria chalumnae* Smith (poisson coelacanthidé). Annal. Sci. Nat., Sér. 12, Sér. B 15: 292–328.

Millot, J., J. Anthony & D. Robineau. 1972. État commenté des captures de *Latimeria chalumnae* Smith (poisson, crossoptérygien, coelacanthidé) effectuées jusqu'au mois d'octobre 1971. Bull. Mus. Natl. Hist. Nat. Paris, Sér. 3, Zool. 39: 533–548.

Millot, J., J. Anthony & D. Robineau. 1978. Anatomie de *Latimeria chalumnae.* Vol. 3, Appareil digestif – appareil respiratoire – appareil urognital – glandes endocrines – appareil circulatoire – téguments – écailles – conclusions générales. C.N.R.S., Paris. 198 pp.

Millot, J. & N. Carasso. 1955. Note préliminaire sur l'oeil de *Latimeria* (Coelacanthidae). C.R. hebd. Séances Acad. Sci. Paris, Sér. D 241: 576–577.

Millot, J., R. Nieuwenhuys & J. Anthony. 1964. Le diencéphale de *Latimeria chalumnae* Smith (poisson coelacanthidé). C.R. hebd. Séances Acad. Sci. Paris, Sér. D 258: 5051–5055.

Millot, J. & A. Policard. 1955. Sur la structure inframiscrocopique du tissu conjonctif du coelacanthe. Bull. Microsc. Appl. Paris 5: 94–95.

Mok, H.-K. 1981. The posterior cardinal veins and kidneys of fishes with notes on their phylogenetic significance. Japan. J. Ichthyol. 27: 281–290.

Munk, O. 1964. The eye of *Calamoichthys calabaricus* Smith, 1865 (Polypteridae, Pisces) compared with the eye of other fishes. Vidensk. Medd. Dansk Naturhist. For. 127: 113–123.

Nelson, G.J. 1968. Gill-arch structure in *Acanthodes.* pp. 130–143. *In:* T. Ørvig (ed.) Current Problems of Lower Vertebrate Phylogeny, Nobel Symp. 4, Almquist & Wiksell, Stockholm.

Nelson, G.J. 1969. Gill arches of the phylogeny of fishes, with notes on the classification of vertebrates. Bull. Amer. Mus. Nat. Hist. 141: 475–552.

Nelson, G.J. 1970. Subcephalic muscles and intracranial joints of sarcopterygian and other fishes. Copeia 1970: 468–471.

Nelson, G.J. 1973. Relationships of clupeomorphs, with remarks on the structure of the lower jaw in fishes. pp. 333–349. *In:* P.H. Greenwood, R.S. Miles & C. Patterson (ed.) Interrelationships of Fishes, Zool. J. Linn. Soc. Lond. 53, Suppl. 1.

Nieuwenhuys, R. 1964. Comparative anatomy of the spinal cord. pp. 1–57. *In:* J.C. Eccles & J.P. Schad (ed.) Organization of the Spinal Cord, Progress in Brain Research, vol. 11, Elsevier, Amsterdam.

Nieuwenhuys, R. 1965. The forebrain of the crossopterygian fish *Latimeria chalumnae* Smith. J. Morphol. 117: 1–24.

Nieuwenhuys, R. 1974. Topological analysis of the brain stem: a general introduction. J. Comp. Neur. 156: 255–276.

Nieuwenhuys, R., J.P.M. Kremers & C. van Huijzen. 1975. The brain of the crossopterygian fish *Latimeria chalumnae.* Acta Morphol. Neerl.-Scand. 13: 306.

Nieuwenhuys, R., J.P.M. Kremers & C. van Huijzen. 1977. The brain of the crossopterygian fish *Latimeria chalumnae:* a survey of its gross structure. Anat. Embryol. 151: 157–169.

Nolf, D. 1985. *Otolithi piscium.* pp. 1–145. *In:* H.-P. Schultze (ed.) Handbook of Paleoichthyology, Vol. 10, Gustav Fischer, Stuttgart.

Northcutt, R.G. 1980. Anatomical evidence of electroreception in the coelacanth (*Latimeria chalumnae*). Zbl. Vet. Med. Comp. Anat. Histol. Embryol. 9: 289–295.

Northcutt, R.G. 1986. Electroreception in nonteleost bony fishes. pp. 257–285. *In:* T.H. Bullock & W. Heiligenberg (ed.) Electroreception, John Wiley & Sons, New York.

Northcutt, R.G. 1987. Lungfish neural characters and their bearing on sarcopterygian phylogeny. pp. 277–297. *In:* W.E.

Bemis, W.W. Burggren & N.E. Kemp (ed.) The Biology and Evolution of Lungfishes, Centennial Suppl. of J. Morphol. 1, Alan R. Liss, New York.

Northcutt, R.G. 1989. The phylogenetic distribution and innervation of craniate mechanoreceptive lateral lines. pp. 17–78. *In:* S. Coombs, P. Gösner & H. Münz (ed.) The Mechanosensory Lateral Line: Neurobiology and Evolution, Springer Verlag, New York.

Northcutt, R.G. & T.J. Neary. 1975. Observations on the optic tectum of the coelacanth, *Latimeria chalumnae.* Amer. Zool. 15: 806.

Northcutt, R.G., T.J. Neary & D.G. Senn. 1978. Observations on the brain of the coelacanth *Latimeria chalumnae.* External anatomy and quantitative analysis. J. Morphol. 155: 181–192.

Ohman, P. 1974. Fine structure of the retinal pigment epithelium of the river lamprey *Lampetra fluviatilis* (Cyclostomi). Acta Zool. (Stockholm) 55: 245–254.

Ørvig, T. 1977. A survey of odontodes ('dermal teeth') from developmental, structural, functional, and phyletic points of view. pp. 53–75. *In:* S.M. Andrews, R.S. Miles & A.D. Walker (ed.) Problems in Vertebrate Evolution, Acad. Press, London.

Pearson, R. & L. Pearson. 1976. The vertebrate brain. Academic Press, London. 744 pp.

Pegueta, V.P. 1968. Enchondral ossification in *Latimeria chalumnae* Smith. Dopov. Akademii Nauk UKR RSR Ser. B, Geol., Geophys., Chem., Biol. 7: 653–656. (In Russian.)

Peyer, B. 1968. Comparative odontology. Univ. Chicago Press, Chicago. 347 pp.

Pfeiffer, W. 1968. Über die Epidermis von *Latimeria chalumnae* J.L.B. Smith (Crossopterygii, Pisces). Z. Morphol. Ökol. Tiere 63: 419–427.

Poplin, C. 1981. Les homologies du pont prootique chez les ostéichthyens. Cybium 3, sér. 5: 3–17.

Reilly, S.M. & G.V. Lauder. 1988. Atavisms and the homology of hyobranchial elements in lower vertebrates. J. Morphol. 195: 237–245.

Robineau, D. 1973. Signification fonctionnelle de l'articulation intra-cranienne chez *Latimeria chalumnae* (poisson crossoptérygien coelacanthidé). C.R. hebd. Séances Sci. Paris, Sér. D 277: 1341–1343.

Robineau, D. 1975. Le système de la veine jugulaire et ses homologies chez *Latimeria chalumnae* (Pisces, Crossopterygii, Coelacanthidae). C.R. hebd. Séances Acad. Sci. Paris, Sér. D 281: 45–48.

Robineau, D. 1976. Les organes régulateurs de la pression artérielle céphalique chez *Latimeria chalumnae* (Crossopterygii, Coelacanthidae). Revue Trav. Inst. (scient. tech.) Pêch. marit. 40: 730–732.

Robineau, D. 1987. Sur la signification phylogénétique de quelques caractères anatomiques remarquables du coelacanthe *Latimeria chalumnae* Smith, 1939. Annal. Sci. nat., Zool., Paris, sér. 13, 8: 43–60.

Robineau, D. & J. Anthony. 1971. Le problème de la veine cave postérieure chez *Latimeria chalumnae* (poisson crossoptéryga

gien). C.R. hebd. Séances Acad. Sci. Paris. Sér. D 273: 689–692.

Robineau, D. & J. Anthony. 1973. Biomécanique du crâne de *Latimeria chalumnae* (poisson, crossoptérygien, coelacanthidé). C.R. hebd. Séances Acad. Sci. Paris, Sér. D 276: 1305–1308.

Rochon-Duvigneaud, A. 1958. L'oeil et la vision. pp. 1099–1140. *In:* P.P. Grassé (ed.) Traité de Zoologie, Vol. 13 (2), Masson, Paris.

Rosen, D.E., P.L. Forey, B.G. Gardiner & C. Patterson. 1981. Lungfishes, tetrapods, paleontology and plesiomorphy. Bull. Amer. Mus. Nat. Hist. 167: 159–276.

Roux, G.H. 1942. The microscopic anatomy of *Latimeria* scale. South Afr. J. Med. Sci., Biol. Suppl. 7: 1–18.

Sasagawa, I., M. Ishiyama & H. Kodera. 1985. Fine structure of the pharyngeal teeth in the coelacanthid fish (*Latimeria chalumnae*). Chikyu Kagaku [Earth Sci.] 39(2): 105–115.

Schaeffer, B. 1941. A revision of *Coelacanthus newarki* and notes on the evolution of the girdles and basal plates of the median fins in the Coelacanthini. Amer. Mus. Novit. 1110: 1–17.

Schaeffer, B. 1948. A study of *Diplurus longicaudatus* with notes on the body form and locomotion of the Coelacanthini. Amer. Mus. Novit. 1378: 1–32.

Schaeffer, B. 1952. The Triassic coelacanth fish *Diplurus,* with observations on the evolution of the Coelacanthini. Bull. Amer. Mus. Nat. Hist. 99(2): 25–78.

Schaeffer, B. 1967. Osteichthyan vertebrae. J. Linn. Soc. Lond., Zool. 47: 185–195.

Schaeffer, B. & J.T. Gregory. 1961. Coelacanth fishes from the continental Triassic of the western United States. Amer. Mus. Novit. 2036: 1–18.

Schultze, H.-P. 1980. Eier legende und lebend gebärende Quastenflosser. Natur und Museum 110: 101–108.

Schultze, J.-P. 1987. Dipnoans as sarcopterygians. pp. 39–74. *In:* W.E. Bemis, W.W. Burggren & N.E. Kemp (ed.) The Biology and Evolution of Lungfishes, Centennial Suppl. of J. Morphol. 1, Alan R. Liss, New York.

Schultze, H.-P. 1988. Notes on the structure and phylogeny of vertebrate otoliths. Copeia 1988: 257–259.

Schultze, H.-P. 1991. CT scan reconstruction of the palate region of *Latimeria chalumnae.* Env. Biol. Fish. 32: 183–192. (this volume)

Shellis, R.P. 1978. The role of the inner dental epithelium in the formation of the teeth in fish. pp. 31–42. *In:* P.M. Butler & K.A. Joysey (ed.) Development, Function and Evolution of Teeth, Academic Press, London.

Shellis, R.P. & D.F.G. Poole. 1978. The structure of the dental hard tissues of the coelacanthid fish *Latimeria chalumnae* Smith. Archs. oral Biol. 23: 1105–1113.

Smith, J.L.B. 1939a. A living fish of Mesozoic type. Nature 143: 455–456.

Smith, J.L.B. 1939b. The living coelacanthid fish from South Africa. Nature 143: 748–750.

Smith, J.L.B. 1940. A living coelacanthid fish from South Africa. Trans. Roy. S. Afr. 28: 1–106.

178

Smith, J.L.B. 1953a. The second coelacanth. Nature 171: 99–101.

Smith, J.L.B. 1953b. Problems of the coelacanth. S. Afr. J. Sci. 49: 279–281.

Smith, M.M. 1978. Enamel in the oral teeth of *Latimeria chalumnae* (Pisces: Actinistia): a scanning electron microscope study. J. Zool. 185: 355–369.

Smith, M.M. 1979a. SEM of the enamel layer in oral teeth of fossil and extant crossopterygian and dipnoan fishes. Scanning Electron Microsc. 1979: 483–490.

Smith, M.M. 1979b. Scanning electron microscopy of odontodes in the scales of a coelacanth embryo, *Latimeria chalumnae* Smith. Arch. oral Biol. 24: 179–183.

Smith, M.M. & M.H. Hobdell. 1973. Comparisons between the microstructure of scales of *Latimeria chalumnae* and extant dipnoan and teleostean scales. J. Dent. Res. 52: 957.

Smith, M.M., M.H. Hobdell & W.A. Miller. 1972. The structure of the scales of *Latimeria chalumnae*. J. Zool. Lond. 167: 501–509.

Stensiö, E. 1947. The sensory lines and dermal bones of the cheek in fishes and amphibians. K. svenska VetenskAkad. Handl., ser. 3, 24(3): 1–195.

Stensiö, E. 1963. The brain and the cranial nerves in fossil, lower craniate vertebrates. Skr. norske Vidensk-Akad. Mat.-naturv. Kl. 1963: 1–120.

Suzuki, N. & T. Hamada. 1990. Examination of the modern coelacanth with X-ray photography and X-ray computed tomography. Scient. Pap. Arts Sci., Univ. Tokyo. (in press.)

Suzuki, N., Y. Suyehiro & T. Hamada. 1985. Initial report of expeditions for coelacanth – Part I – Field Studies in 1981 and 1983. Scient. Pap. Coll. Arts Sci., Univ. Tokyo 35: 37–79.

Tanaka, Y. 1985. An anatomical study on the spleen of archaic fishes 1. Coelacanthiformes and Dipneusti. Acta Haematol. Japan. 48: 710–723.

Thomson, K.S. 1966a. Mobility of the skull and fins in the coelacanth (*Latimeria chalumnae* Smith). Amer. Zool. 6: 565–566.

Thomson, K.S. 1966b. Intracranial mobility in the coelacanth. Science 153: 999–1000.

Thomson, K.S. 1967. Mechanisms of intracranial kinetics in fossil rhipidistian fishes (Crossopterygii) and their relatives. J. Linn. Soc., Zool. 46: 223–253.

Thomson, K.S. 1970. Intracranial movement in the coelacanth *Latimeria chalumnae* Smith (Osteichthyes, Crossopterygii). Postilla 149: 1–12.

Thomson, K.S. 1973. New observations on the coelacanth fish, *Latimeria chalumnae*. Copeia 1973: 813–814.

Trewavas, E. 1958. The coelacanth yields its secrets. Discovery, May 1958: 196–205.

Uyeno, T. 1988. Coelacanths. Newton 8(12): 40–47. Graphic Science Magazine. (In Japanese.)

Valk, J., C. MacLean & P.R. Algra. 1985. Basic principles of nuclear magnetic resonance imaging. Elsevier, Amsterdam. 197 pp.

Vannier, M.W., J.L. Marsh & J.O. Warren. 1984. Three-dimensional CT reconstruction images for craniofacial surgical planning and evaluation. Radiology 150: 179–184.

van Kemenade, J.A.M. 1976. Anatomy and cytology of the pituitary gland of the coelacanth fish *Latimeria chalumnae* Smith. Gen. comp. Endocrinol. 29: 264.

van Kemenade, J.A.M. & J.W. Kremers. 1976. The pituitary gland of the coelacanth fish *Latimeria chalumnae* Smith: general structure and adenohypophysial cell types. Cell Tissues Res. 163: 291–311.

Véran, M. 1988. Les éléments accessoires de l'arc hyoïdien des poissons téléostomes (Acanthodiens et Ostéichthyens) fossiles et actuels. Mém. Mus. natl. Hist. Nat. Paris, Sci. Terre, 54: 1–98.

Wahlert, G. von. 1961. Über einige Skelett-Formen von Fisch-Flossen und ihre phylogenetische Bedeutung. Verh. Deutsch. Zool. Ges. 1961: 498–508.

Wahlert, G. von. 1968a. *Latimeria* und die Geschichte der Wirbeltiere. Eine evolutionsbiologische Untersuchung. Fortschr. Evolutionsforsch. 4: 1–125.

Wahlert, G. von. 1968b. Demonstration von *Latimeria chalumnae* J.L.B. Smith. Zool. Anz., Suppl. 31: 527–529.

Wahlert, G. von & H. von Wahlert. 1962. Funktion und biologische Bedeutung der Quastenflossen. Natur und Museum 92: 7–12.

Wahlert, G. von & H. von Wahlert. 1967. Bau und Funktion der paddelförmigen Unpaarflossen von *Latimeria chalumnae* J.L.B. Smith (Actinistia, Osteichthyes). Stuttgart. Beitr. Naturk. 172: 1–3.

Wake, D.B. 1979. The endoskeleton: the comparative anatomy of the vertebral column and ribs. pp. 192–237. *In:* M.H. Wake (ed.) Hyman's Comparative Vertebrate Anatomy, 3rd ed., Univ. Chicago Press, Chicago.

White, D.N. 1988. Multidimensional imaging in maxillofacial surgery. Facial Plastic Surgery 5: 197–206.

Wiley, E.O. 1979a. Ventral gill arch muscles and the phylogenetic relationships of *Latimeria*. pp. 56–67. *In:* J.E. McCosker & M.D. Lagios (ed.) The Biology and Physiology of the Living Coelacanth, Occ. Pap. Calif. Acad. Sci. 134, San Francisco.

Wiley, E.O. 1979b. Ventral gill arch muscles and the interrelationships of gnathostomes, with a new classification of the Vertebrata. Zool. J. Linn. Soc. 67: 149–179.

Witkowski, A. & W. Szymczak. 1976. Have Crossopterygii been ovoviviparous? Przegląd Zool. 20: 174–178. (In Polish.)

Woodward, A.S. 1940. The surviving crossopterygian fish, *Latimeria*. Nature 146: 53–54.

Woolson, S.T., P. Dev, L.L. Fellingham & A. Vassiliadis. 1986. Three-dimensional imaging of bone from computerized tomography. Clin. Orthopaed. Related Res. 202: 239–248.

Young, S.W. 1984. Nuclear Magnetic Resonance Imaging. Basic Principles. Raven Press, New York. 163 pp.

Zangerl, R. & H.-P. Schultze. 1989. X-radiographic techniques and applications. pp. 165–178. *In:* R.M. Feldmann, R.E. Chapman & J.T. Hannibal (ed.) Paleotechniques, Paleontol. Soc. Spec. Publ. 4.

Appendix 1

Survey of the literature concerning the anatomy (gross morphology and histology) of *Latimeria chalumnae* Smith.

Gross morphological studies:
Osteology

Head: J.L.B. Smith 1939a, b, 1940, Woodward 1940, Schaeffer 1952, Millot & Anthony 1956a, 1958a, c, 1960a, 1965, Trewavas 1958, Jarvik 1960, 1963, 1980, Schaeffer & Gregory 1961, Lehman 1966, Bjerring 1967, 1984, 1985, D.E. McAllister 1968, 1971, Jollie 1972, Adamicka & Ahnelt 1976, Hughes 1976, Millot et al. 1978, Compagno 1979, Lagios 1979, Anthony 1980, 1984, Lauder 1980a, Meunier 1980, Rosen et al. 1981, Robineau 1987, Forey 1988.

Intracranial joint: J.L.B. Smith 1940, Millot & Anthony 1955b, 1958a, c, 1960a, 1965, Nelson 1970, Jollie 1972, Bjerring 1967, 1973, 1977, 1978, Millot et al. 1978, Lagios 1979, Anthony 1980, 1984, Jarvik 1980, Lauder 1980a, Poplin 1981, Robineau 1987, Forey 1988.

Lower jaw: J.L.B. Smith 1940, Woodward 1940, Millot & Anthony 1956a, 1958a, 1960a, Jessen 1966, D.E. McAllister 1971, Jollie 1972, Nelson 1973, Janvier 1974, Adamicka & Ahnelt 1976, Millot et al. 1978, Miller 1979, Jarvik 1980, Forey 1988.

Palate: Millot & Anthony 1958a, Peyer 1968, Schultze 1987, 1991.

Visceral skeleton: J.L.B. Smith 1940, Woodward 1940, Millot & Anthony 1958a, c, 1960a, 1965, Jarvik 1963, 1980, D.E. McAllister 1968, Peyer 1968, Nelson 1968, 1969, Jollie 1972, Adamicka & Ahnelt 1976, Hughes 1976, Bjerring 1977, Millot et al. 1978, Compagno 1979, Lauder 1980a, Rosen et al. 1981, Forey 1988, Reilly & Lauder 1988, Véran 1988.

Shoulder girdle: J.L.B. Smith 1940, Schaeffer 1941, Millot & Anthony 1958a, c, 1965, Lehman 1966, Jollie 1972, Rosen et al. 1981, Anthony 1984.

Axial skeleton: J.L.B. Smith 1940, Millot & Anthony 1956a, 1958a, c, 1960a, 1965, Lehman 1966, Schaeffer 1967, Grassé 1976, Hughes 1976, Andrews 1977, Wake 1979, Laerm 1979, Jarvik 1980, Lauder 1980b, Locket 1980, Rosen et al. 1981, Anthony 1984, Robineau 1987, Balon et al. 1988, Forey 1988, Ahlberg 1989.

Appendicular skeleton: J.L.B. Smith 1939a, b, 1940, Woodward 1940, Millot & Anthony 1958a, c, 1960a, 1965, François 1959, Wahlert 1961, Wahlert & Wahlert 1962, 1967, Lehman 1966, D.E. McAllister 1971, Jollie 1972, Grassé 1976, Jarvik 1980, 1981, Rosen et al. 1981, Suzuki et al. 1985, Balon et al. 1988, Forey 1988.

Scales: J.L.B. Smith 1939a, b, 1940, Roux 1942, Millot & Anthony 1956a, 1958a, c, 1965, D.E. McAllister 1971, Ørvig 1977, Millot et al. 1978, Miller 1979, Locket 1980, Meinke 1982, Suzuki et al. 1985.

Musculature

Cranial musculature: Millot & Anthony 1958a, 1965, Janvier 1974, Adamicka & Ahnelt 1976.

Subcephalic muscles: Millot & Anthony 1958a, 1965, Nelson 1970, Bjerring 1967, 1968, 1971, Jarvik 1980.

Gill arch musculature: Millot & Anthony 1958a, Wiley 1979a, b.

Eye musculature: Millot & Anthony 1958a, 1965.

Fin musculature: Millot & Anthony 1958a, c, 1965, Millot et al. 1978.

Body musculature: Millot & Anthony 1956a, 1958a, c.

Nervous system

Brain and spinal cord: Millot & Anthony 1956a, 1958a, c, 1962, 1965, 1966, Stensiö 1963, Millot 1964, Nieuwenhuys 1964, 1965, 1974, Millot et al. 1964, Lemire 1971, Kremers 1975, Nieuwenhuys et al. 1975, 1977, Northcutt & Neary 1975, Anthony & Robineau 1976a, Grassé 1976, Hughes 1976, Pearson & Pearson, 1976, Northcutt et al. 1978, Kremers & Nieuwenhuys 1979, Lagios 1979, Anthony 1980, 1984, Locket 1980, Northcutt 1987, Robineau 1987.

Cranial nerves: Millot & Anthony 1965, Bjerring 1971, 1972, Hughes 1976, Millot et al. 1978, Northcutt et al. 1978, Kremers & Nieuwenhuys 1979, Locket 1980, Jarvik 1981, Anthony 1984, Robineau 1987, Northcutt 1989, Bemis & Northcutt 1991.

Lateral line system: J.L.B. Smith 1939a, b, 1940, Jarvik 1942, Stensiö 1947, Millot & Anthony 1955a, 1956a, 1958a, c, 1959, 1965, Lehman 1966, Jollie 1972, Jarvik 1980, 1981, Hensel 1986, Northcutt 1986, 1989.

Sense organs

Rostral organ: Millot & Anthony 1954, 1956b, c, 1958a, c, 1965, Lehman 1966, Grassé 1976, Anthony 1980, 1984, Jarvik 1980, Northcutt 1980, 1986, 1989, Rosen et al. 1981, Bemis & Hetherington 1982, Bjerring 1986.

Nasal organ: Millot & Anthony 1958a, c, 1965, Lehman 1966, Grassé 1976, Rosen et al. 1981, Robineau 1987.

Eye: Lenoble & Le Grand 1954, Millot & Carasso 1955, Rochon-Duvigneaud 1958, Millot & Anthony 1958a, c, 1965, Munk 1964, Cole 1968, Locket 1974, 1980, Anthony & Robineau 1976a, Millot et al. 1978.

Inner ear and otolith: Millot & Anthony 1958a, c, 1965, Carlström 1963, Jarvik 1980, Nolf 1985, Fritzsch 1987, Maisey 1987, Fritzsch & Wake 1988, Schultze 1988.

Respiratory system

Gills: Anthony & Robineau 1968, Hughes 1972, 1976, Millot et al. 1978.

Lung: Millot & Anthony 1958c, 1973b, Grossner 1968, Millot et al. 1978, Lagios 1979, Anthony 1984, Robineau 1987.

Circulatory system

Gill and branchial arteries: Millot & Anthony 1958b, Anthony & Robineau 1968, Hughes 1976, Millot et al. 1978, Anthony 1980, Robineau 1987.

Cranial arteries: Millot & Anthony 1965, Anthony & Robineau 1967, Robineau 1976, Millot et al. 1978.

Hypothalamo-hypophysial portal vascular system: Lagios 1972, Millot et al. 1978.

180

Heart and ventral aorta: Millot & Anthony 1958c, 1965, Anthony et al. 1965, Anthony & Robineau 1968, Grassé 1976, Hughes 1976, Millot et al. 1978, Anthony 1980, 1984, Robineau 1987.
Jugular vein: Robineau 1975, Millot et al. 1978, Anthony 1980.
Posterior vena cava: Robineau & Anthony 1971, Millot et al. 1978, Anthony 1980, Robineau 1987.
Posterior cardinal vein: Millot et al. 1978, Anthony 1980, Mok 1981.
Renovascular system: Lagios 1974, Millot et al. 1978, Anthony 1980.

Digestive system
Intestine: Millot & Anthony 1958a, c, 1965, 1973a, Grassé 1976, Millot et al. 1978, J.A. McAllister 1987.
Esophagus: Millot & Anthony 1958c, 1965, Millot et al. 1978.
Liver: Millot & Anthony 1958c, 1965, Grassé 1976, Millot et al. 1978, Anthony 1980, 1984.
Spleen: Millot et al. 1978, Tanaka 1985.
Pancreas: Millot & Anthony 1958c, 1973a, Grossner 1968, Epple & Brinn 1975, Millot et al. 1978, Locket 1980, Anthony 1984, Robineau 1987.
Stomach: Millot & Anthony 1958a, c, 1965, Grassé 1976, Millot et al. 1978.

Excretory and reproductive systems
Interrenal tissues: Millot et al. 1978, Lagios & Staskov-Concannon 1979.
Kidney: Millot & Anthony 1958c, 1965, 1973b, c, d, Grassé 1976, Millot et al. 1978, Anthony 1980, Locket 1980, Mok 1981.
Excretory and reproductive systems: Millot & Anthony 1958c, 1960a, b, c, 1965, 1972, 1973b, c, d, Anthony & Millot 1972, Grassé 1976, Witkowski & Szymczak 1976, Dingerkus et al. 1978, Millot et al. 1978, Lagios 1979, Anthony 1980, Locket 1980, Schultze 1980, J.A. McAllister 1987.

Endocrine system
Pituitary gland: Millot & Anthony 1965, Lagios 1975, 1979, van Kemenade & Kremers 1976, van Kemenade 1976, Millot et al. 1978, Northcutt et al. 1978, Anthony 1980, Locket 1980, Robineau 1987.
Postanal gland: Millot & Anthony 1972, Millot et al. 1978, Lagios 1979, Robineau 1987.
Thymus and thyroid: Millot & Anthony 1956d, 1958c, Chavin 1972, 1976, Millot et al. 1978, Locket 1980.

Histological studies:
Hard tissues
General mineralized tissues: Castanet et al. 1975, Meunier et al. 1974, Miller 1979.

Scales: J.L.B. Smith 1939b, 1940, Roux 1942, Bernhauser 1961, M.M. Smith & Hobdell 1973, M.M. Smith et al. 1972, Ørvig 1977, Fukuda et al. 1978, Shellis & Poole 1978, Giraud et al. 1978a, b, Millot et al. 1978, Miller 1979, M.M. Smith 1979b, Meunier 1980, Meunier & Géraudie 1980, Locket 1980, Meinke 1982.
Teeth: Bernhauser 1961, Isokawa et al. 1968, Miller 1969, 1979, Miller & Hobdell 1968, Peyer 1968, Hobdell & Miller 1969, Grady 1970, Castanet et al. 1975, Millot et al. 1978, Shellis 1978, Shellis & Poole 1978, M.M. Smith 1978, 1979a, Locket 1980, Meinke 1982, Sasagawa et al. 1985.
Bone and cartilage: Pegueta 1968, Francillon et al. 1975, Mathews 1975, Miller 1979, Meunier 1980.
Actinotrichia: Géraudie & Meunier 1980.

Skin
Epidermis and dermis: Pfeiffer 1968, Millot et al. 1978, Locket 1980.
Mucous cells: Millot et al. 1978, Imaki & Chavin 1984.
Pigmentation: Imake & Chavin 1973, Lamer & Chavin 1975, Millot et al. 1978, Miller 1979, Locket 1980.

Nervous system and sense organs
Brain: Millot & Anthony 1956a, 1958a, c, 1965, 1966, Nieuwenhuys 1964, Lemire 1970, Millot et al. 1978, Kremers & Nieuwenhuys 1979.
Eye: Lenoble & Le Grand 1954, Millot & Anthony 1965, Dartnall 1972, Locket 1973a, b, 1974, Ohman 1974, Locket 1980.
Nasal organ: Millot & Anthony 1965.
Rostral organ: Millot & Anthony 1965.
Pineal complex: Hafeez & Merhige 1977.

Endocrine system
Pancreas: Grossner 1968, Millot & Anthony 1972, Melmed & Holt 1975, Millot et al. 1978.
Pituitary gland: van Kemenade & Kremers 1976, van Kemenade 1976.
Postanal gland: Millot & Anthony 1972, Lemire 1976, 1977, Millot et al. 1978, Lagios 1979, Lemire & Lagios 1979.
Thyroid: Chavin 1972.

Miscellaneous
Gills: Hughes 1972, 1976, 1980, Millot et al. 1978.
Conjunctive tissue: Millot & Policard 1955.
Granular juxtaglomerular cell: Lagios 1974.
Hypothalamo-hypophysial portal vascular system: Lagios 1972.
Excretory system: Millot et al. 1978, Lagios & Stako-Concannon 1979, Locket 1980.
Digestive system: Fraschini 1967, Millot et al. 1978.

Abbreviations used in Figures:

ac.d	dorsal arcualia
ac.v	ventral arcualia
Af	anal fin
Ang	angular
ant.p	antotic process
ar.d	dorsal arcual plate
ar.v	ventral arcual plate
Bb	basibranchial
Bb.p	basibranchial tooth plate
b.p D1	basal plate of first dorsal fin
br	brain
brc	braincase
Bs	basisphenoid
ca.p	calcareous pavement
Cb 1	ceratobranchial 1
Cb 2	ceratobranchial 2
Cb 3	ceratobranchial 3
Cb 4	ceratobranchial 4
Cb 5	ceratobranchial 5
Ch	ceratohyal
Cla	clavicle
Cor	coronoid IV
cra	cranial cavity
D2	second dorsal fin
Eb	epibranchial
ect	ectethmoid
Enpt	entopterygoid
Eppt	epipterygoid
es	esophagus
G	lateral gular
h.a.	haemal arch
hp	hypophysis
hrt	heart
hyo	hyomandibular
Ih	interhyal
int	spiral intestine
ioc	infraorbital canal
io.s	interorbital septum
lbr	labyrinth
le.D1	lepidotrichia of first dorsal fin
liv	liver
Lj	lacrimojugal

lu	lung
m.epa	epaxial musculature
m.gen	geniohyoideus muscle
m.hyp	hypaxial musculature
Mk	Meckel's cartilage
m.ob.s	obliquus superior muscle
m.ob.i	obliquus inferior muscle
m.rc.i	rectus internus muscle
m.st	sternohyoideus muscle
m.sub	subcephalic muscle
m.t.v	transversi ventrales 2 muscle
na.c	nasal capsule
na.p	posterior external narial tube
nc	notochord
nc.d	division of notochord
nr.a	neural arch
nr.b	neural arch base
nr.s	neural spine
n.I	cranial nerve I
n.II	cranial nerve II
n.V mx	maxillary branch of cranial nerve V
nV.op.	ophthalmic profundus branch of cranial nerve V
Op	operculum
Pa	parietal
p.g	pelvic girdle
pol	polar cartilage
Pp	postparietal
Pra	prearticular
pro	prootic
Q	quadrate
ro.c	rostral cavity
Ro.l	lateral rostral
ro.t.a	anterior rostral tube
sac	saccular otolith
sc	scales
soc	supraorbital canal
scl	sclerotic plate
sto	stomach
sp.c	spinal cord
t.m.sub	tendon of subcephalic muscle
Uh	urohyal

Environmental Biology of Fishes **32**: 183–192, 1991.

CT scan reconstruction of the palate region of *Latimeria chalumnae*

Hans-Peter Schultze
*Museum of Natural History and Department of Systematics & Ecology, The University of Kansas,
Lawrence, KS 66045-2454, U.S.A.*

Received 2.8.1989   Accepted 2.8.1990

Key words: Sarcopterygii, Actinistia, Coelacanth, Morphology, CT scan, Relationships

Synopsis

The palate of *Latimeria chalumnae* is described based mainly on three-dimensional CT scan reconstruction. It is compared with that of other osteichthyans. The palate of *L. chalumnae* compares best with that of rhipidistians; it is more advanced than that of actinopterygians in having fewer bones. This tendency toward bone reduction in the palate is even more pronounced in dipnoans. The interpretation of features of the Early Devonian genus *Diabolepis* determines if authors consider dipnoans or actinistians more closely related to tetrapods. Both groups are only distant relatives of tetrapods.

Introduction

The osteology of *Latimeria chalumnae* Smith 1939 has been described by Smith (1940) and Millot & Anthony (1958). Skull roof, cheek region and lower jaw have been described quite accurately (see compilation by Schultze & Cloutier 1991), whereas the palate has been presented only incompletely by Smith (1940, fig. 10). Smith (1940) figured the toothed part of the dermal bones, and Millot & Anthony (1958, fig. 2) presented the whole parasphenoid in contact with the endocranium but omitted other dermal bones (figured in medial view, Millot & Anthony 1958, pl. 38) except premaxillae and vomers. Schultze (1987, fig. 7B) compiled a composite palate from figures in Millot & Anthony (1958, fig. 2; pl. 2, 31, 36, 38).

The lack of a satisfactory presentation of the palate can be understood immediately if one realizes that the bones of the medial side of the palatoquadrate and the parasphenoid are not exposed on the palate except for their denticulated areas. The subcephalic muscles (Millot & Anthony 1958, pl.

37) and a pavement of ossifications (Fig. 1; 'calcaires pavant' of Millot & Anthony 1958, pl. 1, 14–18, 25) lie below them so that Smith (1940) even figures a 'suprapterygoid denticulate area'.

Here I will describe the composition of the palate as perceived by CT scan (Computed Tomography) sections and a three-dimensional reconstruction made from them. Palatal morphology of *L. chalumnae* will then be compared with that of other osteichthyans.

Material and methods

The description of the palate of *Latimeria chalumnae* is based on specimen VIMS 8118, a female of 1452 mm standard length deposited in the ichthyological collection of the Virginia Institute of Marine Science, The College of William and Mary, Gloucester Point. The Computed Tomography (CT scan or popular CAT scan) was done on the frozen specimen with a Siemens DR-GH scanner in the radiological unit of the Riverside Hospital in

Newport News, Virginia on January 3, 1988. The 3D-reconstruction from tapes with the CT scan sections were assembled and computed on a CE-MAX-1000 in the Department of Radiology of Johns Hopkins Hospital in Baltimore, Maryland, on August 22 and 23, 1988. The techniques are described in detail by Schultze & Cloutier (1991).

Description

The palate of *Latimeria chalumnae* is composed of premaxillae ('rostral dental plates' of Smith 1940), vomers ('postrostral dental plates' of Smith 1940), two pairs of dermopalatines ('prevomer' and 'autopalatine' of Smith 1940; 'prédermo-palatin' and 'dermo-palatin' of Millot & Anthony 1958), elongated ectopterygoids, broad entopterygoids, an unpaired median parasphenoid and two unpaired arcual plates ('pièce préoccipitale souschordale antérieure, postérieure' of Millot & Anthony 1958).

Premaxillae: As in all actinistians the premaxillae are the only bones of the outer dental arcade; maxillae present in actinopterygians and other crossopterygians are missing. The premaxillae form a pair of bones at the anterior margin of the palate (Millot & Anthony 1958, fig. 2); the three-dimensional reconstruction of the palate (Fig. 2) indicates the same for specimen VIMS 8118. Smith (1940) figured four pairs of separate dental plates; these plates were referred to by other authors as fragmented premaxillae. One pair to many dental plates represents variation of the premaxillae within the species. The teeth increase in size from the symphysis posterolaterally, but they remain small in comparison to the fang-like teeth on the inner dental arcade.

The inner dental arcade is formed by one pair of vomers, two pairs of dermopalatines (one pair less than actinopterygians) and one pair of ectopterygoids; each carrying enlarged fang-like teeth, but small in comparison to other crossopterygians. The inner dental arcade appears to be a continuous arcade with fang-like teeth except for the vomers on the three-dimensional reconstruction (Fig. 2).

Vomers: The vomers form triangular plates which meet each other in the midline. Laterally, they articulate with the anterior dermopalatines, but they have no common articulation with premaxillae or parasphenoid.

Dermopalatines: Two long oval shaped dermopalatines follow the vomer posterolaterally. Labially they carry small teeth and lingually fang-like teeth which appear prominently on the three-dimensional reconstruction (Fig. 2). The anterior dermopalatine articulates with an anterolateral flange of the parasphenoid; the posterior dermopalatine articulates with the anterior dermopalatine and the posteriorly following ectopterygoid.

Ectopterygoid: An elongated bone (five times longer than the posterior dermopalatine) follows the posterior dermopalatine and attaches posteriorly to the lateral margin of the entopterygoid. Large and small teeth are intermixed on the ectopterygoid.

Paired entopterygoids and an unpaired median parasphenoid occupy most of the area of the palate. Two unpaired median arcual plates follow the parasphenoid.

Entopterygoid: The entopterygoid attaches to the medial side of the palatoquadrate (Fig. 1) so that it forms the connection between the ventrally positioned ectopterygoid and the dorsal and median stem of the parasphenoid. Larger teeth compared to those on the main toothed area sit on the margin of the entopterygoid where it attaches to the ectopterygoid. An edentulous triangular portion reaches anteriorly to the posterior dermopalatine. The entopterygoid does not reach the parasphenoid.

Parasphenoid: The parasphenoid underlies the anterior unit of the endocranium, the ethmosphenoid; it reaches from the ventral fissure of the intracranial joint to the anterior autopalatines. The parasphenoid has a narrow shaft or stem, which ends in two posterolateral swallow-tail shaped extensions. The stem diminishes in width from its posterior end anteriorly until about the middle of its length. From there on, the lateral margins flare out anterolaterally to reach the anterior dermopalatine. The parasphenoid is five times as wide in the anterior triangular portion than at the narrowest part of the shaft. The triangular anterior portion carries an elevated toothed area in the form of an

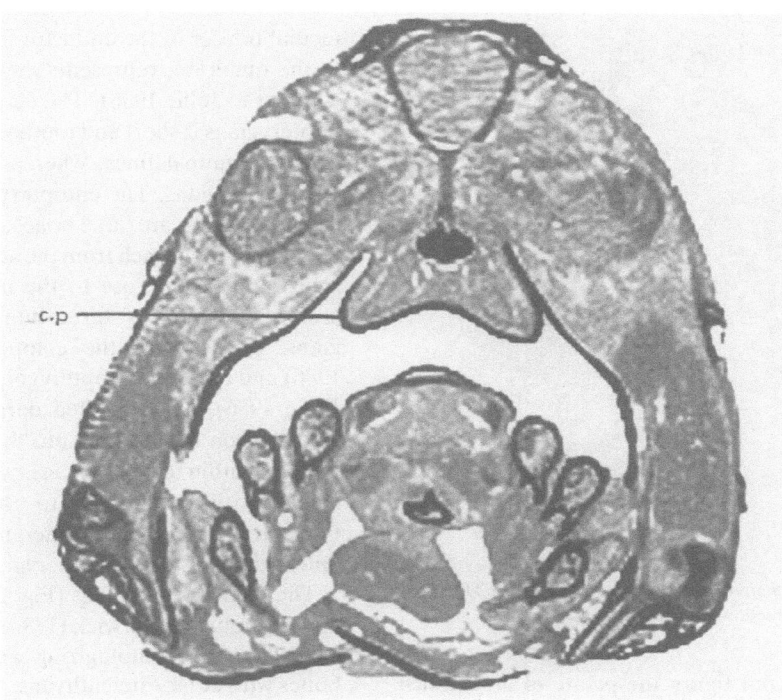

Fig. 1. Latimeria chalumnae, specimen VIMS 8118. CT scan section (C44, scan 181/191) through posteriororbital region. c.p = pavement of ossifications.

elongated oval. This oval extends at its widest extent to the width of the narrowest part of the shaft. The teeth along the margin of the toothed area are larger than the ones in the middle.

Arcual plates: Two arcual plates follow the parasphenoid at about the same level in the mid line (Fig. 2) below the notochord in the otico-occipital region. The anterior one is one-quarter wider and longer than the second one. The posterolateral swallow-tail shaped extensions of the parasphenoid underlie the anterior arcual plate.

The palatal bones are covered deeply by soft tissue (Fig. 1) except for the toothed areas. The superficial portion of the skin is packed with ossifications ('calcaires pavant' of Millot & Anthony 1958; 'small tooth plates' of Nelson 1969, p. 486, pl. 83, fig. 2). Smith (1940, p. 71) spoke of 'dentate area of skin' and figured (Smith 1940, fig. 10) a 'suprapterygoid denticulate area' between entopterygoid and toothed area of parasphenoid. This pavement appears as a complete bony cover of the palate in the CT scan sections (Fig. 1). Therefore, this pavement had to be cut away with the lower jaw in each CT scan section before the palatal bones could be reconstructed (Fig. 2). The toothed areas of dermal bones together with the pavement of ossification form a continuous biting surface of the palate despite large gaps between palatal bones.

Comparison

Gardiner (1984, fig. 75, 76) and Schultze (1987, fig. 7) have compared the palate of osteichthyans including tetrapods. These authors differ in the interpretation of bones anterior and lateral to the entopterygoids in the dipnoan *Griphognathus* and of the pterygoid bones in actinopterygians. Gardin-

186

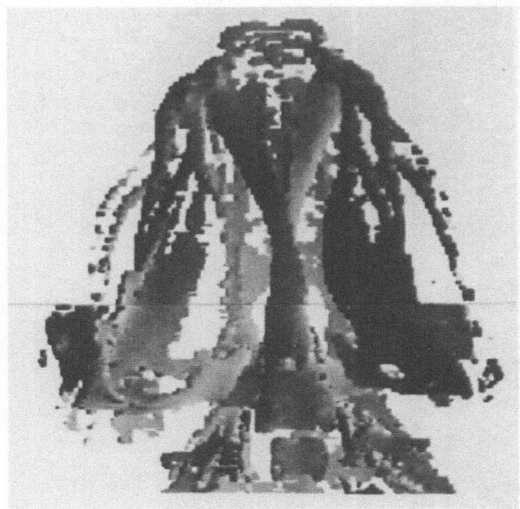

Fig. 2. Latimeria chalumnae, specimen VIMS 8118. Three-dimensional reconstruction of palate from CT scan sections.

er (1984) does not figure the palate of an actinistian.

The palate of *Latimeria* (Fig. 3A) compares closely with that of rhipidistians (Fig. 3D: porolepiforms, Fig. 3E: osteolepiforms) with outer dental arcade (maxilla missing in actinistians), inner dental arcade (paired vomers, two pairs of dermopalatines but one pair in rhipidistians and tetrapods, paired ectopterygoids) and large entopterygoids between inner dental arcade and unpaired median parasphenoid. The parasphenoid lies below the ethmosphenoid and ends in front of the intracranial joint. The broad anterior plate reaching the anterior dermopalatines is a unique feature of the parasphenoid of *L. chalumnae*. The palate is distinct from that in actinopterygians (Fig. 3B) and dipnoans (Fig. 3C). The complete outer dental arcade with premaxilla and maxilla is present in actinopterygians, but missing in dipnoans. The inner dental arcade has a higher number of bones in primitive actinopterygians than in crossopterygians: one pair of vomers, three pairs of dermopalatines and one pair of ectopterygoids. I follow here the homologization of Gardiner (1984) who considered the ectopterygoid as the most posterior bone of the inner dental arcade whereas the bone surrounding the

medial border of the adductor fossa, and reaching to the quadrate, represents the entopterygoid (in contrast to Jollie 1986). The ectopterygoid of actinopterygians is short and toothed, of about the size of the dermopalatines, whereas it is elongated in crossopterygians. The entopterygoids of primitive sarcopterygians are large bones, occupying most of the palate, they reach from the adductor fossa anteromedially to or close to the midline. The same area is occupied by three bones in actinopterygians: posterior (= the ectopterygoid of Jollie 1986) and anterior entopterygoid (the pterygoid of Jollie 1986), and so-called dermometapterygoid. The division of this area into three ossifications is unique within osteichthyans. The parasphenoid has a position anterior to the ventral fissure (= ventral part of the intracranial joint of crossopterygians) as in crossopterygians.

The palate of dipnoans (Fig. 3C) is quite different. Campbell & Barwick (1984) demonstrated the difficulties of homologizing any of the palatal bones with other osteichthyans; they denied homologization by using ingroup comparison (within dipnoans). If we accept *Diabolepis* as the sister-group of dipnoans (Chang & Yu 1984, Maisey 1986), one can support some of Campbell & Barwick's (1984) conclusions even by outgroup comparison. In *Diabolepis* (Fig. 4A), both nasal openings lie outside the outer dental arcade, outside premaxilla and ?maxilla, but ventral of the snout. Here *Diabolepis* is very similar to *Youngolepis* (Fig. 4B). A posterior continuation of the outer dental arcade in form of a maxilla has to be assumed for both forms because the area posterior to the posterior nasal opening and the premaxilla is identical build with overlapped area of lacrymal. Nevertheless the maxilla is not known in *Diabolepis, Youngolepis* or *Powichthys*. The premaxillae of *Diabolepis* are reaching onto the palate. The movement of the nasal openings onto the palate – a scenario (Schultze 1987) necessary to attain the pattern in dipnoans (Fig. 4B–E) – transfers the premaxillae onto the palate. It follows that the margin of the snout in dipnoans cannot be homologized with premaxillae and maxillae of other osteichthyans in contrast to Rosen et al. (1981) and Gardiner (1984). Campbell & Barwick (1984)

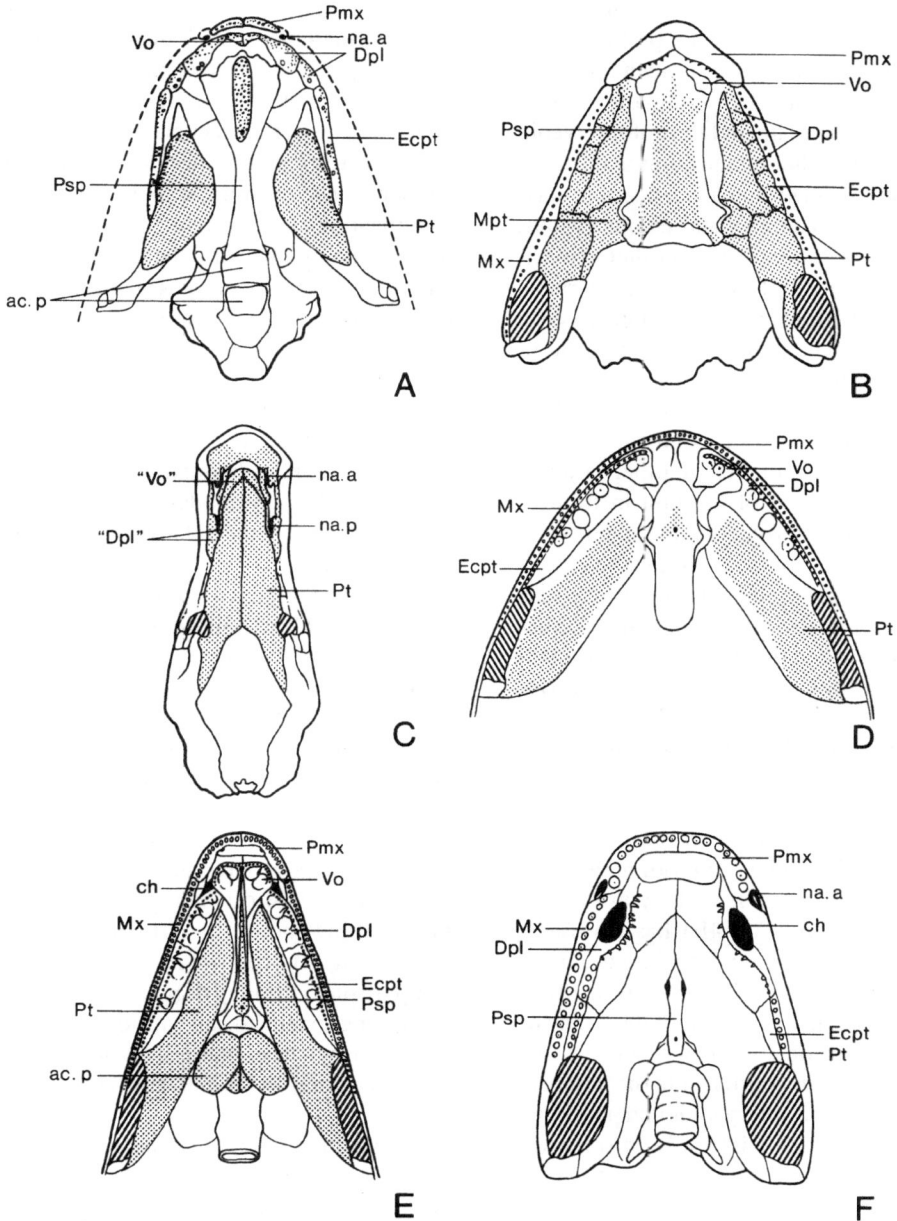

Fig. 3. Comparison of the palate of A– *Latimeria chalumnae* with that of B– a primitive actinopterygian (*Mimia toombsi*), C– a dipnoan (*Griphognathus whitei*), D– a porolepiform (*Glyptolepis groenlandica*), E– an osteolepiform (*Eusthenopteron foordi*) and F– a primitive tetrapod (*Ichthyostega* sp.). B– after Gardiner (1984, fig. 75A), C– after Miles (1977, fig. 6), D– after Jarvik (1972, fig. 3), E– after Jarvik (1980, fig. 124), F– after Jarvik (1980, fig. 171B). ac.p = arcual plate; ch = choana; Dpl = dermopalatine; 'Dpl' = so-called dermopalatine of dipnoans; Ecpt = ectopterygoid; Mpt = so-called dermometapterygoid; Mx = maxilla; na.a, na.p = anterior, posterior nasal opening; Pmx = premaxilla; Psp = parasphenoid; Pt = entopterygoids; s.Psp = swallow-tail shaped extension of parasphenoid; Vo = vomer, 'Vo' = so-called vomer of dipnoans.

reached the same conclusion by referring to the bone mosaic in the upper lip of the primitive dipnoan *Uranolophus* (Denison 1968) which cannot be homologized with premaxillae. Only a few bones occur between the upper lip and the entopterygoid of dipnoans and their homologization is difficult. We could expect to find premaxillae, possible maxillae, vomers, dermopalatines and ectopterygoids by comparison with *Diabolepis* and other osteichthyans. In contrast, we find a reduced and variable number of bones with changing relations to each other (Fig. 4C–G). The pairs of bones in front of the entopterygoids in *Uranolophus* (Fig. 4C), *Dipnorhynchus* (Fig. 4D), *Speonesydrion, Griphognathus* (Fig. 4E) and some other dipnoans may be homologized with the vomers of other osteichthyans using the criterion of position in front of entopterygoids and posteromedial to the posterior nasal opening; still one could not exclude the possibility that these are the premaxillae or premaxillae fused with the vomers, or dermopalatines (Miles 1977). But what are then the three bones in front of the entopterygoids in *Chirodipterus* (Fig. 4F), *Holodipterus* (Miles 1977, fig. 72) and other dipnoans? Most authors interpret the unpaired median bone as a vomer and the pair of bones as dermopalatines. *Gnathorhiza, Ceratodus, Neoceratodus* (Fig. 4G) and other post-Devonian dipnoans have a pair of very anterior bones behind the anterior nasal openings. Are these really vomers as they are always named? *Griphognathus* (Fig. 4E) has a variable number of bones anterolateral to its entopterygoids (Campbell & Barwick 1984, Schultze 1990). The 'extra' dermal bone (Rosen et al. 1981) may be the premaxilla shifted onto the palate; medial to it are the paired vomers and posterior to the latter are the dermopalatines. Entopterygoids and parasphenoid can be homologized with those of other osteichthyans. The parasphenoid is situated behind the ventral fissure in all dipnoans except the primitive genus *Uranolophus* (Denison 1968, Schultze 1991, in contrast to Campbell & Barwick 1989) where it reaches far anterior between the entopterygoids. Thus the upper jaw and palate of dipnoans deviates strongly from that of other osteichthyans and tetrapods (Fig. 3F).

In conclusion, the palate of *Latimeria* compares best with that of rhipidistians. The number of bones of the inner dental arcade and of the pterygoid complex is reduced in both groups over that in actinopterygians. This reduction may be an advanced feature (Rosen et al. 1981, Gardiner 1984), but polarity cannot be decided by outgroup comparison as such an outgroup is missing. A further complication is presented by the palate of dipnoans. The number of palatal bones anterior and lateral to the entopterygoids is further reduced, but it is not clear if that pattern was derived from a crossopterygian pattern (Fig. 5A) most closely related to porolepiforms (Maisey 1986, Chang 1990) or parallel to that of crossopterygians (Fig. 5B: Campbell & Barwick 1987, Schultze 1987, 1990). The position of the dipnoans within the sarcopterygians (including tetrapods) must be resolved in order to determine the relative position of actinistians to tetrapods.

The closest relatives to tetrapods (Fig. 5) are the panderichthyids (Schultze 1987, 1990, Vorobyeva & Schultze 1990), the osteolepiforms being the next closest group (Holmes 1985, Maisey 1986, and others) followed by porolepiforms. The common ancestor of these groups lived in the early Lower Devonian (about 400 million years ago) or in the Late Silurian (as far back as 420 million years ago). About that time, we must also find the common ancestor of all sarcopterygians because there are no earlier records of osteichthyans. This means that the splitting between different sarcopterygian groups (Fig. 5) occurred in a short time period in comparison to the long duration of these groups. In comparing extant forms with each other we may choose to include or ignore a 400 million year history of each group, a temporal distance of at least 800 million years.

The position of dipnoans as the closest relatives of tetrapods (Rosen et al. 1981, Gardiner 1984, Forey 1987) has been debated by many authors (Schultze 1981, 1987, 1990, Holmes, 1985, Maisey 1986, Panchen & Smithson 1987, Campbell & Barwick 1987, Chang 1990); these authors favor osteolepiforms, represented only as fossils, as the closest relatives. Still the dipnoans are the closest living relatives in some of these schemes (Fig. 5A) where the Early Devonian genus *Diabolepis* serves to link

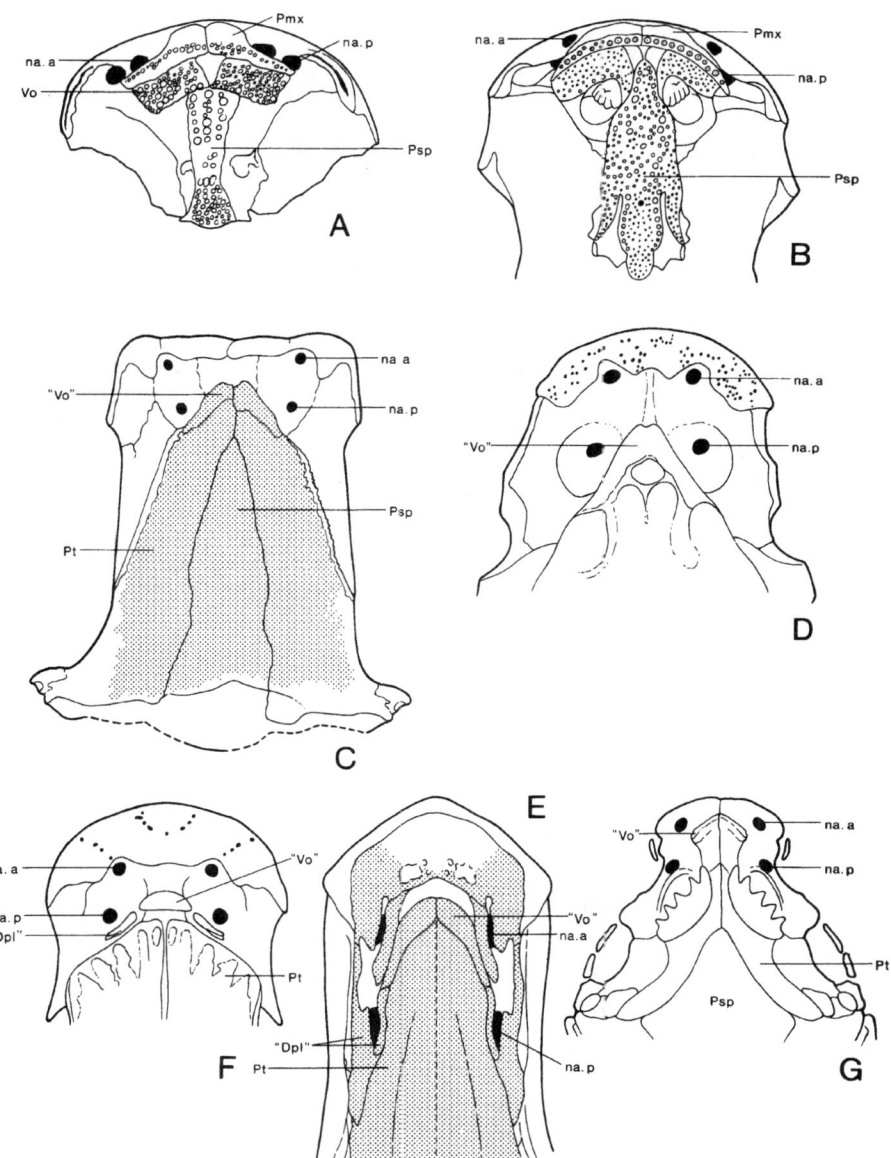

Fig. 4. Anterior palate of A– *Diabolepis speratus,* B– *Youngolepis praecursor,* and of the dipnoans C– *Uranolophus wyomingensis,* D–
Dipnorhynchus suessmilchi, E– *Griphognathus whitei,* F– *Chirodipterus australis* and G– *Neoceratodus forsteri.* A– after Chang & Yu
(1984, fig. 2B), B– after Chang (1982, fig. 10 + 7B), C– original, D– after Thomson & Campbell (1971, fig. 74 + 75), E– after Miles
(1977, fig. 57), F– after Miles (1977, fig. 67), G– many sources. 'Dpl' = so-called dermopalatine of dipnoans; na.a, na.p = anterior,
posterior nasal opening; Pms = premaxilla; Psp = parasphenoid; Pt = entopterygoid; Vo = vomer; 'Vo' = so-called vomer of
dipnoans.

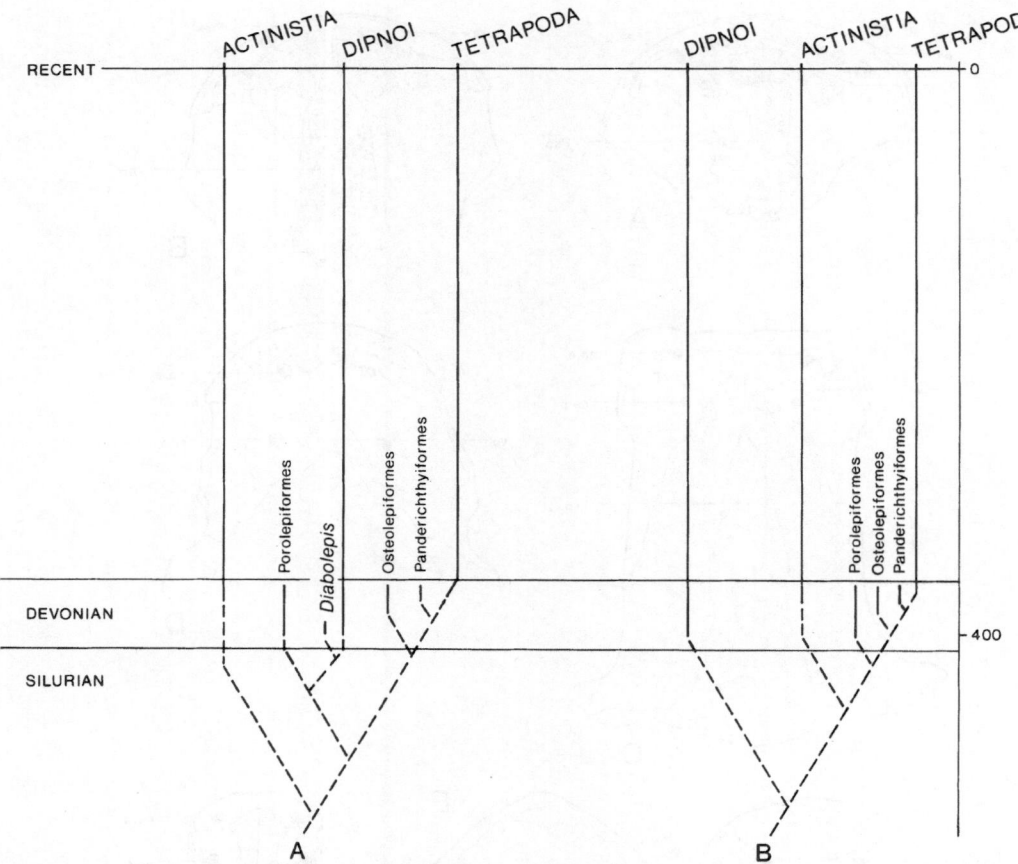

Fig. 5. Temporal position of actinistians in relation to tetrapods: A– dipnoans closer to tetrapods (after Maisey 1986); B– actinistians closer to tetrapods (after Schultze 1987).

Table 1. Terminology of the bones of the palate.

Smith 1939	Millot & Anthony 1958	Schultze 1987/this paper
rostral dental plates (2 pairs)	prémaxillaire	premaxilla
postrostral dental plate	vomer	vomer
prevomer dental plate	prédermo-palatin	dermopalatine
autopalatine dental plate	dermo-palatin	dermopalatine
ectopterygoid	ectoptérygoïde	ectopterygoid
entopterygoid	entoptérygoïde	entopterygoid
suprapterygoid denticulate area	'calcaires pavant'	pavement of ossifications
parasphenoid	parasphénoïde	parasphenoid
–	pièce préoccipitale souschordale	arcual plates

primitive porolepiforms (youngolepiforms) and dipnoans (Maisey 1986). *Diabolepis* closely resembles youngolepiforms (Fig. 4B) in most features, but Chang & Yu (1984) cited features that indicate a closer relationship to dipnoans. These features are very doubtful, and Campbell & Barwick (1987) and Panchen & Smithson (1987) argued for a sistergroup relationship of *Diabolepis* to youngolepiforms or porolepiforms respectively and not to dipnoans. If that is true, actinistians can be considered as the closest living relatives of tetrapods (not in Panchen & Smithson's 1987 scheme). Still there is the other possibility that dipnoans and actinistians together are the sistergroup of tetrapods (Northcutt 1987, Chang 1990, Forey et al. 1990). It appears that the early splitting of the sarcopterygians into different groups within a relatively short time span makes it difficult to resolve the sequence of events and thus the exact interrelationships of all sarcopterygian groups. We will see a continuing discussion over the relationships of sarcopterygians in the future.

Acknowledgements

The author is very thankful to J. Musick, Virginia Institute of Marine Science, for the invitation to take part in the investigation of two *Latimeria chalumnae* specimens donated by the Explorers Club, New York, and for the arrangement of the radiological research. J. Daimler, Radiological Unit, Riverside Hospital, Newport News, Virginia, generously supported our request to make CT scan serial cross sections, and Christie White spent many hours CT scanning the fish. The three-dimensional reconstruction was made possible through a generous invitation by J. Zinreich, Department of Radiology, Johns Hopkins Hospital, Baltimore, Maryland. Cindy Quinn spent two days with the author working on the CEMAX 1000. Photographic prints were prepared by J. Chorn, and the drawings by Ann Musser, both from the Museum of Natural History, The University of Kansas, Lawrence. The manuscript was typed by J. Elder and J. Wiglesworth, Word Processing Center, Division of Biological Sciences, The University of Kansas, Lawrence. The English was improved by J. Chorn, Museum of Natural History, The University of Kansas, Lawrence. I would like to express my thanks to all these people.

References cited

Campbell, K.S.W. & R.E. Barwick. 1984. The choana, maxillae, premaxillae and anterior palatal bones of early dipnoans. Proc. Linn. Soc. New South Wales 107: 147–170.

Campbell, K.S.W. & R.E. Barwick. 1987. Paleozoic lungfishes – a review. J. Morph. Suppl 1: 93–131.

Campbell, K.S.W. & R.E. Barwick. 1989. *Uranolophus:* a reappraisal of a primitive dipnoan. Mem. Ass. Australas. Palaeontols. 7: 87–144.

Chang, M.-M. 1982. The braincase of *Youngolepis,* a Lower Devonian crossopterygian from Yunnan, South-Western China. Dept. Geol., Univ. Stockholm. 113 pp.

Chang, M.-M. 1990. 'Rhipidistians', dipnoans and tetrapods. *In:* H.-P. Schultze & L. Trueb (ed.) Origins of Major Groups of Tetrapods, Controversies and Consensus, Cornell University Press, Ithaca. (in press).

Chang, M.-M. & X.B. Yu. 1984. Structure and phylogenetic significance of *Diabolichthys speratus* gen. et sp. nov., a new dipnoan-like form from the Lower Devonian of eastern Yunnan, China. Proc. Linn. Soc. New South Wales 107: 171–184.

Denison, R.H. 1968. Early Devonian lungfishes from Wyoming, Utah, and Idaho. Fieldiana: Geol. 17: 353–413.

Forey, P.L. 1987. Relationships of lungfishes. J. Morph. Suppl. 1: 75–91.

Forey, P.L., B.G. Gardiner & C. Patterson. 1990. The lungfish, the coelacanth, and the cow revisited. *In:* H.-P. Schultze & L. Trueb (ed.) Origins of Major Groups of Tetrapods, Controversies and Consensus, Cornell University Press, Ithaca. (in press).

Gardiner, B.G. 1984. The relationships of the palaeoniscid fishes, a review based on new specimens of *Mimia* and *Moythomasia* from the Upper Devonian of Western Australia. Bull. Brit. Mus. (Natur. Hist.), Geol., 37: 173–428.

Holmes, E.B. 1985. Are lungfishes the sister group of tetrapods? Biol. J. Linn. Soc. 25: 379–397.

Jarvik, E. 1972. Middle and Upper Devonian Porolepiformes from East Greenland with special references to *Glyptolepis groenlandica* n. sp. Medd. Grønland 187: 1–295.

Jarvik, E. 1980. Basic structure and evolution of vertebrates, Vol. 1. Academic Press, London. 575 pp.

Jollie, M. 1986. A primer of bone names for the understanding of actinopterygian head and pectoral girdle skeletons. Can. J. Zool. 64: 365–379.

Maisey, J.G. 1986. Heads and tails: a chordate phylogeny. Cladistics 2: 201–256.

Miles, R. 1977. Dipnoan (lungfish) skulls and the relationships

192

of the groups: a study based on new species from the Devonian of Australia. Zool. J. Linn. Soc. 61: 1–328.

Millot, J. & J. Anthony. 1958. Anatomie de *Latimeria chalumnae*. Vol. 1: Squelette, muscles et formations de soutien. C.N.R.S., Paris. 122 pp.

Nelson, G. 1969. Gill arches and the phylogeny of fishes, with notes on the classification of vertebrates. Bull. Amer. Mus. Nat. Hist. 141: 475–552.

Northcutt, R.G. 1987. Lungfish neural characters and their bearing on sarcopterygian phylogeny. J. Morph. Suppl. 1: 277–297.

Panchen, A.L. & T.R. Smithson. 1987. Character diagnosis, fossils and the origin of tetrapods. Biol. Rev. 62: 341–438.

Rosen, D.E., P.L. Forey, B.G. Gardiner & C. Patterson. 1981. Lungfishes, tetrapods, paleontology and plesiomorphy. Bull. Amer. Mus. Nat. Hist. 167: 159–276.

Schultze, H.-P. 1981. Hennig und der Ursprung der Tetrapoda. Paläont. Z. 55: 71–86.

Schultze, H.-P. 1987. Dipnoans as sarcopterygians. J. Morph. Suppl. 1: 39–74.

Schultze, H.-P. 1990. A comparison of controversial hypotheses on the origin of tetrapods. *In:* H.-P. Schultze & L. Trueb (ed.) Origins of Major Groups of Tetrapods, Controversies and Consensus, Cornell University Press, Ithaca. (in press).

Schultze, H.-P. 1991. A new long-headed dipnoan (Osteichthyes, Pisces) from the Late Devonian of Iowa, USA. Early Vertebrate Symp., Beijing. (in press).

Schultze, H.-P. & R. Cloutier. 1991. Computed Tomography and Magnetic Resonance Imaging studies of *Latimeria chalumnae*. Env. Biol. Fish. 32: 159–181. (this volume)

Smith, J.L.B. 1940. A living coelacanthid fish from South Africa. Trans. Roy. Soc. S. Afr. 28: 1–106.

Thomson, K.S. & K.S.W. Campbell. 1971. The structure and relationships of the primitive Devonian lungfish – *Dipnorhynchus sussmilchi* (Etheridge). Peabody Mus. Nat. Hist. Bull. 38: VI + 109 pp.

Vorobyeva, E. & H.-P. Schultze. 1990. Panderichthyid fishes and the origin of tetrapods. *In:* H.-P. Schultze & L. Trueb (ed.) Origins of Major Groups of Tetrapods, Controversies and Consensus, Cornell University Press, Ithaca. (in press).

Environmental Biology of Fishes **32**: 193–198, 1991.
© 1991 *Kluwer Academic Publishers.*

Enzymes of the coelacanth *Latimeria chalumnae* evidenced by starch gel electrophoresis

Ann L. Setter[1] & George W. Brown, Jr.[2]
[1] *Aquaculture, University of Idaho, Moscow, ID 83843, U.S.A.*
[2] *School of Fisheries, College of Ocean and Fishery Sciences, University of Washington, Seattle WA 98195, U.S.A.*

Received 21.7.1990   Accepted 21.7.1990

Key words: Primitive fishes, Comparative biochemistry, Enzymology, Evolution, Crossopterygii, Actinistia

Synopsis

The coelacanth, *Latimeria chalumnae*, is often referred to as a living relic. The opportunity to examine its biochemical molecular structure was sought in an effort to define the degree of its genetic variability. The coelacanth is thought to live only in a small area around the Comoro Islands in the Western Indian Ocean. The scenario presented suggests that the coelacanth may have lost genetic variability as a result of genetic drift within a small population. The narrow geographic range of the coelacanth suggests adjustment to a relatively limited environment. The loss of specific alleles through genetic drift can reduce the ability of a fish population to adapt to changes in environmental conditions. The coelacanth needs strong conservation measures to be taken to curtail the capture of specimens and for the protection of its limited natural habitat.

Introduction

Comparative biochemical studies provide information on the nature of physical, chemical, and physiological similarities and differences among extant organisms. In particular, a detailed study of the occurrence and nature of specific enzymes and other proteins in tissues of a large range of chordates can provide insight to the genetic basis and, hence, to the molecular events associated with the evolution from the lower to the higher forms within systematic groups. The continued acquisition of molecular evidence in conjunction with interpretations of findings from the fossil record should lead to an ever clearer picture of evolutionary mechanisms and trends. The epistemological foundation of this approach is succinctly described by Fisher & Whitt (1979).

In 1988, we were fortunate in obtaining tissues from two specimens of the coelacanth (a male and a female) returned to the United States in a frozen condition. We deemed it of great value to apply the electrophoretic technique to extracts of tissues of these two coelacanth specimens in order to document banding patterns which could be compared when more specimens become available. The assembly of a genetic profile from electrophoretic data would provide information on the relatively rare coelacanth itself as well as add to the pool of knowledge of fishes in general. The study would further document the occurrence of given enzymes and would indicate any genetic variants in this crossopterygian fish.

The first enzymes to be assayed in the coelacanth were two liver enzymes of the ornithine-urea cycle, arginase (EC 3.5.3.1) and ornithine carbamoyltransferase (EC 2.1.3.3). Brown & Brown (1967) first disclosed that coelacanth liver contained urea at a level of 1.7 percent (wet weight), comparable to that found in elasmobranchs such as sharks, skates, and rays (see also Pickford & Grant 1967). Since that time various enzymes of the coelacanth have been studied (Kolb & Harris 1972, Hamoir et al. 1973, Goldstein et al. 1973, Kolb et al. 1974, Solomon & Brown 1976, Ferris & Whitt 1978, Fisher & Whitt 1979, Rasmussen 1979, Webb & Brown 1980, Mommsen & Walsh 1989).

In the work described here, we further document the occurrence of various enzymes in tissues of the coelacanth, and report on the observed banding patterns.

Methods

Tissue samples

Tissue samples were obtained from Craig V. Sullivan who participated in the dissection of two frozen coelacanths (a male [VIMS 8117] and a female [VIMS 8118]) in January of 1988 at the Virginia Institute of Marine Science, Gloucester Point. At the University of Washington the samples were held in a freezer at $-80°$ C until required for use. For starch-gel electrophoresis approximately 0.5 g of each tissue was removed from storage and put into iced, individual test tubes with PTP (see Aebersold et al. 1987), a tissue extraction buffer. Test tubes containing tissue were stored at $-80°$ C until used. Advantage was taken of the freeze-thaw cycle for enzyme extraction (passive). Five tissues in common from each coelacanth specimen were studied – kidney, liver, muscle, pancreas, spleen. The urinary bladder of the male specimen was also examined. In the case of the female specimen, conus arteriosus, conus venosus, hemolyzed blood, and notochord were also examined.

Starch-gel electrophoresis

The organic chemicals used were for the most part obtained from Sigma Chemical Co. (St. Louis). Inorganic chemicals were of reagent or analytical grade. The enzymes studied (for Enzyme Commission (EC) Numbers see Florkin & Stotz 1965), enzyme name abbreviations, and tissues examined are given in Table 1.

Banding patterns observed conform to the documentation of Harris & Hopkinson (1977). Starch gels were routinely prepared the day before electrophoretic runs. Gel buffers and electrode buffers, respectively, were as follows: System 1. tris-citrate, pH 8.7, and lithium-borate, pH 8.5 (Ridgway et al. 1970), (E8.5) – pH of electrode modified; System 2. tris-borate, pH 8.7, and tris-borate, pH 8.7 (Brewer & Sing 1970), (TBE); System 3. citric acid, pH 6.5, pH 5.5, + NAD added to gel and to cathodal electrode tray; electrode buffer, citric acid, pH 6.5 or pH 5.5 (Clayton & Tretiak 1972); System 4. tris-citrate, pH 7.0, and tris-citrate, pH 7.0 (Shaw & Prasad 1970); System 5. tris-phosphate, pH 8.2, and tris-phosphate, pH 8.2 (Busack et al. 1979).

Gels were placed on ice packs for cooling before placing paper wicks against a cross-section cut in the starch slab. Test tubes containing tissue were removed from the freezer, the tissues were allowed to thaw, and tubes were then centrifuged at room temperature at 600G for 3 min. A paper wick was dipped into each test tube to absorb a portion of the protein slurry and was then placed across the cut face of the gel. Electric current (65 milliamps) from a Heathkit power supply was passed through the gel for 4–6 h. Marker dye was placed on several wicks so that migration of the proteins through the gel could be monitored.

Reagent solutions (recipes) used for stains and staining procedures were those for the standard electrophoretic methods described by Aebersold et al. (1987). After the electrophoretic run, the gels were sliced, and staining solutions were mixed with agar and applied to bring out the banding patterns specific for given enzymes. The banding patterns were recorded as genotypes, but allele frequencies were not calculated for coelacanth tissues since there was no apparent variation. Banding patterns were recorded according to their migration distance from the origin. While this is not common practice, we felt it was more descriptive than using the standard percent migration with respect to the common allele since all loci are monomorphic and as such would have been labeled 100. Photographs were made of the gels for later reference. Drawings were made of the photos and are displayed in Fig. 1 with the buffer and migration distance of the loci noted.

Results

In CK we saw activity in kidney and pancreas in addition to liver and muscle findings which were reported previously (Fisher & Whitt 1979, Hamoir et al. 1973). We found activity for AK in liver, pancreas, kidney and urinary bladder along with muscle, the latter as previously reported Fisher & Whitt (1979). Hamoir et al. (1973) found three

zones of activity for PGM in muscle. We saw activity for this enzyme in all tissues tested at two isoloci. Only two isoloci for GPI were detected, in contrast to four previously reported by Hamoir et al. (1973). Activity for GPI was seen largely in the male, although in kidney from the female, a pattern unlike that of the male was seen. This discrepancy needs to be resolved, but we have scored the observed genotype here as monomorphic.

For GAP, two isoloci were found. GAP-1 (Locus 1 is nearest to origin) at a migration distance of − 21 mm was apparent in pancreas, urinary bladder, kidney and hemolyzed blood. GAP-2 (Locus 2 (− 14 mm) was further from the origin) was seen only in spleen. Three isoloci were seen in esterase, with several tissues specific for only one of the loci.

Table 1. Enzymes tested for in coelacanth tissues.

EC No.	Name	Abbreviation	Results[1,2,3] (positive indicated)
Activity present			
1.1.1.8	glycerol-3-phosphate dehydrogenase	GPD	Mu
1.1.1.27	lactic dehydrogenase	LDH	Mu, CV, HB, Pa
1.1.1.37	malic dehydrogenase	MDH	Mu, UB, CV, HB, Pa
1.1.1.42	isocitric dehydrogenase	IDH	Ki, Li, Mu, Pa
1.1.1.12	glyceraldehyde-3-phosphate dehydrogenase	GAP	Pa, UB, Bl, Sp
1.15.1.1	superoxide dismutase	SOD	All but Mu
2.7.3.2	creatine kinase	CK	Ki, Mu, Pa
2.7.4.3	adenylate kinase	AK	UB, CA, HB, Ki, Li, Mu, Pa, Sp
3.1.1.1	esterase	EST	Ki, Li, Pa, Sp, UB, CA, CV, No
3.1.3.11	fructose diphosphatase (D-fructose-1, 6-diphosphate 1-phosphatase)	FDP	Mu, Ki, Pa
3.4.1.1	leucine aminopeptidase	LGG	All but No
5.3.1.8	mannose phosphate isomerase	MPI	All but CV
5.3.1.9	glucose phosphate isomerase	GPI	Ki, Li, Mu, Pa, Sp
4.1.2.13	aldolase	ALD	Hb, Ub, Mu, CA, CV
2.7.5.1	phosphoglucomutase	PGM	All
Activity not detected[4]			
1.1.1.44	phosphogluconate dehydrogenase	PGD	
1.1.1.40	malic enzyme	ME	
1.1.1.49	glucose-6-phosphate dehydrogenase	GD	
2.6.1.1	aspartic aminotransferase	AAT (GOT)	
4.2.1.3	aconitase	AH	

[1] Abbreviations used: CA = conus arteriosus, CV = conus venosus, HB = hemolyzed blood, Ki = kidney, Li = liver, Mu = muscle, No = notochord, Pa = pancreas, Sp = spleen, UB = urinary bladder.

[2] The following tissues were tested: in the male: kidney, liver, muscle, pancreas, spleen, urinary bladder; in the female: conus arteriosus, conus venosus, hemolyzed blood, kidney, liver, muscle, notochord, pancreas, spleen.

[3] The results are noted if either the male or female specimen showed activity; a couple of discrepancies between the specimens are noted in the text.

[4] See methods for tissues studied.

196

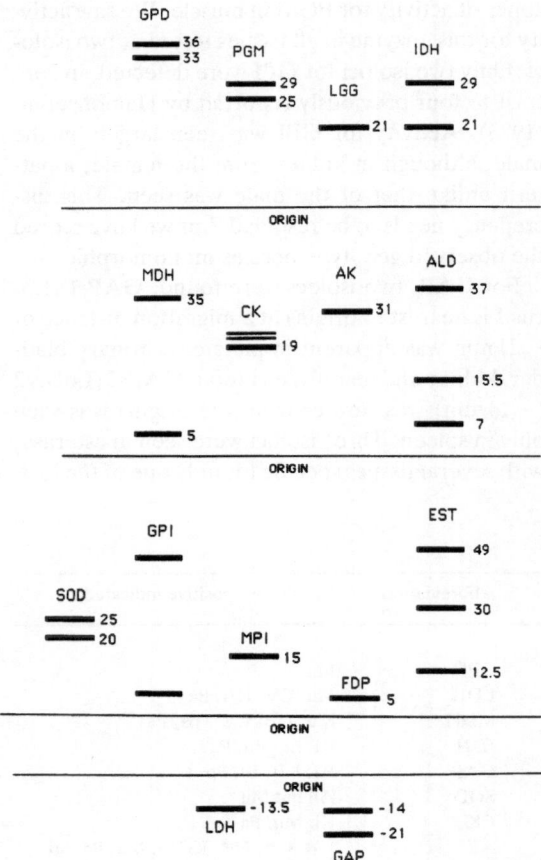

Fig. 1. Enzyme abbreviations given in Table 1 are noted above each of the drawings showing the structure and migration distance of the loci in millimeters. The loci are referred to in the text numerically (e.g. locus 1, locus 2, etc.) and were numbered consecutively with locus 1 always closest to the origin. The buffer systems used for the various enzymes were as follows: System 1 – LGG, GPI; System 2 – GPD; System 3 – CK, MDH, LDH, EST, AK, FDP, MPI, SOD, GAP; System 4 – IDH; System 5 – ALD.

All three were observed in pancreas, liver and kidney. Only EST-3 was seen in spleen (female), conus arteriosus and conus venosus, and urinary bladder. EST-1 by itself was only found in notochordal fluid. IDH shows two loci, the faster migrating band being largely specific to liver tissue, and the slower band being found in muscle and kidney. Pancreas faintly exhibits the liver locus (IDH-2) as well as the locus seen in muscle and kidney (IDH-1), with an intermediate shadow visible between the two loci. Activity for the peptidase (LGG) appeared as one strong locus in all tissues examined except notochord, and possibly with another faster locus specific to spleen and pancreas.

Two loci were observed for MDH. Only in muscle did the activity for MDH overlap between the two specimens for this system. A number of other tissues – conus venosus, hemolyzed blood, pancreas, and urinary bladder – showed activity for this system. One locus was noted for FDP in muscle, liver, and kidney. One isolocus was obtained for MPI from the following tissues: spleen, pancreas, liver, conus venosus, urinary bladder, muscle, kidney, hemolyzed blood. GPD exhibited two loci and was seen only in muscle. SOD also showed two loci but was seen in liver, spleen, pancreas, conus venosus, urinary bladder, kidney, and hemolyzed blood.

The coelacanth, exhibits an enzymic profile lacking in genetic variability as revealed by the two specimens examined here.

Discussion

Coelacanths exhibited a lack of molecular variation in the 14 enzymes demonstrated through electrophoresis. The coelacanth population, as evidenced by the two specimens examined, is shown to be very homogeneous. This is probably due to the seemingly small geographic area where the coelacanth is known to reside. From a fitness standpoint, multiple isoenzymatic forms may be advantageous in a variable environment (Bryant 1974). With no variability presently documented for the coelacanth, its capability for adaptation under changing environmental conditions may be restricted.

Continued adaptation among a limited number of individuals within a specific environment may over time cause a reduction in genetic variability either due to genetic drift or a population bottleneck. Such a situation is relevant to the lack of enzymic variations found in the coelacanth, which currently resides in a relatively limited environment. The long range result becomes predictable.

Because of limited gene flow, the coelacanth population will inevitably be constrained by the effects of inbreeding. The species has found a successful niche, yet it may be or become limited at some point in its ability to adjust to a broader environment if the lack of variability noted from these two individuals accurately describes the genetic structure of the population.

In fishes, horizontal starch gel electrophoresis is used largely as a tool for determining genetic variability between species or stocks, or both, thus providing a definitive genetic identity (Utter et al. 1987). This technique is most applicable for populations separated geographically, spatially, or temporally within an aquatic environment. Although only two specimens were available from the Comoro Islands for this analysis, the allelic structure was worth noting in light of past and anticipated future studies. Precise determinations of banding patterns are usually verified through an adequate sample size (greater than 50 individuals). Other researchers have noted that an accurate heterozygosity estimate can be assembled from a small number of individuals when a large number of loci are examined (Gorman & Renzi 1979). The data obtained here do not show any variation between the two individuals in tissues where the analysis overlapped.

There were shortfalls to this study which may have skewed the results. Some enzyme denaturation may be expected to have occurred during delivery and storage of the original tissues. Also, loss of enzyme activity may have occurred due to oxidation of sulfhydryl groups at active sites or elsewhere on the enzyme. Further studies should include an evaluation of the use of thiol reagents (e.g. mercaptoethanol) in the tissue extraction agent and in enzyme staining solutions.

Furthermore, Rasmussen (1979) reported that peak activity for LDH in assay solutions depended upon the concentration of added urea. Test studies should be done on this and other enzymes of the coelacanth to see what effects varying concentrations of urea may have on enzymes. The entire body space of the coelacanth is permeated by urea (DiJulio & Brown 1975); hence urea may have a profound effect on the aggregation of monomeric units and the activity of a number of enzymes.

The coelacanth lives in a relatively small, circumscribed oceanic area, the Comoro Islands, a geographic region limited by natural physical barriers. The long term scenario for a small population in which all individuals are genetically very similar and exist as an isolated group will be a population that will remain limited to their current narrow geographic range. The gradual sloughing off of genetic variability over generations is characteristic of a small population affected by genetic drift. The situation evidenced here for the coelacanth points to the need for strong conservation measures to curtail the capture of specimens and for the protection of its limited natural habitat.

Acknowledgements

We are indebted to John A. Musick of the Virginia Institute of Marine Science for extending the courtesy to one of our colleagues, Craig V. Sullivan, to participate in the dissection of the two coelacanth specimens and in making tissues available to us. We also thank the Society for the Protection Of Old Fishes for coordinating efforts to obtain tissues and for a transportation allowance for Sullivan. The School of Fisheries also provided additional travel support. School of Fisheries Contribution No. 747.

References cited

Aebersold, P.B., G.A. Winans, D.J., Teel, G.B., Milner & F.M. Utter. 1987. Manual for starch gel electrophoresis: a method for detection of genetic variation. NOAA Tech. Rept., National Marine Fisheries Service No. 61. 19 pp.

Brewer, G.J. & C.F. Sing. 1970. An introduction to isozyme techniques. Academic Press, New York. 186 pp.

Brown, G.W. Jr. & S.G. Brown. 1967. Urea and its formation in coelacanth liver. Science 155: 570–573.

Bryant, E.H. 1974. On the adaptive significance of enzyme polymorphisms in relation to environmental variability. Amer. Nat. 108 (959): 1–19.

Busack, C.A., R. Halliburton & G.A.E. Gall. 1979. Electrophoretic variation and differentiation in four strains of domes-

198

ticated rainbow trout (*Salmo gairdneri*). Can. J. Genet. Cytol. 21: 81–94.

Clayton, J.W. & D.N. Tretiak. 1972. Amine-citrate buffers for pH control in starch gel electrophoresis. J. Fish. Res. Board Can. 29: 1169–1172.

DiJulio, D.H. & G.W. Brown, Jr. 1975. Urea levels in tissues of the coelacanth, *Latimeria chalumnae*. Amer. Zool. 15: 801 (abstract).

Ferris, S.D. & G.S. Whitt. 1978. Genetic and molecular analysis of nonrandom dimer assembly of the creatine kinase isozymes of fishes. Biochem. Genet. 16: 811–829.

Fisher, S.E. & G.S. Whitt. 1979. Evolution of the creatine kinase isozyme system in the primitive vertebrates. pp. 142–159. *In:* J.E. McCosker & M.D. Lagios (ed.) Occasional Papers of the California Academy of Sciences 134, San Francisco.

Florkin, M. & E.H. Stotz (ed.) 1965. Enzyme nomenclature, Vol. 13, 2nd ed. Elsevier Publishing Co., Amsterdam. 219 pp.

Goldstein, L.S., S. Harley-DeWitt & R.P. Forster. 1973. Activities of ornithine-urea cycle enzymes and of trimethylamine oxidase in the coelacanth, *Latimeria chalumnae*. Comp. Biochem. Physiol. 44B: 357–362.

Gorman, G.C. & J. Renzi, Jr. 1979. Genetic distance and heterozygosity estimates in electrophoretic studies: effects of sample size. Copeia 1979: 242–249.

Hamoir, G., A. Pirout & C. Gerday. 1973. Muscle proteins of the coelacanth *Latimeria chalumnae* Smith. J. Mar. Biol. Assoc. U.K. 53: 763–784.

Harris, H. & D.A. Hopkinson. 1977. Handbook of enzyme electrophoresis in human genetics. Elsevier/North Holland Publishing Co., New York. (Unpaginated, with Supplement.)

Kolb, E. & J.I. Harris. 1972. Purification and properties of glycolytic enzymes from coelacanth (*Latimeria chalumnae*) muscle. Biochem. J. 130: 26.

Kolb, E., J.I. Harris & J. Bridgen. 1974. Triose phosphate isomerase from the coelacanth. Biochem. J. 137: 185–197.

Mommsen, T.P. & P.J. Walsh. 1989. Evolution of urea synthesis in vertebrates: the piscine connection. Science 243: 72–75.

Pickford, G.E. & F.B. Grant. 1967. Serum osmolality in the coelacanth, *Latimeria chalumnae*: urea retention and ion regulation. Science 155: 568–570.

Rasmussen, L. 1979. Some biochemical parameters in the coelacanth: ventricular and notochordal fluids. pp. 94–100. *In:* J.E. McCosker & M.D. Lagios (ed.) Occasional Papers of the California Academy of Sciences 134, San Francisco.

Ridgway, G.J., S.W. Sherburne & R.D. Lewis. 1970. Polymorphism in the esterases of Atlantic herring. Trans. Amer. Fish. Soc. 99: 147–151.

Shaw, C.R. & R. Prasad. 1970. Starch gel electrophoresis – a compilation of recipes. Biochem. Genet. 4: 297–320.

Solomon, F.P. & G.W. Brown, Jr. 1976. Enzymes of the coelacanth, *Latimeria chalumnae*. Northwest Section Meeting of the Society for Experimental Biology and Medicine, University of Oregon Medical School, Oct. 30, Portland, Oregon. (abstract).

Utter, F., P. Aebersold & G. Winans. 1987. Interpreting genetic variation detected by electrophoresis. pp. 21–46. *In:* N. Ryman & F. Utter (ed.) Population Genetics & Fishery Management, Washington Sea Grant Program, Seattle.

Webb, J.T. & G.W. Brown, Jr. 1980. Glutamine synthetase: assimilatory role in liver as related to urea retention in marine Chondrichthyes. Science 208: 293–295.

Environmental Biology of Fishes **32**: 199–218, 1991.

Guppies, toadfish, lungfish, coelacanths and frogs: a scenario for the evolution of urea retention in fishes

Robert W. Griffith
Department of Biology, Southeastern Massachusetts University, North Dartmouth, MA 02747, U.S.A.

Received 20.6.1989   Accepted 20.7.1990

Key words: Nitrogen metabolism, Vertebrate origins, Gnathostome evolution, Osmoregulatory strategies, Anadromous fishes, *Latimeria chalumnae*

Synopsis

The question of how (and why) the ureosmotic strategy, characteristic of *Latimeria chalumnae* and the chondrichthians evolved is addressed. There are three requirements for ureosmotic regulation: urea synthesis via the ornithine-urea cycle, urea tolerance involving biochemical and physiological adjustments, and urea retention that requires renal, branchial, metabolic and reproductive adaptations. Several examples of lower vertebrates in which urea plays a physiological role are considered to see whether they might provide insight into the origin of ureosmotic regulation. The guppy shows high urea synthesis and retention during embryonic development, and it is possible that a developmental role of urea is a general phenomenon in fishes. The toadfish, thought to be an enigma with high urea synthesis in the absence of an obvious physiological role of urea, is ureotelic under some conditions. Its urea excretion is likely related to renal function and/or parental care. In lungfish high ureogenesis is associated with estivation in periodically dry habitats. The resultant hyperuremia prevents ammonia toxicity, inhibits water loss and may repress metabolism. *Latimeria* is a classic marine ureosmotic regulator in which urea is used as an osmolyte that allows osmotic equilibrium with sea water while maintaining low ion levels. Adults of the frog, *Rana cancrivora*, are also ureosmotic regulators in brackish water. A scenario is proposed that suggests how ureosmotic regulation could have evolved in *Latimeria* and other fishes. The ornithine-urea cycle (composed of an arginine synthetic pathway and a second pathway that splits arginine into urea) occurred in fossil anadromous agnathans. Here the first pathway functioned in the ammocoete-like larvae for the generation of arginine to supplement a protein-deficient diet of algae, whereas the arginase pathway was important in the embryo for vitellin catabolism. Gnathostome evolution was associated with trends towards large eggs and prolonged development, requiring a complete ornithine-urea cycle for ammonia detoxification in embryos. Retention of a complete ornithine-urea cycle throughout adult life (via paedomorphosis) would preadapt any relatively large, sluggish, euryhaline fish for ureosmotic regulation when it was exposed to sea water. It is suggested that ureosmotic regulators evolved from freshwater or anadromous ancestors that entered the marine habitat. Once early ureosmotic regulators were established in the sea there would have been strong selection for internal fertilization and development, as is seen in *Latimeria* and many elasmobranchs. It is suggested that ureosmotic regulation was a common strategy in Paleozoic marine gnathostomes.

Introduction

Study of the relationships among nitrogen metabolism, habitat, reproduction and phylogeny in vertebrates has proven to be a most productive line of inquiry for physiologists. There is a clear relationship between the principal end product of nitrogen metabolism (ammonia vs. urea vs. uric acid) and the degree of adaptation of various vertebrate classes to terrestrial habitats. Most fish, larvae and permanently aquatic amphibians and certain aquatic reptiles excrete ammonia which requires considerable dilution with water because of its toxicity. Adult amphibians and mammals excrete urea which is less toxic but still entails much water loss because of its solubility and the difficulty of producing urine with high solute concentration. Most reptiles and birds (as well as insects and certain other terrestrial invertebrates) excrete uric acid which is of limited solubility and requires little urinary water loss.

For terrestrial animals there are also relationships among reproduction, development and nitrogen excretion. In reptiles and birds with a cleidoic egg, uric acid is favored as the nitrogenous end product during development because it can be stored in a crystalline form in the allantois and left behind at hatching. In placental mammals the high solubility of urea is an advantage because it can readily cross the placenta and be excreted through the mother's kidneys. In amphibians the aquatic larvae excrete ammonia, and metamorphosis to the terrestrial adult is accompanied by the induction of urea synthesis and ureotelism. A few desert-living anuran species are uricotelic as adults.

In fishes there are also connections among nitrogen metabolism, habitat, and phylogeny. In lungfishes (particularly *Protopterus* spp.) urea accumulates in the tissues during estivation when their habitat dries up. The build-up of urea prevents ammonia toxicity, inhibits desiccation and may play a role in depressing metabolism. An even more striking example of a connection between nitrogen metabolism and habitat is the use of urea as an osmolyte in certain marine fishes. In all marine elasmobranchs, holocephalans and the coelacanth, *Latimeria chalumnae*, urea is maintained at very high levels (around 400 mM) permitting them to be in approximate osmotic equilibrium with seawater while maintaining characteristic low vertebrate ion levels. The ureosmotic regulation of the sharks and their kin seemed so striking and unique that the eminent renal physiologist Homer Smith (1953) suggested, prior to the discovery of ureosmotic regulation in *Latimeria*, that the name of the elasmobranchs or chondrichthians should be modified to the uremichthians. More recently Løvtrup (1977) and Lagios (1979, 1982) appear to concur with Homer Smith by linking *Latimeria* with the chondrichthians in their cladistic phylogenies based, at least partly, upon osmoregulatory characters.

Much discussion has focused on the question of the homology of ureosmotic regulation in the chondrichthians and *Latimeria* (Pang et al. 1977, Griffith & Pang 1979, Griffith 1985, Løvtrup 1977, Compagno 1979, Lagios 1979, 1982, Wiley 1979, Forey 1988). While it is doubtful that definitive evidence for or against the homology of ureosmotic regulation in *Latimeria* and chondrichthians will ever arise, significant differences in the ability of the kidneys to reabsorb urea (Griffith et al. 1974), disparity in mechanisms of internal fertilization (a necessary concomitant of ureosmotic regulation in fishes [Griffith & Pang 1979]), and the occurrence of ureosmotic regulation in the frog *Rana cancrivora* (Gordon et al. 1961) suggest that the feature was surely independently evolved twice, probably thrice and that it is of questionable taxonomic value.

To a physiologist a more intriguing question than that of homology is how (and why) two particular groups of marine fishes, the chondrichthians and the coelacanths, evolved ureosmotic regulation, whereas other marine fishes, the actinopterygians and lampreys, evolved hypo-osmotic regulation. Both strategies are most reasonably explained as consequences of reinvasion of the marine habitat by freshwater (or diadromous) fishes (Morris 1972, Pang et al. 1977, Griffith 1985). In several publications (Griffith & Pang 1979, Griffith 1981, 1985) the question of choice of osmoregulatory strategy was addressed using arguments based on the physiology of contemporary ureosmotic fishes and logical extrapolations therefrom. It was concluded that ureosmotic regulation is advantageous to fishes that are (a) large, (b) sluggish and (c) possess internal fertilization and that (d) constancy of environmental salinity, and, obviously, the abilities to (e) synthesize and (f) tolerate urea are closely correlated with the occurrence of ureosmotic regulation. However, such an analysis and the listing of features that may now be associated with ureosmotic regulation is unlikely to satisfactorily explain those factors and conditions responsible for the initial evolution of urea retention in the early paleozoic ancestor(s) of the extant chondrichthians and *L. chalumnae*. This question may be addressed through another approach: the development of a scenario or model that attempts to explain the functional cause(s) behind the origin of urea retention in early fishes. Gans (1985) has discussed the utility of such scenarios which may be regarded as hypotheses, subject to refutation as new data accrues, and subject to competition with other scenarios that might be proposed. Such scenarios or models can prove very useful in allowing us to direct our attention to vital and productive questions. Thomson (1971) has previously discussed the question of the origins of urea synthesis and retention, but a wealth of new information on the role of urea in animals should make a reconsideration of the topic worthwhile.

In this paper I will (1) review those characteristics that might be regarded as prerequisites for the evolution of urea retention; (2) consider concrete examples or 'case studies' of living fishes that might serve as models for, or provide us with clues as to, the origin of this strategy and (3) develop a scenario or hypothesis that explains, step by step and with justification, the process by which ureosmotic regulation might have evolved in the ancestor(s) of living chondrichthians and *Latimeria chalumnae*.

Characteristics necessary for ureosmotic regulation

Urea synthesis

In animals urea can be derived either by the catabolism of purines, from the breakdown of dietary arginine, or it can be synthesized from ammonia (or glutamine) and carbon dioxide via the ornithine-urea cycle. If it is to be a significant osmolyte and/or the principal form by which nitrogenous waste is excreted, urea is mostly generated by the ornithine-urea cycle. This cycle requires the expenditure of metabolic energy (the equivalent of 4 ATP per urea) and it involves five enzymes: carbamoyl phosphate synthetase = CP Sase; ornithine carbamoyl transferase; arginosuccinate synthetase; arginosuccinate lyase, and arginase. The key initial enzyme CP Sase and other aspects of the ornithine-urea cycle differ among vertebrate classes. Lungfishes and tetrapods possess CP Sase I which can use only ammonia as substrate. Their arginase is located in the cytosol, and mitochondrial uptake favors ornithine rather than arginine. In all fishes (save the lungfish) where CP Sase has been demonstrated it reacts preferentially with glutamine rather than ammonia (CP Sase III), their arginase is mitochondrial and uptake by the mitochondria is specific for arginine (Mommsen & Walsh 1989). Hence, in most fishes urea is generated within the mitochondria, whereas in lungfish and tetrapods urea is formed extramitochondrially.

A complete ornithine-urea cycle has not been shown for most aquatic invertebrates nor for hagfish or lampreys (Read 1968, 1975, Mommsen & Walsh 1989), but it has been demonstrated in representatives of all gnathostome classes except birds, where it presumably has been secondarily lost. It also occurs in some terrestrial invertebrates (land planarians, earthworms and some molluscs). Although a complete cycle would seem to be primitive for gnathostomes, it is not always expressed. Enzymes of the ornithine-urea cycle may be subject to repression on both evolutionary and developmental bases. Although marine elasmobranchs have high urea cycle activity, stingrays of the family Potamotrygonidae have evolved low ureogenic activity, presumably in association with a strictly freshwater habitat and low blood urea levels (Goldstein & Forster 1971). Most teleost fishes have minimal or non-detectable ornithine-urea cycle activities but a few such as toadfish (*Opsanus* spp.) have remarkably high activity (Read 1971). The ornithine-urea cycle is induced during the metamorphosis of amphibians (Brown & Cohen 1960) but it is repressed during the later stages of embryonic development in some fishes (Depeche et al. 1979).

An important factor that will obviously influence urea synthesis is the availability of the substrates ammonia and/or glutamine plus carbon dioxide. Glutamine, the preferred substrate of the CP Sase III of most fishes, is synthesized from ammonia in the liver by an energy-requiring reaction involving the enzyme glutamine synthetase. This is found at high activity levels in ureosmotic fishes (Brown & Brown 1985). Ammonia availability for urea synthesis, either directly or through glutamine, is a complex issue involving rates of protein catabolism and branchial and renal ammonia exchange. Because epithelial ammonia transport can involve direct diffusion of free ammonia, the diffusion of ammonium ion or ionic exchange of ammonium for sodium, and since pH, salinity, blood flow and a host of other variables affect the operation of these exchange mechanisms (Evans & Cameron 1986), it is obvious that the actual rates of urea synthesis in animals may be highly variable, even if the urea synthetic capacity remains constant.

Urea tolerance

Although one normally thinks of it as being relatively harmless since it is a compound involved in ammonia detoxification, urea is a protein denaturant with deleterious effects on biochemical systems that must be offset or withstood if an animal builds up high concentrations in its tissues. Such adaptations can involve special modifications in structure that render proteins resistant to the effects of urea such as occur in the hemoglobin of elasmobranchs (Bonaventura et al. 1974), in the composition of the cartilage proteins of urea retaining fishes (Mathews 1962, 1967) and in some enzyme systems (Yancey & Somero 1978). Yancey & Somero (1979, 1980) have pointed out the important role that 'counteracting' solutes (e.g. trimethylamine oxide, taurine, and some amino acids that occur in a 1 : 2 ratio with urea in the cells of ureosmotic regulators) can play in ameliorating the destabilizing effects of urea on enzymes.

Although urea has many adverse effects upon biochemical systems, urea toxicity is not likely to be a very serious impedent to the gradual evolution of urea retention. In a simple experiment Griffith et al. (1979) exposed the teleost, *Fundulus heteroclitus,* to elevated environmental urea levels. In sea water most fish survived for over a week at environmental and tissue urea concentrations of around 400 mM and a few survived for short periods, when urea levels were gradually increased, at levels as high as 1000 mM. Incidentally, microbial breakdown of urea to ammonia represents one source of toxicity for teleost fishes exposed to high urea (Hugonenq & Florence 1921) and the presence of ureolytic bacteria in elasmobranchs (Knight et al. 1988) suggests that some ureosmotic fishes might, under certain circumstances, be subject to toxic levels of ammonia as an indirect consequence of their osmoregulatory strategy.

Urea retention

Although the synthesis of urea is an obvious prerequisite for ureosmotic regulation, the prevention of excessive urea loss is equally important. Most epithelia of animals are very permeable to urea which rapidly reaches equilibrium between extracellular fluid and tissues, readily diffuses across the respiratory and other surface epithelia of most aquatic animals, and freely passes across the glomerulus to be lost in the urine. Presumably urea also passes across the gut mucosa to the lumen where it can be excreted or broken down by microorganisms. Because urea synthesis requires energy, it is obvious that mechanisms that limit the epithelial, renal and gut loss of urea will be highly advantageous to ureosmotic fishes.

In those species of fish studied (including marine elasmobranchs, freshwater teleosts and lampreys) most urea loss occurs through the gills rather than via the kidneys (Boylan 1967, Payan et al. 1973, Chan & Wong 1977, Vellas & Serfaty 1967, Read 1968). However, in euryhaline elasmobranchs exposed to dilute media, renal urea loss is elevated above that of branchial loss and consequently plasma urea levels drop. Although marine elasmobranchs lose most urea via the gills and their water permeability is quite high, the actual permeability to urea is still, by necessity, very low relative to that of other animals (Payan et al. 1973).

In part, lowered branchial urea loss of ureosmotic fishes would seem to entail an adaptive change in membrane permeability (probably involving relatively thick blood/water barriers and characteristic 'inclusions' [Hughes 1976]), but it also seems to involve a relatively low respiratory surface area. Elasmobranchs and *Latimeria chalumnae* possess extremely low gill surface area to weight ratios relative to comparably sized teleosts (Hughes & Morgan 1973). Presumably, these branchial adaptations to resist urea loss are associated with a relatively low metabolic rate as has been shown for some sharks (Brett & Blackburn 1978, Carey et al. 1982). Because of the familiar relationship between body size and surface/volume ratio and because urea synthesis is volume dependent whereas loss is surface area dependent, ureosmotic regulation is more economical to large than to small fishes. Furthermore, developing embryos face serious problems of urea balance (Griffith & Pang 1979). Consequently, extant ureosmotic fishes possess internal fertilization and undergo development within the mother or within 'cleidoic' egg cases.

Renal urea loss is virtually eliminated in elasmobranchs by tubular urea reabsorption associated with sodium transport and a countercurrent exchange system (Schmidt-Nielsen et al. 1972, Lacy et al. 1985). Tubular urea reabsorption is apparently lacking in *Latimeria chalumnae* and in *Rana cancrivora* (Griffith et al. 1976, Schmidt-Nielsen & Lee 1962), where its absence may be compensated through low glomerular filtration rates or by activity of the urinary bladder.

The possibility of loss of urea through the gut in ureosmotic fishes has been little explored. Analyses of the gut fluid of elasmobranchs suggest that urea readily diffuses into the lumen (Smith 1931), but the degree to which this contributes to overall urea loss has not been investigated. In many mammals, particularly herbivores, urea that diffuses into the lumen of the gut serves as a nitrogen source for microbial protein synthesis. This protein synthesis is of obvious nutritional benefit to the host. Mommsen & Walsh (1989) have suggested that this might also occur in the toadfishes, *Opsanus* spp.

Some case studies

The guppy

Depeche et al. (1979) studied the urea content and urea synthetic capability (measured by the incorporation of radio-labelled bicarbonate into urea) of live-bearing guppies, *Poecilia reticulata*, during various developmental stages. They compared animals adapted to fresh water with those in saline, and they also measured urea content and synthesis in an oviparous teleost, *Oncorhynchus mykiss*, for comparison. A summary of their data is presented in Table 1.

In both fresh and sea water, the urea content of guppy embryos increases during embryonic development until the later stages, when it declines toward relatively low adult levels. The urea concentrations in young embryos (21 mM in freshwater and 47 mM in saline) are, to our knowledge, the highest recorded in teleosts, other than those exposed artificially to elevated urea levels. The activity of the ornithine-urea cycle is clearly correlated with the urea concentrations of embryos; ureogenesis is low in oocytes, peaks during the early gestational stages and declines during the later embryonic stages. Although urea levels in the rainbow trout are considerably lower than in *P. reticulata* and the data on urea synthesis may not be directly comparable quantitatively, it is apparent that rainbow trout also demonstrated an increase in urea concentration and ureogenesis during development that is followed by a decline towards adult levels at hatching. A five-fold increase in non-protein nitrogen during the embryonic development of *O. mykiss* was previously reported by Suyama & Ogino (1958). In both guppies and trout, environmental salinity had only minor effects upon urea synthesis and concentration.

It would seem from Depeche et al. (1979) that urea synthesis is important in the embryonic development of teleosts, particularly those that undergo internal development. This may well be associated with a restricted opportunity to exchange ammonia with the environment, combined with high protein catabolism of ovovitellin. The study of Depeche et al. (1979) should be replicated on other teleost and non-teleost fishes to see how widespread the phenomenon is distributed phylogenetically, and to assess to what degree internal development, egg size and other factors influence its expression.

The toadfish

The toadfishes, *Opsanus tau* and *O. beta,* represent an enigma in regard to nitrogen metabolism. These fishes have extraordinarily high activities of ornithine-urea cycle enzymes in their livers, comparable to or higher than those of ureosmotic elasmobranchs (Read 1971, Mommsen & Walsh 1989). However, *Opsanus* neither builds up urea in its blood plasma (Marshall & Grafflin 1932, Griffith 1981) nor has it been found to excrete significant amounts of urea (Marshall & Grafflin 1932, Read 1971, Mommsen & Walsh 1989). As one possible explanation for the curious possession by toadfish of high urea synthetic capacity with no demonstrated role of urea in osmoregulation or nitrogen excretion, Mommsen & Walsh (1989) suggested that urea might be transported across the gut to serve as a nitrogen source for symbiotic microbial protein synthesis, much like the way nitrogen is cycled in ruminant mammals. Since toadfish are fully carnivorous fish with an ample supply of dietary protein for their own protein synthetic needs,

Table 1. Ontogenic changes in urea concentration and synthesis in guppies and rainbow trout (after Depeche et al. 1979).

Species	Stage or tissue	Salinity	Urea content[1]	Urea synthesis[2]
Trout	Adult muscle	FW	1.4	bk
	Adult liver	FW	–	bk
	Oocytes	FW	3.1	bk
	Ovulated oocytes	FW	5.2	bk
	Stage 7	FW	3.4	bk
	Stage 16	FW	5.2	340
	Stage 25	FW	6.6	85
	Stage 32 (larvae)	FW	3.5	16
	Stage 7	dilute SW	7.1	131
	Stage 16	dilute SW	5.1	540
	Stage 25	dilute SW	6.8	125
	Stage 32 (larvae)	dilute SW	3.4	bk
Guppy	Vitellogenic female	FW	5.6	2850 (liver)
	Small oocytes	FW	8.3	4.0
	Large oocytes	FW	30.1	6.0
	Young embryo	FW	20.7	21.8
	Late embryo	FW	5.6	21.6
	Vitellogenic female	saline	6.7	–
	Small oocytes	saline	8.9	–
	Large oocytes	saline	36.2	–
	Young embryo	saline	46.6	42.3
	Late embryo	saline	10.9	28.7

[1] Values are in mM kg^{-1} H$_2$O.

[2] All values for trout are in pmol g^{-1} h^{-1}; values for guppy tissues are pmol g^{-1} h^{-1}; values for guppy embryos are in pmol per 100 eggs per hour. bk = background.

Mommsen & Walsh's (1989) suggestion is less than convincing.

In some unpublished experiments on nitrogen excretion in *Opsanus tau,* I believe I have found part of the answer to the toadfish enigma. *Opsanus* is ureotelic, at least under some circumstances. Previous studies of nitrogen excretion in toadfish were short term (Read [1971] studied for 2–6 h; Mommsen & Walsh [1989] for 1 h or less) and were conducted in sea water. I also observed in seawater tests (of 24–48 h duration) that most nitrogenous waste excreted is in the form of ammonia (87%). Only one of ten fish excreted a substantial proportion of its nitrogenous waste as urea (40%). However, in five fish transferred to dilute (20%) seawater and followed for a number of days a high proportion (on average over 50%) of the nitrogen excreted is in the form of urea. The pattern seen in individual fish is of a relatively consistent excretion rate of ammonia coupled with irregular 'bursts' of urea excretion occurring every few days. This pattern is logically attributable to the fish sequestering urea in their urinary bladder and micturating at infrequent intervals.

If we accept that *Opsanus* retains urea in the bladder and excretes it irregularly, we must question why they bother to do so when it would be much easier and metabolically cheaper to simply excrete ammonia across the gills. Two possibilities occur to me; one related to the toadfish's reproductive biology and the other to its osmoregulatory physiology. These two functional explanations are not mutually exclusive.

The reproductive explanation suggests that urea excretion might be related to parental care. Toadfish reproduce by laying eggs under stones, in tin cans and in similar hiding places with the male parent guarding the nest until the embryos hatch (Bigelow & Schroeder 1953). These nests may even occur in the intertidal zone. Were the male toadfish to remain with his young in a nest of very restricted size and limited water exchange and were he to excrete only ammonia, it could build up to toxic levels and poison the young. Conversion of ammonia to urea and its retention in the urinary bladder would obviate these problems.

The second possible functional role of urea is as an aid to water balance in the aglomerular toadfish. *Opsanus* is euryhaline, surviving well for months even in markedly hypo-osmotic 5% seawater (Lahlou et al. 1969) and tolerating fresh water for a number of days. Although the toadfish is relatively impermeable to water and ions it still requires some means of water excretion in dilute media which cannot, of course, occur via filtration since toadfish lack glomeruli. Bieter (1931) demonstrated that urea (together with magnesium and sulfate salts but not 'conventional' diuretics, NaCl or glucose), is a diuretic in *Opsanus*. It is reasonable to suggest that the toadfish actively secretes urea via the kidney tubules and/or urinary bladder and that this secretion induces osmotic water flow into the tubule or bladder, particularly in dilute media. Active secretion of urea has been documented in the kidneys of several vertebrates including amphibians (Forster 1954) and hagfish (Read 1975).

The lungfish

The dipnoans are a classic case study demonstrating the relationship between water availability and nitrogen metabolism. Of the three genera of extant lungfishes, the African *Protopterus* is most readily subject to environmental desiccation and regularly estivates in cocoons, the South American *Lepidosiren* also estivates but its burrows tend not to dry up completely, and the Australian *Neoceratodus* is fully aquatic (Carlinski & Barrio 1972, Funkhouser et al. 1972). While estivating, *Protopterus* and *Lepidosiren* build up large quantities of urea in the tissues that circumvent ammonia toxicity and may contribute to adaptive metabolic repression (Smith 1930, Janssens 1964, Carlinsky & Barrio 1972). Accordingly, the capacity for urea synthesis via the ornithine-urea cycle is highest in *Proptopterus*, essentially similar in *Lepidosiren* and quite low in *Neoceratodus* (Janssens & Cohen 1968, Goldstein et al. 1967, Funkhouser et al. 1972, Carlinski & Barrio 1972). Unlike other fishes, the African lungfish, *Protopterus,* possesses carbamoyl phosphate synthetase I and cytosolic arginase (Mommsen & Walsh 1989). This is probably also true for *Lepidosiren* (Carlinski & Barrio 1972). It would be of interest to know whether the aquatic Australian lungfish, *Neoceratodus,* possesses CPS I like other lungfishes or CPS III like other aquatic fishes.

The pattern of aquatic nitrogen excretion in lungfishes generally follows the pattern of urea synthetic capacity. *Neoceratodus* excretes less than 10% of its amine nitrogen as urea (Goldstein et al. 1967), *Lepidosiren* excretes somewhere around 30% of its nitrogen as urea (Funkhouser et al. 1972, Carlinski & Barrio 1972) and *Protopterus,* depending on temperature, also excretes 30% urea or more (Smith 1930, Janssens & Cohen 1968). However, in reviewing the literature it becomes obvious that a great deal of variability in the aquatic excretion of urea and ammonia occurs in lungfishes, some of which is related to environmental factors and some of which is apparently due to high individual variation. It should also be pointed out that the number of specimens documented in the published literature is small for all three genera.

Homer Smith's (1930) classic studies on *Protopterus* demonstrated that temperature has a pronounced effect on aquatic nitrogen excretion. For one specimen the total nitrogen excretion at 30°C was almost double that at 20°C and the urea contribution to total nitrogen excretion was 60% at 30°C but only 30% at 20°C. Carlinski & Barrio (1972) found that feeding can affect the form in which nitrogen is excreted. *Lepidosiren* fed a diet of snails eliminated 40% of their nitrogen as urea whereas starved fish excreted only 21% as urea. In unpublished experiments on *Lepidosiren* and *Protopterus* (R.W. Griffith, P.K.T. Pang, K.S. Thomson & W. Sawyer) we observed the following: (a) there is an inverse relationship between size and total nitrogen excretion (in mgN kg^{-1}h^{-1}) in both *Lepidosiren* and *Protopterus* but there is no apparent correlation between size and the percent nitrogen excreted as urea; (b) low temperature depresses overall nitrogen metabolism in *Protopterus*; (c) lowering the oxygen tension by bubbling nitrogen into the chamber does not influence urea or ammonia excretion in *Protopterus*; (d) exposure to hypertonic media (40% seawater) depresses ammonia and urea excretion in both *Protopterus* and *Lepidosiren* and (e) following return to fresh water from saline media there is a marked but transient increase in urea (but not ammonia) excretion.

Although experimental variables have some influence on nitrogen excretion in lungfish, a more notable observation is the great variability that occurs between individuals of the same species in the ratio of urea to ammonia excreted under identical experimental conditions. Carlinski & Barrio (1972) obtained percentages of urea nitrogen excreted that ranged from 0% to 92% in unfed *Lepidosiren.* While testing two large *Protopterus aethiopicus* of virtually identical history and measured under identical conditions, we found that one specimen was consistently ureotelic (68% urea) while the other was consistently ammonotelic (11% urea). It would seem that whether lungfish are ammonotelic or ureotelic while in water is largely an irrelevant physiological question. The important issue is whether they are capable of converting ammonia to urea during estivation.

Certain teleost fish, such as mudskippers and clingfishes are, like lungfishes, exposed to aerial habitats. In those species tested modest increases in tissue or fluid urea levels accompany emersion (Gordon et al. 1969, Marusik et al. 1981).

Latimeria *and other ureosmotic fishes*

When Pickford & Grant (1967) reported that the coelacanth, *Latimeria chalumnae,* builds up high levels of urea in its body fluids as do the elasmobranchs, it caused zoologists to revise their ideas on the uniqueness of ureosmotic regulation, to question conventional schemes of vertebrate phylogeny and to reassess the notion that the coelacanth would prove to be a useful model for the evolution of tetrapods. What is probably not widely known is that the manuscript for their paper in Science was nearly rejected for publication because one of the anonymous reviewers objected to publishing a paper based on a single individual, lacking measures of variability between coelacanth specimens. Only when the companion paper by Brown & Brown (1967) on ureogenesis in *L. chalumnae* was submitted, was Pickford & Grant's paper considered acceptable. With new data on levels of urea and other solutes in blood and other fluids from the frozen-thawed VIMS specimens combined with published data on other frozen-thawed and living coelacanths (Pickford & Grant 1967, Lutz & Robertson 1971, Rasmussen 1979, Cole 1968, 1973, Griffith et al. 1974, 1975) it is now possible to provide those elusive standard errors to satisfy that overzealous reviewer (Table 2). Data from Griffith (1981) on an elasmobranch, the blue shark *Prionace glauca,* are also presented for comparison.

Although the blood chemistry of *L. chalumnae* appears to be superficially very similar to that of elasmobranchs, there are meaningful differences if one relies upon data obtained from the living specimen studied by Griffith et al. (1974) rather than frozen-thawed specimens which are subject to fluid and solute shifts between body compartments and post-mortem changes. Although urea and trimethylamine oxide levels are similar to those of sharks, *L. chalumnae* has lower monovalent ion levels and is somewhat hypo-osmotic to seawater in contrast to the hyperosmotic elasmobranchs. Consequently, *L. chalumnae* is in negative water balance and

might be expected to exhibit lower renal filtration than elasmobranchs.

The mechanisms of urea synthesis in *L. chalumnae* are, as far as we know, identical to those in sharks and their kin. *L. chalumnae* possesses activities of ornithine-urea cycle enzymes comparable to those of elasmobranchs (Brown & Brown 1967, Goldstein et al. 1973), and their carbamoyl phosphate synthetase is CPS III like that of elasmobranchs and most other fishes (Mommsen & Walsh 1989).

Whether *L. chalumnae* possesses molecular and physiological adaptations like those of sharks to deal with the deleterious biochemical effects of hyperuremia has not yet been fully explored. *Latimeria* has muscle levels of trimethylamine oxide that are around 300 mM (Griffith 1981). Since it has muscle urea levels of approximately 400 mM, the ratio of trimethylamine oxide to urea is even higher than the 1 : 2 value characteristic of elasmobranchs. This suggests that the coelacanth uses the counteracting solute strategy like sharks and their kin (Yancey & Somero 1979, 1980). *L. chalumnae* cartilage protein possesses a high sulfate content, like that of sharks, which is probably an adaptation to

Table 2. Composition of the blood and other body fluids of *Latimeria chalumnae* and the blue shark.

Parameter	Fluid	VIMS specimens	Live specimen[a]	All specimens[b]	Blue shark[c]
Osmolarity	Blood	1031[d]	931	1021 ± 52	1036 ± 24
(mOsm kg⁻¹)	Notochordal	1477[e]	1058	1268	
	aqueous humor	779[d], 911[e]	937	895 ± 40	
Chloride	Blood	165	187	163 ± 29	256 ± 4
(mM)	Notochordal	297	204	250 ± 27	
	aqueous humor	165, 355	194	238 ± 59	
Urea	Blood	365	377	346 ± 20	391 ± 9
(mM)	Notochordal	495	420	442 ± 27	
	aqueous humor	287, 134	367	363 ± 47	
Trimethyl-	Blood	108	122	113 ± 5	99 ± 6
amine oxide	Notochordal	117	154	143	
(mM)	aqueous humor	43, 52	–	46	

[a] Based on the data of Griffith et al. (1974).

[b] Based on data from VIMS specimen(s) and all frozen-thawed and live specimens recorded in Griffith (1980). Data are means plus standard errors when N is 3 or greater.

[c] From Griffith (1981). Data are means plus standard errors.

[d] Larger female specimen in relatively good condition. Other parameters as for osmolarity.

[e] Smaller male specimen with obvious freezer burn.

resist urea-induced disassociation (Mathews 1966, 1967). The hemoglobin of *Latimeria* is urea resistant (Mangum 1991) as in elasmobranchs (Bonaventura et al. 1974) but whether any of the coelacanth's enzymes show urea resistance as do some of those of elasmobranchs (Yancey & Somero 1978) remains a question of considerable interest.

Branchial loss of urea in *L. chalumnae* appears to be restricted by a combination of low respiratory surface area, a relatively thick branchial blood-water barrier and perhaps by the existence of 'inclusions' underneath the outermost epithelial layer. Such inclusions also occur in the ureosmotic elasmobranchs and in lungfish (Hughes 1976). These adaptations to restrict branchial urea loss presumably also decrease the potential for gill oxygen transport and should be associated with low metabolic rate. Although based on few measurements and species, published values for the metabolic rate of elasmobranchs are relatively low (Brett & Blackburn 1978, Carey et al. 1982) and metabolism is believed to be very low in *L. chalumnae* (Balon et al. 1988). *Latimeria's* renal urea loss is likely limited because of a low glomerular filtration rate associated with the coelacanth's hyposmotic state (Griffith et al. 1976). Unlike elasmobranchs and holocephalans, the coelacanth seems to be unable to reabsorb urea in its kidneys (Griffith et al. 1976), presumably because it lacks the structural and physiological mechanisms required (Schmidt-Nielsen et al. 1972, Lacy et al. 1985). The problem of excessive urea loss during development has been circumvented in *Latimeria* by internal fertilization and development (Smith et al. 1975). Chondrichthians, of course, all have internal fertilization and the embryos either develop within the mother or have protective egg cases (Wourms 1977).

L. chalumnae is not exposed to dilute media in its marine habitat (Balon et al. 1988) but many elasmobranchs are euryhaline and occur in dilute media, including freshwater (Pang et al. 1977). The physiological mechanisms involved in adaptation to dilute media in elasmobranchs generally include a substantial increase in glomerular filtration rate that leads to enhanced renal urea excretion and lowered blood urea levels (Goldstein et al. 1968, Payan et al. 1973, Chan & Wong 1977). The elevated glomerular filtration serves to simultaneously excrete excess water and to lower the fish's osmotic gradient. Some euryhaline elasmobranchs also appear to show adaptive changes in hepatic urea synthesis (Payan et al. 1973). However, certain stenohaline elasmobranchs do not adjust urea loss in response to dilution of the media (Haywood 1974). The only extant group of strictly freshwater elasmobranchs, stingrays of the family Potamotrygonidae, have evolved low blood urea levels, urea synthesis occurs at about 5% the rate of marine elasmobranchs, and the kidneys do not reabsorb urea (Thorson et al. 1967, Griffith et al. 1973, Goldstein & Forster 1971).

The crab-eating frog

The crab-eating (or mangrove) frog, *Rana cancrivora*, is a ureosmotic regulator like *Latimeria chalumnae* and the chondrichthians, maintaining high blood urea levels (up to 350 mM) that bring it into approximate osmotic equilibrium with its brackish or seawater habitat (Gordon et al. 1961). Although *R. cancrivora's* hepatic ureogenic system has not yet been fully characterized, other ranid frogs have high ornithine-urea cycle enzymes (Brown & Cohen 1960) with CP Sase I and cytosolic arginase (Mommsen & Walsh 1989). One may presume this is also true of *R. cancrivora*. Ureogenesis is induced by saline acclimation in the crab-eating frog (Balinski et al. 1972). Like the coelacanth and unlike chondrichthians, the crab-eating frog does not reabsorb urea in its renal tubules (Schmidt-Nielsen & Lee 1962), although there may be some urea reabsorption by the urinary bladder. Reduced filtration accounts for most of the reduction of urinary urea loss. The crab-eating frog obviously does not face a problem of branchial urea loss since it breathes with lungs. The problem of excessive urea loss during the early developmental stages where gill respiration is used is 'solved' in *R. cancrivora* by the presence of hypo-osmotic regulation in the tadpoles (Gordon & Tucker 1965). Urea synthesis is not induced until metamorphosis as, of course, is typical of anurans generally. Biochemical mechanisms of dealing with hyperuremia and the possible utilization of the counteracting solute strategy have not yet been investigated in *R. cancrivora*.

Other anuran amphibians, including *Xenopus laevis, Bufo viridis* and *Scaphiopus couchi,* build up urea in their tissues in response to hypertonic media or in association with estivation in a desiccating environment (Gordon 1965, McClanahan 1967, McBean & Goldstein 1970). This adaptive response, however, does not occur in all Amphibia (Schoffeniels & Tercefs 1966). In general, the elevated urea levels in these species, as in *R. cancrivora*, are a consequence of decreased renal filtration and enhanced urea synthesis. Urea retention has also been reported in saline-adapted terrapin, *Malaclemys centrata* (Gilles-Baillien 1973). The phenomenon of hyperuremia in response to increased salinity appears to be widely distributed phylogenetically in vertebrates.

A scenario for the evolution of ureosmotic regulation

What the scenario should explain

Scenarios are hypotheses that attempt to explain the historical process of evolution using functional, ecological and developmental data and arguments. As should be true of other hypotheses, a good scenario should be consonant with all relevant biological and paleontological data, it should be logical and parsimonious and it should be as specific as possible to make it subject to testing and refutation. A scenario that attempts to explain the evolution of ureosmotic regulation in fishes should deal with the following questions: (1) when, how and why did the vertebrate ornithine-urea cycle originate; (2) what was the original role(s) of urea in the ancestor(s) of ureosmotic regulators; (3) if the original role(s) were other than as an osmolyte, how and why did the ureosmotic regulators subsequently come to use urea for osmoregulation. We also should consider whether ureosmotic regulation is likely to be a phenomenon that evolved a number of times in fishes and we might speculate on how widespread the phenomenon was in the fauna of the Paleozoic seas.

There obviously are no direct data on ureosmotic regulation or other possible uses of urea in fossil fish, so that our arguments must be based on the utilization of urea by living fishes and upon logical extrapolations from these extant fishes back to their fossil ancestors (and perhaps to other related fossil fishes). Since most of the relevant vertebrate evolution occurred in the Paleozoic (principally before and during the Devonian, by which time the ureogenic elasmobranchs, coelacanths and the other major vertebrate groups had diverged), most of our attention will be focused on conditions and evolution during that era. Although no physiological data exist, considerable information on the morphology and habitat of Paleozoic fishes is available (see, for example Moy-Thomas & Miles 1971, Thomson 1969, 1971, Campbell & Barwick 1986), that will provide some guidelines for reconstructing the evolution of nitrogen metabolism.

Briefly, my scenario will suggest that: (1) a functional ornithine-urea cycle was absent in early agnathan fishes but all component enzymes were present, albeit at different developmental stages; (2) a complete ornithine-urea cycle evolved in early gnathostomes as a means of detoxifying ammonia during early embryogenesis but the cycle was repressed later in development and was not functional in adults; and (3) the pedogenic retention of the ureogenic pathway by adults that invaded the marine habitat explains the origin of urea retention in the ancestor(s) of the extant ureosmotic fishes. This scenario is amplified and defended below.

Was the vertebrate ornithine-urea cycle primitive, did it evolve in agnathans, or did it originate in gnathostomes?

The ornithine-urea cycle is apparently based upon the joining of a common nutritional pathway for arginine biosynthesis that occurs even in microorganisms with an arginine splitting system useful for gluconeogenic protein catabolism (Campbell 1972). The initial enzyme in the synthetic pathway, CP Sase is also used in purine synthesis, although vertebrates use a distinct form of the enzyme (CP Sase II) for this function. Thus, the ornithine-urea cycle does not involve truly novel enzyme systems, only a novel use of pre-existing ones. However, the co-occurrence of an arginine synthetic system (of utility to organisms deficient in dietary protein) with an arginase system (of obvious adaptive value only to organisms with a dietary excess of the amino acid arginine) would seem to be a very unlikely circumstance. Only when the end product of these two sets of reactions, urea, has some specific and important function would one normally expect to find a complete ornithine-urea cycle.

In invertebrate animals, only land planarians, earthworms and a variety of terrestrial or amphibious molluscs have been shown to possess complete ornithine-urea cycles (Campbell 1972). In these animals urea functions in ammonia detoxification, as an aid to shell deposition in snails, and as an osmolyte during desiccation. A complete cycle is not known to occur in any marine or freshwater invertebrate, nor does it occur in terrestrial insects. The most parsimonious explanation for its occurrence in several distinct phyla of terrestrial invertebrates and in vertebrates (presumably all independently evolved from aquatic invertebrate ancestors) is that a complete functional cycle was independently acquired in these groups. However, the genetic capacity to synthesize all component enzymes was widely distributed in animals including the ancestral lines of vertebrates, molluscs, annelids and flatworms.

The living hagfish, *Bdellostoma cirrhatum* lacks a complete ornithine-urea cycle and individual component enzymes are below detectable levels (Read 1975). A complete cycle is also absent in another living agnathan, the lamprey *Entosphenus tridentatus,* but this species possesses measurable levels of two key enzymes, CP Sase and arginase (Read 1968). Superficially, the most parsimonious interpretation would be that a complete ornithine-urea cycle evolved in the early ancestor of the gnathostomes since living agnathans lack it and since it occurs in representatives of all gnathostome classes (except birds which have secondarily lost it). However, the occurrence of measurable activities of components of both arginine synthetic and arginine splitting pathways in the lamprey suggest that this simple explanation may not be valid. It

would be of particular interest to study the occurrence of enzymes of the ornithine-urea cycle in developmental stages of lampreys (and, indeed, hagfish if they ever are available). I believe it more than coincidental that these two key enzymes co-occur in lampreys which have a nutrition based largely on protein (vitellin) during early embryonic development, switch to a protein-poor diet of algae during their ammocoete larval stages, and then may revert to a protein-rich diet as parasitic or non-feeding adults. If the hypothesis by Griffith (1985, 1987) that the protovertebrate was anadromous with a life style not unlike that of the lamprey, except for the specialized parasitic adult period, is valid then all components of the ornithine-urea cycle would have functional roles in the protovertebrate, albeit during different life-history intervals. A testable corollary of my hypothesis is that the arginine synthetic steps of the ornithine-urea cycle should occur at substantial levels in ammocoete larva and they might be repressed at metamorphosis.

Hence, in my scenario the enzymes of the ornithine-urea cycle were present in the protovertebrate and early agnathans with arginase playing a nutritional role in the embryo and arginine synthesis being nutritionally important in the ammocoete-like larva. These animals were neither ureosmotic nor ureotelic and their capacity to synthesize urea at any given life-history interval was doubtful.

Predation, development and urea synthesis in early gnathostomes

Most scenarios for the origin of jaws in early gnathostomes logically assume some shift in feeding mode from microphagy or deposit feeding in the agnathan ancestors to predation (Thomson 1971, Mallatt 1984, 1985). The evolution of an efficient predatory habitus with effective jaws, major modifications in the branchial apparatus and effective sensory systems would have placed certain ecological and developmental constraints on the earliest gnathostome fishes. The problems are of the relative ineffectiveness and vulnerability of small predators in many habitats, and the time and energy reserves it takes for the embryological development of the requisite morphological adaptations for predation. In predatory vertebrates there are three principal strategies for dealing with these problems: (1) a dramatic metamorphosis from larval non-selective microphagy to adult predation (as in Amphibia and lampreys); (2) producing large numbers of relatively small young in plankton-rich habitats that begin feeding as microphagous predators and modify the size of their prey as they grow (as in many teleosts); and (3) producing small numbers of embryos that are very well supplied with energy reserves from the mother so that they can be effective macrophagous predators when they become independent (as in elasmobranchs, the coelacanth, dipnoans, many actinopterygians and all amniotes). I believe that the very early evolution of many of the gnathostome lines involved channelization into the third strategy and this may well have been the pathway by which the original gnathostome evolved from agnathan ancestors. Certainly, protracted embryonic development is the most widely distributed strategy among non-specialized gnathostomes, and it seems to me that the gradual prolongation of embryonic dependence provides a more plausible scenario for the evolution of jawed fishes from agnathans than do alternatives based on the evolution of precocious predation or dramatic metamorphosis.

Assuming that our view of early gnathostome evolution is correct, a long embryonic development would have some interesting implications on nitrogen metabolism. Presumably the principal source of energy for the embryo was the protein vitellin, as it is in most oviparous vertebrates. We may assume that the eggs were benthic and that lipid reserves would necessarily have been limited for reasons of buoyancy. The developing embryo, needing energy as well as building blocks for its own protein, would produce excess amine nitrogen from protein catabolism. Approximately 50% of the protein reserves of teleost eggs are catabolized rather than incorporated into embryonic protein (Love 1970). Most fish embryos and larvae seem to have limited membrane permeability to water and ions (Tytler & Blaxter 1988), and branchial and renal transport systems are not operative in early

development (Holliday 1969). Consequently, the problem of ammonia toxicity would certainly arise. The arginine synthetic system that was nutritionally functional in agnathan larvae might then be subverted to generate arginine for urea synthesis. The data of Depeche et al. (1979) who found a buildup of urea levels and urea synthetic capacity in guppies and trout embryos provide substantiation for our hypothesis that urea synthesis plays a significant role in the embryonic development of gnathostome fishes. The extension of the study of Depeche et al. (1979) on the ontogenic development of urea levels and ureogenetic capacity to other fishes of diverse phylogeny and developmental type would, of course, serve as a test of our hypothesis.

Hence, according to my scenario many of the early gnathostomes developed the capacity to generate urea and used this capacity to avoid ammonia toxicity during protracted embryonic development. They showed a moderate build up of urea in their tissues during development that later abated when renal and branchial ammonia excretion systems were fully functional.

Did water deprivation or aerial respiration play some role in the origins of ureosmotic regulation in marine fish?

The lepidosirenid lungfishes, as well as some anuran amphibia, have high ornithine-urea cycle enzymes and build up urea in their tissues when they are subjected to desiccation (as when the lungfish's temporary ponds and rivers dry up and they estivate). Terrestrial life and air breathing in certain invertebrates, Amphibia and some amniotes are associated with high rates of urea synthesis and ureotely. Thomson (1971) has suggested that the existence of accessory lung breathing in fishes might have favored the shift from ammonotely towards ureotely early in gnathostome evolution. We should consider whether these phenomena might have any bearing on the origin of ureosmotic regulation in marine fishes.

Neither the extant marine ureosmotic regulating fishes, *Latimeria chalumnae* and the chondrichthians, nor their close fossil relatives show any structural or physiological features suggestive of past association with periodically dry habitats. Although the adaptive response of urea build-up in response to desiccation that is seen in *Protopterus* spp. might be regarded as a useful preadaptation for ureosmotic regulation in the marine environment and we cannot exclude the possibility that it played some role in fossil marine lungfishes, it is improbable that this was of significance to ancestors of the extant ureosmotic regulators.

A terrestrial evolutionary history with ureotelism associated with lung breathing and water conservation certainly were important in preadapting *Rana cancrivora* for ureosmotic regulation in a marine habitat. However, the chondrichthians lack any vestiges of lungs or a swimbladder, and the coelacanths either have ossified 'lungs' or a fat-filled one as in *L. chalumnae*. Furthermore, air-breathing actinoptergians such as *Polypterus*, *Periophthalmus* and *Amia* do not have elevated ornithine-urea cycles (Brown & Brown 1985, Gregory 1977). Although the suggestion by Thomson (1971) that lung breathing was related to nitrogen metabolism in early gnathostomes cannot be totally rejected, it is not a parsimonious general explanation for ureotely or urea retention in aquatic fishes.

I believe it is relevant that the key initial enzyme of the ornithine-urea cycle differs between the dipnoans plus the tetrapods (CPS I) and other fishes (CP Sase III). It would seem that the original enzyme was CP Sase III which we have suggested plays a role during early development in fishes and can be used by adult ureosmotic fishes for osmoregulation. CP Sase I would then be a novel acquisition by the evolutionary line that led to the dipnoans and tetrapods, functioning in adults for ammonia detoxification during water deprivation. It would be of interest whether glutamine-dependent CP Sase III plays any role during the development of dipnoans and tetrapods and it would also be of interest whether CP Sase I shows sequence homology with CP Sase III or is otherwise derived.

Could urea retention have originated in a strictly marine ancestor?

We have not yet fully addressed the question of the habitat in which urea synthesis and urea retention evolved. In an earlier scenario Griffith (1987) proposed that the origin of vertebrates involved anadromy. This scenario provided a coherent framework that accounted for the evolution of vertebrate osmoregulatory features, bone, and the process of cephalization. In the current paper we have seen how an anadromous scenario also provides a plausible explanation for the co-occurrence of arginine biosynthesis and arginase activity in the early jawless vertebrates. I also think that it is feasible that anadromy was maintained from the protovertebrate through the direct line to the earliest gnathostome. Certainly most of the phylogenetic groups of early gnathostomes contained marine, freshwater and euryhaline species during the Devonian (Moy-Thomas & Miles 1971, Thomson 1969, Campbell & Barwick 1986) and anadromy would provide the predatory adult with sufficient large prey in the marine habitat, before the 'explosion' of freshwater invertebrates in the Devonian, while maintaining the developing young in a safe freshwater habitat. Anadromy is also consistent with my scenario for the developmental role of urea. Regardless of whether the early gnathostome ancestor of the elasmobranchs and bony fish were anadromous or not, however, it is almost certain that it was associated with fresh water for reasons discussed below.

Scenarios could, of course, be proposed that have ureosmotic regulation evolving in a strictly marine ancestor without freshwater involvement. One possibility is that ureosmotic regulation could evolve from the ionosmotic (non-)regulatory strategy such as is found in the hagfish and most marine invertebrates which maintain blood ions at seawater levels. Such scenarios are implausible. They require the development of specialized morphological and physiological adaptations to regulate ions (chloride cells, branchial ion pumps, ion-impermeable epithelia etc.) plus adaptations to synthesize, retain and tolerate urea. Furthermore, such scenarios require the incipient ureosmotic regulator to expend a great deal of metabolic energy for urea synthesis and ion transport for 'benefits' of questionable value. The only specific benefits of regulation of ion levels at low levels that have been documented in the literature are that certain enzymes function best at relatively low concentrations of ions. However, these enzymes occur intracellularly and function in a relatively low ion environment even in non-regulating invertebrates. Consequently, most osmoregulatory physiologists believe that ion regulation mechanisms, found in both ureosmotic regulation and hypo-osmotic regulators, are derived from an animal that lived in fresh water where low ion levels are of obvious adaptive benefit (Smith 1953, Morris 1972, Pang et al. 1977, Griffith 1985).

A second possible scenario could have ureosmotic regulation evolving from a marine hypo-osmotic regulator like the teleosts. There is no compelling a priori physiological reason why such a scenario is improbable. In fact, believing there was a good possibility that certain sluggish marine teleosts in a constant marine habitat (e.g. deep sea teleosts) might have evolved ureosmotic regulation, I surveyed 59 species of marine fish in search of ureosmotic teleosts (Griffith 1981). I have concluded (probably a number of years after most other physiologists) that ureosmotic marine teleosts don't exist. Presumably, once locked into the hypo-osmotic strategy, ureosmotic regulation is no longer an option. It might be of interest that the highest teleost blood urea level that I observed in that survey was not in a marine species but in the anadromous white perch, *Morone americanus*. While in brackish water this species had blood urea levels of 8.4 mM in comparison with 1.2 mM in fresh water (Griffith unpublished).

The fresh water to sea water transition

Perhaps the easiest part of developing my scenario is explaining how a fish that is large, sluggish, derives energy from protein catabolism and possesses a full complement of ornithine-urea cycle enzymes can become a ureosmotic regulator. As long as the fish is capable of maintaining blood sodium and chloride levels by active salt secretion (an absolute

necessity for euryhalinity), simply exposing the fish to hypertonic media will do the trick. Transfer from fresh water to high salinity puts the animal in negative water balance, lowering renal nitrogen excretion. Consequently, enhanced urea synthesis occurs and blood and tissue urea levels buildup. This occurs in euryhaline elasmobranchs when transferred from fresh to sea water, it occurs in euryhaline Amphibia and my preliminary data suggest it also occurs in the African lungfish, *Protopterus* spp. It almost certainly would also have occurred in suitably large and metabolically sluggish paleozoic gnathostomes that had retained substantial ureogenic capacity via the ornithine-urea cycle enzymes from their embryonic development rather than repressing the cycle prior to adulthood. Ureosmotic regulation would be a natural consequence of the paedomorphic retention of high ornithine-urea cycle enzymes into the adult marine phase of early anadromous gnathostomes.

Following the initial development of ureosmotic regulation in the adult of an anadromous gnathostome, there would be strong selective pressure to eliminate the freshwater phase, if possible, by the evolution of internal fertilization and development. Adaptations for internal fertilization such as the claspers of chondrichthians or erectile caruncules of coelacanths would have been required, of course. The switch to internal development would simultaneously eliminate the risks associated with anadromous migrations and provide an opportunity for the young to be larger and more effective predators when they first became independent. It also would have made these fish fully-qualified marine ureosmotic regulators like *Latimeria chalumnae*, the elasmobranchs and the holocephalans are today.

Did ureosmotic regulation evolve more than once in paleozoic gnathostomes – is it homologous in Latimeria chalumnae *and chondrichthians?*

I am afraid that creating scenarios provides us with no more insight into the question of the independent vs. common origin of ureosmotic regulation in the chondrichthians and *L. chalumnae* than do conventional approaches (i.e. Griffith & Pang 1979). However, our scenario implies that many early gnathostomes were preadapted for ureosmotic regulation by the possession of protracted development, urea synthesis and retention during development and, possibly, anadromy. Whether the common ancestor of the chondrichthians and *Latimeria* was anadromous with a marine phase is moot, although the occurrence of a gut-associated ion secreting gland in both lines and in *Protopterus* (Lagios & McCosker 1977) is suggestive that this might have been the case. Similarly, whether the ancestor's marine phase, if it existed, was a ureosmotic regulator is also moot. While it is likely that the chondrichthians and *L. chalumnae* achieved fully independent marine ureosmotic status with internal development independently, since the mechanisms for internal fertilization are distinct, the question of whether their common ancestor might have had a ureosmotic marine adult phase is, I think, unanswerable at present.

How widespread was ureosmotic regulation in Paleozoic fishes?

It might be amusing to speculate on how common the ureosmotic strategy was among Paleozoic marine fishes of different phylogenetic lines. I believe that it is doubtful that most fossil agnathan fishes could have been ureosmotic. The living agnathans are neither ureosmotic nor ureogenic and if the fossil agnathans had protracted ammocoete-like larval stages (either in fresh or sea water) it is most unlikely that they would have been either. Only those agnathan lines that developed internal fertilization and development in association with the marine habitat (if any existed) would be viable candidates for ureosmotic regulation.

The earliest gnathostomes to show up in the fossil record were the acanthodians. We normally picture these fishes as relatively small, active predators – occupying a niche like that of pelagic teleosts today (Moy-Thomas & Miles 1971). Such fish, being active and small, are unlikely to have been ureosmotic regulators. I view the acanthodians as an early side branch off the gnathostome line that channelized into the teleost-like strategy of precocious predation. This may well also hold for early actinopterygians.

I believe that ureosmotic regulation was the usual mode of adapting to sea water for most of the remaining groups of gnathostome fishes, all of which contain representatives from marine habitats during the early Paleozoic. Placoderms have been reported to possess claspers (Miles 1967), evidence of internal fertilization and, one presumes, ureosmotic regulation in at least some members of this group. The living lungfishes build up urea in response to desiccation and to saline challenges and while there is no evidence of internal fertilization in dipnoans we must remember that internal development is not an absolute prerequisite of ureosmotic regulation, particularly for anadromous animals. The descendents of the rhipidistians include amphibians which are ureotelic as adults, one of which is a marine ureosmotic regulator. Like lungfishes, there is no evidence of internal fertilization in rhipidistians but one would expect euryhaline species to be ureosmotic regulators while in sea water. The living coelacanth is a ureosmotic regulator and we expect this was the usual state in other marine coelacanths, some of which are known to have had internal development (Watson 1927). The Carboniferous coelacanth *Rhabdoderma* which is known to be oviparous (Schultze 1972) has been reported from both fresh and seawater habitats (Thomson 1969), was likely anadromous and, we suspect, would have been a ureosmotic regulator while in the sea. Since the living marine elasmobranchs and holocephalans are all ureosmotic regulators and since both lines are predominantly marine it is parsimonious to assume that this was the case with all fossil marine chondrichthians also. Strictly freshwater elasmobranchs, like the pleurocanths, would have possessed low urea levels as do the extant freshwater potamotrygonid stingrays.

General conclusions and some suggestions for further study

Ureosmotic regulation is a derived physiological

feature found in extant marine chondrichthians, *Latimeria chalumnae* and in *Rana cancrivora*. Although it is possible that the feature is synapotypic in the coelacanth and chondrichthians, the capacity to synthesize urea appears to be primitive in gnathostomes and urea retention is a natural consequence of exposure to saline media or desiccation in ureogenic lower vertebrates. A reasonable scenario can be proposed for the origin of ureosmotic regulation that includes the following points: (a) all component enzymes of the ornithine-urea cycle occurred in agnathan ancestors of the gnathostomes but arginine synthetic and arginase pathways functioned during different stages of embryonic development; (b) the early evolution of predatory gnathostomes involved trends towards protracted embryonic development which, in turn, favored the occurrence of a functional ornithine-urea cycle in embryos to detoxify ammonia produced by the catabolism of vitellin; and (c) the paedomorphic retention of a complete ornithine-urea cycle into adulthood preadapted suitably large sluggish fishes for ureosmotic regulation whenever they entered the marine habitat.

Several corollaries of the above scenario are testable by physiological and biochemical studies on living fishes. It would be expected during lamprey development that the relative activity of arginase would be highest prior to hatching, but that this enzyme would be repressed and arginine synthetic pathways (including CP Sase) would be enhanced during the ammocoete period. It would also be anticipated that extension of the study of Depeche et al. (1979) to a broad variety of teleost and nonteleost fishes would demonstrate that the role of urea synthesis in ammonia detoxification is a general phenomenon in the early development of fishes. Furthermore, urea synthetic capacity and embryonic urea levels should be directly correlated with egg size and with internal development. I should finally like to point out that perhaps the most promising area of future research in the field of vertebrate nitrogen metabolism centers around the question of how the genes coding for the vertebrate ornithine-urea cycle enzymes are expressed or repressed during development. I believe it likely that relatively small modifications in the epigenetic

control of ureogenic gene expression have led to truly profound evolutionary changes in the physiology of vertebrates. Understanding this epigenetic control and how it varies could provide us with considerable insight into the evolutionary processes by which many of the major groups of vertebrates, including the gnathostomes, tetrapods, amniotes and mammals originated.

Acknowledgements

I should like to thank Jack Musick (VIMS) and the Explorer's Club of New York for access to fluids from two frozen *Latimeria chalumnae* specimens which provided some of the data for this paper. Facilities and support for unpublished studies on nitrogen excretion in toadfish and lungfish that are cited here were provided by Keith Thomson (then of Yale University) and Wilbur Sawyer and Peter Pang (Columbia University).

References cited

Balinski, J.B., S.E. Dicker & A.B. Elliot. 1972. The effect of long-term adaptation to different levels of salinity on urea synthesis and tissue amino acid concentrations in *Rana cancrivora*. Comp. Biochem. Physiol. 43B: 71–82.

Balon, E.K., M.N. Bruton & H. Fricke. 1988. A fiftieth anniversary reflection on the living coelacanth, *Latimeria chalumnae:* some new interpretations of its natural history and conservation status. Env. Biol. Fish. 23: 241–280.

Bieter, R.N. 1931. The action of some diuretics upon the aglomerular kidney. J. Pharmacol. Exp. Therap. 43: 399.

Bigelow, H.B. & W.C. Schroeder, 1953. Fishes of the Gulf of Maine. Fishery Bull. U.S. Fish Wildlife Ser. Vol. 53. 577 pp.

Bonaventura, J., C. Bonaventura & B. Sullivan. 1974. Urea tolerance as a molecular adaptation of elasmobranch hemoglobins. Science 186: 57–59.

Boylan, J.W. 1967. Gill permeability in *Squalus acanthias*. pp. 197–206. *In:* P.W. Gilbert, R.F. Mathewson & D.P. Rall (ed.) Sharks, Skates and Rays, Johns Hopkins Press, Baltimore.

Brett, J.R. & J.M. Blackburn. 1978. Metabolic rate and energy expenditure in the spiny dogfish, *Squalus acanthias*. J. Fish. Res. Board Can. 35: 816–821.

Brown, G.W. & S.G. Brown. 1967. Urea and its formation in coelacanth liver. Science 155: 570–573.

Brown, G.W. & S.G. Brown. 1985. On urea formation in primitive fishes. pp. 321–337. *In:* R.E. Foreman, A. Gorbman,

216

J.M. Dodd & R. Olsson (ed.) Evolutionary Biology of Primitive Fishes, Plenum Press, New York.

Brown, G.W. & P.P. Cohen. 1960. Comparative biochemistry of urea synthesis. 3. Activities of urea-cycle enzymes in various higher and lower vertebrates. Biochem. J. 75: 82–91.

Campbell, J.W. 1972. Nitrogen excretion. pp. 279–316. In: C.L. Prosser (ed.) Comparative Animal Physiology, 3rd. Ed., W.B. Saunders, Philadelphia.

Campbell, K.S.W. & R.E. Barwick. 1986. Paleozoic lungfishes: a review. J. Morphol. Suppl. 1: 93–131.

Carey, F.G., J.W. Kanwisher, O. Brazier, G. Gabrielson, J.C. Casey & H.L. Pratt. 1982. Temperature and activities of a white shark, Carcharodon carcharias. Copeia 1982: 254–260.

Carlinski, N.J. & A. Barrio 1972. Nitrogen metabolism of the South American lungfish, Lepidosiren paradoxa. Comp. Biochem. Physiol. 41B: 857–873.

Chan, D.K.O. & T.M. Wong. 1977. Physiological adjustments to dilution of the external medium in the lip-shark Hemiscyllium plagiosum (Bennett), II. Branchial, renal and rectal gland function. J. Exp. Zool. 200: 85–96.

Cole, D.F. 1968. Anterior chamber of the coelacanth eye. Br. J. Ophthal. 52: 415–418.

Cole, D.F. 1973. Intraocular fluid composition in the coelacanth, Latimeria chalumnae. Exp. Eye Res. 16: 389–395.

Compagno, L.J.V. 1979. Coelacanths: shark relatives or bony fishes? Occ. Pap. Calif. Acad. Sci. 134: 45–52.

Depeche, J., R. Gilles, S. Daufresne & H. Chaipello. 1979. Urea content and urea production via the ornithine-urea cycle during the ontogenic development of two teleost fishes. Comp. Biochem. Physiol. 63A: 51–56.

Evans, D.H. & J.N. Cameron. 1986. Gill ammonia transport. J. Exp. Zool. 239: 17–23.

Forey, P. 1988. Golden jubilee for the coelacanth Latimeria chalumnae. Nature 336: 727–732.

Forster, R.P. 1954. Active cellular transport of urea by frog renal tubules. Amer. J. Physiol. 179: 372–377.

Funkhouser, D., L. Goldstein & R.P. Forster. 1972. Urea biosynthesis in the South American lungfish, Lepidosiren paradoxa: relation to its ecology. Comp. Biochem. Physiol. 41A: 439–443.

Gans, C. 1985. Scenarios: why? pp. 1–9. In: R.E. Foreman, A. Gorbman, J.M. Dodd & R. Olsson (ed.) Evolutionary Biology of Primitive Fishes, Plenum Press, New York.

Gilles-Baillien, M. 1973. Isosmotic regulation in various tissues of the diamondback terrapin, Malaclemys centrata (Latreille). J. Exp. Biol. 59: 39–43.

Goldstein, L. & R.P. Forster. 1971. Urea biosynthesis and excretion in freshwater and marine elasmobranchs. Comp. Biochem. Physiol. 39B: 415–421.

Goldstein, L., P.A. Janssens & R.P. Forster. 1967. Lungfish Neoceratodus forsteri: activities of ornithine-urea cycle and enzymes. Science 157: 316–317.

Goldstein, L., W.W. Oppelt & T.H. Maren. 1968. Osmotic regulation and urea metabolism in the lemon shark Negaprion brevirostris. Amer. J. Physiol. 215: 1493–1497.

Goldstein, L., S. Harley-DeWitt & R.P. Forster. 1973. Activ-

ities of ornithine-urea cycle enzymes and trimethylamine oxidase in the coelacanth, Latimeria chalumnae. Comp. Biochem. Physiol. 44B: 357–362.

Gordon, M.S. 1965. Intracellular osmoregulation in skeletal muscle during salinity adaptation in two species of toads. Biol. Bull. (Woods Hole) 128: 218–229.

Gordon, M.S., K. Schmidt-Nielsen & H.M. Kelly. 1961. Osmotic regulation in the crab-eating frog (Rana cancrivora). J. Exp. Biol. 38: 659–678.

Gordon, M.S., I. Boetius, D.H. Evans, R. McCarthy & L.C. Oglesby. 1969. Aspects of the physiology of terrestrial life in amphibious fishes. I. The mudskipper, Periophthalmus sobrinus. J. Exp. Biol. 50: 141–149.

Gordon, M.S. & V.E. Tucker. 1965. Osmotic regulation in tadpoles of the crab-eating frog (Rana cancrivora). J. Exp. Biol. 42: 437–445.

Gregory, R.B. 1977. Synthesis and total excretion of waste nitrogen by fish of the Periophthalmus (mudskipper) and Scartelsaas families. Comp. Biochem. Physiol. 57A: 33–36.

Griffith, R.W. 1980. Chemistry of the body fluids of the coelacanth, Latimeria chalumnae. Proc. Roy. Soc. (Lond.) B. 208: 329–347.

Griffith, R.W. 1981. Compositon of the blood serum of deep-sea fishes. Biol. Bull. (Woods Hole) 160: 250–264.

Griffith, R.W. 1985. Habitat, phylogeny and the evolution of osmoregulatory strategies in primitive fishes. pp. 69–80. In: R.E. Foreman, A. Gorbman, J.M. Dodd & R. Olsson (ed.) Evolutionary Biology of Primitive Fishes, Plenum Press, New York.

Griffith, R.W. 1987. Freshwater or marine origin of the vertebrates? Comp. Biochem. Physiol. 87A: 523–531.

Griffith, R.W., M.B. Mathews, B.L. Umminger, B.F. Grant, P.K.T. Pang, K.S. Thomson & G.E. Pickford. 1975. Composition of fluid from the notochordal canal of the coelacanth, Latimeria chalumnae. J. Exp. Zool. 192: 165–172.

Griffith, R.W. & P.K.T. Pang. 1979. Mechanisms of osmoregulation in the coelacanth: evolutionary implications. Occ. Pap. Calif. Acad. Sci. 134: 79–93.

Griffith, R.W., P.K.T. Pang & L.A. Benedetto. 1979. Urea tolerance in the killifish, Fundulus heteroclitus. Comp. Biochem. Physiol. 62A: 327–330.

Griffith, R.W., P.K.T. Pang, A.K. Srivastava & G.E. Pickford. 1973. Serum composition of freshwater stingrays (Potamotrygonidae) adapted to fresh and dilute sea water. Biol. Bull. (Woods Hole) 144: 304–320.

Griffith, R.W., B.L. Umminger, B.F. Grant, P.K.T. Pang, L. Goldstein & G.E. Pickford. 1976. Composition of bladder urine of the coelacanth, Latimeria chalumnae. J. Exp. Zool. 196: 371–380.

Griffith, R.W., B.L. Umminger, B.F. Grant, P.K.T. Pang & G.E. Pickford. 1974. Serum composition of the coelacanth, Latimeria chalumnae Smith. J. Exp. Zool. 187: 87–102.

Haywood, G.P. 1974. The exchangeable ionic space, and salinity effects upon ion, water and urea turnover rates in the dogfish, Poroderma africanum. Mar. Biol. 29: 267–276.

Holliday, F.G.T. 1969. The effects of salinity on the eggs and larvae of teleosts. pp. 293–311. *In:* W.S. Hoar & D.J. Randall (ed.) Fish Physiology, Vol. 1, Academic Press, New York.

Hughes, G.M. 1976. On the respiration of *Latimeria chalumnae*. Zool. J. Linn. Soc. 59: 195–208.

Hughes, G.M. & M. Morgan. 1973. The structure of fish gills in relation to their respiratory function. Biol. Rev. 48: 419–475.

Hugonenq, L. & G. Florence. 1921. Experience du cours se reapportant a l'azotemia. Bull. Soc. Chim. Biol. 3: 174–175.

Janssens, P.A. 1964. The metabolism of the aestivating African lungfish. Comp. Biochem. Physiol. 11: 105–117.

Janssens, P.A. & P.P. Cohen, 1968. Nitrogen metabolism in the African lungfish. Comp. Biochem. Physiol. 24: 879–886.

Knight, I.T., D.J. Grimes & R.R. Colwell. 1988. Bacterial hydrolysis of urea in the tissues of carcharhinid sharks. Can. J. Fish. Aq. Sci. 45: 357–360.

Lacy, E.R., E. Reale, D. Schlusselberg, W.K, Smith & D.J. Woodward. 1985. A renal countercurrent system in marine elasmobranch fish: a computer assisted reconstruction. Science 227: 1351–1354.

Lagios, M.D. 1979. The coelacanth and Chondrichthyes as sister groups: a review of shared apomorph characters and a cladistic analysis and reinterpretation. Occ. Pap. Calif. Acad. Sci. 134: 25–44.

Lagios, M.D. 1982. *Latimeria* and the Chondrichthyes as sister groups: a rebuttal to recent attempts at refutation. Copeia 1982: 942–948.

Lagios, M.D. & J.E. McCosker. 1977. A cloacal excretory gland in the lungfish *Protopterus*. Copeia 1977: 176–178.

Lahlou, B., I.W. Henderson & W.H. Sawyer. 1969. Renal adaptations by *Opsanus tau*, a euryhaline aglomerular teleost to dilute media. Amer. J. Physiol. 216: 1266–1272.

Love, R.M. 1970. The chemical biology of fishes. Academic Press, New York. 547 pp.

Løvtrup, S. 1977. The phylogeny of vertebrata. John Wiley and Sons, London. 330 pp.

Lutz, P.L. & J.D. Robertson. 1971. Osmotic constituents of the coelacanth, *Latimeria chalumnae* Smith. Biol. Bull. (Woods Hole) 141: 553–560.

Mallatt, J. 1984. Early vertebrate evolution: pharyngeal structure and the origin of gnathostomes. J. Zool. (Lond.) 204: 169–183.

Mallatt, J. 1985. Reconstructing the life cycle and the feeding of ancestral vertebrates. pp. 59–68. *In:* R.E. Foreman, A. Gorbman, J.M. Dodd & R. Olsson (ed.) Evolutionary Biology of Primitive Fishes, Plenum Press, New York.

Marshall, E.K. & A.L. Grafflin. 1932. The function of the proximal convoluted segment of the renal tubule. J. Cell. Comp. Physiol. 1: 161–176.

Marusik, E.T., F. Balbontin, S.M. Galli-Gallardo, M. Garreton, P.K.T. Pang & R.W. Griffith. 1981. Osmotic adaptations of the Chilean clingfish, *Sicyases sanguineus*, during emersion. Comp. Biochem. Physiol. 68A: 123–126.

Mathews, M.B. 1962. Sodium chondroitin sulfate-protein complexes of cartilage, III. Preparations from shark. Biochim. Biophys. Acta 58: 92–101.

Mathews, M.B. 1966. The molecular evolution of cartilage. Clin. Orthop. 48: 267–283.

Mathews, M.B. 1967. Macromolecular evolution of connective tissue. Biol. Rev. 42: 499–551.

McBean, R.L. & L. Goldstein. 1970. Renal function during osmotic stress in the aquatic toad, *Xenopus laevis*. Amer. J. Physiol. 219: 1115–1123.

McClanahan, L. 1967. Adaptations of the spadefoot toad, *Scaphiopus couchi*, to desert environments. Comp. Biochem. Physiol. 20: 73–99.

Miles, R.S. 1967. Observations on the ptyctodont fish, *Rhamphodopsis* Watson. J. Linn. Soc. Lond. (Zool.) 47: 99–120.

Mommsen, T.P. & P.J. Walsh. 1989. Evolution of urea synthesis in vertebrates: the piscine connection. Science 243: 72–75.

Morris, R. 1972. Osmoregulation. pp. 193–239. *In:* M.W. Hardisty & I.C. Potter (ed.) The Biology of Lampreys, Vol. 2, Academic Press, London.

Moy-Thomas, J.A. & R.S. Miles. 1971. Paleozoic fishes. 2nd Ed., W.B. Saunders, Philadelphia. 259 pp.

Pang, P.K.T., R.W. Griffith & J.W. Atz. 1977. Osmoregulation in elasmobranchs. Amer. Zool. 17: 365–377.

Payan, P., L. Goldstein & R.P. Forster. 1973. Gills and kidney in ureosmotic regulation in euryhaline skates. Amer. J. Physiol. 224: 367–372.

Pickford, G.E & F.B. Grant. 1967. Serum osmolarity in the coelacanth, *Latimeria chalumnae:* urea retention and ion regulation. Science 155: 568–570.

Rasmussen, L.E. 1979. Some biochemical parameters in the coelacanth, *Latimeria chalumnae*, ventricular and notochordal fluid. Occ. Pap. Calif. Acad. Sci. 134: 94–110.

Read, L.J. 1968. A study of ammonia and urea production and excretion in the freshwater-adapted form of the Pacific lamprey, *Entosphenus tridentatus*. Comp. Biochem. Physiol. 26: 455–466.

Read, L.J. 1971. The presence of high ornithine-urea cycle enzyme activity in the teleost, *Opsanus tau*. Comp. Biochem. Physiol. 39B: 409–413.

Read, L.J. 1975. Absence of ureogenic pathways in the liver of hagfish, *Bdellostoma cirrhatum*. Comp. Biochem. Physiol. 51B: 139–141.

Schmidt-Nielsen, B., B. Truniger & L. Rabinowitz. 1972. Sodium-linked urea transport by the renal tubule of the spiny dogfish, *Squalus acanthias*. Comp. Biochem. Physiol. 42A: 13–25.

Schmidt-Nielsen, K. & P. Lee. 1962. Kidney function in the crab-eating frog (*Rana cancrivora*). J. Exp. Biol. 39: 167–177.

Schoffeniels, E. & R.R. Tercefs. 1966. L'osmoregulation chez les batraciens. Ann. Soc. R. Zool. Belg. 96: 23–39.

Schultze, H.-P. 1972. Early growth stages in coelacanth fishes. Nature (New Biology) 236: 90–91.

Smith, C.L., C.S. Rand, B. Schaeffer & J.W. Atz. 1975. *Latimeria*, the living coelacanth, is ovoviviparous. Science 190: 1105–1106.

Smith, H.W. 1930. Metabolism of the lungfish, *Protopterus aethiopicus*. J. Biol. Chem. 86: 97–130.

Smith, H.W. 1931. The absorption and excretion of water and

218

salts by the elasmobranch fishes, II. Marine elasmobranchs. Amer. J. Physiol. 98: 296–310.

Smith, H.W. 1953. From fish to philosopher. Little, Brown and Co., Boston. 264 pp.

Suyama, M. & C. Ogino. 1958. Changes in chemical composition during development of rainbow trout eggs. Bull. Jap. Soc. Sci. Fish. 23: 785–788.

Thomson, K.S. 1969. The environment and distribution of paleozoic sarcopterygian fish. Amer. J. Sci. 267: 457–464.

Thomson, K.S. 1971. The adaptation and evolution of early fishes. Q. Rev. Biol. 46: 139–166.

Thorson, T.B., C.M. Cowan & D.E. Watson. 1967. *Potamotrygon* spp.: elasmobranchs with low urea content. Science 158: 375–377.

Tytler, P. & J.H.S. Blaxter. 1988. The effects of external salinity on the drinking rates of larvae of herring, plaice and cod. J. Exp. Biol. 138: 1–15.

Vellas, F. & A. Serfaty. 1967. Sur l'excretion ureique de la carpe (*Cyprinus carpio* L.). Arch. Sci. Physiol. 21: 185–192.

Watson, D.M.S. 1927. The reproduction of the coelacanth fish, *Undina*. Proc. Zool. Soc. Lond. 1927: 453–458.

Wiley, E.O. 1979. Ventral gill arch muscles and the phylogenetic relationships of *Latimeria*. Occ. Pap. Calif. Acad. Sci. 134: 56–67.

Wourms, J.P. 1977. Reproduction and development in chondrichthyan fishes. Amer. Zool. 17: 379–410.

Yancey, P.H. & G.N. Somero. 1978. Urea-requiring lactate dehydrogenase of marine elasmobranch fishes. J. Comp. Physiol. 125: 135–141.

Yancey, P.H. & G.N. Somero. 1979. Counteraction of urea destabilization of protein structure by methylamine osmoregulatory compounds of elasmobranch fishes. Biochem. J. 183: 317–323.

Yancey, P.H. & G.N. Somero. 1980. Methylamine osmoregulatory solutes of elasmobranch fishes counteract urea inhibition of enzymes. J. Exp. Zool. 212: 205–213.

Environmental Biology of Fishes **32**: 219–222, 1991.

Urea and chloride sensitivities of coelacanth hemoglobin

Charlotte P. Mangum
Department of Biology, College of William and Mary, Williamsburg, VA 23185, U.S.A.

Received 1.8.1989   Accepted 24.7.1990

Key words: Oxygen transport, Oxygen equilibrium, *Latimeria chalumnae,* Actinistia

Synopsis

At pH 6.96–6.98, 20° C and in the absence of inorganic ions, the O_2 affinity of thawed *Latimeria chalumnae* hemoglobin was 1.53–1.86 mmHg; cooperativity was 1.00–1.13. These values are essentially the same as those in the literature for samples that had never been frozen. There was no clear effect of either urea (up to > 3M) or KCl (up to > 1M) on O_2 binding. Thus the hemoglobins of the coelacanth, as well as those of most of the elasmobranchs examined, are insensitive to urea, a major intracellular osmolyte in these groups and a denaturing agent in higher vertebrates. However, the absence of comparable information on more primitive hemoglobins and also on teleost hemoglobins precludes a clear evolutionary interpretation of the origin of urea sensitivity of the hemoglobins in higher vertebrates.

Introduction

In 1963 Manwell mentioned that the hemoglobin (Hb) of the dogfish *Squalus acanthias* (= *suckleyi*) is not sensitive to high levels of urea, although he showed no data. Later, Bonaventura et al. (1974a) reported in detail the virtual insensitivity of skate, ray and shark Hb to urea, which denatures many proteins including the Hb of higher vertebrates. They interpreted the low urea sensitivity of these Hb as a property that had evolved specifically as an adaptation to the high urea levels found in the elasmobranch red blood cell (RBC). Subsequently the Hb of additional elasmobranchs were shown to be insensitive or almost so (Mumm et al. 1978, Martin et al. 1979, Scholnick 1989). However, the evolutionary interpretation of the urea insensitivity of the elasmobranch Hb was questioned by Edelstein et al. (1976) who showed that, in contrast to mammalian Hb, the Hb of a teleost resists dissociation by urea. Thus structural insensitivity to urea does not require its presence within the RBC.

More recently Weber et al. (1983) reported contrary findings, viz. that O_2 binding of *S. acanthias* Hb is sensitive to urea concentrations within the physiological range. In their original experiments the urea sensitivity could have been in fact a response to the cyanates that form spontaneously in aged urea solutions (Stark et al. 1960, Marier & Rose 1964). Cyanates carbamylate proteins, which can modify their functional properties. Using deionized solutions of urea, however, Weber (1983) reported a sensitivity that actually exceeded the original. Still different results were obtained by Scholnick (1989), who found that the urea sensitivity of stripped Hb and RBC of North American members of *S. acanthias* is so small that it is barely detectable. Thus present information indicates that only one of eight elasmobranchs examined contains a Hb with appreciable urea sensitivity and the response within this species appears to vary. Although the Hb of amphibians and higher vertebrates are clearly sensitive to urea, its effect on

HbO$_2$ binding in other groups of fishes and in lower taxa has not, to my knowledge, been examined.

The chloride sensitivity of Hb is also of evolutionary interest. NaCl lowers the O$_2$ affinity of most vertebrate Hb, including those from at least one sciaenid teleost (Bonaventura et al. 1976). My unpublished data on two species of euryhaline sciaenids suggest that the Cl$^-$ effect should be appreciable within the physiological range of Cl$^-$ in the blood. In contrast, NaCl raises the O$_2$ affinity of the elasmobranch Hb examined by Bonaventura et al. (1974a), but not those investigated by Mumm et al. (1978) and Martin et al. (1979). Moreover, NaCl can lower the O$_2$ affinity of other elasmobranch Hb (C.P. Mangum unpublished observation; J. Bonaventura personal communication). While the Cl$^-$ effect should be detectable within the physiological range, it is unlikely to be very great in the stenohaline elasmobranchs examined thus far.

The present contribution reports the effects of urea and KCl (because it is more relevant physiologically than NaCl) on the Hb of *Latimeria chalumnae*. The goal of investigating KCl as well as urea sensitivity is not to identify responses of physiological importance to this presumably stenohaline species. Instead the aim is to enlarge the phylogenetic scope of the database on effector sensitivity of Hb and thus to permit more confident phylogenetic inferences.

Previous workers have demonstrated the following properties of coelacanth Hb: (1) In the physiological pH range coelacanth RBC have a very high O$_2$ affinity relative to even elasmobranch Hb. Coelacanth RBC almost lack cooperativity, a feature also found among elasmobranchs, but the pH dependence of coelacanth Hb resembles that of higher vertebrate Hb more than elasmobranch Hb (Wood et al. 1972). (2) Purified coelacanth Hb has a decidedly higher O$_2$ affinity (P$_{50}$) than RBC (though about the same cooperativity), implicating sensitivity to organic phosphates (Wood et al. 1972); the existence of organic phosphate sensitivity is variable among elasmobranchs but not teleosts. (3) In spite of its low cooperativity, the O$_2$ affinity of purified coelacanth Hb is sensitive to inorganic phosphate as well as ATP (Wood et al. 1972, Bonaventura et al. 1974b). (4) Unlike some

but not all of the elasmobranch Hb (Fyhn & Sullivan 1975), coelacanth Hb is tetrameric in the (presumably) oxy-conformation (Weber et al. 1973, Bonaventura et al. 1974b).

Materials and methods

Hemoglobin in the best condition was obtained from an immature female (VIMS 8118) while the carcass was thawing, by allowing lysed material and a few RBC to drip from the gills into a container. The material was centrifuged, dialyzed at 4°C overnight against 0.05 M Tris maleate buffer at a ratio of about 1:1000 and concentrated by membrane centrifugation. The filtrate that passed through the Centricon (AMICON) membrane (molecular weight cutoff 30000) was virtually colorless, indicating that little or none of the material had dissociated. Because the retentate clearly contained metHb, I decided to make one series of O$_2$ binding measurements immediately, using the nonoptical cell respiration method (Mangum & Lykkeboe 1979). At the end of this series, the metHb was reduced with Na borohydride, the preparation dialyzed again and the experiments repeated. In both series the urea solutions were prepared within minutes of use and kept on ice to prevent cyanate formation, as recommended by Stark et al. (1960) and Marier & Rose (1964).

Results

The value for O$_2$ affinity obtained in the absence of urea or KCl is very similar to that reported by Bonaventura et al. (1974b) in the absence of phosphate and at the same pH (Fig. 1). When all of the data in Table 1 are combined, the 95% confidence interval (1.04–1.20) around the mean value (1.12) does not overlap 1.0. Though significant, cooperativity is very slight, as also found by Bonaventura et al. (1974b). The present sample must have been in a respiratory condition similar to that examined by Bonaventura et al. (1974b), even though their sample had not been frozen and the present one had. The purified coelacanth Hb examined by

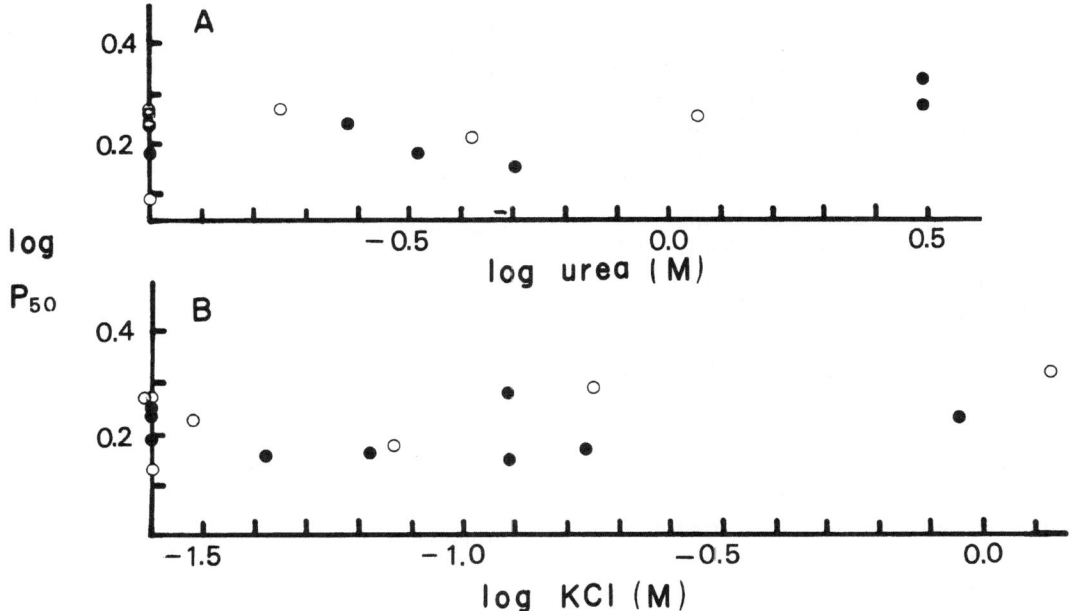

Fig. 1. Response of *Latimeria chalumnae* HbO₂ affinity (P₅₀) and cooperativity (n) to urea (panel A) and KCl (panel B), (●) before and after (○) reduction with Na borohydride. pH 6.96–6.98, 20°C, 0.05 M Tris maleate buffer.

Wood et al. (1972) had a lower O₂ affinity, very likely due to the presence of inorganic PO₄.

Neither KCl nor urea, added in concentrations well above the physiological range found in animal bloods, influences O₂ binding (Fig. 1, Table 1). The 95% confidence intervals around the slopes of the regression lines describing each set of data in Figure 1 overlap zero. Reduction with Na borohydride also had little or no effect.

Discussion

Like all but one of the elasmobranch Hb previously

examined, *L. chalumnae* Hb is insensitive to urea. This property may protect it from denaturation in vivo, as also inferred earlier. The effect of urea on O₂ equilibrium properties of teleost Hb and those of more primitive animals is not known. Thus the available information does not permit a distinction between two alternative evolutionary hypotheses: (1) Urea insensitivity evolved specifically in the elasmobranchs and coelacanths to protect the Hb in RBC of animals that utilize it as an osmolyte, or (2) urea sensitivity evolved in the teleosts when high levels of urea were no longer a problem and the elasmobranchs and coelacanths merely retained the insensitivity inherited from their urea-free ancestors. It will be interesting to examine the sensitivity of a wider spectrum of Hb.

Unlike almost all tetrameric vertebrate Hb examined, the exceptions including a few elasmobranch Hb (Martin et al. 1979), *L. chalumnae* Hb is also insensitive to Cl⁻. Given its large response to PO₄, which binds at the same site (Dickerson & Geis 1983), this finding is surprising. Although

Table 1. Cooperativity of HbO₂ binding in *Latimeria chalumnae*. pH 6.96–6.98, 20°C, 0.05 M Tris maleate buffer.

Control	1.12 + 0.05 (8)*
0.12 –3.15 M urea	1.15 + 0.07 (9)*
0.025–1.35 M KCl	1.14 + 0.05 (10)*

* Mean + S.E. (n).

222

Wood et al. (1972) reported little ATP sensitivity at pH 7.47, the decrease in O_2 affinity at 6.65 was dramatic. Bonaventura et al. (1974b) found that inorganic PO_4 sensitivity also diminishes at high pH but an effect is still very clear at the pH employed in the present investigation. The absence of Cl^- sensitivity of *L. chalumnae* Hb clearly cannot be interpreted as a specific adaptation to wide variability in the RBC because the species is not believed to be highly euryhaline. It is more likely to be an inherited feature. Unfortunately, the response of more primitive Hb is poorly known, though one would expect them to be Cl^- insensitive because they are not sensitive to organic PO_4 and they rarely exhibit appreciable cooperativity. We recently found no Cl^- sensitivity of an annelid RBC (Mangum et al. 1989), to my knowledge the only one examined outside of the vertebrates. It will be interesting to ascertain the relationship between Cl^- and phosphate sensitivity in a wide variety of RBC Hb.

Acknowledgement

Supported by NSF DCB 84–14856 (Regulatory Biology) and 88–16172 (Physiological Processes). I thank the Explorer's Club and the Virginia Institute of Marine Science Coelacanth Program (J.A. Musick, Principal Investigator) for access to the material.

References cited

Bonaventura, C., B. Sullivan & J. Bonaventura. 1976. Spot hemoglobin. Studies on the Root effect hemoglobin of a marine teleost. J. Biol. Chem. 251: 1871–1876.

Bonaventura, J., C. Bonaventura & B. Sullivan. 1974a. Urea tolerance as a molecular adaptation of elasmobranch hemoglobins. Science 196: 57–59.

Bonaventura, J., R.G. Gillen & A. Riggs. 1974b. The hemoglobin of the crossopterygian fish, *Latimeria chalumnae* (Smith). Subunit structure and oxygen equilibria. Arch. Biochem. Biophys. 163: 728–734.

Dickerson, R.E. & I. Geis. 1983. Hemoglobin. Benjamin Cummings Publ. Co., Menlo Park. 176 pp.

Edelstein, S.J., B. McEwen & Q.H. Gibson. 1976. Subunit dissociation in fish hemoglobins. J. Biol. Chem. 251: 7632–7637.

Fyhn, U.E.H. & B. Sullivan. 1975. Elasmobranch hemoglobins: dimerization and polymerization in various species. Comp. Biochem. Physiol. 50B: 119–130.

Mangum, C.P. & G. Lykkeboe. 1979. The influence of inorganic ions and pH on the oxygenation properties of the blood in the gastropod mollusc *Busycon canaliculatum*. J. Exp. Zool. 207: 417–430.

Mangum, C.P., J.A. Colacino & T.L. Vandergon. 1989. Oxygen binding by single red blood cells of the annelid bloodworm *Glycera dibranchiata*. J. Exp. Zool. 249: 144–149.

Manwell, C. 1963. Foetal and adult hemoglobins of the spiny dogfish, *Squalus suckleyi*. Arch Biochem. Biophys. 101: 504–511.

Marier, J.R. & D. Rose. 1964. Determination of cyanate, and a study of its accumulation in aqueous solutions of urea. Anal. Biochem. 7: 304–309.

Martin, J.P., J. Bonaventura, H.J. Fyhn, U.E.H. Fyhn, R.L. Garlick & D.A. Powers. 1979. Structural and functional studies of the hemoglobin isolated from Amazon stingrays of the genus *Potamotrygon*. Comp. Biochem. Physiol. 60B: 189–194.

Mumm, D.P., D.H. Atha & A. Riggs. 1978. The hemoglobin of the common stingray *Dasyatis sabina*: structural and functional properties. Comp. Biochem. Physiol. 60B: 189–194.

Scholnick, D.A. 1989. Anion and urea sensitivity of three elasmobranch hemoglobins. M.A. Thesis, College of William and Mary, Williamsburg. 54 pp.

Stark, G.R., W.H. Stein & S. Moore. 1960. Reactions of the cyanate present in aqueous urea with amino acids and proteins. J. Biol. Chem. 235: 3177–3182.

Weber, R.E. 1983. TMAO (trimethylamine oxide)-independence of oxygen affinity and its urea and ATP sensitivities in an elasmobranch hemoglobin. J. Exp. Zool. 228: 551–554.

Weber, R.E., J.F. Bol, K. Johansen & S.C. Wood. 1973. Physicochemical properties of the hemoglobin of the coelacanth *Latimeria chalumnae*. Arch. Biochem. Biophys. 154: 96–105.

Weber, R.E., R.M.B. Wells & J.E. Rosetti. 1983. Allosteric interactions governing oxygen equilibria in the haemoglobin system of the spiny dogfish, *Squalus acanthias*. J. Exp. Biol. 103: 109–120.

Wood, S.C., K. Johansen & R.E. Weber. 1972. Haemoglobin of the coelacanth. Nature 239: 283–285.

Reproduction, feeding and parasites

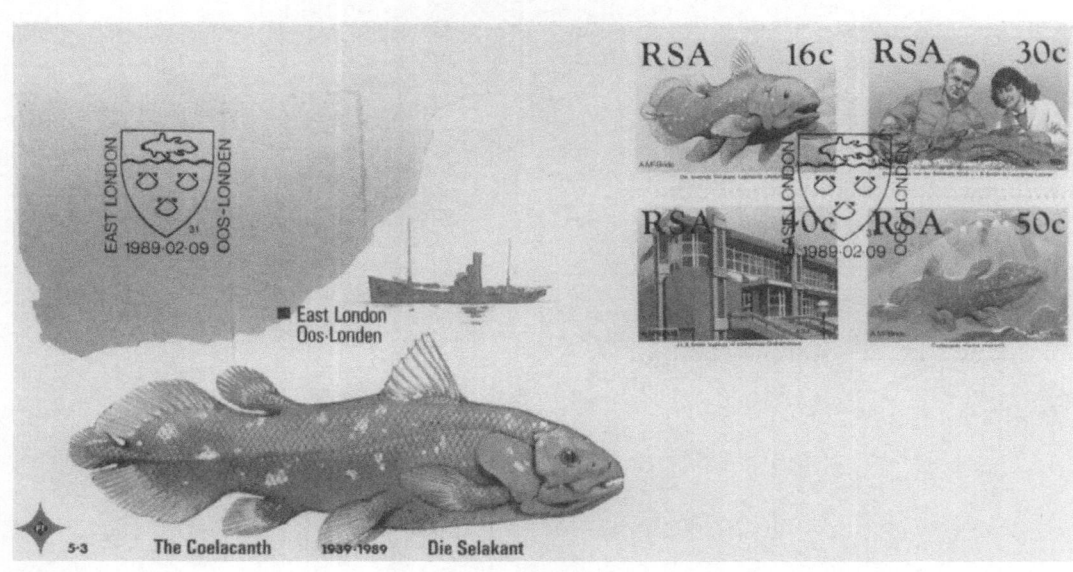

Environmental Biology of Fishes **32**: 225–248, 1991.
© 1991 *Kluwer Academic Publishers.*

Viviparity and the maternal-embryonic relationship in the coelacanth *Latimeria chalumnae*

John P. Wourms[1], James W. Atz[2] & M. Dean Stribling[1]
[1] *Department of Biological Sciences, Clemson University, Clemson, SC 29634, U.S.A.*
[2] *American Museum of Natural History, Department of Herpetology and Ichthyology, New York, NY 10024, U.S.A.*

Received 1.9.1989   Accepted 8.7.1990

Key words: Embryo, Fetal nutrition, Lecithotrophy, Matrotrophy, Oophagy, Oviduct, Placenta, Uterus, Yolk, Yolksac

Synopsis

Embryos of *Latimeria chalumnae* develop in well-vascularized compartments in the uterine region of the right oviduct. Compartments conform to the shape of their embryos and yolksacs; they represent a stable, gestation-induced oviductal modification. Late-term pups possess large, flaccid, vascular yolksacs almost devoid of yolk. The sac is in close contact with, but does not adhere to, the lumenal uterine surface. A massive vascular plexus occurs in the wall of the compartment at the site of contact with the yolksac; together they constitute a non-adherent, transposable placenta. The exterior surface of the yolksac is bounded by an attenuated, single-layered, squamous epithelium that surrounds an intercommunicating bed of cortical sinuses. The cortex of the sac is composed mostly of connective tissue stroma. The inner surface is bounded by a layer of yolk-digesting merocytes. Residual yolk occurs as yolk platelets that include yolk crystals. The interior surface of the sac is invested by an uniquely specialized vitelline circulation; no connection seems to exist between the interior of the yolksac and gut. The uterine wall consists of: (1) a lumenal surface composed of an anastomosing network of capillaries with a layer of attenuated, very thin, squamous epithelium, (2) a well-vascularized connective tissue stroma, (3) alternating transverse and longitudinal layers of smooth muscle, also well-vascularized, and (4) an external epithelial layer. Comparison of egg dry weight (184 g) with the estimated dry weights of a late-term pup (171 to 239 g) and a neonate (200 to 280 g) reveals a weight change of −7 to +30% and +9 to +52%, respectively. This is indicative of matrotrophy. In one female specimen, 19 remarkably large ovulated eggs were found and in another about 30 somewhat smaller ovarian ones. These are many more than ever could be accommodated in the uterine space. During the early and middle phases of development, embryos must be lecithotrophic, using their yolk reserves, with oophagy of fragmented supernumerary eggs as the most probable source of additional nutrients. The well-developed embryonic gut contains brown, amorphous yolk-like material. The limited amount of metachromatic secretory product of the uterine glands can play little or no role in embryonic nutrition.

Introduction

Latimeria chalumnae, the only living coelacanth, is a relict of a group of lobe-finned fishes long be-

lieved to have become extinct in the late Cretaceous. Since the first exemplar was captured in 1938, about 200 specimens have been caught of which about half have become available for scientific study (Balon et al. 1988). *L. chalumnae* inhabits the waters around the volcanic islands of Comoros in the western Indian Ocean at depths of 115 m or more (Fricke et al. 1987, Balon et al. 1988). Most of what is known about coelacanth biology is based on a few hand-lined individuals that survived for a short time after capture and a limited number of frozen or formalin-fixed specimens (McCosker & Lagios 1979, Locket 1980, Suzuki et al. 1985, Balon et al. 1988). The first direct observations of free-ranging fish in their natural environment were made only recently by Fricke et al. (1987). Not until 1975 was *L. chalumnae* discovered to be viviparous (Smith et al. 1975), and information about coelacanth reproduction and development is still very limited (reviewed in Locket 1980, Schultze 1980, Wourms 1981, Balon et al. 1988, Wourms et al. 1988).

The circumstances that led to the discovery of five well-developed young inside the specimen of *L.chalumnae* at the American Museum of Natural History have been recounted by Atz (1976), but the disposition of these unique yolksac pups has yet to be put on record. As is shown in Figure 1 of Smith et al. (1975), they were located more or less evenly along the oviduct and for identification purposes were numbered consecutively commencing with the one nearest the infundibulum. Decisions as to how they might best be put to use were made in the Department of Ichthyology over a period of about a year. The first decision was to sequester pup 5 in order to make it available, undisturbed, should some unanticipated relationship between mother and offspring subsequently become apparent. Pup 2 was selected for photography because its body was straighter and its fins more naturally held than in the other three. It was then decided that this specimen would be maintained intact in the ichthyological collections of the AMNH. Pup 3 was chosen for clearing and staining and it, too, now resides in the museum's collections. As time passed, the view that specimens ought to be made more accessible to foreign investigators gained acceptance; the British Museum (Natural History) and the Muséum National d'Histoire Naturelle appeared to be the institutions most able to discharge

this responsibility. By February 1977, pups 4 and 5 had been sent to London and Paris, respectively. In the meantime, Michael D. Lagios of the Children's Hospital of San Francisco proposed that one of the pups be serially sectioned in order to produce a permanent histological atlas that could provide more anatomical data for a larger number of workers than any other type of preparation. Eventually pup 1 was turned over to Lagios; unfortunately, circumstances beyond his control prevented the completion of the project. Nevertheless, the entire head and the rectal-urinary-genital region were embedded in nitrocellulose, serially sectioned at 50 microns, and stained with Gomori trichrome stain (Lagios in litt. 1989). The 1235 51-by-76 mm slides have now been catalogued and are available for study at the AMNH (Bemis & Northcutt 1991). Sections from these series were used in a comparative study of the coelacanth ear by Fritzsch (1987).

Photographs of the pups are scattered through the literature: No. 1 (Myking 1977), No. 2 (Smith et al. 1975, Atz 1976, McAllister & Smith 1978, Schultze 1980, Tanaka 1989), No. 4 (Locket 1976, 1980), and No. 5 (Balon 1984, 1985, Hensel 1986). Only those in Myking (1977) and Tanaka (1989) are reproduced in color.

The purpose of this communication is to examine the maternal-embryonic relationship in the coelacanth, *Latimeria chalumnae*. Morphological investigations of maternal and fetal tissues, weight determinations, and information from the literature have been used to accomplish this. Preliminary reports of our research have appeared in Wourms et al. (1980, 1988).

Materials and methods

The principal material used in this study was derived from a 1.6 m, 65 kg female specimen in the American Museum of Natural History (CCC no. 29) and four of the five pups that it contained. The female had been landed near Mutsamudu, Anjouan Island, Comoros in 1962 (Smith et al. 1975). It has been injected carefully with formaldehyde shortly after it died but before decomposition had

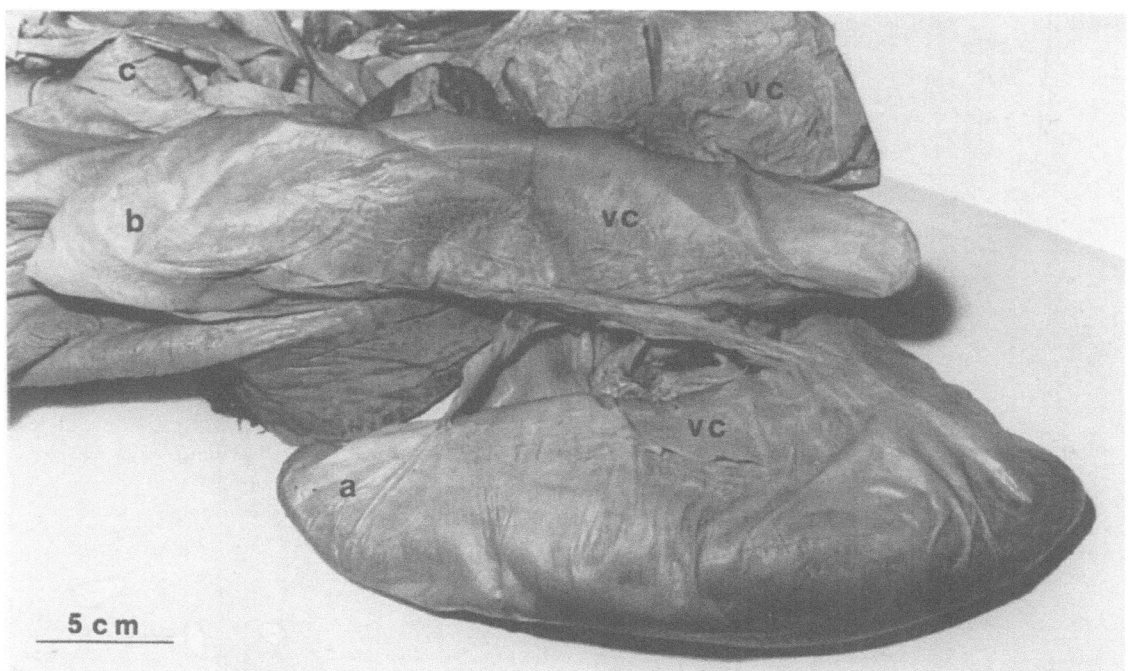

Fig. 1. Within the uterus of the gravid female, embryos are enclosed in compartments. Compartment a contains pup 5. Pups 4 and 3 have been removed from compartments b and c. Each compartment has its own well-developed vascular complex (vc). Photographed in December 1976 at the American Museum of Natural History.

set in. Some years later, it was stored in 50% iso-propyl alcohol. The quality of tissue preservation at the anatomical, histological, and ultrastructural (SEM) levels proved to be good. Pup 5 was used to ascertain compartmentalization and the topographic relationship of the yolksac to the uterine wall; pup 2 provided anatomical information on the sac; scanning electron microscopy and light-microscopy histology were carried out on the yolksac of pup 3; and transverse serial sections of pup 1 were examined.

Standard methods of dissection and macrophotography were used to examine pup 5 in situ and to remove it for examination. Uterine and yolksac tissues were routinely processed for scanning electron microscopy (SEM), i.e., osmicated, critical-point dried, sputter-coated with gold, and then examined with an ETEC scanning electron microscope at 10–30 kv. Yolksac tissue was gently son-icated before processing to remove loose yolk particles. Previous studies (Wourms & Cohen 1975, Knight et al. 1985) have demonstrated that significant morphological information can be obtained when adult or embryonic fish tissue, obtained from museum specimens, is processed and examined with SEM. For light microscopic, histological observation, tissues were dehydrated in a graded series of ethanol and embedded in glycol methacrylate (JB-4 embedding medium, Polysciences, Inc.). Embedded tissues were sectioned at 2–3 μm with a glass knife on a Sorvall JB-4 microtome, and sections were mounted on glass slides and stained with toluidine blue. Photomicrographs were taken through a Wild compound microscope equipped with flat field fluorite objectives and a Wild automatic camera system, images being recorded on 35 mm Technical Panchromatic film 2415 (Kodak, Inc.).

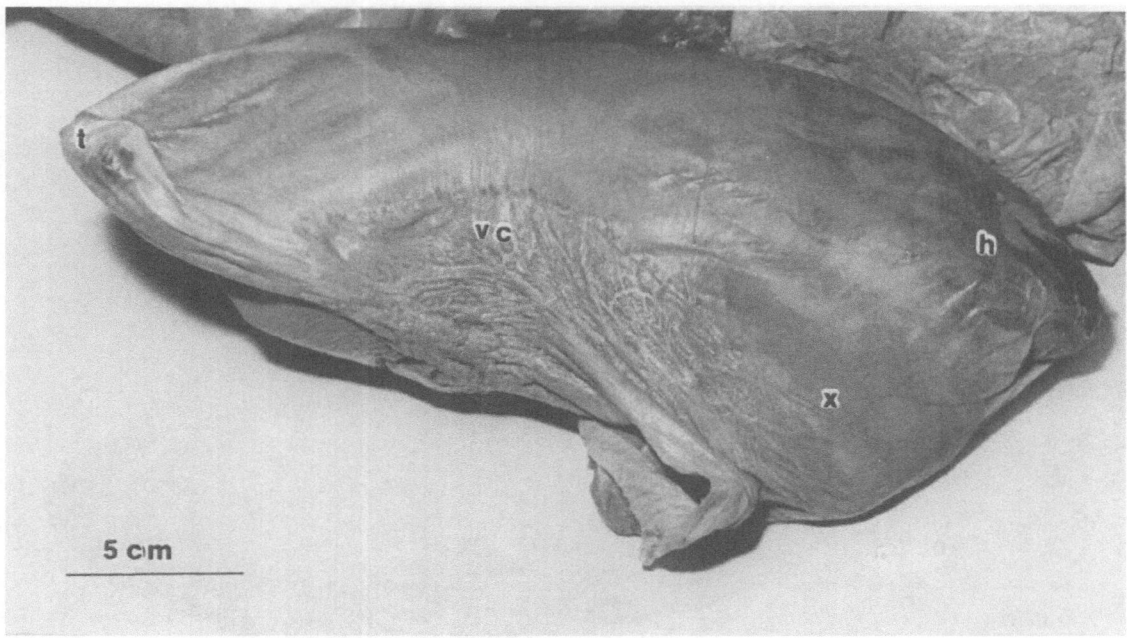

Fig. 2. The compartment that encloses pup 5 has been rotated 180° to highlight its vascular complex (vc). The head (h) is to the right and is visible through the somewhat translucent uterine wall while the tail (t) is to the left. The uterine wall and its well-developed vascular complex (vc) conform to the shape of the underlying yolksac (x). cf. Figure 3.

Results

Oviduct-uterus

Only the right oviduct is functional in *L.chalumnae,* the left being reduced to a mere cord with an obliterated lumen (Millot et al. 1978). In the gravid AMNH specimen, the five well-developed young were enclosed in dilated regions of the structure, all with their heads facing toward the infundibulum, i.e., away from the urogenital orifice (Smith et al. 1975). In the ensuing account, we shall refer to the region of the oviduct that bears the developing young as the uterus. When referring to the advanced developmental stages of the livebearing coelacanth, we have elected to use the term pup. This is appropriate for a number of reasons. First, such individuals are far too advanced (definitive phenotypes) to be called embryos inasmuch as they exhibit a preponderance of juvenile characteristics. Because of its mammalian connotations, fetus

seems inappropriate. Finally, there is precedence for the term pup, which is used regularly to describe the late-term and neonatal stages of sharks.

Each of the dilated uterine regions that enclose a developing embryo or pup constitutes a compartment. The wall of the uterus takes on the shape of the enclosed developing fish (Fig. 1, 2), and adjacent compartments are connected by constricted regions or isthmuses that permit direct communication between them (Fig. 1). The oviduct of mature, non-gravid females shows neither compartments nor constrictions (Millot et al. 1978). The process of compartmentalization of developing embryos in the coelacanth is analogous to what occurs in many species of sharks (Wourms et al. 1988), but the situation appears less complex in the coelacanth. In sharks the space in the uterine portion of the oviduct is partitioned by means of apparent growth of the wall, and each embryo is enclosed in a compartment not in direct communication with any other. In contrast, simple hypertrophy of al-

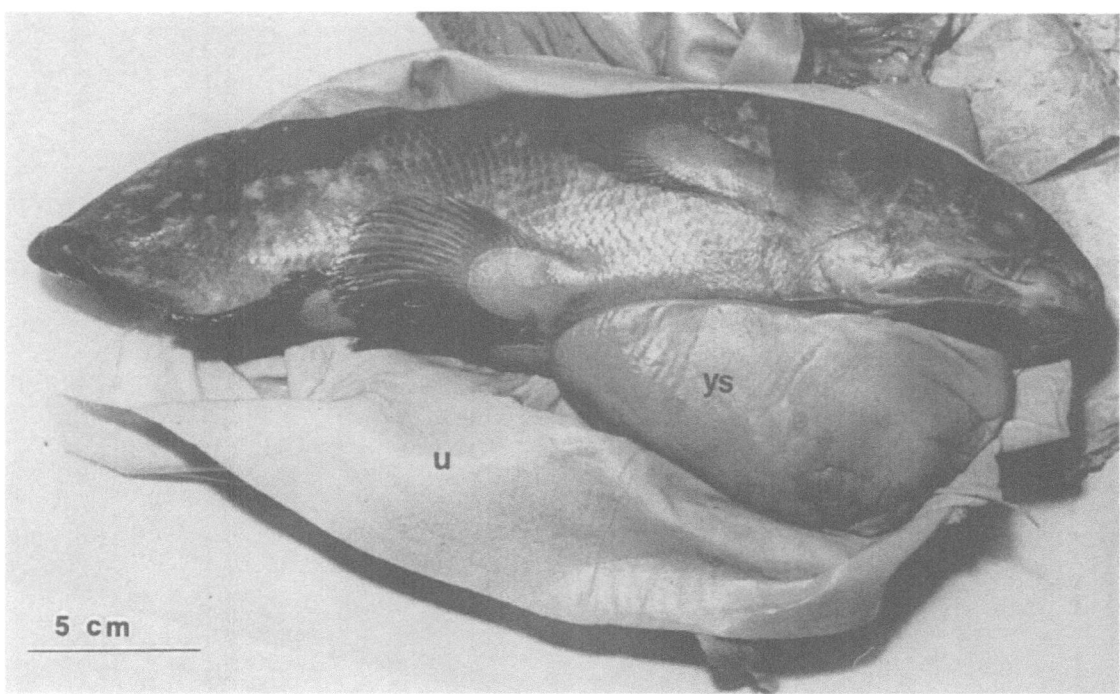

Fig. 3. After opening the compartment, the uterine wall was reflected to expose pup 5 lying on its left side, the same position as in Figure 2. The yolksac (ys) is in intimate contact with the vascularized region of the uterine wall (u).

ternate regions of the coelacanth uterus has sufficed to achieve a sequential, linear separation of the pups in a manner similar to some mammals such as the rabbit and pig.

Smith et al. (1975) removed four pups from the female, leaving pup 5 in utero. This pup and its compartment were examined both before and after its removal (Fig. 2, 3, 8). The two outstanding features of the compartment are its shape and its vascularization. The thin but tough uterine wall conformed closely to the shape of the enclosed pup and its yolksac, the head, yolksac, and folded tail all being clearly discernible (Fig. 2,3). This conformity of shape persisted even after the pup had been removed and appears to result from the coordinated growth of uterine tissue to accomodate the young rather than any elastic deformation. When the uterus was opened and reflected, the yolksac was seen to be in intimate contact with the lumenal surface of the uterine wall. Nevertheless, the two

did not adhere to each other nor was there any kind of permanent connection. Because the shape of the uterine wall conforms to that of the yolksac, the topographic relationship between the two tissues is inherently stable, although they seem able to undergo a limited translocation relative to each other. The compartment is heavily vascularized (Fig. 1, 2), and a massive vascular plexus is closely apposed to the ventral surface of the yolksac and the ventral, post-anal region of the pup (Fig. 2, 3, 4). The parts of the uterine wall that are in contact with the dorsal and lateral surfaces of the pup lack a vascular plexus (compare the dorsal aspect of the compartment in Figure 1 with its ventral aspect in Figure 2). The conclusion that each embryo-containing compartment has its own plentiful supply of blood seems inescapable. Vascular plexuses of this type and magnitude do not occur in the mature, non-gravid oviduct (Millot et al. 1978).

After pup 5 has been removed, an examination

230

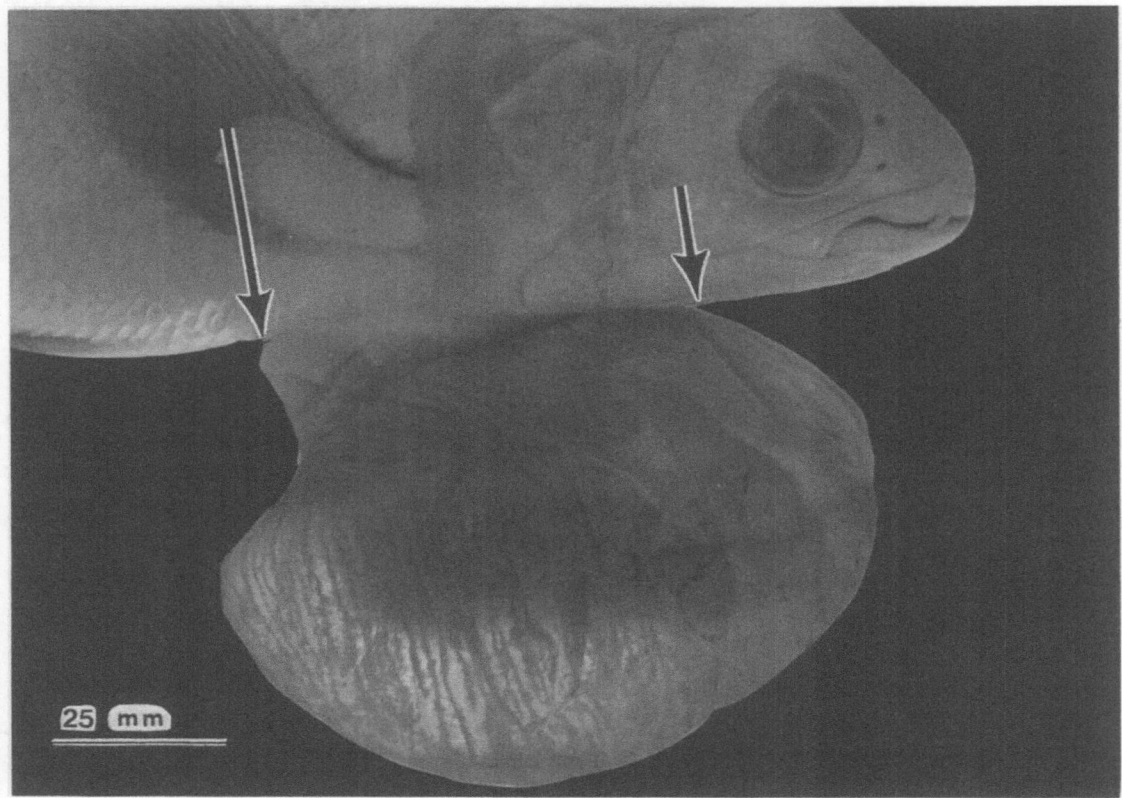

Fig. 4. The yolksac of pup 2 was photographed while illuminated by strong transmitted light. Although many blood vessels are obscured by residual yolk within the sac, several of the vertically oriented, more or less parallel ones are visible. Arrows indicate the anterior and posterior limits of the long attachment zone between the ventral body wall and the yolksac.

of the lumenal surface of the uterine epithelium revealed that it is smooth-surfaced. It exhibits none of the macroscopic surface modifications described, for example, in the matrotrophic sharks and rays (Wourms et al. 1988). Superficially, its morphology resembles that of *Squalus acanthias,* thus presenting the appearance of a type Ia uterine environment, according to the classification of Ranzi (1934) (cf. Needham 1942, Hoar 1969).

Preserved uterine tissue is buff-colored, fibrous in texture, and thin (0.5–1.0 mm) but tough. When examined in transverse section by light microscopy (Fig. 5, 6d), the uterine wall exhibits the following organization: (1) a lumenal epithelium that is associated with a capillary network in most regions, (2) a fibrous connective tissue stroma, (3) a region of transversely oriented smooth muscle, (4) a region of longitudinally oriented smooth muscle in a connective tissue matrix, (5) a superficial region of longitudinally oriented smooth muscle, and (6) an external oviductal epithelium. The uterine wall is well vascularized with large blood vessels in both the connective tissue stroma and smooth muscle layers. Low-magnification scanning electron microscopy confirmed this observation.

Scanning electron micrographic surveys of the lumenal surface (Fig. 5b, 6a) depict a slightly convoluted but relatively flat surface organized into distinct, lobular domains. The boundaries of adjacent domains are formed by deep infolding of the surface (cf. surface view of folds in Figure 5b with the cross-section through a fold in Figure 5a). Most

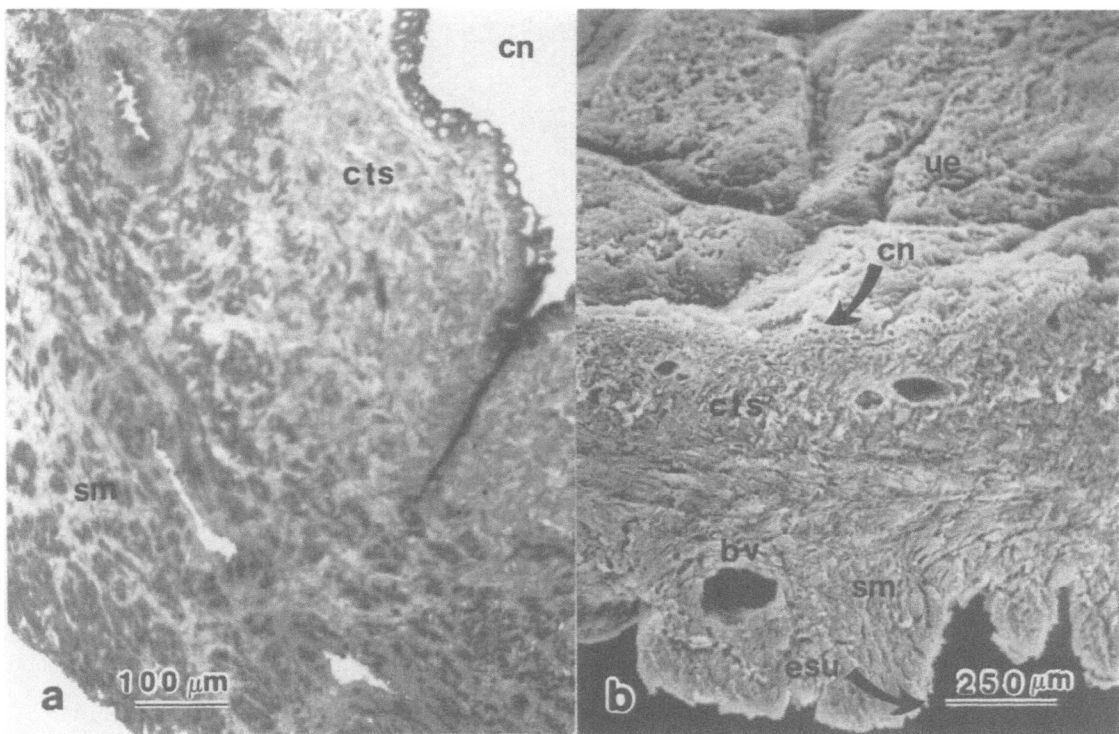

Fig. 5. a – The capillary network (cn) lies along the lumenal surface at the upper right in this light micrograph of a section through the uterine wall. Beneath it are a connective tissue stroma (cts) and an exterior zone of smooth muscle (sm). At the far right, the surface is thrown into a deep fold that separates adjacent domains. b – In this scanning electron micrograph of a section through the uterine wall, the lumenal surface of the uterine epithelium is at the top. Circular profiles of the capillary network (cn) are exposed along the upper margin of the cut surface. Blood vessels (bv) lie within the connective tissue stroma (cts) and smooth musculature (sm) of the wall.

of the lumenal surface consists of an anastomosing network that looks like a web of interlocking cables (Fig. 6b, c, 7a). In some places, the cable-like capillary network takes on the appearance of hemispherical acinar glands (Fig. 6c, 7b). Extra-cellular material, possibly a product of secretion, occurs in the interstices between adjacent capillary cables and in the lumina of acinar structures. This material exhibits intense red-violet metachromasia when stained with toluidine blue. In other areas, lumenal surfaces that lack the capillary network show patches of ciliated epithelium (Fig. 7d). Such patches occur as discrete domains surrounded by ones with capillary networks.

The anastomosing capillary network forms the major feature of the uterine lumenal surface, both in the region juxtaposed to the vascular plexus of the embryonic yolksac as well as in other regions. It is composed of capillaries about 10–15 μm in diameter surmounted by an extremely thin, attenuated, simple squamous epithelium (Fig. 6c,d, 7a). Beneath the capillary network lies a subcortical space or sinus that appears to be lined with an endothelium (Fig. 6d). The epithelium covering the capillary network displays at least two morphological states. In the first, the lumenal epithelium is thrown into a series of pleats or folds (Fig. 7b). The question might be raised as to whether this is an artifact of preservation or tissue processing; if it were, the condition should be present uniformly within a particular specimen, but this is not the case. There are, for example, several areas in Fig-

232

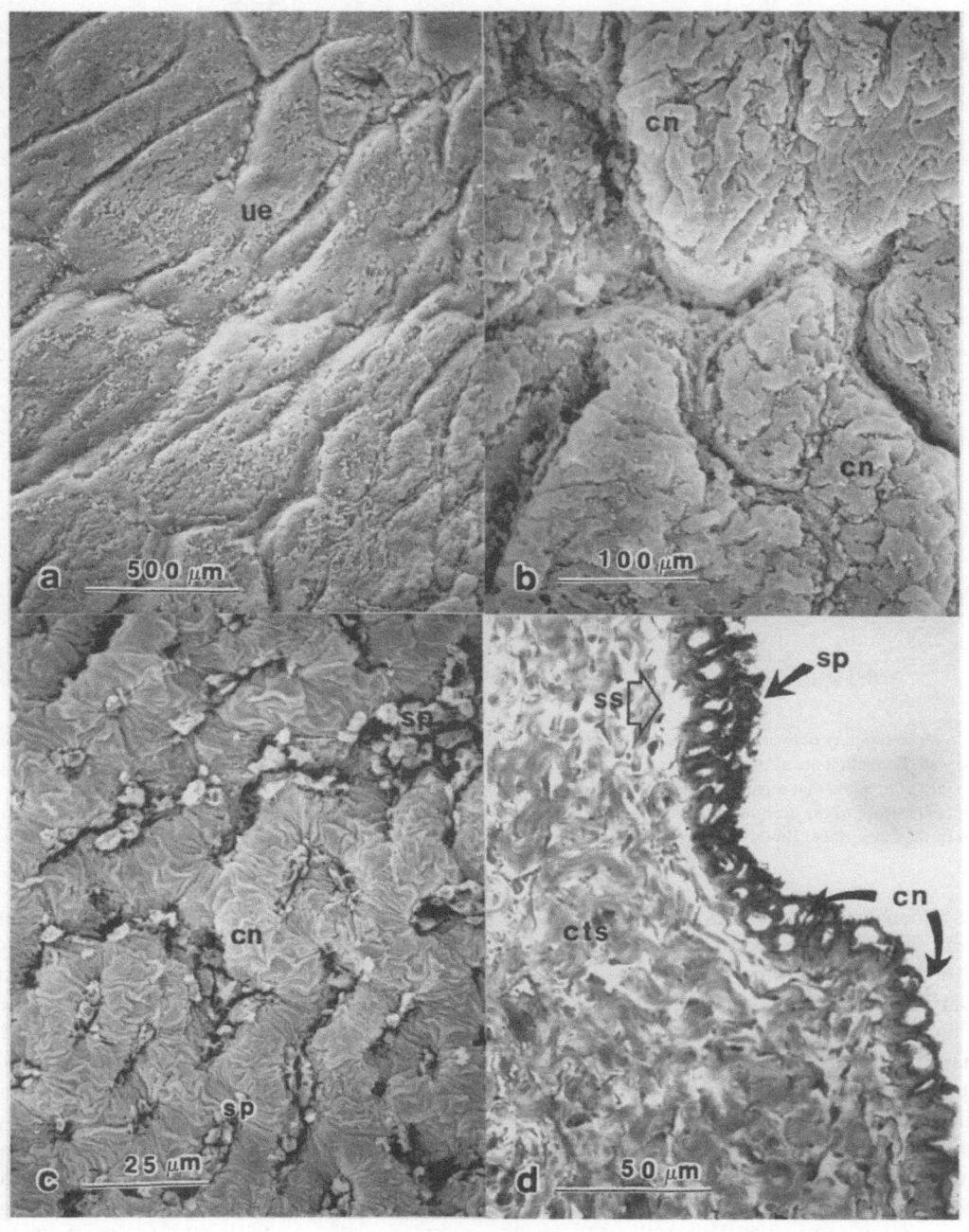

ure 7a (see labels ue and cn) in which the epithelium is relatively smooth. Moreover, the folds or wrinkles are not encountered in the ciliated areas (Fig. 7d), nor in samples of yolksac. In the second, putative state, the lumenal epithelial cells are organized either as cellular processes or pedicles, similar to those of mammalian kidney podocytes, or as a branching series of deeply incised ridges (Fig. 7c). Given the limitations of tissue preservation, however, it is not possible to distinguish between these two alternatives. Nevertheless, the major conclusion that can be reached is that the surface area of the uterine lumenal epithelium is extensively amplified, either by pleats, ridges, or branching processes.

Late-term pup and its yolksac

Only four of the five pups were removed from the female *L.chalumnae* at the time of their discovery. When it was subsequently removed, in December 1976, pup 5 measured approximately 303 mm TL (285 mm SL according to McAllister & Smith 1978). The other four pups ranged from 301 to 327 mm in total length at the time of their removal in September 1975 (Smith et al. 1975). Except for the presence of the large yolksac, late-term pups generally resemble miniature adults. The yolksac

←

*Fig. 6.*a – The lumenal surface of the uterine epithelium (ue) depicted in this scanning electron micrograph is organized into distinct lobular domains.

b – At higher magnification, many but not all regions of the lumenal uterine epithelium are organized into capillary networks (cn).

c – The lumenal surface forms an anastomosing capillary network (cn) that gives the appearance of a web of interlocking cables. Extra-cellular material, possibly a secretory product (sp), is found in the interstices between cables and lumina of acinar structures.

d – Light microscopy of sections through the uterine wall reveals that the capillary network (cn) and associated secretory product (sp) display intense metachromatic staining with toluidine blue while the connective tissue stroma (cts) stains less intensely except for its fibrous elements. Transverse sections of the capillaries reveal that they are encased with an attenuated squamous epithelium. A subcortical space (ss) lies immediately beneath the capillary network.

of pup 5 is shaped like an elongated ellipsoid hemisected along its long axis. It measures about 130 mm by 70 mm along its axes. The attachment of the sac to the belly originates between the posterior margins of the pair of gular plates, and in pup 2 it continues posteriorly for 55 mm and represents 18% of its total length. (NB: the total length of pup 2 at present is 308 mm; it has shrunk from the 322 mm it measured in 1975.) Although flaccid, the sac still contains yolk. Except for its vascularization, it exhibits no obvious anatomical modifications.

The embryonic gut, which was removed from pup 3, is well developed but simple. A short pharynx leads into the dorsal portion of a large pyramid-shaped stomach, the anterior portion of which is broad and the posterior narrowing to a blunt tip. Rugae are present on its inner surface. The intestine originates from the ventral aspect of the anterior end of the stomach, executes a loop to the right, and extends rearward as a simple, tubular structure. The mid-region of the intestine is expanded but it narrows posteriorly before terminating in the cloaca. No evidence could be found of any direct connection between the digestive system and the yolksac -- such as the vitelline duct of elasmobranch embryos. A spleen, pancreas, and bilobed liver are associated with the gut. Stomach and intestinal contents consisted mostly of a gelatinous substance associated with some dark-brownish-black amorphous material. Preliminary examination of stained sections through the intestine of pup 1 support these observations. On the basis of this anatomical evidence, it is reasonable to conclude that the digestive system of late-term pups is functional.

In late-term *L.chalumnae,* the only exposed surface that retains its embryonic characteristics is the yolksac (Fig. 8, 9). This structure comprises a yolksac cortex of modest thickness (about 0.5–1.0 mm), yolk granules associated with the inner surface of the cortex as well as the associated blood vessels, and the vascular elements of a vitelline circulation that ramifies over the surface of the sac (Fig. 4, 10a, b).

The extent and pattern of organization of the vitelline circulation are most unusual. When the intact yolksac of pup 2 was examined with strong

234

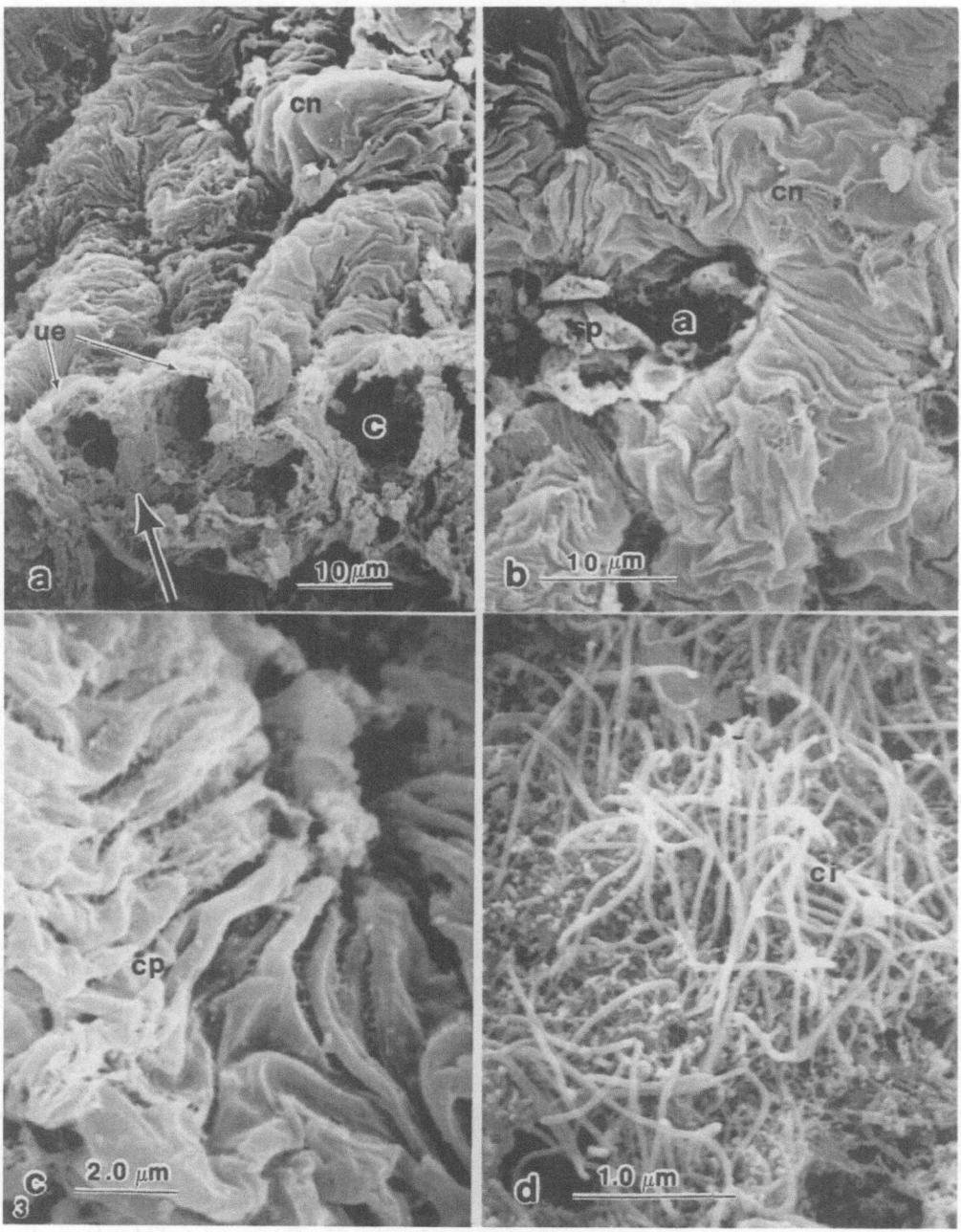

*Fig. 7.*a – Scanning electron microscopy of the uterine wall reveals the relationship between the surface epithelium (ue) and underlying capillaries (c) of the anastomosing capillary network (cn) that were exposed by cutting through the uterine wall. An arrow indicates the cut surface. b – In some areas of the lumenal epithelium, illustrated in this scanning electron micrograph, the capillary network (cn) resembles an acinar structure. Extracellular material, possibly a secretory product (sp), is associated with the lumen of the acinus (a). The squamous epithelium that comprises the lumenal surface often is much plicated. c – A scanning electron micrograph in which the squamous epithelium that forms the outer surface of the capillary network displays cell processes (cp) and pedicels similar to those of kidney podocytes. d – In some regions of the lumenal uterine epithelium, as depicted in this scanning electron micrograph, the capillary network is replaced by a ciliated epithelium (cn).

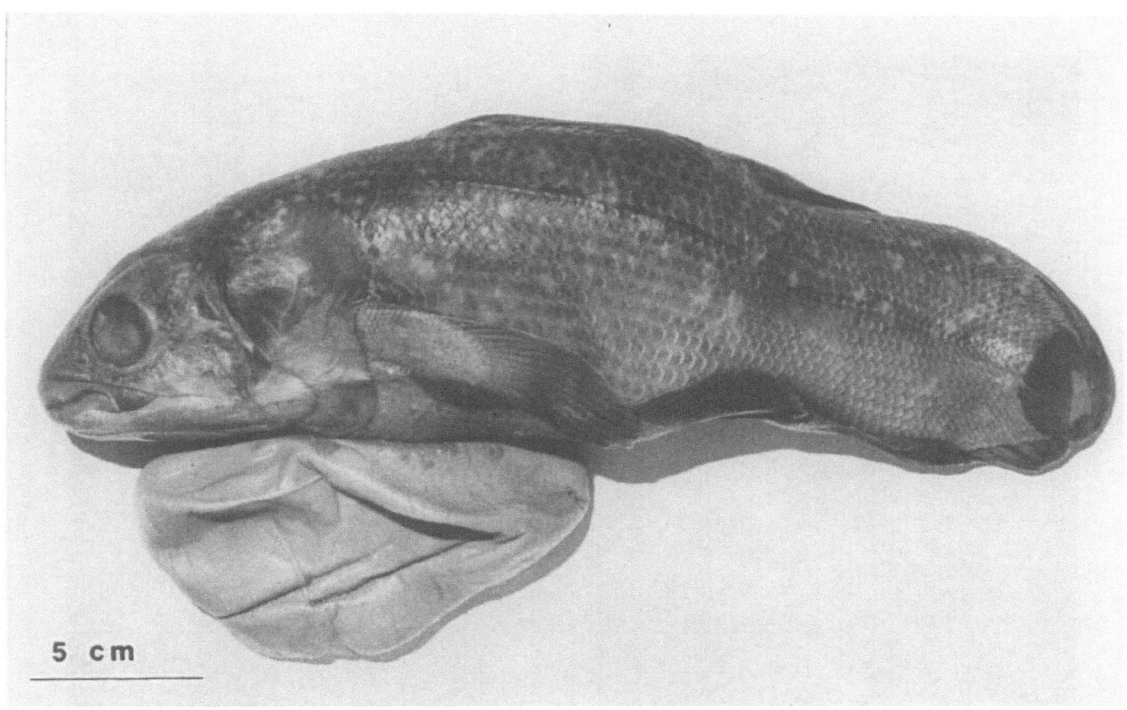

Fig. 8. Pup 5 lying on its right side after removal from its compartment. The yolksac is flaccid with a long junction between its proximal margin and the ventral surface of the pup.

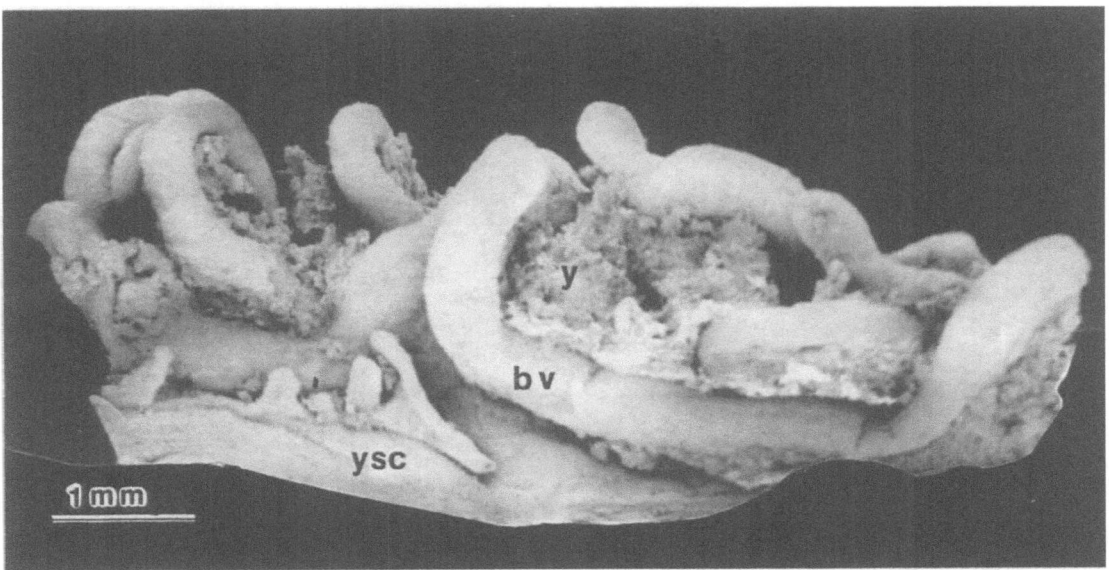

Fig. 9. Low magnification scanning electron micrographs reveal extensive vascularization (bv) along the interior surface of the yolksac. Yolk (y) is associated with the blood vessels. The cortex (ysc) of the yolksac is at the bottom.

236

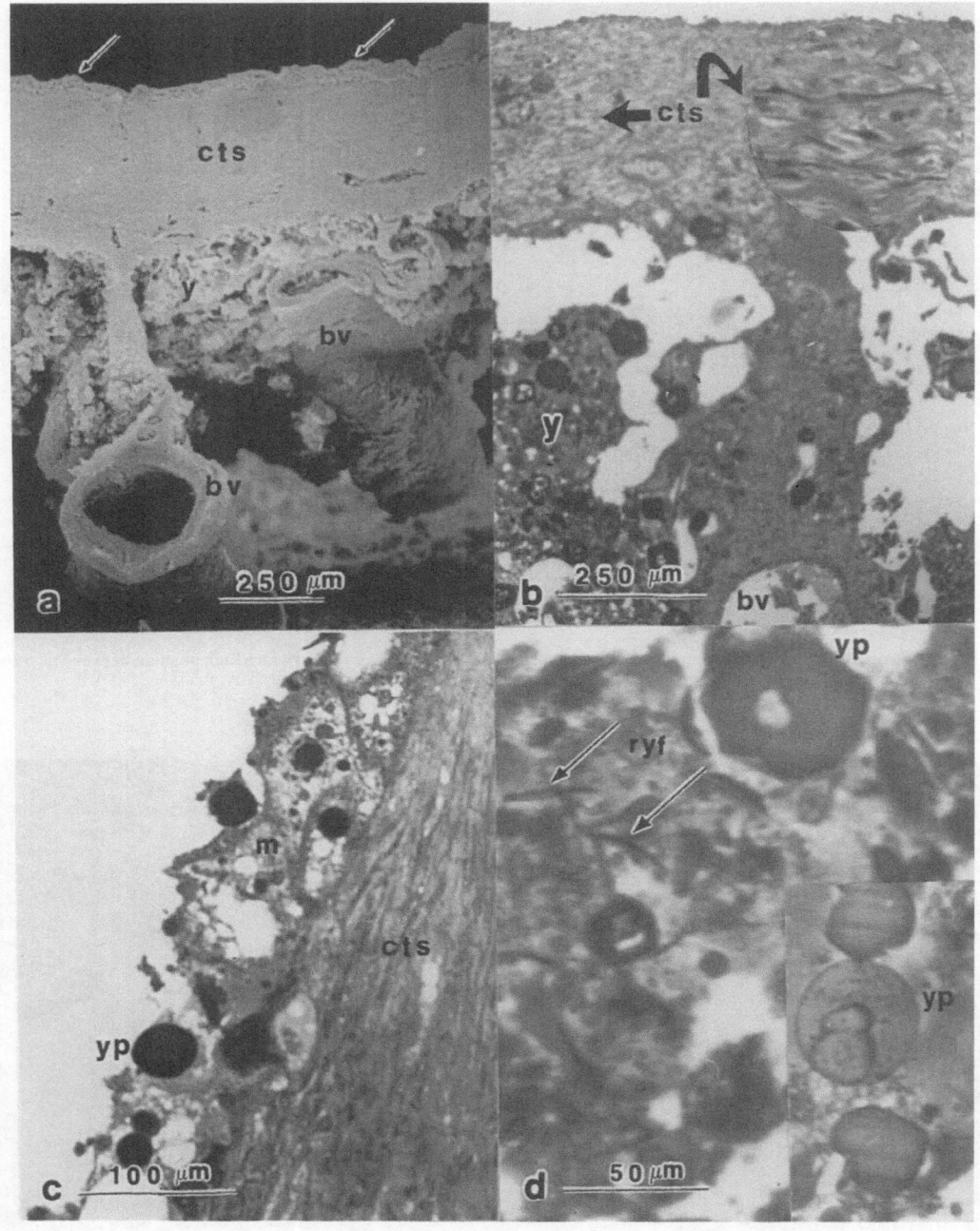

transmitted light, vessels could be seen in some detail (Fig. 4). On each side, 65–70 blood vessels of differing widths lie more or less parallel to one another and about equally spaced. They are oriented at right angles to a line formed by the juncture of the sac and the body wall. It would appear that the distal ends of some of these vessels may converge on some circular locus within the anterior, distal aspect of the sac. Further details of the yolksac circulation in coelacanths will be provided in a subsequent publication.

Light microscopic and scanning-electron micrographs provide complementary information on histological organization (Fig. 10a, b). Most of the yolksac cortex consists of a dense fibrous connective tissue stroma populated by fibroblasts. The exterior surface of the sac is a single-layered squamous or low-cuboidal epithelium of surprising structural complexity that surmounts the underlying connective tissue stroma. The interior surface of the yolksac cortex is lined with a layer of cells, the most abundant type of which tentatively is identified as a merocyte. (Fig. 10c). Elements of the vitelline circulation are associated with the interior surface of the yolksac cortex and are joined to it either by the connective tissue stroma or by densely-staining, yolk-rich cytoplasmic struts (Fig. 10a, b). The lumena of the vitelline vessels contain blood cells and coagulated serum. In some instances, it appears that the yolk-rich struts actually are part of the yolk syncytium. Yolk, in the form of aggregated yolk platelets (Fig. 10d), adheres to the inner surface of the yolksac cortex and the vitelline blood vessels and also is present within the cytoplasm of the vascular struts and in the yolk syncytium.

Late-term pups still carry an appreciable amount of un-utilized yolk. In the yolksac of pup 3, the particulate yolk was estimated to occupy a volume of about 25–35 ml. Yolk occurs in the form of spherical platelets about 45 micra in diameter (Fig. 10d), similar to those in amphibian eggs. Platelets consist of an outer and an inner region. The outer one is formed of an amorphous matrix that exhibits a blue-violet metachromasia when stained with toluidine blue. Crystalline yolk, usually in the form of polyhedrons is embedded within the matrix and exhibits a blue-green metachromasia. Yolk crystals are identical to those found in small (10–15 mm) ovarian eggs of *Latimeria* by Grodziński (1972). Based on their crystalline structure, Lange (1983) concluded that the platelets are composed of lipovitellin and phosvitin. What appear to be stages in the utilization of yolk are readily observable; a putative sequence is outlined here. During the first stages, yolk platelets are phagocytosed by the merocytes that line the inner surface of the yolksac cortex (Fig. 10c). Yolk platelets also come to lie within the cytoplasm of the vascular struts (Fig. 10b). Yolk platelets with vacuoles inside their polyhedral yolk crystals have been observed; rod-like structures, which appear to be linear aggregates of small granules, occur within these vacuoles. The next putative stage in the processing of yolk is exemplified by platelets in which the space formerly occupied by blue-green yolk crystals has become occluded by purple metachromatic rods and granules. Subsequently, the rod-like aggregates are found to have dispersed within the cytoplasm.

The outer surface of the yolksac was examined for specialized structures that might serve a trophic or transport function. When visualized only in surface views, the exterior yolksac epithelium consists of an irregularly ordered sheet of cells that present a much folded and convoluted surface (Fig. 11a). When examined with SEM over a range of magnifi-

←

Fig. 10. a – In this scanning electron micrograph of a transverse section through the yolksac, the cortex in the upper portion of the micrograph consists primarily of a connective tissue stroma (cts). A thin epithelial layer (arrows) forms its external surface. Yolk (y) and blood vessels (bv) of the vitelline circulation lie beneath the cortex. b – This light micrograph of a transverse section through the cortex of the yolksac corresponds to the preceding figure. The connective tissue stroma (cts and inset) contains fibrous and cellular elements. Densely staining cytoplasm contains yolk (y) and supports elements of the blood vasculature system (bv). c – In this light micrograph, fusiform merocytes (m), i.e. cells that contain yolk platelets (yp), line the interior surface of the yolksac adjacent to the connective tissue stroma (cts). d – This light micrograph depicts various stages in yolk utilization. Yolk platelets (yp and inset) exhibit a violet metachromatic cortex and a green metachromatic core. During yolk utilization, filamentous structures, the residual yolk filaments (ryf) are produced.

238

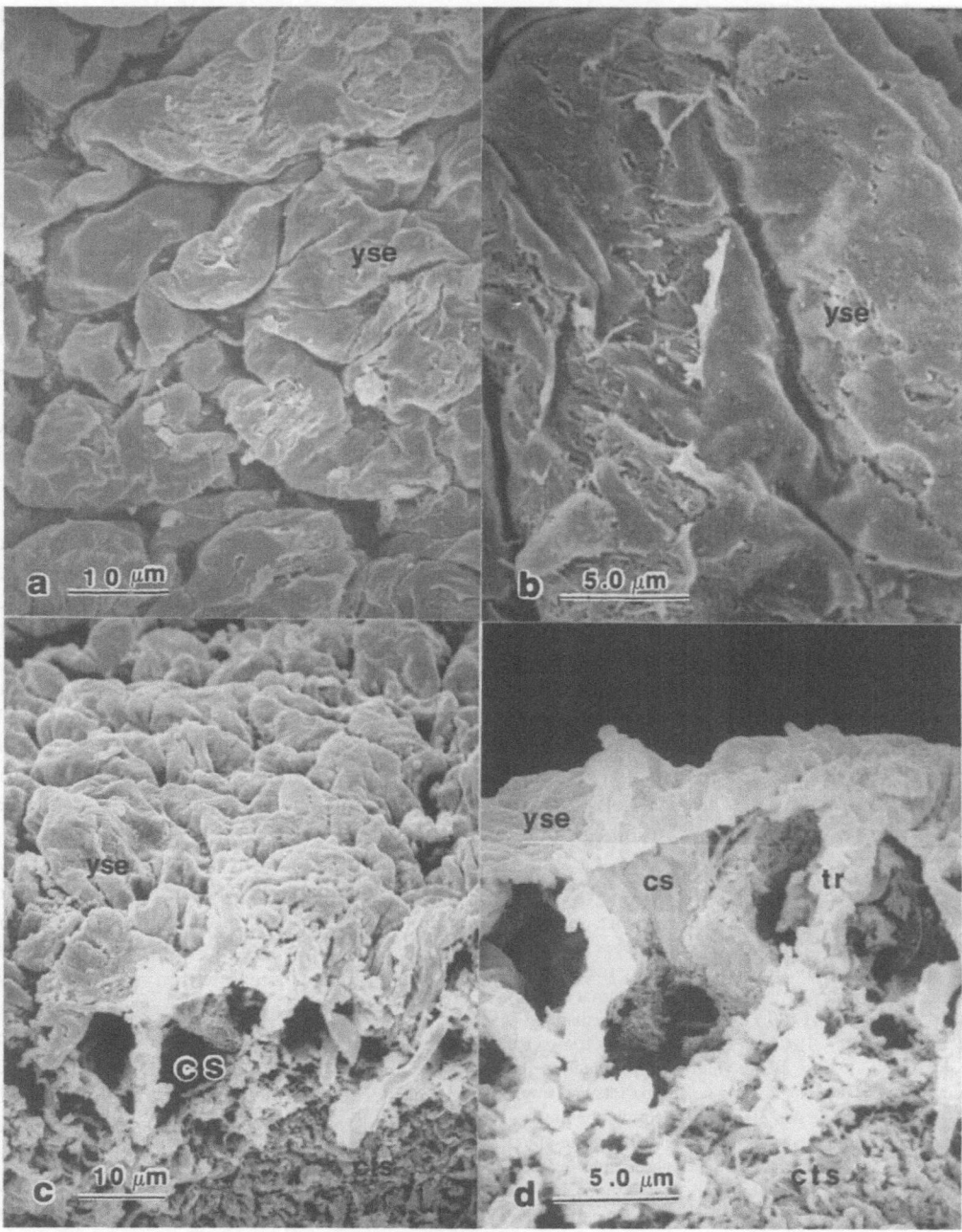

Fig. 11. a – This scanning electron micrograph presents an overview of the irregularly arranged cells that form the convoluted exterior surface (yse) of the yolksac. b – A higher magnification scanning electron micrograph reveals that the cells on the exterior of the yolk sac (yse) are essentially smooth-surfaced and lack any amplification of the cell surface. In some areas, e.g. at lower left, the cell membrane has been torn, revealing the underlying cytoskeleton. c – This scanning micrograph depicts both the exterior surface of the yolksac epithelium (yse) and its interior, the latter in transverse section. The yolksac epithelium encloses a series of cortical sinuses (cs). d – In this scanning electron micrograph of a transverse section through the yolksac cortex, the thin yolksac epithelium at the top of the figure encloses a well developed system of cortical sinuses (cs). Columnar trabeculae (tr) extend from the under surface of the epithelium to a loosely organized boundary at the upper surface of the connective tissue stroma.

cations (4X to 10000X), however, cells that form the epithelial sheet prove to be essentially smooth-surfaced (Fig. 11b). They lack amplifications of cell surfaces such as microvilli, microplicae, and cilia. The small pits visible on the surface may serve a transport function, but considering the method of tissue preservation that was used, they more likely represent artifacts. En face views of transverse sections through the yolksac cortex are more enlightening (Fig. 11c, d); here the epithelium is seen to enclose a well-developed system of inter-communicating cortical sinuses with columnar trabeculae that extend from the undersurface of the epithelium across the sinuses to terminate in a loosely organized boundary on the upper surface of the connective tissue stroma. These cortical sinuses appear to be associated with capillaries of the vitelline circulation, but the precise relationship and whether or not they are contiguous with the vitelline circulation has yet to be resolved. This pattern of cellular organization differs considerably from the anatomosing capillary network that occurs on the lumenal surface of the uterus. The system of cortical sinuses superficially resembles the vascular bulbs found on the surface of the pericardial trophoderm of *Anableps* (Knight et al. 1985) as well as the subcapsular sinuses of mammalian lymph nodes (Kessel & Kardon 1979).

Table 1. Estimated weight change during gestation of *Latimeria chalumnae.*

	Wet weight	Solids %	Dry weight	Change %
Egg	319 g(1)	58%(1)	184(1)	–
Late-term embryo	684 g(2)	25%(3)	171(3)	– 7%
		30%(3)	205(3)	+ 11%
		35%(3)	239(3)	+ 30%
Term embryo or juvenile	800 g(4)	25%(3)	200(3)	+ 9%
		30%(3)	240(3)	+ 30%
		35%(3)	280(3)	+ 52%

(1) Devys et al. (1972).
(2) Corrected on the basis that weight in alcohol (547 g) is 80% of total wet weight.
(3) Estimated % solids and corresponding dry weights based on values for full-term shark pups from Ranzi (1932, 1934) and Needham (1942).
(4) Anthony & Robineau (1976).

Weight determination

To distinguish between the lecithotrophic and matrotrophic forms of embryonic nutrition, it is necessary to determine the per cent change in dry weight that may take place between the mature egg and the term embryo (or pup) during gestation. According to Devys et al. (1972), the dry weight of the single coelacanth egg that has been chemically analyzed totaled 184 g. Inasmuch as none of the pups could be destroyed, it became necessary to obtain an estimate of the dry weight of pup 2, the only one that seems to have weighed. Its wet weight was 547 g, but it had been kept in alcohol, and so a correction factor to account for water loss through dehydration had to be applied. Accordingly, 547 g was considered to represent 80% of its total wet weight, that is, 684 g. Ranzi (1932, 1934) and Needham (1942) have shown that the dry weight of shark pups at term is about 30% of their wet weight. Using this factor, we estimate the dry weight of pup 2 to be 205 g, which represents an increase of 11% in dry weight during gestation. If we accept 25% and 35% as reasonable variations in the conversion factor of Ranzi and of Needham, the estimated dry weight of pup 2 would range from 171 to 239 g, corresponding to changes in dry weight during gestation of from − 7 to + 30%. When the same correction factors are used for the neonate or juvenile described by Anthony & Robineau (1976), the increases in dry weight range from 9 to 52% (Table 1).

Discussion

The maternal-embryonic trophic relationship

Following the report by Smith et al. (1975) that *L. chalumnae* is ovoviviparous, it was generally inferred that coelacanths exhibit a simple lecithotrophic form of viviparity. In lecithotrophy, the embryo depends entirely on its own yolk reserves during gestation, whereas in matrotrophy, the embryo receives an additional supply of maternal nutrients to supplement its initial supply of yolk (Wourms 1981). Coelacanths, as well as other animals that develop from yolk-rich eggs, undoubtedly are lecithotrophic during the earlier phases of their development. To be able to make a critical distinction between lecithotrophy and matrotrophy in livebearing forms, however, it is necessary to determine the change in dry weight that takes place during gestation by comparing the dry weight of the mature egg and term embryo. During the development of oviparous elasmobranchs, which by definition are lecithotrophic, there is at least a 21% loss of dry weight. Paradoxically, the loss of weight in some lecithotrophic, viviparous species may be even more extensive, e.g. − 40% in the spiny dogfish *Squalus* and − 54% in the European brown dogfish *Centrophorus*. In matrotrophic, viviparous sharks, rays, and bony fishes, there is either no net change in weight or an increase that can exceed 1 000 000%. The increase in weight is proportional to the extent of maternal nutrient transfer (reviewed in Wourms 1981, Wourms et al. 1988).

Three stages in the development of *L. chalumnae* presently are available for study: (1) about 19 mature, ovulated eggs (Anthony & Millot 1972, Millot & Anthony 1974), (2) five late-term pups (Smith et al. 1975), and (3) a single 420 mm free-living specimen (Anthony & Robineau 1976). We concur with Balon et al. (1988) that the latter is a neonate inasmuch as it still contains yolk in its body cavity. (N.B. The interior of the coelacanth yolksac is continuous with the body cavity, and the presence of yolk there is thus good evidence of the neonatal state.) By weighing one of the pups and using published data on the dry weight of a mature egg and the wet weight of the neonate, we have calculated the changes in estimated weight that occurred during gestation: from − 7 to + 30% for the late-term pup and from + 9 to + 52% for the neonate (Table 1). It is reasonable to conclude that some degree of matrotrophy characterizes the development of *L. chalumnae*.

Having established that coelacanth embryos receive a supplemental supply of maternal nutrients during their gestation, it then becomes necessary to determine the origin of the nutrients and the pathway of nutrient transfer. Four possible sources will be considered: (1) yolk reserves, (2) ingestion of histotrophe produced by the uterine mucosa, (3) uptake of uterine histotrophe by the yolksac epithelium, and (4) oophagy or adelphophagy, i.e. the ingestion of eggs or embryos (cf. Wourms 1981, Balon 1986, Wourms et al. 1988 for extensive discussions of patterns of embryonic nutrition). Based on the depletion of yolk and comparative studies of other fishes, it appears obvious that yolk reserves are the major source of nutrition during the earlier stages of development in *L. chalumnae*. It is surprising that a significant amount of yolk still remains in the yolksac of the late-term pup and that yolk has been detected in the body cavity of a neonate (Balon et al. 1988). It is tempting to speculate that a dual source of nutrition is maintained during the period immediately after parturition. Unused yolk reserves could supplement ingested food or serve as a substitute for ingested food, were that supply interrupted. The advantage of this nutritional strategy is that it would reduce neonatal mortality and allow growth to proceed at a constant rate during the first phase of juvenile life (cf. Balon 1985, 1986).

There is little evidence that uterine histotrophe plays a significant role in the nutrition of coelacanth embryos. It may be entirely possible, however, that the molecular constituents of histotrophe may serve some vital immunological or endocrinological function. Examination of the compartments in which the five pups resided did not reveal the presence of large amounts of histotrophe; in fact, the small quantities of extra-cellular material presumed to be histotrophe could be detected only at the microscopic level. Inasmuch as each compartment in the preserved, gravid uterus was isolated from the adjacent ones, presumably by post-mortem contraction of muscular inter-compartmental isthmuses, it seems unlikely that any histotrophe would have been lost by leakage. Moreover, small quantities of putative histotrophe were preserved in situ; if a copious amount had been present in the living fish, it would have been retained during preservation. Of course, it is entirely possible that during earlier phases of gestation, histotrophe may have been secreted in larger volume and contributed to nutrition. By comparison, specimens of fishes like matrotrophic rays and sharks, in which histotrophe is a vital element in embryonic nutrition, are characterized by the presence of large amounts of protein- or lipid-rich material throughout gestation (Wourms 1977, 1981, Wourms et al. 1988). It seems reasonable to conclude that only limited amounts of histotrophe are produced during coelacanth gestation. The structure of the uterine lumenal epithelium provides additional support for this conclusion. In the classification of piscine uterine environments set forth by Ranzi (1934), L. chalumnae would belong in group Ia in which the uterine epithelium is either smooth, as in Squalus, or forms short villi, as in Centrophorus, and secretion or fluid transport are minimal or absent. Jollie & Jollie (1967) have described an extensive system of juxta-epithelial capillaries in Squalus acanthias. Unfortunately, direct comparison between the observations made on the uterus of Squalus with transmission electron microscopy and on that of Latimeria with scanning electron microscopy is impossible. Nevertheless, extensive capillary beds undoubtedly are present in both cases – with capillaries that are in very close proximity to the uterine lumen.

A striking feature of the lumenal uterine epithelium in L. chalumnae is its close association with an underlying anastomosing capillary network. The SEM structure of this network is remarkably similar to what has been described in the trophonemata, i.e. the villous extensions of the uterine wall in certain matrotrophic rays (Hamlett et al. 1985, Wourms & Bodine 1984, Wourms et al. 1988). In these rays, secretion from glandular regions, as well as the transport of low- and high-molecular substances from the blood vascular system, appear to take place. Arguing by analogy, the transport of components with low molecular weight, such as amino acids, sugars, and fatty acids, across the capillary network of the coelacanth uterus and their subsequent release into its lumen cannot be ruled out – nor can it be verified on the basis of morphology alone. Thus, the presence or absence of histotrophe that contains small molecules must remain problematic. On the other hand, histotrophe that contains substances with high molecular weights, such as proteins and protein-carbohydrate complexes, seems to be present only in very limited amounts, if at all. As far as the pups are concerned, in their advanced state of development they could ingest and digest any kind of histotrophe. The transport of components of low molecular weight across the yolksac epithelium also is possible, but the situation with large molecules presents more problems inasmuch as microvilli, which are often associated with epithelial cell surfaces known to take up macromolecules, are absent. Their absence, however, does not necessarily preclude such uptake. Definitive evidence of macromolecular endocytosis can be obtained only from transmission electron micrographs or experiments using tracer molecules.

Oophagy, the ingestion of supernumerary eggs by developing young, may well be the major source of supplemental nutrients for coelacanth pups. The eggs are enormous: up to 9 cm in diameter and 334 g in weight. Anthony & Millot (1972) found 19 such eggs in the body cavity of a 1.63 m female, and a specimen dissected more recently in Japan contained about 30 ovarian eggs up to 4.5 cm in diameter (Suyehiro et al. 1982). The number of pups in a brood is limited by the size of the female, especially since only one oviduct is functional. The five pups in the 1.6 m female took up nearly all the available space in the oviduct. What would have been the fate of the 19 ovulated eggs in the other female of the same size? At the very most, such a fish could accommodate seven or eight developing embryos, and 11 or 12 eggs would than be superfluous-eggs, it might be pointed out, that represent a considerable outlay of energy in their production. We have suggested that these eggs serve as nutrients for the embryos that survive to term (Wourms et al. 1980). Late-term pups have a well-developed digestive system with a large stomach, and the stomach and intestine of pup 3 contained ingested material. Preliminary microscopic examination of the stomach contents and of histological sections through the gut of pup 1 suggest that the ingested material is yolk. Further studies may permit more definite conclusions.

Unresolved also is the mechanism by which the pups are exposed to and consume the egg material. Do all the ovulated eggs enter the oviduct in which some then survive and develop while the rest eventually become fragmented and ingested? Alternately, is the number of ovulated eggs that enters the oviduct limited to the number that can be supported to term? In the latter case, the supernumerary eggs would become fragmented in the body cavity and the fragments passed into the oviduct and conveyed to the developing young. On the basis of what is known about oophagy in sharks (Wourms et al. 1988), the first model appears to be more likely.

Oophagy probably is a relatively primitive specialization which may have evolved to take advantage of the wastage of gametes that is not uncommon in viviparous animals with large broods. It could have had its origin in the comsumption of moribund eggs by surviving members of a brood. A transition from casual scavenging to developmentally-programmed oophagy would seem to require rather small changes in maternal and embryonic physiology and morphology. Oophagy and adelphophagy (embryonic cannibalism) have evolved independently a number of times among various chondrichthyans and osteichthyans (Wourms et al. 1988). Among chondrichthyans, both are characteristic reproductive modes in sharks of the order Lamniformes (Compagno 1984). The evolution of oophagy in *L. chalumnae* must differ from that in other fishes, however, because of the enormous size of the coelacanth egg. In oophagous lamniform sharks, e.g. *Eugomphodus, Alopias,* and *Cetorhinus,* the acquisition of oophagy apparently is correlated with a reduction in egg size (reviewed in Wourms et al. 1988). Moreover, at least in sharks, oophagy appears to be a simple, efficient strategy for attaining neonatal gigantism, a feature with obvious survival value in a predominantly fish-eating group. Late-term pups and neonates of *L. chalumnae,* although of considerable size as compared with the offspring of other bony fishes, are significantly smaller than the pups of most oophagous sharks (Wourms et al. 1988); in fact, they are not appreciably larger than the pups of *Centrophorus* and *Chlamydoselachus,* both of which also develop from usually large eggs. We conclude that oophagy in *L. chalumnae* is only marginally developed. Perhaps an evolutionary stasis has occurred in which the ingestion of supernumerary eggs evolved but without a reduction in egg size. Some clues toward

the resolution of this paradox have been offered by Balon (1991) who points out that the egg of the Carboniferous coelacanth *Rhabdoderma,* considered by Schultze (1980, 1985) to be oviparous, is relatively large (36 mm in diameter) whereas the Jurassic coelacanth *Holophagus* is viviparous (Watson 1927). Viviparity thus evolved perhaps 200 million or more years ago in this group of fishes and has been retained. Did the evolution of large eggs predispose the coelacanths toward viviparity? Did viviparity lead to the evolution of even larger eggs? According to this scheme, oophagy evolved subsequently. In this connection, the fact that the primitive viviparous sharks *Chlamydoselachus* and *Ginglymostoma* also have very large eggs requires consideration.

We suggest two scenarios to help explain the retention of enormous egg size along with a modest degree of oophagy. The first scenario is based on the concept of developmental constraints in which the highly buffered genetic-epigenetic developmental program is recognized as being able either to resist developmental changes, i.e. show negative constraint, or to favor such changes, i.e. show positive constraint. An egg of enormous size represents a highly specialized state of the oocyte, the evolution of which must have involved both qualitative and quantitative changes in this type of cell as well as its interactions with other tissue systems such as liver, blood vascular, and endocrine. Although the developmental program of oogenesis may be quite flexible, once altered to produce such a highly specialized egg, it may have proved refractory to further evolutionary change. This would seem to apply to *L. chalumnae,* in which a reduction in egg size, the predicted state, does not occur. The second scenario recognizes that *L. chalumnae* represents a relict group of animals and, in common with other such representatives, it probably exhibits an exceedingly slow rate of evolutionary change. The enormous size of the coelacanth egg may have become established about 200 million years ago, at a time of more rapid evolutionary change, as indicated by the much greater taxonomic diversity of coelacanths at that time. Along with many other coelacanth characters, egg size has changed little since then. This putative evolutionary status of the reproductive mode of *L. chalumnae*, in which a primitive form of oophagy (matrotrophy) is coupled with eggs of enormous size (a primitive but specialized form of lecithotrophy), may be one of the factors that helps account for the relative lack of success of the coelacanths and the current status of *L. chalumnae* as a relict species.

The coelacanth yolksac placenta

Can the yolksac of the late-term pup of *L. chalumnae* properly be called a yolksac placenta? Amoroso (1981) suggested that it can. Wourms et al. (1988) considered the question and accepted Amoroso's interpretation but with reservations. Here we review this issue in the light of our more detailed description and comprehensive analysis of the structure of both the uterus of the gravid female and the yolksac of the late-term pup.

The yolksac of the late-term pup of *L. chalumnae* is indeed a non-adherent, transposable placenta; Amoroso's (1981) insight was correct but for the wrong reasons, a situation that subsequently affected his interpretation of the function of the yolksac. According to Mossman (1937), an animal placenta is an intimate apposition or fusion of fetal organs to maternal or paternal tissue for physiological exchange. We believe that our functional analysis has shown that the yolksac serves as the embryonic (fetal) portion of the yolksac placenta of which the maternal portion consists of the the highly vascularized zone of contact of the uterine lumenal epithelium that lies within the embryo's uterine compartment.

These conclusions are based on the following observations and interpretations. In the only known brood of *L. chalumnae*, five advanced young were developing in constricted, compartment-like regions of the uterus (oviduct). Each compartment has an asymmetric shape that conforms to that of the pup it shelters. All the pups are oriented with the head facing toward the anterior end of the compartment. The isthmus connecting adjacent compartments is muscular with the muscles in a contracted state. We conclude that the conforming shape of each compartment is a stable

configuration during the period of gestation and not an artifact of contraction produced during the course of preservation. The pups possess large, flaccid, heavily vascularized yolksacs which are nearly empty of yolk. The sac is in intimate contact with the uterine wall but it does not adhere to it, nor is there any direct connection or interdigitation with the wall. The yolksac is free to move, i.e. it is transposable. Where the uterine tissue conforms to the yolksac, a distinct zone of contact is formed. A single major vascular plexus occupies the area of intimate contact. Vascular plexuses of this type have not been seen in the mature, non-gravid oviduct-uterus (Millot et al. 1978). Thus, the zone of contact between the yolksac and the uterine wall with its vascular plexus forms a stable, structurally specialized maternal-embryonic tissue relationship that develops only during gestation.

The yolksac of the late-term pup, in a sense, is incongruous. It is the only external part of the pup fully to retain its embryonic characteristics. The yolksac cortex can be regarded as an extension of the body wall, and its lumen is continuous with the body cavity. An extensive zone of attachment between body and yolksac originates between the gular plates and continues posteriorly for nearly a fifth of the total length of the fish. In contrast the chondrichthyan yolksac is connected to the body by a narrow yolk stalk which contains the vitelline duct and the vitelline artery and vein (Wourms 1977). The extensive zone of attachment, the absence of a vitelline duct, and the organization of tissues distinguish the coelacanth yolksac from the chondrichthyan yolksac (cf. Mossman 1987).

Retention of a large yolksac in the late stages of development in viviparous fishes is a specialized condition (Balon 1985). For example, it does not occur in lecithotrophic, viviparous sharks like *Squalus acanthias* in which there is a progressive diminution and resorption of the yolksac as gestation proceeds. A similar process of dimunition and resorption occurs during the initial lecithotrophic phase in the development of nonplacental, matrotrophic rays (Hamlett et al. 1985). It is only in the placental sharks that the yolksac is retained after the partial or nearly complete depletion of the yolk reserves and before implantation. The latter condi-

244

tion is approximated in the coelacanth. The distal portion of the selachian yolksac establishes contact with the uterine wall to form the site of placental attachment; in the coelacanth, however, the entire surface of the yolksac remains in contact with uterine tissue and serves as a site of transfer. Other similarities exist between the coelacanth yolksac placenta and the yolksac placenta of certain sharks. Based on the morphology of the maternal-fetal junction, Wourms et al. (1988) recognized six classes of shark placentae. The embryonic placenta of sharks of class I, e.g. *Carcharhinus falciformis*, the silky shark (Gilbert & Schlernitzauer 1966), closely resembles the embryonic portion of the coelacanth yolksac placenta. In the sharks of this group, the exterior epithelium of the sac is reduced to a thin layer of flattened cells, and many capillaries lie directly beneath the epithelium. Maternal and embryonic tissues do not interdigitate; instead the embryonic placental tissue rests on the maternal portion of the placenta. An egg envelope, which is a type of collagen-rich extra-cellular matrix, actually intervenes between embryonic and maternal tissues, the latter remaining unchanged at the site of contact except for an extensive development of subcortical capillary networks.

We have already considered the role of the uterine wall and the yolksac in maternal-embryonic nutrient transfer. An anatomosing capillary network characterizes the lumenal epithelium of the uterine wall. The network is similar to that found in the trophonemata of matrotrophic rays (Hamlett et al. 1985), the pneumocyte-capillary arrangement in the alveolar sac of the lung, the choroid plexus (Kessel & Kardon 1979), and the pecten of the avian eye (Kokkala personal communication). These structures all participate either in molecular or in gas exchange. Similarly, the exterior surface of the yolksac consists of a thin, single-layered epithelium that covers an intercommunicating system of cortical sinuses associated with capillaries of the vitelline circulation. In the late-term pup, the vitelline circulation of the yolksac is massively developed. The arrangement of many parallel, more or less equally spaced blood vessels entering the embryonic circulation along the entire junction between the sac and belly wall never seems to have

been reported in the embryo of any other fish. Thus, although there are features of the yolksac that need further study, we do recognize that it is structurally specialized for molecular transport and gas exchange.

On a basis of the extensive vascularity of the maternal and embryonic tissues as well as the attenuated, single-layer epithelia that form the placental interface, we conclude that the primary functions of the coelacanth yolksac placenta are gas exchange and the exchange of materials of low molecular weight such as metabolites and metabolic wastes.

Acknowledgements

Research has been supported in part by National Science Foundation Grants DCB-82085 25 and DCB 8609690, National Institute of Health Biomedical Research Support Grant 2-507-RRO7180, and a Guggenheim Fellowship to J.P.W. One of us (J.P.W) takes this occasion to acknowledge the long-standing cooperation of the Department of Herpetology and Ichthyology at the American Museum of Natural History in permitting access to the fish collections and the Bashford Dean Library. At the museum, C. Lavett Smith, Carl J. Ferraris, and Tom Trombone were extremely helpful with photography. Irene Kokkala kindly made her unpublished scanning electron micrographs of vascularization in the pecten of the avian eye available for comparison.

References cited

Amoroso, E.C. 1981. Viviparity. pp. 3–25. *In:* S.R. Glasser & D.W. Bullock (ed.) Cellular and Molecular Aspects of Implantation, Plenum Press, News York.

Anthony, J. & J. Millot. 1972. Première capture d'une femelle de coelacanthe en état de maturité sexuelle. C.R. Acad. Sc. Paris Sér. D 274: 1925–1926.

Anthony, J. & D. Robineau. 1976. Sur quelques caractères juvéniles de *Latimeria chalumnae* Smith (Pisces, Crossopterygii, Coelacanthidae). C.R. Acad. Sc. Paris Sér. D 283: 1739–1742.

Atz, J.W. 1976. *Latimeria* babies are born, not hatched. Underwater Naturalist 9(4): 4–7.

Balon, E.K. 1984. Patterns in the evolution of reproductive styles in fishes. pp. 35–53. *In:* G.W. Potts & R.J. Wootton (ed.) Fish Reproduction: Strategies and Tactics, Academic Press, London.

Balon, E.K. 1985. The theory of saltatory ontogeny and life history models revisited. pp. 13–30. *In:* E.K. Balon (ed.) Early Life History of Fishes: New Developmental, Ecological and Evolutionary Perspectives, Developments in Env. Biol. Fish. 5, Dr W. Junk Publishers, Dordrecht.

Balon, E.K. 1986. Types of feeding in the ontogeny of fishes and the life-history model. Env. Biol. Fish. 16: 11–24.

Balon, E.K. 1991. Probable evolution of the coelacanth's reproductive style: lecithotrophy and orally feeding embryos in cichlid fishes and in *Latimeria chalumnae*. Env. Biol. Fish. 32: 249–265. (this volume)

Balon, E.K., M.N. Bruton & H. Fricke. 1988. A fiftieth anniversary reflection on the living coelacanth, *Latimeria chalumnae:* some new interpretations of its natural history and conservation status. Env. Biol. Fish. 23: 241–280.

Bemis, W.E. & R.G. Northcutt. 1991. Innervation of the basicranial muscle of *Latimeria chalumnae*. Env. Biol. Fish. 32: 147–158. (this volume)

Compagno, L.J.V. 1984. Sharks of the world. An annotated and illustrated catalogue of shark species known to date. FAO Fish. Synop. No. 125 (4, Pt. 1) viii + 249.

Devys, M., A. Thierry, M. Barbier & M.M. Janot. 1972. Premières observations sur les lipides de l'ovocyte du coelacanthe (*Latimeria chalumnae*). C.R. Acad. Sc. Paris Sér. D 275: 2085–2087.

Fricke, H., O. Reinicke, H. Hofer & W. Nachtigall. 1987. Locomotion of the coelacanth *Latimeria chalumnae* in its natural environment. Nature 329: 331–333.

Fritzsch, B. 1987. Inner ear of the coelacanth fish *Latimeria* has tetrapod affinities. Nature 327: 153–154.

Gilbert, P.W. & D.A. Schlernitzauer. 1966. The placenta and gravid uterus of *Carcharhinus falciformis*. Copeia 1966: 451–457.

Grodziński, Z. 1972. The yolk of *Latimeria chalumnae* Smith. Folia histochem. cytochem. 10: 11–18.

Hamlett, W.C., J.P. Wourms & J.W. Smith. 1985. Sting-ray placental analogue: structure of the trophonemata in *Rhinoptera bonasus*. J. submicrosc. Cytol. 17: 31–40.

Hensel, K. 1986. Morphologie et interprétation des canaux et canalicules sensoriels céphaliques de *Latimeria chalumnae* Smith, 1939 (Osteichthyes, Crossopterygii, Coelacanthiformes). Bull. Mus. natn. Hist. nat. Paris Sér. 4, 8: 379–407.

Hoar, W.S. 1969. Reproduction. pp. 1–72. *In:* W.S. Hoar & D.J. Randall (ed.) Fish Physiology, Volume 3, Academic Press, New York.

Jollie, W.P. & L.G. Jollie. 1967. Electron microscopic observations on accommodations to pregnancy in the uterus of the spiny dogfish, *Squalus acanthias*. J. Ultrastruct. Res. 20: 161–178.

Kessel, R.G. & R.H. Kardon. 1979. Tissues and organs. W.H. Freeman and Co., San Francisco. 317 pp.

Knight, F.M., J. Lombardi, J.P. Wourms & J.R. Burns. 1985.

Follicular placenta and embryonic growth of the viviparous four-eyed fish (*Anableps*). J. Morph. 185: 131–142.

Lange, R.H. 1983. Les cristaux de lipovitelline-phosvitine dans l'ovocyte de *Latimeria chalumnae* Smith 1939 (Coelacanthidae, Pisces). Étude comparative. C.R. Acad. Sc. Paris Sér. III 297: 393–396.

Locket, N.A. 1976. A future for the coelacanth? New Scientist 70: 456–458.

Locket, N.A. 1980. Some advances in coelacanth biology. Proc. R. Soc. Lond. B208: 265–307.

McAllister, D.E. & C.L. Smith. 1978. Mensurations morphologiques, dénombrements méristiques et taxonomie du coelacanth, *Latimeria chalumnae*. Naturaliste can. 105: 63–76.

McCosker, J.E. & M.D. Lagios (ed.) 1979. The biology and physiology of the living coelacanth. Occ. Pap. Calif. Acad. Sci. 134. 175 pp.

Millot, J. & J. Anthony. 1974. Les oeufs du coelacanthe. Science et Nature 121: 3–4 (+ color cover photograph).

Millot, J., J. Anthony & D. Robineau. 1978. Anatomie de *Latimeria chalumnae*, Vol. 3, Appareil digestif, appareil respiratoire, appareil urogénital, glandes endocrines, appareil circulatoire, téguments, écailles, conclusions générales. C.N.R.S., Paris. 198 pp.

Mossman, H.W. 1937. Comparative morphogenesis of the fetal membranes and accessory uterine structures. Contrib. Embryol. Carnegie Inst. 26: 129–246.

Mossman, H.W. 1987. Vertebrate fetal membranes. Rutgers University Press, New Brunswick. 383 pp.

Myking, L.M. 1977. Old four legs: the living fossil. Sea Frontiers 23: 334–341.

Needham, J. 1942. Biochemistry and morphogenesis. Cambridge University Press, London. 785 pp.

Ranzi, S. 1932. The physio-morphological basis of embryonic development in sharks. (Le basi fisio-morfologische dello sviluppo embrionale dei selaci.) Parti I. Pubbl. St. Zool. Napoli 13: 209–290. (In Italian).

Ranzi, S. 1934. The physio-morphological basis of embryonic development in sharks. (Le basi fisio-morfologische dello sviluppo embrionale dei selaci.) Parti II, III. Pubbl. St. Zool. Napoli 13: 331–437. (In Italian).

Schultze, H.-P. 1980. Eier legende und lebend gebärende Quastenflosser. Natur und Museum 110: 101–108. (English translation from the Department of Geology, Field Museum of Natural History, Chicago).

Schultze, H.-P. 1985. Reproduction and spawning sites of *Rhabdoderma* (Pisces, Osteichthyes, Actinistia) in Pennsylvanian deposits of Illinois, USA. Neuvième Congr. Internat. Stratigr. Geol. Carbonifère, Washington, Compte Rendu 5: 326–330.

Smith, C.L., C.S. Rand, B. Schaeffer & J.W. Atz. 1975. *Latimeria*, the living coelacanth, is ovoviviparous. Science 190: 1105–1106.

Suyehiro, Y., T. Uyeno & N. Suzuki. 1982. Coelacanth: dissecting a living fossil. Newton graphic Science Magazine 2(8): 82–93. (In Japanese).

Suzuki, N., Y. Suyehiro & T. Hamada. 1985. Initial report of

246

expeditions for coelacanth - Part I- Field studies in 1981 and 1983. Sci. Pap. Coll. Arts Sci. Univ. Tokyo 35: 37–79.

Tanaka, S. 1989. Extant frilled sharks. Collecting and Breeding 51: 50, 61–63. (In Japanese).

Watson, D.M.S. 1927. The reproduction of the coelacanth fish, *Undina*. Proc. Zool. Soc. Lond. 1927: 453–457.

Wourms, J.P. 1977. Reproduction and development of chondrichthyan fishes. Amer. Zool. 17: 379–410.

Wourms, J.P. 1981. Viviparity: the maternal-fetal relationship in fishes. Amer. Zool. 21: 473–515.

Wourms, J.P. & A.B. Bodine. 1984. Structure and function of trophonemata, a placental analogue, during early gestation of the butterfly ray. p. 407. *In:* S. Seno & Y. Orada (ed.) International Cell Biology 1984, Academic Press, Orlando. (Abstract)

Wourms, J.P. & D.M. Cohen. 1975. Trophotaeniae, embryonic adaptations in the viviparous ophidioid fish, *Oligopus longhursti*: a study of museum specimens. J. Morph. 147: 385–401.

Wourms, J.P., B.D. Grove & J. Lombardi. 1988. The maternal-embryonic relationship in viviparous fishes. pp. 1–134. *In:* W.S. Hoar & D.J. Randall (ed.) Fish Physiology, Volume 11B, Academic Press, San Diego.

Wourms, J.P., M.D. Stribling & J.W. Atz. 1980. Maternal fetal nutrient relationships in the coelacanth, *Latimeria*. Amer. Zool. 20: 962. (Abstract)

Addendum

In a recent publication, Heemstra & Compagno (1989) have challenged the hypotheses of oophagy and placental viviparity in *Latimeria chalumnae* that were set forth by Wourms et al. (1980, 1988) as well as Balon et al. (1988). Although many of the issues raised by the two authors have been addressed in the present paper, several points require further amplification. Moreover, several factual errors ought to be corrected. The hostile tone of their article is puzzling inasmuch as the authors must be aware of the tentative nature of several of our arguments, as well as their own, this being the result of the paucity of information on the reproduction of *Latimeria* and the difficulty in obtaining any additional data on the subject. Contrary to their assessment, all of our publications have been based primarily on original research on *Latimeria* and oophagous and placental sharks.

Several points in the criticism of the oophagy hypothesis are addressed below:

(1) Heemstra & Compagno indicate that evidence bearing on the relationship of the uterine young of *Latimeria* to their mother was presented in Smith et al. (1975) and Atz (1976); in fact, such evidence was specifically avoided inasmuch as it was being reserved for future use by Atz and his collaborator, Wourms.

(2) In the discussion of embryonic weight gain, much is made of the fact that the dry weight of the pup was extrapolated from its wet weight. It was necessary to extrapolate the dry weight because the pup was too valuable to destroy by desiccation. The process of extrapolation is confused with a sentence taken out of context, viz. '*this approach* is useful as a first approximation' (Wourms et al. 1988 p. 36) in which 'approach' refers to the comparison of egg and term fetal dry weights and not to extrapolation. Moreover, in the original publication, the next sentence points out that the approximation errs on the conservative side by underestimating weight increase.

(3) In the next paragraph Heemstra & Compagno ask if the presence of supernumerary eggs is evidence for oophagy and conclude that it is not. We agree that supernumerary eggs in and of themselves are not evidence of oophagy, but the occurrence in *L. chalumnae* of as many as 10–12 supernumerary eggs of such a large size is unusual in any organism and represents a substantial investment of maternal energy. Our discussions have been based on the documented number of 19 ovulated eggs. Heemstra & Compagno suggest that it would be possible for a coelacanth to bring 19 pups to term since the limiting factor is space in the oviduct and not in the body activity. On the basis of 320 mm pups and allowing 50 mm between adjacent compartments, we calculated that 19 embryos would occupy 7.0 meters of uterine space in a 2.0 meter fish.

(4) In their discussion of uterine cannibalism in sharks, a subject also reviewed in Wourms et al. (1988), Heemstra & Compagno call attention to the retention of large yolksacs in the late-stage embryos of some sharks. While large yolksacs are associated with late developmental phases of some sharks, especially those with large eggs (such as *Chlamydoselachus* and *Centrophorus*), we have observed that the yolksacs are turgid and filled with unused yolk. Moreover, as the yolk is consumed,

their yolksacs diminish in size. This is not the case in *Latimeria*.

(5) In any discussion of oophagy, it is important to remember that oophagy has evolved independently among teleosts and among sharks. Among sharks, oophagy is not confined to Lamniformes as Heemstra & Compagno state but also occurs in at least one species within the Carcharhiniformes, namely *Pseudotriakis microdon* (Yano 1989). Oophagy, like any other reproductive process, displays evolutionary diversity. In its simplest form, it appears to have evolved as a means by which energy present in unfertilized or moribund eggs could be utilized by the viable members of a brood. From a relatively simple situation such as this, various patterns of oophagy presumably evolved. In the sand tiger shark *Carcharias (Eugomphodus) taurus,* oophagy is specialized and presumably evolved from a less complex reproductive pattern; the packaging of many (10–17) small, 10 mm eggs in a single egg case is a situation seldom seen in sharks. Heemstra & Compagno err in considering the specialized characteristic of small egg size as a general characteristic of oophagy in fishes.

(6) The state of activity of the ovary of *L. chalumnae* is most likely not relevant. The amount of yolk reserves contained in the supernumerary eggs is more than adequate to account for the modest increase in weight that appears to take place in *Latimeria*. Fragmented eggs would produce a mixture of yolk platelets and fluid components constituting an embryotrophe that could be easily ingested by developing pups. There is no reason to discard the notion that the ingestion of egg-derived embryotrophe could take place during early-mid development even when yolk reserves are still mostly intact; in fact, Yano (1989, and unpublished) has pointed out a somewhat analogous situation in the oophagous carcharhinid shark, *Pseudotriakis microdon,* in which oophagy complements lecithotrophy. Heemstra & Compagno are probably right in stating that the narrow diameter of the passageway between adjacent uterine compartments would prevent the transit of intact supernumerary eggs, but it would not prevent the transit of fragmented egg-derived embryotrophe. Finally, it should be pointed out that Wourms et al. (1988)

did not attribute adelphophagy, i.e. cannibalization of similarly sized siblings, to the coelacanth.

(7) Wourms et al. (1988) noted that the well developed gut of pup 3 contained ingested brown, amorphous material. Examination of histological sections through the gut of pup 1 confirmed this. Based on preliminary studies, which require confirmation, the ingested material appears to be yolk. Should this be the case, it would be positive proof of oophagy.

The crux of the argument by Heemstra & Compagno concerning the lack of any placenta in *L. chalumnae* is contained in the following sentence: 'In *Latimeria* there is no maternal/fetal interface; the yolk sac is not attached to the oviduct, hence there is no attachment site; and consequently, *Latimeria* does not have a placenta'. In the present paper, we have shown that there is indeed a well-defined maternal/fetal interface. Moreover, the yolksac is actually in apposition with a specialized region of the uterus. Both here and previously (Wourms et al. 1988, p. 53), it is clearly stated that the yolksac is actually in apposition to the uterine lining, *not* attached to or fused with it. The word 'apposition' completely satisfies the first of the two alternative criteria of a maternal/fetal placental interface in Mossman's (1937) definition of a placenta (cf p. 36 in Wourms et al. 1988). Heemstra & Compagno have failed to follow the argument involving the modern concept of a placenta, an essential point of which is that fusion or attachment is not the sole criterion for evaluating the maternal-fetal placental interface. Their expressed inference of specious argumentation on our part arises, apparently, from their insistence on using what is to us an outmoded, narrow definition of the placenta. A matter of definition has been turned into an implication of deception.

Our characterization of the maternal-fetal relationship in *L. chalumnae* as a form of placental viviparity was not arrived at lightly. As we have pointed out here and elsewhere, the late E.C. Amoroso (1981), an acknowledged authority on placentation, was the first to suggest such a placental relationship. Amoroso's insight was correct but his reasoning and conclusions were wrong. Subsequently we evaluated his suggestion and adopted

248

it, based on the evidence of our own research. This evidence and our interpretation of it are presented at length in the present paper. We reiterate our conclusion that *Latimeria* has a placental form of viviparity.

We cannot understand Heemstra & Compagno's belief that 'the poor quality of the tissue preservation in the only known gravid female precludes any substantive evaluation of nutrient transfer based on histology or cytology'. To the best of our knowledge, no attempt was ever made to evaluate first-hand the state of preservation of the gravid female and its offspring. Moreover, our paper demonstrates that the quality of preservation is certainly adequate for the purposes of light microscopic histology and scanning electron microscopy.

Finally, we find it rather curious that Heemstra & Compagno have reverted to using the term ovoviviparity. Currently, a distinction is made only between oviparity and viviparity; it is recognized that ovoviviparity represents an obfuscating, artificial category (reviewed in Wourms et al. 1988, p. 30).

Heemstra, P.C. & L.J.V. Compagno. 1989. Uterine cannibalism and placental viviparity in the coelacanth? A skeptical view. S. Afr. J. Sci. 85: 485–486.

Yano, K. 1989. Comments on the reproductive mode of the false cat shark *Pseudotriakis microdon*. Program and Abstracts. 69th Annual Meeting Amer. Soc. Ichthyol. Herpetol: 164.

Environmental Biology of Fishes **32**: 249–265, 1991.

Probable evolution of the coelacanth's reproductive style: lecithotrophy and orally feeding embryos in cichlid fishes and in *Latimeria chalumnae*

Eugene K. Balon
Institute of Ichthyology, Department of Zoology, College of Biological Science, University of Guelph, Guelph, Ontario N1G 2W1, Canada

Received 20.6.1989   Accepted 8.7.1990

Key words: Viviparity, Vitellogenesis, Direct development, Parallel evolution, Egg size, Size at release, Alprehost, Speciation, Fossils, Lake Tanganyika, *Cyphotilapia frontosa*, Crossopterygians, Actinistia, *Rhabdoderma exiguum*

Synopsis

The living coelacanth is a livebearer. Yolk seems to be the main source of nutrients and of oxygen to the embryo (fetus). Long before birth, young may also possibly feed orally on histotrophe secretion and egg debris. This type of reproduction evolved, as in most other fishes, from oviparity. The Carboniferous coelacanth *Rhabdoderma exiguum* had eggs of much lesser yolk volume and may represent an earlier form of oviparity with hiding, guarding or brooding type of parental care. The Jurassic coelacanth *Holophagus (Undina)* and the Cretaceous *Axelrodichthys* appear to have already evolved the internal-bearing style. Much of this evolutionary sequence is similar to that in cichlids. Ancestral cichlids are substrate tenders and nesters, with small eggs, little yolk and a feeding larva with indirect development. Mouthbrooding cichlids evolved a few, large eggs with denser yolk, direct development and, ultimately, orally feeding embryos while yolk is still in ample supply. Mixed feeding from yolk and orally ingested food in cichlids and in coelacanths is shown to be an enhanced mode of food delivery to the embryos over that from each source separately, in order to produce directly a better developed or larger young at the time of release, i.e. independence. Increase in egg size is regarded as an environmentally induced, altered pattern of yolk synthesis and an initial component of the epigenetic mechanism leading towards greater specialization. Carotenoids are incorporated within the yolk to assist the oxidative metabolism of the developing embryo.

Introduction

> ...each new study brings to light fascinating information on the ways in which the coelacanth has found unique answers to evolutionary and ecological problems.
>
> M.N. Bruton (1989a)
> in *The Living Coelacanth Fifty Years Later*

Wourms et al. (1980, 1988, 1991) concluded, for well documented reasons, that the living coela-canth is a livebearer that utilizes both lecithotrophy and matrotrophy. The embryos (later fetuses = young = pups) of *Latimeria chalumnae* appear to depend mostly on their large yolk supply, but they increase their weight to such an extent that oophagy and possibly the digestion of histotrophe through the embryonic gut and especially placentotrophy are likely. The latter is supported by the presence of some histotrophe in the oviduct and apposition of the yolksac with the uterus, the former by space limitations for the development of all

(19) ovulated eggs found in one female. The living coelacanth's eggs are very large – 90 mm in diameter and 334 g in weight. Among known fishes only the gulper shark, *Centrophorus granulosus*, has eggs of similar size. Fetuses of this shark, however, lose weight during gestation in spite of maternal nutrient transfer via histotrophe secretion (Wourms et al. 1991). Because young of *L. chalumnae* gain about 9 to 43% in weight during gestation, a denser yolk and more effective transfer of nutrients are strongly suggested. In the absence of other absorptive structures than the weakly alligned capillary plexuses of the yolk and uterine walls (Wourms et al. 1991), oral ingestion and intestinal digestion are the main nutrient delivery routes during most of the latter part of gestation. The oophagy is supported further by the fact that 'all of the young were situated with their heads directed away from the urogenital orifice' (Smith et al. 1975, p. 1106), thus facilitating feeding on egg fragments which are passed down the oviduct.

While I can add little information about the maternal-fetal relationship explained by Wourms et al. (1991), I present here a hypothesis on the evolutionary trajectory for the production of such an extremely large egg. En passant, I intend to contribute an alternative explanation for the embryonic oxygen supply over and above the capillary plexuses of the yolksac and uterine walls found by Wourms et al. (1991).

The living coelacanth, first thought to be oviparous, was relatively recently determined to be a livebearer, primarily as a result of the discovery of three individual females, one with 19 mature eggs (Anthony & Millot 1972, Millot & Anthony 1974), another with 30 ovarian eggs (Suyehiro et al. 1982, 1983) and the third with five yolksac fetuses (Smith et al. 1975). Decisive data for the correct interpretation were, however, accumulating before the latter find: the evidence for internal fertilization appeared with the description of the male's urogenital papilla and erectile folds surrounding the cloaca (Millot & Anthony 1960a, b, 1973a, b); whereas Griffith & Thomson (1973) pointed out that embryos cannot retain urea for osmoregulation in sea water and must, therefore, develop in the female's

body and be released only after they have achieved this ability because the eggs are not enclosed in impermeable cases as in oviparous sharks. As is often the case, however, it took the undisputable fact of the five fetuses in the oviduct to finally convince everybody. Unfortunately, some still insist on the long-abandoned term ovoviviparity (e.g. Schultze 1985, Smith 1986) which was replaced, at Wourms' (1981) suggestion, by the more fitting term 'obligate lecithotrophic livebearing' (Balon 1981a). Some others insist on ovoviviparity simply because they cannot take progress in science without irritation (e.g. Heemstra & Compagno 1989) in spite of expert opinions (e.g. Wourms et al. 1988, 1991). It now seems that *L. chalumnae* belongs to an even more advanced matrotrophic reproductive guild of internal bearers than the simple lecithotrophic viviparity, a new guild (Balon 1981b, 1990b) which combines lecithotrophy, oophagy, trophodermy and placentotrophy (Wourms et al. 1988, 1991).

Detailed reviews and interpretations were recently published in some of the jubilee compilations (e.g. Balon et al. 1988, Forey 1988, Bruton 1989a, b, c, Greenwood 1988, 1989, Stobbs 1989a, b, Balon 1990a), on most other problems of coelacanth biology, so it is not necessary here to repeat this information (see also Courtenay-Latimer 1988. These data support the interpretation of *L. chalumnae* reproduction accepted here.

Trends in the evolution of reproductive styles

I have devoted at least two separate studies to trends in the evolution of reproductive styles (Balon 1981a, 1984a). The recognition of such trends initially led to the classification of reproductive styles into discrete guilds (Balon 1975a, 1981b, 1990b). For the purpose of this study, therefore, only the relevant aspects will be summarized (Balon 1986a).

Most fishes broadcast gametes into the external environment. Their eggs are small and, more importantly, of low yolk volume and density but in large numbers, often hundred of thousands and as many as 40 million per female (e.g. Topp & Girardin 1971). The resultant ontogeny is indirect with the larva as an additional food gathering interval to complement the yolk, which is insufficient to form the definitive phenotype alone. The latter is remodelled from a tissue-rich larva via a more or less drastic metamorphosis (e.g. Flegler-Balon 1989).

A changing internal and external environment, especially an increase in complexity and diversity, accelerates specialization, which is normally an irreversible tendency within the sequence of generations (Balon 1983, 1986b, 1988a, b). This acceleration of specialization is in part due to the demand for isolation in order to avoid competition. The most conspicuous results of specialization are an increase in egg size, or more correctly, yolk volume and density, and a decrease in egg numbers, followed by hiding, nesting, brooding or bearing the offspring (e.g. Blumer 1979, 1982, Baylis 1981, Barlow 1981, Balon 1977, 1984a). These breeding styles were interpreted as security arrangements to guard the ever increasing investment in the yolk (Balon 1978, 1981b).

Matsuda (1987, p. 28) believed that the production of larger eggs is a direct response to novel environments by an endocrine mechanism, specifically that producing 17 β-estradiol, the hormone which induces the production of vitellogenin, and the hormones which regulate the final endocytosis. In his words, 'It appears most probable that the functioning of the endocrine mechanism has been altered by novel environmental stimuli in the new environments, and that the consequently altered pattern of vitellogenesis has resulted in the entry of a larger amount of yolk (or yolk precursor) into more limited numbers of eggs, so that the egg becomes larger'. This hormonally mediated, altered pattern of yolk synthesis is considered by Matsuda as the proximate cause of egg-size increase.

The ultimate cause of the increase in egg size is, according to Matsuda (1987), natural selection – which is supposed to hone the results of the proximate cause (and fix it by genetic assimilation); in my opinion, however, epigenetic mechanisms should rather be seen as this ultimate cause (see also Løvtrup 1982, 1989). As explained elsewhere in detail, the mechanisms which cause specialization and the trends responsible, among others, for egg-size increases, are the same as those responsible for the constant production of more generalized or more specialized forms – the saltatory altricial-precocial homeorhetic states (alprehost, Balon 1988b, 1989). The altered pattern of yolk synthesis and uptake is only a part or a consequence of the epigenetic mechanisms responsible for both the proximate and ultimate process. Non-Darwinian selection – the final mortality – is at most the ecological mechanism of the process (Løvtrup 1982, 1987).

As a result of the larger yolk supply the entire ontogeny is altered: it becomes direct, without a larva. A larger young is produced at first oral feeding and especially at the release from parental guarding. Furthermore, the acceleration of oral feeding while a large yolk supply is still available delivers nutrients from two sources simultaneously, thus creating even larger and better formed independent young (contrary to the interpretation given by Yaganisawa & Sato 1990 that it serves mainly as training for exogenous feeding).

Ultimately, even the exposure of gametes is eliminated by internal fertilization, and a large yolk and overlapping oral feeding are replaced by other maternal-fetal nutrient transfer mechanisms which are facilitated by the proximity of various relevant structures within the mother's body (e.g. Balon 1981b, Wourms et al. 1988).

The above patterns of egg-size increase, and to a lesser extent the mechanisms, are well documented for a variety of invertebrates (e.g. Matsuda 1987), agnathans (e.g. Gorbman 1983), and among fishes for chondrichthyians (Compagno 1990) and most Arctic, Antarctic and deep-sea taxa (Marshall 1953), and also for sibling species of sculpins (Goto 1980, 1990), salmonins (e.g. Balon 1980, 1984b), amphibians (Duellman & Trueb 1986, Duellman 1989) and birds (e.g. Nice 1962, Ricklefs 1979, 1983, O'Connor 1984). It is, therefore, obvious that the trend from general to special and from simple to complex, reflected, among others, in egg-size increase and in successively more complex reproductive styles, is one of the more general patterns of evolution.

a. The case of cichlid fishes

To illustrate the above trends in the evolution of reproductive styles and to create a parallel for the interpretation of the reproduction in the living coelacanth, I have selected several cichlid fishes from Lake Tanganyika. Of course, the same trends are known among cichlids from other lakes, including tilapines which range from the substrate clutch-guarding *Tilapia* with small eggs to mouthbrooding *Oreochromis* with large eggs (e.g. Peters 1963, Noakes & Balon 1982). A group of endemic cichlids of common ancestry may provide more compelling evidence.

Among the more than 40 genera and over 130 species of cichlid fishes in Lake Tanganyika only four species are not endemic (e.g. Poll 1956, Brichard 1978, Coulter et al. 1986). The lake was formed in the Pliocene, some 3 to 5 million years ago, and all the labroid cichlid fishes there are thought to have evolved from 4 to 5 founder riverine taxa (Greenwood 1964, 1981, Fryer & Iles 1972). In contrast to the other rift-valley lakes of

Africa, Lake Tanganyikan cichlids are the most diverse with respect to reproductive styles. Many tend the clutch on the substrate, in cavities, in empty shells or in specially constructed nests, while others perform various modes of mouthbrooding (e.g. Brichard 1978).

Boulengerochromis microlepis, a representative guarding sand nester produces 8 to 12 thousand small, 2 mm long eggs (Table 1). Because this cichlid is one of the largest in the lake, its eggs are largest among the substrate tenders and nesters, but are still smaller than those of any mouthbrooder (Kuwamura 1986). The vigorously guarded embryos within the nest pit swim up and begin first oral feeding as larvae when about 8.4 mm long (Kuwamura & Mihigo 1988).

Another cichlid, *Haplotaxodon microlepis*, which is less than half the size of *B. microlepis* when mature, guards its clutch in a sand-pit nest, but after hatching collects the embryos into its buccal pouch and broods them (Kuwamura 1988). Its clutch size is only about 100, the eggs are somewhat larger (2.4 × 1.8 vs. 2.0 × 1.5 mm) and the young at first swimming and oral feeding are over 9.2 mm long.

The very small *Tanganicodus irsacae* is a typical biparental mouthbrooder that collects eggs imme-

Table 1. Some life-history variables (in numbers and mm) of nesting and mouthbrooding cichlids from Lake Tanganyika.

	Clutch size	Egg dimension	Size at hatching	Size at first swimming	Size at first oral feeding	Size at release	Time at release in days	Source
Guarding nester								
Boulengerochromis microlepis	8000	2.0 × 1.5	5.2	8.4	8.5	–	–	Kuwamura & Mihigo (1988)
Guarding nester and mouthbrooder								
Haplotaxodon microlepis	100	2.4 × 1.8	4.7	9.2	9.4	–	–	Kuwamura & Mihigo (1988)
Nonfeeding mouthbrooder								
Tanganicodus irsacae	12	3.5 × 2.5	5.1	–	11.2	11.0	14	Kuwamura et al. (1989)
Feeding mouthbrooder								
Tropheus duboisi	8	6.1	5.5	–	10.0	18.0	31	Yamagisawa & Sato (1989)
Tropheus moorii	12	6.3	5.5	–	12.0	18.0	33	Yamagisawa & Sato (1989)
Cyphotilapia frontosa	17	6.0 × 4.0	6.0	–	11.0	23.0	54	Balon (1985)

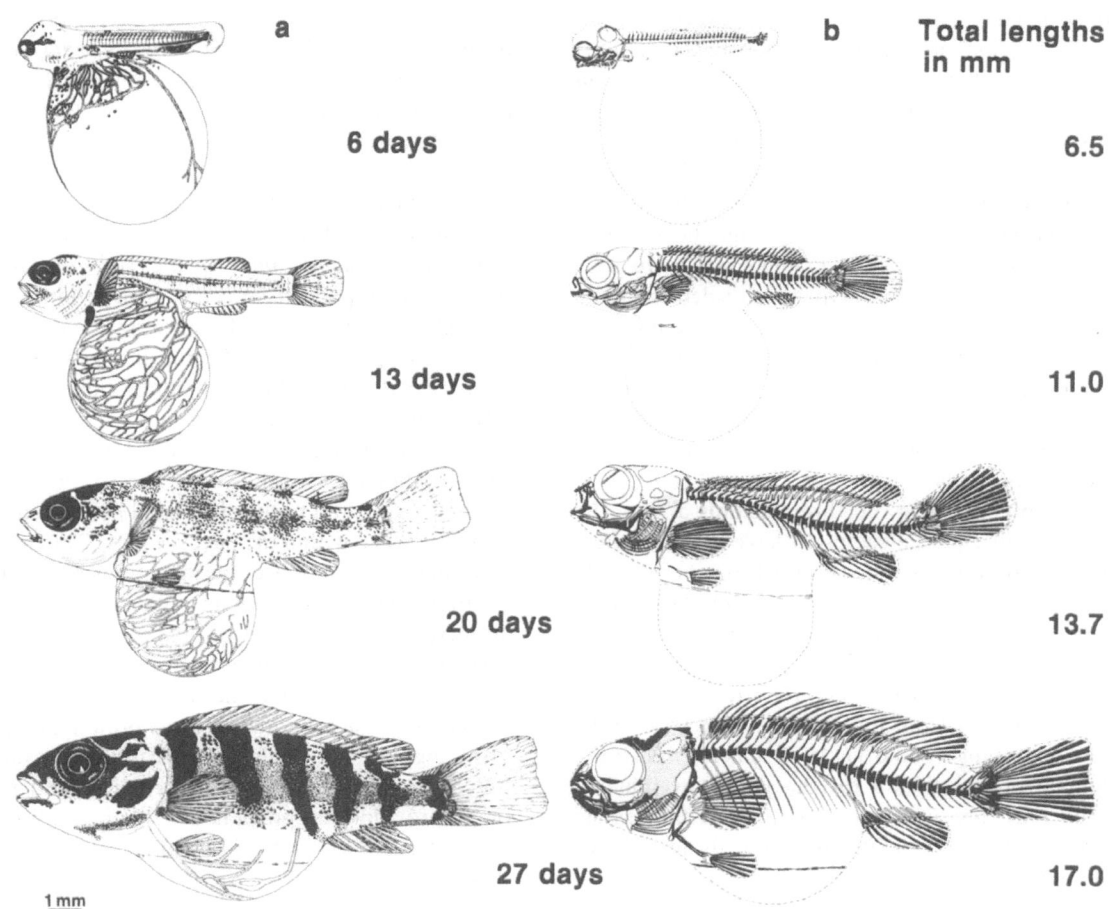

a

b

Total lengths in mm

6 days

6.5

13 days

11.0

20 days

13.7

27 days

17.0

1 mm

Fig. 1. Four decisive stages of *Cyphotilapia frontosa* based on live (a) and cleared and stained (b) specimens: At the time of mating, fertilized eggs (6.0 × 4.0 mm) are picked up by the female of this deepwater cichlid from Lake Tanganyika, and mouthbrooded until a few large juveniles (23 mm TL) are released 54 days later. In the buccal cavity of the female the embryos hatch on the fifth day atop a large yolk (a) with the first elements of the axial skeleton already cartilaginous (stippled in b). They start feeding orally when 13 days old. By that time fins have differentiated and many skeletal elements are calcified (black in the respective b). When 27 days old they become fully formed definitive phenotypes which still retain much of the yolk but have a completely calcified skeleton (lowermost b), although at that time they are only half way through the brooding interval and feed on particles inhaled by the mother and on cellular debris (egg envelopes, the breakdown particles of dead siblings).

diately upon deposition and retains them in the buccal cavity for about 14 days at which time fully formed young 11 mm long are released and begin oral feeding. Only about 12 eggs 3.5 mm long constitute a clutch (Kuwamura et al. 1989).

In *Cyphotilapia frontosa* mouthbrooding has been further improved by nearly doubling the egg size (6.0 × 4.0 vs. 3.5 × 2.5 mm in *T. irsacae*) and

by orally feeding embryos during their 54 day retention time in the buccal pouch (Balon 1985). Out of the large eggs numbering on average 17 per clutch (Table 1) embryos hatched after 5 to 6 days of incubation (Fig. 1). When 13 days old and about 11 mm long the embryos, still with a large yolksac, had differentiated fins, a much calcified skeleton and a formed gut. At that time, while still in the buccal pouch, they began oral ingestion and intestine digestion of particles (detritus, plankton) that were inhaled by the female and, possibly, on debris from egg envelopes and dead siblings. The brooding female took in food particles more often over time, and whenever a cloud of brine-shrimp nauplii was introduced, moved immediately into it, inhaling vigorously. Half way through the brooding interval the young, now 17 mm long, were fully formed definitive phenotypes with calcified skeletons but still possessed a considerable remnant of yolk. They were finally released as 23 mm long young directly into the adult habitat, between depths of 10 to 50 m. Most other mouthbrooding cichlids release their young into inshore, shallow nursery grounds.

When comparing the three cichlids above, the increase in egg size seems to have a lesser effect than feeding in the buccal pouch on the ultimate size of the young at release (Fig. 2). Oral feeding of yolksac juveniles in the buccal pouch doubles their size at release.

Yanagisawa & Sato (1990) reported on two more cichlid species from Lake Tanganyika with orally feeding young (Table 1). In *Tropheus moorii* the female provides food for the young by inhaling food particles, whereas in *T. duboisi* the female provides food for the young and feeds herself. Although much smaller than *C. frontosa* (Table 2) when mature, both *Tropheus* spp. have eggs larger than those of *C. frontosa*; their clutch size is smaller, however, and the young are only 18 mm long at release (vs. 23 mm in *C. frontosa*).

In summary (Fig. 3), based on available data for other cichlid species, egg size increases steadily from nest tenders which fan the clutch, to those which fan, and after hatching, move the young by mouth to various pits, to those which mouthbrood and finally provide food for oral feeding of the young in the buccal pouch. In Figure 3 tenders are represented by *Cichlasoma nigrofasciatum*, a guarding species which tends and fans the clutch on a substrate (Balon 1960). When the embryos hatch, the parents transfer them several times, using the mouth (huddling), into new pits excavated in the substrate. The two nest-guarding tilapias (*Tilapia sparrmanii* and *T. rendalli*) do the same but retain the free embryos longer in their buccal pouches during such transfers. *Aequidens paraguayensis* and *Gymnogeophagus balzanii*, similar to *Haplotaxodon microlepis*, guard and fan the clutch in a nest until hatching, when the embryos are collected into the mouth and brooded continuously in the buccal cavity until the final release which coincides with the first oral feeding. *Oreochromis niloticus, Sarotherodon melanotheron*, and *Labeotropheus trewavasae* represent species with continuous mouthbrooding but of different duration depending on the yolk supply reflected in egg size (left scale). *Cyphotilapia frontosa* represents the most advanced style with exogenous feeding starting in early young during mouthbrooding, but for most of the brooding interval the young survive on mixed feeding – yolk and particles inhaled by the female and/or foraged within the buccal cavity. The major consequence of all these specializations is the production of more viable young, which is reflected in an increase in size and in completion of definitive form at the time of release from the protection of the parental body (bold line). Hatching (dashed line) is at first accelerated with the prolongation of brooding time but the larger yolk ultimately facilitates the incorporation of more carotenoid pigments providing an endogenous oxygen supply (see the last section), and so hatching can again be delayed.

In spite of the very large size differences among the adult females, relative increase in egg diameter, lowest in the substrate guarders (Table 2), remains within 3 to 8% in the mouthbrooders and the living coelacanth. A similar relationship applies to the size at release.

b. The similar case of the coelacanths

The fragments of data on reproduction of the coelacanths can be better interpreted if more complete cases of the cichlids and other fishes are considered. Coelacanths, however, evolved their specialized reproductive styles, from oviparity to internal bearing, during an enormous time span of over 380 million years. Cichlids had at most 6 million years to evolve from oviparity to the most specialized

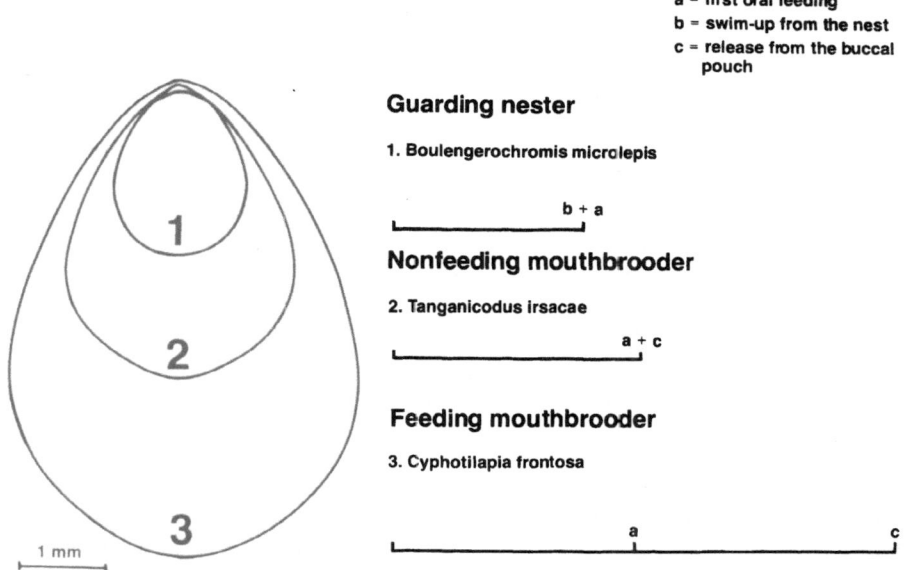

Fig. 2. Comparison of egg size and size at first oral feeding (a), swim-up from the nest (b) and at release from the buccal pouch (c) in three cichlids from Lake Tanganyika: (1) a guarding nester, (2) nonfeeding mouthbrooder and (3) feeding mouthbrooder (after Kuwamura & Mihigo 1988 and Balon 1985).

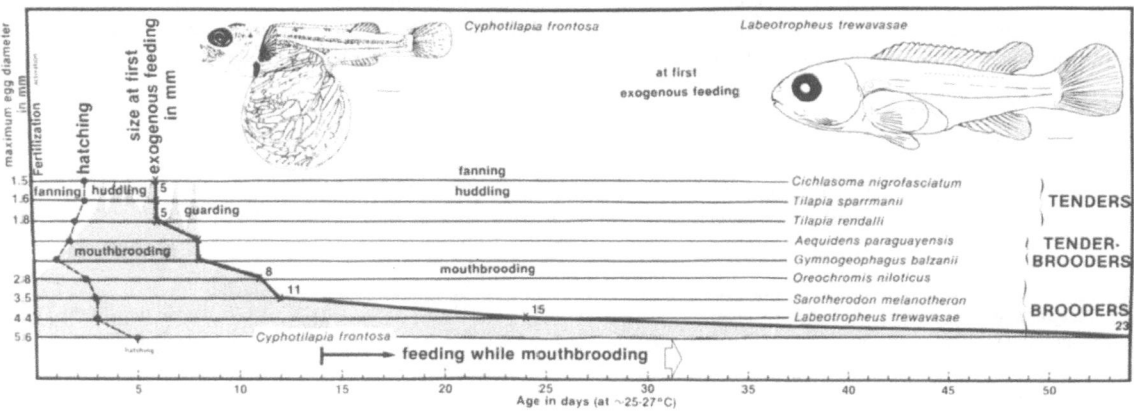

Fig. 3. The evolution of reproductive styles in cichlid fishes from tenders of the clutch on a substratum, via intermediate initial substrate tenders and subsequent mouthbrooders, to mouthbrooders with the immediate uptake of ova after or prior to fertilization and, finally, to orally feeding young while mouthbrooded. Fish figures represent the size and forms of *Cyphotilapia frontosa* (left) and *Labeotropheus trewavasae* (right) at the time of first exogenous feeding. Buccal uptake and brooding until first release are represented by the dotted screen. The ordinate gives maximum egg diameter, the abscissa time from fertilization, the dashed line hatching and the bold line size at first oral feeding.

Table 2. Some relative life-history variables of cichlids from Lake Tanganyika as compared to those of the coelacanth.

	In % of adult female TL			Mean TL of adult female (mm)
	egg diameter	size at release	size of yolksac juvenile*	
Boulengerochromis microlepis	0.4	1.7	1.8	480
Haplotaxodon microlepis	1.0	4.7	4.7	200
Tanganicodus irsacae	6.7	21.1	21.5	52
Tropheus duboisi	7.3	21.4	11.9	84
Tropheus moorii	7.8	22.2	14.8	81
Cyphotilapia frontosa	2.8	10.7	5.1	215
Latimeria chalumnae	5.6	26.2	19.9	1600

* Approximately comparable to the 5 yolksac fetuses in the 1.6 m TL female of *L. chalumnae* (Smith et al. 1975).

case of external bearing. Speculatively, given time, the evolution of livebearing cichlids can be expected (Balon 1981a).

Schultze (1980, 1985) documented oviparity in the lower Carboniferous coelacanth *Rhabdoderma exiguum*. Caseless eggs of this species were found in estuarine deposits, thus implying freshwater or at least brackish spawning grounds. Their relatively large size (Table 3) may have been accompanied by some hiding or even guarding, akin to recent salmonins or cichlids (Balon 1977, 1980). This we

may never know, but the environmental relationships presented by Schultze (1985) hint at some estuarine migration, also suggested from the data for the contemporary *Caridosuctor populosum* (Lund & Lund 1985, Lund et al. 1985).

The yolksac-bearing young of *R. exiguum* described by Schultze (1972, 1985) display an advanced shape and well developed skeletal elements and should be considered juveniles (not larvae) with specialized direct development (Fig. 4). Were the coelacanths of the Devonian the ancestral non-

Table 3. Some life-history variables of feeding mouthbrooder, and oviparous and viviparous coelacanths.

	Egg diameter (mm)	Size at release (mm)	Size of yolksac juvenile* (mm)	Maximum yolksac Ø of juvenile* (mm)	Mean TL of adult female (cm)	Source
Cyphotilapia frontosa	6	23	11	9	21.5	Balon unpublished
Rhabdoderma exiguum	36	53	48	36	60.0	Schultze (1985)
Latimeria chalumnae	90	420	318	105	160.0	Smith et al. (1975) Balon et al. (1988)

	In % of adult female TL			TL of yolksac juvenile* in their maximum yolksac diameter	Clutch size
	egg diameter	size at release	size of yolksac juvenile*		
Cyphotilapia frontosa	2.8	10.7	5.1	1.2	17
Rhabdoderma exiguum	6.0	8.8	8.0	1.3	–
Latimeria chalumnae	5.6	26.2	19.9	3.0	19/5

* Comparable to the 5 yolksac juveniles in the 1.6 m TL female of *L. chalumnae* (Smith et al. 1975).

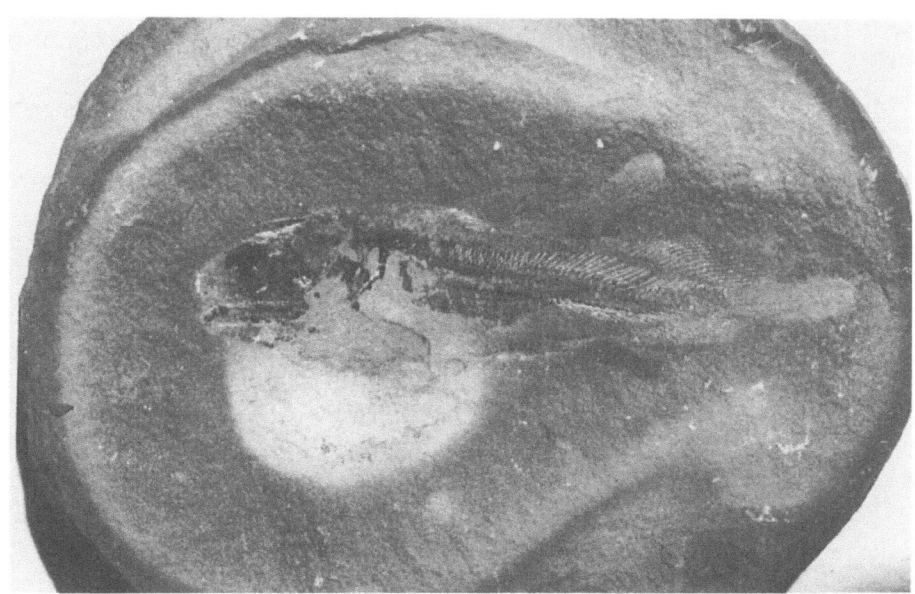

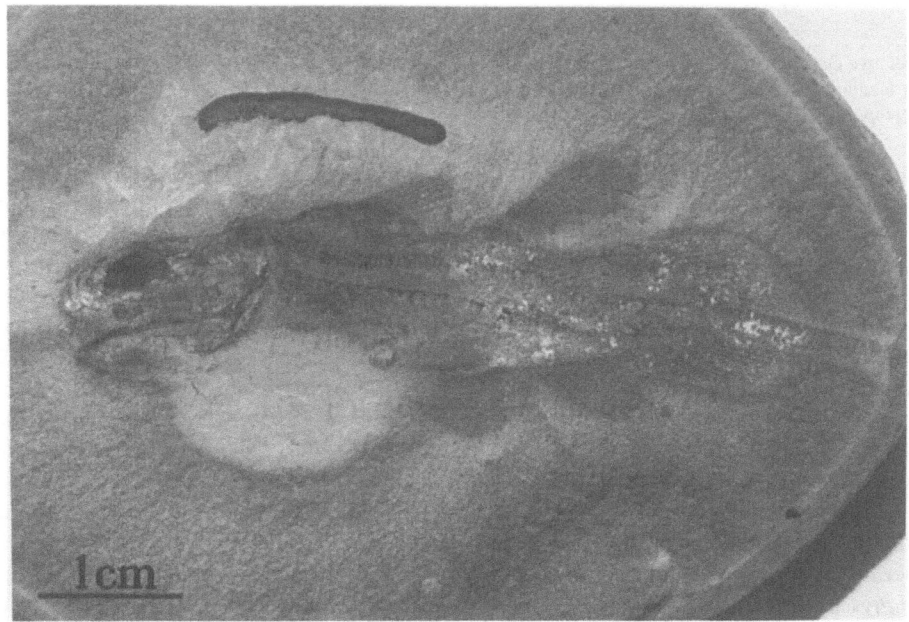

Fig. 4. Two 5 cm long 'yolksac juveniles' (ROM # 43565, 47481) of *Rhabdoderma exiguum* from the Mazon Creek, Illinois (Essex fauna of the Francis Creek Shale, pit 11, Upper Carboniferous). Courtesy of Desmond Collins, Royal Ontario Museum, Toronto.

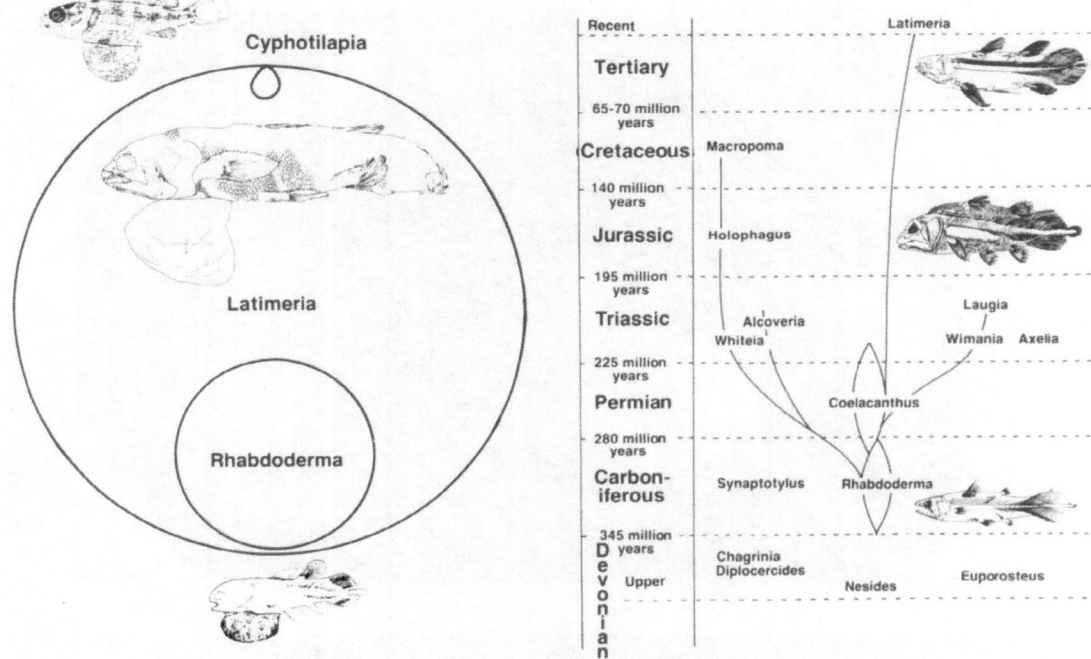

Fig. 5. The relative egg size of the coelacanths *Rhabdoderma* and *Latimeria* (circles) versus *Cyphotilapia* (upper pole) with the body shape at first feeding against the coelacanths in various geological times (right) with the figures of oviparous *Rhabdoderma* and the viviparous *Holophagus* (= *Undina*) and *Latimeria*.

guarding gamete scatterers with small eggs and high fecundity? Moreover, was *R. exiguum* by any chance a mouthbrooder? Some marine mouth-brooding catfishes may serve as another model for this possibility. The eggs of ariid catfishes only about 10 to 14 mm in diameter and 15 to 80 in number, are mouthbrooded by the male (e.g. Jaya-ram 1978, Coates 1988).

The Jurassic coelacanth *Holophagus*, as pointed out by Watson (1927), is ultimately considered to be viviparous (Schultze 1980), in spite of interim skepticism by ever present disbelievers that the young in the body cavity are an artefact of a fossil-ization overlap. It is possible that the holotype of *Axelrodichthys araripensis* is also a pregnant fe-male with 3 embryos instead of the calcified swim-bladder (Maisey 1986, 1987); after all it would be easier to imagine eggs or embryos in the female's body cavity shifted during fossilization. Internal bearing in coelacanths, therefore, might have evolved 200 to 110 million years ago. No wonder

then that the egg size of *Latimeria chalumnae* is so enormous. Matsuda's (1987) proximate cause of egg-size increase, mediated hormonally by stimuli from the frequently changing environment, would suffice by itself to explain this, without invoking the amplification of the epigenetic mechanisms (Fig. 5). Absolute egg size increased only 2.5 fold be-tween *R. exiguum* and *L. chalumnae*, while that between *Boulengerochromis microlepis* and *Tro-pheus duboisi*, *T. moorii* or *Cyphotilapia frontosa* is about 3 fold (Table 1, 3).

As yolk density seems to be the same in *C. fron-tosa* and *L. chalumnae* (Table 4), the difference in size at first oral feeding must be determined by the yolk volume alone (Fig. 6). As the increase in size from first oral feeding to release is about 60 to 70% in *L. chalumnae* but only 48% in *C. frontosa*, oral feeding of young in the buccal cavity must be less efficient than feeding, assisted by the 'placental apposition' of yolksac and uterine capillary plexus-es (Wourms et al. 1991), on histotrophe and/or on

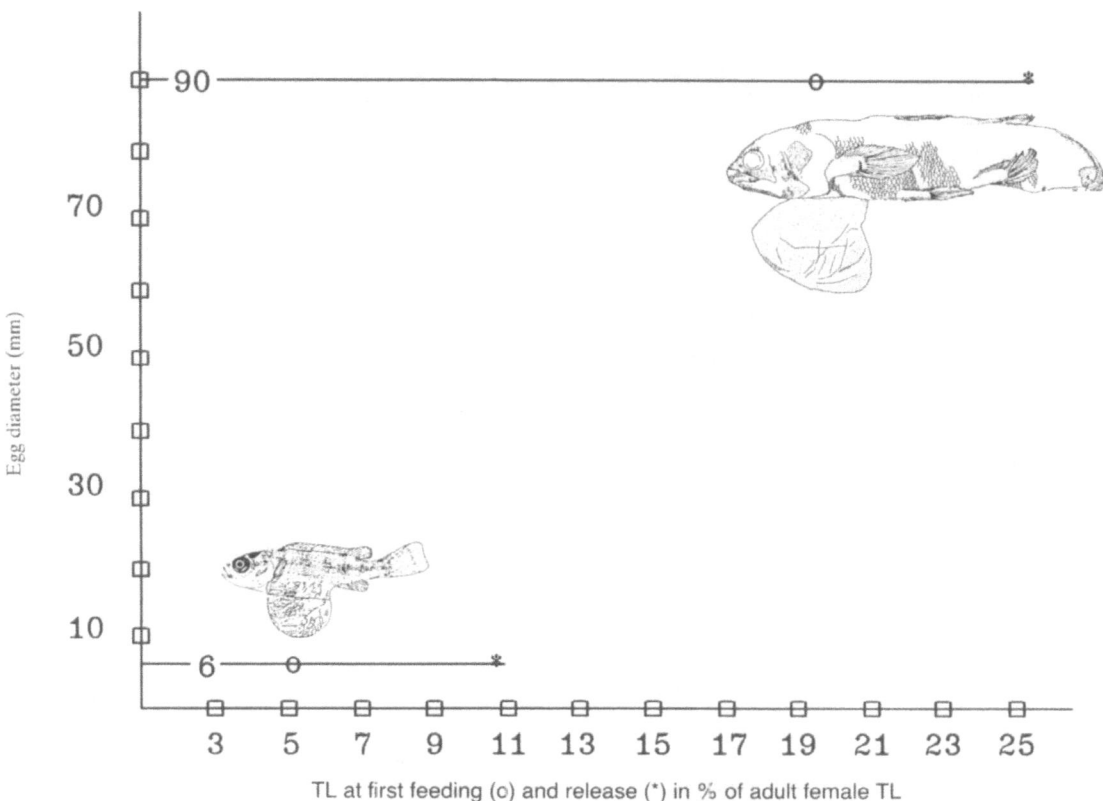

Fig. 6. Yolk density being equal, the yolk volume alone probably determines the size of *C. frontosa* and *L. chalumnae* at first oral feeding.

Table 4. Average composition of an egg in percent of moisture, lipids, proteins and carotenoids.

	Moisture	Lipids	Proteins	Carotenoids
Indirect development				
Lota lota	81	2	16	0+
Gadus morhua	76	2	20	2
Perca fluviatilis	74	6	18	2
Direct development				
Neogobius melanostomus	62	6	27	5
Cyphotilapia frontosa	44	35	20	0.016
Latimeria chalumnae	42	37	7	14

Fig. 7. Three eggs from the 163 cm TL *L. chalumnae* female captured near Anjouan in 1972 (CCC no. 79) compared to an orange fruit (bottom right). Photograph courtesy of J. Anthony.

the debris from sibling eggs practiced by *L. chalumnae*.

The enormous egg size of the living coelacanth (Fig. 7) is unique among fishes, but for the slightly smaller egg of the gulper shark mentioned earlier or a few other less documented elasmobranchs. This difference in egg size among most fishes and the living coelacanth is no larger than that between living agnathans – lampreys and hagfishes (Fig. 8). Of course, in both instances the creatures with larger eggs have direct development without a larva, labeled by some as cases of embryonization (Sharov 1966, Matsuda 1987). I prefer to consider them as the directly developing precocial alternatives of the epigenetic dichotomy, the specialized members, which through isolation avoid competition (Balon 1975b, 1985, 1989, Flegler-Balon 1989, Løvtrup 1989).

Carotenoids in the yolk: suggestion on their significance

The overwhelming circumstantial evidence that carotenoids are implicated in the storage and supply of metabolic oxygen (or equivalent) has been reviewed more than once (e.g. Soin 1956, 1962, 1967, Mikulin & Soin 1975, Balon 1977, 1981b, c) since the time attention was drawn to it for the first time by Smirnov (1950). The mechanism was finally explained by Karnaukhov (1979). Most respiratory physiologists, however, have regrettably not yet addressed this problem.

Working on filtrating molluscs, Karnaukhov (e.g. 1971, 1979, 1988 and cited herewith) explains the role of carotenoids in oxidative metabolism as follows: '... carotenoids may act as electron acceptors, and, together with haemoproteins, form a system of intracellular oxygen reserve (accumu-

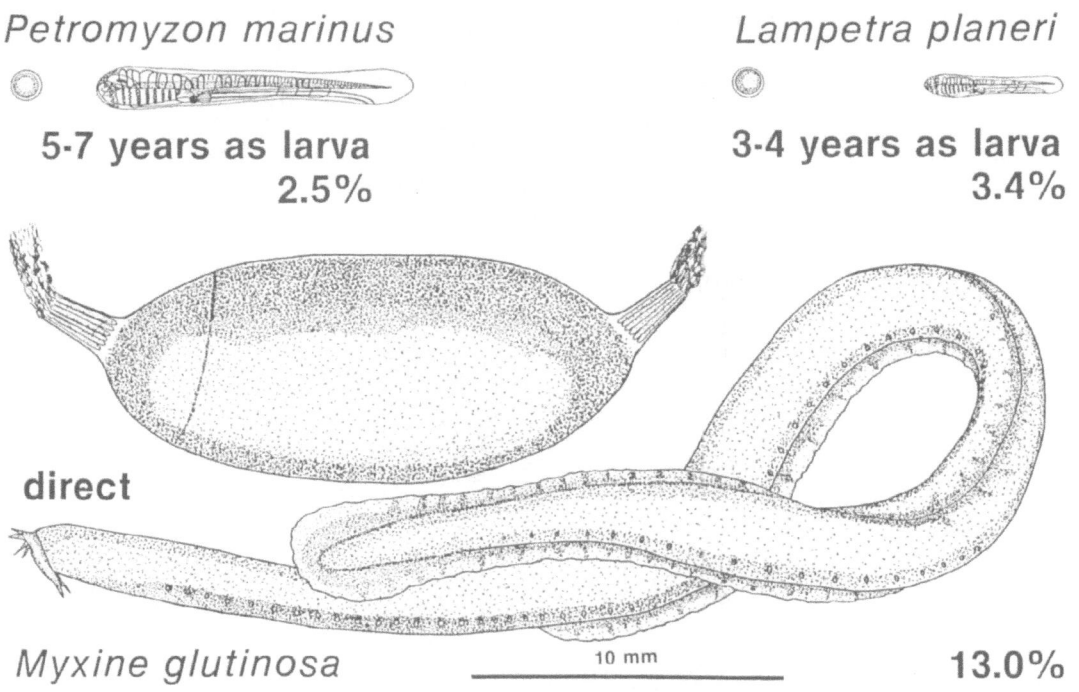

Petromyzon marinus

5-7 years as larva
2.5%

Lampetra planeri

3-4 years as larva
3.4%

direct

Myxine glutinosa

10 mm

13.0%

Fig. 8. The relationship between egg size and mode of development in living agnathans. Eggs of sea lamprey, *Petromyzon marinus*, brook lamprey, *Lampetra planeri*, and the hagfish, *Myxine glutinosa*, ammocetes larvae (for lampreys) and a juvenile (for hagfish) at the onset of exogenous feeding, are drawn to scale. Their sizes (10, 4.4 and 65 mm) at first oral feeding are given in per cent of the respective average adult female size. While *P. marinus* and *L. planeri* with their small eggs (1.0 and 1.1 mm in diameter) develop indirectly, the hagfish produces only a few large eggs (17 mm long) and develops directly (from Balon 1986b).

lator) in the cytosomes [Fig. 9]. Thus, cytosomes can provide energy for the cell when the rate of oxygen penetration into the tissue is low' (Karnaukhov 1979, p. 156).

At the request of a reviewer I also take the liberty to copy the next paragraph from Karnaukhov's paper: 'The electron-acceptor and electron-donor properties of the conjugated double-bond chain of carotenoids (. . .) allow it to connect an oxygen molecule in place of a (central) unsaturated double bond with the help of haemoprotein. (. . .) The colourless, oxygenated carotenoid may serve as an electron-acceptor equivalent of molecular oxygen and can be considered analogous with the oxidized form of the well-known cytochrome oxidase ($a + a_3$) of the mitochondrial respiratory chain. Carotenoids can accumulate (. . .) oxygen (or electron-acceptor equivalents of oxygen) in

resting cells under conditions of low-rate oxygen penetration (in hypoxic condition). When the cells enter an active phase under these conditions, the rate of their oxygen consumption (. . .) becomes higher than the rate of oxygen penetration from the environment (. . .). The oxygen deficiency arising in this situation is relieved by oxygen released from the carotenoid intracellular accumulator (. . .) and carotenoids once more assume their colour. When the cell returns to rest, the oxygen reserve in the carotenoids intracellular accumulator can be restored'.

Fishes with small eggs and indirect development have a yolk with a high moisture content and a low lipid and carotenoid content (Table 4). The yolk of fishes with pelagic larvae like burbot, cod or perch contains over 75% water and less than 2% carotenoids. The yolk of fishes with large eggs and direct

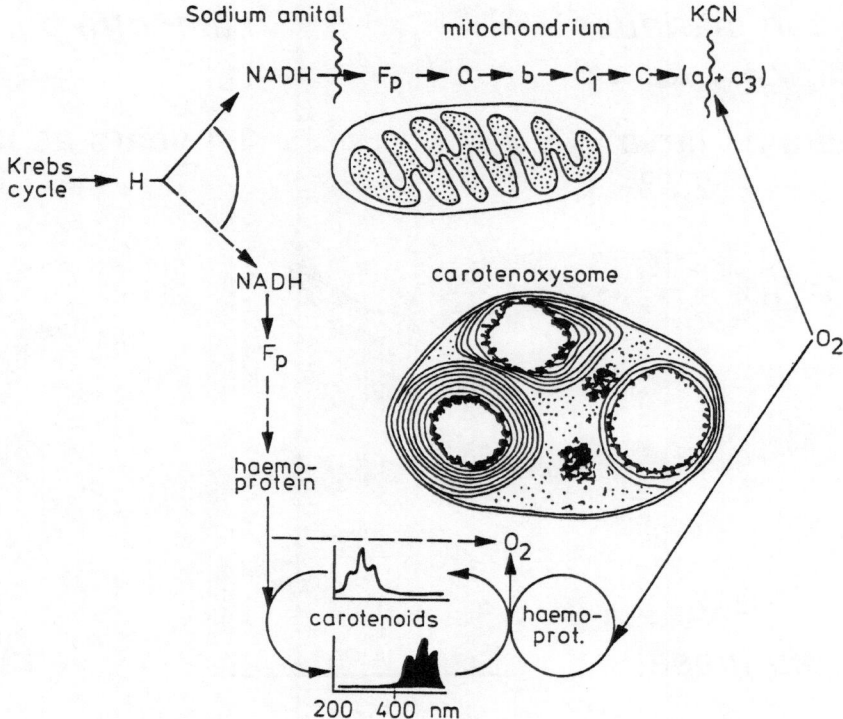

Fig. 9. Scheme of respiratory relationship between mitochondria and the carotenoxysomes (from Karnaukhov 1979).

development contains substantially less water, and more lipids, proteins and carotenoids. For example, in the goby *Neogobius melanostomus*, a guarding nester (see Balon 1981b), the yolk contains 62% water, 6% lipids, 27% proteins and 5% carotenoids. In contrast, the yolk of *C. frontosa* and *L. chalumnae* contains 44 and 42% moisture, 35 and 37% lipids, 20 and 7% proteins, and 0.02 and at least 14% of carotenoids, respectively (Table 4). In the yolk of the living coelacanth the proteins seem to be replaced by carotenoids, which are reported by Devys et al. (1972) to be as high as 21% (corrected here for the effect of preservative). If these differences are confirmed by better carotenoid estimates in the future, *L. chalumnae* yolk is likely to be implicated not only in the nutrient supply but also as the source of molecular oxygen for the developing embryo and juvenile.

The small eggs and planktonic free embryos of indirectly developing fishes have unrestricted ac-

cess to dissolved oxygen in the water. In *C. frontosa*, because of the constant churning by the female, the eggs, embryos and juveniles in the buccal pouch are assured of a steady supply of oxygen with the water. Fanning of the clutch in the nest serves the same function in *N. melanostomus*.

Compared to the incubation time of oviparous scorpaenids, the viviparous *Sebastes* have 5–10 times longer gestation period. One of the explanations given for this prolongation of development is anoxia encountered in the ovarian cavity by the embryos of live-bearing rockfishes (Wourms 1991). The quantity of carotenoids in yolk of these highly fecund and primitive livebearers is very low.

The eggs and embryos of *L. chalumnae* may also encounter an oxygen deficit during the close confinement in the oviduct compartments, in spite of the juxtaposition of capillary plexuses on the yolk and on the uterine wall (Wourms et al. 1991). The carotenoids may release oxygen molecules upon demand (Karnaukhov 1979). However, A.Y. Mikulin (in litt. 1.10.1990) considers Karnaukhov's interpretation to be incorrect and the problem still unresolved. This aspect alone warrants further investigation, if only to fulfill the prophecy of the opening quotation.

Acknowledgements

John Wourms facilitated much of this study by giving me in advance the abstract and an early draft of his contribution. As in the past we often come to the same conclusion. Mike Bruton enabled my participation in the current quests for the coelacanths and with David Noakes and Christine Flegler-Balon also deserves thanks for the hard labor on the improvement of the first draft, and for making it less drafty. Marie Rush and Steve Crawford struggled with idiosyncratic computers to complete the tables, figures and slides in time, and equally deserve my gratitude.

References cited

Anthony, J. & J. Millot. 1972. Première capture d'une femelle de coelacanthe en estat de maturité sexuelle. C.R. Acad. Sci. Paris, Ser. D 224: 1925–1927.

Balon, E.K. 1960. Development of *Cichlasoma nigrofasciatum* (Guenther) during the embryo period of life. Věst. Čs. spol. zool. 24: 199–214. (In Slovak.)

Balon, E.K. 1975a. Reproductive guilds of fishes: a proposal and definition. J. Fish. Res. Board Can. 32: 821–864.

Balon, E.K. 1975b. Terminology of intervals in fish development. J. Fish. Res. Board Can. 32: 1663–1670.

Balon, E.K. 1977. Early ontogeny of *Labeotropheus* Ahl, 1927 (Mbuna, Cichlidae, Lake Malawi), with a discussion on advanced protective styles in fish reproduction and development. Env. Biol. Fish. 2: 147–176.

Balon, E.K. 1978. Reproductive guilds and the ultimate structure of fish taxocenes: amended contribution to the discussion presented at the mini-symposium. Env. Biol. Fish. 3: 149–152.

Balon, E.K. 1980. Early ontogeny of the lake charr, *Salvelinus (Cristivomer) namaycush*. pp. 485–562. *In:* E.K. Balon (ed.) Charrs: Salmonid Fishes of the Genus *Salvelinus*, Dr W. Junk Publishers, The Hague.

Balon, E.K. 1981a. About processes which cause the evolution of guilds and species. Env. Biol. Fish. 6: 129–138.

Balon, E.K. 1981b. Additions and amendments to the classifica-

tion of reproductive styles in fishes. Env. Biol. Fish. 6: 377–389.

Balon, E.K. 1981c. Saltatory processes and altricial to precocial forms in the ontogeny of fishes. Amer. Zool. 21: 567–590.

Balon, E.K. 1983. Epigenetic mechanisms: reflections on evolutionary processes. Can. J. Fish. Aquat. Sci. 40: 2045–2058.

Balon, E.K. 1984a. Patterns in the evolution of reproductive styles in fishes. pp. 35–53. *In:* C.W. Potts & R.J. Wootton (ed.) Fish Reproduction: Strategies and Tactics, Academic Press, London.

Balon, E.K. 1984b. Life histories of Arctic charrs: an epigenetic explanation of their invading ability and evolution. pp. 109–141. *In:* L. Johnson & B.L. Burns (ed.) Biology of the Arctic Charr, University of Manitoba Press, Winnipeg.

Balon, E.K. 1985. Early life histories of fishes: new developmental, ecological and evolutionary perspectives. Developments in Env. Biol. Fish. 5, Dr W. Junk Publishers, Dordrecht. 280 pp.

Balon, E.K. 1986a. Types of feeding in the ontogeny of fishes and the life-history model. Env. Biol. Fish. 16: 11–24.

Balon, E.K. 1986b. Saltatory ontogeny and evolution. Rivista di Biologia/Biology Forum 79: 151–190.

Balon, E.K. 1988a. Tao of life: universality of dichotomy in biology. 1. The mystic awareness. Rivista di Biologia/Biology Forum 81: 185–231.

Balon, E.K. 1988b. Tao of life: universality of dichotomy in biology. 2. The epigenetic mechanisms. Rivista di Biologia/Biology Forum 81: 339–380.

Balon, E.K. 1989. The epigenetic mechanisms of bifurcation and alternative life-history styles. pp. 467–501. *In:* M.N. Bruton (ed.) Alternative Life-History Styles of Animals, Kluwer Academic Publishers, Dordrecht.

Balon, E.K. 1990a. The living coelacanth endangered: a personalized tale. Tropical Fish Hobbyist 39: 117–129.

Balon, E.K. 1990b. Epigenesis of an epigeneticist: the development of some alternative concepts on the early ontogeny and evolution of fishes. Guelph Ichthyol. Rev. 1: 1–48.

Balon, E.K., M.N. Bruton & H. Fricke. 1988. A fiftieth anniversary reflection on the living coelacanth, *Latimeria chalumnae*: some new interpretations of its natural history and conservation status. Env. Biol. Fish. 23: 241–280.

Barlow, G.W. 1981. Patterns of parental investment, dispersal and size among coral-reef fishes. Env. Biol. Fish. 6: 65–85.

Baylis, J.R. 1981. The evolution of parental care in fishes, with reference to Darwin's rule of male sexual selection. Env. Biol. Fish. 6: 223–251.

Blumer, L.S. 1979. Male parental care in the bony fishes. Quart. Rev. Biol. 54: 149–161.

Blumer, L.S. 1982. A bibliography and categorization of bony fishes exhibiting parental care. Zool. J. Linn. Soc. 76: 1–22.

Brichard, P. 1978. Fishes of Lake Tanganyika. T.F.H. Publications, Neptune City. 448 pp.

Bruton, M.N. 1989a. The living coelacanth fifty years later. Trans. Roy. Soc. S. Afr. 47: 19–28.

Bruton, M.N. 1989b. The coelacanth – can we save it from extinction? WWF Reports (October-November): 10–12.

264

Bruton, M. 1989c. Does the coelacanth occur in the Eastern Cape? The Naturalist (Port Elizabeth) 33: 5–13.

Coates, D. 1988. Length-dependent changes in egg size and fecundity in females, and brooded embryo size in males of forktailed catfishes (Pisces: Ariidae) from the Sepik River, Papua New Guinea, with some implications for stock assessment. J. Fish Biol. 33: 455–464.

Compagno, L.J.V. 1990. Alternative life-history styles of cartilaginous fishes in time and space. Env. Biol. Fish. 28: 33–75.

Coulter, G.W., B.R. Allanson, M.N. Bruton, P.H. Greenwood, R.C. Hart, P.B.N. Jackson & A.J. Ribbink. 1986. Unique qualities and special problems of the African Great Lakes. Env. Biol. Fish. 17: 161–183.

Courtenay-Latimer, M. 1988. Odd amusing bits of the coelacanth discovery. A retrospective in 1988. The Coelacanth, the Journal of the Border Historical Society 26 (2): 19–24.

Devys, M., S. Thierry, M. Barbier & M.M. Jonot. 1972. Premières observations du coelacanthe (*Latimeria chalumnae*). C.R. Acad. Sci. Paris 275 (Ser. D): 2085–2087.

Duellman, W.E. 1989. Alternative life-history styles in anuran amphibians: evolutionary and ecological implications. pp. 101–126. *In:* M.N. Bruton (ed.) Alternative Life-History Styles of Animals, Kluwer Academic Publishers, Dordrecht.

Duellman, W.E. & L. Trueb. 1986. Biology of amphibians. McGraw-Hill Book Company, New York. 670 pp.

Flegler-Balon, C. 1989. Direct and indirect development in fishes – examples of alternative life-history styles. pp. 71–100. *In:* M.N. Bruton (ed.) Alternative Life-History Styles of Animals, Kluwer Academic Publishers, Dordrecht.

Forey, P.L. 1988. Golden jubilee for the coelacanth *Latimeria chalumnae*. Nature 336 (6201): 727–732.

Fryer, G. & T.D. Iles. 1972. The cichlid fishes of the Great Lakes of Africa. T.F.H. Publications, Neptune City. 641 pp.

Gorbman, A. 1983. Reproduction in cyclostome fishes and its regulation. pp. 1–29. *In:* W.S. Hoar, D.J. Randall & E.M. Donaldson (ed.) Fish Physiology, Vol. 9a, Academic Press, San Diego.

Goto, A. 1980. Geographical distribution and variation of two types of *Cottus nozawae* in Hokkaido, and morphological characteristics of *C. amblystomopsis* from Sakhalin. Japan. J. Ichthyol. 27: 97–105.

Goto, A. 1990. Alternative life-history styles of Japanese freshwater sculpins revisited. Env. Biol. Fish. 28: 101–112.

Greenwood, P.H. 1964. Explosive speciation in African lakes. Proc. R. Inst. G. Brit. 40: 256–269.

Greenwood, P.H. 1981. The haplochromine fishes of the east African lakes. Kraus International Publications, Munich. 839 pp.

Greenwood, P.H. 1988. A living fossil fish the coelacanth *Latimeria chalumnae*. British Museum (Natural History), London. 16 pp.

Greenwood, P.H. 1989. Fifty years a 'living fossil' – the coelacanth fish *Latimeria chalumnae*. Biologist 36: 15–19.

Griffith, R.W. & K.S. Thomson. 1973. *Latimeria chalumnae*: reproduction and conservation. Nature 242: 617–618.

Heemstra, P.C. & L.J.V. Compagno. 1989. Uterine cannibalism and placental viviparity in the coelacanth? A skeptical view. S. Afr. J. Sci. 85: 485–486.

Jayaram, K.C. 1978. Curious cat fishes of India. Zoologiana 1: 9–17.

Karnaukhov, V.N. 1971. Carotenoids in oxidative metabolism of molluscoid neurons. Exp. Cell. Res. 64: 301–306.

Karnaukhov, V.N. 1979. The role of filtrator molluscs rich in carotenoid in the self-cleaning of fresh waters. Symp. Biol. Hung. 19: 151–167.

Karnaukhov, V.N. 1988. Biological functions of carotenoids. Nauka Press, Moscow. 241 pp. (In Russian).

Kuwamura, T. 1986. Substratum spawning and biparental guarding of the Tanganyikan cichlid *Boulengerochromis microlepis*, with notes on its life history. Physiol. Ecol. Japan 23: 31–43.

Kuwamura, T. 1988. Biparental mouthbrooding and guarding in a Tanganyikan cichlid *Haplotaxodon microlepis*. Japan. J. Ichthyol. 35: 62–68.

Kuwamura, T. & N.K. Mihigo. 1988. Early ontogeny of a substrate-brooding cichlid, *Boulengerochromis microlepis*, compared with mouthbrooding species in Lake Tanganyika. Physiol. Ecol. Japan 25: 19–25.

Kuwamura, T., M. Nagishi & T. Sato. 1989. Female-to-male shift of mouthbrooding in a cichlid fish, *Tanganicodus irsacae*, with notes on breeding habits of two related species in Lake Tanganyika. Env. Biol. Fish. 24: 187–198.

Løvtrup, S. 1982. The four theories of evolution. Rivista di Biologia 75: 53–66, 231–272, 385–409.

Løvtrup, S. 1987. Darwinism: the refutation of a myth. Croom Helm, London. 469 pp.

Løvtrup, S. 1989. On divergent and progressive evolution. pp. 55–69. *In:* M.N. Bruton (ed.) Alternative Life-History Styles of Animals, Kluwer Academic Publishers, Dordrecht.

Lund, R. & W.L. Lund. 1985. Coelacanths from the Bear Gulch limestone (Namurian) of Montana and the evolution of the Coelacanthiformes. Bull. Carnegie Nat. Hist. 25: 1–74.

Lund, W.L., R. Lund & G.A. Klein. 1985. Coelacanth feeding mechanisms and ecology of the Bear Gulch coelacanths. Compte Rendu Neuvième Congr. Internat. Stratigr. Geol. Carbonifere (Washington & Champaign-Urbana 1979) 5: 492–500.

Maisey, J.G. 1986. Coelacanths from the Lower Cretaceous of Brazil. Amer. Mus. Novit. 2866: 1–30.

Maisey, J.G. 1987. New fossil coelacanth named after Dr. Herbert R. Axelrod. Tropical Fish Hobbyist 36: 76–84.

Marshall, N.B. 1953. Egg size in Arctic, Antarctic and deep-sea fishes. Evolution 7: 328–341.

Matsuda, R. 1987. Animal evolution in changing environments with special reference to abnormal metamorphosis. John Wiley & Sons, New York. 355 pp.

Mikulin, A.Y. & S.G. Soin. 1975. The functional significance of carotenoids in the embryonic development of teleosts. J. of Ichthyol. 15: 749–759.

Millot, J. & J. Anthony. 1960a. Le cloaque chez les coelacanthes. Bull. Mus. natn. Hist. nat. Paris 32: 287–289.

Millot, J. & J. Anthony. 1960b. Appareil génital et reproduction des coelacanthes. C.r. Acad. Sci. Paris 251 (Ser. D): 442–443.

Millot, J. & J. Anthony. 1973a. L'appareil excréteur de *Latimeria chalumnae* Smith (Poisson, Coelacanthidé). Ann. Sci. nat. B 15: 293–328.

Millot, J. & J. Anthony. 1973b. Variations sexuelles de l'appareil excréteur du coelacanthe, *Latimeria chalumnae*. Connexions avec l'appareil génital male. C.r. Acad. Sci. Paris 276 (Ser. D): 2447–2448.

Millot, J. & J. Anthony. 1974. Les oeufs du coelacanthe. Science et Nature 121: 3–4.

Nice, M.M. 1962. Development of behavior in precocial birds. Trans. Linn. Soc. New York, Vol. 8. 211 pp.

Noakes, D.L.G. & E.K. Balon. 1982. Life histories of tilapias: an evolutionary perspective. pp. 61–82. *In:* R.S.V. Pullin & R.H. Lowe-McConnell (ed.) The Biology and Culture of Tilapias, ICLARM Conf. Proc. 7, Manila.

O'Connor, R.J. 1984. The growth and development of birds. John Wiley and Sons, Chichester. 326 pp.

Peters, H.M. 1963. Eizahl, Eigewicht und Gelegeentwicklung in der Gattung *Tilapia* (Cichlidae, Teleostei). Int. Rev. ges. Hydrobiol. Hydrogr. 48: 547–576.

Poll, M. 1956. Poissons Cichlidae. Rés. sci. Explor. hydrobiol. lac Tanganika (1946–1947) 3 (5b): 1–619.

Ricklefs, R.E. 1979. Adaptation, constraint, and compromise in avian postnatal development. Biol. Rev. 54: 269–290.

Ricklefs, R.E. 1983. Avian postnatal development. pp. 1–83. *In:* D.S. Farner, J.R. King & K.C. Parkes (ed.) Avian Biology, Vol. 7, Academic Press, New York.

Schultze, H-P. 1972. Early growth stages in coelacanth fishes. Nature New Biol. 236 (64): 90–91.

Schultze, H-P. 1980. Eier legende und lebend gebärende Quastenflosser. Natur und Museum 110: 101–108. (English translation from the Department of Geology, Field Museum of Natural History, Chicago.)

Schultze, H-P. 1985. Reproduction and spawning sites of *Rhabdoderma* (Pisces, Osteichthyes, Actinistia) in Pennsylvanian deposits of Illinois, USA. Compte Rendu Neuvième Congr. Internat. Stratigr. Geol. Carbonifère (Washington & Champaign-Urbana 1979) 5: 326–330.

Sharov, A.G. 1966. Basic arthropodan stock with special reference to insects. Pergamon Press, Oxford. 271 pp.

Smirnov, A.I. 1950. The significance of carotenoid pigmentation in the embryonic-larval stages of cyprinid fishes (Pisces, Cyprinidae). Dokl. Acad. Sci. USSR 73: 609–612. (In Russian.)

Smith, C.L., C.S. Rand, B. Schaeffer & J.W. Atz. 1975. *Latimeria*, the living coelacanth, is ovoviviparous. Science 190: 1105–1106.

Smith, M.M. 1986. Latimeriidae. pp. 152–153. *In:* M.M. Smith & P.C. Heemstra (ed.) Smiths' Sea Fishes, Macmillan, Johannesburg.

Soin, S.G. 1956. On the respiratory significance of carotenoid pigment in salmonid fish eggs and in other representatives of the order Clupeiformes. Zool. Zh. 35: 1362–1369. (In Russian.)

Soin, S.G. 1962. Embryonic respiratory adaptations in fishes and their development in the Baikal sculpins (Cottoidei). Voprosy Ichtiologii 2: 127–139. (In Russian.)

Soin, S.G. 1967. Ecomorphological data on the relationship of carotenoids with the embryonic respiration in fishes. pp. 340–350. *In:* Metabolism and Biochemistry of Fishes, Nauka Press, Moscow. (In Russian.)

Stobbs, R. 1989a. The coelacanth enigma. The Phoenix, Magazine of the Albany Museum 2 (2): 8–15.

Stobbs, R.E. 1989b. Laxative lipids and the survival of the living coelacanth. S. Afr. J. Sci. 85: 557–558.

Suyehiro, Y., T. Uyeno & N. Suzuki. 1982. Coelacanth, dissecting a living fossil. Newton (Graphic Science Magazine) 2 (8): 82–93. (In Japanese.)

Suyehiro, Y., T. Hamada & N. Suzuki. 1983. Coelacanths, mystery of living fossil revealed. Newton (Graphic Science Magazine) 3 (5): 78–93. (In Japanese.)

Topp, R.W. & D.L. Girardin. 1971. An adult louvar, *Luvarus imperialis* (Pisces, Luvaridae), from the Gulf of Mexico. Copeia 1971: 181–182.

Watson, D.M.S. 1927. The reproduction of the coelacanth fish *Undina*. Proc. Zool. Soc. Lond. 1: 453–457.

Wourms, J.P. 1981. Viviparity: the maternal-fetal relationship in fishes. Amer. Zool. 21: 473–515.

Wourms, J.P. 1991. Reproduction and development of *Sebastes* in the context of the evolution of piscine viviparity. Env. Biol. Fish. 30: 111–126.

Wourms, J.P., J.W. Atz & M.D. Stribling. 1991. Viviparity and the maternal-embryonic relationship in the coelacanth *Latimeria chalumnae*. Env. Biol. Fish. 32: 225–248. (this volume)

Wourms, J.P., M.D. Stribling & J.W. Atz. 1980. Maternal-fetal nutrient relationship in the coelacanth, *Latimeria*. Amer. Zool. 20: 962. (Abstract)

Wourms, J.P., B.D. Grove & J. Lombardi. 1988. The maternal-embryonic relationship in viviparous fishes. pp. 1–134. *In:* W.S. Hoar & D.J. Randall (ed.) Fish Physiology, Vol. 11b, Academic Press, San Diego.

Yanagisawa, Y. & T. Sato. 1990. Active browsing by mouth-brooding females of *Tropheus duboisi* and *Tropheus moorii* (Cichlidae) to feed the young and/or themselves. Env. Biol. Fish. 27: 43–50.

Environmental Biology of Fishes **32**: 267–273, 1991.

Observations on locomotion and feeding of released coelacanths, *Latimeria chalumnae*

Teruya Uyeno
Division of Vertebrate Paleontology, National Science Museum, 3-23-1 Hyakunin-cho Shinjuku-ku, Tokyo 160, Japan

Received 7.9.1990   Accepted 27.10.1990

Key words: Caudal fin, Buoyancy, Colouration, Feeding, Median fins

Synopsis

Two line-caught coelacanths, *Latimeria chalumnae*, were released for observation of their free-swimming movements in the open sea at a depth of about 50 m. Observations were made of their movement and use of the fins, interaction with humans (scuba diver), and colouration (white patches on dark background) as a possible means of camouflage. A new nomenclature is proposed for the caudal fin region and finnage.

Introduction

The natural history and behaviour of *Latimeria chalumnae* Smith, the living coelacanth, are like a jigsaw puzzle. On fitting a key part in place one realizes that other important pieces are still missing. A rare chance to observe and record on film and videotape two adult specimens swimming in the open sea, albeit not at their normal depth, was afforded the third Japan-Comoros team in July, 1986.

This joint project between the Japan Scientific Expedition of the Coelacanth (JASEC) [Mission des Recherches Scientifiques du Coelacanthe au Japon (MRSCJ)] and the Federal Islamic Republic of the Comoros (République Féderale Islamique des Comores) provided the opportunity for on-going investigations of *L. chalumnae* (see Suzuki et al. 1985). Another aspect of this project is the continuing education of the native fisherman on handling live coelacanths for scientific study. The results of this, as they affected the second specimen reported herein, have already substantially increased our knowledge of the coelacanth.

The specimens

According to information provided by Christian Olhagaray, the first specimen (CCC no. 135) was hooked by Mohammed Youssof Kari, 400–500 m from shore at a depth of 250 m on 4 July 1986 at around 0130 h, using a fist-size piece of tuna as bait. Towed at a depth of 3–5 m, the fish was brought to shore in front of Hotel Coelacanthe, Moroni, at 1030 h, and at about 1630 h was placed in a $2 \times 1 \times 1$ m metal cage anchored at 50 m depth, 50 m offshore. As the fishermen were afraid the fish would escape, the tow rope was kept attached, almost certainly making the movements recorded on cine-film unnatural. At 1030 h the next day this specimen died and was brought to shore. It measured 125 cm in total length and was estimated to weigh around 60 kg.

A second specimen (CCC no. 136), the video-tape of which provides the basis for this report, was likewise hooked on a hand-line from a pirogue (outrigger dugout) by a native fisherman, Mohammed Islam, on 17 July 1986 at 0100 h, about 400 m offshore, around 200 m deep using 'roudi', a gempylid fish, as bait. This coelacanth, 140 cm long and weighing about 65 kg, was brought to shore at the same location as the first specimen and placed in the aforementioned cage around 1000 h. It was in far better condition than the first one, possibly due to the fact that it stopped fighting the line after a relatively short time. Once caged, the hook was removed from the fish's mouth, and the specimen was released so that a videotape of free-swimming locomotion could be taken.

Although with the initial specimen the natives were reluctant to remove the tether, by the time of its death they became aware that the coelacanth is both slow and tame, and so they readily released the second specimen without fear of it swimming away quickly. One point immediately became apparent; the coelacanth is a very slow swimmer, as its ancestor probably was some 350 million years ago; it had changed its hydrodynamic structures little since the Palaeozoic (e.g. Balon et al. 1988). Initially, the fish swam extremely slowly but, as it descended (to around 50 m), its speed very gradually increased and, although still moving relatively slowly, it appeared generally more active. After videotaping was completed it was herded back to the cage where, under observation, it remained constantly in motion until it died the next day at 1930 h.

Hydrodynamic structures and movements

The locomotion of the coelacanth can be classified as an extreme example of the tetraodontiform mode (Breder's nomenclature as modified by Lindsey 1978). The second dorsal (D2) and first anal (A1) fins act as the sole propulsive force by flapping from side to side in unison. This mode closely resembles that of the ocean sunfish (family Molidae), where the tail is virtually absent (Lighthill 1969). In both cases, the stiff anterior rays serve to keep these fins stretched out like paddles. As previous authors (e.g. Millot & Anthony 1958, McAllister & Smith 1978) have not developed a uniform nomenclature (and abbreviations) for the fins of *L. chalumnae*, I propose an interim one in Fig. 1. This fin nomenclature is different because the fins of the

caudal region (CR), posterior to the anus, are treated here in a new way. The tripartite finnage has been collectively referred to as the 'caudal fin' by previous authors, but when one scrutinizes carefully the skeletal structures of this area, it is apparent that the fin rays of the dorsal and ventral lobes extend well into the body musculature until they articulate with and are supported by pterygiophores adjoining the neural and haemal spines of the axial skeleton. Mainly for this reason, these lobes could better be called the third dorsal (D3) and second anal (A2) fin, respectively. These fins undulate like those of the balistid fishes, although in the coelacanth they do not apparently serve for propulsion.

The actual caudal peduncle (Hubbs & Lagler 1964) is miniscule, extending only from where the last fin ray of A2 exits the scalation to where the most anterioventral ray of the small true caudal fin (CF) disappears under the scalation. Only the rays of this most posterior lobe are not supported by pterygiophores. Thus, this lobe (known to vary greatly in size among different individuals) is the only fin which should be called the caudal fin, and yet is serves almost no function in locomotion. Earlier researchers have named similar structures in fossil forms, such as *Coelacanthus*, the 'axial (caudal) lobe' (axl) (Goodrich 1930, Moy-Thomas & Westoll 1935). This caudal fin proper of the living coelacanth has also been called the 'accessory lobe' (Locket 1980, Locket & Griffith 1972), the 'axial supplement' (Smith 1939a, b), the 'supplementary caudal' (Smith 1939c), the 'caudal tuft' (Forey 1988) and, most mistakenly, the 'epicaudal' (Fricke et al. 1987, Balon et al. 1988).

It was interesting to observe that *L. chalumnae* appeared to prefer swimming forward inclined slightly head downward, rather than completely horizontal. All movements during the periods of observation were deliberately and smoothly executed with no sudden stops, starts, or changes of course. Locomotion and fin maneuvers, except for the second dorsal fin (D2) and first anal fin (A1), were extremely sluggish. The underwater script writer J. Stevens observed first in 1965 that 'the pectorals had a stabilizer role, while the anal and posterior dorsal fins assured active forward movement in conjunction with the tail' (Millot et al. 1978). Independently, Locket & Griffith (1972) confirmed that 'forward propulsion was achieved by the concerted action of the second dorsal and anal fins'. They described this unusual side-to-side synchronous sculling action perfectly, so only a few additional notes need to be made here. Even though the W-shaped configuration of the myomeres in the living coelacanth is consistent with that of most other fish, there are probably no or very little alternating serial contractions of the segments because no sinusoidal body wave-like action (Harris 1937, Hertel 1966) was observed. Contrary to J. Stevens' observation about movement being 'in conjunction with the tail', my analysis agrees with Locket & Griffith (1972) that '(no) active movement of the tail were seen'. The entire caudal region (rather than tail) acts passively, somewhat bending slightly with the water currents.

Although Locket & Griffith (1972) went on to conjecture that 'the abundant body musculature suggests, however, that powerful tail strokes are possible', my dissections on several preserved specimens show that the strong body musculature (especially in the caudal region) functions mainly to help keep the body rigid and on an even keel, as it moves through the water. Fricke et al. (1987) reported on 'fast starts' (Hertel 1966) where powerful tail strokes and subsequent sinusoidal body movement occurred. Regrettably, the movement following this activity of short duration was not observed by Fricke et al. (1987) to see if it could be sustained. In the two captured individuals no such activity was seen.

Most astonishing is the sinusoidal movement of the second dorsal fin and its peduncle moving like a ball-and-socket jointed arm through a full 330° three dimensionally, impeded only by the body below it. If one considers the second dorsal and first anal fins to be 0° when held erect on a vertical plane aligned with the body, the D2 and A1 move down and up respectively on the same side of the body more than 90°, past the horizontal, until the distal ends almost touch (Fig. 2). This activity serves to compress the water between the fins much as a flattened oar does as it moves down into the water. In this way the coelacanth is pushed for-

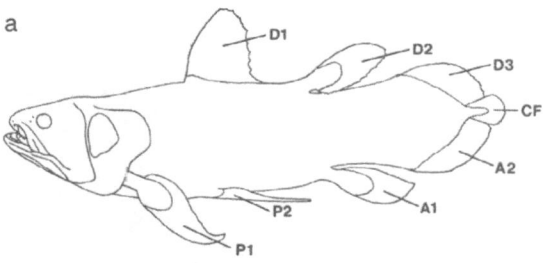

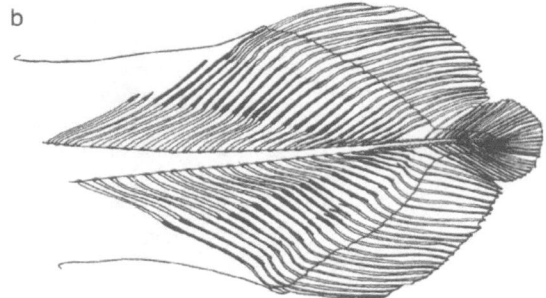

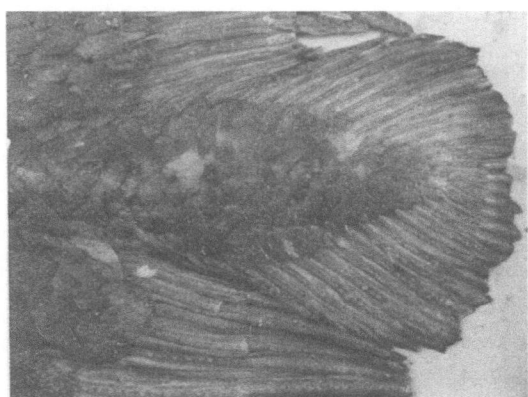

Fig. 1. a – Proposed coelacanth fin nomenclature: D1 – first dorsal, D2 – second dorsal, D3 – third dorsal, CF – caudal fin, P1 – Pectoral, P2 – pelvic, A1 – first anal, A2 – second anal (modified from Millot & Anthony 1958). b – Skeletal structure of the caudal region. Note soft rays of D3 and A2 articulate with the pterygiophores (shown in solid black) while those of CF connect directly to the notochord. c – Photograph showing external separation of CF from D3 and A2.

ward. As in single oar sculling where the oar is moved through the water first on one side of the boat and then the other, the two fins alternate this compression action, first on one side of the body

270

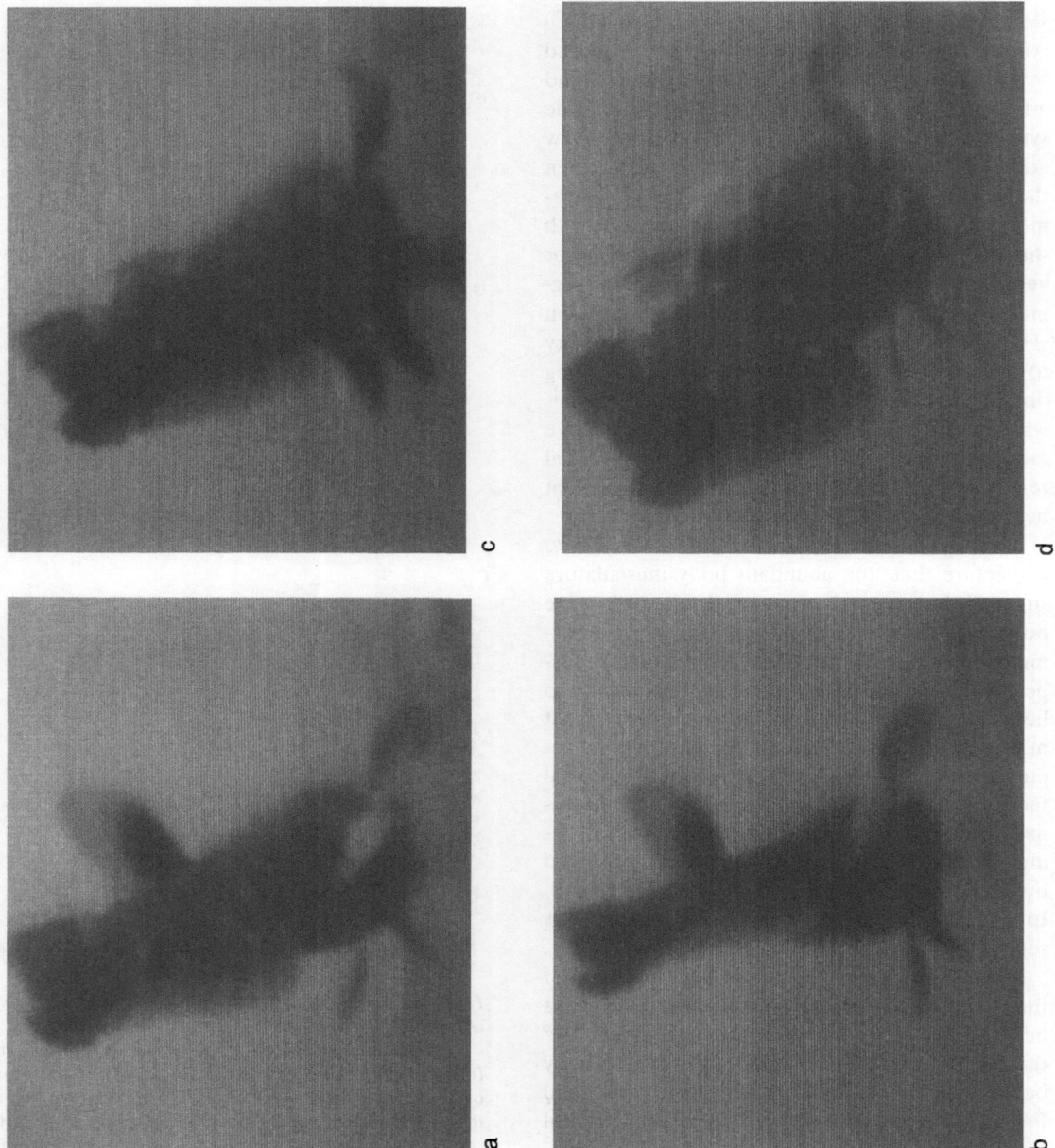

Fig. 2. Sequence from a posteroventral view showing the movement of the coelacanth's fins, especially that of the second dorsal (D2) and first anal (A1). From left to right: a – D2 is on the animal's left hidden by the caudal region, while A1 is seen on the left just below this region; b – D2 has moved to the right and down as A1 (whitish) moved to the right; c – A1 (whitish) moved up further; d – D2 has moved back to the left and turned (only a wisp is seen) while A1 is moving left, appearing as a blur just below the caudal.

and then the other. During straight forward motion, the amplitude of alternating joint fin movement is roughly equal on both sides of the body. When gradually turning, this fin movement is usually repeated only to the side away from the turn. In both cases, speed appears to be controlled by how many compressions are made in a given time, the greater the number, the faster the forward progression, and how close the two fins are brought together, the closer, the faster. While being observed in these video records, *L. chalumnae* at its fastest appeared to swim forward at less than 10 m min⁻¹, executing about 75 to 80 compressions (alternating right and left side compressions counted separately) during this time.

In general, observations on locomotion at shallow depths no greater than 50 m appear to agree with those Fricke et al. (1987) recorded at 117 and 198 m, however, no extended drifting activities (except for occasional headstands) were observed in video records perused by me; the anal and second dorsal fins moved faster and through a much greater arc. It is too early to conjecture about the correlation of depth and current movements to these differences in fin movement and locomotion, or to accept the explanation given in the commentary to 'The Story of the Coelacanth' SABC video that the faster movements are caused by respiratory stress of the dying fish.

The pectoral fins (P1), as noted by earlier authors, serve mainly as stabilizers in conjunction with the pelvics (P2) and the D3 and A2 to prevent the fish from rolling. Movements of the paired fins were unequal, independent, and roughly one-quarter out of phase with the second dorsal and first anal fins' compression action. Rather than calling it a 'sculling movement' (Locket & Griffith 1972), it is more akin to hand movements during an aquatic crawl, as the fins also twist while sculling. The first dorsal fin seems to serve as a rudder to the fish's forward movement.

Throughout its forward swimming, the coelacanth's pelvic fins remained spread at approximately 30° to the longitudinal axis, as already noted by Locket & Griffith (1972). Only very gentle, minimal movements of these fins were observed. The first dorsal fin remained fully erect most of the time, while at other times it was folded about halfway, but never completely closed.

One theory voiced by many authors, including the popular natural history writer David Attenborough (1979), is that the pectoral and pelvic fins could have been used to help the coelacanth move over the rocky sea floor and would be of real assistance for crawling out of water. These paired fins are quite different from those of the lungfishes and mudskippers which do move along the substrate both in and out of the water; the extinct *Eusthenopteron* has been assumed to have done likewise. Furthermore, the position of the pelvics in relation to the anal fin 1 and 2 would prove to be a severe hindrance to this type of locomotion in *L. chalumnae*. Most importantly though, after observing *L. chalumnae* swimming in the open sea, it is apparent that the lack of sustained caudal mobility would make such paired fin assisted crawl virtually impossible. Rather, the fish with its slightly negative buoyancy (Nevenzel et al. 1966) is well suited for near-bottom or midwater swimming and dwelling. The fact that most catches have been along the rather steep volcanic slopes descending to 300 m within 1 km from the western shore of the Grand Comoro seems to support this view.

Feeding

Based on this fish's body shape, fin size and location, McCosker's (1979) conjecture that the coelacanth is not unlike large groupers, perching on reef platforms, lunging short distances to capture prey seems reasonable. Current observations seem to support part of this idea, except that the fish probably does not 'lunge', but rather drifts and executes a sudden, deliberate bite within a short distance using its powerful jaws when prey moves within reach. Such a movement was seen when the second caged specimen readily took a piece of cuttlefish offered by the diver. Just as the coelacanth's movements can be related to those of the ocean sunfish, *Mola mola*, it also appears that some feeding habits may be similar to that species. A group of some four young of ocean sunfish was observed near Ito, Izu Peninsula, in a headstand position feeding off

272

the ocean floor, at about 40 m depth (H. Masuda personal communication). Judging from the stomach contents of three *L. chalumnae* collected during our expeditions, the coelacanth probably feeds at least in part on the ocean floor. So far I have removed unmasticated, slightly digested whole specimens of a deep sea witch-eel, *Ilyophis brunneus* (1500 m) (Suyehiro et al. 1984), a swell shark, *Cephaloscyllium sufflans* (40–440 m), and *Beryx decadactylus* (200–600 m) (Smith 1986) from the stomach. The first two are known to be bottom or near-bottom dwellers, while the last is mainly epibenthic. Their occurrence in these stomachs supports an earlier report (McCosker 1979) on food habits. The headstanding, drift feeding orientation of the coelacanth may thus offer some further supportive evidence for an electroreceptive function of the rostral organ (Bemis & Hetherington 1982).

Acknowledgements

This research has been supported by the Japan Scientific Expedition of the Coelacanth (JASEC), whose director, Kimihei Shinonoi, has not only been instrumental in providing research materials, but also in constantly furnishing invaluable assistance and encouragement throughout the project. Christian Olhagaray, Jean-Louis Geraud, and Jean Pierre Rebondy have assisted by providing much of the data including the film, videotape, and photographs. Also, I acknowledge the devoted interest and support of the joint project by the late Ahmed Abdallah, President of the Federal Islamic Republic of the Comoros, as well as all the fishermen and people of the Grand Comoro who have willingly given of their time and efforts. Neal M. Teitler assisted in the preparation of this manuscript. John E. McCosker, David Noakes and Christine Flegler-Balon kindly reviewed and offered corrections which helped to improve this paper.

References cited

Attenborough, D. 1979. Life on earth. Collins/BBC, London. 319 pp.

Balon, E.K., M.N. Bruton & H. Fricke. 1988. A fiftieth anniversary reflection on the living coelacanth, *Latimeria chalumnae*: some new interpretations of its natural history and conservation status. Env. Biol. Fish. 23: 241–280.

Bemis, W.E. & T.E. Hetherington. 1982. The rost[r]al organ of *Latimeria chalumnae*: morphological evidence of an electroreceptive function. Copeia 1982: 467–471.

Forey, P.L. 1988. Golden jubilee for the coelacanth *Latimeria chalumnae*. Nature 336: 727–732.

Fricke, H., O. Reinicke, H. Hofer & W. Nachtigall. 1987. Locomotion of the coelacanth *Latimeria chalumnae* in its natural environment. Nature 329: 331–333.

Goodrich, E.S. 1930. Studies on the structure and development of vertebrates. Macmillan, London. 837 pp.

Harris, J.E. 1937. The mechanical significance of the position and movements of the paired fins in the Teleostei. Carnegie Inst. Wash. Publ. 475: 171–189.

Hertel, H. 1966. Structure, form, movement. Reinhold, New York. 251 pp.

Hubbs, C.L. & K.F. Lagler. 1964. Fishes of the Great Lakes region. University of Michigan Press, Ann Arbor. 213 pp.

Lindsey, C.C. 1978. Form, function and locomotory habits in fish. pp. 1–100. *In:* W.S. Hoar & D.J. Randall (ed.) Fish Physiology, Vol. 7, Locomotion, Academic Press, New York.

Lighthill, M.J. 1969. Hydromechanics of aquatic animal propulsion. Ann. Rev. Fluid Mech. 1: 413–446.

Locket, N.A. 1980. Some advances in coelacanth biology. Proc. R. Soc. London. B 208: 265–307.

Locket, N.A. & R.W. Griffith. 1972. Observations on a living coelacanth. Nature 237: 175.

McAllister, D.E. & C.L. Smith. 1978. Mensurations morphologiques, dénombrements méristiques et taxonomie du coelacanthe, *Latimeria chalumnae*. Naturaliste canadien 105: 63–76.

McCosker, J.E. 1979. Inferred natural history of the living coelacanth. Occ. Pap. Calif. Acad. Sci. 134: 17–24.

Millot, J. 1955. First observation on a living coelacanth. Nature 175: 362–363.

Millot, J. & J. Anthony. 1958. Anatomie de *Latimeria chalumnae*. Tome 1: squelette, muscles et formations de soutien. C.N.R.S., Paris. 122 pp.

Millot, J., J. Anthony & D. Robineau. 1978. Anatomie de *Latimeria chalumnae*. Tome 3: Appareil digestif, appareil respiratoire, appareil urogénital, glandes endocrines, appareil circulatoire, téguments, écailles, conclusions générales. C.N.R.S., Paris. 198 pp.

Moy-Thomas, J.A. & T.S. Westoll. 1935. On the Permian coelacanth, *Coelacanthus granulatus* Ag. Geol. Mag. 72: 446–457.

Nevenzel, J.C., W. Rodegker, J.F. Mead & M.S. Gordon. 1966. Lipids of the living coelacanth, *Latimeria chalumnae*. Science 152: 1753–1755.

Smith, J.L.B. 1939a. A living fish of Mesozoic type. Nature 143: 455–456.

Smith, J.L.B. 1939b. A surviving fish of the order Actinistia. Trans. R. Soc. S. Afr. 27: 47–50.

Smith, J.L.B. 1939c. A living coelacanthid fish from South Africa. Trans. R. Soc. S. Afr. 28: 1–106.

Smith, M.M. 1986. Latimeriidae. pp. 152–153. *In:* M.M. Smith & P.C. Heemstra (ed.) Smiths' Sea Fishes, Macmillan South Africa, Johannesburg.

Suyehiro, Y., T. Tsutsumi & T. Uyeno. 1984. On dissection of coelacanth, *Latimeria chalumnae*. pp. 12–13. *In:* Proceedings of the First Symposium on Coelacanth Studies, Tokyo.

Suzuki, N., Y. Suyehiro & T. Hamada. 1985. Initial report of expeditions for coelacanth – Part 1 – Field studies in 1981 and 1983. Sci. Pap. Coll. Arts Sci., Univ. of Tokyo 35: 37–79.

Mmadi younow PKyan.
محمد يونس

Environmental Biology of Fishes **32**: 275–279, 1991.
© 1991 *Kluwer Academic Publishers.*

Stomach contents of *Latimeria chalumnae* and further notes on its feeding habits

Teruya Uyeno[1] & Toshio Tsutsumi[2]
[1] *National Science Museum, Tokyo, 169 Japan*
[2] *Keikyu Aburatsubo Marine Park, Miura, 238-02 Japan*

Received 30.10.1990   Accepted 23.11.1990

Key words: Prey, Benthic, Epibenthic, Nocturnal, Piscivorous, Drift feeder, Headstand, Bottom feeder, Coelacanth, Crossopterygii

Synopsis

Stomach contents of three *Latimeria chalumnae* dissected in Japan support the hypothesis that the coelacanth is a predominantly nocturnal bottom or near-bottom drift feeder. Prey items identified in this study (*Ilyophis brunneus, Cephaloscyllium sufflans* and *Beryx decadactylus*) and those reported previously are mostly benthic or epibenthic dwellers. The drifting of the coelacanth in a headstanding posture would account for the easy capture of prey that moves within or just above the bottom strata at night.

Introduction

Of the over 172 specimens of the living coelacanth, *Latimeria chalumnae,* that have been recorded as captured to date (Bruton & Coutouvidis 1991), gut contents have been reported in very few cases (McCosker 1979). In 1967, Japan received its first coelacanth as a gift from French president Charles DeGaulle, and this specimen preserved in formalin was then given to the Yomiuriland Aquarium for display. The senior author has performed gross dissections on this fixed specimen (which remains in formalin) and on four of the six frozen ones presented by the Government of the Federal Islamic Republic of the Comoros to the Japan Scientific Expedition of the Coelacanth (JASEC). Of these fish, three held identifiable stomach contents that the authors consider to be of significance in interpreting the coelacanth's feeding habits, and form the basis for this report.

Coelacanth prey

Smith (1953), Millot (1954) and Millot & Anthony (1958) were the first to identify and record prey found in the stomach of *L. chalumnae*. This and later published records were reviewed by McCosker (1979). These records are summarized in the top half of Table 1.

Upon dissection of JASEC no. 1 and 2 (CCC no. 123, 121), only a cephalopod beak of unidentifiable species was found within the stomach of the former, while that of the latter was empty. Prey identified from JASEC no. 3, 4 and 6 (CCC no. 124, n and 149) were *Ilyophis brunneus* (Fig. 1a), *Cephaloscyllium sufflans* (Fig. 1b), and *Beryx decadactylus* (Fig. 1c, d), respectively (Suyehiro et al. 1984). These records are summarized in the bottom half of Table 1.

Inferred feeding habits

Evidence from the prey suggests that *Latimeria chalumnae* is a predominantly piscivorous benthic or epibenthic feeder. The coelacanth's headstand position observed and photographed by Fricke et al. (1991) in the natural habitat and by the JASEC team in waters about 50 m depth on released specimens, is indicative of drift feeding fishes (Uyeno 1991). This vertical, head-down position has been observed in a number of other species including the leaf fish *Monocirrhus polyacanthus*, snipefish *Macrorhamphosus scolopax*, shrimpfishes *Acoliscus strigatus* and *Centriscus scutatus,* young sunfish *Mola mola,* and young snooks *Lates japonicus.*

Interestingly, a preliminary list compiled by Suzuki et al. (1985) of the marine ichthyofauna of the Comoros and local names derived from observation and discussion with local fishermen about their catch does not include most of the prey identified from the stomachs of the coelacanths as shown in Table 1. These species probably are absent from surveys because most of these prey are deepwater fish seldom caught or seen. Although *Coranthus polyacanthus* migrates up to 150 m depth (Fraser 1972, Hayashi 1984) and the swell shark *Cephaloscillium sufflans* can be found along shallow bottoms of 40 m depth (Bass 1986), these and other prey species usually dwell quite deep: lanternfish *Diaphus metopoclampus* – down to 850 m (Hulley 1986), the stout beardfish *Polymixia nobilis* – 640 m (Heemstra 1986b), deepwater snappers *Symphysanadon* sp. – 500 m (Katayama 1984), *Cephaloscillium sufflans* – 440 m (Bass 1986), *Beryx decadactylus* – 600 m (Heemstra 1986a, Nielsen 1973) and the rather rare deep sea witch-eel, *Ilyophis brunneus,* has been caught at depths from 600 to 2668 m (Castle 1986, Machida 1984, Blache et al. 1973). From the prey ingested by the specimens shown in Table 1, and because these and other coelacanths have taken bait including the gempylid *Promethichthys prometheus,* tuna, octopus and *Decapterus* sp.

Table 1. Prey items found in the stomach and intestine of the coelacanth. The 'CCC no.' refers to the Coelacanth Conservation Council inventory number given by Bruton & Coutouvidis (1991).

CCC no.	Prey items	Prey size, depth of natural occurrence and notes	Reference and older specimen identification
?	*Diaphus metopoclampus* lanternfish	TL 20 cm whole; 375–850 m daytime; 90–850 m at night	Millot & Anthony (1958, p. 2592)
2	Scales and eyeballs	in intestine	Smith (1953, p. 101); RUSI 614
57	*Polymixia nobilis* stout beardfish	SL 17–18 cm (gill arches, hyoid, dentary, cleithrum, several vertebrae); near bottom dweller, 183–640 m	McCosker (1979, p. 20); USNM 20587
69	*Coranthus polyacanthus* (two) deepwater cardinal fish	SL 12–14 cm, partly digested; 150 m	McCosker (1979; p. 21); CAS 24862
73	Entire cuttlefish		McAllister (1971, p. 17)
88	Scales and otoliths	in intestine	McCosker (1979, p. 21); SIO 75-347
90	*Symphysanadon* sp. (two) deepwater snappers	SL 7–8 cm, intact and partially digested; some species 119–500 m on or near bottom	McCosker (1979, p. 20); CAS 33111
123	Cephalopod beak		Uyeno & Tsutsumi (this paper); JASEC no. 1
135	*Ilyophis brunneus* deepsea witcheel	TL 4.9 cm, whole, partially digested; 600–2668 m	Suyehiro et al. (1984, p. 12, Uyeno 1989, p. 20); JASEC no. 3
136	*Cephaloscyllium sufflans* swell shark	TL 48 cm, whole, partially digested; 40–440 m	Uyeno (1990, p. 20); JASEC no. 2
149	*Beryx decadactylus*	SL 34.5 cm, slightly digested; 200–600 m	Uyeno (1990, p. 21); JASEC no. 6

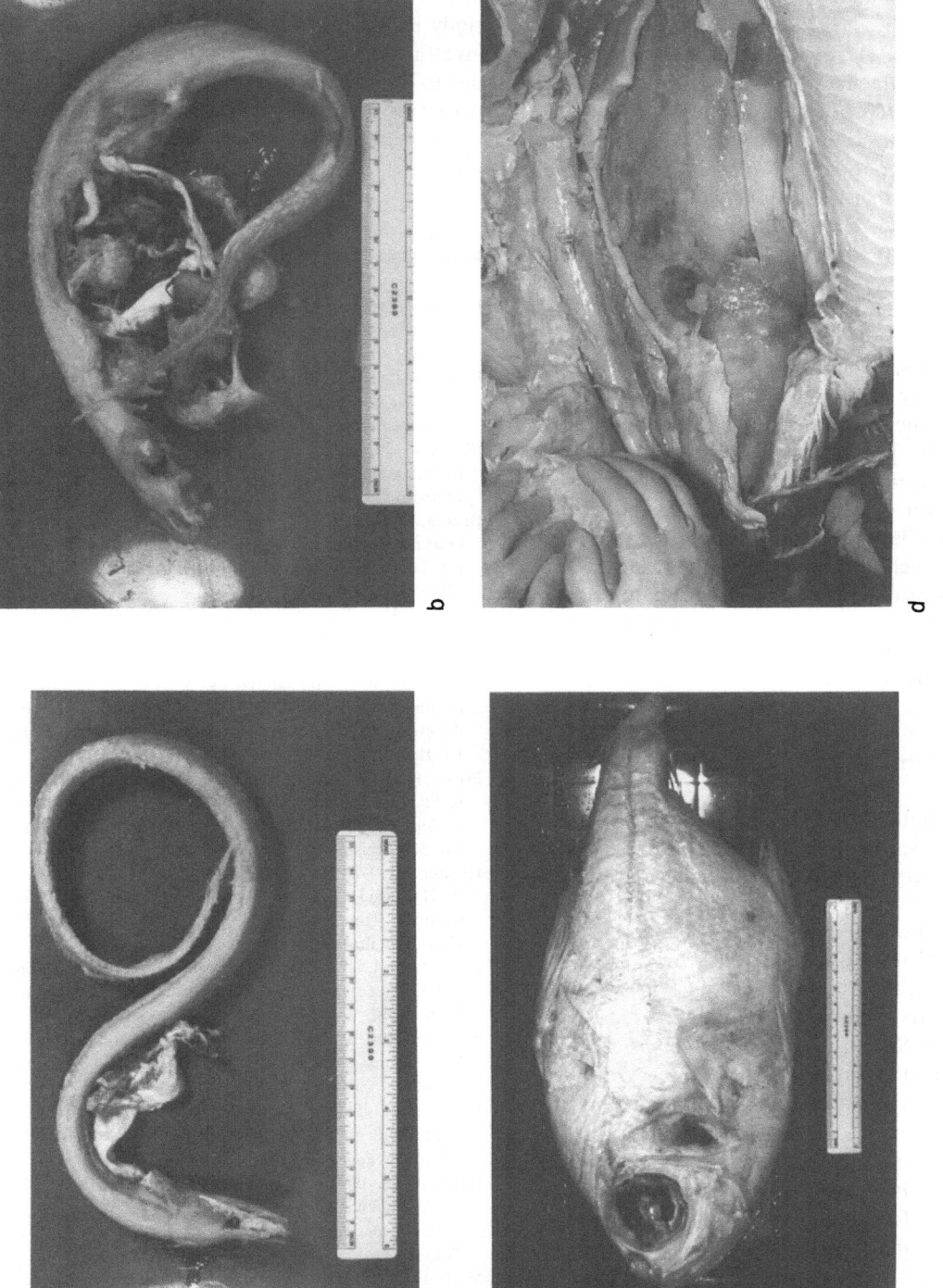

Fig. 1. a – Deepsea witch-eel *Ilyophis brunneus*, 49 cm TL, removed from the stomach of JASEC specimen no. 3. (CCC no. 124). b – Swell shark *Cephaloscyllium sufflans*. 48 cm TL, recovered from the gut of JASEC specimen no. 4. (CCC n). c – Whole, very slightly digested *Beryx decadactylus*, 34.5 cm SL, removed from JASEC specimen no. 6. (CCC no. 149). d – The *Beryx* (same as in Fig. 1c) in situ, shown occupying practically the entire stomach of the coelacanth.

(Bruton & Coutouvidis 1991), *Latimeria* is clearly not prey specific.

Fricke et al. (1991) have provided convincing evidence that the natural habitat of *L. chalumnae* lies within the 180–210 m depth range and that nocturnal migration occurs between 182–253 m. Growing evidence from photographs and stomach contents indicate that *L. chalumnae* is a slow-swimming solitary feeder well adapted to drifting in a vertical, headstand position just above the ocean floor, even at depths of 600 m or more. Although Bemis & Hetherington (1982) believe that the rostral organ is electroreceptive and may serve in sensing prey movement, this function has not been unequivocally proven. When the coelacanth is directly over the prey, the mouth is probably quickly opened, the prey grasped and then sucked in whole. An action somewhat like this was observed when a diver offered a piece of cephalopod to a captured coelacanth. Once this food was brought to almost touching the mouth, it was grasped so fast by the powerful jaws with the head simultaneously whipping to one side that the videotaped action was blurred beyond clear resolution, even when viewed in a single frame mode.

Acknowledgements

We are grateful to JASEC director Kimihei Shinonoi whose dedication and administrative support from the very beginning have made our team's research and reports possible. Ongoing thanks are also due to the Government of the Federal Islamic Republic of the Comoros as they have extended continuous cooperation in making our joint research efforts not only feasible, but a pleasure as well. Naoki Suzuki of Jikei University School of Medicine and Masayoshi Hayashi of the Yokosuka City Museum of Natural History provided invaluable information necessary for compiling this report, for which we extend our sincerest appreciation. Tomio Iwamoto kindly took time to personally arrange for the senior author to examine the coelacanths and stomach contents that are part of the fish collections of the California Academy of Sciences. We wish to acknowledge our indebtedness to the late Yasuo Suyehiro who originated this study as well as organized Japan's first scientific investigation of the coelacanth. Also, thanks are due to Neal M. Teitler who greatly assisted in preparing the manuscript and taking the photographs, and to Eugene Balon for help in editing its contents.

References cited

Bass, A.J. 1986. Scyliorhinidae. pp. 89–95. *In:* M.M. Smith & P.C. Heemstra (ed.) Smiths' Sea Fishes, Macmillan, Johannesburg.

Bemis, W.E. & T.E. Hetherington. 1982. The rost[r]al organ of *Latimeria chalumnae:* morphological evidence of an electroreceptive function. Copeia 1980: 467–470.

Blache, J., M.L. Bauchot & L. Saldanha. 1973. Synaphobranchidae. pp. 252–253. *In:* J.C. Hureau & Th. Monod (ed.) Clofnam I, UNESCO, Paris.

Bruton, M.N. & S.E. Coutouvidis. 1991. An inventory of all known specimens of the coelacanth *Latimeria chalumnae,* with comments on trends in the catches. Env. Biol. Fish. 32: 371–390. (this volume)

Castle, P.H.J. 1986. Synaphobranchidae. pp. 188–191. *In:* M.M. Smith & P.C. Heemstra (ed.) Smiths' Sea Fishes, Macmillan, Johannesburg.

Fraser, T.H. 1972. Comparative osteology of the shallow water cardinal fishes (Perciformes: Apogonidae) with referene to the systematics and evolution of the family. Ichthyol. Bull. J.L.B. Smith Inst. Ichthyol. 34: 1–105.

Fricke, H., K. Hissmann, J. Schauer, O. Reinicke, L. Kasang & R. Plante. 1991. Habitat and population size of the coelacanth *Latimeria chalumnae* at Grand Comoro. Env. Biol. Fish. 32: 287–300. (this volume)

Hayashi, M. 1984. *Coranthus polyacanthus* (Vaillant). p. 145. *In:* H. Masuda, K. Amaoka, C. Araga, T. Uyeno & T. Yoshino (ed.). The Fishes of the Japanese Archipelago, Tokai University Press, Tokyo.

Heemstra, P.C. 1986a. Berycidae. pp. 409–410. *In:* M.M. Smith & P.C. Heemstra (ed.) Smiths' Sea Fishes, Macmillan, Johannesburg.

Heemstra, P.C. 1986b. Polymixiidae. p. 432. *In:* M.M. Smith & P.C. Heemstra (ed.) Smiths' Sea Fishes, Macmillan, Johannesburg.

Hulley, P.A. 1986. Order Myctophiformes. pp. 282–322. *In:* M.M. Smith & P.C. Heemstra (ed.) Smiths' Sea Fishes, Macmillan, Johannesburg.

Katayama, K. 1984. Symphysanodontidae. p. 138. *In:* H. Masuda, K. Amaoka, C. Araga, T. Uyeno & T. Yoshino (ed.) The Fishes of the Japanese Archipelago, Tokai University Press, Tokyo.

Machida, Y. 1984. *Ilyophis brunneus* Gilbert. p. 26. *In:* H. Masuda, K. Amaoka, C. Araga, T. Uyeno & T. Yoshino (ed.)

The Fishes of the Japanese Archipelago, Tokai University Press, Tokyo.

McAllister, D.E. 1971. Old four legs: a 'living fossil'. Mus. Natl. Sci. Nat. Coll. Odyssey Ser. 1: 1–25.

McCosker, J.E. 1979. Inferred natural history of the living coelacanth. pp. 17–24. *In:* J.E. McCosker & M.D. Lagios (ed.) The Biology and Physiology of the Living Coelacanth, Occ. Pap. Calif. Acad. Sci. 134, San Francisco.

Millot, J. 1954. New facts about coelacanths. Nature 174: 426–427.

Millot, J. & J. Anthony. 1958. Crossoptérygiens actuels *Latimeria chalumnae.* pp. 2553–2597. *In:* P.-P. Grassé (ed.) Traité de Zoologie, Vol. 13, Agnathes et Poissons, Anatomie, Ethologie, Systématique, Masson, Paris.

Nielsen, J. 1973. Berycidae. p. 337. *In:* J.C. Hureau & Th. Monod (ed.) Clofnam I, UNESCO, Paris.

Smith, J.L.B. 1953. The second coelacanth. Nature 171: 99–107.

Smith, J.L.B. 1958. Old fourlegs. Second Edition. Pan Books Ltd., London. 284 pp.

Suyehiro, Y., T. Tsutsumi & T. Uyeno. 1984. On dissection of coelacanth, *Latimeria chalumnae.* pp. 12–13. *In:* Proc. 1st Symp. on Coelacanth Studies, Tokyo, 1984, Japan Sci. Comm. of Coelacanth Studies, Tokyo.

Suzuki, N., Y. Suyehiro & T. Hamada. 1985. Initial report of expeditions for coelacanth – Part I – Field studies in 1981 and 1983 –. Sci. Papers Coll. Arts Sci. Univ. Tokyo 35: 37–79.

Uyeno, T. 1989. Investigating the ecology of the coelacanth based on feeding habits. Newton Graphic Science Magazine 9(13): 20–21. (In Japanese).

Uyeno, T. 1991. Observation on locomotion and feeding of released coelacanths, *Latimeria chalumnae.* Env. Biol. Fish. 32: 267–273. (this volume)

Note

C.C. Baldwin (personal communication) kindly provided the following description of the stomach content of CCC no. 141 (CCC no. 140 had no identifiable remains in the stomach): 'The stomach contents of the large female coelacanth, VIMS 08118 (CCC no. 141), have been given the catalog number VIMS 08118A. The stomach contained a single recognizable item: a vertebral column without head or tail, comprised of 89 vertebrae and measuring ca. 38 cm in length (specimen is coiled, and accurate measurement difficult to make). The size of the vertebrae decreases considerably posteriorly, the most anterior vertebral centrum measuring 5.6 mm in length, the most posterior, 2.5 mm. The small size of the last vertebra suggests that the posterior end of the column may be close to complete. Most vertebrae lack prominent projections (neural spines, haemal spines, etc.). Vertebral centra 10–22 each bear a well-developed pair of wing-like parapophyses ventrally, and vertebrae 72–76 have a ventral projection that appears to be a haemal spine. The parapophyses are robust structures that would not seem to be easily detached from the centra; thus 13 pairs of parapophyses is probably an accurate count and may indicate the presence in the predigested specimen of 13 pairs of ribs. Alternatively, the five haemal spines on the posterior vertebrae are very fragile and tear easily when handled, and the predigested specimen probably had many more of these structures.

Based on the morphology of the vertebral column, particularly the small size of the most posterior vertebrae and the presence of nine vertebrae preceding the first centrum bearing parapophyses, it seems likely that the total number of vertebrae is not much greater than the observed 89. The number of vertebrae, length and general morphology of the specimen suggest that it may be the remains of an anguilliform eel, and a comparison of the remains to a skeleton of *Anguilla rostrata* reveals many similarities. Although many anguilliforms have vertebral counts greatly exceeding 100, representatives of several families found in the western Indian Ocean (e.g., Anguillidae, Xenocongridae, Moringuidae, Cyemidae) have between 90 and 110 vertebrae, a range that likely includes the number of vertebrae in the predigested specimen.

It should be noted, however, that other teleost taxa occurring in the western Indian Ocean, including gadiforms, zoarciforms and ophidiiforms, contain members with vertebral counts similar to that of the digested specimen, but comparative osteological material is lacking for those taxa'.

J.A. Musick

Environmental Biology of Fishes **32**: 281–283, 1991.
© 1991 *Kluwer Academic Publishers.*

Juvenile anisakine parasites from the coelacanth *Latimeria chalumnae*

Dennis A. Thoney[1] & William J. Hargis, Jr.
Virginia Institute of Marine Science, School of Marine Science, College of William and Mary, Gloucester Point, VA 23062, U.S.A.
[1] *Present address: New York Aquarium, Boardwalk and West 8th St., Brooklyn, NY 11224, U.S.A.*

Received 15.6.1989   Accepted 20.7.1990

Key words: Co-evolution, Nematode, Anisakidae, Host

Synopsis

Two coelacanths captured in waters off of the Comoro Islands were examined for parasites. Eight third-stage anisakines of the genus *Terranova* or *Pulchrascaris* were found in the spiral valve and rectum of the two coelacanths examined. The poor condition of these defrosted specimens prevented further identification. The depauperate parasitofauna in coelacanths may result from their unique physiology and morphology and because they are a relict fish that has survived millions of years beyond other relatives and potential intermediate hosts.

Introduction

The coelacanth, *Latimeria chalumnae* Smith, is the only known extant species of the Coelacanthiformes, an order that was relatively abundant over a hundred million years ago. If parasite phylogeny reflects host phylogeny (Farenholz' Rule), and the more primitive the host, the more primitive the parasites it harbors (Szidat's Rule), then relict hosts such as the coelacanth would be expected to have parasites with primitive features. Just as the coelacanth provided valuable information regarding the evolution of fishes, primitive characters of its parasites should provide information on the evolution of parasites.

Thus far, the few coelacanths examined have had very few parasites. Monod (1954) described larvae of the gnathiid isopod *Praniza milloti* from the gills of three individuals he examined. Dollfus & Campana-Rouget (1956) reported the occurrence of one specimen of the tetraphyllidean metacestode *Scolex polymorphus unilocularis*, two specimens of a trypanorhynch metacestode, three specimens of various stages of the nematode *Ascarophis* sp., and about thirty specimens of *Anisakis* from four coelacanths they examined. Hargis (1958) also found a number of *P. milloti* on the pair of formalin-preserved gill arches from a specimen captured 'in good condition' near the landing of Itsandra on Grand Comoro. Kamegai (1971) found four larval nematodes, one of which was identified as *Anisakis* (Type II) and four trypanorhynch metacestodes of the genus *Tentacularia*. He also described the gill monogenean *Dactylodiscus latimeris* Kamegai 1971, the only member of the family Dactylodiscidae.

The purpose of this study was to determine and evaluate the parasites of two coelacanths captured in waters off the Comoro Islands.

Materials and methods

The two coelacanths studied were captured by handline off the Comoro Islands. The smaller specimen (VIMS 8117) was 972 mm TL and weighed 13.5 kg. Date of capture of this specimen is unknown, but it was severely freezer-burned, indicating that it had been frozen for considerable time. The second specimen (VIMS 8118) captured in 1986 was 1452 mm SL and weighed 53.8 kg. This specimen appeared to be in good condition when defrosted. Both specimens were necropsied. The skin, buccal cavity, coelom, liver, kidney, swim bladder, and mesenteries were examined macroscopically for parasites. The esophagus, stomach, duodenum, spiral valve, and rectum were examined using a steromicroscope at 4–25x. Gill filaments were removed and examined from the four arches on one side of the head of specimen # 8118. The remaining four left gill arches with filaments were kept intact for detailed anatomical study and only examined macroscopically for parasites. The gill filaments of specimen # 8117 were badly freezer-burned and were not examined.

Nematodes collected were cleared in glycerin for identification. Paraffin sections of one nematode were stained with hematoxylin and eosin. Two anisakine juveniles have been deposited at the USNM Helminthological Collection, Beltsville, Maryland (No. 81038). One specimen of *Anisakis* sp. (No. 16700) from a coelacanth examined by Kamegai (1971) was obtained from the Meguro Parasitological Museum, Tokyo, for comparison with our specimens.

Results

The smaller coelacanth (VIMS 8117) had one third-stage anisakine nematode in the spiral valve and the larger one (VIMS 8118) had seven third-stage anisakines in the spiral valve and rectum.

Anisakine juveniles

Description of third-stage juveniles (based on 6 specimens) (measurements in mm): Body 4.25–10.31 long by 0.18–0.45 wide at greatest width. Lacked visible alae. Esophagus 0.72–1.06 long. Ventriculus 0.35–0.53 long by 0.08–0.11 wide; ven-

tricular appendage absent. Intestinal cecum 0.38–0.50 long by 0.02 wide. Nerve ring located between 21–25% of anterior extremity of worm. Excretory pore not visible. Four pyriform rectal glands. Tail conically shaped, 0.13–0.19 long, with numerous minute spines at tip.

Remarks: Our specimens closely resemble the *Anisakis* juveniles described by Dollfus & Campana-Rouget (1956) in size and morphology, except that ours have an intestinal cecum. Deardorff (1987) discussed the difficulties of identifying anisakines lacking adult characters, but the presence of an intestinal cecum in our specimens places them in either the genus *Terranova* or *Pulchrascaris*. Both genera have been reported from the western Indian Ocean, where our coelacanths were collected. All specimens were in such poor condition because of prolonged freezing that they could not be identified further. In most of our parasites the intestinal cecum was difficult to see, but it was distinct in two specimens. Its presence also was verified in paraffin sections. The lack of an intestinal caecum in the specimen collected by Kamegai (1971) distinguish it from our specimens. The excretory canal was not observed.

Discussion

The coelacanth specimens examined in this study had extremely depauperate parasite faunas. Isopods, cestodes, and monogeneans found previously in coelacanths were not found in these individuals. In fact, the only parasites collected were larval stages of anisakine nematodes probably belonging to the genus *Terranova* or *Pulchrascaris*. Unless sharks feed on coelacanths, the presence of these particular juvenile nematodes in coelacanths probably represent accidental infections. Coelacanths probably become infected with these anisakines by feeding on infected fishes and invertebrates.

Examination of stomach contents of our coelacanths indicated that they fed mostly on fishes. According to Holmes (1990), marine piscivores do not necessarily have greater parasite-species richness than benthic feeders or planktivores, but they do generally have greater parasite-species richness than most freshwater fishes. Also, large coelacanths are thought to be quite old, which should have allowed accumulation of parasites over time. Why then have the coelacanths examined thus far had such sparse parasite faunas?

High urea content and differences in intestinal morphology are but a few of the features distinguishing coelacanths from teleosts. Parasites of teleosts may not be able to tolerate a host with such a unique physiology, thus preventing cross-infection by many otherwise suitable parasitic species. Although coelacanths may be physiologically more similar to sharks than teleosts, there must be sufficient differences to prevent cross infection of parasites.

Except for the gill monogenean D. latimeris, few known parasites are restricted to coelacanths. Dactylodiscus latimeris has many primitive characters for the Dactylogyroidea (D.L. Kritsky, personal communication) suggesting that it may be a relict parasite that has co-evolved with the coelacanth over a long period of time. Because the coelacanth is a relict fish that has managed to survive for millions of years beyond other known relatives, those heteroxenous parasites that may have coevolved with the coelacanth could have become extinct because of the extinction of their intermediate hosts. Williams et al. (1987) noted that other relict hosts such as the holocephalans also have presumed relict parasites such as the copepod Vanbenedenia kroyeri Malm, the monogeneans Chimaericola leptogaster (Leuckart) and Calicotyle affinis Scott, the gyrocotylideans of the genus Gyrocotyle, and the aspidocotylean Multicalyx elegans (Olsson). Although, the life cycles of M. elegans and Gyrocotyle are not completely known, most of these parasites, including D. latimeris from the coelacanth, are monoxenous, which would allow completion of

their life cycles even had an ancient intermediate host become extinct.

Acknowledgements

We wish to thank J.A. Musick and all other colleagues involved with our two coelacanth hosts. Without their cooperation this study would not have been possible. The loan of a specimen by S. Kamegai, Meguro Parasitology Museum, Tokyo, Japan is greatly appreciated. We also wish to thank E.M. Burreson and T.L. Deardorff for reviewing earlier drafts of this manuscript. VIMS Contribution No. 1615.

References cited

Deardorff, T.L. 1987. Redescription of *Pulchrascaris chiloscyllii* (Johnston and Mawson, 1951) (Nematoda: Anisakidae), with comments on species in *Pulchrascaris* and *Terranova*. Proc. Helminthol. Soc. Wash. 54: 28–39.

Dollfus, R.P. & Y. Campana-Rouget. 1956. Helminthes trouves dans le tube digestif de coelacanthes. Mem. Inst. Sci. Madagascar 11: 33–41.

Hargis, W.J., Jr. 1958. Coelacanths and monogeneids. J. Parasitol. 44: 191.

Holmes, J.C. 1990. Helminth communities in marine fishes. pp. 101–130. *In:* G.W. Esch, A.O. Bush & J.M. Aho (ed.) Parasite Communities: Patterns and Processes, Chapman Hall, London.

Kamegai, S. 1971. On some parasites of a coelacanth (*Latimeria chalumnae*): a new Monogenea, *Dactylodiscus latimeris* n. g., n. sp. (Dactylodiscidae n. fam.) and two larval helminths. Res. Bull. Meguro Parasitol. Mus. 5: 1–5.

Monod, T. 1954. Sur une larve de gnathiidé (*Praniza milloti* nov. sp.) parasite du coelacanthe. Mem. Inst. Sci. Madagascar, Ser. A, 9: 91–94.

Williams, H.H., J.A. Colin & O. Halvorsen. 1987. Biology of gyrocotylideans with emphasis on reproduction, population ecology and phylogeny. Parasitology 95: 173–207.

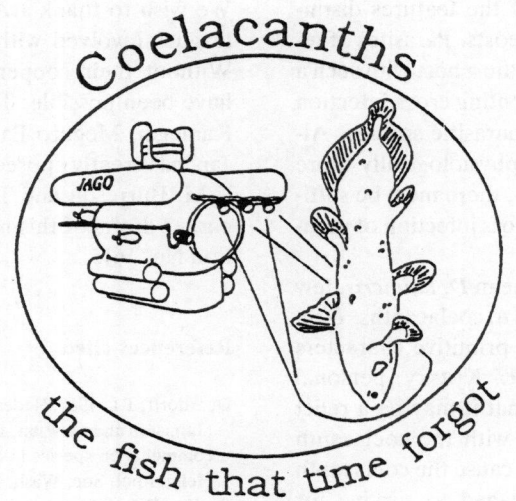

Coelacanths

the fish that time forgot

Ecology and conservation

Environmental Biology of Fishes **32**: 287–300, 1991.
© 1991 *Kluwer Academic Publishers.*

Habitat and population size of the coelacanth *Latimeria chalumnae* at Grand Comoro

Hans Fricke[1], Karen Hissmann[1], Jürgen Schauer[1], Olaf Reinicke[1], Lutz Kasang[1] & Raphael Plante[2]
[1] *Max-Planck-Institut für Verhaltensphysiologie, 8130 Seewiesen, Germany*
[2] *Centre d'Océanologie de Marseille, Station Marine d'Endoume, 13007 Marseille, France*

Received 8.8.1990   Accepted 25.10.1990

Key words: Living fossil, Crossopterygian, Social behaviour, Cave dwelling, Submersible study, Home range, Migration

Synopsis

In 1987 and 1989 coelacanths were observed for the first time in their natural habitat with the help of submersibles. Coelacanths were found between 150–253 m depth, their preferential depth seems to be around 200 m; the water temperature ranged between 16.5–22.8° C. During the day coelacanths aggregate in small non-aggressive groups in sheltered lava-caves. Caves might be a limiting factor for distribution. At night they leave the caves for hunting by drifting singly along the steep lava slopes. They migrate between different caves located within a large home range covering more than 8 km coastline. Coelacanths are site-attached, some for a period of at least 2 years. Our own observations and earlier catch records show that only the west coast of Grand Comoro is a suitable coelacanth habitat with more structural complexity and prey fish abundance than other coastlines of the island. From our survey we estimated a total coelacanth population off Grand Comoro to be 150–210 individuals; a saturated population would be 370–510 individuals. This small relict population seems to be stable. International protection of coelacanths against commercial interests is needed.

Introduction

In 1987 we observed, for the first time, living coelacanths in their natural habitat along the west coast of Grand Comoro, Western Indian Ocean (Fricke et al. 1987, Fricke & Schauer 1987, Fricke & Plante 1988, Fricke 1988). We encountered six coelacanths at depths between 117–198 m only at night. They drifted singly with perfect buoyancy close to the substrate and did not use their fins for locomotion along the bottom (Fricke et al. 1987). We suspected that coelacanths lived during the day in water deeper than 200 m which was the depth limit for our former research submersible GEO. To follow coelacanths down to their proposed lower depth range thought to be at around 300 m (Millot et al. 1972, McCosker 1979) we built the new submersible JAGO with a depth capacity down to 400 m. In fall 1989 we continued our diving operation with the new submersible and discovered that during the day coelacanths are cave dwellers and aggregate in small groups of up to 10 individuals (Fricke & Hissmann 1990, Fricke et al. 1991). They shelter in protected deep lava caves and hunt for fish singly at night.

A first account of the coelacanth's ecology was published by Forster (1974) and McCosker (1979). McCosker emphasized that the natural habitat can only be inferred because of the inaccessibility of the fish's environment. A decade ago, he gave a remarkable summary of our ecological knowledge of coelacanths. Meanwhile the use of small submersibles opened a new era for field studies and, thanks to modern technology, our knowledge on behaviour and natural history of the beast expands rapidly. Here we summarize our observations from 1987 and 1989 on the coelacanth's habitat. We report, for the first time, estimates of their population size at Grand Comoro – an issue of particular relevance for conservation efforts. In the last few years, coelacanths have become prominent in the world press because of the various threats to them by humans (see Bruton 1989a, b). The results of our study should help to evaluate the chances of the coelacanth's survival within the near future.

Materials and methods

Time table

The research was carried out during three cruises: cruise 1 from December 1986 to January 1987, cruise 2 from April to May 1987 with MV METOKA as mothership, and cruise 3 from October to December 1989 with MV SEA EAGLE as supportship. The two-man submersible GEO, depth capacity 200 m, was used during cruises 1 and 2; the two-man submersible JAGO, depth capacity 400 m, during cruise 3. During cruise 1 we performed a total of 29 dives, 10 dives during cruise 2, and 71 dives during the 3rd cruise.

Physical measurements

We used for measuring temperature: Testotherm 7300 with probe PT 100, resolution 0.1°C, Lenzkirchen, Germany; for salinity: WTW, Salinometer LF 191, resolution 0.1%, Weilheim, Germany; for the slope angle: inclinometer, type "Expert", with resolution 1°; and for continuous measurements of temperature, salinity and depth the CTD probe "Seabird", U.S.A. used in connection with a Toshiba computer Laptop 1600-40 and the computer program "Seasoft".

Observational techniques

Observations were recorded on a small Panasonic tape recorder and with still cameras (Nikon, Pentax, Minolta), on video (Sony V 50 E) and 16 mm cinefilm (Bolex H 16, Arriflex SR).

The locations of coelacanth caves were marked with beer bottles painted white. A 3 mm guiderope, starting between 100–150 m depth and laid perpendicular to the coast, aided submersible navigation to the caves. Because caves looked very similar, drawings of each cave were made which highlighted prominent topographical features of the surroundings.

Environmental surveys

Depth, salinity, temperature, slope angle, structural complexity of the bottom, sand-rock-coverage, number of potential prey species and prey fish abundance were measured at different stations along the east and west coast of Grand Comoro. Each station consisted of at least 10 spot checks. To guarantee independence of measurements, each spot check was separated by at least 30 sec travel time of the submersible. All data were pooled in 3 depth classes of 80–120 m, 140–160 m and 180–200 m and were presented in per cent of total observation.

For estimating the fish abundance the submersible was stationed approximately 2 m above the bottom and perpendicular to the coastline. The numbers of visible fish on the bottom area, illuminated by the submersible's main lights, were estimated in classes: n = 0, 1–3, 4–10, 11–30 and > 30 individuals.

Structural complexity of bottom was estimated in 5 classes from flat bottom without macrostructure (class 0) to highest complexity with many crevices and holes (class 4).

Sand-rock-coverage was estimated in 4 classes: class 0 (sand only), class 1 (predominant sand), class 2 (predominant rock), class 3 (rock only). The data between east and west coast or between different depths along the same coast were treated with

Fig. 1. Coelacanths have individual white body marks, different on both sides of the body, which were used for individual identification.

the non-parametric Kolmogorov-Smirnov test (Siegel 1956). Other statistical tests are indicated in the text.

Identification of individuals

Coelacanths have white body marks (Fig. 1) which are different on both sides of the body in each individual. In the field it is impossible to identify coelacanths according to these individual patterns. We made black-and-white prints from all slides, videos and cinepictures where body marks were clearly visible. Additionally, the 16 mm films (which were coded for dive- and cave-numbers) and videos were screened in order to facilitate easier identification of both body sides. A catalogue of

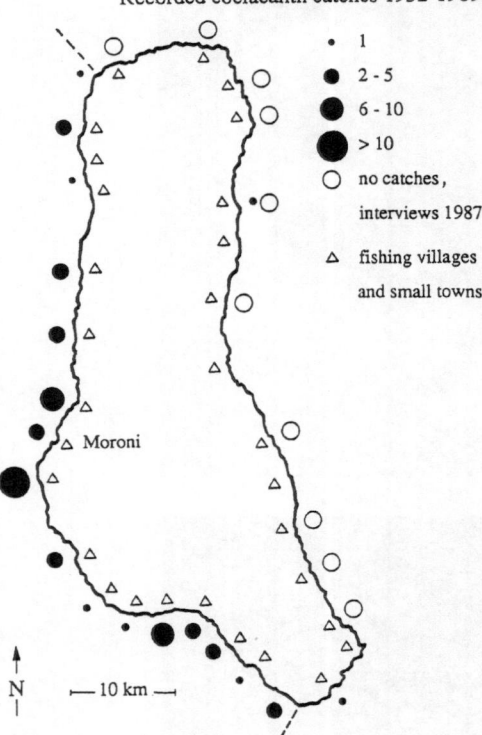

Recorded coelacanth catches 1952-1989

- 1
- 2 - 5
- 6 - 10
- > 10
○ no catches, interviews 1987
△ fishing villages and small towns

Moroni

N ⟶ 10 km ⟶

Fig. 2. Map of Grand Comoro (capital Moroni) with locations of coelacanth catches (filled circles) and distribution of coastal villages and small towns (open triangles). The dotted lines indicate the boundaries of the coelacanth habitat along the west coast.

individual body patterns was obtained which can also be used for further field studies.

Results and observations

Distribution of coelacanth catches

Figure 2 shows the locations of coelacanth catches from 1954 to 1989 along the entire coastline of Grand Comoro; Figure 3 summarizes the annual catch rate. The average annual catch rate is 2.6 ± 1.5 individuals ($x \pm$ SD, n = 69). There is no decrease or increase in catch rate over the years (χ^2, p > 0.5). Millot et al. (1972) did not report coelacanth catches from the east coast. This might be explained by the inaccessibility of the east coast at that time. A recently built road provides better access now. In 1987 we interviewed fishermen along the entire coast about coelacanth catches. Although active traditional fishing from outrigger canoes takes place along the entire coast, there are still no oral reports of coelacanth catches on the east coast, as if the east coast was unsuitable as coelacanth habitat. The two earlier records on coelacanth catches on the east coast off Bouni and the south coast off Chindini may, therefore, be based on missinformation. According to all available information and taking the two doubtful records into account, 98.6% of all coelacanth catches occurred

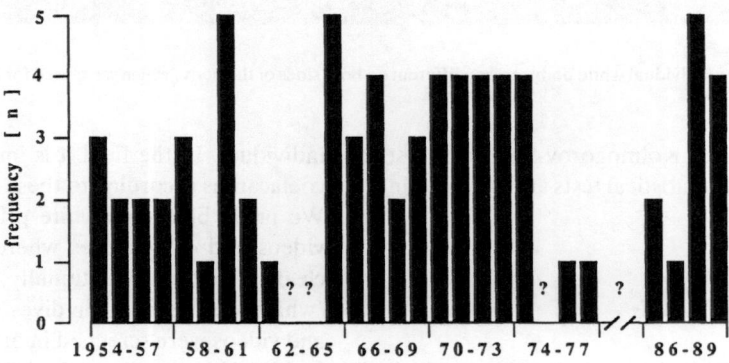

Fig. 3. Annual catch rate of coelacanths at Grand Comoro. Data base: Millot et al. 1972, McCosker 1979, Bruton, de San personal communication and own observation.

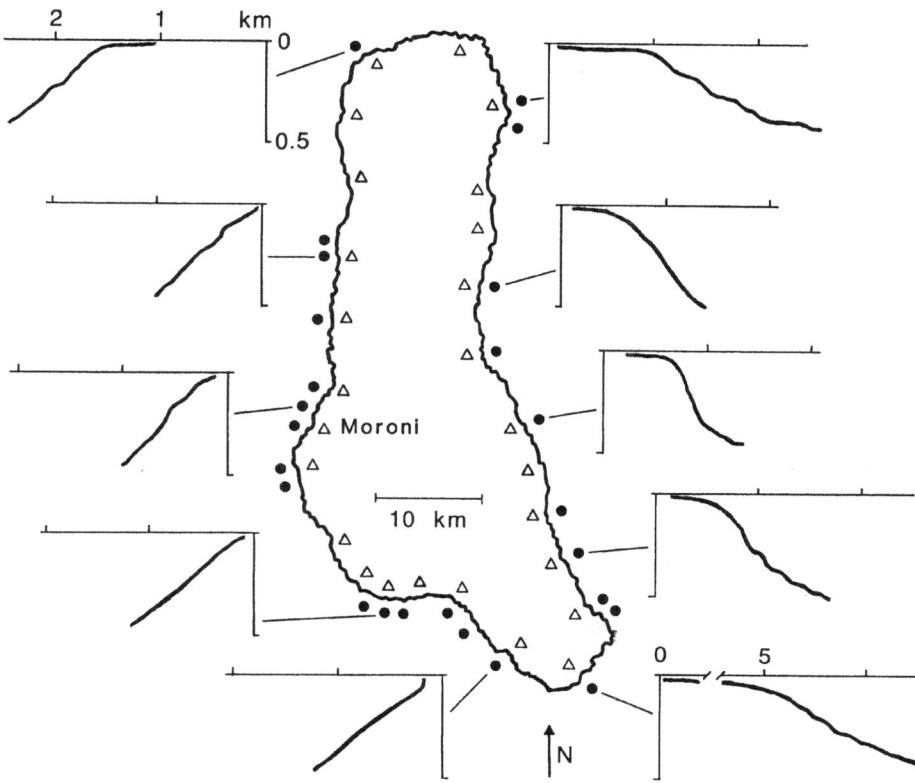

Fig. 4. Map of Grand Comoro with location of bathymetric profiles and stations of the environmental survey (filled circles; open triangles = coastal villages and small towns).

on the west coast. The west coast seems to provide a suitable habitat for coelacanths. Therefore, the coastline between Mitsamiuli and Chamboini is defined as coelacanth habitat. The differential distribution gives the unique chance to study ecological parameters suitable for coelacanth settlement along one but not along the other coast.

Environmental survey

A bathymetric survey off the entire coast of Grand Comoro shows large sandy fans spreading up to 7 km offshore at the north and south end of the island (Fig. 4). The west coast is characterized by steep drop offs close to the shore with an average slope angle of 45° ± 14° (n = 122). A large sub-

marine fan extends off Iconi at a depth of 200–300 m and continues in deeper water. Aerial photographs of the west coast indicate many recent lava flows (Fig. 5). The latest eruption took place in 1978 at Singani. We followed the flow down to 400 m depth. The east coast has very similar slope angles (44° ± 15, n = 33). However, flat sandy coralline terraces are formed in shallower water. The east coast is geologically older and more eroded, particularly at the northern end. From our bathymetric survey we can conclude that the coelacanth habitat is formed by steep lava-covered slopes close to the shoreline. This topographical situation is different from all other coastlines at Grand Comoro.

We performed a submersible survey down to 200 m depth in order to study further environmen-

292

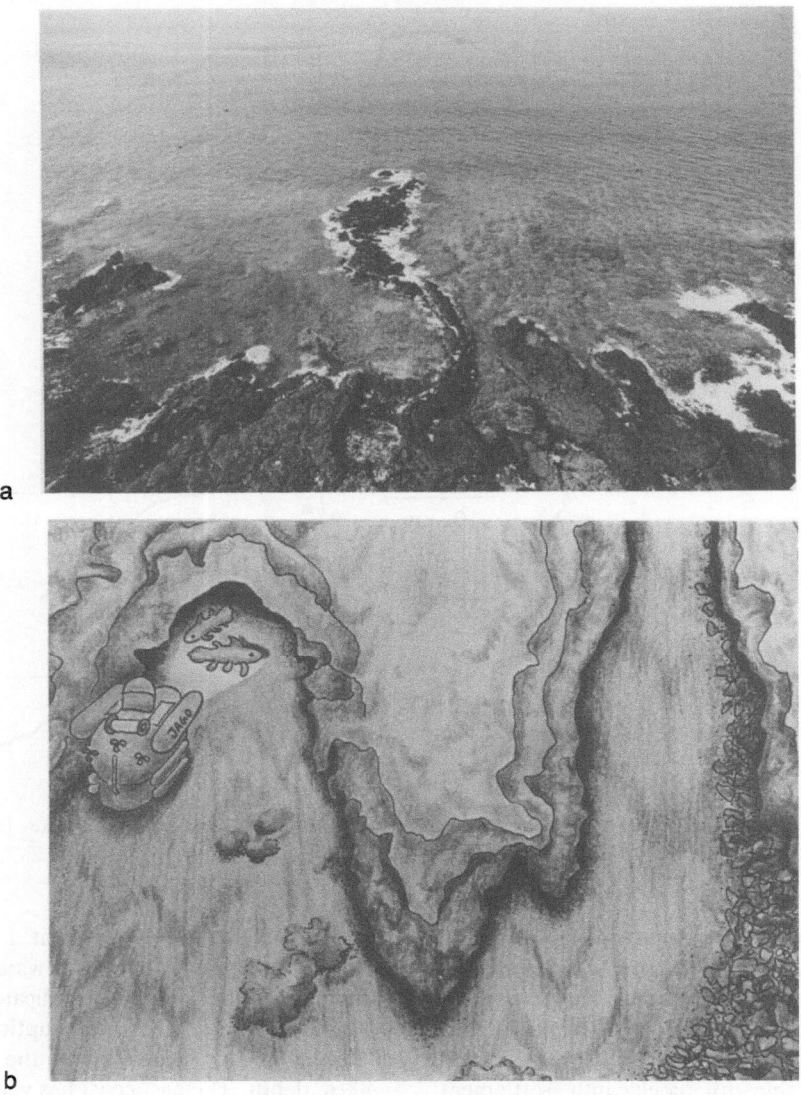

Fig. 5. a – Aerial photograph of the south-west coast of Grand Comoro. b – Drawing of the lava flows which continue under water. Coelacanth caves are located on the lower end of such flows.

tal differences between the east and the west coasts to determine which factors might be responsible for the presence or absence of coelacanths.

Figure 4 shows stations of submersible dives devoted to data collecting. We measured angle of slope, temperature, sand-rock-coverage, structural complexity of the bottom, number of prey species, prey fish abundance and largest prey size. The data have been partially published elsewhere (Fricke & Plante 1988). Figure 6 shows prey fish abundance, sand-rock-coverage and structural complexity at both coasts at different depths. Potential prey fish species observed by us were: *Etelis* sp., *Epinephalus* sp., *Synagrops* sp., *Anthias* sp.,

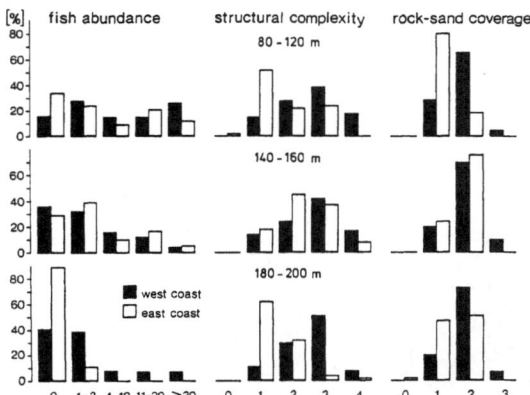

Fig. 6. Data from environmental survey of prey fish abundance, structural complexity of the bottom and sand-rock-coverage of the east and west coast.

Symphysanodon sp., a variety of unidentified apogonids, a common unidentified chaetodontid, *Pristigenus niphonia, Histiopterus typus.*

The main results of the surveys were: (1) in January to May the water temperature of the west coast is significantly colder than the east coast with the 18° thermocline being located at 173 ± 10 m (n = 16) and 193 ± 9 m (n = 6) respectively (t-test, p < 0.01); (2) the east coast has significantly more sand coverage (p < 0.001) and less structural complexity (p < 0.001); (3) prey-fish abundance decreases significantly along both coasts with depth, east coast p < 0.01 and west coast p < 0.001); (4) although there are no differences in number of prey species and species composition, the east coast has less fish than the west coast (p < 0.002).

Earlier it was speculated that coelacanths live at the mouth of submarine freshwater outlets (Forster 1974, McCosker 1979), which do indeed occur in shallow water. We could not confirm such outlets

Table 1. Depth of the 18° C isotherm during different months of the year measured on the south-west coast of Grand Comoro.

	depth (m)	SD	min	max	n
November	228	6	217	243	530
January	187	8	178	199	6
May	174	4	167	181	6

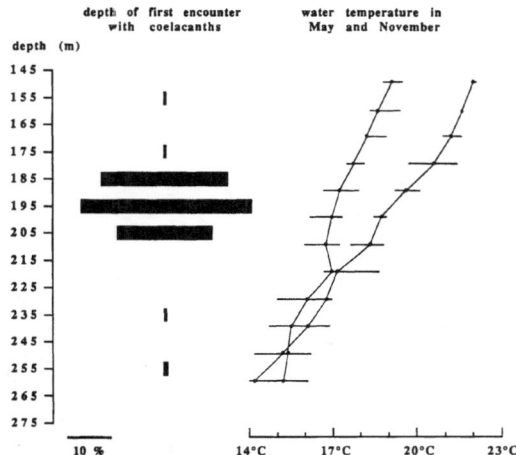

Fig. 7. Depth distribution of coelacanth encounters and temperature profiles during different seasons: a – total of all first day and night encounters in 1987 and 1989; first encounters are pooled in depth classes of 10 m each and given in the middle of each class, e.g. class 180–189 m at 185 m, b – average temperatures during cold (May) and warm underwater seasons (November) with minimum and maximum values.

by salinity measurements in deeper water below 80 m along both coasts.

Depth distribution of encountered coelacanths

In 1987, we encountered six coelacanths only at night. The first encounter took place between 150–198 m depth on the south-west coast. One individual was followed with the submersible over a period of almost 5 hours. It migrated vertically from maximum 198 m to minimum 117 m and remained 56% of the observation time in water temperature of less than 18° C. It should be noted that this individual was under constant illumination of the submersible lights which could have influenced its behaviour.

In 1989, we found a total of 38 individuals during night and day; at night between 195–250 m, during day between 182–253 m. Figure 7a summarizes the depth of all first coelacanth encounters during day and night. Ninety-seven percent of all first sightings were between 180–210 m depth.

Oxygen saturation of the coelacanth blood is highest at around 15–20° C (Hughes & Itazawa

294

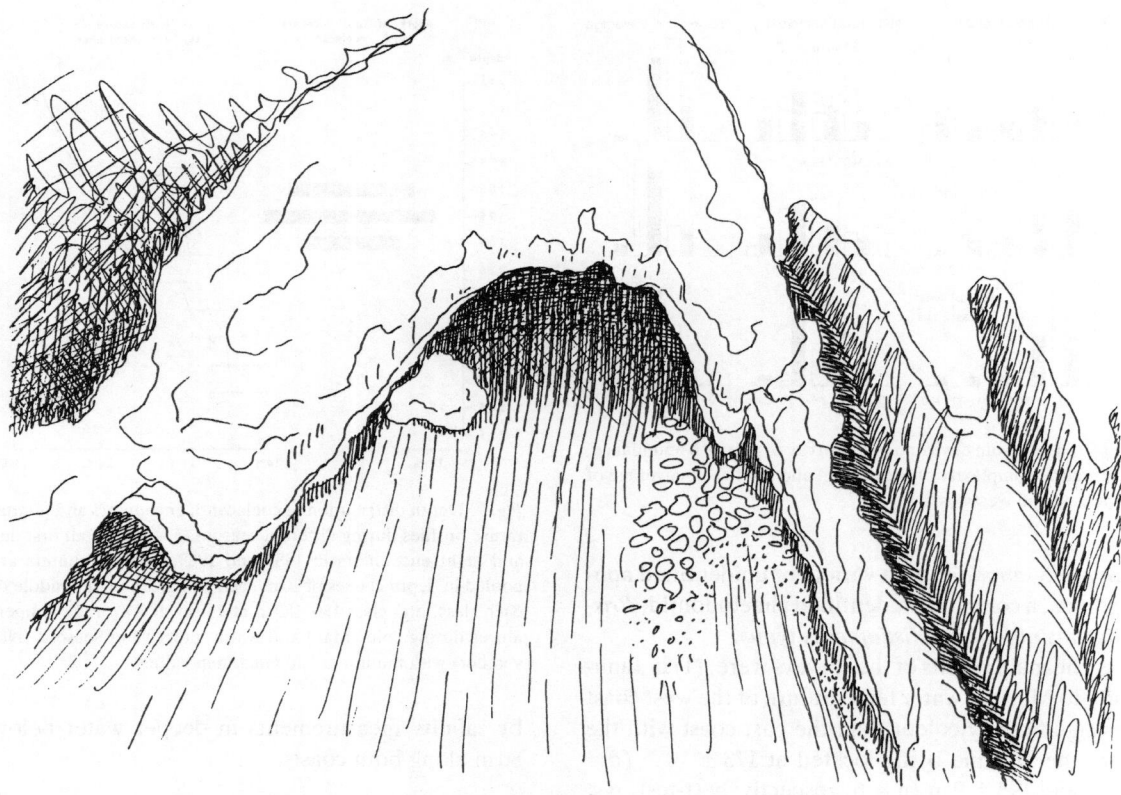

Fig. 8. Sketch of cave entrance (cave 6) at 185 m depth. The occupied caves were always located in still water below lava overhangs (drawing by K. Hissmann).

1972), and could determine the preferential depth. Table 1 shows the 18° C thermocline of the southwest coast in November, January and May. Clearly, in November the thermocline is located in much deeper water than during the other months. The situation in November is caused by stable warm surface water that is not mixed with colder deep water. The water bodies are mixed by the monsoon winds starting in mid December. Figure 7b shows temperature profiles in May and November. The depth of most encounters with coelacanths (180–210 m) has an average water temperature of 16.5–18° C in May and 18.5–20.5° C in November.

Cave dwelling by coelacanths

During the day all coelacanths were found in deep lava caves which penetrated 3–5 m into the lava-rock (Fig. 5b, 8, 9). We routinely visited six occupied caves at depths between 182–253 m along a stretch of 8 km coastline. The caves were not occupied every day and the number of individuals changed from day to day. Table 2 shows depth, occupation rate, size of aggregations and mean/min/max temperature at the cave entrance. The highest observed temperature of an occupied cave was 22.8° C. This is so far the highest observed temperature tolerance of a coelacanth in its natural habitat. We frequently observed shimmering water above the cave entrance due to mixing of cold water with warm water trapped under the ceiling of

Fig. 9. A group of 10 coelacanths aggregating in cave No. 2 off Grand Comoro during the daytime.

the cave. Figure 10 illustrates one example; temperature difference was 4.8° C. This phenomenon is linked with the occurrence of internal waves (Piton et al. 1990). Continuous temperature records over 22 hours from December 6 to 7 close to a cave at 192 m depth showed temperature variations between 18.7–22.3° C which changed with the tide; higher temperatures were recorded during low tide. Salinity was on average 35.1‰. A detailed description of the coelacanth's physical environment is published elsewhere (Piton et al. 1990).

During deep water surveys below 220 m depth we found only very few caves. At this depth the lava slopes were more eroded and more covered with sand than the slopes in shallower waters. The deepest cave at 253 m was occupied only once (Table 2). Another deep cave at 259 m was visited repeatedly over 18 days but no coelacanths were found. On the other hand we found caves in shallower waters above 180 m which were also unoccupied. The occupied caves were inhabited also by crustaceans and benthic fish. A *Lysmata*-type shrimp with long white antennae touched the coelacanths frequently but no actual cleaning of the fish was observed. All caves were occupied by the flashlightfish *Photoblepharon* sp. and a variety of unidentified apogonid fish.

Table 2. Field data of six caves, occupied by coelacanths, collected during November 5 to December 4, 1989.

Cave	Depth (m)	Temperature range °C	Number of controls	Occupation rate	Number of occupants	Mean number of occupants
1	193	17.1–22.3	31	86%	1– 6	3.0
2	205	16.4–22.7	17	69%	1–10	3.1
3	182	20.6–23.0	12	70%	1– 4	1.5
4	188	19.2–23.4	12	55%	1– 6	1.3
5	253	15.4–19.1	8	14%	1	–
6	185	18.2–21.6	5	100%	4– 6	4.5

296

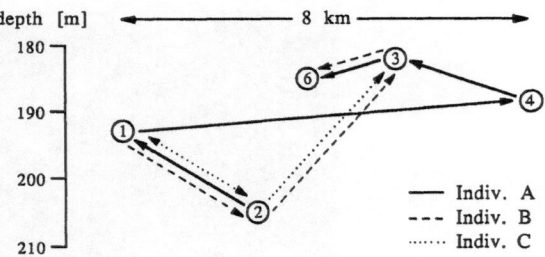

Fig. 11. Position of five caves (see Table 1) with intercave migration of 3 individuals. Individual A started from cave 2 on 28 November and was finally observed in cave 6 on 10 December; individual B started from cave 1 on 5 November and was finally observed in cave 6 on 14 December, individual C started from cave 2 on 28 November and was finally found in cave 3 on 13 December. The intercave migrations do not occur on a straight course.

Fig. 10. Schematical section through a cave to illustrate the formation of warm water pockets with shimmering interfaces. Largest observed temperature difference of 4.8°C on 30 November 1989 (drawing by K. Hissmann).

Coelacanth movements and site attachment

During dusk, coelacanths leave their day-time caves and drift singly close to the bottom of the lava fields. During drifting they hunt for fish. One feeding attempt was filmed. We could not detect preferences for certain drifting directions, depth or areas. Before sunrise, but still in total darkness, coelacanths probably return to the nearest cave. This can be inferred from data of intercave migrations. Figure 11 illustrates intercave migrations of three individuals. One large female (individual A) was found in all caves between 182–205 m depth. She occupied a large home range over at least 8 km coastline (Fricke et al. 1991). Nineteen individuals were repeatedly observed and are considered as residents. They have at least a temporary site attachment to their daytime resting habitat and their nocturnal feeding grounds. We rediscovered in the

same area three of the six individuals photographed in 1987. Therefore a long-term site attachment is very likely. Nineteen individuals were only seen once; they are considered as non-residents.

Size distribution and population estimates

The sizes of coelacanths were estimated. We divided the 38 encountered individuals in two size classes of smaller and larger than 100 cm total length. Two individuals, i.e. 5%, were smaller than 100 cm. The smallest individuals, approximately 60–80 cm long, were found together with a large female (Fig. 12); the smallest individual and the female were observed over 5 days only and did not change their resting cave.

The size distribution of coelacanths landed at Grand Comoro were similar (Millot et al. 1972, McCosker 1979, de San personal communication, Bruton 1989a). Three out of 43 individuals (≈ 7%) were smaller than 100 cm. Our observation area with the occupied caves covered approximately 9% of the entire coelacanth habitat (Fig. 2), starting south of Mitsamiuli and north of Chamboini. If we assume that our observation area is representative of the coelacanth habitat, estimations of the total population at Grand Comoro are possible (Table 3). We estimated 210 individuals if the number of

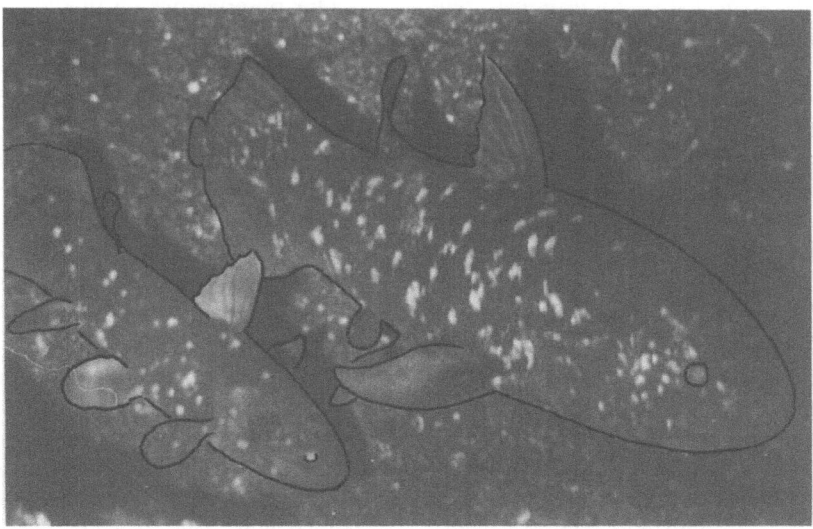

Fig. 12. The smallest individual together with a large female in cave No. 6, well camouflaged by their white dots against the cave background covered with small white sponges and oyster shells.

the observed residents is taken as the basis for the calculation; 150 individuals from the observed mean cave occupants. If we take the total of all identified coelacanths into account, an estimate of 420 individuals is obtained. The maximum number of occupants observed in each cave probably indicates the level of cave saturation; these account approximately for 370 individuals (Table 3). We

Table 3. Estimations of population size at Grand Comoro. The limits of the coelacanth habitat are shown in Fig. 3. The observation area is for conservation reasons not indicated on the map.

	Number of coelacanths along 8 km coastline	Calculated total population for entire coastline
Residents	19	210
Non-residents	19	210
Total of identified coelacanths	38	420
Total of mean cave occupants	13.5	150
Total of maximum cave occupants	33	370

interpreted overcrowding effects with increasing flight readiness of the inhabitants.

Discussion

Habitat and distribution

The entire habitat of the living coelacanth comprises daytime-resting caves which are located within large nocturnal feeding grounds extending over several kilometers. Our 1987 habitat survey revealed differences in structural complexity and sand-rock-coverage between the west and east coast; along the latter no coelacanths were caught. In light of the new field observations in 1989, scarcity of caves due to higher sand coverage and less structural complexity of the bottom (caused probably by erosion of the older lava flows) might be the most significant environmental factor for absence of coelacanths along the east coast. The caves are used for shelter during inactive periods, probably for various reasons: one might be protection against large predatory fish. We encountered sharks; one coelacanth had a damaged pectoral fin with bite-marks of a shark.

Spacious caves along the west coast occur mainly at the end of submarine lava flows. Although we have no quantitative data on the depth distribution of such caves, they are rare below 220 m due to heavier sand coverage of the eroded deeper slopes. Scarcity of caves might be one of the limiting factors for the lower depth distribution of coelacanths on the west coast.

The temperature-dependent oxygen saturation of coelacanth blood (Hughes & Itazawa 1972, Hughes 1975) suggests that coelacanths prefer a depth corresponding to water temperatures of 15–20°C with an upper threshold of probably 22–23°C. The water temperature changes considerably during the seasons (Table 1). In November, the 18°C isotherm is located more than 50 m deeper than in January and May. This might have an effect on the fish's nocturnal feeding migrations. The depth of our first encounters with coelacanths differed during the seasons and were shallower during the cold period. This might be due to vertical nocturnal feeding migrations in the cold season. We observed one vertical movement during the cold but none during the warm season.

Most coelacanths were encountered between 180–210 m depth which correlates with the varying depth of the 18–20°C isotherm (Fig. 7). Obviously water temperature determines the upper distributional limits of the fish. Although the colder temperature of the deeper water would be favourable for respirational demands, the scarcity of caves might prevent the occupation of waters below 220 m depth.

Furthermore, there are very few prey fish in deeper water (Fig. 6) which might be an additional limiting factor for coelacanth residence. Coelacanths may be forced to take advantage of the higher prey abundance in shallow water of less than 120 m depth which could be achieved by vertical migrations. There is no proof yet of our earlier hypothesis that resting during inactive periods in colder water might save metabolic costs. In colder water metabolism is slower and less food is needed, an advantage in areas of low prey densities (Fricke & Plante 1988).

Our observations on the coelacanths' environment suggest that the breeding stock at Grand Comoro has a very narrow habitat range which may limit the total population size. Limiting environmental constraints for the lower depth distribution could be scarcity of suitable caves below 220 m and the decrease of prey fish abundance. Respirational demands probably prevent the occupation of shallower zones with higher prey abundance. But it has to be noted that also at suitable depth, the number of 'good' caves is in short supply. We found only a very limited number of occupied caves. This might be due to the peculiar location of the caves mainly at the end of submarine lava flows. The characteristics of such caves are a certain slope angle (Fig. 10), large internal space with further subterranean ramifications and a certain position to the prevailing currents. The caves are always located in still water, otherwise warm water pockets could not be trapped along the roof of the caves (Fig. 10). Retreat in the calm water of caves during inactive periods probably saves energy. It might be a key factor for the survival of coelacanths as a large predatory fish in a marginal habitat of low prey density.

Population size

The boundaries of the coelacanth habitat at Grand Comoro were initially determined by the locations of coelacanth catches which are limited to the west coast (Fig. 2). They can now be more precisely defined by the submarine topography and its characteristics (Fig. 6). We counted the number of coelacanths for population estimates by repeated visits to an area at a certain distance. The data were not easily obtained and cannot compete with similar terrestrial studies. Nevertheless, they are the only available data on coelacanth population size presently known. According to our bathymetric soundings and environmental data the submarine topography seems to be uniform throughout the coelacanth habitat. Therefore we assume that our sample is representative when expanded to the entire coelacanth habitat.

Table 3 shows relatively small differences in total

population size between estimates based on residents (210 individuals) and number of mean cave occupants (150 individuals). The differences could be due to observation errors, i.e. not all caves were found by us. If we take the total of all encountered coelacanths (residents and non-residents) into account, the total population size would double to 420 individuals. This seems to be less correct because it includes all occasional invaders from neighbouring home ranges and possible floaters. The social system of coelacanths is only incompletely known and no information exists on the origin of those occasional invaders and floaters. Therefore we consider a total of 200 individuals as the most reliable figure. If we consider the highest number of observed individuals in each cave as a measure of cave saturation, we obtain a maximum population size of 370 individuals, i.e. the carrying capacity of the coelacanth habitat at Grand Comoro. If observational errors are considered and the proportion between mean cave occupants and total residents is taken into account, then we get 520 individuals for the saturation of carrying capacity. Thus, the estimated population size of approximately 200 individuals would suggest a habitat saturation of 40–55% at Grand Comoro. Both figures for total population size and carrying capacity give indeed very small numbers and the question arises whether such small populations are able to survive.

Coelacanths are ovoviviparous[1] (Smith et al. 1975, Atz 1976, Heemstra & Compagno 1989), their fecundity is low and gestation period more than 12 months. No information exists on longevity, growth or age at sexual maturity or on age structure or sex ratios of populations. However, the scanty knowledge of life-history parameters suggests already that coelacanths are highly specialized and precocial (Balon et al. 1988). The low proportion of juveniles in the two independent samples (5% and 7% of the total) hints at a low recruitment and an apparently high longevity. In the future life-history parameters should be more

[1] An old term now substituted by lecithotrophic viviparity and not applicable to *L. chalumnae* as shown elsewhere in this volume (see e.g. Wourms et al.) [editors].

thoroughly studied in order to detect and predict population changes.

Population size and survival of coelacanths

The low number of coelacanths at Grand Comoro raises concern about their survival. Annually 2–5 individuals are eliminated by traditional human fishing efforts (Fig. 3). This would be 1–1.5% of the estimated population. Coelacanths are caught at night by local fishermen as an accidental bycatch of fishing for the oilfish *Ruvettus pretiosus* with deep water lines from outrigger canoes (Millot et al. 1972). Recent commercial interests of public aquariums led to estimates on speed of extinction thought to be less than two decades (Bruton 1989a). Another reason for concern is the increasing numbers of fishing canoes and consequently increased fishing activity (Bruton 1989b). Are coelacanths really threatened with extinction? At a first glance the answer would be clearly yes. However, a critical view reveals that there are no signs yet for increased night time fishing which is locally considered dangerous (Stobbs 1989). Present rapid developments are targeting pelagic offshore fishing grounds beyond the coelacanths' habitat (de San personal communication). So, in reality fishing pressure on coelacanths has not increased and might even decrease in the near future. Figure 4 indicates that the annual fishing rate has not changed over the last three decades. The stable catch rate might suggest that traditional fishing does not harm the coelacanth population at Grand Comoro and that coelacanths have been caught by traditional means probably for a very long time. This might immediately change if new commercial interests (e.g. from public aquaria) develop or if rumours are substantiated that coelacanths can be used for pharmaceutical purposes. For the time being it seems that the low population size might have persisted there for centuries or longer.

Furthermore, the size distribution of captured coelacanths indicates that all size classes have been equally effected by line fishing and no selective fishing for certain size classes exists. This should have the least disturbing effect on population structure. Nevertheless, with our present technical means we should concentrate on monitoring the population size of the coelacanths at Grand Comoro and studying the stability of the small relict population.

In the past few years the necessity for international efforts to protect coelacanths has been discussed many times and is treated in a separate paper in this book. Although the coelacanth population is small indeed, we have reasons to believe that coelacanths might survive if afforded international protection.

Acknowledgements

The study was generously funded by grants to Hans Fricke by the German Research Community (DFG, AZ Fr 369/7-1, 7-2), by the Frankfurt Zoological Society – Help for Threatened Wildlife – (ZGF-Project-No. 1036/87) and the World Wildlife Fund Germany. The study was also supported by the Max-Planck-Society. We thank particularly Wolfgang Wickler for constant support of our field work, and also the administration and the workshops of the Max-Planck-Institute in Seewiesen. We are especially thankful to N. Schuch and numerous German companies who helped us to build the new submersible JAGO although they could never understand that JAGO has been constructed for a single fish species only. We thank the crews of METOKA and SEA EAGLE for their skills and friendship to make our diving operations in often adverse weather conditions possible. In Moroni we thank J.-L. Geraud, R. Rossi and M. de San for much help and friendly hospitality. The CNDRS and Mission Française assisted us in many ways. We are grateful that the Government of the Federal Islamic Republic of the Comoros permitted our diving operations in their territorial waters.

References cited

Atz, J. 1976. *Latimeria* babies are born, not hatched. Underwater Naturalist 9: 4–7.

Balon, E.K., M.N. Bruton & H. Fricke. 1988. A fiftieth anniversary reflection on the living coelacanth, *Latimeria chalumnae*: some new interpretations of its natural history and conservation status. Env. Biol. Fish. 23: 241–280.

Bruton, M.N. 1989a. The living coelacanth fifty years later. Trans. Roy. Soc. S. Afr. 47: 19–28.

Bruton, M.N. 1989b. The coelacanth – can we save it from extinction? WWF-Report Oct/Nov: 10–12.

Forster, G.R. 1974. The ecology of *Latimeria chalumnae* Smith: results of field studies from Grande Comore. Proc. R. Soc. Lond. B 186: 291–296.

Fricke, H. 1988. Coelacanths. The fish that time forgot. Nat. Geogr. Mag. 175: 824–838.

Fricke, H. & K. Hissmann. 1990. Natural habitat of coelacanths. Nature 346: 323–324.

Fricke, H. & R. Plante. 1988. Habitat requirements of the living coelacanth *Latimeria chalumnae* at Grande Comoro, Indian Ocean. Naturwissenschaften 75: 149–151.

Fricke, H., O. Reinicke, H. Hofer & W. Nachtigall. 1987. Locomotion of the coelacanth *Latimeria chalumnae* in its natural environment. Nature 329: 331–333.

Fricke, H. & J. Schauer. 1987. Im Reich der lebenden Fossilien. GEO 10: 14–34.

Fricke, H., J. Schauer, K. Hissmann, L. Kasang & R. Plante. 1991. Coelacanth *Latimeria chalumnae* aggregates in caves: first observations on their resting habitat and social behaviour. Env. Biol. Fish. 30: 281–285.

Heemstra, P.C. & L.J.V. Compagno. 1989. Uterine cannibalism and placental viviparity in the coelacanth? A skeptical view. S. Afr. J. Sci. 85: 485–486.

Hughes, G.M. & Y. Itazawa. 1972. The effect of temperature on the respiratory function of coelacanth blood. Experientia 28: 1247.

McCosker, J.E. 1979. Inferred natural history of the living coelacanth. pp. 17–24. *In:* J.E. McCosker & M.D. Lagios (ed.) The Biology and Physiology of the Living Coelacanth, Occ. Pap. Calif. Acad. Sci. 134, San Francisco.

Millot, J., J. Anthony & D. Robineau. 1972. État commenté des captures de *Latimeria chalumnae* Smith (Poisson, Crossoptérygien, Coelacanthidé) effectuées jusqu'au mois d'octobre 1971. Bull. Mus. Natn. d'Hist. Nat. Zool. 39: 533–548.

Piton, B., L. Kasang, M. Marsac & R. Plante. 1990. L'habitat du coelacanth aux Comores: quelques données d'environnement physique. Document Scientific ORSTOM (in press).

Siegel, S. 1956. Nonparametric statistics for the behavioral science. International Student Edition, Tokyo.

Smith, C.L., C.S. Rand, B. Schaeffer & J.W. Atz. 1975. Latimeria, the living coelacanth, is ovoviviparous. Science 190: 1105–1106.

Stobbs, R.E. 1989. Laxative lipids and the survival of the living coelacanth. S. Afr. J. Sci. 85: 557–558.

Environmental Biology of Fishes **32**: 301–311, 1991.
© 1991 *Kluwer Academic Publishers.*

The demography of the coelacanth *Latimeria chalumnae*

Michael N. Bruton[1] & Michael J. Armstrong[2]
[1] *J.L.B. Smith Institute of Ichthyology, Private Bag 1015, Grahamstown, 6140 South Africa*
[2] *Aquatic Sciences Research Division, Newforge Lane, Belfast BT9 5PX, Northern Ireland*

Received 1.10.1990   Accepted 16.1.1991

Key words: Length, Weight, Sex ratio, Growth, Mortality, Exploitation, Predation, Conservation, Comoros

Synopsis

The very sparse data that are available on the abundance, population structure and biology of the coelacanth *Latimeria chalumnae* off Grand Comoro are summarised, and some simple numerical analyses are carried out to explore certain aspects of the population dynamics, particularly the age-profile of the population. The object has not been to provide estimates of key demographic parameters, such as mortality rates, but to propose various scenarios that are useful for comparison with real data as they become available. The analysis also makes it possible to reach some preliminary conclusions that are relevant to the management of the coelacanth population. For instance, it appears that the catch rate of coelacanths by artisanal fishermen may have a negligible effect on coelacanth survivorship, and it is more likely that population size and structure are determined by natural mortality rates and birth rates. It is suggested that predation is the main cause of natural mortality and that the main predators of coelacanths are likely to be large sharks. Interference with the traditional patterns of the Comoran artisanal fishery may threaten the coelacanth. Several important gaps in our knowledge of coelacanth demography are identified.

Introduction

During the last five years research on the coelacanth has entered a new phase with the first observations on the living animal in its natural environment by Hans Fricke and his colleagues (Fricke 1988a, b, Fricke & Plante 1988, Fricke & Hissmann 1990, Fricke et al. 1991). This research has revealed that the coelacanth is mainly concentrated in a narrow depth zone (about 180–220 m) within its distributional range along some shores of Grand Comoro and Anjouan islands (Fricke & Hissmann 1990). Within this narrow range the coelacanth is threatened by a variety of socio-economic and biological factors (see Bruton 1987, 1988, 1989, Balon et al. 1988, Bruton & Stobbs 1991).

There is now worldwide concern that this unique fish may be threatened with extinction (Forey 1988, Bruton 1989, Greenwood 1989, Balon 1990, Bruton & Stobbs 1991). In an effort to contribute towards the campaign to conserve the coelacanth, we have synthesized all available data on the demography of *Latimeria chalumnae* and have performed some simple numerical analyses of these data in order to explore some aspects of the dynamics of the population of coelacanths. The demographic data available on the coelacanth are insufficient to allow an accurate assessment of population characteristics, but the numerical analyses allow us to make some predictions and, perhaps more importantly, to identify gaps in our knowledge.

302

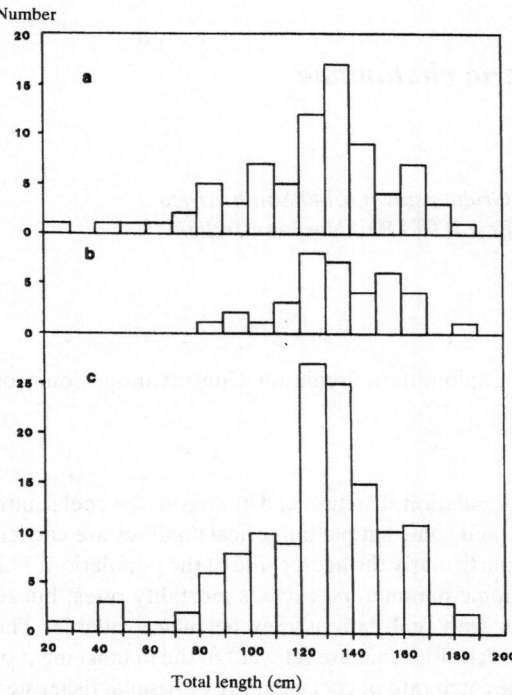

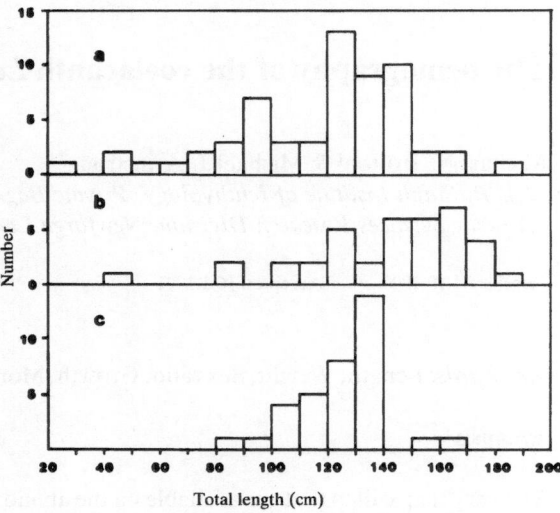

Fig. 1. The length frequency of coelacanths known to have been caught off Grand Comoro (a) and Anjouan (b) and the combined length frequency (c). The sample sizes for the different graphs are as follows: a = 81, b = 38, c = 134 (which is the total number of specimens for which length measurements were available at the time of preparation of this paper; the island of origin of some specimens is unknown).

Fig. 2. The length frequency of all known unsexed (a), female (b) and male (c) coelacanths. The sample sizes for the different graphs are as follows: a = 63, b = 36, c = 35.

Size frequency

According to Bruton & Coutouvidis (1991), 158 coelacanth specimens are definitely known to have been caught since 1938 in that we have a record of their capture and/or their current location. We also have information on an additional fourteen specimens that have been caught but for which little data are available.

The length frequency of all specimens for which lengths are available (at the time of preparation of this paper, Fig. 1) reveals that the modal length for the whole sample is 120 to 129 cm within a length range from 25 to 183 cm. This sample includes 60 specimens that have not been sexed, including several juveniles and young males, which probably explains the relatively low modal value obtained

for the whole sample relative to the modes for males and females (Fig. 2). The length frequency of specimens that have been sexed indicates that the modal length of females is greater than that of males (160 vs. 130 cm). Females also reach a larger maximum size; the largest male measures 155 cm TL whereas the largest females reach 183 cm in length and 95 kg in weight (Table 1). A 164 cm specimen was recorded by Millot et al. (1972) as a 'male (?)' which, if confirmed, would be the largest male known. We have received unconfirmed re-

Table 1. Details on the largest coelacanths caught to date (nk = not known).

Year captured	Capture location	Total length (cm)	Weight (kg)	Sex
1983	Mutsamudu, Anjouan	183.1	nk	F
1986	Iconi, Grand Comoro	182.0	nk	nk
1960	Itsandra, Grand Comoro	180.0	95	F
1989	Dzouhatsu, Grand Comoro	176.5	nk	?F
1981	Mitsamiuli, Grand Comoro	175.4	84	F
1981	Grand Comoro	170.0	nk	nk

ports of specimens measuring over 200 cm and 100 kg, presumably females, being landed at Grand Comoro. M. de San, Head of the European Economic Community's fisheries development programme in the Comoros, and the official responsible for recording coelacanth catches at Grand Comoro, informed us on 13.11.1987 that he had personally seen two coelacanths that exceeded 2 m in length.

Only 17 of 143 coelacanths (12%) were smaller than 100 cm TL (Bruton & Coutouvidis 1991). It is interesting to note that Fricke & Hissmann (1990) report that, on the basis of their observations from a research submersible, fishes less than 100 cm TL represented about 5% of the total number observed.

The weight distribution of a sample of 65 coelacanths indicates that the modal weight of males is 30–39 kg and that of females 70–79 kg (Fig. 3).

Length : weight relationship

The statistical data used in the calculation of the relationship of weight to total length are given in Table 2. One specimen was excluded from these calculations as its weight measurement was clearly incorrect: a 125 cm fish of unknown sex 'was estimated to weigh about 60 kg' (Uyeno 1991) but probably weighed about 25–30 kg. The accuracy of the length : weight calculations is reduced by the fact that some of the specimens were weighed on inaccurate scales in deep freeze plants or airports in the Comoros. Furthermore, some of the specimens were weighed when fresh whereas others were weighed after they had been frozen or preserved in formalin, sometimes several months after capture. The low r² values given in Table 2 reflect these inaccuracies as well as the inherent variation of the length : weight relationship among individuals in the sampled population. The exponent 'b' is higher for females than for males, which indicates that females are more robust at a given size than males (Fig. 4).

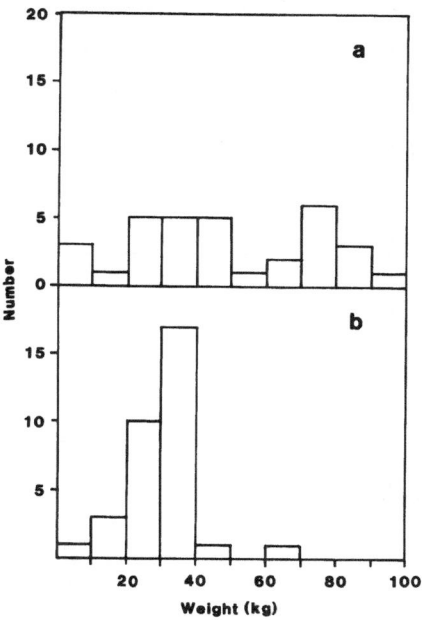

Fig. 3. The weight frequency of female (a) and male (b) coelacanths from Grand Comoro and Anjouan for which measurements are available. The sample sizes for the two graphs are: a = 32, b = 33.

Sex ratio

As coelacanths are not sexually dimorphic, they have to be dissected to be sexed. The sex ratio of the 75 specimens whose sex is known is 1.00 : 0.97 in favour of males. As shown in Fig. 5, the proportion of females in that part of the catch that has been sexed has increased in recent years. However, a large part of the catch in recent years has not been sexed and the correct sex ratio is therefore unknown.

Growth rate

Hureau & Ozouf (1977) examined the scales of 12 coelacanths ranging in size from a 31 cm fullterm embryo to a 180 cm adult female, and found 2 to 23 rings on their scales. They suggested that two rings are laid down in the scales each year in association with the seasonal change in growth rate during

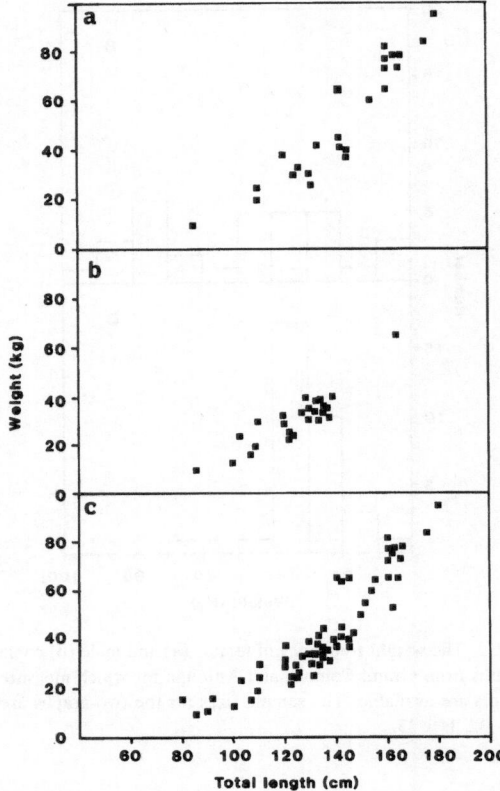

Weight (kg)

Total length (cm)

Fig. 4. The length : weight relationship of all known specimens of female (a) and male (b) coelacanths and of the total sample of unsexed and sexed fishes (c). The sample sizes for the different graphs are as follows: a = 34, b = 28, c = 85.

January/February and August/September. Their age estimates range from one month before parturition for the embryo to 10 years 11 months for the 180 cm female. They estimated the length at birth to be 30.4 cm and the average annual size to be 47.1, 61.5, 75.5, 89.0, 103.4, 113.7, 120.6, 135.5, 150.0 and 168.6 cm (Fig. 6). The average annual length increment was estimated to be 13.0 cm. They predicted that the gestation period lasts more than 13 months. Their growth curve is linear and not asymptotic, and is probably not valid.

Likewise, Suyehiro (1983) estimated that the natural life span of the coelacanth is about 10 years, but provided no details on how he had made that calculation. Uyeno (1984) determined the growth

rate of a coelacanth using scale rings and estimated that an individual reaches a length of about 170 cm TL after about 7–8 years. These low estimates for the longevity of the coelacanth seem unlikely as other large reef-dwelling fishes, such as some snappers and groupers, reach ages greater than 20 years (Manooch 1987). Our opinion is that the coelacanth reaches an age well in excess of 20 years, possibly 40 or 50 years.

Balon et al. (1988) disputed these growth estimates and suggest that one ring may be laid down each year, as is the case with many tropical fishes whose scales record an internal circa-annual rhythm in spite of two distinct rainy seasons in their natural habitat (see Balon 1974). It seems likely that the rings are formed annually after the yolksac juvenile has been born. The length at birth is uncertain, but Balon et al. (1988) reported the capture of a 42.3 cm individual with fragments of yolk still in the body cavity, which indicates that it had recently been born. Following Balon et al. (1988), it has been assumed that the length at birth is about 40 cm (see also Balon 1977), and that only one ring is formed annually from birth. The annual growth increment up to about 20 years of age would then be about 6.5 cm, as opposed to 13.0 cm according to Hureau

Table 2. Estimates of the slopes and intercepts of the relationship between the logarithm of fish weight (g) and the logarithm of total length (cm) for the coelacanth *L. chalumnae* caught off Grand Comoro and Anjouan. The calculations for the whole population include data for coelacanths whose island of capture is unknown (n = sample size, b = slope).

n	b	std error of b	intercept	r^2
Whole population				
Females				
28	3.233250	0.069932	− 5.28937	0.9731
Males				
34	2.654537	0.063765	− 4.09376	0.8222
Both sexes				
84	3.044642	0.0814	− 4.8964	0.9211
Anjouan population				
Both sexes				
26	3.014186	0.053978	− 4.80791	0.9357
Grand Comoro population				
Both sexes				
59	2.939564	0.097213	− 4.69122	0.9031

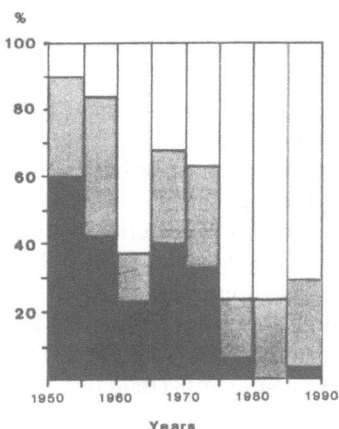

Fig. 5. The ratio of unsexed (open hatching), female (shaded) and male (solid block) coelacanths caught in different years (n = 151).

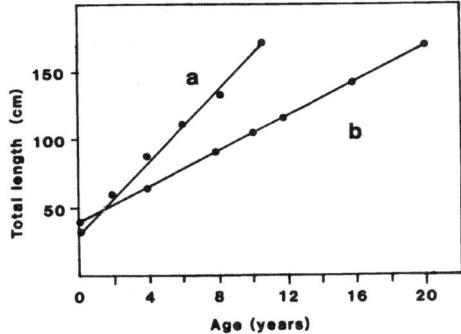

Fig. 6. The annual growth rate of the coelacanth as estimated by Hureau & Ozouf (1977) (a) and Balon et al. (1988) (b) based on readings of scale rings.

& Ozouf (1977). The annual increment of 6.5 cm corresponds to an annual increase in fresh weight of 30 to 50% over the first five years of life, as opposed to figures of 60 to 200% for Hureau & Ozouf's growth rate. The latter figures appear to be excessive for a species whose life style is typically precocial rather than altricial (Bruton & Stobbs 1991). The revised growth rate of the coelacanth following Balon et al. (1988) is plotted in Fig. 6.

Population estimation

No estimate of the absolute population size of coelacanths was possible before Fricke's dives off the Comoros. Previously, estimates of relative abundance had been made from the number of coelacanth catches along different shores (e.g. Stobbs 1989). Catch records revealed that coelacanths appeared to be most abundant on the west and southwest coasts of Grand Comoro with smaller numbers along the northern, western and southern shores of Anjouan. No coelacanths are known to have been caught off Moheli (but see Bruton & Coutouvidis 1991).

Fricke & Plante (1988) conducted diving surveys along the western and eastern shores of Grand Comoro using a research submersible and found

that the preferred habitat of the coelacanth – rocky substrata without sand cascades – did not occur along the eastern shore. In general, they found that the location of coelacanth catches coincided with the observed distribution of coelacanths in their natural habitat. More recently, Fricke & Hissmann (1990) and Fricke et al. (1991) have found that coelacanths are cave-dwellers during the day, and they suggest that the numbers of caves available as daytime refuges may limit their distribution. Along the western shores of Grand Comoro they found several occupied caves at depths from 180 to 210 m, whereas below 220 m caves were rare and coelacanths were rarely encountered. No rocky caves were found during a brief survey of the northeastern shore of Anjouan and no coelacanths were seen (H. Fricke personal communication), although they must occur there since at least 42 specimens have been caught at Anjouan (Bruton & Coutouvidis 1991).

Fricke & Hissmann (1990), using two different methods involving observations from a submersible, estimated that the number of coelacanths detectable by this means along the west coast of Grand Comoro was between 150 and 210, comprising mostly fish longer than 100 cm TL. When they used the highest number of observed coelacanths per cave as an indication of cave saturation, they estimated that 370 to 520 individuals > 100 cm TL would be the habitat's carrying capacity for the whole island. Considering that Grand Comoro appears to be the main habitat of the coelacanth, this suggests that the total population is very small. No estimates of variance or likely levels of bias are given by Fricke & Hissmann (1990) because of the small sample sizes.

Population structure

In our calculations we have used a figure of 500 individuals as the present population size within the entire habitat of coelacanths off Grand Comoro and Anjouan. This estimate refers to coelacanths > 100 cm total length (TL), which is the length range most frequently seen (95% of observed fishes, Fricke & Hissmann 1990) and caught by Comoran fishermen (88% of coelacanths caught, Bruton & Coutouvidis 1991). A comparatively high figure has been used as it is possible that Fricke's preliminary population size calculation is an underestimate due to our limited knowledge of the depth distribution of the coelacanth.

By varying the rate of natural mortality, we reconstructed a population with a stationary age distribution of 500 individuals > 100 cm TL, and in which the number of animals at age zero (i.e. the number at birth) was equal to the sum of the numbers of females longer than the age at first maturity (according to the assumed growth parameters) multiplied by the average number of young born per female. As the latter is poorly known, we have used values of three (a more conservative estimate) and five per female, which is the number reported by Smith et al. (1975). We have also assumed a sex ratio of 1 : 1, which is the likely condition in the wild. Also, in order to derive some understanding of the age-dependent pattern of exploitation, the average catch-at-age over the last 36 years was inferred from the data on fish length and was incorporated in the analysis as the most likely age distribution of an average annual catch (Table 3). The values in Table 3 appear as fractions of a fish per age class as only about 3.5 fish were known to have been caught on average each year since 1952 at the time that this analysis was carried out.

The age distributions were calculated in accordance with the growth rate for *L. chalumnae* suggested by Balon et al. (1988), which is half that calculated by Hureau & Ozouf (1977, Fig. 6). Clearly this is a major weakness of the analysis as the accuracy of the growth estimate by Balon et al. (1988) has not yet been tested. We have nevertheless carried out analyses based on this growth estimate in the hope that they will stimulate further research leading to the determination of more accurate parameters for a future model.

Assuming that one ring is formed per year, the average annual growth rate is about 6.5 cm year^{-1} from birth, and the fish attain about 170 cm in their twentieth year after birth (Balon et al. 1988). Other estimated parameters of the coelacanth's life cycle are: age above which most coelacanths are caught by Comoran fishermen (80 cm TL) = 6 years, age at 100 cm when they are first seen by Fricke = 9 years, and age at first maturity at 125 cm = 13 years.

Using the growth rate estimate of Balon et al. (1988), ages were allocated to fish that had been caught over the last 36 years and compiled into an aggregate age frequency that was then converted

Table 3. Estimated catches at age for 107 coelacanths, for which lengths were measured, at Grand Comoro and Anjouan over the past 36 years. The average catch per year is determined assuming a total catch of 126 fish over this period (the number known to have been caught at the time of this analysis), and using the growth estimate for the coelacanth of Balon et al. (1988).

Age	Total catch	Average catch
6	3	0.1
7	6	0.2
8	0	0.0
9	5	0.16
10	8	0.26
11	2	0.07
12	12	0.39
13	12	0.39
14	17	0.56
15	9	0.29
16	6	0.20
17	5	0.16
18	11	0.36
19	5	0.16
20	2	0.07
21+	4	0.13

into an average annual catch-at-age. Table 3 includes fish from 6 to 21 years old, with the 21-year category also including fishes older than 21 years. There will undoubtedly be errors in these initial age allocations, especially at higher ages, because of variable growth, the difficulty of reading rings on the scales, and the small annual sample sizes. Beamish & McFarlane (1983) have pointed out that scales may provide very inaccurate estimates of the growth rates of fishes beyond ages of 4 to 10 years.

The average age composition of the population over time is a function of the number of fish born per year and the age-specific natural and fishing mortality. In our analysis, the overall population size was scaled up or down so that the number older than 9 years (> 100 cm) was equal to 500 individuals, whilst retaining the stationary age distribution referred to earlier. The mortality rates of juveniles are likely to be greater than those of adults (unless cave-dwelling provides a high degree of protection from predators), but these data are not available and cannot as yet be incorporated into the analysis. Population parameters were also estimated for a population size of 1000 individuals > 100 cm TL for comparison.

Table 4. Annual instantaneous rate of natural mortality (M), and ratios of catch to population numbers of five-year old and older coelacanths (F), for different 'survey' estimates of population size of fish > 100 cm TL. The natural losses of coelacanths of age 6 and more years (i.e. in the age range exploited by man) are given in parentheses. Separate estimates are given for assumed birth rates of 3 and 5 young per female per spawning. The first rank (500 coelacanths > 100 cm TL) most closely approximates Fricke & Hissmann's (1990) estimate of the saturation population size at Grand Comoro and therefore probably gives the more accurate estimation of natural and fishing mortality rate. The figures for a population of 1000 fishes > 100 cm TL are given for comparison. Note that the fishing mortality rate is substantially less than the natural mortality rate in all instances.

No. of coelacanths > 100+ cm TL	Birth rate = 3		Birth rate = 5	
	M	F	M	F
500	0.165 (137)	0.0042	0.192 (174)	0.0039
1000	0.166 (276)	0.0021	0.194 (349)	0.0019

Given values of the number of fish at age zero, the annual rate of natural mortality and the annual catch from each age class, the numbers attaining each age can be computed as follows:

$$N_{i+1} = [N_i \times \exp(-M/2) - C_i] \times \exp(-M/2),$$

where N_i = the number of fish attaining age i, M = the annual instantaneous rate of natural mortality, and C_i is the annual catch of fish of age i, for convenience assumed to be taken as a single pulse midway through the year. In the present case, the total mortality rate $Z [= Ln (N_{i+1}/N_i)]$ in each age class is very similar to the natural mortality rate because of the very small annual catches.

In order to calculate the total population size, we computed the average number of fish in each age class as the number of fish dying at that age divided by the total mortality rate at that age and summed these values over the appropriate range of ages. For fish of 13 years and older, the expected number of young being born was half the total number of adults multiplied by the birth rate. The number of fish at age zero (N_0) and the natural mortality rate (M) were iteratively adjusted until the following conditions were met:

(1) The number of fish of 9 years (about 100 cm) and older was 500, and

(2) the total expected births equalled the calculated number of fish at age zero.

The natural mortality rate and the overall rate of fishing mortality, i.e. the ratio of the average annual catch of coelacanths (3.5 fish) to the average number of coelacanths aged 6 years or over (the exploited part of the population) for a reconstructed population that meets the above conditions, are given in Table 4. Natural mortalities of 0.166 and 0.193 (about 85 to 82% annual survival) provided stationary age distributions satisfying the birth rate conditions. To obtain a population of 500 individuals of 9 years and over, very low fishing mortality rates were required, the catches of 6-year old and older coelacanths representing only 0.4% of the population in this age range. The fishermen's catches are, by these estimates, almost insignificant, particularly when compared with the annual losses to predators of 137 to 174 six-year old and

308

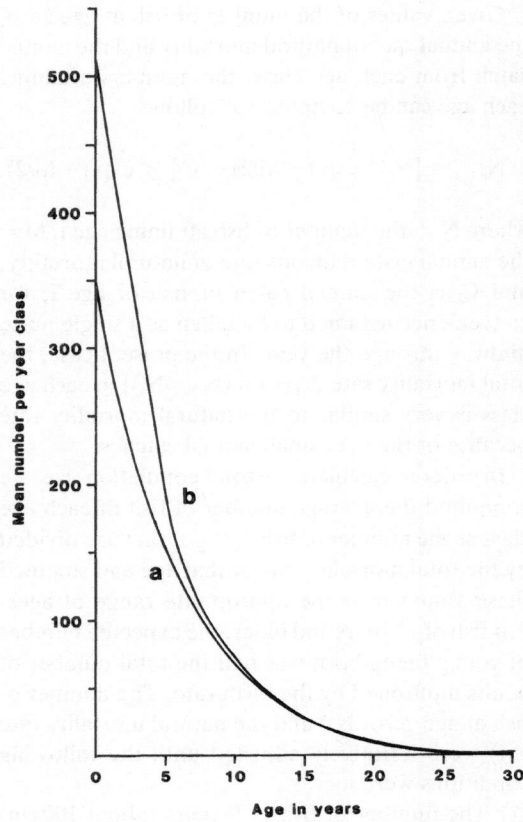

Fig. 7. The mean numbers of coelacanths in different age classes for 3 births (a) and 5 births (b) per female, showing the survivorship curve predicted by the model for a population of approximately 500 individuals > 100 cm TL.

older animals. It would appear that the natural mortality rate is driving the age composition of the

population, according to this analysis. This will be true even for population sizes of 150 to 210 as estimated by Fricke & Hissmann (1990) for the west coast of Grand Comoro only.

Factors affecting the abundance of large predators, such as a fishery for sharks, may thus be of greater significance to the survival of the coelacanth than the direct impact of native fishermen on coelacanths. However, overfishing of prey species, which may well have occurred off the western coast of Grand Comoro (Stobbs 1987), where most coelacanths are found (Fricke & Hissmann 1990), could reduce the prey available to coelacanths and affect their individual growth and survivorship rates.

The numbers of coelacanths in different length ranges in the reconstructed population for stationary age distributions which meet the population size and birth rate conditions described above, are given in Table 5 and Fig. 7. It is interesting to note that there are three times more fish in the 100–150 cm length group than in the 150+ cm group, if the average number of young born is three; and four times more, if the average number of young born is five. For the population estimate of 500 fish > 100 cm TL, the percentage of mature fish (> 125 cm TL) in the total population is estimated to be 11% if there are three births per female and about 7% if there are five births. The low representation of fishes < 100 cm TL from both catches as well as observed records still remains to be explained, but may be related to their habitation of small caves in relatively shallow water where they are both difficult to observe and difficult to catch.

Table 5. Numbers of coelacanths in different total length ranges in the population for stationary age distributions meeting the population size and birth rate conditions described in the text. The number of mature male and female fish (i.e. > 125 cm TL) is given as N(mat). Separate estimates are given for assumed birth rates of 3 and 5 young per female. The total number of births = N(mat)/2 × birth rate. The first rank (500 fishes > 100 cm TL) most closely approximates Fricke & Hissmann's (1990) estimate of the number of coelacanths off Grand Comoro and probably gives the best estimates of stationary age distribution for the different birth rates. The estimates for 1000 fishes > 100 cm TL are given for comparison.

No. of coelacanths > 100 cm TL	Birth rate = 3				Birth rate = 5			
	< 100	100–150	150+	N(mat)	< 100	100–150	150+	N(mat
500	1770	375	126	251	2406	401	100	224
1000	3536	743	255	505	4808	795	204	452

Discussion

These analyses clearly have several limitations due to our inadequate knowledge of coelacanth biology, but they have indicated that fishing mortality is unlikely to be a major determinant of the population size of the coelacanth unless catch rates increase markedly. The catch rates of coelacanths were initially low (1.9 on average each year in the 1950s) but stabilised at a higher level in subsequent decades (4.9 in the 1960s, 4.8 in the 1970s and 4.0 in the 1980s, Bruton & Coutouvidis 1991) despite the expansion of the artisanal fishery throughout these decades (Stobbs 1987, Stobbs & Bruton 1991). This trend could indicate that the coelacanth population is already in decline. However, the increased fishing effort has not necessarily increased the pressure on the coelacanth (Fricke & Hissmann 1990) as much of the current effort using fibreglass boats (see Stobbs & Bruton 1991) is directed at offshore fishing for gamefish that have aggregated at fish aggregation devices (M. de San personal communication 1990). Coelacanths, on the other hand, are caught almost entirely from traditional paddling canoes at night (Stobbs 1987, Stobbs & Bruton 1991). The deployment of fish aggregation devices, and the availability of more modern, motorised boats, may therefore have decreased the risk of catching coelacanths.

At present levels of fishing, it appears therefore that the artisanal fishery does not have a marked impact on the coelacanth. This conclusion is supported by the observation by Fricke et al. (1991) of a group of coelacanths in a cave off one of the busiest artisanal fishing ports in the Comoros.

Gaps in our knowledge

Our analyses have identified several important gaps in our knowledge of the dynamics of the coelacanth population. These gaps, as well as some deduced from other studies, are listed below:

(1) More accurate age and growth estimates are needed, particularly to calculate the age at first maturity and at the beginning of senescence (if any). The use of a scanning electron emission spectrophotometer to determine the daily deposition of calcium on the scales would be a major advance. It is also important to determine the shape of the growth curve as the linear relationship predicted by Hureau & Ozouf (1977) may only hold over a limited range of ages. The growth rate is more likely to be relatively rapid in the early years and to reach an asymptote in later years, as is typical in fishes. The longevity of the coelacanth also needs to be better established.

(2) Further assessments of the abundance and size composition of the population are needed if the status of the coelacanth is to be accurately determined.

(3) Is there a dependence of birth rate on population size? Dynamic simulation modelling is required to investigate the potential influence of density-dependence on the resilience of the population.

(4) What is the predation rate on coelacanths? Several large shark species, including the bumpytail ragged tooth *Odontaspis ferox,* have been seen by Fricke and his colleagues in the habitat of the coelacanth (personal communication, and video of their November 1989 expedition to the Comoros), and it is likely that deepsea sharks such as *Hexanchus griseus,* as well as oceanic and shallow water sharks, also occur there (L.J.V. Compagno personal communication 1989). Several coelacanth specimens have missing fins, probably as a result of predation. Furthermore, we need to know whether the mortality rate differs for males and females and for juvenile and adult coelacanths.

(5) Do all mature adults participate in breeding or are there harems and supernumerary males and/or females?

(6) What is the fecundity of the coelacanth, both in terms of the numbers of eggs produced and the number of juveniles that are born? All intact coelacanths in museum or university collections that have not as yet been x-rayed or dissected should be examined as a priority. A better estimate of the length of the gestation period is also needed.

(7) Does the coelacanth exhibit parental care and

310

thereby reduce predation risk to juveniles? Do juvenile coelacanths shelter in caves that are too small for large predators to enter, and thereby reduce predation risk? On the basis of Fricke & Hissmann's (1990) recent observations, it appears that access to caves for shelter during the day is important for both small and large coelacanths.

(8) Better estimates of the length : weight relationship could be used to calculate the condition factors of coelacanths and used to determine whether components of the population are underfed due to overfishing of their potential prey.

(9) The location and relative abundance of small coelacanths (< 100 cm TL) needs to be determined.

(10) From the conservation point of view, it would also be useful to establish whether the coelacanth populations at Grand Comoro and Anjouan represent different gene pools and need to be managed separately, or whether they represent a single gene pool and can be managed together.

Acknowledgements

Our research on the coelacanth has been supported by grants from the Foundation for Research Development, the Sea Fisheries Research Institute, the United Building Society and the J.L.B. Smith Institute of Ichthyology. We are grateful to the government of the Federal Islamic Republic of the Comoros for permission to work in their beautiful country, and to the Director of the Comoran National Museum in Moroni, Damir ben Ali, for his generous assistance. Hans Fricke, Eugene Balon, Phil Heemstra, Colin Buxton and Robin Stobbs kindly commented on a draft of this paper, and Nick James assisted with the length : weight computations. We are particularly grateful to Robin Stobbs for sharing his knowledge on the Comoran fishery. Sheila Coutouvidis collated information from the inventory and drew the figures.

References cited

Balon, E.K. 1974. Fish production of a tropical ecosystem. pp. 249–676. *In:* E.K. Balon & A.G. Coche (ed.) Lake Kariba: A Man-Made Tropical Ecosystem in Central Africa, Monographiae Biologicae 24, Dr W. Junk Publishers, The Hague.

Balon, E.K. 1977. Early ontogeny of *Labeotropheus* Ahl, 1927 (Mbuna, Cichlidae, Lake Malawi), with a discussion on advanced protective styles in fish reproduction and development. Env. Biol. Fish. 2: 147–176.

Balon, E.K. 1990. The living coelacanth endangered: a personalized tale. Tropical Fish Hobbyist 38 (February): 117–129.

Balon, E.K., M.N. Bruton & H. Fricke. 1988. A fiftieth anniversary reflection on the living coelacanth, *Latimeria chalumnae:* some new interpretations of its natural history and conservation status. Env. Biol. Fish. 23: 241–280.

Beamish, R.J. & G.A. McFarlane. 1983. The forgotten requirement for age validation in fisheries biology. Trans. Amer. Fish. Soc. 112: 735–743.

Bruton, M.N. 1987. Is the coelacanth endangered? Ichthos 13: 1.

Bruton, M.N. 1988. 'Fossil' fish trade causes concern. Traffic (U.S.A.) 8: 8.

Bruton, M.N. 1989a. The coelacanth – can we save it from extinction? WWF Reports, October/November 1989: 10–12.

Bruton, M.N. 1989b. The living coelacanth fifty years later. Trans. Roy. Soc. S. Afr. 47: 19–28.

Bruton, M.N. & S.E. Coutouvidis. 1991. An inventory of all known specimens of the coelacanth *Latimeria chalumnae*, with comments on trends in the catches. Env. Biol. Fish. 32: 371–390. (this volume)

Bruton, M.N. & R.E. Stobbs. 1991. The ecology and conservation of the coelacanth *Latimeria chalumnae*. Env. Biol. Fish. 32: 313–339. (this volume)

Forey, P.L. 1988. Golden jubilee for the coelacanth *Latimeria chalumnae*. Nature 336: 727–732.

Fricke, H. 1988a. Coelacanths: the fish that time forgot. Nat. Geog. Mag. 173: 824–838.

Fricke, H. 1988b. J'ai recontre le fossile vivant. Geo 116: 56–75.

Fricke, H. & K. Hissmann. 1990. Natural habitats of coelacanths. Nature 346: 323–324.

Fricke, H., K. Hissmann, J. Schauer, O. Reinicke, L. Kasang & R. Plante. 1991. Habitat and population size of the coelacanth *Latimeria chalumnae* at Grand Comoro. Env. Biol. Fish. 32: 287–300. (this volume)

Fricke, H. & R. Plante. 1988. Habitat requirements of the living coelacanth *Latimeria chalumnae* at Grande Comore, Indian Ocean. Naturwissenschaften 75: 149–151.

Greenwood, P.H. 1989. Fifty years a 'living fossil' – the coelacanth fish *Latimeria chalumnae*. Biologist 36: 15–19.

Hureau, J.-C. & C. Ozouf. 1977. Determination de l'age et croissance du coelacanthe *Latimeria chalumnae* Smith, 1939 (Poisson, Crossopterygien, Coelacanthide). Cybium 3e série, 2: 129–137.

Manooch, C.S. 1987. Age and growth of snappers and groupers. pp. 329–373. *In:* J.J. Polovina & S. Ralston (ed.) Tropical

Snappers and Groupers, Biology and Fisheries Management, Westview Press, London.

Millot, J., J. Anthony & D. Robineau. 1972. État commenté des captures de *Latimeria chalumnae* Smith (Poisson, Crossopterygien, Coelacanthidé) effectuées jusqu'au mois d'Octobre 1971. Bull. Mus. Nat. Hist. Nat., Paris, 3e série, no. 53, Zoologie 39: 533–548.

Smith, C.L., C.S. Rand, B. Schaeffer & J.W. Atz. 1975. *Latimeria*, the living coelacanth, is ovoviviparous. Science 190: 1105–1106.

Stobbs, R.E. 1987. The 'galawas' of the Federal Islamic Republic of the Comores, with notes on artisanal fishing and recommendations for a maritime museum. Invest. Rep., J.L.B. Smith Inst. Ichthyol. 26: 1–21.

Stobbs, R.E. 1989. The coelacanth enigma. The Phoenix 2(2): 8–15.

Stobbs, R.E. & M.N. Bruton. 1991. The fishery of the Comoros, with comments on its possible impact on coelacanth survival. Env. Biol. Fish. 32: 341–359. (this volume)

Suyehiro, Y. 1983. Some views on the dissected specimens of coelacanth. Biennial Report of 1983 of the Keikyu Aburatsubo Marine Park Aquarium 12: 12.

Suzuki, N., Y. Suyehiro & T. Hamada. 1985. Initial report of expeditions for coelacanth – Part 1 – Field studies in 1981 and 1983. Sci. Pap. Coll. Arts Sci., University of Tokyo 35: 37–79.

Uyeno, T. 1984. Age estimation of coelacanth by scale and otolith. pp. 28–29. *In:* Proceedings of the First Symposium on Coelacanth Studies, Tokyo. (In Japanese).

Uyeno, T. 1991. Observations on locomotion and feeding of released coelacanths, *Latimeria chalumnae*. Env. Biol. Fish. 32: 267–273. (this volume)

Environmental Biology of Fishes **32**: 313–339, 1991.
© 1991 *Kluwer Academic Publishers.*

The ecology and conservation of the coelacanth *Latimeria chalumnae*

Michael N. Bruton & Robin E. Stobbs
J.L.B. Smith Institute of Ichthyology, Private Bag 1015, Grahamstown, 6140 South Africa

Received 7.11.1990   Accepted 24.12.1990

Key words: Habitat preferences, Home range, Seasonal activity, Behaviour, Predator-prey relationships, Fishery, Management, Longlining, CITES, Endangered, Coelacanth Conservation Council

Synopsis

Studies on the ecology of the living coelacanth, *Latimeria chalumnae,* are reviewed and assessed. Early predictions on the life history of the coelacanth have proved to be accurate but recent findings have improved our understanding of its habitat and feeding preferences, diel activity patterns and social behaviour. A history of coelacanth conservation reveals that there has been a sustained concern for the survival of this species which has eventually culminated in several effective conservation actions in recent years. The coelacanth is threatened by a number of socio-economic and biological factors, but international action directed at managing the fishery in the Comoros should ensure that the species survives. Recent observations on living coelacanths in their natural environment have greatly improved our knowledge of the behaviour and relative abundance of adults. Important priorities for future research include studies on the distribution and abundance of juveniles and breeding adults, both off the Comoros and elsewhere. The coelacanth is a highly specialised, precocial fish which occupies a unique place in biology. Co-ordinated international efforts should continue to be made to understand and conserve this remarkable fish.

Introduction

The enormous importance of the coelacanth and the obvious vulnerability of the single known breeding population in the Comoros have ensured that the conservation of this species has received attention over the past 50 years. These efforts have, however, been poorly co-ordinated and underfunded, and, until recently, have proved to be largely ineffective. The continued survival of the coelacanth can probably be ascribed more to its inaccessibility and consequent difficulty of capture, and to the relative inefficiency of native fishing methods, than to organised efforts by westernised man. In contrast, efforts to conserve other widely-publicised endangered taxa, such as whales, pandas, turtles and rhinoceroses, have been relatively

well co-ordinated, generously funded and largely successful. Several factors have contributed to this situation, including the extreme value of coelacanth specimens for scientific study (and prestige), which brought about strong competition for specimens, and the almost complete lack of a conservational infrastructure in the Comoros (even today), which has sometimes been abused by organisations seeking coelacanth specimens.

The establishment of the Coelacanth Conservation Council (CCC) in 1987, and the inclusion of the coelacanth in Appendix I of CITES in October 1989, have provided a recent impetus to coelacanth conservation. The CCC has created a useful network of collaborating coelacanth researchers around the world, with this volume as one of the products of that collaboration. In addition, an in-

314

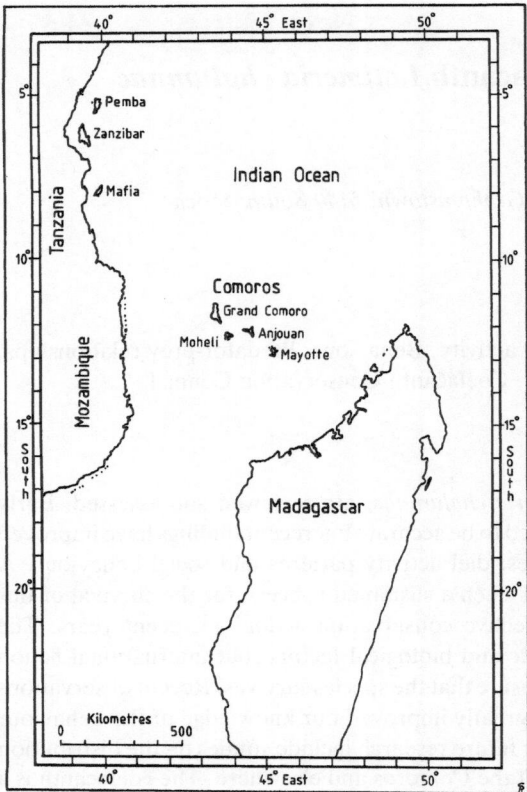

Fig. 1. The Comoros archipelago showing its location in the western Indian Ocean at the northern end of the Mozambique Channel.

ventory and bibliography on the living coelacanth have been compiled (Bruton & Coutouvidis 1991, Bruton et al. 1991) and will continue to be updated by the CCC.

One of the most significant accomplishments in coelacanth conservation has been the development of a pelagic fishery in the Comoros through the deployment of fish aggregation devices and motorised boats. This new fishery has relieved pressure on the inshore reefs and, if sustained, may lead to a recovery of the biota of these systems.

In this paper we briefly discuss the habitat of the coelacanth and review the state of our knowledge on its ecology. We then discuss coelacanth conservation efforts over the past 50 years and evaluate the factors that threaten the survival of the species.

A holistic plan for the conservation of the coelacanth, and a new inferred natural history, are then presented.

The Comoros – the known habitat of the coelacanth

The Comoro archipelago is situated in the northern entrance to the Mozambique Channel, almost equidistant from the African mainland and the northern tip of Madagascar, about 300 km from each (Fig. 1). The islands lie between 11° 20″ and 13° 04″ south latitude and 43° 11″ and 45° 19″ east longitude. In 1976 the three islands of Grand Comoro (known locally as Ngazidja), Anjouan (Nzwani) and Moheli (Moili) united to form the Federal Islamic Republic of the Comoros. The fourth island, Mayotte (Maore), has retained its status as an overseas territory of France. The total area of the four islands is only 2033 km².

The Comoro islands have a hot, wet tropical climate. The 'cool' season lasts from May to October with shade temperatures between 20–22° C, and the hot season from November to April (25–35° C) (Gould 1985). The islands are subject to seasonal, rain-bearing winds during the northerly monsoon, which usually lasts from November to March.

The four islands differ markedly in their topography and age. The youngest, Grand Comoro, is dominated by a huge active volcano, Kartala, on the south side and a smaller extinct volcano, La Grille, on the north side. Kartala has erupted several times during the past 100 years, most recently in 1952, 1965 and 1977; the barren, black lava flows are conspicuous on the western and southern shores. Grand Comoro has steep, densely forested slopes and a steep submarine profile. The other islands do not have active volcanoes, but Anjouan is also steep and mountainous whereas Moheli and Mayotte are older, more eroded islands with less steep profiles. Mayotte has an extensive fringing reef and there are extensive coral reefs to the south of Moheli. The Comoros are separated by ocean depths of 3500 m from Africa and Madagascar. There is a channel about 3000 m deep separating Mayotte from the other islands, which are themselves separated by channels 1500 to 2000 m deep. The south equatorial current flows at a speed of 0.5–3.0 knots past the islands from east to west, and splits into the north-flowing Somali current and the south-flowing Mozambique current on reaching the African coast.

There is considerable pressure on the natural environment of the islands from the rapidly increasing human population, which is very dependent on natural resources for food, medicines, fuel and house- and boatbuilding materials. There are about 420 000 people on the three islands of the Islamic Republic and the rate of population growth is reported to be as high as 4.2% (Newitt 1984). At present growth rates the population is expected to increase to over 1 000 000 by the year 2010 (I. Thorpe personal communication). The economy of the Comoros is one of the poorest in the world with a large foreign debt. Foreign aid represents about 45% of the GNP, agriculture and fishing about 45%, and industry and other activities the remaining 10%. Further details on the Comoros are given by McCosker & McCosker (1976), Newitt (1984) and Gould (1985), and the fishery is described by de San (1983), Suzuki & Tanauma (1984), Suzuki et al. (1985), Stobbs (1987, 1989a) and Stobbs & Bruton (1991).

The submarine topography and fauna of the islands has been described by Millot (1954a), McCosker (1979), Suzuki et al. (1985), Heemstra & Smale (1987), Fricke & Plante (1988), Balon et al. (1988), Bruton et al. (1988), Fricke & Hissmann (1990) and Uyeno et al. (1990). The underwater profile slopes steeply on most shores of Grand Comoro and Anjouan and there is little fringing reef. The reefs have moderate coverings of coral and sponges and support a wide diversity of invertebrates and vertebrates. Corals on the reefs include species of *Acropora* and *Pachyseris,* with gorgonian corals such as *Echinogorgia rigida* and *Junceella juncea* and giant sponges (Fig. 2) in deeper water (Suzuki et al. 1985). At depths of 70 to 120 m fishes from a variety of families are found, including Gobiidae, Eleotridae, Mullidae, Alutenidae, Chaetodontidae, Synodontidae, Cirrhitidae, Labridae, Balistidae and Scorpaenidae (Suzuki et al. 1985).

Fig. 2. A SCUBA diver with a giant sponge at a depth of about 30 m off the west shore of Grand Comoro opposite the Hotel Coelacanthe (November 1987). Photograph by J.-L. Geraud.

Review of studies on coelacanth ecology

Previous accounts of the ecology of the coelacanth are given by Millot (1954b), Smith (1956a), McCosker (1979), Uyeno et al. (1984), Suzuki et al. (1985), Balon et al. (1988), Suyehiro (1988), Bruton (1989a, b), Fricke & Hissmann (1990) and Fricke et al. (1991). The following brief review serves to summarise recent findings as a background to the discussion on the conservation of the species. It therefore deals only with 'whole animal' ecology rather than with the diverse and fascinating literature on physiological and morphological adaptations to the environment.

Smith (1939a, b, 1940, 1950, 1953, 1956a), supported by Millot (1954b), correctly predicted that the coelacanth inhabits rocky reefs in moderately deep water off the East African coast or Madagascar, in a manner similar to large serranids. White (1939) of the British Museum (Natural History) suggested rather that *L. chalumnae* is 'almost certainly a wanderer from deeper parts of the sea to which its kind have retreated in the face of fierce competition with the more active modern types of fishes'. Smith (1956a) also predicted that the coelacanth is a piscivore.

Millot (1954a) provides the first observations on coelacanth ecology based on studies in the Comoros. He states that 'Coelacanths inhabit, around the Comoro Islands, basalt rocky bottoms which slope very steeply (25°–45°). They seem to live normally on the actual bottom or quite close to it, between 200 and 400 metres depth at least...'. Elsewhere he states that the coelacanth normally frequents rocky bottoms between 300 and 500 m, rising occasionally to 200 or 150 m (Millot 1954b). He dismisses a suggestion by Bertin (1953, in Millot 1954b) that coelacanths might venture into shallower water to mate.

Millot (1954b) also reports that coelacanths are always caught as a bycatch at night as the deepsea fishes which Comoran fishermen seek to catch – *Ruvettus pretiosus* and species of *Promethichthys, Priacanthus* and *Muraenesox* – 'only bite at night'. He reports further that the night fishermen leave after sunset and return at about 0300 h. On the basis of interviews with Comoran fishermen by a colleague R. Fourmanoir, he further states that coelacanths are mainly caught between October and April, and that the fishermen mainly use the 'roudi', *Promethichthys prometheus,* as bait for *Ruvettus* (and hence coelacanths). The bait is typically hung 8 to 10 m above the bottom to avoid being snagged. He reports that the second and third coelacanths were caught 100 and 30 m above the substrate, which now seems unlikely given our current knowledge of coelacanth behaviour. Millot (1954b), on the basis of observations made by an oceanographer, M. Menache, describes the coelacanth habitat off Anjouan as a steep slope with a general water circulation from east to west and a marked thermocline between 100 and 200 to 250 m. A diagrammatic representation of *Ruvettus* fishing in this habitat is given in Figure 3.

Thomson (1973, p. 63) predicted that '*Latimeria* is not a truly deepsea fish. We might reconstruct, therefore, that *Latimeria* is a fish that lives a relatively sedentary existence near the bottom. At night it may come toward the surface to follow prey species, and migrate back down the submarine slopes of the islands during the day. If this is correct, then apart from detailed ecological preferences that we do not know of, the species could be widely dispersed within the area immediately around the Comoros. But to move beyond this region, the fish would probably have to swim in open water, away from the bottom, which may explain its restricted distribution. As to population size, we cannot guess, but common sense tells us that to be safe we should avoid a catch rate higher than the present one'.

More recent findings on coelacanth ecology are summarised below:

Habitat preferences. – According to the catch records of Comoran fishermen, coelacanths occur at depths from 40 to 1000 m but have a modal depth range from 100 to 400 m (Fig. 4). Although the accuracy of the fishermen's depth estimates has been questioned (McCosker 1979, Locket 1980, Suzuki et al. 1985), Millot (1954a), Fricke (1988) and Stobbs (1989b) have found that they may be remarkably accurate.

McCosker (1979) reports a depth range for coelacanths of 70 to 600 m with the majority of fishes captured between 150 and 300 m, which probably reflects the depth of the oilfish-catching effort. McCosker & McCosker (1976) consider that these estimates are exaggerated towards greater depths. Locket (1980) states that the most common depths at which coelacanths are caught are 100 to 400 m; a specimen caught off Iconi which he examined (CCC no. 80) was caught at a depth of 165 m about 600 m from shore. Other estimates of the depth range of coelacanths are given by Newman (1971, about 70–660 m), Diamond (1985, 100–300 m) and Thomson (1986, 100–300 m).

Fricke et al. (1987), on the basis of observations from a research submersible, found coelacanths

Fig. 3. Diagrammatic representation of a fisherman in a dugout canoe handlining for the oilfish *Ruvettus pretiosus*, with the possibility of catching a coelacanth as a bycatch. Original illustration by D.P. Voorvelt.

with a length range of circa 120 to 180 cm at depths from 117 to 198 m off Grand Comoro. The depth range of coelacanths during a longer series of dives off Grand Comoro in 1988 was 117 to 198 m (Fricke

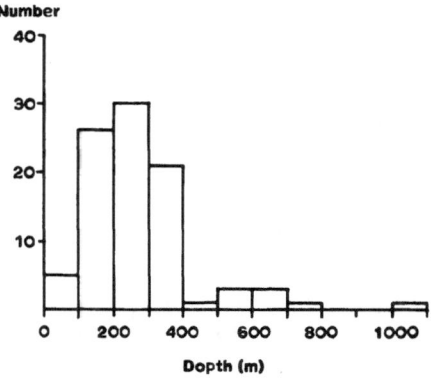

Fig. 4. The estimated depth range at which coelacanths have been caught off Grand Comoro and Anjouan according to the data presented by Bruton & Coutouvidis (1991). n = 91.

& Plante 1988). Subsequently, Fricke & Hissmann (1990) and Fricke et al. (1991) confirmed that coelacanths > 100 cm TL have a modal depth range of 180 to 210 m off the west shore of Grand Comoro near Hahaya, and suggested that the rarity of caves and prey fish below 220 m may limit their distribution. The maximum depths at which the fishermen claim to have caught coelacanths are therefore greater than the maximum depths at which Fricke and his colleagues have observed coelacanths. It is possible, therefore, that the deeper stocks have not as yet been located using the submersible.

The depth preferences of small coelacanths are poorly known and the data are contradictory. Catch records indicate that specimens < 100 cm TL were caught from a wide range of depths (100–600 m, Bruton & Coutouvidis 1991). Suzuki et al. (1985) report that a 25 cm juvenile coelacanth was seen swimming weakly near the water surface off Anjouan by a Comoran fisherman. This fish, which had completely resorbed its yolksac at a smaller size than the other full term juveniles (32 cm, Smith et al. 1975), was apparently 'in a poor state of health at the time of observation though it was still alive' (p. 56). Elsewhere (p. 57) they report that Comoran fishermen claim that small coelacanths are occasionally seen swimming close to the sandy bottom during the day, and that the Comorans have a separate name for this shallow water form

318

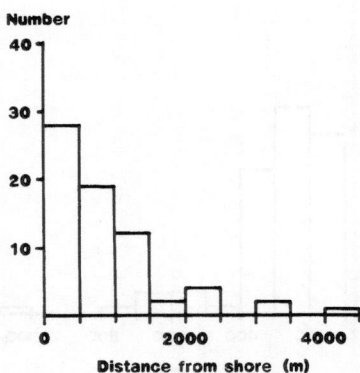

Fig. 5. The estimated distance from shore at which coelacanths have been caught off Grand Comoro and Anjouan according to the data presented by Bruton & Coutouvidis (1991). n = 69.

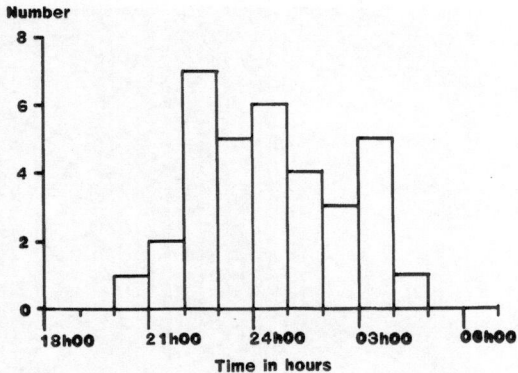

Fig. 6. The time of capture of coelacanths off Grand Comoro and Anjouan according to the data presented by Bruton & Coutouvidis (1991). n = 34.

('mamme') to distinguish it from 'gombessa', the deep water coelacanth.

Coelacanths typically inhabit steeply sloping shores (Fricke & Hissmann 1990). Jean-Louis Geraud, an experienced diver in the Comoros who has an intimate knowledge of the offshore environment there, has hypothesized that coelacanths seek out these steep slopes for three reasons: (a) there are distinct up- and downwelling currents on which the coelacanth can drift to facilitate vertical movements, (b) there are many caves for shelter and prey capture, and (c) outflows of freshwater occur off these shores (personal communication May 1990).

The suggestion that coelacanths live near the outflows of freshwater aquifers under the sea (Forster 1974, McCosker 1979) now seems to be unlikely. While freshwater outflow may occur in shallow water at several localities around Grand Comoro (M.N. Bruton personal observation) and even at depths of 30 and 70 m (McCosker 1979), Fricke has found no evidence of these outflows at the depths at which the coelacanth lives. Millot (1954b) also discussed this interesting theory but found it to be untenable.

Menache (1954) and McCosker (1979) identified thermoclines between 150 to 200 m off Anjouan and 115 and 125 m off Grand Comoro respectively, with water temperatures of 15 to 17° C below the thermocline. In contrast, Newman (1971) found

that water temperatures were fairly uniform from the surface to about 200 m depth at a location offshore of Maludja where a coelacanth was caught in 1958. Fricke et al. (1991) found that coelacanths are normally encountered below the 18° C isotherm, although the water temperatures in caves which they occupy may be as high as 22.8° C.

The distance from shore at which coelacanths are caught ranges from < 100 m to over 4000 m, with the majority of captures less than 1500 m from shore (Fig. 5). Newman (1971) reports a distance from shore of coelacanth fishing of 300 to 2200 m.

Home range. – The only accurate information available is Fricke & Hissmann's (1990) observation that the coelacanth has a home range of at least 8 km of coastline along the western shores of Grand Comoro.

Diel activity. – The time of capture of 34 coelacanths is known; all were caught between 2000 h and 0500 h, i.e. at night (Fig. 6). This result is partly an artifact of the time at which the fishermen are most active (Stobbs 1989a, Stobbs & Bruton 1991) but may also reflect coelacanth activity patterns as Fricke & Plante (1988), Fricke & Hissmann (1990) and Fricke et al. (1991) have found that coelacanths are more active at night and that they may shelter in caves during the day. McCosker (1979) and Fricke & Plante (1988) further suggest that the coelacanth undertakes vertical migrations into shallower water during the night as prey organisms are believed to be more abundant there, but Fricke et al. (1991) found no marked differences in the depth at which they encountered coelacanths during the day and night.

Seasonal activity. – There is a seasonal pattern in the rate at which coelacanths are caught, with higher catches from November to March, the hottest months of the year (Fig. 7). This result is, however, also likely to be an artifact of the seasonal fishing habits of Comoran fishermen whose fishing effort is highest from November to March during the northerly monsoons (kaskazi) and inter-monsoon calms, when sea conditions are more favourable for fishing (Stobbs 1987). The kaskazi blows from November to March with a Beaufort wind force of 2 to 4. In contrast, the southerly monsoon (kuzi) lasts from July to September/October when the winds average force 5 to 6, and fishing activities are severely limited. There may therefore be no seasonal pattern of activity by the coelacanth. The seasonality and lunar periodicity of catches was ascribed by McCosker (1979) partly to the presence in shallow water of rain-fed submarine aquifers, but this now seems unlikely.

Diet and feeding behaviour. – Studies on the feeding preferences of the coelacanth have relied on the examination of the stomach contents of a small number of specimens, many of which had empty stomachs (e.g. those examined by Smith 1939a, Millot 1954b, Thomson 1986). Coelacanths feed on a variety of epibenthic and benthic fishes associated with rocky reefs at moderate depths, including representatives of the families Scyliorhinidae, Synaphobranchidae, Berycidae, Apogonidae, Lutjanidae, Myctophidae and Polymixidae (Uyeno & Tsutsumi 1991). Squids have also been reported in their diet (McAllister 1971), and coelacanths have been caught on *Octopus* bait (CCC no. 93, Bruton & Coutouvidis 1991). The prey taken ranges in size from <5 cm to >60 cm in length. The coelacanth seems to swallow whole prey, and is also able to take baits 15 to 20 cm long (Millot 1954b).

Prior to the research by Fricke et al. (1987), the

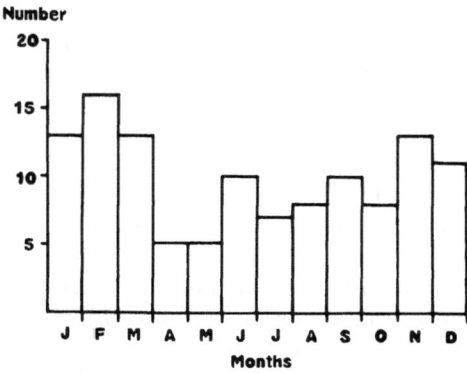

Fig. 7. The month of capture of coelacanths off Grand Comoro and Anjouan according to the data presented by Bruton & Coutouvidis (1991). n = 118.

pattern of locomotion of the coelacanth was variously described as crawling with the paired fins on rocky substrates, stalking like a large piscivorous grouper or even fast swimming in open water (Smith 1956a, Millot & Anthony 1958, von Wahlert 1968, McCosker 1979, Diamond 1985). The locomotory pattern found by Fricke during extensive dives with his submersibles has not confirmed these predictions. He found that the coelacanth is a nocturnal, piscivorous drift-hunter that moves very slowly in up- and downwelling currents, using its paired fins to stabilise the drift motion. The paired and the unpaired lobed fins generate thrust and the large caudal fin is used for fast starts. The paired fins are used synchronously in a pattern that is common in tetrapod locomotion, but they are not used for crawling along the substrate. The six coelacanths which Fricke et al. (1987) observed all performed a curious headstand which lasted for up to 2 min, the purpose of which is unknown, but which may be related to prey detection using the rostral organ (see Balon 1990a). The normally slow movements of the coelacanth were confirmed in subsequent observations by Fricke & Plante (1988), Fricke et al. (1991) and Uyeno (1991). Hughes (1976) concluded from studies on the respiration of the coelacanth that its resting oxygen consumption ($10 \, ml \, kg^{-1} \, h^{-1}$) accords well with an inactive mode of life.

A film sequence taken by H. Fricke of a coelacanth attacking its prey (October 1989) indicates that it used its caudal fin to propel it rapidly forward. It is likely to use the sink-suck method for ingesting prey whereby the mouth is opened rapidly and a negative pressure is created which causes the prey to be sucked in with the surrounding water. Uyeno (1991) confirmed these observations in J.-L. Geraud's video and found that the coelacanth executes a sudden, deliberate bite within a short distance using its powerful jaws, as observed when a captive fish was fed a piece of cuttlefish offered by a diver. Thomson (1966, 1970) noted that the intracranial joint may allow the coelacanth to lift the upper jaw by 15° and thereby widen the gape and increase the strength of the bite.

Parasites. – The coelacanth has a depauperate known parasite fauna, consisting mainly of nematodes, cestodes and isopods (Thoney & Hargis 1991). Kamegai (1971) discovered a new species and genus of monogenean parasite, *Dactylodiscus latimeris,* on the gills of *L. chalumnae* (CCC no. 53). This parasite is apparently restricted to *L. chalumnae* and may be a relict that has evolved with the living coelacanth over a long period of time (Thoney & Hargis 1991).

Demography. – The demography of the coelacanth is reviewed by Bruton & Armstrong (1991).

A brief history of coelacanth conservation

The first coelacanth was caught off South Africa in December 1938 and described by Smith (1939a, b) as a new species, genus, family and suborder of fishes. Because the soft anatomy of this specimen was not saved, Smith resolved to find an additional specimen. His knowledge of fishes led him to predict that coelacanths live in moderately deep water off the East African coast, and he distributed thousands of pamphlets offering a £ 100 reward for the capture of a specimen. The Second World War intervened and it was not until 1952 that another specimen was found, this time in French waters off the Comoros archipelago. Undeterred, Smith flew to the Comoros to collect 'his' fish, and returned to

South Africa with the prized specimen. He described it as a new species *Malania anjouanae* (Smith 1953), but it was later found to be conspecific with *L. chalumnae* (Millot 1955a).

During the 1940s trawlers scoured the south-east coast of South Africa looking for more coelacanths but without success. In 1950 a research vessel from the National Institute of Oceanography in England also searched for the coelacanth off the East London coast but to no avail. In 1953 J.L.B. Smith attempted to organise an expedition to catch a coelacanth but eventually abandoned the plan due to the non-availability of a suitable research vessel.

In 1953, after the capture of the second specimen, the French prohibited foreign scientists from searching for coelacanths in their waters and proposed instead that an international expedition under French leadership should continue to conduct research on coelacanths under the auspices of the Scientific Council for Africa South of the Sahara. Meanwhile J. Millot, later of the Museum National d'Histoire Naturelle in Paris, turned his attention from spiders to coelacanths, and designated the Scientific Research Institute of Madagascar, of which he was then the Director, as the institution responsible for coelacanths taken in French territories. The Madagascar Institute also duplicated Smith's reward of £ 100 for a dead fish but doubled the reward for a live fish, and set up 'fish-embalming stations' at strategic ports in the Comoros (Dugan 1955). At that time the American government even offered a reward of $ 5000 for a coelacanth (Dugan 1955).

The third coelacanth was caught on 24 September 1953 and reached Millot (1954b) in perfect condition. By the end of 1956 the French had 12 well-preserved specimens. In 1954 J. Millot collaborated with Jacques-Yves Cousteau and his 14 man diving team on the 'Calypso' in an investigation of the submarine habitat of the coelacanth in the Comoros using remotely controlled equipment, but they did not attempt to catch a live specimen (Millot 1954, Dugan 1955). Also in 1954 a group of Italian divers under the name 'Spedizione Zoologica Italiana', led by Franco Prosperi, visited Dzaoudzi near Mayotte on a charter vessel and claimed to have photographed a coelacanth in shallow water (Dugan 1955, Prosperi 1957). The photograph, which was published in the popular press in Italy, Britain and France, was regarded as a fake by Millot (1954b).

It was at this point that the first conservation debate broke out. Apparently frustrated by the fact that the international coelacanth expedition had never materialised, Smith (1956b) wrote a caustic letter to The Times of London, suggesting that the French had sufficient specimens for their studies and that they should cease offering a reward for the capture of further coelacanths. He recommended rather that there should be a severe penalty for killing a coelacanth as 'there may well be only a few hundred of them in all', and suggested that 'the present policy is debasing a once important scientific quest to the level of senseless slaughter of one of our most precious heritages in biology' (Smith 1956b). Millot (1956), supported by de Beer (1956) of the British Museum (Natural History), objected strongly to Smith's accusations and stated that every specimen caught had been fully utilised and that more specimens were required for their detailed anatomical studies. Millot (1956) states further that it was 'but elementary justice to pay to the local authorities in the Comoro Islands the warm tribute which they deserve for their sense of responsibility, cooperation in difficult conditions, and devotion to science'.

In 1963, on the occasion of the 25th anniversary of the capture of the first coelacanth, J.L.B. Smith gave a rousing speech during which he again called for the conservation of *L. chalumnae* (Smith 1963). In a popular article dramatically entitled 'The atomic bomb and the coelacanth', Smith states 'In my view there is a very real danger that this priceless heritage from the past may suffer extermination unless steps are taken to prevent it. And to that end I suggest the foundation of an International Society for the Preservation of Coelacanths . . .'. Unfortunately his plea was not heeded and it was some years before internationally coordinated efforts were mounted to conserve the coelacanth.

By 1970 the French had obtained 34 specimens, some of which had been donated or loaned to museums in other countries. A series of expeditions in the late 1960s and early 1970s provided additional specimens for study. These expeditions included the Royal Society Indian Ocean Deep Slope Fishing Expedition of 1969 to Aldabra and Cosmoledo (Forster et al. 1970, McCosker & Lagios 1979), the Royal Society French/British/American Comoros expedition of 1972 (Griffith 1973, Thomson 1973, Anthony 1976, Locket 1980) and the California Academy of Sciences expedition of 1975 to Grand Comoro (McCosker & McCosker 1976, McCosker & Lagios 1979). The latter expeditions attempted to capture live coelacanths but failed to do so.

Little was known about coelacanth reproduction until the 1970s when it became clear that Watson (1927) had been correct in concluding that an Upper Jurassic freshwater coelacanth *Holophagus* (as *Undina*) had two unborn embryos in its body cavity and probably gave birth to live young. Schultze (1972), on the other hand, discovered yolksac juveniles of a saltwater coelacanth (*Rhabdoderma* sp.) from Carboniferous deposits that were not associated with an adult and proposed that this species, and by analogy other coelacanths, was egg-laying like most other fishes and that Watson's specimen was an example of cannibalism. Griffith & Thomson (1973a), on the other hand, suggest that *Rhabdoderma* may be ovoviviparous but that the yolksac juveniles might have been prematurely released from a dead or dying female. Anthony & Millot (1972) and Millot & Anthony (1974) also initially concluded that the coelacanth is egg-laying and has external fertilisation. They also implied that the survival of a species with such small numbers of unprotected eggs would require a degree of parental care of the eggs once they had been laid (Greenwood 1975, Anon 1976, Locket 1976).

Millot (1955b) reports on a 166 cm coelacanth caught off Anjouan on 12.3.1955 (CCC no. 10) that contained eggs of various sizes, with the largest 22 mm in diameter. A 163 cm female coelacanth (CCC no. 79) was caught off Anjouan during the 1972 expedition and found to contain 19 ripe but seemingly unfertilised eggs, each about 9 cm in diameter and 325 g in weight (Anthony & Millot 1972, Millot & Anthony 1974). The discovery that the coelacanth has a low fecundity led Anthony & Millot (1972) to suggest that coelacanth fishing in the

Comoros should be banned between 15 December and 15 March each year during the hypothesized breeding season. Male coelacanths had been found to be in active spermatogenesis from November to February, and large eggs had been found in a female in January (Millot & Anthony 1958).

Griffith & Thomson (1973a) shared Anthony & Millot's (1972) concern for the protection of *Latimeria,* but regarded the year-end ban on coelacanth fishing as premature until more was known about reproductive biology and stock densities. They point out that, if the coelacanth is oviparous, then a year-end fishing ban is sensible, but if the species is ovoviviparous (or viviparous, as has now been found), then a year-end ban makes little sense. They recommend instead that 'the time during which the embryos are retained inside the mother (which could extend for many months as in elasmobranchs) would be the phase most deserving of protection'.

Griffith & Thomson (1973a) further stated that they were not convinced that the native Comoran fishery threatens the survival of the coelacanth as the catch rate (3 to 4 per year) had not changed markedly over the years for which data were available. They did not regard *Latimeria* as being in danger of extinction but emphasized that studies to determine the population levels in the Comoros are critical for the establishment of a sound conservation policy which would both protect *Latimeria* from possible overfishing and preserve the livelihood of the local fishermen.

Griffith & Thomson (1973a) reconsidered the reproductive physiology of the coelacanth because of its implications for conservation and concluded, from the size of the eggs and the similarity of the coelacanth to urea-retaining elasmobranchs, that *L. chalumnae* is ovoviviparous and that the developing embryo uses its yolk for nourishment rather than obtaining nutrients directly from the maternal circulation. Their conclusions were disputed by Millot & Anthony (1974). Griffith & Thomson (1973a) also supported the idea that the coelacanth is likely to show a high degree of parental care, involving either ovoviviparity or guarding a cluster of eggs. The discovery of five well-developed, 30 cm yolksac juveniles in a 65 kg female coela-

canth (CCC no. 29) housed in the American Museum of Natural History, by Smith et al. (1975, see also Anon 1976, Atz 1976, Smith 1976, 1977) finally proved that the coelacanth is livebearing and necessitated a revision of Anthony & Millot's (1972) ideas on oviparity and a closed season to conserve the fish.

The protection of the gravid females of a species is a well established conservation method, but with a gestation period of about 13 months (Locket 1976), it would clearly be impossible to impose a year-round closed season for the coelacanth without interfering with the subsistence fishery. Furthermore, any closed season would be very difficult to enforce in the Comoros. Locket (1976) agreed with Griffith & Thomson (1973a) that the local artisanal fishery probably has little effect on the coelacanth stock and suggested that the coelacanth is not in danger provided that the intensity and methods of fishing do not change.

A better understanding of the reproductive mode of the coelacanth is an essential prerequisite for the formulation of more effective conservation methods. For this reason (and for other less altruistic reasons) repeated attempts have been made to catch a live coelacanth and take it to an aquarium where it could be studied. French attempts to obtain a live coelacanth in the 1950s and 1960s did not come to fruition although observations were made on dying coelacanths by Millot (1955b) and later by Griffith & Locket (1972), Griffith & Thomson (1973b) and Attenborough (1978). In 1969 Margaret Smith attended a meeting with the Director of the Steinhart Aquarium at the California Academy of Sciences to discuss plans to catch and display a live coelacanth (archives of the J.L.B. Smith Institute of Ichthyology). Later in the same year Alan Heydorn, then Director of the Oceanographic Research Institute in Durban, South Africa, announced plans to capture a live coelacanth and display it in Durban (Anon 1964, Anon 1970). Neither plan was successful.

In recent years efforts to catch and display a live coelacanth have been intensified. Expeditions by the Explorers Club and the New York Aquarium in 1986 drew both criticism (Browne 1988) and support (Hamlin 1988). The Japanese Scientific Expedition of the Coelacanth (JASEC) attempted to catch live coelacanths off Grand Comoro in 1981/2 and 1983, but without success (Suzuki et al. 1985). A further Japanese expedition to Grand Comoro in late 1989 mounted a determined attempt to catch a live coelacanth using a generous budget provided by the Mitsubishi Corporation, but also failed in its efforts.

Meanwhile research on the coelacanth was re-initiated by the J.L.B. Smith Institute of Ichthyology in Grahamstown. Alarmed by reports that the coelacanth was threatened with extinction, the Institute mounted a series of expeditions to the Comoros in April 1986, March and November 1987 and May 1990. The aim of these expeditions was to determine the conservation status of the coelacanth and the socio-economic and biological factors which threaten its survival. A series of recommendations was made to the Comoran authorities on the implementation of marine conservation measures, including the establishment of marine reserves. The findings of these expeditions were published in scientific papers and reports (Heemstra & Smale 1986, Balon et al. 1988, Bruton 1989a, Bruton et al. 1989, Stobbs 1987, 1989b), in popular articles (e.g. Bruton 1987, 1988a, 1989a, b, Stobbs 1989a, Balon 1990a, b), and in a video which has been shown throughout the world. In addition, Stobbs (1987, 1989a) and Stobbs & Bruton (1991) examined the artisanal fishery in the Comoros and recommended ways in which this important cultural heritage could be preserved.

One of the most significant recent developments in coelacanth conservation has been the initiation of research by Hans Fricke of the Max-Planck-Institut für Verhaltensphysiologie on the behaviour and abundance of the living coelacanth using a research submersible. Fricke has been able to penetrate into the natural habitat of the coelacanth for the first time and has provided invaluable information, inter alia, on their habitat requirements (Fricke & Plante 1988, Fricke & Hissmann 1990), relative abundance and conservation status (Fricke & Hissmann 1990, Fricke et al. 1991). Fricke's popular articles (e.g. Fricke 1988) and his videos have also rekindled the public's interest in coelacanths and their conservation.

In March 1987 the members of Fricke's expedition from Germany (H. Fricke and R. Plante) and the members of Bruton's expedition from South Africa and Canada (E.K. Balon, M.N. Bruton, C. Flegler-Balon) met in a small restaurant in Moroni on Grand Comoro to discuss the conservation status of the coelacanth. Evidence from their different investigations had revealed that increased incentives were being provided by foreign organisations for the capture and illegal trade of coelacanths. In addition, Fricke's initial diving observations had indicated that coelacanths are not abundant in the Comoros. The group therefore resolved to establish an international organisation to coordinate efforts to study and conserve the coelacanth. The Coelacanth Conservation Council (CCC) was formally announced at a conference in Grahamstown, South Africa, in July 1987 (Bruton 1988a, b, c). The CCC has its headquarters in Moroni and the Secretariat is currently in Grahamstown with agencies in Canada, the United Kingdom, the U.S.A., Germany and Japan. In addition to its general objectives of promoting coelacanth research and conservation, the CCC has initiated the establishment of an international registry of coelacanth researchers and the compilation of a coelacanth inventory (Bruton & Coutouvidis 1991) and bibliography (Bruton et al. 1991), which are published for the first time in this volume.

A review of the natural history and conservation status of the coelacanth published on the fiftieth anniversary of its discovery (Balon et al. 1988) provided information on the trade in coelacanths and made a number of recommendations on their conservation. Other anniversary publications (Forey 1988, 1989, Fricke & Plante 1988, Bruton 1989a, Greenwood 1989, van Bruggen 1989, Balon 1990a, b) also called for stricter conservation measures.

The current status of the coelacanth

Coelacanth capture records: 1938–1990

The inventory of coelacanth catches lists 172 specimens caught between 1938 and 1990 (Bruton &

Coutouvidis 1991). In addition, an unknown number of coelacanths has been traded illegally. The average annual catch rate since 1952 is 4.39 specimens with the number caught per year ranging from 1 in the early 1950s to 11 in 1986.

It is imperative for the coelacanth conservation effort that this detailed inventory of catches continues to be kept and that those specimens that are caught are made available for scientific study. At present, coelacanth catches in Grand Comoro are documented by M. de San, Directeur du Projet Pêche Artisanale, and the specimens are kept in the SOCOVIA freezer in Moroni. Coelacanths caught on Anjouan are recorded by Abdallah Massonde in Mutsamudu and the specimens are kept in the freezer at Le Centre de Formation des Pêcheurs, Ecole Nationale de la Pêche, near Mutsamudu.

The coelacanth was placed on Appendix II of CITES (Convention on International Trade in Endangered Species of Wild Fauna and Flora) in 1976 to provide it with partial protection. In 1987 a motion was proposed to have the coelacanth deleted from Appendix II on the basis that there was insufficient evidence that trade threatened its survival (CITES 1987), but this motion was defeated. The founder members of the CCC campaigned successfully in 1988 for *L. chalumnae* to be upgraded from Appendix II to Appendix I (Balon et al. 1988, Bruton 1988c, Fricke 1988) so as to afford full protection to the species. This decision was made by the CITES assembly in October 1989 and came into effect in January 1990, and requires that all signatories to CITES should refrain from trading commercially in coelacanths. In addition, all transfers of coelacanth specimens for bona fide scientific study should have full documentation from both the exporting and the importing countries, and all transfers of coelacanths should be reported to CITES.

In September 1989, in anticipation of the CITES decision, the then President of the Comoros, the late Ahmed Abdallah, decreed that all trade in the coelacanth should be strictly controlled. This policy has been continued by the current President, Ahmed Johar. However, the Comoros is not as yet a signatory to CITES and is therefore not obliged to heed its recommendations.

Previously Comoran fishermen were paid generously by the government for coelacanths, the amount ranging from CFA 1400 to 50 000 (50 CFA = 1 French Franc). Coelacanths were sold to foreign researchers for about CFA 50 000 to 60 000. The prices reported for various private deals between fishermen and scientists range from five shillings! (Greenwood 1988) to US $280 (Dugan 1955) and Canadian $200 (Newman 1971). At least two research groups (Newman 1971, McCosker & McCosker 1976) even offered a fisherman who caught a live, healthy coelacanth an all-expenses-paid trip to Mecca. The price paid to Sidi Bakari, the official government taxidermist in Mutsamudu, for a preserved and mounted coelacanth is CFA 50 000 (S. Bakari personal communication May 1990).

The CCC also campaigned for *L. chalumnae* to be upgraded from category 'K' in the International Red List of Threatened Species (IUCN 1986) (meaning that insufficient is known for the species to be properly categorised) to 'V' (vulnerable to extinction if the present factors threatening its existence continue to operate). This campaign has also been successful (A. Collares Pereira personal communication July 1990).

As far as the international community is concerned, the coelacanth therefore has adequate protection in terms of conventions and lists, but the challenge is to translate these abstract concepts into meaningful actions in the field. Before discussing various conservation measures that have been proposed, it is necessary first to consider the socio-economic and biological factors that threaten the survival of the coelacanth.

Factors threatening the survival of the coelacanth

Socio-economic factors

Socio-economic factors threatening the survival of the coelacanth are summarised in Figure 8 and discussed below:

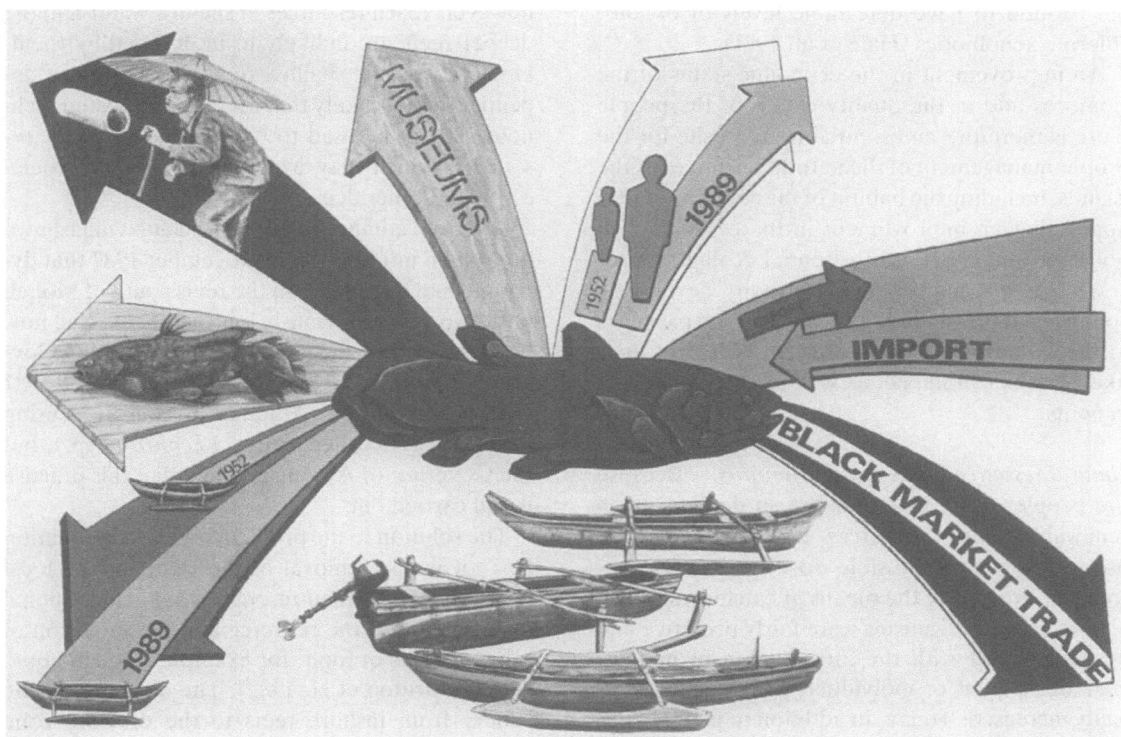

Fig. 8. Diagrammatic representation of some socio-economic factors that may threaten the survival of the coelacanth (clockwise from top right): the rapidly expanding human population in the Comoros, a large trade imbalance and weak economy, illegal trade in the coelacanth, increased fishing efficiency, increased fishing effort, poor methods of preservation and taxidermy which render specimens that are caught as useless for further study, a rumour among some Far Eastern people that the fluid from the notochord of the coelacanth prolongs human life, and a trade in coelacanths to museums and other research institutions. Drawing by D.P. Voorvelt.

Economic status of the Comoros. – The Federal Islamic Republic of the Comoros is one of the poorest countries in the world. In 1980 the value of imports exceeded that of exports by 295% (Gaspart 1983, Newitt 1984). The gross national product was about US $120 million in 1982, which was equivalent to about US $285 per capita at the time (Anon 1987). The real growth rate of the economy in the period 1973 to 1982 was 0.0%. Only 40% of the population is economically active and virtually all manufactured goods as well as large quantities of food, fuel and building materials have to be imported (personal communications with Comoran officials).

These factors, combined with the rapidly in-creasing human population (Newitt 1984), mean that many Comoran people are forced to satisfy their short term survival needs and cannot afford the 'luxury' of long term planning in the use of renewable natural resources. The problems associated with this short term view are already obvious in the reduced ability of many terrestrial and aquatic ecosystems in the Comoros to sustain present levels of utilisation. For instance, the collection of coral for making lime is likely to have damaged the life-support system of many reef-dwelling animals, as have the use of poisons and explosives for catching fish. The increased use of herbicides and pesticides on land is likely to have a cumulative effect in the sea, where coelacanth tissues have already been found to have detectable levels of organo-chlorine xenobiotics (Hale et al. 1991).

An improvement in the economic status of the Comoros and in the quality of life of the people there is therefore an essential prerequisite for the proper management of the natural resources of the islands, including the habitat of the coelacanth. It is hoped that an improvement in the economy will not be at the expense of essential ecological processes, as has happened in so many developing countries. It is essential that economic aid packages to the Comoros are therefore scrutinised for their likely ecological impact as well as their economic benefits.

Natural resource needs in the Comoros. – Because the people of the Comoros are so dependent on renewable natural resources, they have tended to use every means possible to exploit these resources. Previously the means of catching and collecting marine organisms were fairly primitive and inefficient, but with the introduction of modern gear the impact of individuals has been dramatically increased. Today, in addition to paddled dugouts, they have available outboard and inboard powered 'fibreglass' boats as well as nylon lines, steel hooks, gill nets, seine nets, diving equipment and spearguns. The impact of fishing activities on coelacanth survival is considered separately by Stobbs & Bruton (1991).

In May 1990 one of us (M.N.B.) observed young Comoran spearfishermen operating off an outrigger canoe opposite the site of the old Maloudja Hotel on the north shore of Grand Comoro. During one hour the two individuals shot at least 40 fish with their two spearguns on the fringing reef at a depth of 5 to 10 m. The fishes killed included chaetodons, *Acanthurus* sp., small rockcods, cardinalfish, soldierfish and lizardfish. Another worrying development is the use of small seine nets to catch intertidal organisms at low tide. Teams of women were seen to collect lizardfish, gobies, soles, chaetodons and pipefish in the intertidal zone in front of the Al Amal Hotel on the north shore of Anjouan in May 1990, and this practise is prevalent on all the islands. Divers and intertidal foragers can only,

however, reach resources in shallow water and the deeper reefs are unlikely to be harmed by them. However, as the shallow reefs become more depauperate, it is likely that more sophisticated technologies will be used to tap the deeper water resources, which may eventually affect the coelacanth and other deep reef species.

A retired administrative official at Nyumashuwa on Moheli informed us in November 1987 that dynamite had been used on the reefs south of Moheli until two years previously but that its use had now been stopped by the gendarme. The Deputy Governor of Moheli, Ali Said, informed us on 6.11.1987 that fishermen on Anjouan poison fish using crushed leaves called 'uruva' (*Tephrosia* sp.), but the Governor of Anjouan denied that this practise is still carried out.

The solution to the problem of over-exploitation lies not in the removal of the rights of the local people to exploit their resources, but in the rational management of the resource and the provision of other sources of food, for example through aquaculture (Bruton et al. 1989). The diversion of the fishery from inshore reefs to the offshore zone through the deployment of fish aggregation devices in the open ocean (Stobbs & Bruton 1991) is another classical example of how to resolve an over-exploitation problem without infringing on the traditional rights of the people.

Lack of conservation legislation in the Comoros. – In the Comoros environmental issues are dealt with by the Ministry of Equipment, Environment and Urbanisation and the Ministry of Production, Industry and Art. Agriculture, forestry and fisheries and the management of fauna and flora fall under the latter Ministry through a Central Federal Directorate for Rural Development (SODOPEC). In 1979 a national museum, the Centre National de Documentation et de la Recherche Scientifique (CNDRS), was established in Moroni. The museum has a small research section concerned with biology.

As far as we could ascertain, there is no current legislation in the Comoros concerned with the conservation of living natural resources except in the broadest terms, although coral and sand are apparently protected by presidential edicts. It also appears that the Comoros is not a signatory to any international conservation conventions, such as the Ramsar Convention and the World Heritage Site Convention. There is a presidential decree that turtles should not be killed and that all coelacanths caught must be sold to the state, but these decrees are not always observed. It is therefore apparent that new legislation will have to be introduced in order to provide for the protection and long term sustainable use of marine resources.

On their arrival at the international airport at Hahaya, Grand Comoro, visitors are handed a note that reads as follows:

'Important information
Wishing to protect the natural environment of our country we inform you that:
* Underwater fishing using air tanks is prohibited in the country;
* Underwater fishing with or without air tanks is not authorised in the region of the Noumachea islets and of the southern coast of Moheli Island;
* Collection and sale of (the following) shells are prohibited: *Charonia tritonis, Cypraecassis rufa* and *Cassis cornuta;*
* Sale of sea turtles and their shells is not authorised.'

Despite this injunction, the taxidermist near Mutsamudu routinely receives, and prepares for resale, various turtles, sea shells and coelacanths, and the airport gift shop at Hahaya has all three banned shells for sale.

Lack of a conservational infrastructure in the Comoros. – Even when appropriate conservation legislation is introduced, the Comoros lacks the management infrastructure to enforce the legislation. There are currently no conservation officers in the Islamic Republic.

Vulnerability to longline fishing. – The coelacanth is likely to be vulnerable to commercial multi-hook longlining. Deepset longlines, whether deployed on the sea floor or a short distance above the sub-

strate, have the potential to catch large numbers of coelacanths, as well as other demersal fishes, and the use of such fishing gear on the inshore slopes of any of the Comoros islands should not be allowed.

Neither the fishing school founded by the Japanese on Anjouan nor the Projet FED on Grand Comoro encourage commercial multi-hook longline fishing. There seems to be no justification for deepset longlines in the Comoran context since more than adequate catches of pelagic fishes are possible at the fish aggregation devices and using the trolling capabilities of japawa and fedawa boats (Stobbs & Bruton 1991).

Illegal trade. – During the period of French authority in the Comoros the trade in coelacanth specimens was strictly controlled. After the Comoros became independent in 1975, and especially during the Soilih regime from 1976 to 1978, an unknown number of coelacanths changed hands illegally without any official record being kept. In addition, most of the official files in the Comoros were destroyed during the Soilih regime.

Until October 1989 (when the CITES decision was made), coelacanths were freely available to anyone who had sufficient money to buy a specimen. All coelacanths caught were required, by Presidential edict, to be sold to the government, but there was no restriction on their resale by the government. Coelacanths were advertised for sale and sold to biologists, medical researchers, anglers and tourists as well as donated to dignitaries. Each legally sold coelacanth was usually issued with an export permit, but the other requirements for a CITES Appendix II species, e.g. that a scientific authority may advise the exporting state if there is evidence that the species may be threatened, were not observed. Some purchasers appear to have acquired their specimens without official permission and documentation and to have exported them illegally, probably to bypass complicated bureaucratic procedures. If the importing country is a CITES signatory, then the authorities in that country are at fault for allowing the specimen to be imported without a permit.

Since the reclassification of *L. chalumnae* on

Appendix I, trade in the coelacanth has been banned, and there is now a threat that an illegal trade may again develop, especially considering the economically depressed state of the Comoros and the continuing interest of museums and aquaria in obtaining specimens.

Increased incentives to land coelacanths. – Before the rescheduling of *L. chalumnae* on CITES, several museums and aquaria offered incentives to fishermen to catch coelacanths (see above). It is now illegal to offer such incentives, and it is hoped that this threat will no longer have to be considered.

Scientific trade in coelacanths. – In terms of Appendix I of CITES, coelacanths may still be donated or sold to countries for bona fide scientific research, but each application should be screened by a scientific authority. A proposal that the Coelacanth Conservation Council should be the screening body has been made to the Comoran government. While there is considerable merit in attempting to capture and translocate a live coelacanth to an aquarium in a western country for research, such an action should not be embarked upon lightly. It is extremely important that the coelacanth should be de-commercialised. Competitive attempts by rival aquaria to obtain the first live specimen will not be in the interest of coelacanth conservation.

Increased fishing effort. – Stobbs & Bruton (1991) have outlined the changes in the Comoran fishery in recent years. During the period 1962 to 1987 the total number of paddled outrigger canoes (galawas) increased from 1409 (Boulinier-Giraud 1973) to over 4000 (Stobbs 1987). Since 1986 a number of more modern craft have been introduced into the fishery through Japanese and European Economic Community aid programmes, including outboard and inboard motor driven 'fibreglass' boats (japawas and fedawas, Stobbs & Bruton 1991). During the past three years many of the traditional fishing practices have fallen away and large numbers of traditional canoes are no longer in use. As the traditional canoes are used for fishing over inshore reefs whereas the more modern vessels are primarily intended for pelagic fishing, these recent developments may relieve pressure on inshore reefs, and therefore on the coelacanth and its prey. The main danger to the biota of the inshore reefs would therefore be the use of motorised vessels for shallow water fishing, which clearly needs to be avoided.

Increased fishing efficiency. – The more efficient motor-driven boats that have been introduced recently into the Comoran fishery do not constitute a direct threat to the coelacanth. The coelacanth would, however, be severely threatened if their greater seaworthiness, larger size, larger crews and more efficient tackle were applied to *Ruvettus* fishing, with the likelihood of catching coelacanths as a bycatch.

Loss of specimens for study. – Until recently, most of the coelacanths caught off Anjouan were sent to a taxidermist, S. Bakari, near Mutsamudu for preservation and preparation. Bakari removes and discards the soft anatomy (including eggs), supports the specimen internally with wooden struts and coconut fibre, and fits a '*Ruvettus*' eye. The specimen is then allowed to dry out, after which it is virtually useless for scientific study. It is essential that all specimens caught are made available for study. For this reason the Comoran government's present policy is to deepfreeze all specimens.

Elixir. – There is a rumour among some Far Eastern people that the fluid in the notochord of the coelacanth acts as a life-prolonging elixir (Balon et al. 1988, Stobbs 1989b). This dangerous rumour needs to be dispelled as it could threaten the coelacanth in the same way as the alleged aphrodisiac properties of the rhinoceros horn have threatened the rhinoceros.

Increased trade to museums. – Because of the importance of the coelacanth in biology there is still a considerable demand for specimens for display in museums. More recently the demand has extended to the acquisition of live coelacanths for display in aquaria. While it is important that those specimens

Fig. 9. Diagrammatic representation of some biological factors that make the coelacanth vulnerable to extinction (clockwise from top right): narrow geographical range, narrow habitat preferences, height in the food chain, slow breeding rate, and membership of a complex community with possible mutualisms and commensalisms. Drawing by D.P. Voorvelt.

that are accidentally caught should be fully utilized, it is not desirable for the museum trade to be an incentive to catch or land more coelacanths.

Biological factors

The biological factors that make the coelacanth vulnerable to extinction are summarised in Figure 9 and discussed below:

Narrow geographical range. – Except for the first specimen, which was caught off South Africa, all known specimens of the coelacanth have been caught off Grand Comoro and Anjouan in the Comoros (Bruton & Coutouvidis 1991). The coelacanth therefore has a very narrow known range,

although it may occur off Madagascar or the mainland African coast, or elsewhere. The narrow known range means that the species is vulnerable to habitat degradation in the Comoros, which is a real possibility considering the economic plight of that country. Conservation actions therefore need to be directed at managing the coelacanth population in the Comoros as well as determining whether the species occurs elsewhere. The limited geographical range of *L. chalumnae* is ascribed by McCosker (1979) to the reduced vagility which accompanies its livebearing reproductive mode.

Narrow habitat preferences. – Even within the Comoros the coelacanth has narrow known habitat preferences. The depth range from which catches have been made ranges from <100 to >1000 m

(Fig. 4) with 86% of the specimens recorded between 100 m and 400 m. Fricke & Hissmann (1990) and Fricke et al. (1991) found that coelacanths prefer heterogenous, rocky habitats in which caves and overhangs are found (as occur off the west coast of Grand Comoro) rather than more homogenous, sandy substrata (as on the east coast of Grand Comoro and the shores of Moheli). The major fishing effort in the Comoros is exerted on rocky shores (Stobbs & Bruton 1991).

Height in the food chain. – The coelacanth is a predator of fishes and squid (Uyeno & Tsutsumi 1991) and is thus vulnerable to decreases in the stocks or availability of these species or their prey. The data available suggest, however, that the coelacanth is euryphagous and may switch its feeding to whichever prey is most abundant.

Slow breeding rate. – The slow breeding rate of the coelacanth may cause it to be particularly vulnerable to density-independent mortality risks with which it did not co-evolve, such as overfishing. The removal of female fish from the population before they breed would have a particularly deleterious effect (Bruton & Armstrong 1991).

Membership of a complex community. – The coelacanth is a member of a highly complex community and may be involved in commensalisms and/or mutualisms which would be disrupted by overfishing or habitat degradation within its range.

Vulnerability to longline fishing. – The coelacanth, being a large-mouthed piscivore, may be particularly vulnerable to baited longlines or handlines. Furthermore, *L. chalumnae* occurs within a similar depth range to the main target species of the Comoran night handline fishery, the oilfish *Ruvettus pretiosus,* and is of a similar size and is therefore likely to take bait from the same size hook.

Vulnerability to changed water temperatures and pressures on capture. – Many of the coelacanths caught by artisanal fishermen have remained alive for several hours after capture but eventually succumb to physiological stresses and die. Their death has been ascribed to several causes, such as hypoxia (Hughes 1976, 1980), temperature-related respiratory problems or the effects of decompression (Locket 1976, Balon et al. 1988, Fricke & Plante 1988, Bruton 1989a). Hughes (1980) suggests that death is due to hypoxic stresses which may lead to acidiotic conditions in the blood and tissue that would affect the dissociation characteristics of the blood (Hughes & Itasawa 1972) so as to reduce the efficiency with which oxygen is transported from the gills to the tissues. Temperature stresses at the surface, where water temperatures may be 24°C (8 to 10°C higher than in the habitat of the coelacanth, McCosker 1979), would aggravate the situation by moving the dissociation curve further to the right, as the oxygen affinity of coelacanth haemoglobin achieves its maximum level between 15 and 20°C (Hughes & Itasawa 1972, Hughes 1976). Respiratory stress would be exacerbated by an increased metabolic rate at the elevated temperatures. Furthermore, Hughes (1976) found that the coelacanth cannot significantly increase its oxygen absorption rate since their gill area/body mass ratio was an order of magnitude lower than that of any other fish examined at that time.

McCosker (1979) and Fricke & Plante (1988) report that coelacanths may make vertical migrations into shallower water where prey organisms are more abundant. These vertical migrations may extend through temperature ranges of 6 to 7°C, possibly as wide as 10°C (Forey 1984), and pressure ranges of 15 atmospheres. It is also known that the coelacanth has no gas-filled swimbladder; its almost neutral buoyancy is maintained by quantities of low density wax-ester lipids in its tissues, including the swimbladder (Spark 1982). Rapid decompression, such as would be experienced by a coelacanth caught on a hook and pulled to the surface by an artisanal fisherman, would thus not result in the usual effects of physical barotrauma where these are due to gas expansion. Death is more likely to be due to one of several other side-effects of decompression and/or muscular exertion, such as hypoxia, lowered pH of the blood or intracellular tissues and consequent acidosis, a Bohr

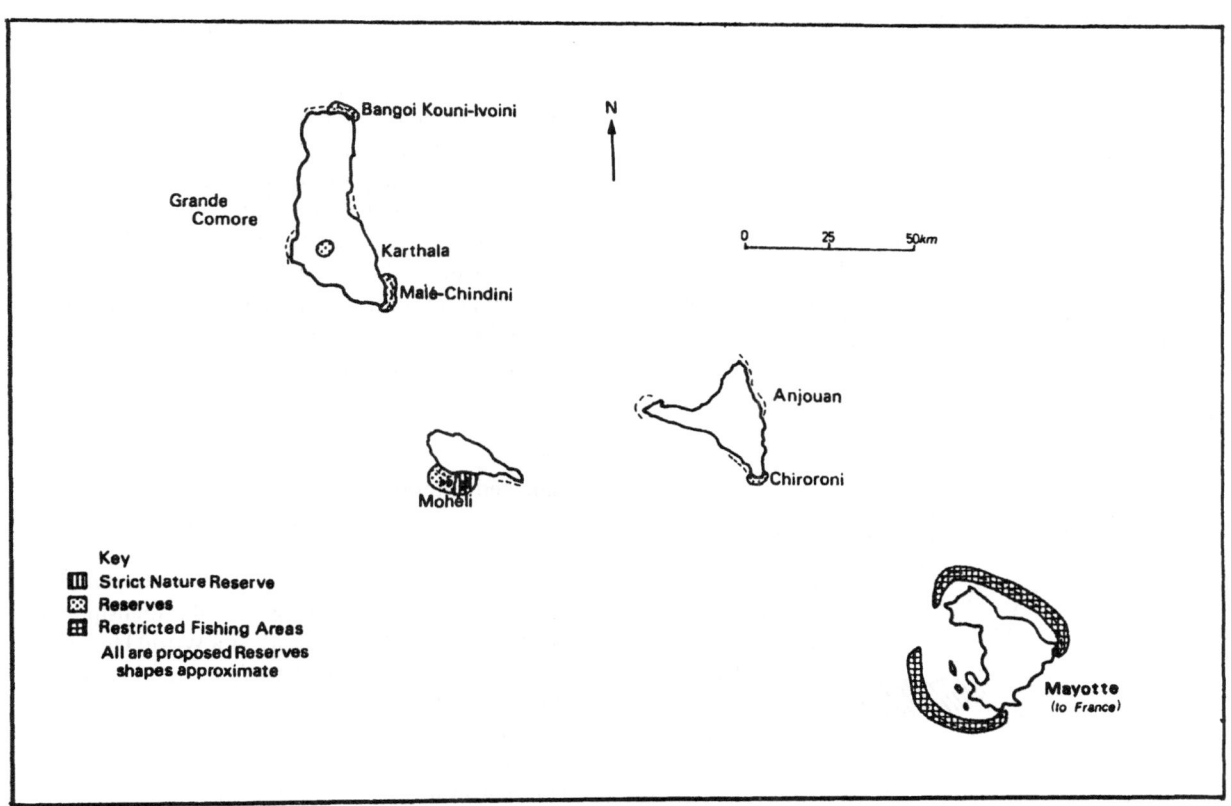

Fig. 10. The marine reserves in the Comoros proposed by IUCN/UNEP (1987).

shift, lactic acid accumulation, and/or stress-induced disruption of the electrolyte balance in the blood (hyperkalaemia).

Proposed conservation measures

A variety of conservation measures have been proposed for the coelacanth taking into account the above threats:

Decompression experiments. – As the majority of coelacanths that have been caught, and are likely to be caught in future, are accidental bycatches, there is merit in attempting to return these specimens to their natural habitat in an undamaged state. At least three attempts at recompression have been

made, but all failed (J.-L. Geraud and J. Hamlin personal communications 1988, 1989, 1990). Coelacanths that have been brought to the surface have suffered from a variety of decompression stresses (see above).

Catch limits. – All coelacanths caught in the Comoros have been landed as an accidental bycatch of nighttime *Ruvettus* fishing. It would thus be impractical to introduce a catch limit on coelacanths, whether by size or number. It would also be difficult to attempt any regulation of the deep hand-lining fishery for *Ruvettus* unless suitable alternatives could be found (see Stobbs 1989b and below). The popularity of the fish aggregation devices deployed in the Comoros by Projet FED could contribute significantly towards providing for the nu-

332

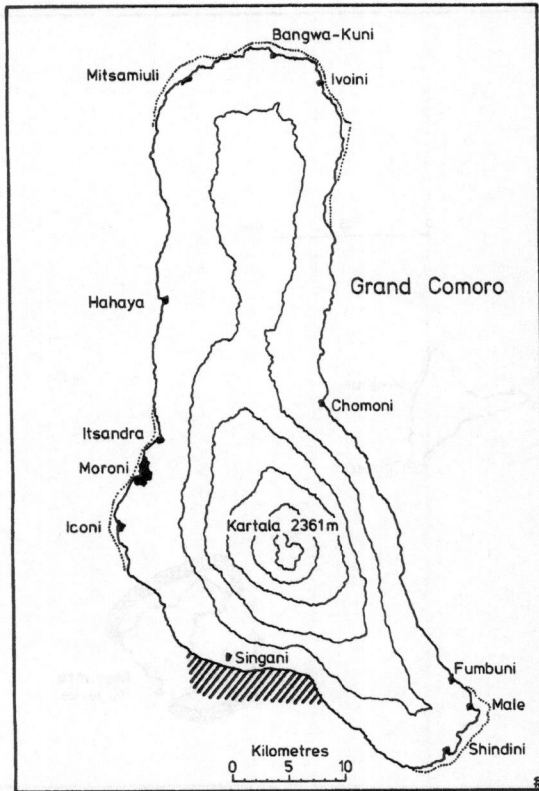

Fig. 11. The proposed Coelacanth National Park (shaded) along the south shore of Grand Comoro. After Bruton et al. (1989).

tritional requirements of subsistence fishermen and their families, but it must not be overlooked that *Ruvettus,* and to a lesser extent *Latimeria,* have also been used traditionally for medicinal purposes.

Marine reserves. – The establishment of marine reserves in the Comoros has a strong likelihood of success as the Comoran authorities are strongly in favour of this conservation measure (A. Mroudjae personal communication 1990). Several proposals for marine reserves in the Comoros have been made (e.g. H.A. Kay personal communication 1986, IUCN/UNEP 1987, Bruton et al. 1989, Dahl 1989) but none has as yet been implemented due to lack of funds. Dahl (1989), in an appendix to his report, presents draft legislation for the establish-

ment of a marine reserve to the south of Moheli. In the 'IUCN Directory of Afrotropical Protected Areas' (IUCN/UNEP 1987) it is stated that 'SODO-PEC intends to create marine parks: at Moheli on the islands of Nisoumachona (5 ha) (a strict nature reserve with fishing prohibited and only open to scientists, surrounded by a buffer zone where artisanal fishing is permitted); marine areas between Bangwa Kuni and Ivoini, and Shindini and Male (in the north and south of Grand Comoro respectively); and Chiroroni on Anjouan' (Fig. 10).

Detailed proposals for marine conservation in the Comoros are presented by Bruton et al. (1989), including the establishment of several strict nature reserves, marine national parks, resource reserves, managed nature reserves, anthropological reserves, natural landmarks and a world heritage site that were selected according to the criteria of Salm & Clarke (1984). The recommendation that has the most relevance here is the proposed establishment of a Coelacanth National Park as a World Heritage Site along the southern shores of Grand Comoro (Fig. 11). It is recommended that this reserve should extend from near Singani eastwards to a point west of Dembeni, and about 4 km out to sea. The reason for the establishment of this National Park is that it is one of the few known habitats of the coelacanth where high densities of these fishes have been found in an area that is lightly fished by artisanal fishermen. The protection of the site would therefore not unduly interfere with traditional fishing activities.

The Coelacanth National Park would be managed as a strict nature reserve (sensu Salm & Clarke 1984) with no public access and no removal of fauna and flora except for pelagic fishing by locally based artisanal fishermen using paddled galawas. It would, however, be useful to retain local fishermen in the area so that they could become involved in policing the reserve. No motorised boats should be allowed into the area except for monitoring or research purposes. An essential aspect of the management of this Park would be to obtain the support of the local people so that the reserve becomes a source of pride to them. A small educational centre could be established at Singani for this purpose, and research in the Park should be encouraged. Consideration should also be given to establishing a buffer zone around the proposed Park where fishing levels are controlled by only allowing traditional fishing methods and prohibiting commercial or recreational fishing.

Modern conservation theory emphasizes that nature reserves are useless if they are preserved enclaves in a broader area of severely degraded habitat. It is essential therefore that the shoreward boundary of the reserve is also managed to ensure that soil erosion and pollution originating from outside the reserve do not interfere with essential ecological processes within the reserve.

A conserved area for coelacanths should also be considered on the west coast of Grand Comoro in the light of Fricke et al.'s (1991) recent findings.

Closed seasons. – As the coelacanth is a livebearer with a gestation period that may approximate 13 months, and apparently has an extended breeding season over the austral summer, it would be impractical to implement a closed season even if coelacanths could be specifically targeted. However, the fact that the coelacanth is a bycatch of the oilfish fishery, and that Comoran fishermen depend on oilfish and other species for their livelihood, makes this conservation measure impractical.

Reduced incentives and trade. – The upgrading of the coelacanth to Appendix I of CITES has theoretically ended legal trade that may detrimentally affect the species' survival. It will, however, be necessary to control the illegal trade and also to ensure that the Comoros and its major trading partners sign the CITES. It is also imperative that all countries that are signatories of CITES should comply with the convention by reporting all trade in coelacanths to the relevant authorities and by abiding by the decisions of the screening committee appointed to approve coelacanth trade for scientific purposes.

Alternative laxatives. – A novel conservation proposal was made by Stobbs (1989b) who noted that the main reason why Comorans catch oilfish (and hence coelacanths) is because of the laxative and apparent anti-malarial properties of the flesh of these fishes. The oilfish is highly valued in subsistence fisheries throughout the Indo-Pacific for its use as a natural purgative. *Ruvettus* oil is also regarded in the Comoros as an effective mosquito repellant, which is a valuable asset as over 90% of the population suffers from malaria.

Oilfish lipid is predominantly an indigestible wax ester that is present in muscle tissue to the extent of 30–71% dry mass (Nevenzel et al. 1966, Spark & de Wit 1981). *Latimeria* muscle lipid is composed of wax esters of similar composition to those of *Ruvettus* (Nevenzel et al. 1966). Because *Latimeria* is caught as a bycatch of the *Ruvettus* fishery, and *Ruvettus* is primarily valued for its laxative qualities, it would seem that the availability of an alternative laxative might lead to reduced fishing pressure on the coelacanth, as the need to fish at night might then be diminished.

Almost without exception fishermen on Grand Comoro interviewed by Stobbs (1989b) indicated their willingness to abandon the dangerous and tiring operation of nightfishing for *Ruvettus,* if a culturally acceptable substitute for nesa oil could be found. Eventually nightfishing may stop completely as most of the fishes that are caught for food are caught during the day. As coelacanths are almost always caught at night, a switch from night to day fishing may reduce catches without interfering with traditional fishing rights.

A holistic plan for the conservation of the coelacanth. – The conservation of the coelacanth in the Comoros should not be seen as a single species conservation problem that can be solved by high technology modern techniques, such as captive propagation or the establishment of feral colonies elsewhere. The problem should rather be seen in an holistic context in which a heightened awareness is created among the local Comoran population of the need to conserve renewable natural resources for their own good. If the inshore marine ecosystem in the Comoros is properly managed, and high technology gear is not brought to bear on the coelacanth and other inshore reef species, then it will represent a renewable natural resource to the Comoran people in perpetuity.

The Comoros is a classical example of a country whose capacity to support people at an adequate standard of living has been severely reduced in recent years. Clearly there is a need to match short-term goals of feeding and sheltering people with longterm goals of maintaining essential ecological processes and life-support systems. We must accept, however, that conservation will only become a way of life to the Comoran people if they perceive that their economic status and quality of life will improve as a result of conservation actions. Conservation initiatives should therefore be a combination of legislation, enforcement and education, with as much involvement of the local people as possible. Above all, conservation should not be seen as an activity that is mainly for the benefit of foreign tourists.

The most important force in conservation is the development of a conservation ethic in the people whose livelihood is threatened by the depletion of the resources. The Comoran people, through their Islamic beliefs and their close empathy with nature, have an intrinsic conservation ethic, but there is a danger, as human demographic pressures mount, that the realities of survival will overrule the ethics of conservation (especially if the local people are encouraged to carry out irrational acts by immoral financial incentives from abroad).

Against the above background, the following specific recommendations are made on the conservation of the coelacanth:

(1) A Coelacanth National Park should be established along the south shore of Grand Comoro as a component of the Comoran marine reserves plan.
(2) Nature conservation legislation which protects marine resources outside the formally proclaimed marine reserves should also be promulgated and enforced, not only to protect the coelacanth but also to protect other marine life so that a variety of species and ecosystems can continue to sustain the natural resource-based economy of the Comoros.
(3) The incentive to catch or trade in the coelacanth should be removed entirely. The capture of live coelacanths should be for scientific purposes only and all attempts at live capture should be approved by a scientific advisory body such as the Coelacanth Conservation Council. Consideration should be given to the establishment of a suitable aquarium in the Comoros for initial studies on live coelacanths so as to reduce the risk of mortality due to the stress of transport, and to ensure that the Comoros continues to play a central role in coelacanth research.
(4) Those specimens that are caught incidently should be properly curated and preserved by qualified people and should be made available for study by the scientists who are best able to carry out the necessary research. Both frozen as well as chemically preserved material is needed for a variety of histological and cytological investigations.
(5) An educational campaign should be carried out in the Comoros to inform the local populace of the importance of the coelacanth both to their nation and to mankind as a whole.
(6) Good quality fibre-reinforced plastic replicas of the coelacanth (such as those currently sold by the East London Museum in South Africa) should be made available for display in museums as an alternative to real specimens.
(7) Further research should be carried out to establish whether a coelacanth that has been caught and brought to the surface can be released and survive recompression to 200 m. If this is the case, an incentive should be provided to fishermen to release coelacanths that are caught.
(8) Further research should be carried out from manned and remotely-controlled research submersibles on the habitat preferences, abundance, demography and conservation status of juvenile and adult coelacanths. This research should also be extended to other western Indian Ocean islands and to the coast of mainland Africa. A fulltime researcher should be placed in the Comoros to monitor catches and facilitate the establishment of the marine reserves.
(9) The oilfish fishery should be studied further in order to devise ways in which oilfish can be caught without catching coelacanths. Possible substitutes for oilfish should also be sought as this species seems to be caught mainly for its purgative and anti-malarial properties rather than for food.
(10) The offshore fishery should be further developed in order to reduce fishing pressure on the over-exploited inshore reefs. Protein production could also be increased through fishfarming in

ponds on land. The hot, wet climate of the Comoros, combined with the abundance of inexpensive
labour and a ready market for fish products, would
suggest that the culture of African tilapias and catfishes could be successfully undertaken.

(11) The small socio-economic units of the rural
fishing villages should be retained as far as possible
as they provide stability and a sustained income in a
volatile society. The threats posed by urbanisation
and its concomitant problems of poverty, disease,
loss of traditions and the breakup of family units,
need to be recognised and countered without
interfering with the opportunities for the Comoran
people to reach their full potential. The dangers of
replacing labour-intensive, dispersed fisheries with
capital-intensive, localised fisheries also need to be
recognised.

(12) The international community needs to assist
the Comoros to realise the full potential of its rich
natural resources.

Conclusion

The early reconstructions of the natural history of
the coelacanth by Millot (1954b) and Smith (1956a)
were remarkably accurate considering that they
only had dead material to work on. Although
McCosker (1979) wrote 'It is patently presump-
tuous to describe the natural history and behavior
of a living fish which (. . .) has never been observed
in a healthy living condition', he also made a significant contribution to coelacanth ecology. The writ-
ings of K.S. Thomson, E.K. Balon, R.W. Griffith,
H.-P. Schultze, H. Fricke, P.L. Forey, D.E. McAllister, T. Uyeno, N. Suzuki and many others have
all elucidated coelacanth ecology from different
angles. The recent observations on coelacanths
from research submersibles by Hans Fricke and his
colleagues represent a most significant advance
which allows us to comment for the first time on the
behaviour and interspecies relationships of living
coelacanths in their natural habitat.

The coelacanth is now hypothesized to be a long-
lived, slow growing, epibenthic drift-predator inhabiting deep rocky reefs. They have a slow metabolic rate and feed singly on reef-associated fishes and

Fig. 12. Diagrammatic representation of the putative breeding
cycle of the coelacanth. After internal fertilisation has occurred,
a few large eggs develop in the female which embryos hatch and
develop into large yolksac juveniles that feed endogenously
from the yolk, as well as exogenously on yolk debris and placental transfer (according to Balon 1991). The juvenile has resorbed
its yolksac when it is born at a length of about 32 cm TL.
Drawing by D.P. Voorvelt.

squids, possibly during diel migrations into shallower water at night. The coelacanth may return to
deeper water during the day where it aggregates in
caves, possibly to escape predation from sharks
and/or to seek refuge from the strong currents that
are characteristic of its habitat. There is evidence
that coelacanths may return repeatedly to the same
caves and that they have a home range. When in
the caves they hover in midwater, rarely touching
one another or the substrate, although the possibility of pheromonal or electrosensitive communication cannot be ruled out. Prey species are ap-
proached slowly and may then be ingested by a combination of a rapid forward burst and negative water
pressure created by the suddenly opened mouth.

Coelacanths are social animals and may have a
prolonged courtship. They are livebearers which
produce a few, very large young that are well-
endowed with yolk and may even feed exogenously
before birth (Balon 1991). The young develop directly from embryos into the definitive phenotype

336

(juvenile) without an intervening larval period. Although the young are large and capable predators when they are born, only a few can be produced in each breeding season. The newly-born young are likely to be predators of small fishes and invertebrates and may seek refuge from their many predators in small caves, or with the parent/s or other adults in large caves. Adult coelacanths are probably preyed on by large sharks.

Coelacanths thus have many of the characteristics of precocial animals – few, large eggs, dense yolk, no larvae, large size at first feeding, high parental investment per individual young, high individual fitness of young, and a low reproductive rate. They can also be surmised to be highly specialised members of a diverse community that are mainly subject to density-independent mortality, and would be vulnerable to man-induced increases in mortality rates.

Coelacanths are first recruited into the *Ruvettus* fishery as a bycatch at about 6 years and 80 cm TL, but the number caught per year by artisanal fishermen is probably insignificant. The coelacanth nevertheless has many of the attributes which Norton (1986) regards as typical of animals that are threatened with extinction: rarity, large individual size, height in the trophic pyramid, biotically controlled evolution, low dispersibility, few offspring, greater individual longevity, specialised nature, and habitation of ecosystems with high biodiversity. Such animals have typically evolved in response to pressures resulting from the complex interactions between species (and are therefore more dependent for their survival on the continued existence of these other species) rather than in response to abiotic factors. The loss of some species from such an ecosystem is therefore likely to have a cascading effect on the other species. It is for this reason that indirect impacts on the coelacanth, for instance overfishing of their wide array of prey species, may have a greater impact on their survival than the minimal direct impact of catches by artisanal fishermen (Bruton & Armstrong 1991).

The lower depth limit of the coelacanth is still unknown and may be greater than present findings indicate. Likewise, the coelacanth may be widespread along the eastern shores of Madagascar, or along the coasts of northern Zululand or the southeast coast of South Africa where deep rocky reefs are found and the southbound Mozambique current flows close to the coast. The possibility that coelacanths have an even wider range can also not be dismissed as the Indian Ocean has been poorly explored using appropriate fishing gear.

Despite its ancient origin, the coelacanth is a highly specialised animal that has found solutions to evolutionary and ecological problems that are different from those of any other animal. Many aspects of their morphology, physiology and behaviour are unique or have tetrapod affinities, and it is likely that many surprises await future coelacanth researchers. It could be argued that the coelacanth has become overspecialised and may therefore not be able to compensate for changes that are occurring in its habitat through the intervention of man, such as increased mortality rates and decreased abundance of prey. If this is the case, the present generation has an awesome responsibility to ensure that this remarkable animal, which has provided such a unique window into the past, persists as a living component of our biota.

Acknowledgements

We are grateful to the Minister of the Interior in the Comoros, M. Ali Mroudjae, as well as the Director of the Cabinet, Abderemane Mohamed-Sidi, for their assistance and generosity during our May 1990 expedition to the Comoros. We are also grateful to Damir ben Ali and Abdu Shakur Aboud of the Centre National de Documentation et de la Recherche Scientifique in Moroni, and Jean-Louis Geraud, for their continued help, and to E.K. Balon, H. Fricke and J. Schauer for fruitful discussions. Our May 1990 expedition to the Comoros was kindly sponsored by the United Building Society in South Africa. We are grateful to the other members of our expedition (E.K. Balon, R. Cloutier and R. Reynolds) for their assistance in the field. We are also grateful to Sheila Coutouvidis and Jean Pote for background research and help with the manuscript, and to J.A. Musick, E.K. Balon and P.H. Greenwood for comments on the manuscript. Margaret Crampton and Eve Cambray assisted with literature retrieval. Research on this project was supported by grants from the Foundation for Research Development, the Southern African Nature Foundation and the J.L.B. Smith Institute of Ichthyology.

References cited

Anon. 1964. Coelacanth gets a blue label. Peabody Museum News 1: 3–6.

Anon. 1970. No holds barred in quest for elusive coelacanths. African Aquarist 3: 13.

Anon. 1976. So the coelacanth does bear live young. Nature 259: 81–82.

Anon. 1987. The Comores. The South African Connexion 2: 46–47.

Anthony, J. 1976. Operation coelacanthe. Arthaud, Paris. 201 pp.

Anthony, J. & J. Millot. 1972. Premiere capture d'une femelle de Coelacanthe en stat de maturite sexuelle. C.R. Acad. Sci. Paris, Sér. D 224: 1925–1927.

Attenborough, D. 1978. Life on Earth. Collins, London. 226 pp.

Atz, J.W. 1976. *Latimeria* babies are born, not hatched. Underwater Naturalist 9(4): 4–7.

Balon, E.K. 1990a. The living coelacanth endangered: a personalized tale. Tropical Fish Hobbyist 38: 117–129.

Balon, E.K. 1990b. Tracking the coelacanth: a follow-up tale. Tropical Fish Hobbyist 38: 122–131.

Balon, E.K. 1991. Probable evolution of the coelacanth's reproductive style: lecithotrophy and orally feeding embryos in cichlid fishes and in *Latimeria chalumnae*. Env. Biol. Fish. 32: 249–265. (this volume)

Balon, E.K., M.N. Bruton & H. Fricke. 1988. A fiftieth anniversary reflection on the living coelacanth, *Latimeria chalumnae:* some new interpretations of its natural history and conservation status. Env. Biol. Fish. 23: 241–280.

Boulinier-Geraud, G. 1973. Etude morphologique de la pirogue a balancier aux Comores et dans l'ouest de l'Ocean Indien. Memoire de Maitrise Specialisee d'Ethnologie, Dissertation at the Université Paris 10. 126 pp.

Browne, M.W. 1988. Do scientists pose a threat to rare 'fossil fish'? New York Times, March 22, pp. C1, C8.

Bruton, M.N. 1987. Is the coelacanth endangered? Ichthos 13: 1.

Bruton, M.N. 1988a. Coelacanth Conservation Council. Quagga 24: 12–13.

Bruton, M.N. 1988b. Coelacanth Conservation Council. Env. Biol. Fish. 23: 315–319.

Bruton, M.N. 1988c. 'Fossil' fish trade causes concern. Traffic (U.S.A.) 8: 8.

Bruton, M.N 1989a. The living coelacanth fifty years later. Trans. R. Soc. S. Afr. 47: 19–28.

Bruton, M.N. 1989b. Does the coelacanth occur in the Eastern Cape? The Naturalist 33: 5–13.

Bruton, M.N. & M.J. Armstrong. 1991. The demography of the coelacanth *Latimeria chalumnae*. Env. Biol. Fish. 32: 301–311. (this volume)

Bruton, M.N , C.D. Buxton, G.R. Hughes & R.E. Stobbs. 1989. Recommendations on marine conservation in the Federal Islamic Republic of the Comoros. Invest. Rep., J.L.B. Smith Inst. Ichthyol. 34: 1–104.

Bruton, M.N. & S.E. Coutouvidis. 1991. An inventory of all known specimens of the coelacanth *Latimeria chalumnae*, with comments on trends in the catches. Env. Biol. Fish. 32: 371–390. (this volume)

Bruton, M.N., S.E. Coutouvidis & J. Pote. 1991. Bibliography of the living coelacanth *Latimeria chalumnae*, with comments on publication trends. Env. Biol. Fish. 32: 403–433. (this volume)

CITES. 1987. Convention on International Trade in Endangered Species of Wild Fauna and Flora. Amendment to Appendices I & II of the Convention. Proposal for deletion of *Latimeria chalumnae* from Appendix II: 1–2.

Dahl, A.L. 1989. Ressources cotieres, reserves marines et tourisme aux Comores. Rapport d'une Mission OMT/PNUD a la Republique Federale Islamique des Comores. 20 pp.

de Beer, G. 1956. Conservation of coelacanths. The Times, London, 5 June 1956.

de San, M. Profil de la pêche artisanale aux Comores. SWIOP document RAF/79/065. 21 pp.

Diamond, J.M. 1985. In quest of the wild and weird. Discover (Los Angeles) 6: 34–36, 38–42.

Dollfus, R.P. & G. Campana-Rouget. 1956. Helmenthes trouves dans le tube digestif de coelacanthes. Mem. Inst. scient. Madagascar A 11: 33–41.

Dugan, J. 1955. The fish. Colliers, September 1955: 64–69.

Forey, P.L. 1984. The coelacanth as a living fossil. pp. 166–169. *In:* N. Eldredge & S.M. Stanley (ed.) Living Fossils, Springer-Verlag, New York.

Forey, P. 1988. Golden jubilee for the coelacanth *Latimeria chalumnae*. Nature 336: 727–732.

Forey, P. 1989. Le coelacanthe. La Recherche 20: 1318–1326.

Forster, G.R. 1974. The ecology of *Latimeria chalumnae* Smith: results of field studies from Grande Comore. Proc. R. Soc. Lond. B186: 291–296.

Forster, G.R., J.R. Badcock, M.R. Longbottom, N.R. Merrett & K.S. Thomson. 1970. Results of the Royal Society Indian Ocean deep slope fishing expedition 1969. Proc R. Soc. Lond. B175: 367–404.

Fricke, H. 1988. Coelacanths. The fish that time forgot. Nat. Geogr. Mag. 175: 824–838.

Fricke, H. & K. Hissmann. 1990. Natural habitat of coelacanths. Nature 346: 323–324.

Fricke, H., K. Hissmann, J. Schauer, O. Reinicke, L. Kasang & R. Plante. 1991. Habitat and population size of the coelacanth

338

Latimeria chalumnae at Grand Comoro. Env. Biol. Fish. 32: 287–300. (this volume)

Fricke, H. & R. Plante. 1988. Habitat requirements of the living coelacanth *Latimeria chalumnae* at Grande Comore, Indian Ocean. Naturwissenschaften 75: 149–151.

Fricke, H., O. Reinicke, H. Hofer & W. Nachtigall. 1987. Locomotion of the coelacanth *Latimeria chalumnae* in its natural environment. Nature 329: 331–333.

Gaspart, C. 1983. The Comoro Islands since independence: an economic appraisal. Proc. ICIOS Conference, 1979, Section 2, Perth. 56 pp.

Gould, D.E. 1985. Let's visit the Comores. Burke, London. 96 pp.

Greenwood, P.H. 1975. Norman's 'A history of fishes'. Ernest Benn, London. 236 pp.

Greenwood, P.H. 1988. The 'sixpenny coelacanth', and some notes on *Latimeria* in a social context. Ichthos, special edition 2: 5–6.

Greenwood, P.H. 1989. Fifty years a 'living fossil' – the coelacanth fish *Latimeria chalumnae*. Biologist 36: 15–19.

Griffith, R.W. 1973. A live coelacanth in the Comoro Islands. Discovery (New Haven) 9: 27–33.

Griffith, R.W. & N.A. Locket. 1972. Observations on a living coelacanth. Nature 237: 175.

Griffith, R.W. & K.S. Thomson. 1973a. *Latimeria chalumnae*: reproduction and conservation. Nature 242: 617–618.

Griffith, R.W. & K.S. Thomson. 1973b. Observations on a dying coelacanth. Amer. Zool. 12: 730.

Hale, R.C., J. Greaves, J.L. Gundersen & R.F. Mothershead II. 1991. Occurrence of organochlorine contaminants in tissues of the coelacanth *Latimeria chalumnae*. Env. Biol. Fish. 32: 361–367. (this volume)

Hamlin, J. 1988. Capture of 'fossil fish' promotes its ultimate survival. The New York Times, 9 April 1988: A30.

Heemstra, P.C. & M.J. Smale. 1987. Fisheries resources of the Islamic Republic of the Comores, with recommendations for the wise use and conservation of the marine fauna of these islands. Invest. Rep., J.L.B. Smith Inst. Ichthyol. 23: 1–34.

Hughes, G.M. 1976. On the respiration of *Latimeria chalumnae*. Zool. J. Linn. Soc. 59: 195–208.

Hughes, G.M. 1980. Ultrastructure and morphometry of the gills of *Latimeria chalumnae*, and a comparison with the gills of associated fishes. Proc. R. Soc. Lond. B 208: 309–328.

Hughes, G.M. & Y. Itasawa. 1972. The effect of temperature on the respiratory function of coelacanth blood. Experientia 28: 1247.

Hureau, J.-C. & C. Ozouf. 1977. Determination de l'age et croissance du coelacanthe *Latimeria chalumnae* Smith, 1939 (Poisson, Crossopterygien, Coelacanthide). Cybium 2: 129–137.

IUCN. 1986. Red list of threatened animals. The IUCN Conservation Monitoring Centre, Cambridge, U.K. International Union for the Conservation of Nature and Natural Resources. 105 pp.

IUCN/UNEP. 1987. The IUCN directory of Afrotropical protected areas. IUCN, Gland. 1034 pp.

Kamegai, S. 1971. On some parasites of a coelacanth *Latimeria chalumnae*: a new monogenea *Dactylodiscus latimeris* new genus, new species (Dactylodiscidae new family) and two larval helminths. Res. Bull. Meguro parasitol. Mus. 5: 1–5.

Locket, N.A. 1976. A future for the coelacanth? New Scientist 58: 456–458.

Locket, N.A. 1980. Some advances in coelacanth biology. Proc R. Soc. Lond. B208: 265–307.

Locket, N.A. & R.W. Griffith. 1972. Observations on a living coelacanth. Nature 237: 175.

McAllister, D.E. 1971. Old four legs: a 'living fossil'. Mus. natl. Sci. nat. Coll. Odyssey Ser. 1: 1–25.

McCosker, J.E. 1979. Inferred natural history of the living coelacanth. pp. 17–24. *In*: J.E. McCosker & M.D. Lagios (ed.) The Biology and Physiology of the Living Coelacanth, Occ. Pap. Calif. Acad. Sci. 134, San Francisco.

McCosker, J.E. & M.D. Lagios (ed.). 1979. The biology and physiology of the living coelacanth. Occ. Pap. Calif. Acad. Sci. 134: 1–175.

McCosker, S. & J.E. McCosker. 1976. To the islands of the moon. Pacific Discovery 29: 19–32.

Menache, M. 1954. Étude hydrologique sommaire de la région d'Anjouan en rapport avec la pêche de trois des coelacanthes. Mém. Inst. scient. Madagascar A9: 151–185.

Millot, J. 1954a. New facts about coelacanths. Nature 174: 426–427.

Millot, J. 1954b. Le troisième coelacanthe, historique, éléments d'écologie morphologie externe, documents divers. Naturaliste malagache, suppl. 1: 1–26.

Millot, J. 1955a. Unité specifique des coelacanthes actuels. Nature (Paris) 83(3238): 58–59.

Millot, J. 1955b. First observations on a living coelacanth. Nature 175: 362–363.

Millot, J. 1956. Conservation of coelacanths. The Times, London. 14 June 1956.

Millot, J. & J. Anthony. 1958. Anatomie de *Latimeria chalumnae*. Tome 1. Squelette et muscles. C.N.R.S., Paris. 122 pp.

Millot, J. & J. Anthony. 1974. Les oeufs du coelacanthe. Science et Nature 121: 3–4.

Millot, J., J. Anthony & D. Robineau. 1972. État commente des captures de *Latimeria chalumnae* Smith (Poisson, Crossoptérygien, Coelacanthide) effectuées jusqu'au mois d'Octobre 1971. Bull. Mus. Natn. d'Hist. Nat. Zool. 39: 533–548.

Monod, T. 1954. Sur une larve de gnathiide (*Praniza milloti* nov. sp.) parasite du *Latimeria chalumnae* (coelacanthe). Mem. Inst. scient. Madagascar Sér. A 9: 91–94.

Myking, L.M. 1977. Old four legs. The living fossil. Sea Frontiers 23: 334–341.

Nevenzel, J.C., W. Rodegker, J.F. Mead & M.S. Gordon. 1966. Lipids of the living coelacanth, *Latimeria chalumnae*. Science 152: 1753–1755.

Newman, M. 1971. Search for a coelacanth. Vancouver Public Aquarium Newsletter 15: 1–4.

Newitt, M. 1984. The Comoro islands. Struggle against dependency in the Indian Ocean. Gower, London. 144 pp.

Norton, B.G. 1986. On the inherent danger of undervaluing

species. pp. 110–137. *In:* B.G. Norton (ed.) The Preservation of Species, the Value of Biological Diversity, Princeton University Press, Princeton.

Northcutt, R.G. 1980. Anatomical evidence of electroreception in the coelacanth (*Latimeria chalumnae*). Zentralbl. Vet. Med., Reihe C, 9: 289–295.

Prosperi, F. 1957. Vanished continent. An Italian expedition to the Comoro Islands (transl. D. Moore). Hutchinson, London. 233 pp.

Randall, J.E. 1980. Conserving marine fishes. J. Fauna Pres. Soc. 15: 287–291.

Salm, R.V. & J.R. Clarke. 1984. Marine and coastal protected areas: a guide for planners and managers. IUCN, Gland. 302 pp.

Schulze, H.-P. 1972. Early growth stages in coelacanth fishes. Nature New Biol. 236(64): 90–91.

Smith, C.L., C.S. Rand, B. Schaeffer & J.W. Atz. 1975. *Latimeria,* the living coelacanth, is ovoviviparous. Science 190: 1105–1106.

Smith, J.L.B. 1939a. A living fish of Mesozoic type. Nature 143: 455.

Smith, J.L.B. 1939b. A surviving fish of the order Actinistia. Trans. R. Soc. S. Afr. 27: 47–50.

Smith, J.L.B. 1940. A living fossil (*Latimeria chalumnae*). Smithsonian Report, 1940: 321–328.

Smith, J.L.B. 1950. *Latimeria chalumnae.* pp. 79–80. *In:* J.L.B. Smith (ed.) The Sea Fishes of Southern Africa, 2nd ed., Central News Agency, Cape Town.

Smith, J.L.B. 1953. The second coelacanth. Nature 171: 99–107.

Smith, J.L.B. 1956a. Old fourlegs. The story of the coelacanth. Longmans, Green & Co., London. 260 pp.

Smith, J.L.B. 1956b. Conservation of coelacanths. The Times, London, 4 June 1956.

Smith, J.L.B. 1963. The atomic bomb and the coelacanth. The Daily Dispatch, 10 December 1963.

Smith, M.M. 1976. Surprise – a baby. The Eastern Cape Naturalist 59: 4–6.

Smith, M.M. 1977. Surprise, surprise – a baby! Afr. Wildl. 31: 30–31.

Spark, A.A. 1982. New light on marine waxes, fats and oils. S. Afr. J. Sci. 78: 303–305.

Spark, A.A. & A.A. de Wit. 1981. Lipids of a new species of stingray *Hexatrygon bickelli.* Ann. Rep. Fish. Ind. Res. Inst. 24: 48–50.

Stobbs, R.E. 1987. The 'galawas' of the Federal Islamic Republic of the Comores, with notes on artisanal fishing and recommendations for a maritime museum. Invest. Rep., J.L.B. Smith Inst. Ichthyol. 26: 1–21.

Stobbs, R.E. 1989a. The coelacanth enigma. The Phoenix 2: 8–15.

Stobbs, R.E. 1989b. Laxative lipids and the survival of the living coelacanth. S. Afr. J. Sci. 85: 557–558.

Stobbs, R.E. & M.N. Bruton. 1991. The fishery of the Comoros, with comments on its possible impact on coelacanth survival. Env. Biol. Fish. 32: 341–359. (this volume)

Suyehiro, Y. 1988. Gombessa. Be-Pal Books, Tokyo. 254 pp. (In Japanese).

Suzuki, N. & K. Tanauma. 1984. Coelacanth fishing by native fishermen. pp. 8–9. *In:* Proceedings of the First Symposium on Coelacanth Studies, Tokyo. (In Japanese).

Suzuki, N., Y. Suyehiro & K. Hamada. 1985. Initial report of expeditions for coelacanth – Part 1 – Field studies in 1981 and 1983. Sci. Pap. Coll. Arts Sci., Univ. Tokyo 35: 37–79.

Thomson, K.S. 1966. Intracranial mobility in the coelacanth. Science 153: 999–1000.

Thomson, K.S. 1970. Intracranial movement in the coelacanth *Latimeria chalumnae* Smith (Osteichthyes, Crossopterygii). Postilla 149: 1–12.

Thomson, K.S. 1973. Secrets of the coelacanth. Nat. Hist. 82: 58–65.

Thomson, K.S. 1986. A fishy story. Amer. Sci. 74: 169–171.

Thoney, D.A. & W.J. Hargis, Jr. 1991. Juvenile anisakine parasites from the coelacanth *Latimeria chalumnae.* Env. Biol. Fish. 32: 281–283. (this volume)

Uyeno, T. 1984. Age estimation of coelacanth by scale and otolith. pp. 28–29. *In:* Proceedings of the First Symposium on Coelacanth Studies, Tokyo. (In Japanese).

Uyeno, T. 1989. Dissecting a coelacanth. National Science Museum News 244: 6–9.

Uyeno, T. 1990. Investigating the lifestyle of *Latimeria* on the basis of its feeding habits. Newton Graphic Science Magazine 9(13): 20–21. (In Japanese).

Uyeno, T. 1991. Observations on locomotion and feeding of released coelacanths, *Latimeria chalumnae.* Env. Biol. Fish. 32: 267–273. (this volume)

Uyeno, T., T. Hamada, S. Murakami, K. Takahashi, C. Olhagaray & N. Suzuki. 1990. Coelacanth: Japanese research to date. Newton Graphic Science Magazine 9(13): 10–31. (In Japanese).

Uyeno, T. & T. Tsutsumi. 1991. Stomach contents of *Latimeria chalumnae* and further notes on its feeding habits. Env. Biol. Fish. 32: 275–279. (this volume)

Uyeno, T., T. Tsutsumi, N. Suzuki & T. Hamada. 1984. Distribution and ecology. p. 34. *In:* Proceedings of the First Symposium on Coelacanth Studies, Tokyo. (In Japanese).

van Bruggen, A.C. 1989. Fifty year coelacanth. Biovisie Mag. Vakblad voor Biologen 69: 5–6. (In Dutch).

von Wahlert, G. 1968. *Latimeria* und die Geschichte der Wirbeltiere. Eine evolutionsbiologische Untersuchung. G. Fischer Verlag, Stuttgart. 125 pp.

Watson, D.M.S. 1927. The reproduction of the coelacanth fish *Undina.* Proc. Zool. Soc. Lond. 1: 453–457.

White, E.I. 1939. One of the most amazing events in the realm of natural history in the twentieth century: the discovery of a living fish of the coelacanth group, thought to have been extinct 50 million years off South Africa. The Illustrated London News Suppl., March 1939.

Wourms, J.P., M.D. Stribling & J.W. Atz. 1980. Maternal-fetal nutrient relationship in the coelacanth, *Latimeria.* Amer. Zool. 20: 962. (abstract).

It is no longer justified or legal to acquire a live or preserved coelacanth for museum or aquarium display. Good plastic replicas are available from several institutions. A rubber coelacanth as a hydraulic mobile can even substitute for a living fish in public aquarium exhibits. This one, held by Les Kaufman, was designed and used at the New England Aquarium, Boston. Photo by E.K. Balon, 28.12.1989.

Environmental Biology of Fishes **32**: 341–359, 1991.

The fishery of the Comoros, with comments on its possible impact on coelacanth survival

Robin E. Stobbs & Michael N. Bruton
J.L.B. Smith Institute of Ichthyology, Private Bag 1015, Grahamstown, 6140 South Africa

Received 7.11.1990   Accepted 10.1.1991

Key words: Artisanal fishing, Canoes, Conservation, Fishing methods, *Ruvettus*

Synopsis

The traditional methods of deepsea handline fishing in the Comoros are described. The main target species is the oilfish *Ruvettus pretiosus,* and the coelacanth *Latimeria chalumnae* is caught as a bycatch. In recent years motorised dugout canoes as well as outboard- and inboard-powered boats have been introduced into the fishery, and more efficient fishing tackle has become available. The more modern gear is intended for use on pelagic fishes and has added a new dimension to the fishery. Traditional handline fishermen are not considered to be a threat to the coelacanth, but the fishermen equipped with motorised boats and modern tackle would constitute a real threat if they direct their efforts on inshore reefs.

Introduction

Subsistence fishermen are universally considered to occupy the lowest levels of any social system, yet recent FAO studies have shown that an estimated 10 million subsistence fishermen support about 100 million of the world's population (Thomson 1980). Subsistence fishermen throughout the world are facing increasing socio-economic pressures as, without the benefit of official aid, they are expected to meet the protein needs of a rapidly expanding population. Nearly 50% of the world's fisheries landings come directly from the activities of subsistence fishermen, which is a community that contributes more to employment, and uses far less energy, than the often heavily subsidised large-scale fisheries (Thomson 1980).

For centuries the fishermen of small islands operated in harmony with their island's biota while rationally exploiting those areas available to them. The fishermen met their needs for protein with equipment and techniques that were adequate to

the task, though not always efficient by modern standards (Reeves 1985).

In the Comoros there has always been a traditional ethic of rational exploitation. Until recently, cultural taboos and tradition prevented overexploitation while allowing everyone a fair share of marine resources. Marine resource sharing was the traditionally accepted rule, and a managed and adequate fishery developed to meet the needs of the islands' growing population. The Comoran population has, however, now outstripped the capabilities of the fishing community to supply its needs. The rate of population increase is reportedly one of the highest in Africa (Newitt 1984).

Furthermore, the Comoran fishing community is recruiting fewer younger men from the family unit as a result of social re-adjustment in a society that is striving after the materialism of the Western world (Grand Comoro fishermen of Dzahadjou personal communication 1987, 1988). However, the picture is not entirely negative. The Comoro Islands lie directly in the path of the warm westerly South

Equatorial Current that brings abundant shoals of pelagic fishes to the region. The recent development of an offshore pelagic fishery is resulting in an increasing supply of fishes, especially tuna, to the marketplace and indirectly leading to some recovery of the nearshore resources (M. de San personal communication 1990).

Weather and oceanographic features of the Comoros

The islands of the Comoro archipelago are of recent volcanic origin. Exact age determinations of volcanic islands are not easy to make but the ages of the Comoro Islands have been reliably placed at about 1 to 1.3 million years for Grand Comoro, 3 to 5 million for Anjouan, 5 to 7 million for Moheli and 7 to 12 million for Mayotte. East of Mayotte there is a submerged dome, Geyser Reef, which is all that remains of an older island between the Comoros and Madagascar.

The islands rise steeply from the ocean floor of 3000 to 3500 m depth. Grand Comoro has a steep underwater profile with some slopes in excess of 45 degrees (Fricke & Plante 1988), and the only shallow shelf areas are at the north and south extremities with a lesser westward projection from Iconi towards Mwamba Raya (Recif Vailheu). Anjouan is steeply sloped along its eastern aspect with a narrow reef-topped shelf extending around its north-western, western and south-western shores. Moheli has a wide to very wide continental shelf, especially on its southern side, and there are extensive coral reefs. Mayotte is completely surrounded by a well-developed reef that encloses one of the largest lagoons in the world.

The south equatorial current flows at a rate of 0.5 to 3 knots (0.9 to 5.5 km per hour) past the Comoro Islands. During the months of the 'kuzi' (southerly monsoon), and under the combined influence of the south-east trades and the southerly monsoon winds, the current splits into the north-flowing Somali current and the south-flowing Mozambique current where it meets the African coast at Ras Kongo (Cape Delgado). During the 'kaskazi' season (northerly monsoon) the Somali current is turned southward and becomes part of the Mozambique current (Anon. 1980, Couper 1989).

Sea surface temperatures vary from 24.5° C in August to over 28° C in February (Anon. 1980), but there is a well marked thermocline at 100 to 200 m (150–200 m on Anjouan; 115–125 m on Grand Comoro) below which the water temperature remains an almost constant 15°–17° C. Nearshore salinities vary little throughout the year although there have been reports of strong fresh water streams issuing from Grand Comoro during the 'kaskazi' rains (McCosker 1979).

The general weather pattern is considerably influenced by monsoon seasons. During the 'kaskazi', the winds blow from the north to north-east and bring hot wet weather. The 'kuzi' season is characterised by strong south to south-easterly winds and cooler, dry weather. Throughout the year there is a predominant easterly wind component which, together with a strong westerly current, combine to make the eastern shores of the islands inhospitable and dangerous (Stobbs 1987). Mean wind directions and velocities are: north to north-east, force 3 to 4 in January; south-east, force 4 in April; south to south-south-east, force 4 in July and back to north-east to east-north-east, force 3 to 4 in October (Anon. 1980).

The Comoros lie within a region where there are thunderstorms on 60 to 100 days a year. These heavy rain storms are accompanied by strong gusty winds and a choppy sea. Over the sea these storms tend to be most frequent at night and can arise with alarming rapidity (Couper 1989, Stobbs personal observation 1956). The region is also subject to the passage of tropical cyclones during the period December to March and, although few cyclones pass directly through the archipelago, the effects of their nearby passage are often devastating (Alexander 1976, Newitt 1984, Couper 1989).

The east and west coasts of Grand Comoro, the east and south coasts of Anjouan and the northeast coast of Moheli are all relatively steep. The Comoros lie within a region where the general wave height is less than 3.5 metres (Couper 1989). Ocean swells and storm-generated waves, which meet these coasts without the tempering effect of offshore shoals or reefs, rear up suddenly and without warning. This makes the task of launching and recovering small canoes a most hazardous procedure. The inhospitability of these east coasts is the principal reason why there has traditionally been so little settlement there. Furthermore, there is no ancient history of canoe development along the eastern shores of Madagascar (Hornell 1944).

The principal fishing season for pelagic fishes corresponds largely with the warmer 'kaskazi' season, from November to March, but many fish species are present throughout the year. The apparent seasonality in pelagic fishing activities may well be a result of weather patterns affecting the fishermen rather than the fish (Cullen & Hemphill 1971, de San 1983). There appears to be no seasonality in the species composition or numbers of nearshore benthic and epibenthic fishes and, here again, the apparent seasonality is probably an artificial manifestation of the fisherman's preference for good weather conditions. Grand Comoro fishermen (1987, 1988) are of the opinion that there is a seasonal pattern to the presence or absence of coelacanths, as there is with *Ruvettus* (Millot 1954). They refuted any suggestion that there might be any connection with tides or the phase of the moon other than those affecting launching and recovery of their canoes and the acknowledged preference for *Ruvettus* to rise to shallow depths on dark moonless nights (Millot 1954, McCosker & McCosker 1976, Suzuki et al. 1985).

Cultural traditions and taboos

Comoran traditions ensure that each fishing village has its fair share of the available natural resources. It is, for example, taboo for a fisherman to fish in the waters of an adjacent village and he is thus unable to follow good weather around the island. Seasonal changes in the monsoon winds allow fishermen on one side of the islands to fish in relative safety while those on the other side are unable to fish except during calm spells; with a change in the season, the one group of fishermen lay up their gear while the others are able to fish.

The vagaries of a monsoon-dominated climate, local weather conditions and cultural taboos result in a situation in which not every fisherman is able or willing to put to sea every day. At best a subsistence fisherman is able to make no more than 180 fishing excursions a year in his paddling canoe (de San 1983).

Millot (1954), Griffith & Thomson (1973) and Locket (1976) were of the opinion that the Comoran artisanal fishery posed no severe threat to the coelacanth provided that the intensity and methods of fishing did not change. Fishing methods and materials are changing, and it seems that there is cause for concern for the biota of the nearshore reefs of the Comoros. Traditions and taboos are being subjected to undermining pressures with the result that the ethic of self-control is being replaced by a desire to exploit at all costs. An intensification of fishing activities will inevitably lead to an increase in the potential threat to all nearshore fish species, including coelacanths and their prey.

Traditionally, the use of explosives and *Tephrosia* ichthyocide is banned, and in 1988 it seemed likely that the use of these illegal fishing methods had ceased in the Comoros (de San 1983, Stobbs personal observation 1988). Some fishermen from Grand Comoro were adamant that *Tephrosia* was still in use, at least on Anjouan, and this opinion is supported by Bourgois (1989). B. Lamarck (personal communication 1988) noted that the use of explosives was on the decline and, in fact, had ceased in the region south of Nyumashua on Moheli. A confidential communication (1990) made the allegation that explosives were again being extensively used on the reefs east of Bangwa-Kuni on Grand Comoro. It was suggested that the source of these explosives was the developers of the harbour facilities in Moroni.

Certain villages once banned the use of fishing nets and traps (de San 1983) and underwater spearguns (Grand Comoro fisherman personal communication 1987). Today, however, fishermen are again using these methods. The use of seine nets is increasing off the southern shores of Moheli and in Mutsamudu bay, Anjouan. Snorkel diving equipment and spearguns are commonplace in many villages along the west coast of Grand Comoro. Traditional wicker rocklobster traps and modern traps made from chicken wire can be found in

344

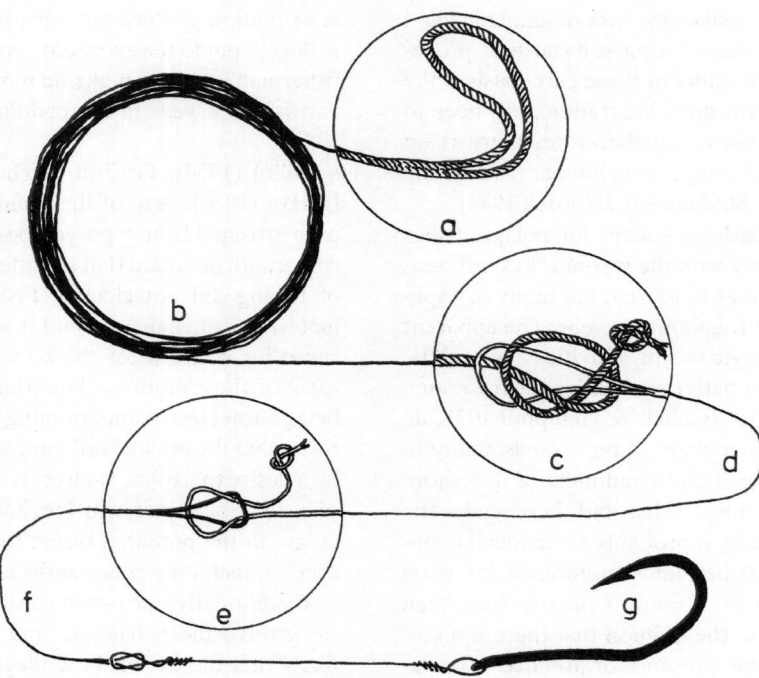

Fig. 1. Traditional Comoran fishing line as used to catch *Ruvettus* and, by accident, coelacanths: a – The proximal terminal eye splice, b – main body of the line 'dingo ya sissi', c – the becket hitch attachment between sections of line, d – the 'mdigo mbeli' leader attached through, e – a becket hitch or sheet bend to, f – a nylon and steel wire trace, g – a 6/0 or larger steel hook (all diagrams, maps and photographs by R.E. Stobbs unless otherwise indicated).

almost all fishing villages but are not extensively used.

Illegal and unethical fishing practices have the potential for profoundly affecting the biota of the limited nearshore reefs and slopes of the Comoros. Explosives and *Tephrosia* are not selective for food fishes and they have a widespread physical and biological impact. Seine nets and traps are also non-selective while the uncontrolled use of spear-guns can result in serious depletion of a wide variety of reef fishes (Bruton & Stobbs 1991). Large-scale and widespread destruction of reef habitats must eventually seriously affect all the islands' nearshore biotas. The known coelacanth communities might also be affected through the eventual reduction in prey species abundance.

Traditional fishing gear

Fishing lines were previously made of coir or other locally grown vegetable fibres painstakingly extracted and twisted into strong, if somewhat crude, twines. In more recent times use was made of cotton fibres and, until very recently, all deepwater fishing lines were made from this fibre. Hand-made line is now being replaced by various grades of monofilament nylon line. The dying art of traditional fishing line making is now practiced by only a very few older men.

The 'tsissi', as traditional cotton deepwater fishing line is known in the Comoros, is a three-strand cotton line twisted and retwisted into a strong cord (Fig. 1). Some of this line is made from standard cod line that is unravelled and then retwisted. The resulting twist is tighter and stronger than the original. Since this line is unique to the Comoros, and is

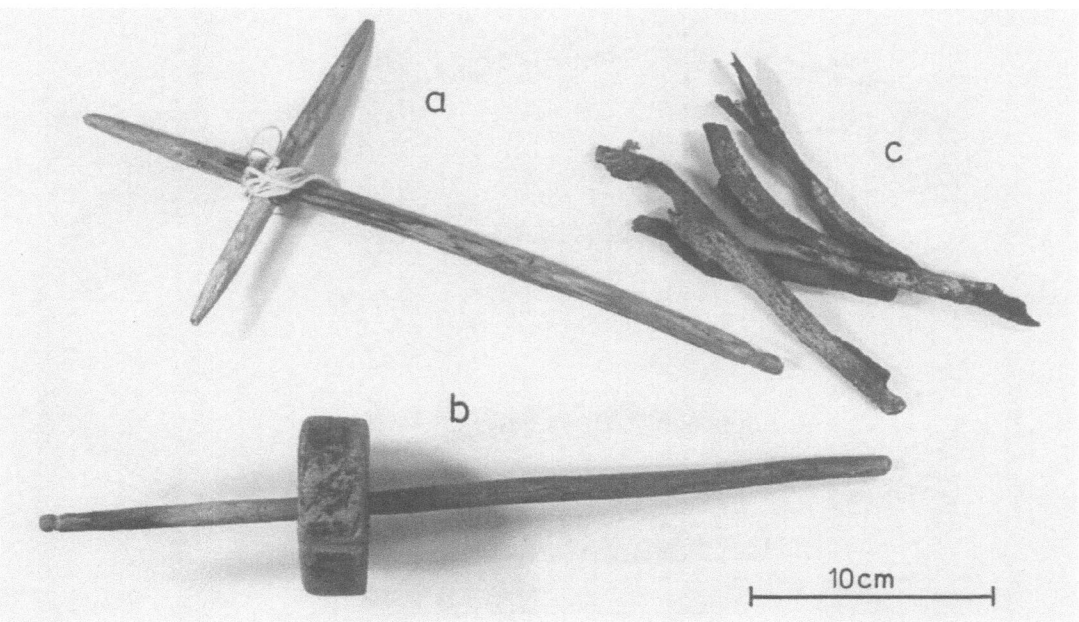

Fig. 2. Equipment used in making a Comoran cotton fishing line: a – the 'teleba' used to hold strands of cotton twist, b – the 'tedje' distaff used to twist the cotton strands, c – pieces of 'm'bessi' bark used to treat the finished fishing line.

the line used when coelacanths are caught (Millot 1954, Suzuki & Tanauma 1984, Suzuki et al. 1985, de San 1983, D. Rotsaert and Grand Comoro fishermen personal communication 1987, 1988 and 1989), its fabrication is described in detail.

Raw cotton or unravelled strands from hanks of ordinary cotton cod line are twisted to a tight, fine pitch by means of a distaff called a 'tedge' (Fig. 2). The loosely stranded line is tied to the shaft of the tedge and this is spun by flicking the shaft between the index finger and thumb. As each section of line is twisted to the required degree, it is collected by winding it upon the 'teleba', where it is kept until there is sufficient line to twist together. When sufficient line has been accumulated on the two teleba, and there is a third length ready from the tedge, the three strands are twisted together to form a strong cord 1.5 to 2.5 mm in diameter. The process is then repeated, 4 to 6 metres at a time, until the required section is complete. This makes the main part of the fishing line, called a 'dingo ya sissi'. On completion of a length of about 150 m an eyesplice is made in one end while the other end is simply knotted

with an overhand knot to keep it from untwisting. There are some line makers who do not splice together short sections to make one continuous 150 metre length but simply make a series of 5 to 8 metre lengths each joined to the other by means of an eyesplice and becket hitch (Day 1964), not a sheet bend as suggested by Millot (1954) (Fig. 1).

The hank of line is then tanned and rotproofed by rubbing it with the wet, green bark of a tree known as 'm'bessi' (Millot 1954, did not offer further identification, neither have we been able to identify this tree which is reported to grow in the forests on Kartala above Moroni). The line is thouroughly impregnated with the resinous juice from the 'm'bessi' bark and is then soaked in sea water for three days. This treatment is repeated until a thick black protective layer is built up. By this time the line is 2 to 3 mm in diameter, rotproof and very strong. It is also easy to hold against the pull of a heavy fish owing to a combination of its large diameter and a tendency to resist slipping even when wet.

According to E. Rotsaert (personal communi-

346

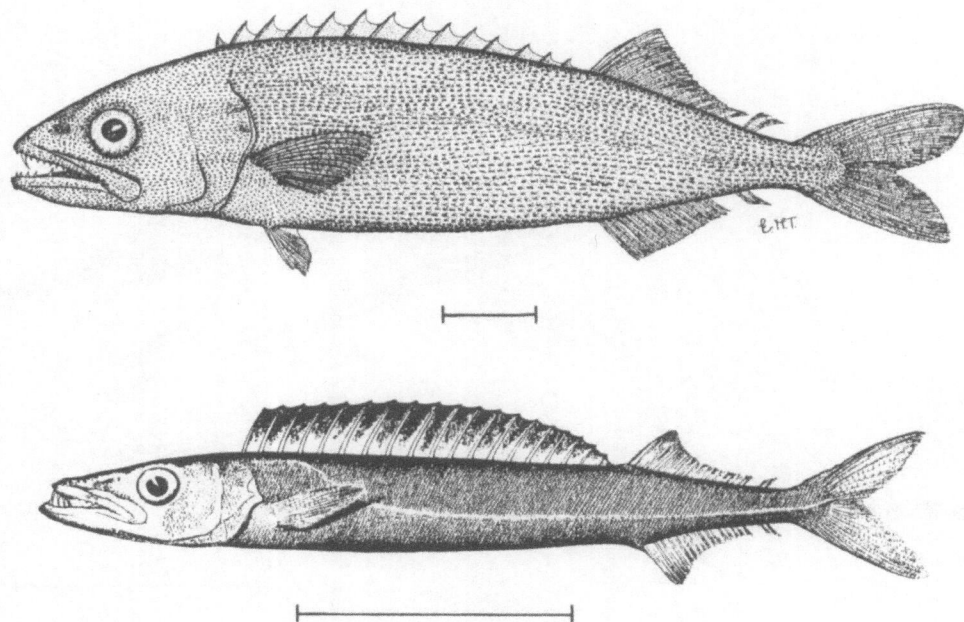

Fig. 3. The oilfish *Ruvettus pretiosus* (above) which is the main target species of deepsea handline fishermen in the Comoros, and the gempylid *Promethichthys prometheus* (below) which is most frequently used for bait when fishing for *Ruvettus*. Scale bar = 10 cm.

cation 1989) it takes about one month of daily dawn to dusk toil to make a 150 m length of 'dingo ya sissi'. One of the remaining makers of traditional cotton line at Iconi, south of Moroni on Grand Comoro, assured one of us (R.E.S. 1990) that it takes at least one to two months to make this length of line.

When in use, each length of 'dingo ya sissi' is kept in a coil, of about 300 mm diameter, with its eyesplice at the inner (proximal) end. The coils of line are kept suspended over a short, thick wooden rod which is stowed athwartships over the canoe gunwhales forward of the fore outrigger boom.

The fishing line is assembled as depicted in Figure 1. The size 6/0, or larger, hook may or may not have a steel trace attached; examples examined had a 100 to 300 mm length of approximately 0.75 mm diameter single strand steel between the hook and monofilament nylon. The leader, 'mdigo mbeli', a 3 to 4 m length of 1.5 to 2.5 mm diameter line made identically to the 'dingo ya sissi', also serves to attach the sinker pebble when deep fishing for 'ne-

sa', *Ruvettus* (Fig. 3). When the line is used for shallow reef fishing, and it is required to be anchored, a short cord is attached to the junction of the trace nylon and the 'mdigo mbeli', and this is firmly tied to a fist-sized rounded sinker pebble (Fig. 4).

Traditional cotton fishing line is today largely replaced by heavy-gauge monofilament nylon line, not because the latter is better but because it is easier and cheaper to obtain on demand and can be purchased in long lengths without joins (Grand Comoro fishermen personal communication 1990). It is further claimed by these fishermen that traditional cotton line is superior to nylon line since the former is easier to handle when wet, stronger and less readily displaced by deepwater currents than the nylon monofilament line.

The traditional fishery

'Nesa' fishing is practiced only on the darkest of

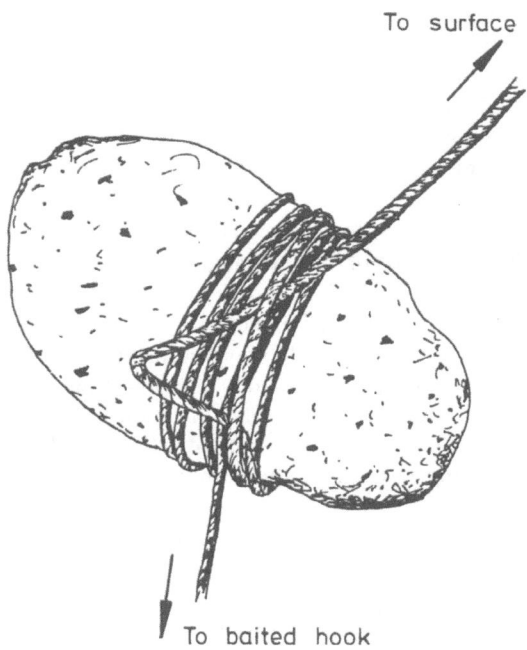

Fig. 4. A Comoran fishing sinker showing how the line is secured with an undertuck.

Fig. 5. Three examples of the forged steel gaffs ('ulowo') used by Comoran fishermen to tow large fish, such as the coelacanth, ashore.

nights when there is no moon and when weather conditions are sufficiently calm that there is little risk of poor weather occurring during the night. Night fishing involves the fisherman and his mate (if he is fortunate enough to find one) in considerable risk since, once he sets out, he cannot return until first light the following day: it would be suicidal to attempt a landing without light on the jagged lava rock shores that are a feature of the Comoro Islands. Only those fortunate fishermen who put out from quiet bays and the infrequent sandy beach are able to return to shore in comparative safety during the night.

'Nesa' fishing is not without considerable risk to the fisherman or, at best, his canoe. The occupation is becoming less and less attractive to artisanal fishermen since they are finding it increasingly difficult to find young partners to accompany them – the calls of city lights, discos and video shows are attracting increasing numbers of potential young fishermen away from their traditional village life and the ranks of learner fishermen.

Grand Comoro fishermen have consistently stated that they consider night fishing to be a dangerous occupation and that, providing they could find an easier way to obtain their daily requirements, they would discontinue this form of fishing (personal communications 1987, 1988). Many fishermen in the villages visited in 1987 and 1988 told of friends that had vanished while fishing. Fisherman I. Mdoionhoma told how two of his close neighbours had vanished without trace while night fishing during the previous three years. It was also communicated to us that, during the previous year, a night fisherman had disappeared but survived his ordeal and was washed ashore on the Kenya coast. The Anjouan fishermen who caught the first mature female coelacanth known to science (CCC no.

10 in Bruton & Coutouvidis 1991), capsized their galawa over deep water while attempting to secure their fish.

Night fishermen set out at last light. In his canoe the fisherman might have with him a pressure paraffin lamp, two paddles, his fishing lines, trace, hooks, a supply of bait (preferably 'roudi', a gempylid fish, *Promethichthys prometheus,* Fig. 3, but any fish, squid, octopus and even crabs are used), a number of fist-sized pebbles (Fig. 4) and one or two forged steel gaffs, 'ulowo' (Fig. 5), tethered to the forward outrigger boom of his 'galawa' and with which he will tow ashore any fish too large to lift into his canoe.

Traditions of *Ruvettus* fishing exist in many parts of the Pacific Ocean but only in the Comoro Islands in the Indian Ocean. It is the deepest artisanal handlining in the world; in parts of the Pacific Ocean depths of 800 m are regularly fished (Gudger 1927). The baited hook must reach the desired depth as rapidly as possible (to avoid the possibility of the bait being taken before it reaches its target depth) and in order to achieve this the line weights are as heavy as practical. Once the baited hook reaches the sea bottom the line weights, or sinkers, are released automatically (in the case of certain Pacific Ocean fishermen, Nordhoff 1928) or by the determined actions of the fishermen (in the Comoros), as described below.

The hook, baited with fillets of 'roudi', is sandwiched between two sinker pebbles although some fishermen prefer to use only one large sinker. The sinker pebbles, 'mbize', are bound together using the 'mdigo mbeli' and secured with an undertuck pulled up so that it is secure enough to hold the rig together for deployment but not so tight that it cannot be freed by a sharp pull on the line (Fig. 4). The fisherman deploys the line rapidly, uncoiling the line and joining additional sections as required. Additional sections of line are joined by means of a becket hitch or becket bend, a type of sheet bend (Day 1964). When the sinkers touch bottom, a depth that the fishermen usually knows with great accuracy, he frees the sinker pebbles. Although nights for 'nesa' fishing are chosen with care, there are almost always some surface or subsurface currents to cope with and the fisherman's task is now to keep his line as vertical as possible and, at the same time, ensure that he is not driven onshore, offshore or too far alongshore. This correction of position is achieved by the fisherman paddling with the loom of his paddle tucked under his armpit and sculling with one hand while the other hand tends the line. With unerring accuracy, even on the darkest of nights, the fisherman is able to keep at least the upper part of his line hanging vertically.

Fishermen who have previously caught a coelacanth claim to know immediately they hook a second by the actions of the hooked fish (Grand Comoro fishermen personal communication 1987, 1988, Hamlin personal communication 1990). Unlike a 'nesa' or shark, both of which fight well according to their size, the 'gombessa', by comparison, is slow but dogged, somewhat like a grouper, 'sahali', and may take as much as 2 hours to bring to boat (but Hamlin personal communication 1990 is of the opinion that this time period is an overestimate). Once on the surface the 'gombessa', or any other large fish, is gaffed through the jaw or gills using the tethered steel gaff; this keeps the fish captive and nominally alive until first light the following day when the fisherman is able to paddle ashore towing the fish alongside or behind the canoe.

'Gombessa' here refers to the coelacanth, *Latimeria chalumnae,* and should not be confused with other fish also named 'gombessa' by Comoran fishermen. These include some of the large labrids such as *Epibulus insidiator, Bolbometopon muricatum* and *Cheilinus undulatus* (Millot 1954, Stobbs 1990).

Canoes and other fishing craft

In the Comoros artisanal fishing is conducted from one of four basic sea craft types: traditional wooden dugout paddling canoes with either single or double outriggers (Fig. 6); outboard-engined wooden dugout canoes with single or double outriggers (Figs 7, 8); small outboard-engined fibreglass (GRP) boats used with or without double outriggers (Fig. 9); and large GRP inboard dieselengined boats used with or without double outriggers (Fig. 10).

Fig. 6. A group of artisanal fishermen with their typical Grand Comoro paddling galawa on the beach at Bouni in April 1987.

Traditional dugout paddling canoes, 'galawa', used by the Comoran fishermen of Grand Comoro, differ little in basic design from all other double outrigger dugout canoes of East Africa. There are, however, some characteristics that make the Grand Comoro 'galawa' unique. High end-pieces give this 'galawa' an unmistakeable, pronounced gunwhale sheer; a configuration not found elsewhere. The outriggers also differ in detail and method of attachment from some of those of central East Africa (Hornell 1944).

It might be considered, from the appearance of the double outrigger configuration, that these canoes have an inherent stability. This is an illusion; the floats are mere planks and are balancers rather than floats. They have little static buoyancy but the angle they make to the water surface provides a measure of dynamic lift. A considerable balancing ability is required from the occupants (M.N. Bruton personal observation 1987; see Fig. 21 in Balon

et al. 1988), especially at night. As in all dugout canoes of the region, the occupants are unable to sit inside the canoe body but must perch upon the thwarts or outrigger booms.

Until recently all dugout canoes of the Comoros archipelago had double outrigger booms and floats similar to, but differing in detail from, the canoes of the African mainland coast. The double outrigger canoes of Madagascar had nearly disappeared, and had been replaced by single outrigger forms, by the beginning of the sixteenth century. Malagasy canoe design greatly influenced trends within the Comoros until today, only on Grand Comoro does the double outrigger configuration persist (Hornell 1944, Stobbs 1987).

The usual dugout paddling canoe is about 4.5 to 5 m overall length (4.00 to 5.12 m, n = 20, on Grand Comoro, 3.36 to 4.50 m, n = 11, on Anjouan, 4.45 to 5.92 m, n = 9, on Moheli). The two smallest canoes seen (on Anjouan and Mayotte in 1988)

Fig. 7. A line-up of Anjouan-type outboard-engined galawa on the sea front at Mutsamudu in April 1987.

were 2.5 m and the largest over 7 m long (Mayotte and Moheli 1987, 1988).

Many canoes are made from a single hollowed log of a number of tree species, especially kapok (*Ceiba pentandra*), jackfruit (*Artocarpus incisa*) and breadfruit (*Artocarpus actilis*) (Stobbs 1987). With suitable trees becoming scarce, there is a growing trend for the canoe underbody to be made from a small diameter log and for this to have carefully inserted gunwhale strakes to increase the freeboard. More detailed accounts on the dugout canoes of the Comoros are given by Boulinier-Giraud (1973) and Stobbs (1987).

Within the past 10 years a number of dugout paddling canoes have been adapted to take a small outboard engine by the addition of a braced plank fitted across the stern. A more recent trend has seen the development of an entirely new type of dugout canoe made purposely to accept a small outboard engine on a flat transomed stern. Each

island has developed its own style and size of hull shape. These new-generation 'galawa' are substantially larger than the paddling canoes (4.59 to 6.33 m, n = 13) and have a hull that is capable of holding a considerable fish catch and all but the very largest fishes.

The outboards used on these newer craft are between 2 and 15 hp (1.5 to 11.2 Kw) which is sufficient to attain speeds of 6 knots or more, thus allowing fishermen to reach fishing grounds that were previously beyond the capabilities of all but the largest sailing or paddling canoes. With the introduction of the outboard engine, the rich fishing grounds of Mwamba Raya (Recif Vailheu, a shallow reef top 30 km WSW of Moroni) and on the south side of Moheli became within reach of large numbers of fishermen. Coastal nearshore reefs and slopes, previously too far for the average paddling fishermen, became well within operational range. With this newfound mobility also came the poten-

Fig. 8. A typical new generation Grand Comoro motorised galawa in Moroni harbour in May 1990.

tial to poach from the traditional fishing grounds of other villages and to escape with impunity. Fishermen could also fish in marginal weather conditions knowing that they could run for shelter should foul weather set in.

A new craft that is replacing the wooden dugout canoe, and proving to be most popular, is the E.E.C. 'fedawa' (Fig. 9), a small (5 to 6 m) outboard-powered GRP boat with seaworthiness and handling characteristics not very different from traditional wooden craft. The outrigger booms are made from semi-rigid PVC piping that is able to flex under the severe sea conditions of the Comoros. This semi-rigidity, like the lashed wooden counterpart of the 'galawa', allows the unit to absorb the knocks of launching and recovery from rocky coves without suffering damage. A vessel with rigidly fixed outriggers might easily be damaged, even completely ripped apart, on these inhospitable shores. This fate has befallen many of the Japanese 'japawa' (see below).

The 'fedawa', of which there are currently three models differing mainly in their hull design, are also produced without outriggers but have sections of polyurethane foam-filled PVC piping affixed to each side of the hull where they function as efficient close-coupled stabilizers. The outboard engines used, 2.5 to 15 hp (1.8 to 11.2 Kw) are of a similar size to those fitted to the new-generation, powered 'galawa'.

There has been complete acceptance of the 'fedawa' by Comoran fishermen throughout the archipelago. So popular have these boats become that at one time the Projet Pêche Artisanale factory was producing 17 boats a month and had advance orders for 15 months production.

The project has sold 300 outboard engines to date. About 250 of these units have been sold to 'fedawa' owners whereas the balance are used on

352

Fig. 9. Fedawa, with outriggers removed, at Iconi in May 1990.

modified 'galawa'. It has been estimated that an additional 300 outboard engines have been purchased from other sources so that there are now over 500 outboard-engined fishing craft in the islands.

The largest and heaviest vessel used in the Comoran fishery is the somewhat unmanageable, single cylinder inboard diesel-engined Japanese GRP boat known locally as a 'japawa' (Fig. 10). These boats have only proved to be successful when used from safe harbours such as Moroni and Shindini on Grand Comoro and Mutsamudu on Anjouan. Many of these craft lie damaged in the smaller fishing villages owing to the lack of adequate harbour or slipway facilities for their safe deployment. 'Japawa' outrigger booms are made from rigid aluminium tubing fixed inflexibly onto both boat and float. So many outriggers have been damaged through lack of suitable harbour facilities that most

of the craft seen today are operated without these floats.

In accepting the larger, outboard powered and inboard diesel engined boats into Comoran fishing, it was hoped that the use of these craft would not seriously affect the nearshore fishery and that they would only be used for offshore fishing for pelagic fishes. This has largely been the case; few fishermen, having paid a considerable sum of money for a boat, are prepared to use that boat to fish in unproductive nearshore waters and reefs; fewer still are prepared to have their investment potentially jeopardised by fishing close to such inhospitable shores as those around Grand Comoro and Anjouan.

Fishing effort

The Comoros are steep-sloping islands where the

Fig. 10. A 9 m japawa, with outriggers intact, on the mangrove flats at Male, Grand Comoro in April 1987.

human population is generally concentrated around a narrow coastal plain. The exception to this generalization is the large inland agricultural communities of Anjouan and, to a lesser extent, Moheli (Fig. 11, 12 and 13). Eastern shores are inhabited less than other shores due to the difficulty of establishing fishing villages on the steep slopes. The western sides of Grand Comoro and Anjouan are, however, less steep and have a moderately wide coastal plain.

Fishing villages are widely, and almost uniformly, dispersed around each of the islands (Fricke & Hissmann 1990), but there is a markedly skewed population distribution in favour of the more protected western shores (Figs 14, 15 and 16). There is, therefore, a heavy fishing pressure exerted upon the western shore of Grand Comoro with only minimal exploitation of the exposed eastern shores. It has been suggested (Stobbs 1989a) that this skewed population distribution is one of the factors re-

sponsible for the apparent greater abundance of coelacanths along the western side of Grand Comoro and not along the eastern shores, according to catch statistics (Bruton & Coutouvidis 1991).

In 1962 33% of the Grand Comoro population lived in coastal villages, and there were 335 permanent and 1209 temporary fishermen. On Moheli, 85% of the population was coastal dwelling with 25 permanent and 130 temporary fishermen. On Anjouan, 32% of the population was coastal dwelling with 19 permanent and 297 temporary fishermen (Boulinier-Giraud 1973). In 1972 the major fishing towns were Mitsamiuli and Vanambuni on Grand Comoro and Mirontsi on Anjouan, each with over 100 galawa. Fourteen villages had over 50 galawa and there were 122 villages with between 25 and 50 galawa (de San 1983).

We do not have up-to-date figures for the numbers of fishermen permanently or temporarily engaged in fishing. It is estimated, however, that

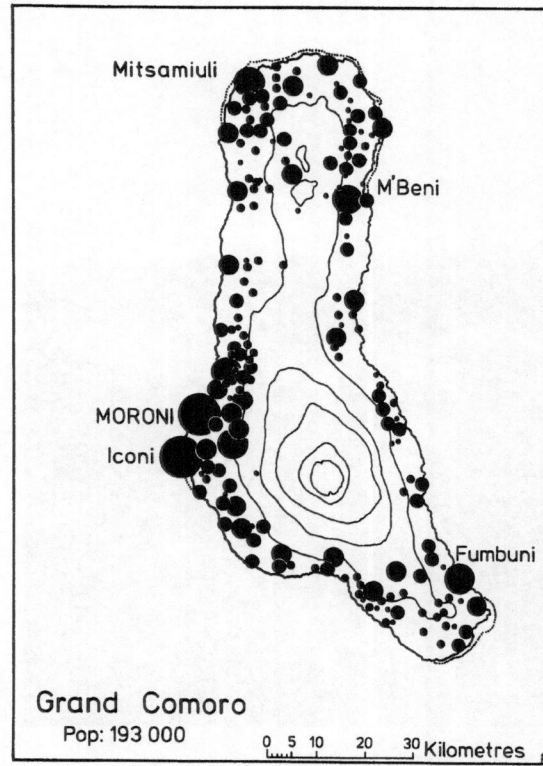

Fig. 11. Map of Grand Comoro showing the human population distribution. Data from Battistini & Vérin (1984).

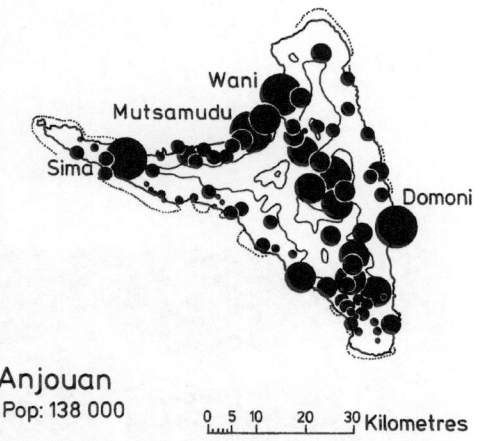

Fig. 12. Map of Anjouan showing the human population distribution. Data from Battistini & Vérin (1984).

there has been a decline in the overall number of fishermen but an increase in the number of villages with concentrations of fishing 'galawa'. Moheli has a low total population and a correspondingly low number of resident fishermen. This island has extensive coral reefs and a wide, shallow continental shelf that is heavily poached by fishermen from the other islands.

Attempts at quantifying artisanal fishing effort in the Comoro Islands have previously been made using an estimate of the population engaged in fishing activities. In 1953 there were only 700 fishermen on Grand Comoro and 250 on Anjouan (Millot 1954). By 1962 the total number of permanent and temporarily employed fishermen had risen to 1544 and 316 respectively (Boulinier-Giraud 1973).

Surveys of numbers of fishing canoes and their

distribution have also been made. A 1962 census, conducted by the Société Centrale pour l'Equipement du Territoire, found that there were then 1409 galawa in the archipelago. A decade later, in 1972, the number of galawa had risen to 3144 (Boulinier-Giraud 1973). De San (1983) estimated that there were 3000 galawa in 1982 and, five years later, it was estimated that there were, at most, 4000 in use in the three islands of the Islamic Republic (Stobbs 1987).

During the May 1990 expedition to the Comoros from the J.L.B. Smith Institute of Ichthyology we found distinct indications that the number of galawa in use was lower than the number estimated in 1987; numbers of derelict galawa were in evidence at most of the fishing villages visited on Grand Comoro and Anjouan. Figure 17 shows the probable decline in numbers of traditional wooden dugout galawa compared with the exponential rise in human population.

While the proportion of traditional galawa appears to have decreased, the numbers of recently introduced outboard-powered fishing boats (modernised galawa as well as fedawa and japawa) have increased sharply. In 1987, there were about 25 modified wooden galawa adapted to carry an outboard engine. Today there are between 500 and 600 motorised galawa and fedawa in use (de San

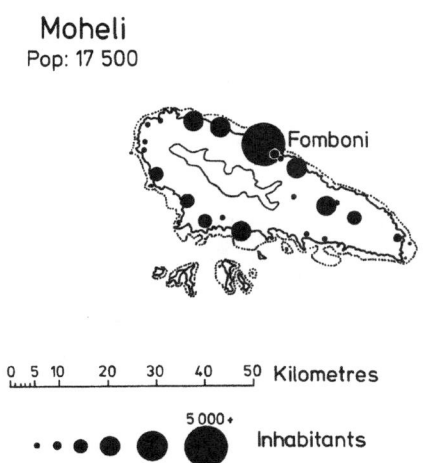

Moheli
Pop: 17 500

Fomboni

0 5 10   20   30   40   50 Kilometres

5 000+   Inhabitants

Fig. 13. Map of Moheli showing the human population distribution. Data from Battistini & Vérin (1984).

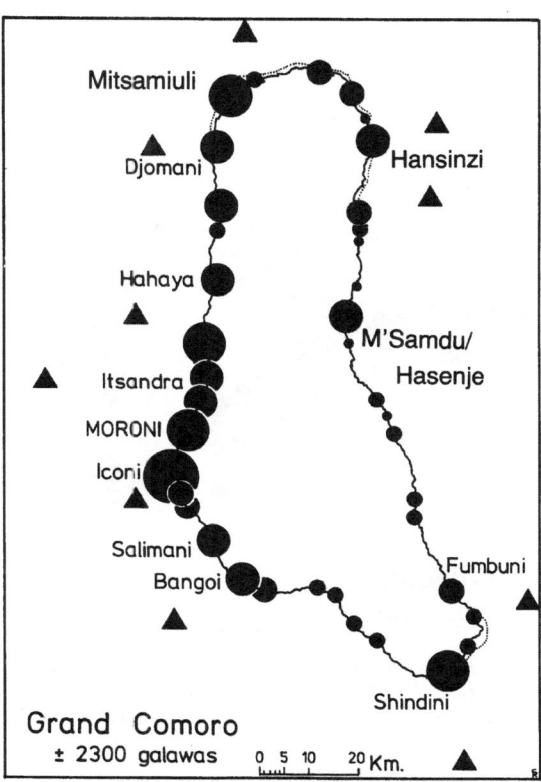

Mitsamiuli

Djomani   Hansinzi

Hahaya

M'Samdu/
Hasenje

Itsandra

MORONI

Iconi

Salimani   Fumbuni

Bangoi

Shindini

Grand Comoro
± 2300 galawas   0 5 10   20 Km.

Fig. 14. Map of Grand Comoro showing fishing canoe distribution and the positions of FADs (triangles). Data from Boulinier-Giraud (1973) and M. de San (personal communication 1990).

personal communication 1990). In addition, about 70 to 80 diesel engine-powered japawa have been imported into the Comoros from Japan.

Despite all the efforts at redirecting fishing effort away from the nearshore reefs to offshore fish aggregating devices (FADs) and trolling for pelagic fishes, i.e. away from the fishing activities that might result in an accidental catch of a coelacanth, there still remains a potential incentive for fishermen to actively try to catch a 'gombessa'. This incentive lies in the monetary reward paid by the Comoran Government for every coelacanth brought to the refrigeration facilities in Moroni. Although payments are usually late, the sum received (CFA 50 000) is well above the average annual earnings of an artisanal fisherman and enough to make the remote chance of success attractive.

We can perhaps draw comfort from the words of M. de San: 'Definitely, in my mind, a higher percentage of boats and fishermen are going offshore to catch tuna around the fish aggregating devices. That, from your point of view, is relieving the pressure on the coelacanth because they (the fishermen) probably do not go as frequently to fish inshore in 200 to 300 meters' (personal communication 1990).

Increased fishing efficiency

There is no reason to believe that modern nylon monofilament fishing line is any more (or less) efficient than traditionally prepared cotton line, and most fishermen prefer the handling characteristics of the latter. The use of modern hooks, steel trace, swivels and artificial lures has not significantly changed traditional fishing practices that might result in the capture of a coelacanth.

Hand forged hooks, made from iron nails as described by Suzuki & Tanauma (1984) and Suzuki et al. (1985), are not as strong as modern steel hooks but appear to have been adequate for all but the largest fishes (personal observation 1988). Steel hooks are cheap and easily available, and we were not able to find a single traditional hookmaker still practicing his craft in 1988.

356

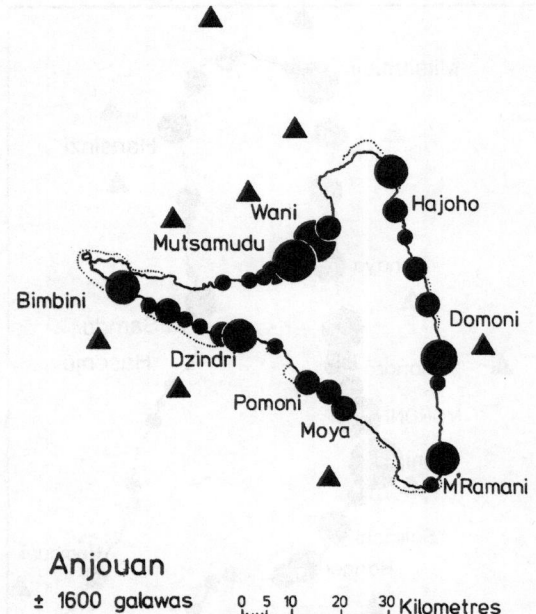

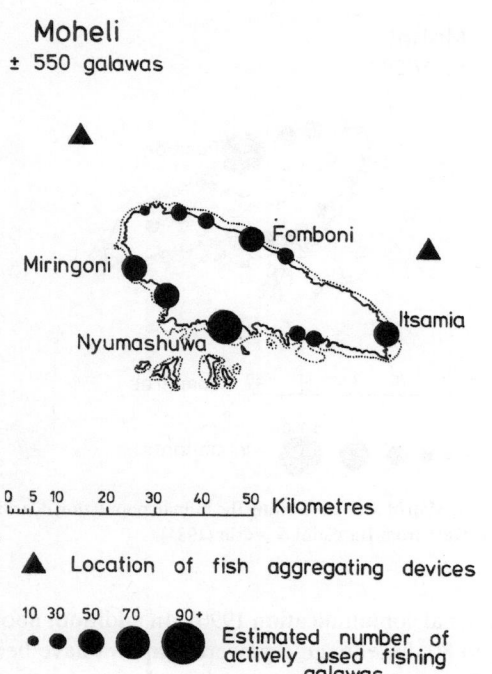

Moheli
± 550 galawas

Anjouan
± 1600 galawas

0 5 10   20   30 Kilometres

Fig. 15. Map of Anjouan showing fishing canoe distribution and the positions of FADs (triangles). Data from Boulinier-Giraud (1973) and M. de San (personal communication 1990).

0 5 10   20   30   40   50 Kilometres

▲ Location of fish aggregating devices

10 30 50 70   90+
● ● ● ● ⬤ Estimated number of actively used fishing galawas

Fig. 16. Map of Moheli showing fishing canoe distribution and the positions of FADs. Data from Boulinier-Giraud (1973) and M. de San (personal communication 1990).

It is in the development of motorised fishing craft that a potential threat exists to the nearshore ecology and, in particular, to the coelacanth. In 1987 the European Economic Community established a fisheries development centre in Moroni under the directorship of a tuna fishing consultant, M. de San. Projet Pêche Artisànale was to serve four main functions, and its outstanding achievements can be measured by the results:

(1) to design and build small outboard-engined GRP boats suitable for local conditions and acceptable to Comoran fishermen; more than 220 'fedawa' have been made and sold.

(2) to sell and repair outboard engines and to train local artisans in the basics of repair and maintenance of outboard engines; more than 40 artisan mechanics have received basic training in the essentials of outboard engine maintenance. In the larger centres there are well-trained mechanics capable of more sophisticated repair work, and major repairs are undertaken at the workshops in Moroni. Some factory workers

have also been well trained in the techniques and complexities of GRP moulding and layup.

(3) to establish a small shop where local fishermen can purchase tuna fishing equipment and outboard engine spares. The small but well-stocked shop at the Projet FED factory in Moroni sells a variety of fishing lures, monofilament fishing line, hooks and other fishing gear as well as outboard engine spares and ancillaries for all the makes and models of engines sold by the project.

(4) to design, prepare and deploy FADs around the islands. At present there are 21 FADs deployed around the islands of the Islamic Republic – 10 around Grand Comoro, 9 around Anjouan and two off Moheli (Figs 14, 15 and 16). All but the furtherest offshore FADs are close enough to be reached by paddling galawa and there is strong competition between the fishermen in these canoes and those in outboard-powered boats. A measure of the effec-

tiveness and popularity of the FADs can be gauged by the annual tuna catches, which have increased from 5000 tons before the deployment of FADs to over 8000 tons today (M. de San personal communication 1990).

Powered GRP boats, such as the heavy inboard-powered japawa and the smaller, lighter fedawa, offer a measure of safety for the fisherman. These boats are inherently more seaworthy than galawa owing to their larger size and greater stability, and they offer the capability of running rapidly for distant safety in the face of deteriorating weather. Despite the comparative comfort and safety offered by these new-generation boats, *Ruvettus* fishermen still prefer the traditional paddling galawa. The reason for this is that a paddling canoe is easier and quicker to manoeuver against currents and keep on station with the fishing line hanging vertical (personal communications with Comoran fisherman, 1987, 1988 and 1990). J. Hamlin (personal communication 1990) attempted to fish for coelacanths from a japawa but soon found that it was almost impossible to keep this large and heavy boat on station.

There has been one reported coelacanth catch from a japawa (Balon et al. 1988, CCC no. 147 in Bruton & Coutouvidis 1991). Even though such catches are not usual, they are certainly possible, even probable. Although outboard-powered craft were introduced to the Comoros in order to enable fishermen to harvest pelagic fishes, they also have the potential to facilitate inshore fishing in previously inaccessible areas.

It has been suggested that fishermen might be discouraged from fishing in places where there is an increased chance of catching a coelacanth. If such a ruling were to interfere with their traditional fishing rights, it could lead to ill feelings and a disregard for the law. If, however, a cheap and culturally acceptable substitute were to be found for the food and medicinal value of *Ruvettus,* as suggested by Stobbs (1989a), then night fishing might eventually cease and coelacanths would no longer be caught accidentally.

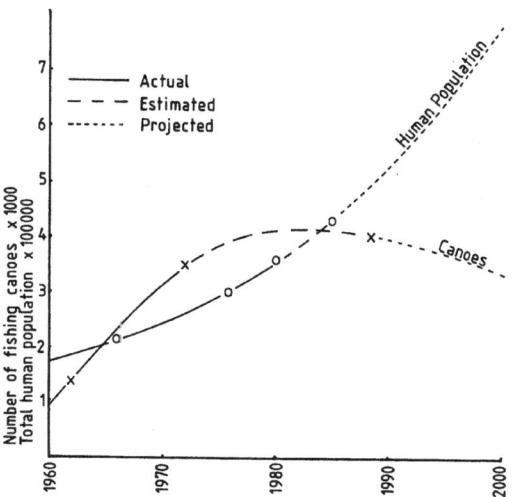

Fig. 17. Graph showing actual and relative decline in numbers of fishing canoes compared to the exponential rise in the human population in the Federal Islamic Republic of the Comoros.

Conclusion

It is apparent that coelacanths cannot be directly targetted. There is no known fishing gear (bait preference or presentation) that might be specific for coelacanths. All known captures, including the East London specimen of 1938, have been taken by accident and not by design. No Comoran fisherman can, or will, predict the capture of a coelacanth, and determined attempts at such capture have thus far proved to be fruitless (J. Hamlin personal communication 1990). Past expeditions have offered considerable rewards for the capture of a coelacanth yet none has ever had a specimen delivered on request within the time alotted.

We have no proof that there has been a sustained increase in fishing effort for the capture of a coelacanth. There has, on the contrary, been a definite, though unmeasured, shift in fishing effort from nearshore benthic fishing to offshore trolling and FAD fishing for pelagic species.

Changes in fishing lines and hooks used in the *Ruvettus* fishery cannot reasonably be seen as contributing to an increase in the possibility of catching a coelacanth. Furthermore, techniques of deep handlining have not changed. It is in the potential

358

for improved sea craft to travel further, faster and with increased safety that there might be a threat to the coelacanth. The attractions of a financial reward for such a catch are also very appealing, especially to the average artisanal fisherman who stands to double, even triple, his income. In an economic predicament such as prevails in the Comoros, where public servants have received their salaries four to six months late over the past two years, the potential of earning a quick reward cannot be overlooked. It is essential, therefore, that the Comoran government should stop paying a high price for coelacanths.

It is our belief that, providing modern equipment is not used in the nearshore fishery, the coelacanth has little to fear from the normal practices of the Comoran artisanal fisherman. The Comoran authorities should, therefore, make every effort to ensure that modern boats are not used for benthic fishing. The entire nearshore region should be made off-limits to mechanised gear and commercial operations and should remain the sole precinct of village-based artisanal fishermen. Furthermore, artisanal fishermen should be encouraged to pursue their calling in a traditional manner while also having the opportunity to improve their quality of life.

Acknowledgements

Our thanks are due to the many fishermen of Grand Comoro, Anjouan and Moheli who spoke to us on traditional fishing practices, especially mvuvi Ibrahima Mdoionhoma of Zahaju Hambu. We also extend our grateful thanks to J. Hamlin, Mr. and Mrs. Tony Kay, D. Rotsaert and M. de San for their help in obtaining information and samples of traditional fishing gear. Financial and logistical support was provided by Damir ben Ali, R. Winterbottom, Marco Boni and Rodger Harding, the South African Department of Foreign Affairs and the United Building Society. We are grateful to E.K. Balon, P.H. Greenwood, J.A. Musick, J.W. Atz and H. Fricke for comments on the manuscript.

References cited

Alexander, D. 1976. Holiday in the Islands, a guide to Comores, Madagascar, Reunion and Rodrigues. Purnell, Cape Town. 251 pp.

Anon. 1980. Africa Pilot Vol. III., Thirteenth edition. Hydrographer of the Navy, Ministry of Defence, Taunton, Somerset. 294 pp.

Balon, E.K., M.N. Bruton & H. Fricke. 1988. A fiftieth anniversary reflection on the living coelacanth, *Latimeria chalumnae:* some new interpretations of its natural history and conservation status. Env. Biol. Fish. 23: 241–280.

Battistini, R. & P. Vérin. 1984. Les Comores. Agence de Coopération Culturelle et Technique. Nathan, Paris. 144 pp.

Boulinier-Giraud, G. 1973. Étude morphologique de la pirogue à balancier aux Comores et dans l'ouest de l'Océan Indien. Mémoire de Maîtrise Spécialisée d'Ethnologie, Dissertation at the Université Paris 10, Paris.

Bourgois, J.-J. 1989. *Tephrosia vogelii,* une plante ichthyotoxique utilisée pour la pêche aux Comores. Bull. Soc. R. Bot. Belg. 133: 81–86.

Bruton, M.N., C.D. Buxton, G.R. Hughes & R.E. Stobbs. 1989. Recommendations on marine conservation in the Federal Islamic Republic of the Comores/Recommandations sur la conservation marine dans la République Fédérale Islamique des Comores. Invest. Rep., J.L.B. Smith Inst. Ichthyol. 34: 1–104.

Bruton, M.N. & S.E. Coutouvidis. 1991. An inventory of all known specimens of the coelacanth *Latimeria chalumnae,* with comments on trends in the catches. Env. Biol. Fish. 32: 371–390. (this volume)

Bruton, M.N. & R.E. Stobbs. 1991. The ecology and conservation of the coelacanth *Latimeria chalumnae.* Env. Biol. Fish. 32: 313–339 (this volume)

Couper, A.D. (ed.). 1989. The Times atlas and encyclopaedia of the sea. Times Books, London. 272 pp.

Cullen, A. & P. Hemphill. 1971. Crash strike. East African Publishing House, Nairobi. 175 pp.

Day, C.L. 1964. The art of knotting and splicing. Adlard Coles Ltd., London. 225 pp.

de San, M. 1983. Profil de la pêche artisanale aux Comores. SWIOP document RAF/79/065. 21 pp.

Fricke, H. & K. Hissmann. 1990. Natural habitat of coelacanths. Nature 346: 323–324.

Fricke, H. & R. Plante. 1988. Habitat requirements of the living coelacanth *Latimeria chalumnae* at Grande Comore, Indian Ocean. Naturwissenschaften 75: 149–151.

Griffith, R.W. & K.S. Thomson. 1973. *Latimeria chalumnae:* reproduction and conservation. Nature 242: 617–618.

Gudger, E.W. 1927. Wooden hooks used for catching sharks and *Ruvettus* in the South Seas; a study of their variation and distribution. Amer. Mus. Nat. Hist. 28: 200–348.

Hornell, J. 1944. The outrigger canoes of Madagascar, East Africa and the Comoro Islands. Mariner's Mirror 30: 3–18, 170–185.

Locket, N.A. 1976. A future for the coelacanth? New Scientist 58: 456–458.

McCosker, J.E. 1979. Inferred natural history of the coelacanth. pp. 17–24. *In:* J.E. McCosker & M.D. Lagios (ed.) The Biology and Physiology of the Living Coelacanth, Occ. Pap. Calif. Acad. Sci. 134, San Francisco.

McCosker, J.E. & S. McCosker. 1976. To the islands of the moon. Pacific Discovery 29: 19–28.

Millot, J. 1954. Le troisième coelacanthe, historique, élements d'écologie, morphologie externe, documents divers. Naturaliste Malgache, suppl. 1: 1–26.

Newitt, M. 1984. The Comoro Islands, struggle against dependence in the Indian Ocean. Gower Publishing Co., Aldershot. 144 pp.

Nordhoff, C.B. 1928. Fishing for the oilfish. Nat. Hist. 28: 40–45.

Reeves, P. 1985. Fish slips from the poor man's table. ICLARM 8(2): 5–6.

Stobbs, R.E. 1987. The 'galawas' of the Federal Islamic Republic of the Comores, with notes on artisanal fishing and recommendations for a maritime museum. Invest. Rep., J.L.B. Smith Inst. Ichthyol. 26: 1–21.

Stobbs, R.E. 1989a. Laxative lipids and the survival of the living coelacanth. S. Afr. J. Sci. 85: 557–558.

Stobbs, R.E. 1989b. The coelacanth enigma. The Phoenix 2(2): 8–15.

Stobbs, R.E. 1990. Comorian fish names with a preliminary list of Malagasy common names. Invest. Rep., J.L.B. Smith Inst. Ichthyol. 35: 1–34.

Suzuki, N., Y. Suyehiro & T. Hamada. 1985. Initial reports of expeditions for coelacanth – Part 1 – Field studies in 1981 and 1983. Sci. Pap. Coll. Arts Sci., Univ. Tokyo. 35: 37–79.

Suzuki, N. & K. Tanauma. 1984. Coelacanth fishing by native fishermen. Proc. First Symposium on Coelacanth Studies, Tokyo (In Japanese).

Thomson, D. 1980. Conflict within the fishing industry. ICLARM 3(3): 13–14.

Environmental Biology of Fishes **32**: 361–367, 1991.
© 1991 *Kluwer Academic Publishers.*

Occurrence of organochlorine contaminants in tissues of the coelacanth *Latimeria chalumnae*

Robert C. Hale, John Greaves, Jennifer L. Gundersen & Robert F. Mothershead II
Virginia Institute of Marine Science, College of William and Mary, Gloucester Point, VA 23062, U.S.A.

Received 1.8.1989   Accepted 10.8.1990

Key words: Polychlorinated biphenyl, DDT, Gas chromatography, Mass spectrometry, Crossopterygii

Synopsis

Gas chromatography coupled with electrolytic conductivity detection and electron capture negative chemical ionization mass spectrometry have been used to identify and quantify organochlorine xenobiotics in tissues from two specimens of the living coelacanth *Latimeria chalumnae*. Compounds identified include polychlorinated biphenyls (PCB), 4,4'-DDT and its metabolites 4,4'-DDD and 4,4'-DDE. Levels of these compounds in the specimens were observed to be generally related to tissue lipid content. Highest concentrations of the xenobiotics were present in the lipid-rich swim bladder, followed by adipose tissue and liver. Levels ranged from 89 to 510 μg kg^{-1} for PCB and 210 to 840 μg kg^{-1} for $\sum$-DDT (including DDD and DDE) on a wet-weight basis. Organochlorine concentrations in relatively lipid-poor tissues, i.e. muscle and kidney, were lower. Parent DDT contributed significantly to the $\sum$-DDT burden. PCB congeners containing five to seven chlorines were most prevalent. Component distributions did not match those present in common commercial PCB formulations. However, major congeners observed were similar to those reported for other fishes.

Introduction

Organochlorine xenobiotics are ubiquitous pollutants of the ecosphere (Hutzinger et al. 1974, Gregor & Gummer 1989). Examples include polychlorinated biphenyls (PCB) and various pesticides, e.g. DDT (4,4'-dichloro-diphenyl-trichloroethane) and its major breakdown products DDD (4,4'-dichloro-diphenyl-dichloroethane) and DDE (4,4'-dichloro-diphenyl-dichloroethylene). To a lesser extent chlordane and toxaphene have also been found (Musial & Uthe 1983, Muir et al. 1988). The oceans may represent the ultimate sink for many organochlorine compounds. These xenobiotics occur at all depths, including the deep ocean (Schulz et al. 1988).

Organochlorines are generally lipophilic and resistant to metabolism and are thus subject to bioaccumulation directly from the water or via the food chain (van der Oost et al. 1988). In biota there are reports of organochlorines in fishes from a number of locales, including: the Atlantic (Harvey et al. 1973, Kramer et al. 1984), Bering Sea (Takagi et al. 1975), southeastern Pacific (Tanabe et al. 1980), Antarctic (Subramanian et al. 1983) and Arctic (Muir et al. 1988) oceans. Results have generally centered on residues in muscle or liver tissue. Concentrations normally observed have been in the μg kg^{-1} to mg kg^{-1} range. Typically levels were higher in samples from the northern hemisphere (Kramer et al. 1984). This distribution reflects the greater degree of industrialization in this region. In addition to fishes that inhabit shallow waters, a few studies have examined residues in deepsea species, such as *Antimora rostrata* (Barber & Warlen 1979). In this particular species, $\sum$-DDT concentrations in the liver were in the 1000–2700 μg kg^{-1} range, on a wet-weight basis.

The coelacanth, *Latimeria chalumnae*, is the only known extant crossopterygian. All representatives of this group were believed to have been extinct for 70 million years, until a live coelacanth was captured in 1938. Although rare, a number of specimens have been captured in the last 50 years. The range of the coelacanth appears to be restricted to the vicinity of the Comoro Islands, northwest of Madagascar. The species is demersal, frequenting water depths of 100 to 800 m, and piscivorous (Thomson 1973). Two specimens were obtained from the area of the Comoro Islands and provided an opportunity to investigate the tissue burdens of xenobiotics in this rare species.

Materials and methods

Sample preparation

Glassware used in this study was subjected to a rigorous cleaning procedure. Initially, it was soaked in an alkaline cleaning agent (Contrad 70, Curtin Matheson Scientific, Inc., Houston, TX) overnight. Glassware was then washed in detergent and water, rinsed with 3N HCl and deionized water. Nonvolumetric glassware was baked at 500° C overnight, volumetric glassware at 100° C. It was subsequently sealed with solvent rinsed aluminum foil and stored. Immediately prior to use glassware was solvent rinsed. High purity solvents used were toluene, methanol, acetone and dichloromethane (Burdick and Jackson Division, Baxter Healthcare Corp., Muskegon, MI).

A male and female coelacanth (both immature, VIMS collection numbers 8117 and 8118) were examined. Liver, kidney, swimbladder, mesenteric adipose tissue and skeletal muscle were removed from these fish and analyzed for organochlorine contaminants. Each tissue examined was chemically dried by mixing with a 1 : 9 mixture of synthetic amorphous precipitated silica (QUSO G 35, Degussa Corp., Teterboro, NJ) and pre-extracted

sodium sulfate (reagent grade, Fisher Scientific, Fairlawn, NJ). The desiccated samples received a surrogate standard containing decachlorobiphenyl and were subsequently soxhlet extracted (48 h) with dichloromethane. Blanks were analyzed in parallel with the samples as a check on possible laboratory contamination. The dichloromethane was removed by rotary evaporation and replaced with hexane. Removal of lipids was accomplished by treatment with concentrated sulfuric acid (Veierov & Aharonson 1978). Extracts were further purified by gel permeation chromatography (Hale 1988) or by passage through Florisil (Fisher Scientific, Fairlawn, NJ).

Analysis

A gas chromatograph (GC, Model 3300, Varian Associates, Walnut Creek, CA) equipped with an electrolytic conductivity detector (ELCD, OI Corporation, College Station, TX) and a GC-quadrupole mass spectrometer (GC-MS, Model ELQ400-2, Extrel Corporation, Pittsburgh, PA) were employed to separate, identify and quantify organochlorine xenobiotics in the coelacanth tissues. The GC-ELCD was utilized in the halogen selective mode and the GC-MS was used in the electron capture negative chemical ionization (NCI) mode. Methane served as the moderator gas. The gas chromatographic column was a fused silica capillary column coated with DB-5 (95% methyl – 5% phenyl-silicone phase, 30 m long × 0.33 mm i.d. × 0.25 μm film thickness, J & W Scientific Inc., Folsom, CA). After injection of the sample in the splitless mode, the GC column temperature was maintained at 60° C for two minutes. The temperature was then increased at a rate of 4° min^{-1} to a final temperature of 310° C, followed by a 10 min hold at this temperature. The carrier gas was helium with a head pressure of 1.0 kg cm^{-2}. The ELCD reaction-tube temperature was set to 950° C. The electrolyte was n-propanol. For the GC-MS analyses, the capillary column was connected directly to the ion source of the MS. The interface was maintained at 260° C and the source temperature was held at 100° C.

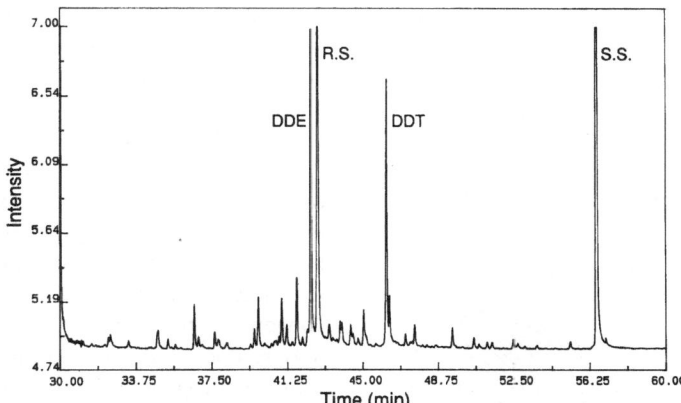

Fig. 1. GC-ELCD chromatogram of an extract of the kidney from the female coelacanth. The majority of the unlabeled peaks were identified as PCB. R.S. = retention standard (2,4'-DDD). S.S. = surrogate standard (decachlorobiphenyl).

A retention standard mixture was added to, or coinjected with, the purified extracts to assist in analyte identification. The components of the mixture were: 2-chloronaphthalene, alpha-hexachlorocyclohexane, 2,4'-DDD and pentachlorobenzene. Quantitation of organochlorines was based on comparison of the chromatographic peak areas detected with that of pentachlorobenzene. Corrections were made for analytical losses based on recovery of the surrogate standard. Appropriate factors to correct for differences in detector response between the internal standard and the sample components were also applied.

Lipid contents of the tissues were estimated by gravimetric measurement of an aliquot of the dichloromethane extract after evaporation of the solvent at room temperature.

Results and discussion

PCB

PCB were identified in all tissues examined. A GC-ELCD profile of the extract from kidney is shown in Fig. 1. A histogram depicting the contributions of the various chlorinated homolog groups to the total PCB content of the swim bladder, from the immature male fish, is provided in Fig. 2. Patterns were similar for the remaining tissues in both

coelacanths. Table 1 presents the PCB concentration in each tissue on a wet-weight basis. The highest PCB levels occurred in swim bladder and adipose tissues. These tissues in the coelacanth contained over 90% extractable lipoidal material. Relative PCB concentrations in the other tissues were liver > kidney > muscle. Extractable lipid contents of these tissues were: liver 32%, muscle 16% and kidney 8.8%. The high lipid and wax content of the coelacanth has been previously noted (Benson & Lee 1975). The composition of the lipid pool itself has been reported to be important in the partitioning of fat soluble pollutants (Schneider 1982). No attempt to distinguish between the classes of lipids was made in the current study.

The PCB congener patterns observed differ significantly from commercial PCB formulations. As noted previously (Fig. 2), the most common homologs in the coelacanth tissues were those containing five to seven chlorine atoms per molecule, in agreement with reports of PCB residues in other fish species (e.g. Kramer et al. 1984). This is similar, in terms of percent chlorination, to Aroclor 1254 and 1260. These products were originally manufactured in the U.S. by Monsanto. Similar products were marketed under different names (e.g. Clophen and Kaneclor) by manufacturers in Europe and Japan. In contrast to the distribution of homologs in fishes, it has been estimated that the majority of PCB released to the environment contained

364

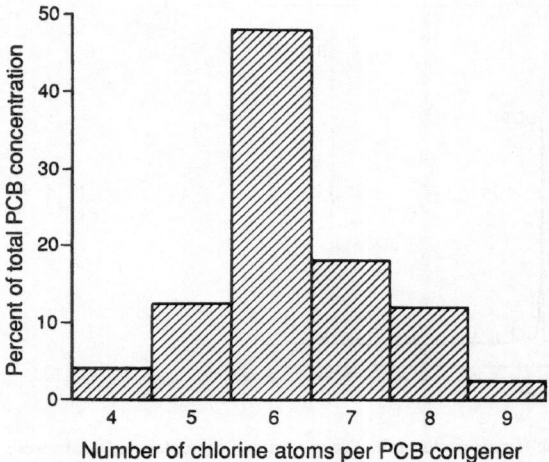

Fig. 2. Distribution of PCB homologs in an extract from the swimbladder of the male coelacanth. The pattern observed was typical for all tissues examined.

less than five chlorines (National Research Council 1979). This preferential accumulation of the higher chlorinated congeners in these organisms is attributed to their high lipophilicity. Major congeners detected in the coelacanth were identified as: 101 (2,2′,4,5,5′-pentachlorobiphenyl), 138 (2,2′, 3,4,4′,5′-hexachlorobiphenyl), 153 (2,2′,4,4′,5,5′-hexachlorobiphenyl) and 180 (2,2′,3,4,4′,5,5′-heptachlorobiphenyl), as determined by the retention data of Mullin et al. (1984). This agrees well with published data on other fishes (Maack & Son-

zogni 1988, Niimi & Oliver 1989). Identifications are tentative since some co-elution of these PCB congeners with other PCB standards was observed and has been reported by other researchers (Duinker et al. 1988, Roos et al. 1989).

Detailed consideration of the overall distribution of congeners suggests significant alteration/fractionation of the original PCB formulations via biological uptake processes, metabolism and abiotic or physical degradation. As a specific example, congener 174 (2,2′3,3′,4,5,6′-heptachlorobiphenyl) is one of the less recalcitrant PCB. It has vicinyl hydrogens at the meta and para positions of one of the aromatic rings and is, therefore, a candidate for metabolic alteration (Niimi & Oliver 1983). In Aroclor 1260 the ratio of congeners 180 : 174 is 10 : 4, whereas in the mesenteric fatty tissue from the coelacanth the ratio is 10 : 1. This suggests possible metabolic alteration of the mixture. Consideration of ratios with respect to congener 180 is valid because this congener is a diagnostic peak for Aroclor 1260 (Zell & Ballschmitter 1980). Any contribution of congener 174 by other PCB formulations would reduce the 180 : 174 ratio. The 10 : 1 ratio obtained for this tissue is, therefore, a conservative estimator of the change in the PCB mixture. It is not, however, possible to determine at what stage the alteration occurred, i.e. by metabolism within the coelacanth or prior to accumulation. It does seem reasonable to expect that, in common with other fishes (Niimi & Oliver 1983), the coela-

Table 1. Concentrations of PCB, Σ-DDT, DDE, DDD and DDT in selected tissues of a female and male coelacanth. Both specimens examined were immature. Values are in μg kg^{-1} and are expressed on a wet-weight basis. NQ = compound detected, but concentration too low to quantify.

	PCB	Σ-DDT	DDE	DDD	DDT
Female					
Kidney	77	77	46	2.3	29
Liver	89	210	120	12	79
Swimbladder	330	840	520	42	280
Mesenteric adipose	280	750	490	23	240
Muscle	38	9.4	6.0	0.27	3.1
Male					
Swimbladder	510	620	440	19	160
Mesenteric adipose	190	300	240	NQ	65
Muscle	14	17	13	0.68	3.7

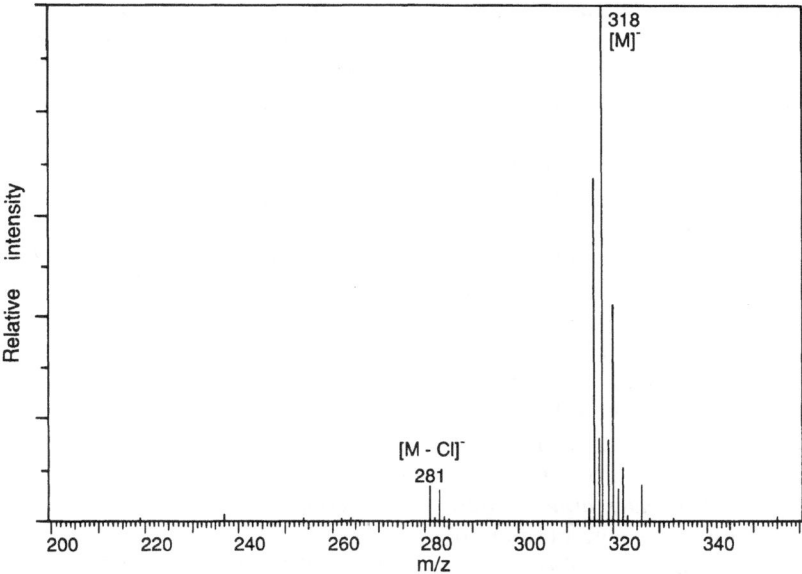

Fig. 3. NCI mass spectrum of 4,4'-DDE from an extract of the kidney from the female coelacanth.

canth is capable of metabolizing PCB to some degree.

DDT

DDT and its metabolites, DDD and DDE, were identified in all tissues (Table 1). Levels of $\sum$-DDT were typically higher than those of PCB in the tissues examined. In common with PCB, the highest $\sum$-DDT concentrations were found in tissues with the greatest lipid content. An NCI-mass spectrum of 4,4'-DDE is shown in Fig. 3. This spectrum, obtained from an extract of the kidney from the female coelacanth, compares exactly with that of an authentic standard. Although the predominant component was DDE, the DDT : DDE ratio of 1 : 2 in the coelacanth is elevated compared to recent data for shallow water fishes. Ratios for these latter organisms are typically 1 : 10 or more (Kramer et al. 1984, Oliver & Niimi 1988). Studies on levels of organochlorines in deep water fishes have shown elevated concentrations of DDT relative to its breakdown products (Barber & Warlen 1979, Kramer et al. 1984). The tissue burdens of

DDD were much lower in the coelacanths, generally representing less than 5% of the $\sum$-DDT.

It is noteworthy that parent DDT has been identified as a major contaminant in the coelacanth tissue. The half-life of DDT in biological systems is limited. Since DDT use was banned in many developed countries in the early 1970's, environmental residues of the parent compound in these countries have generally decreased; although some data suggest a recent increase attributable to aerial input from countries where the insecticide is still in use (Rapaport et al. 1985). The presence of parent DDT in the coelacanth tissue may, therefore, be indicative of the current use of this insecticide in countries bordering the Indian Ocean. Such use of DDT has been documented in Zimbabwe (Thompson 1983), South Africa (Davies & Randall 1990) and India (Rapaport et al. 1985). In addition to DDT, its derivatives and PCB there was evidence of other chlorinated compounds in the coelacanth tissues. At present additional efforts are being made to determine the identities of these compounds.

Insufficient data are available to determine if any of these pollutants are present at levels that would threaten the coelacanth population. Samples ana-

366

lyzed in the present study were obtained from immature specimens. Tissue burdens of many organochlorines have been observed to increase with the age and size of the fish (Subramanian et al. 1983). Concentrations in adult coelacanths may be higher than those reported here. Since the female specimen was immature, no ova were available for analysis. Eggs of coelacanths are very large and lipid rich (Thomson 1973). Consequently, elevated concentrations of lipophilic pollutants would be expected to occur in this tissue. This may have effects on the breeding success of the species. Reproductive impairment of fishes, attributed to $\sum$-DDT burdens of ovaries, has recently been reported (Hose et al. 1989). Stegeman et al. (1986) have reported that levels of cytochrome P-450E in the deep-sea fish *Coryphaenoides armatus* were positively correlated with tissue concentrations of PCB. This suggests that some physiological processes may already have been impacted by the presence of pollutants in the deep ocean. As issues such as ocean disposal of wastes are debated, the occurrence and impact of xenobiotics on the deep-sea and open-ocean environments must be considered.

In conclusion, lipophilic organochlorine pollutants were detected in the tissues of both coelacanths examined. The remoteness of their habitat has not provided a sufficient buffer to prevent exposure of the coelacanths to PCB and DDT. This finding was not unexpected because atmospheric transport has resulted in global distribution of recalcitrant organochlorines (Manchester-Neesvig and Andren 1989). Movement of these xenobiotics to depth in the oceans has been facilitated by various phenonomena, e.g. sinking fecal pellets (Elder & Fowler 1977). Particularly noteworthy was the large contribution parent DDT made to the total organochlorine tissue burden of the fish. Factors contributing to the accumulation of the pollutants are the high lipid content of the coelacanth and the position of the fish in the trophic ladder.

At present the ability of the coelacanth to transform and eliminate xenobiotics is unknown. It is interesting to note that if current technology had been available to analyse for PCB and DDT at the time of the capture of the first coelacanth in 1938, it is unlikely these pollutants would have been detected. The insecticidal properties of DDT were not identified until 1939; and although commercial production of PCB began in 1929, their use escalated greatly only after the second world war.

Acknowledgements

The Explorer's Club and Virginia Institute of Marine Science Coelacanth Program are gratefully acknowledged for access to study material. Technical assistance was provided by G. Vadas, E. Harvey, R. Edstrom and P. Mason. VIMS contribution number 1585.

References cited

Barber, R.T. & S.M. Warlen. 1979. Organochlorine insecticide residues in deep sea fish from 2500 m in the Atlantic Ocean. Environ. Sci. Techol. 13: 1146–1148.

Benson, A.A. & R.F. Lee. 1975. The role of waxes in oceanic food chains. Sci. Amer. 232: 76–86.

Davies, R. & R.M. Randall. 1990. Historical and geographical patterns in eggshell thickness of African fish eagles, *Haliaeetus vocifer*, in relation to pesticide use within southern Africa. *In:* R.D. Chancellor (ed.) Proceedings Third World Conference on Birds of Prey, World Working Group on Birds of Prey. (in press.)

Duinker, J.C., D.E. Schultz & G. Petrick. 1988. Selection of chlorinated biphenyl congeners for analysis in environmental samples. Mar. Poll. Bull. 19: 19–25.

Elder, D.L. & S.W. Fowler. 1977. Polychlorinated biphenyls: penetration into the deep ocean by zooplankton. Fecal pellet transport. Science 197: 359–361.

Gregor, D.J. & W.D. Gummer. 1989. Evidence of atmospheric transport and deposition of organochlorine pesticides and polychlorinated biphenyls in Canadian Artic snow. Environ. Sci. Technol. 23: 561–565.

Hale, R.C. 1988. Disposition of polycyclic aromatic compounds in blue crabs, *Callinectes sapidus*, from the southern Chesapeake Bay. Estuaries 11: 255–263.

Harvey, G.R., H.P. Miklas, V.T. Bowen & W.G. Steinhauer. 1973. Observations on the distribution of chlorinated hydrocarbons in Atlantic Ocean organisms. J. Mar. Res. 32: 103–118.

Hose, J.E., J.N. Cross, S.G. Smith & D. Diehl. 1989. Reproductive impairment in a fish inhabiting a contaminated coastal environment off southern California. Environ. Poll. 57: 139–148.

Hutzinger, O.S., S. Safe & V. Zitko. 1974. The chemistry of PCBs. CRC Press, Boca Raton. 269 pp.

Kramer, W., H. Buchert, U. Reuter, M. Biscoito, D.G. Maul, G. Le Grand & K. Ballschmiter. 1984. Global baseline pollution studies IX: C_6–C_{14} organochlorine compounds in surface-water and deep-sea fish from the eastern North Atlantic. Chemosphere 13: 1255–1267.

Maack, L. & W.C. Sonzogni. 1988. Analysis of polychlorobiphenyl congeners in Wisconsin fish. Arch. Environ. Contam. Toxicol. 17: 711–719.

Manchester-Neesvig, J.B. & A.W. Andren. 1989. Seasonal variation in the atmospheric concentration of polychlorinated biphenyl congeners. Environ. Sci. Technol. 23: 1138–1148.

Muir, D.C.G., R.J. Norstrom & M. Simon. 1988. Organochlorine contaminants in Arctic marine food chains: accumulation of specific polychlorinated biphenyls and chlordane-related compounds. Environ. Sci. Technol. 22: 1071–1079.

Mullin, M.D., C.M. Pochini, S. McCrindle, M. Romkes, S.H. Safe & L.M. Safe. 1984. High-resolution PCB analysis: synthesis and chromatographic properties of all 209 PCB congeners. Environ. Sci. Technol. 18: 468–476.

Musial, C.J. & J.F. Uthe. 1983. Widespread occurrence of the pesticide toxaphene in Canadian east coast marine fish. Intern. J. Environ. Anal. Chem. 14: 117–126.

National Research Council. 1979. Polychlorinated biphenyls. National Academy of Sciences, Washington D.C. 182 pp.

Niimi, A.J. & B.G. Oliver. 1983. Biological half-lives of polychlorinated biphenyl (PCB) congeners in whole fish and muscle of rainbow trout (*Salmo gairdneri*). Can. J. Fish. Aquat. Sci. 40: 1388–1394.

Niimi, A.J. & B.G. Oliver. 1989. Distribution of polychlorinated biphenyl congeners and other halocarbons in whole fish and muscle among Lake Ontario salmonids. Environ. Sci. Technol. 23: 83–88.

Oliver, B.G. & A.J. Niimi. 1988. Trophodynamic analysis of polychlorinated biphenyl congeners and other chlorinated hydrocarbons in the Lake Ontario ecosystem. Environ. Sci. Technol. 22: 388–397.

Rapaport, R.A., N.R. Urban, P.D. Capel, J.E. Baker, B.B. Looney, S.J. Eisenreich & E. Gorham. 1985. 'New' DDT inputs to North America: atmospheric deposition. Chemosphere 14: 1167–1173.

Roos, A.H., P.G.M. Kienhuis, W.A. Traag & L.G.M.Th. Tuinstra. 1989. Problems encountered in the determination of 2,3,4-2′,4′,5′-hexachlorobiphenyl (CB 138) in environmental samples. Intern. J. Environ. Anal. Chem. 36: 155–161.

Schneider, R. 1982. Polychlorinated biphenyls (PCBs) in cod tissues from the Western Baltic: significance of equilibrium partitioning and lipid composition in the bioaccumulation of lipophilic pollutants in gill-breathing animals. Meeresforsch. 29: 69–79.

Schulz, D.E., G. Petrick & J.C. Duinker. 1988. Chlorinated biphenyls in North Atlantic surface and deep water. Mar. Poll. Bull. 19: 526–531.

Stegeman, J.J., P.J. Kloepper-Sams & J.W. Farrington. 1986. Monooxygenase induction and chlorobiphenyls in the deep-sea fish *Coryphaenoides armatus*. Science 231: 1287–1289.

Subramanian, B.R., S. Tanabe, H. Hidaka & R. Tatsukawa. 1983. DDTs and PCB isomers and congeners in Antarctic fish. Arch. Environ. Contam. Toxicol. 12: 621–626.

Takagi, M., H. Murayama & S. Soma. 1975. PCB contents in several species of flatfish collected in the Eastern Bering Sea. Bull. Jap. Soc. Sci. Fish. 41: 685–690.

Tanabe, S., H. Tanaka, R. Tatsukawa & I. Nakamura. 1980. Organochlorine compound residues in several species of fishes caught off Chile. Bull. Jap. Soc. Sci. Fish. 46: 763–769.

Thompson, W.R. 1983. DDT in Zimbabwe. World Working Group on Birds of Prey 1: 41–47.

Thomson, K.S. 1973. Secrets of the coelacanth. Nat. Hist. 82: 58–65.

Van der Oost, R., H. Heida & A. Opperhuizen. 1988. Polychlorinated biphenyl congeners in sediments, plankton, molluscs, crustaceans, and eel in a freshwater lake: implications of using reference chemicals and indicator organisms in bioaccumulation studies. Arch. Envir. Contam. Toxicol. 17: 721–729.

Veierov, D. & N. Aharonson. 1978. Simplified fat extraction with sulfuric acid as cleanup procedure for residue determination of chlorinated hydrocarbons in butter. J. Assoc. Off. Anal. Chem. 61: 253–260.

Zell, M. & K. Ballschmitter. 1980. Baseline studies of the global pollution. III. Trace analysis of polychlorinated biphenyls (PCB) by ECD glass capillary gas chromatography in environmental samples of different trophic levels. Fresenius Z. Anal. Chem. 304: 337–349.

Inventory and bibliography

Environmental Biology of Fishes **32**: 371–390, 1991.
© 1991 *Kluwer Academic Publishers.*

An inventory of all known specimens of the coelacanth *Latimeria chalumnae,* with comments on trends in the catches

Michael N. Bruton & Sheila E. Coutouvidis
J.L.B. Smith Institute of Ichthyology, Private Bag 1015, Grahamstown, 6140 South Africa

Received 15.10.1990   Accepted 3.11.1990

Key words: Fish distribution, Comoros, Fishery, Longlining, Museums, Curation

Synopsis

A list is presented of all known specimens of the coelacanth *Latimeria chalumnae* based on a survey of the literature and of museum, aquarium and university holdings. Details are given of the date, place, time and depth of capture, the name and age of the fisherman, the length, weight and sex of the fish, the first literature record, the method of preservation and the present location of specimens, if known. A new number is assigned to each specimen. At least 172 coelacanths are known to have been caught since 1938. The first coelacanth was caught off South Africa but all properly documented, subsequent specimens have been caught off the islands of Grand Comoro and Anjouan in the Comoros. An appeal is made to the Comoran authorities for each specimen that is caught to be made available for scientific study. Museum authorities are also encouraged to allow their specimens to be X-rayed or dissected so that vital information can be obtained on fecundity, foetal nutrition and dietary preferences. It is essential for the coelacanth conservation effort that this inventory is maintained by the Coelacanth Conservation Council.

Introduction

Lists of the coelacanth specimens known to science have been published previously in the literature. Millot et al. (1972) provided information on the time, date, place, depth and distance from shore of capture, bait used, fisherman's name, weight, length and sex of the fish, and present location of 68 specimens caught between December 1938 and October 1971. Their list was based on a previous list published in a little known Comoran report (Martin 1970). Millot et al. (1972) initiated a numbering system for coelacanth specimens and listed specimens 1 to 66 (an additional two specimens were numbered 14bis and 32bis). Similar information was provided by McCosker (1979) for an additional 19 specimens caught from 1972 to 1977, and for a further one specimen seen in the Maloudja Hotel

but for which no data were available, thus extending his list to specimen number 88. Suzuki & Tanauma (1984) and Suzuki et al. (1985), on the basis of interviews with Comoran fishermen, listed an additional 22 specimens caught between about 1959 and 1977. They also provided details of three coelacanth specimens obtained by the Japanese expeditions to the Comoros in 1981 and 1983. Other partial lists of coelacanths are provided by McAllister & Smith (1978), Thys van den Audenaerde (1984), Ogiso (1986) and Uyeno (1991).

Previous to this inventory, the scattered literature on the coelacanths caught to date made it difficult to review available information as well as to determine catch rates and trends. The founder members of the Coelacanth Conservation Council (CCC) therefore resolved in 1987 to produce a list of coelacanth specimens that could be published

372

and regularly updated (Bruton 1988). It was also recommended that all known specimens should be assigned a CCC number so that all museum, aquarium, university and private holdings of the coelacanth could be cross-referenced to the CCC list.

The compilation of this inventory has proved to be a daunting task as the data available on many specimens are sparse or inaccurate and are scattered in a wide variety of publications or catalogues. Furthermore, some specimens have been transferred from one museum to another, or are in private ownership, and can no longer be traced. The inventory has been compiled after extensive correspondence with museums and individual researchers, and by examining specimens in museums in Europe and North America and during a series of expeditions from the J.L.B. Smith Institute of Ichthyology to the Comoros between 1986 and 1990. In some cases information from different sources on different specimens seemed to apply to the same fish and was collated in the inventory. The draft entry in the inventory for each specimen was sent to every institution or individual known to have a coelacanth. In some cases we received no reply; inadequacies or inaccuracies associated with these entries are therefore unavoidable.

The accuracy of the information in the inventory requires comment. Information of dubious accuracy is preceded by ? in Appendix 1. The accuracy of the estimations by Comoran fishermen of the distance from shore and the depth of capture of coelacanths has been questioned (Stobbs 1987). Fricke (1988), however, found that their estimates of water depth are reasonably accurate. While the length measurements of the specimens are accurate, some of the weights are or were measured on inaccurate scales in the Comoros. Nevertheless, a length : weight regression for the coelacanth provides a reasonably close fit (Bruton & Armstrong 1991). In some cases, the information on a given specimen differs according to the source, e.g. McCosker (1979) gives the length and weight of specimen C83 (CCC no. 96) as 180 cm and 45 kg, whereas the Natural History Museum in Vienna, where the specimen is housed, gives the measurements of the same specimen as 140 cm and 40 kg, which is a more likely combination.

The catch records for the coelacanth during the period 1952 to 1975, while the Comoros was under French control, are likely to be reasonably accurate as the French authorities went to considerable lengths to record and secure all specimens that were landed. After the Comoros declared independence in 1975 catch records became less reliable. This was especially so during the Soilih regime from 1976 to 1978 when most public records were destroyed and western foreign aid agencies were virtually absent from the Comoros. The re-accession of President Abderemane in 1978 heralded a new era during which many western countries, especially from Europe, re-established ties with the Comoros and field research in marine biology and fisheries was re-initiated, especially by the Japanese. During the period 1983 to 1986 the monitoring of coelacanth catches was restarted but unfortunately an illegal trade in coelacanths also developed with the result that many specimens left the Comoros without being properly recorded. At this time lucrative incentives were offered by the staff of aquaria in foreign countries to Comoran fishermen to land coelacanths, and coelacanths were also offered for sale to non-scientific visitors to the Comoros. Some of the specimens purchased by non-scientists fortunately found their way into aquarium or museum collections (e.g. specimens CCC no. 53 and 118). In addition, the Comoran government donated coelacanths to foreign governments (such as France, China, Iraq, Japan, Kuwait), as well as to prominent individuals, such as the Secretary-General of the United Nations, Dr Perez de Cuellar (on 25.10.1985), and to the President of France, F. Mitterand (in July 1990), but full records were not kept of these specimens.

Since the establishment of a European Economic Community Fisheries Development Project in the Comoros in about 1985, coelacanth catches have been monitored by Michel de San on Grand Comoro. The appointment of an official (Abdallah Massonde) to monitor catches on Anjouan has also improved the record there. Furthermore, all coelacanths that are landed in the Comoros have to be sold to the Comoran government and none may now be offered for private sale or barter. The accession of President Johar in 1989 and the interest of his Minister of the Interior, Ali Mroudjae, in the conservation of the coelacanth, have further increased public awareness of the importance of coelacanth conservation.

Structure of the inventory

Information in the inventory is listed under the following headings:

(1) *Number:* The newly assigned Coelacanth Conservation Council/Conseil pour la Conservation du Coelacanthe (CCC) number, replaces the number given to specimens by previous authors, such as Millot et al. (1972), McCosker (1979) and Suzuki et al. (1985). The latter numbers are given under the heading 'First literature reference'. Coelacanth embryos, removed from pregnant females, have been listed as decimal numbers of the CCC number of the parent fish.

(2) *Date of capture:* The day, month and year of capture are given, if known. In some cases the only information available on the capture date is an estimate by a fisherman of how many years previously a specimen had been caught (e.g. see list in Suzuki et al. 1985), which is likely to result in some inaccuracies. The time of capture is also given for some specimens.

(3) *Site of capture:* The fishing village at which the coelacanth was landed is given, or the island off which the fish was caught. As the same place name often occurs on more than one island, the village name and its island are given when known. In some cases it is only known that the specimen was caught in the Comoros. The location and valid names of the most important fishing villages in the Comoros are given in Figure 1. Some Comoran place names are spelt in different ways by different authors, e.g. Shindini on Grand Comoro = Chindini, Wani on Anjouan = Ouani, Mitsoudje on Grand Comoro = Mitsudje or Mitsuje (R.E. Stobbs personal communication). Most coelacanths have been

caught within 2 km of the coast (Stobbs 1987, Appendix 1).

(4) *Name of fisherman:* The name of the fisherman/fishermen who caught the coelacanth as well as his/their age/s are given, if available.

(5) *Distance from shore at capture:* The distance from the nearest shore of the point at which the coelacanth was caught, as estimated by the fisherman.

(6) *Depth of capture:* The depth of water at the point at which the coelacanth was caught, as estimated by the fisherman.

(7) *Bait used:* The species of fish or squid used as bait. Often the Comoran name only was available, but the scientific name could usually be determined using the list of Comoran fish names compiled by Stobbs (1990a).

(8) *Weight:* The weight of the fish in kilograms as given in the first literature reference or in personal correspondence. Some of the weight estimates are likely to be inaccurate.

(9) *Length:* The total length in centimetres from the anteriormost point of the head to the posteriormost point of the caudal fin, measured in a straight line. These measurements were made on fresh or frozen as well as formalin-preserved specimens, and are therefore likely to be of variable accuracy.

(10) *Sex:* The sex of the specimens could only be determined by dissection as the coelacanth is not obviously sexually dimorphic.

(11) *Condition on capture:* Whether or not the specimen was alive when landed, and how long it stayed alive.

(12) *Method of preservation:* The original and current methods of preservation of the specimen at its final destination (as at September 1990).

(13) *Condition:* The condition of the specimen after capture (good or poor), the state of the eyes and fins etc. Information on subsequent dissection.

(14) *First literature reference:* The first reference to the specimen in an authoritative list in the literature or in personal communications or unpublished reports. Any previous list number assigned to the specimen is also given. Information on a specimen

374

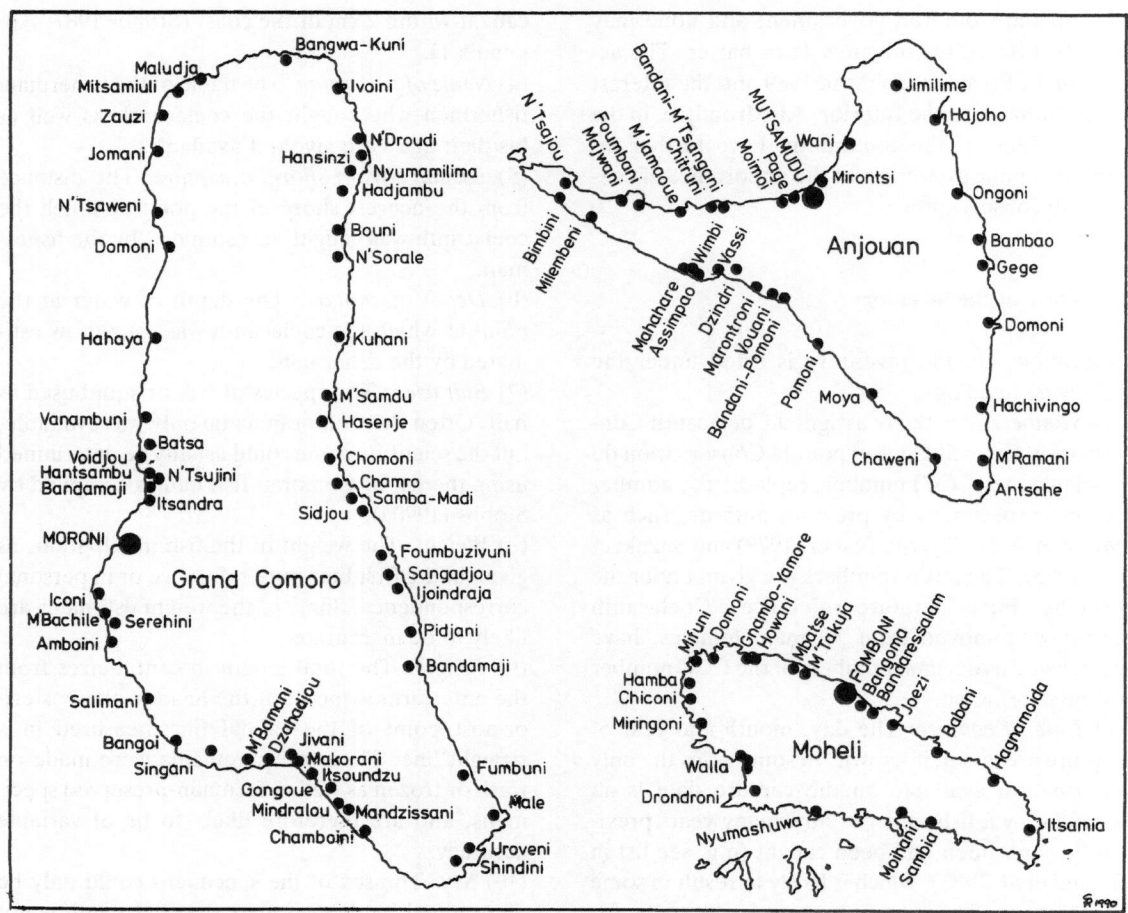

Fig. 1. Map of the three islands in the Federal Islamic Republic of the Comoros showing the most important fishing villages. The position of the islands relative to one another has been altered to facilitate publication. Drawing by R.E. Stobbs.

does not always derive from the original reference but from a variety of sources.

(15) *Current holding:* Previous and present location of the specimen in a museum, aquarium, university or private collection.

(16) *Additional comments:* Any additional comments of significance on the specimen.

Results

The inventory of coelacanth specimens is given in Appendix 1. A total of 172 coelacanth specimens is listed, twice as many as in the previous lists of Millot et al. (1972) and McCosker (1979).

Analyses of trends in the length, weight, geographical distribution, depth of capture and other aspects of the demography and ecology of the coelacanth, based partly on the information in this inventory, have been made separately by Bruton & Armstrong (1991) and Bruton & Stobbs (1991).

Number of specimens and date of capture

The first coelacanth was caught in 1938, the second

in 1952 and the third in 1953 (Fig. 2), and at least one specimen has been recorded each year since 1952 (Fig. 3).

The number of coelacanths caught per year increased from 1.9 in the 1950s to 4.9 in the 1960s, 4.8 in the 1970s and 4.0 in the 1980s, and averaged 4.39 for the period 1952 to 1990 (Fig. 4). This total includes 14 coelacanths (CCC no. a–n) which are known to have been caught since 1952 but for which the year of capture is not known. The years in which most coelacanth catches were recorded were 1986 (11 specimens, when incentives were being provided by foreigners), 1965 (10) and 1973 and 1974 (8 each). Only four coelacanths are known to have been caught in 1989, after various conservation measures had been introduced (Bruton & Stobbs 1991), and two were recorded between January and September 1990.

The number of coelacanths recorded may have some relationship to the presence of incentives offered to Comoran fishermen as well as to the level of interest by western scientists in the Comoros at a given time and the demand by museums and research institutes for specimens. These factors may increase the likelihood of a coelacanth being brought to the attention of western scientists but are unlikely to increase the catch rate as almost all coelacanths have been caught as an accidental by-catch of the oilfish fishery (Stobbs 1987, 1989, Balon et al. 1988, Bruton & Stobbs 1991). According to Stobbs (1987), there is no evidence that the fishermen can adjust their techniques or gear to increase the likelihood of catching a coelacanth. The coelacanth cannot be captured on demand (J.F. Hamlin in litt. 1990 to E.K. Balon).

Place of capture

The first specimen was caught off the south-east coast of South Africa but all subsequent specimens have been caught off the islands of Grand Comoro and Anjouan in the Federal Islamic Republic of the Comoros in the western Indian Ocean. No reliable records are available for coelacanths that might have been caught elsewhere. Bruton (1989a, b) reviews the information available on possible coe-lacanth catches elsewhere and concludes that the most likely localities other than the Comoros are the east coast of Madagascar near Tamatave and the south-east coast of South Africa. Smith (1953a) speculated that coelacanths may occur off Kenya.

Of the 171 coelacanths known to have been caught in the Comoros, 98 were caught off Grand Comoro and 41 off Anjouan. The island of capture of 32 specimens is unknown, and one specimen (the first) was caught off South Africa. Nearly half the coelacanths recorded from Grand Comoro were landed at Iconi, the largest traditional fishing village on the west shore of this island (Table 1, Fig. 1). All but two of the coelacanths caught off Grand Comoro were landed on the north-west, west and south-west coasts (specimen CCC no. 43 was landed at Bouni on the north-east coast, and CCC no. n is recorded from the 'east coast of Grand Comoro'). The majority of specimens caught off Anjouan were landed at Mutsamudu, the largest port. The first specimen to be recorded in the Comoros was landed at Domoni on the western shore of Anjouan in December 1952. No coelacanths have been reliably reported to have been caught off Moheli in the Islamic Republic, or off the French-controlled island of Mayotte. A specimen listed by Osigo (1986) from Chiconi, a village on the west shore of Moheli island in the Comoros, remains to be confirmed but this record is more likely to refer to a locality off the River Chiconi (Mro Chiconi) near Moimoi on Anjouan island. Further information on coelacanth capture sites, catch rates and fishing effort is given by Stobbs (1987, 1989, 1990b) and Bruton & Stobbs (1991).

Current location of coelacanth specimens

Of the 140 coelacanths whose current whereabouts are known, 35 (25%) are in France, 19 (14%) are in the U.S.A., about 18 (11%) are in the Comoros and 8 (6%) are in Japan (Table 2). One specimen, which was confiscated by Japanese customs officials because it was imported without prior permission and documentation, is not included in the inventory as no catch data are available to verify that it is not one of the specimens already recorded.

Fig. 2. The first three coelacanths caught: a – CCC no. 1, a 140 cm adult caught at a depth of 72–100 m off the South African coast by Hendrik Goosen on 22.12.1938, b – CCC no. 2, a 135 cm male caught at a depth of 160 m off Domoni on Anjouan island in the Comoros by Achmed Hussein on 20.12.1952, c – CCC no. 3, a 129 cm male caught at a depth of 200 m off Mutsamudu on Anjouan island in the Comoros by Houmadi Hassani on 24.9.1953 (after Smith 1956).

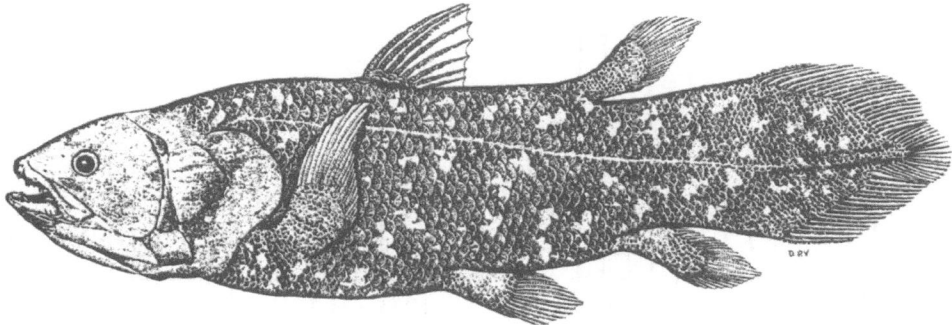

Fig. 3. Drawing of coelacanth CCC no. 118, a 155 cm male caught off Grand Comoro in May 1980 and now on display in the J.L.B. Smith Institute of Ichthyology in Grahamstown. Drawing by D.P. Voorvelt.

Coelacanth specimens are currently housed in collections in 24 countries (Table 2). The museum with the largest holdings of coelacanths is the Museum National d'Histoire Naturelle in Paris, which currently has 18 specimens but which previously distributed at least 37 additional specimens to other museums, research institutes and aquaria in different parts of the world. There are currently six specimens at the American Museum of Natural History in New York (plus 3 embryos) and five at the Kanazawa Aquarium in Japan, four at the British Museum (Natural History) in London and at the Zoological Museum in Turin, and three each at the J.L.B. Smith Institute of Ichthyology in Grahamstown, and the Centre National de Documentation et de la Recherche Scientifique in Moroni.

Dissections

A number of coelacanth specimens have been dissected and exist only as partial specimens. These include a number of specimens listed by Millot et al. (1972) which were not donated to other museums. The tissues of specimens CCC no. 79 and 80, inter alia, were sent to scientists throughout the world for analyses and the results of these studies were published in the volume edited by McCosker & Lagios (1979). Another review that arose out of the collaborative examination of coelacanth tissues was contributed to by Locket (1980). Many of the morphological papers published in this volume

arose from the dissection of coelacanth CCC no. 140 and 141.

Discussion

This inventory attempts to provide the most up-to-date listing of coelacanth specimens presently available but has many shortcomings due to the inadequate data sets for many specimens. It is our hope that coelacanth researchers and curators of museum, aquarium and university collections will inform us of additional data on the specimens in their care so that the inventory can be updated in the future. Further additions and amendments to the inventory will be published in the occasional newsletter of the CCC in the scientific journal Environmental Biology of Fishes. This newsletter is distributed free to all CCC members and other interested parties, who should write to the authors for further information about the CCC.

A further extension of this inventory project is to produce labels giving the CCC number of every coelacanth specimen listed. These labels will then be sent to all the individuals or institutions who have coelacanth specimens in their care with the request that the labels are attached to the specimens.

Latimeria chalumnae was placed on Appendix II of the Convention on International Trade in Endangered Species (CITES) in 1987, which afforded the species partial protection and required member

378

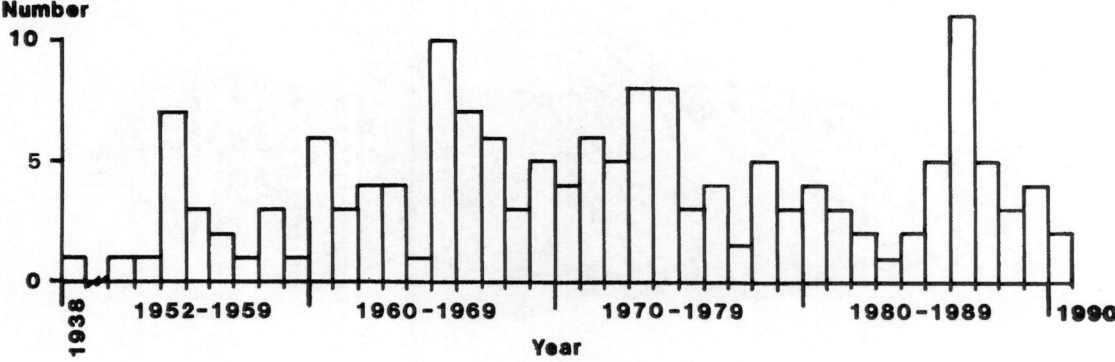

Fig. 4. The catches of coelacanths, *Latimeria chalumnae*, in yearly intervals based on the information in this inventory. n = 148.

states to keep a record of catches of and trade in the coelacanth. Few member states have complied with this recommendation (J. Bernay, CITES, personal communication 1990). *L. chalumnae* was placed on Appendix I of CITES in 1989 (effective January 1990), which affords the species full protection and requires member states to issue permits for and maintain full records of all translocations of coelacanth specimens, and to transfer this information to the CITES office in Lausanne, Switzerland. Once again, few member states have complied (J. Bernay, personal communication 1990). It is hoped that the existence of this official listing of coelacanth specimens will encourage all countries to document the movement and location of coelacanth specimens more fully in future.

Efforts to produce a demographic model of the

coelacanth population have been hampered by a lack of information on the number of young produced (Bruton & Armstrong 1991). It would therefore be highly desirable if all coelacanth specimens which have not as yet been dissected or X-rayed could be investigated as soon as possible so that this vital information can be obtained. Specimen CCC no. 29 (originally no. 26 in Millot et al. 1972) was initially offered to J.L.B. Smith, but he arranged for it to be sent to the American Museum of Natural History where it was dissected some years later and found to contain five yolksac juveniles (Smith et al. 1975). This was the first evidence that the coelacanth is livebearing and put a new perspective

Table 1. The most important fishing villages at which coelacanths have been landed in the Comoros.

Fishing village	No. of coelacanths landed
Grand Comoro	
Iconi	30
Itsoundzu, Itsandra, Salimani	6
M'Bamani	5
Vanambuni	4
Jomani, Hahaya, Mindralou	2
Anjouan	
Mutsamudu	13
Domoni	4
Hajoho, Pomoni	2

Table 2. Known holdings of the coelacanth *Latimeria chalumnae* in museum, university and other collections in different countries.

Country	No. of coelacanths
France	34
Comoros	?19
U.S.A.	21
Japan	9
England	8
Canada, China, Italy, South Africa	5
Austria	4
Australia, Germany, Madagascar, ?U.S.S.R.	2
Algeria, Belgium, Denmark, The Netherlands, Northern Ireland, Scotland, Sweden, Switzerland, Zimbabwe	1

on their life-history style. Since then, no other specimens with juveniles in utero have been found and only three specimens with eggs have been reported, of which only two are available for study (CCC no. 10 and 79). The other specimen with eggs came to the attention of the government taxidermist in the Comoros, Sidi Bakari of Mutsamudu, but regrettably the eggs were not preserved for study. It is therefore imperative that every specimen that is available for study is carefully examined to determine whether it contains young or eggs. Information on, inter alia, the mechanism of foetal nutrition and on the dietary preferences of adult fishes could also be obtained from dissected fishes.

Acknowledgements

We are extremely grateful to the following individuals who assisted us with the compilation of this inventory: Damir ben Ali (Centre Nationale du Recherche Scientifique, Comoros), J.W. Atz and Norma Feinberg (American Museum of Natural History, U.S.A.), E.K. Balon (University of Guelph, Canada), J. Bernay (C.I.T.E.S., Switzerland), M. Burridge (Royal Ontario Museum, Canada), R. Cloutier (British Museum [Natural History], U.K.), P.L. Forey (British Museum [Natural History], U.K.), H. Fricke (Max-Planck-Institut für Verhaltensphysiologie, Germany), Jean-Louis Geraud (Comoros), P.H. Greenwood (British Museum [Natural History], U.K.), W.D. Haacke (Transvaal Museum, South Africa), B. Hersig (Naturhistorisches Museum, Austria), D. Herbert (Natal Museum), I. Illich (Haus der Natur, Austria), J.E. McCosker (California Academy of Sciences, U.S.A.), J.L. Minshull (National Museum, Zimbabwe), J.A. Musick (Virginia Institute of Marine Science, U.S.A.), D.L.G. Noakes (University of Guelph, Canada), D. Robineau (Museum National d'Histoire Naturelle, France), R. Rosenblatt (Scripps Institute of Oceanography, U.S.A.), P.H. Skelton (J.L.B. Smith Institute of Ichthyology, South Africa), R.E. Stobbs (J.L.B. Smith Institute of Ichthyology, South Africa), J.-P. Sylvestre (France), G. Thompson (Ulster Museum, Northern Ireland), N.M. Teitler (Japan), T. Uyeno (National Science Museum, Japan). We are especially grateful to R.E. Stobbs for his assistance with Comoran place names and for drawing Figure 1.

Appendix 1

List of all known specimens of the coelacanth, *Latimeria chalumnae*. For explanation of parenthetical numbers, see text (pp. 373–374)

CCC no. 1: (2) 22.12.1938, (3) off the Chalumna River mouth south-west of East London, South Africa, (4) Hendrik Goosen, (5) 3.5–10 km, (6) 72–100 m, (8) 37.5 kg, (9) 140 cm, (11) alive for 3 h, (12) skinned and dry mounted, (13) near complete, excluding soft anatomy; left side of head dissected, fixed in formalin, (14) Smith (1939), (15) East London Museum, South Africa, (16) Plate III in Smith (1939); Fig. 2a, this paper.

CCC no. 2: (2) 20.12.1952, (3) south of Domoni, Anjouan, (4) Achmed Hussein, (5) 800 m, (6) 160 m, (8) ?37 kg, (9) 135 cm, (10) male, (11) alive when brought to the surface, then killed to bring onto boat, (12) formalin, later 70% propanol, (13) first dorsal fin missing, otherwise complete: cut along dorsal surface, (14) Smith (1953), originally described as *Malania anjouanae*, now synonymised with *Latimeria chalumnae*, (15) on public display in the J.L.B. Smith Institute of Ichthyology in Grahamstown (RUSI 000614); two scales of this specimen are lodged in the Department of Palaeontology in the British Museum (Natural History) (BMNH P34360, P34361), (16) Fig. 2b, this paper.

CCC no. 3: (2) 24.9.1953, 2300 h, (3) Mutsamudu, Anjouan, (4) Houmadi Hassani, (5) 800 m, (6) 200 m, (7) *Promethichthys prometheus* (Pisces, Gempylidae) (Comoran name: roudi), (8) 39.5 kg, (9) 129 cm, (10) male, (11) formalin, (13) dissected, (14) Millot et al. (1972) no. 3, (15) Muséum National d'Histoire Naturelle, Paris, France, (16) Fig. 2c, this paper.

CCC no. 4: (2) 29.1.1954, 0100 h, (3) Iconi, Grand Comoro, (4) Madi M'ze, (5) 600 m, (6) 390 m, (7) roudi, (8) 19.5 kg, (9) 109 cm, (10) female, (12) formalin, (13) good, (14) Millot et al. (1972) no. 4, (15) Antananarivo, Madagascar.

CCC no. 5: (2) 29.1.1954, 2400 h, (3) Mandzissani, Grand Comoro, (4) Ahmada Mrotmana, (5) 280 m, (7) roudi, (8) 34 kg, (9) 127 cm, (10) male, (12) formalin, then 70% alcohol, (13) good condition, subsequently dissected, (14) Millot et al. (1972) no. 5, (15) Muséum National d'Histoire Naturelle, Paris, France.

CCC no. 6: (2) 11.2.1954, (3) Itsandra, Grand Comoro, (6) 150 m, (8) 33 kg, (9) 126 cm, (10) male, (12) formalin, (13) good condition, subsequently dissected, (14) Millot et al. (1972) no. 6, (15) Muséum National d'Histoire Naturelle, Paris, France.

CCC no. 7: (2) 5.9.1954, 2400 h, (3) Anjouan, (5) 700 m, (6) 160 m, (8) 30 kg, (9) 120 cm, (10) male, (11) formalin, (13) good condition, subsequently dissected, (14) Millot et al. (1972) no. 7,

380

(15) Muséum National d'Histoire Naturelle, Paris, France, (16) mounted skeleton on display in exhibition.

CCC no. 8: 12.11.1954, 2400 h, (3) opposite Mutsamudu jetty on Anjouan, (4) Zema ben Said Mohamed & Madi Bacari, (5) 1000 m, (6) 255 m, (7) roudi, (8) 41 kg, (9) 142 cm, (10) immature female with maturing eggs, (11) towed back to the Mutsamudu jetty by dugout canoe and remained alive for 24 h, (12) formalin, (13) good, immediately dissected, (14) Millot (1955), Millot et al. (1972) no. 8, (15) Muséum National d'Histoire Naturelle, Paris, France, (16) this specimen was the first live coelacanth to be observed by scientists.

CCC no. 9: (2) 1954, (4) Sule Sankashi, aged 90 y, (9) 100 cm, (14) Suzuki et al. (1985) no. 18.

CCC no. 10: (2) 12.3.1955, 2000 h, (3) Anjouan, (5) 1500 m, (6) 300 m, (8) 78.5 kg, (9) 166 cm, (10) female, (12) formalin, (13) good, first maturing female discovered with eggs during dissection, (14) Millot et al. (1972) no. 9, Dugan (1955), (15) Muséum National d'Histoire Naturelle, Paris, France.

CCC no. 11: (2) 18.3.1955, 0200 h, (3) M'Bamani, Grand Comoro, (5) 500 m, (6) 250 m, (8) 26 kg, (9) 131 cm, (10) female, (12) formalin, (13) partly dissected, (14) Millot et al. (1972) no. 10, (15) Muséum National d'Histoire Naturelle, Paris, France.

CCC no. 12: (2) 15.4.1955, 0400 h, (3) Iconi, Grand Comoro, (4) Mahada, (5) 600 m, (6) 300 m, (7) roudi, (8) 22.5 kg, (9) 122 cm, (10) male, (12) formalin, (13) dissected, (14) Millot et al. (1972) no. 11, (15) Muséum National d'Histoire Naturelle, Paris, France.

CCC no. 13: (2) 3/4.5.1956, 2400 h, (3) Itsoundzu, Grand Comoro, (5) 300 m, (6) 200 m, (8) 60 kg, (9) 154 cm, (10) female, (12) formalin, (13) fairly good condition, subsequently dissected, (14) Millot et al. (1972) no. 12, (15) on display in the public gallery of the British Museum (Natural History), London, England, (16) donated by J. Millot and J. Anthony from the Muséum de la Homme, Paris, France.

CCC no. 14: (2) 9 or 27.5.1956, 0300 h, (3) Vanambuni, Grand Comoro, (5) 150 m, (6) ?150 m, (8) 39 kg, (9) 134 cm, (10) male, (13) dissected, (12) formalin, (14) Millot et al. (1972) no. 13, (15) Muséum National d'Histoire Naturelle, Paris, France.

CCC no. 15: (2) 27.12.1957, (3) Grand Comoro, (5) 200 m, (6) 200 m, (8) 25 kg, (9) 110 cm, (10) female, (12) formalin, (14) Millot et al. (1972) no. 14, (15) British Museum (Natural History), London, England.

CCC no. 16: (2) February 1958, 0100 h, (3) Grand Comoro, (10) female, (14) Millot et al. (1972) no. 14 bis, (15) Air Comores, Moroni, Comoros.

CCC no. 17: (2) 3.9.1958, (3) Salimani, Grand Comoro, (8) 35 kg, (9) 135 cm, (10) male, (12) formalin, (13) partly dissected, (14) Millot et al. (1972) no. 15, (15) Muséum National d'Histoire Naturelle, Paris, France.

CCC no. 18: (2) 19.11.1958, (3) Iconi, Grand Comoro, (8) 36 kg, (9) 135 cm, (14) Millot et al. (1972) no. 16.

CCC no. 19: (2) 30.10.1959, (3) M'Bamani, Grand Comoro, (6) 180 m, (8) 35 kg, (9) 132 cm, (10) male, (12) formalin, (13) dissected, (14) Millot et al. (1972) no. 17, (15) Muséum National d'Histoire Naturelle, Paris, France.

CCC no. 20: (2) 1.1.1960, (3) Itsandra, Grand Comoro, (6) 600 m, (8) 95 kg, (9) 180 cm, (10) female, (11) alive for short period after landing, (12) formalin, (13) poor, (14) Millot et al. (1972) no. 18, (15) Muséum National d'Histoire Naturelle, Paris, France.

CCC no. 21: (2) 21.2.1960, (3) Itsandra, Grand Comoro, (6) 600 m, (8) 40 kg, (9) 145 cm, (10) female, (11) alive for short period after capture, (12) formalin, (13) good but poorly fixed, (14) Millot et al. (1972) no. 19, (15) Muséum National d'Histoire Naturelle, Paris, France.

CCC no. 22: (2) 19.6.1960, 2200 h, (3) Itsoundzu, Grand Comoro, (5) 1000 m, (6) 300 m, (8) 31 kg, (9) 130 cm, (10) male, (12) formalin, (13) good, (14) Millot et al. (1972) no. 20, (15) Muséum National d'Histoire Naturelle, Paris, France.

CCC no. 23: (2) 23.6.1960, (3) between Iconi and Moroni, Grand Comoro, (5) 700 m, (6) 250 m, (8) 33–35 kg, (9) 125–130 cm, (10) male, (14) Millot et al. (1972) no. 21, (15) University Zoological Museum, Copenhagen, Denmark.

CCC no. 24: (2) 20.7.1960, (3) between Iconi and Moroni, Grand Comoro, (5) 1000 m, (6) 250 m, (8) 64 kg, (9) 140–145 cm, (10) female, (13) partly dissected, (12) formalin, (14) Millot et al. (1972) no. 22, (15) Muséum National d'Histoire Naturelle, Paris, France.

CCC no. 25: (2) 1960, (3) Mizinjaju, at Iconi, Grand Comoro, (4) Bakkari Issa, aged 65 y, (5) 100 m, (6) 210 m, (7) roudi, (9) 150 cm, (11) dead on landing, (14) Suzuki et al. (1985) no. 13.

CCC no. 26: (2) 8.4.1961, (3) Mindralou, Grand Comoro, (6) 250 m, (8) 33 kg, (9) 135 cm, (10) male, (14) Millot et al. (1972) no. 23, (15) American Museum of Natural History, New York. Exchanged in February 1977 from the Muséum National d'Histoire Naturelle, Paris, France.

CCC no. 27: (2) 4.8.1961, (3) Grand Comoro, (8) 38 kg, (9) 132 cm, (10) male, (12) formalin, (13) good, (14) Millot et al. (1972) no. 24, (15) Muséum National d'Histoire Naturelle, Paris, France.

CCC no. 28: (2) 10.10.1961, (3) Anjouan, (8) 34.5 kg, (9) 130 cm, (10) male, (12) formalin, (13) dissected, (14) Millot et al. (1972) no. 25, (15) American Museum of Natural History, New York; this specimen was obtained from the Muséum National d'Histoire Naturelle in Paris in exchange for embryo CCC no. 29.5, on 2.2.1977.

CCC no. 29: (2) 8.1.1962, (3) Mutsamudu, Anjouan, (8) 65 kg, (9) 160 cm, (10) female, (13) contained five embryos about 30–32 cm TL with yolksacs 8 × 13 cm in maximum diameter (yolksac juveniles), (14) Millot et al. (1972) no. 26, Smith et al. (1975), (15) American Museum of Natural History, New York (AMNH 32949; 19.4.1965), (16) originally offered to J.L.B. Smith by G. Garrouste; x-rays of each embryo are on file at the American Museum of Natural History; information on each embryo is listed below in caudad-rostral order:

CCC no. 29.1: (3) most caudad in oviduct, (9) 30.3 cm on removal, (16) histological lantern slides prepared by M.D. Lagios between 1976 and 1988 (see Bemis & Northcutt 1991); available for study at the American Museum of Natural History, New York.

CCC no. 29.2: (9) 32.2 cm on removal, (12) in alcohol, (13) half absorbed yolksac, (15) American Museum of Natural History,

(16) this embryo measured 30.8 cm on 21.2.1980 and is preserved whole.

CCC no. 29.3: (9) 32.1 cm on removal, (12) cleared and stained with alizarin red in 1976; counterstained in 1980, (16) TL measured in glycerin on 21.2.1989 was 34.1 cm.

CCC no. 29.4: (9) 32.7 cm on removal, (15) sent to British Museum (Natural History) as a gift on 3.12.1975, (16) dissected on one side.

CCC no. 29.5: (9) 30.3 cm when removed in December 1976, (15) sent to Muséum National d'Histoire Naturelle, Paris, France, in exchange for CCC no. 28.

CCC no. 30: (2) 28.2.1962, (3) Mutsamudu, Anjouan, (4) Zema Mohamed, (8) 30 kg, (9) 124 cm, (10) male, (12) formalin, (13) good, (14) Millot et al. (1972) no. 27, (15) Muséum National d'Histoire Naturelle, Paris, France.

CCC no. 31: (2) 15.3.1962, (3) between Domoni and N'Tsaweni, Grand Comoro, (4) Moindjie Soilihi, (6) 200 m, (7) roudi, (8) 45 kg, (9) 142 cm, (10) female, (12) formalin, (13) good, (14) Millot et al. (1972) no. 28, (15) Muséum National d'Histoire Naturelle, Paris, France.

CCC no. 32: (2) 1962, (14) Millot et al. (1972) no. 29.

CCC no. 33: (2) 21.9.1963, (3) Anjouan, (8) 45 kg, (9) about 135 cm, (14) Millot et al. (1972) no. 29, (15) on display in the British Museum (Natural History), London.

CCC no. 34: (2) 1963, (9) 114 cm, (14) Millot et al. (1972) no. 31, (15) Institute of Paleontology, Pavie, Italy.

CCC no. 35: (2) 1963, (9) 108 cm, (14) Millot et al. (1972) no. 32, (15) Museum of Natural History, Besancon, France.

CCC no. 36: (2) 1963, (14) Millot et al. (1972) no. 32 bis, (15) Cambridge Museum, Cambridge, England.

CCC no. 37: (2) 11.12.1964, (3) Mutsamudu, Anjouan, (4) Ahmed Hussein, (5) 1000 m, (6) 150 m, (7) *Octopus* sp., (8) 35.5 kg, (9) 137 cm, (10) male, (12) 10% formalin, liver, tissue an scales sent to D.I. Fox and G.F. Crozier, (14) Millot et al. (1972) no. 33, (15) University of California, Los Angeles.

CCC no. 38: (2) 1965, (3) off Shezani, near Moroni, Grand Comoro, (4) Madi Yussuf, aged 73 y, (5) 500 m, (6) 600 m, (9) 80 cm, (11) dead when landed, (14) Suzuki et al. (1985) no. 3.

CCC no. 39: (2) 1.1.1965, 2400 h, (3) Itsandra, Grand Comoro, (8) 43 kg, (9) 147 cm, (14) Millot et al. (1972) no. 34, (15) Museum of Oceanography, Quimper, France.

CCC no. 40: (2) early 1965, (15) sold to the American Exploration Society.

CCC no. 41: (2) 20.1.1965, (3) Mutsamudu, Anjouan, (4) Zema Mohamed, (9) 150 cm, (14) Millot et al. (1972) no. 35, (15) Science Faculty, Tananarive University, Madagascar.

CCC no. 42: (2) 21.1.1965, 2330 h, (3) Mutsamudu, Anjouan, (4) Houmadi Mderemane & Abdh de M'Djihari, (9) 139 cm, (14) Millot et al. (1972) no. 36, (15) Université, Paris VII, Paris.

CCC no. 43: (2) 21.3.1965, 0200 h, (3) Bouni, Grand Comoro, (4) Youssouf Ali, (6) 300 m, (8) 31 kg, (9) 131 cm, (14) Millot et al. (1972) no. 37, (15) Australian Museum, Sydney.

CCC no. 44: (2) April 1965, (3) west coast, Grand Comoro, (8) 25 kg, (9) 123 cm, (14) Millot et al. (1972) no. 38, (15) Los Angeles County Museum of Natural History, Los Angeles.

CCC no. 45: (2) 12.6.1965, (3) Jimilime, Anjouan, (6) 280 m, (8) 55 kg, (9) 152 cm, (14) Millot et al. (1972) no. 39, (15) Swedish Museum of Natural History, Stockholm.

CCC no. 46: (2) 1.8.1965, (3) Shindini, Grand Comoro, (6) 100 m, (8) 75 kg, (9) 162 cm, (14) Millot et al. (1972) no. 40, (15) Natural History Museum, Leiden, Netherlands.

CCC no. 47: (2) 18.8.1965, (3) Jomani, Grand Comoro, (8) 25 kg, (9) 124 cm, (10) male, (12) originally in formalin, now in 75% ethanol, (13) fair, tail slightly damaged, on public display, (14) Millot et al. (1972) no. 41, (15) Museum of Comparative Zoology, Harvard University, Cambridge, U.S.A.

CCC no. 48: (2) 9.2.1966, 2300 h, (3) Hahaya, Grand Comoro, (4) Hassani M'zima, (6) 300 m, (9) 160 cm, (10) female, (14) Millot et al. (1972) no. 42, (15) Staatliches Museum für Naturkunde, Stuttgart, Germany.

CCC no. 49: (2) 25.2.1966, 2200 h, (3) Mutsamudu, Anjouan, (6) 350 m, (7) roudi, (9) 124 cm, (12) frozen, (14) Millot et al. (1972) no. 43, (15) Museum of Natural History, Geneva, Switzerland.

CCC no. 50: (2) 14.3.1966, (3) Iconi, Grand Comoro, (4) Ali M'voura, (6) 150–200 m, (7) *Cypselurus bahiensis* (Pisces, Exocoetidae), (8) 15.87 kg, (9) 107 cm, (10) male, (14) Millot et al. (1972) no. 44, (15) Yale University, New Haven, U.S.A.

CCC no. 51: (2) 18.9.1966, (3) Jomani, Grand Comoro, (8) 73.9 kg, (9) 165 cm, (10) female, (14) Millot et al. (1972) no. 45, (15) Royal Institute of Natural Sciences, Brussels, Belgium.

CCC no. 52: (2) 20.11.1966, 2400 h, (3) Joumbi, Anjouan, (5) 400 m, (6) 250 m, (8) 65 kg, (9) 155 cm, (14) Millot et al. (1972) no. 46, (16) N. Teitler on behalf of T. Uyeno in litt. 1990 states that this specimen is not in the Yomiuri-Land Aquarium, Tokyo, Japan, as stated by Millot et al. 1972.

CCC no. 53: (2) 20.12.1966, 2400 h, (3) off north shore of Anjouan, (5) 400 m, (6) 250 m, (7) gempylid, (8) 54 kg, (9) 150 cm, (10) female, (12) body and viscera separated, both in 6% formalin, (13) dissected by Japanese research team, (14) obtained by Satoru Kamegai, reported to us by T. Uyeno (in litt. 1990), (15) gift to M. Shoriki, Head of the Yomiuri Newspaper Co. from the French government in recognition of his cultural contributions to France and Japan (Kamegai 1971); now housed at the Yomiuri-Land Aquarium, Japan.

CCC no. 54: (2) 1966, (4) Sule Sankashi, aged 90 y, (9) 50 cm, (14) Suzuki et al. (1985) no. 19.

CCC no. 55: (2) 12.2.1967, (3) Mutsamudu, (4) Zema Houmadi, (5) 450 m, (6) 300 m, (7) roudi, (8) 65 kg, (9) 145 cm, (12) formalin, (13) good, (14) Millot et al. (1972) no. 47, (15) Natural History Museum, La Rochelle, France.

CCC no. 56: (2) 17.2.1967, (3) Singani, Grand Comoro, (8) 45 kg, (9) 118 cm, (10) male, (14) Millot et al. (1972) no. 48, (15) Senckenberg Museum, Frankfurt, Germany.

CCC no. 57: (2) 1.3.1967, (3) Grand Comoro, (4) Said Mehezi, (5) 450 m, (6) 300 m, (7) roudi, (8) 73 kg, (9) 165 cm, (10) female, (14) Millot et al. (1972) no. 49, (15) Birmingham Medical School, Birmingham, Alabama, U.S.A., (16) later sent to the Smithsonian Institution, Washington, D.C; USNM 20587.

CCC no. 58: (2) 13.6.1967, (3) Salimani (syn. Hambou), Grand Comoro, (4) Soilihi Foumou, (5) 200 m, (6) 150 m, (7) roudi, (8) 30 8 kg, (9) 130 cm, (10) female, (14) Millot et al.

(1972) no. 50, (15) Musée Zoologique de Strasbourg, Strasbourg, France.

CCC no. 59: (2) 25.8.1967, (3) Iconi at Chezani, Grand Comoro, (4) Hamidi Oissoule & Ali M'sa Ali, (5) 300 m, (6) 76 m, (7) roudi, (8) 15.1 kg, (9) 107 cm, (12) frozen, (13) tissue in poor state for histological investigation, (14) Grady (1970), Millot et al. (1972) no. 51, (15) Field Museum of Natural History, Chicago, U.S.A. (FMNH 76057), (16) there is some confusion about the information on this specimen, and there is a possibility that data on two specimens, caught on the same date, have been combined in this account.

CCC no. 60: (2) 1967, (4) Sule Sankashi, aged 90 y, (9) 150 cm, (14) Suzuki et al. (1985) no. 20.

CCC no. 61: (2) 21.1.1968, (3) Mutsamudu, Anjouan, (4) Cheikn Ahmed Affondi, (6) 160 m, (7) roudi, (8) 50 kg, (9) 150 cm, (10) female, (14) Millot et al. (1972) no. 52, (15) Field Museum of Natural History, Chicago, U.S.A.

CCC no. 62: (2) 26.9.1968, (3) M'Bamani, Grand Comoro, (4) M'Saidie Madi Abd, (5) 1000 m, (6) 150 m, (7) roudi, (8) 33.8 kg, (9) 132 cm, (10) male, (12) formalin, (14) Millot et al. (1972) no. 53, (15) Natural History Museum, Nantes, France.

CCC no. 63: (2) 31.10.1968, (3) Iconi, Grand Comoro, (4) Ali M'Dahoma, (5) 1000 m, (6) 200 m, (7) roudi, (8) 13 kg, (9) 100 cm, (10) male, (14) Millot et al. (1972) no. 54, (15) Haut-Commissaire, Moroni, Grand Comoro.

CCC no. 64: (2) 16.2.1969, (3) Mutsamudu, Anjouan, (4) Zema Mohamed, (6) 200 m, (7) roudi, (8) 42 kg, (9) 133 cm, (10) female, (12) frozen, (14) Millot et al. (1972) no. 55, (15) National Museum of Canada, Ottawa, (16) a cast of this specimen is on permanent loan to the Musée d'Histoire naturelle de Miguasha, Quebéc, Canada.

CCC no. 65: (2) 13.3.1969, (3) Dzahadjou, Grand Comoro, (4) Tadjiri Himidi, (5) 1000 m, (6) 150 m, (7) roudi, (8) 31.5 kg, (9) 138 cm, (10) male, (14) Millot et al. (1972) no. 56, (15) Musée de la Pêche, Concarneau, France.

CCC no. 66: (2) 24.3.1969, (3) Vanambuni, Grand Comoro, (4) Aboudou Hamedi, (5) 1000 m, (6) 60 m, (7) *Decapterus* sp. (Pisces, Carangidae) (Comoran name: hanale), (8) 32 kg, (9) 137 cm, (10) male, (12) frozen, (14) Millot et al. (1972) no. 57, (15) Royal Scottish Museum, Edinburgh, Scotland.

CCC no. 67: (2) 1969, (3) Mizinjaju, at Iconi, Grand Comoro, (4) Adam Ally, aged 76 y, (5) 500 m, (6) 260 m, (7) roudi, (9) 90 cm, (11) alive when landed, (14) Suzuki et al. (1985) no. 11.

CCC no. 68: (2) 15.8.1969, 2200 h, (3) between Moroni and Itsandra, Grand Comoro, (4) Nahouza M'Dahoma, (5) 1000 m, (6) 200 m, (7) *Tylosurus choram* (Pisces, Scomberesocidae) (Comoran name: M'Tsoumboui), (8) 25 kg, (9) 124 cm, (10) male, (14) Millot et al. (1972) no. 58, (15) Museum of the Reunion, Paris, France.

CCC no. 69: (2) 1.1.1970, 2200 h, (3) Itsoundzu, Grand Comoro, (4) Oussoufa M'Latamou, (5) 1000 m, (6) 300 m, (7) roudi, (8) 26 kg, (9) 122 cm, (10) male, (12) formalin, (14) Millot et al. (1972) no. 59, (15) Steinhart Aquarium, California Academy of Sciences, San Francisco, U.S.A.; CAS 24862.

CCC no. 70: (2) 23.7.1970, 2200 h, (3) Mindralou, near Dembeni, Grand Comoro, (4) Abdou Moiramboini & Issa Mkou-

found, (5) 500 m, (6) 180–200 m, (7) roudi, (8) 28.5 kg, (9) 120 cm, (10) male, (12) frozen, (13) dissected, (14) Millot et al. (1972) no. 60, (15) Laboratory of Chemical Biology, Science Faculty, Paris, France.

CCC no. 71: (2) 20.11.1970, (3) M'Bamani (syn. Hambou), Grand Comoro, (4) M'Saidie Soilihi, (5) 800 m, (6) 70 m, (7) 'djadge' (Millot et al. 1972), (8) 73 kg, (9) 160 cm, (10) female, (12) received frozen, then fixed in formalin, (13) dissected, (14) Millot et al. (1972) no. 61, (15) Muséum National d'Histoire Naturelle, Paris, France.

CCC no. 72: (2) 21.11.1970, (3) Maludja, Grand Comoro, (6) 17 m, (9) 121 cm, (10) male, (12) frozen, (14) McCosker (1979) states that this specimen is not included in the list of Millot et al. (1972), (15) Los Angeles County Museum of Natural History, Los Angeles, U.S.A.

CCC no. 73: (2) 27.2.1971, (3) Assimpao, Anjouan, (4) Abdou Charif, (5) 1000 m, (6) 300 m, (8) 77 kg, (9) 160 cm, (10) female, (12) transported in formalin and later exhibited in a tank in propanol, (14) Millot et al. (1972) no. 62, (15) MacMillan Tropical Gallery, Public Aquarium of Vancouver, Canada.

CCC no. 74: (2) 5.3.1971, 0300 h, (3) Itsandra, Grand Comoro, (4) Mohamed Soilih, (5) 1500 m, (6) 100 m, (7) *Lutjanus* sp. (Pisces, Lutjanidae) (Comoran name: hazi), (8) 38 kg, (9) 133 cm, (10) male, (14) Millot et al. (1972) no. 63, (15) Hôtel Dieu, Lyon, France.

CCC no. 75: (2) March 1971, (3) off Shezani, near Moroni, Grand Comoro, (6) 350 m, (7) roudi, (9) > 100 cm, (11) alive when landed, (14) Suzuki et al. (1985) no. 9.

CCC no. 76: (2) 3.4.1971, (3) Moroni, Grand Comoro, (4) Youssouf Abdou, (5) 3000 m, (6) 250 m, (7) roudi, (8) 10 kg, (9) 85 cm, (10) female, (14) Millot et al. (1972) no. 64, (15) The Royal Society, London, England.

CCC no. 77: (2) 28.6.1971, (3) Vanambuni, Grand Comoro, (4) Mlaraha Adame, (5) 2000 m, (6) 250 m, (7) hanale, (8) 30 kg, (9) 133 cm, (10) male, (12) formalin, (14) Millot et al. (1972) no. 65, (15) according to Smith et al. (1975), this specimen was not sent to the American Museum of Natural History, New York, U.S.A., as stated by Millot et al. (1972).

CCC no. 78: (2) 16.9.1971, (3) M'Bamani, Grand Comoro, (4) Msaidie Mohamadi, (5) 2000 m, (6) 100 m, (7) roudi, (8) 65 kg, (9) 164 cm, (10) ?male, (14) Millot et al. (1972) no. 66, (15) Shirshov Institute of Oceanology, Moscow, U.S.S.R., (16) possibly the same specimen as CCC no. 98.

CCC no. 79: (2) 5.1.1972, 0100 h, (3) Domoni, Anjouan, (5) 2000 m, (6) 400 m, (7) hanale, (8) 78 kg, (9) 163 cm, (10) female with 19 mature eggs (see Fig. 7 in Balon 1991), (11) alive for 9 h after landing (Locket 1980), (13) good, (14) Millot et al. (1978) no. C 67, McCosker (1979) no. C70, (15) dissected tissues sent worldwide.

CCC no. 80: (2) 22.3.1972, 0200 h, (3) Iconi, Grand Comoro, (4) Madi Youssouf Kaar, (5) 100 m, (6) 165 m, (8) 10 kg, (9) 85 cm, (10) female, (11) alive for 6 h after landing (Locket 1980), (12) dissected, then frozen, (14) McCosker (1979) no. C71, (15) dissected tissues and organs sent to 54 scientists worldwide.

CCC no. 81: (2) 12.5.1972, 2300 h, (3) Iconi, Grand Comoro, (4) Said Ali Kundji, (5) 800 m, (6) 90 m, (8) 38 kg, (9) 120 cm,

Fig. 5. Latimeria chalumnae (CCC no. 91) exhibited at the Haus der Natur, Salzburg, Austria. Photo by E.K. Balon, 30.9.1989.

(10) ?female, (12) formalin, (14) McCosker (1979) no. C72, (15) Aquarium, Le Croisic, France.

CCC no. 82: (2) 12.8.1972, 0300 h, (3) Iconi, Grand Comoro, (4) Mhoumadi Aboudou, (5) 400 m, (6) 100 m, (7) roudi, (9) 95 cm, (12) 70% alcohol, (13) there is an incision into the abdomen but soft anatomy still intact, (14) McCosker (1979) no. C73, (15) Natal Museum, Pietermaritzburg, South Africa, (16) purchased in 1973 for SA R1400 which included the cost of transport; accession number 1527; on public display.

CCC no. 83: (2) 16.10.1972, (3) Dzindri, Anjouan, (4) Tsounou Bacar, (5) 1000 m, (6) 350 m, (8) 30 kg, (9) 120 cm, (10) ?male, (11) alive for 4 h after landing, (14) McCosker (1979) no. C74, (15) Ulster Museum, Ulster, Northern Ireland.

CCC no. 84: (2) February 1973, (3) Mizinjaju, Iconi, Grand Comoro, (4) Yussuf Abdou, aged 54 y, (5) < 500 m, (6) 280 m, (9) 50 cm, (11) alive when landed, (14) Suzuki et al. (1985) no. 6.

CCC no. 85: (2) 6.7.1973, (3) Mitzoudje, Grand Comoro, (5) 500 m, (8) 35 kg, (9) 132 cm, (10) male, (14) McCosker (1979) no. C75.

CCC no. 86: (2) 27.7.1973, (3) Iconi, Grand Comoro, (5) 100 m, (6) 100 m, (8) 10 kg, (9) 86 cm, (10) male, (12) formalin, (14) McCosker (1979) no. C76, (15) M. Nerat, Vienna, Austria.

CCC no. 87: (2) 6.11.1973, 0200 h, (3) Vouani, Anjouan, (6) 175 m, (8) 32 kg, (9) 120 cm, (10) male, (14) McCosker (1979) no. C77, (15) M. Nerat, Vienna, Austria.

CCC no. 88: (2) 22.11.1973, 2100 h, (3) Iconi, Grand Comoro, (4) Ibada Mbelizi, (5) 800 m, (6) 180 m, (8) 24 kg, (9) 103 cm, (10) male, (12) frozen, injected with formalin in 1975, (14) McCosker (1979) no. C78, (15) Marine Vertebrate Collection, Scripps Institute of Oceanography, University of California, La Jolla; S10 75–347.

CCC no. 89: (2) 1973, (3) Milini, at Iconi, Grand Comoro, (4)

Ally Musa Ally, aged 60 y, (5) 250 m, (6) 350 m, (7) roudi, (9) 130 cm, (11) alive when landed, (14) Suzuki et al. (1985) no. 14.

CCC no. 90: (2) 27.11.1973, 0330 h, (3) M'Bachile, Grand Comoro, (5) 400 m, (6) 225 m, (8) 30 kg, (9) 110 cm, (10) male, (12) frozen at − 20 to − 10° C, (14) McCosker (1979) no. C79, (15) California Academy of Sciences, San Francisco, U.S.A.; CAS 33111.

CCC no. 91: (2) 1973, (3) Baco Selemani (probably Salimani), Grand Comoro, (6) 175 m, (9) 120 cm, (10) female, (15) Haus der Natur, Salzburg, Austria, (16) received by Haus der Natur in March 1974; Figure 5 in this paper.

CCC no. 92: (2) 14.2.1974, 0100 h, (3) Mirontsi, Anjouan, (6) 220 m, (8) 40 kg, (9) 139 cm, (10) male, (12) formalin, (13) very good, (14) McCosker (1979) no. C80, (15) Natural History Museum, Vienna, Austria (NMW-76041), (16) the skeleton is on exhibition in the museum.

CCC no. 93: (2) 17.5.1974, 2100 h, (3) Vanambuni, Grand Comoro, (4) Ali M'Dahoma, (5) 300 m, (6) 150 m, (7) *Octopus* sp., (8) 40 kg, (9) 139 cm, (14) McCosker (1979) no. C81, (15) M. Lebret, Paris, France.

CCC no. 94: (2) 17.8.1974, 1645 h, (3) Iconi, Grand Comoro, (4) Said Ahamada, (6) 180 m, (7) *Thalassoma* sp. (Pisces, Labridae) (Comoran name: kakatzi), (8) 0.8 kg, (9) 42.5 cm, (10) female, (12) frozen, then fixed in formalin, (14) McCosker (1979) no. C82, (15) Muséum National d'Histoire Naturelle, Paris, France, (16) the smallest coelacanth caught on a line to date; Fig. 25 in Balon (1977).

CCC no. 95: (2) August 1974, (9) 122 cm, (14) D. Robineau (in litt. 1990), (15) Château de la Bubrese, France.

CCC no. 96: (2) 18.10.1974, (3) Salimani, Grand Comoro, (4) Itsa Monssa, (5) 400 m, (6) 240 m, (7) roudi, (8) 40 kg, (9) 140 cm, (10) female, (13) good, undamaged specimen with intestinal tract intact, (14) McCosker (1979) no. C83, (15) Natural

384

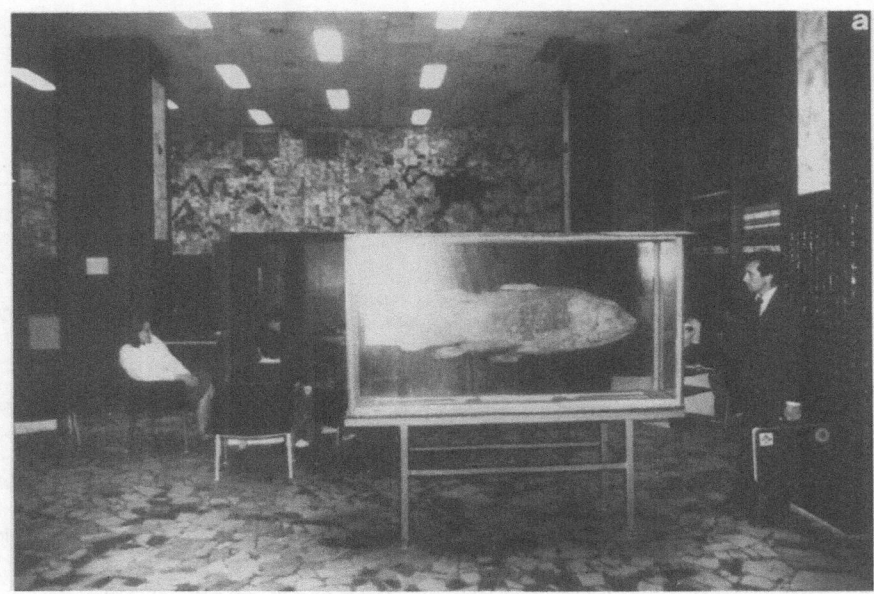

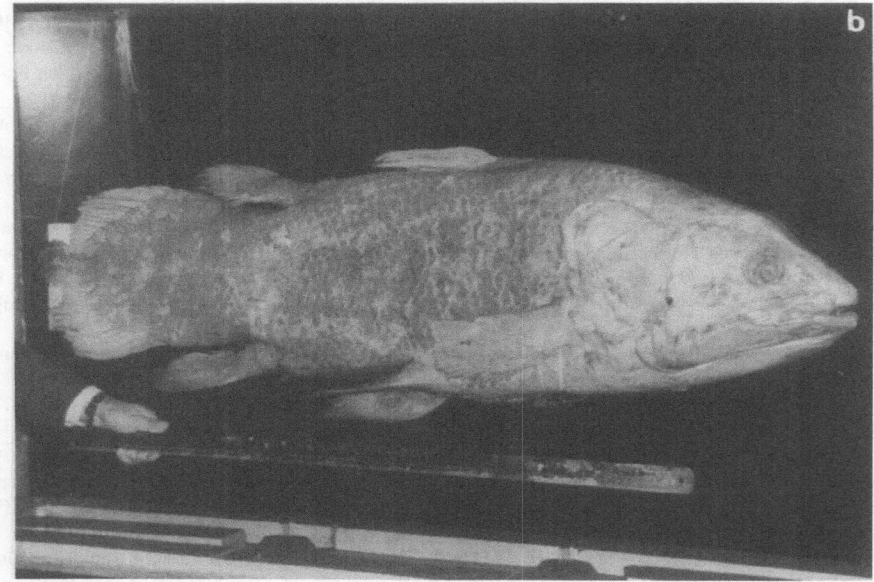

Fig. 6. The only specimen of *Latimeria chalumnae* (CCC no. 98) traced in the U.S.S.R. is in the Institute of Oceanology, Moscow. Photo by E.K. Balon, 17.5.1990.

History Museum, Vienna, Austria (NMW-76040), (16) this specimen was dissected by P. Adamicka; McCosker (1979) gives the length and weight of this specimen as 180 cm and 45 kg.

CCC no. 97: (2) 9.11.1974, (3) Iconi, Grand Comoro, (4) Ali Saadi, (5) 2000 m, (6) 250 m, (7) roudi, (8) 37 kg, (9) 145 cm, (10) female, (12) formalin, (14) McCosker (1979) no. C84.

CCC no. 98: (2) 1974, (3) Grand Comoro, (9) ~ 157 cm, (12) formalin, (13) fair, (14) recorded by E.K. Balon on 21.6.1990, (15) in a sealed glass aquarium in the centre of the cafeteria (Fig. 6a) of the Institute of Oceanology, Moscow, U.S.S.R., (16) this specimen could be the same as CCC no. 78; rediscovered in 1984 encrusted in rust at the Zoological Museum in Moscow and subsequently restored; Figure 6b in this paper.

CCC no. 99: (2) April-May 1974, (3) Mizinjaju at Iconi, Grand Comoro, (4) Athumi Mbelizi, aged 55 y, (5) 250 m, (6) 370 m, (9) 150 cm, (11) alive when landed, (14) Suzuki et al. (1985) no. 8.

CCC no. 100: (2) 22.1.1975, (3) Mromouhouli, Anjouan, (5) 3000 m, (6) 300 m, (7) roudi, (9) 165 cm, (10) female, (12) formalin, (14) McCosker (1979) no. C85, (16) J.W. Atz (personal communication) suggests that one of CCC numbers 100 or 103 went to the Scripps Institute of Oceanography and the other to the Steinhart Aquarium in California.

CCC no. 101: (2) 1975, (3) off Mt N'Gouni at Iconi, Grand Comoro, (4) Madi Yussuf, aged 73 y, (8) 15–16 kg, (9) 80 cm, (11) alive until it reached the coast, (14) Suzuki et al. (1985) no. 4.

CCC no. 102: (2) March 1975, (3) off Mt N'Gouni at Iconi, Grand Comoro, (4) Yussuf Abdou, aged 54 y, (5) 500 m, (7) roudi, (9) 150 cm, (11) alive when landed, (14) Suzuki et al. (1985) no. 7.

CCC no. 103: (2) 27.1.1976, (3) Iconi, Grand Comoro, (14) McCosker (1979) no. C86.

CCC no. 104: (2) 5.4.1976, (3) Grand Comoro, (8) 65 kg, (9) 165 cm, (12) ?formalin, (13) good, (14) Zhu Min (in litt. 1990), (15) Specimen House, Institute of Vertebrate Paleontology and Paleoanthropology, Academia Sinica, Beijing, China.

CCC no. 105: (2) 1.9.1976, (3) Domoni, Anjouan, (8) 11 kg, (9) 90 cm, (12) formalin, (14) McCosker (1979) no. C87, (15) Centre de Recherches Océanographiques et de Pêche, Algeria.

CCC no. 106: (2) 1976, (3) Iconi, Grand Comoro, (4) Said Ahmed Mbae, aged 50 y, (9) circa 25 cm, (11) swimming slowly near the water surface when caught, (14) Suzuki et al. (1985) no. 1, (16) this size estimate describes the smallest free-swimming coelacanth caught to date, possibly a premature birth.

CCC no. 107: (2) January 1977, (3) Iconi, Grand Comoro, (4) Said Ahmed Mbae, aged 50 y, (5) 300–400 m, (6) 300 m, (7) roudi, (9) 120 cm, (11) alive until it reached the coast, (14) Suzuki et al. (1985) no. 2.

CCC no. 108: (2) 1977, (3) Mizinjaju at Iconi, Grand Comoro, (4) Hassana Melinji, (5) 300 m, (6) 250 m, (7) roudi, (9) 150 cm, (11) alive until it reached the coast, (14) Suzuki et al. (1985) no. 12.

CCC no. 109: (2) February 1978, (3) Mizinjaju at Iconi, Grand Comoro, (4) Madi Yussuf, aged 73 y, (5) 500 m, (6) 300 m, (7) roudi, (8) 10 kg, (9) 70 cm, (11) dead when it reached the surface, (14) Suzuki et al. (1985) no. 5.

CCC no. 110: (2) July 1978, (3) Anjouan, (8) 30 kg, (9) 125 cm, (10) ?female, (12) frozen, later injected with formalin, then placed in 75% isopropyl solution, (13) good, (14) J. Minshull (in litt. 1989), (15) Queen Victoria Museum, Harare, Zimbabwe, (16) received from Captain Jack Malloch, chief pilot of an air freight company flying to the Comoros (AFRAIR).

CCC no. 111: (2) December 1978, (9) 129 cm, (12) frozen, (14) D. Robineau (in litt. 1990), (15) Zoological Museum, Turin, Italy.

CCC no. 112: (2) 1978, (3) off Mt Dengu, Grand Comoro, (4) Msakarani Muvura, aged 80 y, (5) 200 m, (6) 100 m, (7) roudi, (9) 100 cm, (11) alive when landed, (14) Suzuki et al. (1985) no. 17.

CCC no. 113: (2) December 1978, (9) 135 cm, (12) frozen, (14) D. Robineau (in litt. 1990), (15) Zoological Museum, Turin, Italy.

CCC no. 114: (2) August 1979, (9) 122 cm, (12) frozen, (14) D. Robineau (in litt. 1990), (15) Zoological Museum, Turin, Italy.

CCC no. 115: (2) September 1979, (9) 100 cm, (13) frozen, (14) D. Robineau (in litt. 1990), (15) Zoological Museum, Turin, Italy.

CCC no. 116: (2) January-February 1979, (3) Iconi, Grand Comoro, (4) Athumi Mbelizi, aged 55 y, (5) 390 m, (6) 500 m, (7) roudi, (9) 60 cm, (11) alive when landed, (14) Suzuki et al. (1985) no. 10.

CCC no. 117: (2) 24.1.1980, (3) Iconi, Grand Comoro, (4) Athumi Mbelizi, aged 56 y, (5) 800 m, (6) 160 m, (7) Exocoetidae, (8) 20 kg, (9) 110 cm, (10) female, (12) formalin, (13) good, (14) D. Robineau (in litt. 1990), (15) O.R.S.T.O.M., Paris, then to the Muséum National d'Histoire Naturelle, Paris, in June 1989.

CCC no. 118: (2) mid-May, 1980, (3) Grand Comoro, (9) 155 cm, (10) male, (12) frozen, later fixed in formalin and displayed in 60% propanol, (13) good, (14) Margaret Smith, (15) on public display in the J.L.B. Smith Institute of Ichthyology, Grahamstown (RUSI 000613), (16) the specimen was dissected on the left side to display the internal organs; Fig. 3 in this paper.

CCC no. 119: (2) 1980, (3) Beni, Grand Comoro, (4) Yusuf Ali, (9) 160–170 cm, (14) Suzuki et al. (1985) no. 16.

CCC no. 120: (2) ?late 1980, (3) Grand Comoro, (9) 100 cm, (12) fixed in formalin, later preserved in propanol, (13) good, (14) Margaret Smith, (15) Transvaal Museum, Pretoria, South Africa.

CCC no. 121: (2) January 1981, (3) Garawani, at Iconi, Grand Comoro, (4) Bakari Isuram, (5) 200 m, (6) 250 m, (7) gempylid, (8) 18.5 kg, (9) 109 cm, (10) female, (12) frozen, (13) good, but with partly damaged fins; after study made into a taxidermic display with artificial eyes prepared by H. Taguchi, (14) Suzuki et al. (1985) no. 2, (15) smallest specimen in Japan, tissues removed for study by JASEC research team at Keikyu Aburatsubo Marine Park Aquarium; currently exhibited at the National Science Museum, Tokyo, Japan.

CCC no. 122: (2) August-October, 1981, (3) Zinzazu, near Itsoundzu, Grand Comoro, (4) Ahamada Isilahi, (5) 500–

1000 m, (6) 500 m, (7) roudi, (9) 170 cm, (11) alive when landed, (14) Suzuki et al. (1985) no. 22.

CCC no. 123: (2) 30.12.1981, (3) Mitsamiuli, Grand Comoro, (4) Soule Assoumani, (5) 400 m, (6) 200 m, (7) gempylid, (8) 84 kg, (9) 175.4 cm, (10) female, (12) frozen, (13) good, slight contusions on head and fins, display specimen with artificial eyes prepared by H. Taguchi, (14) Suzuki et al. (1985) JASEC no. 1, (15) tissues removed for study at the University of Tokyo, Japan; currently housed at the Kanazawa Aquarium, Japan.

CCC no. 124: (2) 1.9.1982, (3) Itsoundzu, Grand Comoro, (4) Ahamada Isilahi, (5) 800 m, (6) 500 m, (7) gempylid, (8) 53 kg, (9) 164 cm, (10) female, (12) frozen, (13) damaged, injury to abdomen; display specimen prepared, with artificial eyes by H. Taguchi, (14) Suzuki et al. (1985) JASEC no. 3, (15) tissues removed for study at the University of Tokyo, Japan, currently housed at the Kanazawa Aquarium, Japan.

CCC no. 125: (2) November 1982, (3) Sahda, ?Grand Comoro, (4) Mwazie Kawa, aged 75 y, (5) 340 m, (6) 150 m, (7) roudi, (14) Suzuki et al. (1985) no. 15.

CCC no. 126: (2) early 1983, (3) Mutsamudu, Anjouan (9) 183.1 cm, (10) female, (12) expertly mounted in Belgium and displayed in a hand-carved wooden cabinet, (13) excellent, all fins and scales intact, artificial eyes, (14) E.K. Balon and M.N. Bruton, 1.5.1987, (15) foyer of President's residence, Moroni, Grand Comoro, (16) Fig. 9 in Balon et al. (1988).

CCC no. 127: (2) 26.11.1984, 0200 h, (3) Chiconi, Anjouan, (6) 250 m, (9) 130 cm, (14) observed by T. Ogiso (1986) in the freezer of the Japanese fishery training school (now the École National de Pêche), Mutsamudu, Anjouan.

CCC no. 128: (2) 16.12.1984, 2200 h, (3) Boukouni, ?Anjouan, (4) Sandani, (6) 200 m, (9) 150 cm, (14) observed by T. Ogiso (1986) in the freezer of the École National de Pêche, Mutsamudu, Anjouan.

CCC no. 129: (2) 28.1.1985, 2200 h, (3) Pomoni, Anjouan, (4) Hassni, (6) 200 m, (14) observed by T. Ogiso (1986) in the freezer of the École National de Pêche, Mutsamudu, Anjouan.

CCC no. 130: (2) 4.6.1985, 0300 h, (3) Wani, Anjouan, (6) 250 m, (9) 130 cm, (14) observed by T. Ogiso (1986) in the freezer of the École National de Pêche, Mutsamudu, Anjouan.

CCC no. 131: (2) 20.9.1985, (3) between Ubeni and Salimani, Grand Comoro, (4) Hama Muhammed, (5) 500 m, (6) 200 m, (9) about 150 cm, (12) frozen, to be dissected in 1991, (13) good, (14) T. Uyeno (in litt. 1990), JASEC no. 5, (15) Kanazawa Aquarium, Japan.

CCC no. 132: (2) 22.11.1985, 2300 h, (3) Vassi, Anjouan, (6) 300 m, (14) observed by T. Ogiso (1986) in the freezer of the École National de Pêche, Mutsamudu, Anjouan.

CCC no. 133: (2) November 1985, (3) Anjouan, (9) 134.3 cm, (12) mounted with coconut fibre and wood, (13) disembowled, injected with formalin, (14) M.N. Bruton, 23.4.1987, (15) on display in Karima restaurant, Domoni, Anjouan. Owned by Habane Said Ali Abdallah Abderemane, (16) Fig. 4 in Balon et al. (1988) and appears on 4 photographs in Balon (1990b, p. 127).

CCC no. 134: (2) early 1986, (3) near Mutsamudu, Anjouan, (9) 159.5 cm, (12) frozen, then fixed in formalin and dried; dry mount prepared by taxidermist, S. Bakari, on Anjouan; stuffed with coconut fibres, supported with wood, internal organs removed, fins complete, (14) M.N. Bruton, 28.4.1987, (15) Centre National de Documentation et de Recherche Scientifique, Moroni, Comoros.

CCC no. 135: (2) 4.7.1986, 0130 h, (3) Grand Comoro, (4) Mohammed Youssof Kari, (5) 400–500 m, (6) 250 m, (7) tuna, (8) about 60 kg, (9) 125 cm, (11) lived for 33 h after capture, (14) Uyeno (1991); JASEC no. 3.

CCC no. 136: (2) 17.7.1986, 0100 h, (3) Grand Comoro, (4) Mohammed Islam, (5) 400 m, (6) 200 m, (7) roudi, (8) about 65 kg, (9) 140 cm, (11) alive for 42 h after capture, (14) Uyeno (1990); JASEC no. 2, (16) a videotape was made of this specimen after its release in shallow water by J.-L. Geraud working for the JASEC team.

CCC no. 137: (2) July 1986, (3) Anjouan, (5) 200 m, (6) 300 m, (9) 124.8 cm, (12) initially frozen, later dried and stuffed with coconut fibre; the eyes were replaced, (13) poor, (14) M.N. Bruton, 1.5.1987, (15) seen in S. Bakari's taxidermy workshop near Mutsamudu, Anjouan, (16) held by S. Bakari in Fig. 24 of Balon et al. (1988).

CCC no. 138: (2) 3 December 1986, (3) Itsandra, Grand Comoro, (9) 132.1 cm, (12) specimen dissected on the left side; tissue samples were taken and placed in alcohol and in a dry collection, (14) Mary Burridge (in litt. 1990), (15) obtained by Peter Stevens, a member of the Explorer's Club expedition in November 1986, and donated to the Royal Ontario Museum, Toronto, Canada (ROM 51809).

CCC no. 139: (2) 23 November 1986, (3) Itsoundzu, Grand Comoro, (4) Madi Issala, (8) 34.6 kg, (9) 109 cm, (14) Musick (in litt. 1990), (15) American Museum of Natural History, New York (AMNH 59196), (16) brought to the shore alive and acquired by the Explorer's Club during their November 1986 expedition.

CCC no. 140: (2) ?1986, (3) Grand Comoro, (8) 13.5 kg, (9) 97.2 cm, (10) immature male, (12) severely freezer-burned indicating that it had been frozen for some time, (13) Hale et al. (1991), (14) obtained by Explorer's Club personnel for the Virginia Institute of Marine Science (VIMS 8117), (15) some of the results arising from the dissection of this specimen are published in this volume; on permanent loan to University of Kansas.

CCC no. 141: (2) November 1986, (3) west coast of Grand Comoro, (8) 53.75 kg, (9) 145.2 cm standard length, (10) immature female, (12) maintained frozen at − 30° C until dissection on 5.1.1988, (13) Cloutier et al. (1988), Schultze & Cloutier (1991), (14) acquired by personnel of the Explorer's Club; airfreighted to the New York Aquarium and transferred to the Virginia Institute of Marine Science (VIMS 8118), (15) this specimen has been subjected to intensive study through dissection, X-ray, RNA sequencing, gel electrophoresis, computed tomography and magnetic resonance imaging; some of the results of these studies are reported in this volume, Figure on p. 17, this volume.

CCC no. 142: (2) December 1986, (3) near Pomoni, Anjouan, (9) 101 cm, (12) dry mount prepared by S. Bakari, Anjouan,

Fig. 7. Three specimens of *Latimeria chalumnae* (CCC no. 153, 154 and 155) in the SOCOVIA cold storage room, Moroni. Photo by E.K. Balon, 15.5.1990.

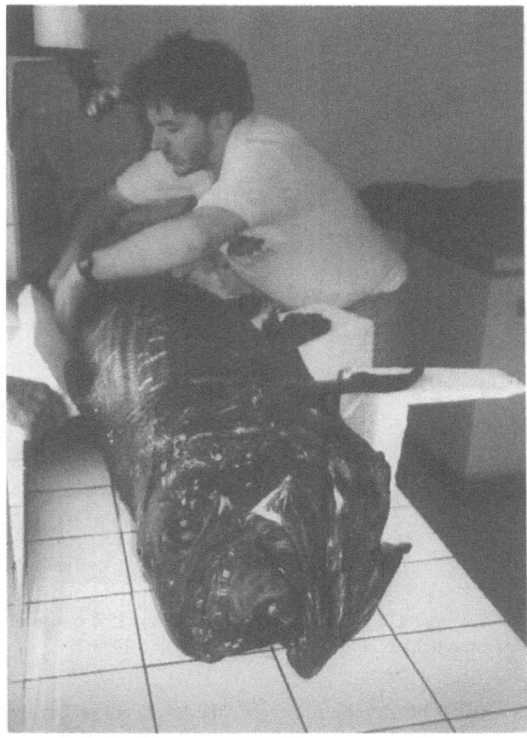

Fig. 8. The large female *Latimeria chalumnae* (CCC no. 154) now at the University of Guelph, Canada, during the initial preparation for gyotaku by R. Cloutier at the Centre National de Documentation et de la Recherche Scientifique, Moroni. Photo by E.K. Balon, 17.5.1990.

(13) poor, many scales missing, (14) M.N. Bruton, 1.5.1987, (15) S. Bakari's taxidermy workshop near Mutsamudu, Anjouan, (16) Fig. 24 in Balon et al. (1988).

CCC no. 143: (2) December 1986, (3) Iconi, Grand Comoro, (9) 182 cm, (12) deep frozen with internal organs intact, (13) good, but caudal fin bent and damaged, (14) M.N. Bruton, 1.5.87, (15) in freezer of CODOPEC at Moroni harbour, Grand Comoro, (16) Fig. 10 in Balon et al. (1988).

CCC no. 144: (2) 9.12.1986, 2100 h, (3) M'Jamaoue, Anjouan, (4) Antoissi Halifa, (8) 27.3 kg, (9) 125 cm, (15) Minorii Shiyata, Japan, (16) on a magnificent gyotaku seen at the Japanese fishing school near Mutsamudu, Anjouan; Fig. 22 in Balon et al. (1988) and on p. 126 in Balon (1990a).

CCC no. 145: (2) 1.2.1987, (3) Anjouan, (9) 100 cm, (12) sewn crudely with string ventrally before being prepared as a dry mount, (13) very poor, flaccid with many scales missing, (14) M.N. Bruton, 11.11.1987, (15) in S. Bakari's taxidermy workshop, near Mutsamudu, Anjouan.

CCC no. 146: (2) June 1987, (3) Anjouan, (8) 82 kg, (9) 161 cm, (10) female, (11) contained about 30 eggs, (12) dry mount, (13) moderate, four scales missing, (14) M.N. Bruton,

11.11.1987, (15) in S. Bakari's taxidermy workshop near Mutsamudu, Anjouan, (16) this was the first coelacanth with eggs that he had treated; the eggs were discarded.

CCC no. 147: (2) 13.6.1987, (3) Hajoho, Anjouan, (4) M'djanga Oussein, Ahamadi Boko & M'Sa M'Kolo, (5) 250 m, (6) 150 m, (8) 81.8 kg, (9) 160 cm, (10) female, (11) landed alive from a japawa, (12) frozen, (14) M.N. Bruton, 11.11.1987, (15) brought to École National de Pêche near Mutsamudu, Anjouan, (16) a video was made of this specimen while it was alive by the Japanese fishing instructors.

CCC no. 148: (2) 1.9.1987, (3) Salimani, Grand Comoro, (4) Abderemane, (9) 120 cm, (12) frozen, (13) excellent, all scales intact, (14) M.N. Bruton, 1.5.1987, (15) in deepfreeze of Electrofood, Moroni, (16) for sale on 1.5.1987.

CCC no. 149: (2) 24.10.1987, (3) Hajoho, Anjouan, (4) Ahmed Zoubert, (6) 250 m, (8) 16.3 kg, (9) 92 cm, (12) frozen, (14) M.N. Bruton, 1.5.1987, (15) seen at the École National de Pêche near Mutsamudu, Anjouan.

CCC no. 150: (2) January 1988, (9) 123 cm, (12) partially

Fig. 9. Large *Latimeria chalumnae* (CCC no. 155) outside the SOCOVIA freezer in Moroni held by Mike Bruton flanked by two Comoran helpers. Photo by E.K. Balon, 15.5.1990.

dissected half of body preserved; the head, gill arches, half pectoral girdle and pelvic girdle were skeletonized, (13) poor, a preliminary dissection was made to remove the gut to reduce the weight for shipping, (14) J.A. Musick (in litt. 1990), (15) American Museum of Natural History, New York (AMNH 56150), (16) W.E. Bemis bought this specimen from the Comoran government freezer in January 1988.

CCC no. 151: (2) January 1988, (9) 141 cm, (10) male, (14) M.N. Feinberg (in litt. 1990), J.A. Musick (in litt. 1990), (15) American Museum of Natural History, New York (AMNH 56150), (16) W.E. Bemis bought this specimen from the Comoran government freezer in January 1988, it was half rotten and partially skeletonized.

CCC no. 152: (2) 13.6.1988, (3) Iconi, Grand Comoro, (4) Ali Mubai, (5) 400 m, (6) 200 m, (7) gempylid, (9) 149.3 cm, (10) female, (12) initially frozen then prepared as a taxidermy display specimen, with artificial eyes, by H. Taguchi, (13) very good, (14) T. Uyeno (in litt. 1990), JASEC no. 6, (15) currently housed at the Kanazawa Aquarium, Japan.

CCC no. 153: (2) 1.2.1989, (3) Grand Comoro, (8) 26 kg, (9) 125.5 cm, (12) deep frozen, (14) Michel de San of Development Pêche Artisanale aux Comores, (15) SOCOVIA, Moroni, bought from Development Pêche Artisanale aux Comores by E.K. Balon on behalf of H.R. Axelrod for the University of Guelph, Canada, on 25.5.1990; kept frozen for research at the J.L.B. Smith Institute of Ichthyology, Grahamstown; Fig. 7 in this paper.

CCC no. 154: (2) 24.6.1989, (3) Hahaya, Grand Comoro, (5)

100 m, (6) 200 m, (9) 168 cm, (10) female, (11) alive during attempts to recompress, then died, (12) deep frozen, (13) excellent, partly defrosted for examination and photography of the posterior parts of oviduct and rectum by E.K. Balon, M.N. Bruton and R. Cloutier in May 1990, (14) photographed by Jean-Louis Geraud when alive during attempts to recompress, (15) donated to the J.L.B. Smith Institute of Ichthyology by the Comoran government in May 1990, obtained from SOCOVIA freezer in Moroni, (16) seven gyotakus were made of this specimen, which was transported frozen to Grahamstown and later fixed in formalin and sent on 12.9.1990 to the University of Guelph in Canada for dissection; Figures 7 and 8 in this paper.

CCC no. 155: (2) 31.10.1989, (3) Dzahadjou, Grand Comoro, (9) 176.5 cm, (10) ?female, (12) deep frozen in SOCOVIA freezer, May 1990, (13) excellent, gills red and firm, (14) M.N. Bruton & E.K. Balon, (15) SOCOVIA freezer, Moroni; belongs to Centre National de Documentation et de la Recherche, Moroni, intended for dissection by Japanese scientists, (16) this specimen was examined alive by H. Fricke on 31.10.1989; a blood sample was taken for haemoglobin analysis; donated to President Mitterand of France when he visited the Comoros in July 1990; Figure 9 in this paper.

CCC no. 156: (2) late 1989, (9) 122 cm, (12) frozen, (14) M.N. Bruton & E.K. Balon, May 1990, (15) SOCOVIA, Moroni, Grand Comoro.

CCC no. 157: (2) 13.11.1990, (3) Salimani, Grand Comoro, (11) alive for 14 h, (14) Michel de San (in litt. 16.11.1990).

CCC no. 158: (2) 1990, (3) Anjouan, (14) Michel de San (in litt. 16.11.1990).

Personal communications on specimens of unknown date of capture

CCC no. a: (8) 80 kg, (9) 170 cm, (14) P.C. Heemstra, (15) Ylang Ylang Hotel, Moroni, Grand Comoro.

CCC no. b: (13) freeze-dried head, a large number of scales and other miscellaneous tissues, (14) S.N. Norris (in litt. 1989), (15) Arizona State University Museum of Natural History, (16) this specimen was acquired by an Arizona State University team collecting marine organisms to test various tissue extracts for cancer-fighting compounds.

CCC no. c: (14) Khalaf (1987), (15) Science and Natural History Museum, State of Kuwait.

CCC no. d: (14) J.-P. Sylvestre (in litt. 1990), (15) Shanghai Museum, China.

CCC no. e: (3) Grand Comoro, (6) 100–200 m, (8) 26 kg, (9) 124 cm, (12) 10% formalin, (13) good, (14) J.-P. Sylvestre (in litt. 1990), Zhu Min (in litt. 1990), (15) Specimen House of Ichthyology, Institute of Hydrobiology, Academica Sinica, Wuhan, China.

CCC no. f: (3) Grand Comoro, (14) J.-P. Sylvestre (in litt. 1990), Zhu Min (in litt. 1990), (15) Museum of Natural History, Beijing, China.

CCC no. g: (14) J.-P. Sylvestre (in litt. 1990), (15) unknown holding, China.

CCC no. h: (14) J.-P. Sylvestre (in litt. 1990), (15) Beijing, China.

CCC no. i: (14) H. Fricke (in litt. 1988), (15) private house, Mutsamudu, Comoros.

CCC no. j: (14) H. Fricke (in litt. 1988), (15) private house, Mutsamudu, Comoros.

CCC no. k: (14) H. Fricke (in litt. 1988), (15) private house, Mutsamudu, Comoros.

CCC no. l: (14) H. Fricke (in litt. 1988), (15) private house, Moroni, Comoros.

CCC no. m: (14) H. Fricke, P. Ansfield & R. Plante (in litt. 1988), (15) private property of a French captain, oral communication by port captain of Mutsamudu, Anjouan.

CCC no. n: (3) east coast of Grand Comoro, (9) 155 cm, (10) female, (12) frozen, (13) good, (14) T. Uyeno (in litt. 1990), JASEC no. 4, (15) Kanazawa Aquarium, Japan.

References cited

Balon, E.K. 1977. Early ontogeny of *Labeotropheus* Ahl, 1927 (Mbuna, Cichlidae, Lake Malawi), with a discussion on advanced protective styles in fish reproduction and development. Env. Biol. Fish. 2: 147–176.

Balon, E.K. 1990a. The living coelacanth endangered: a personalized tale. Tropical Fish Hobbyist 38 (February): 117–129.

Balon, E.K. 1990b. Tracking the coelacanth: a follow-up tale. Tropical Fish Hobbyist 38 (March): 122–131.

Balon, E.K. 1991. Probable evolution of the coelacanth's reproductive style: lecithotrophy and orally feeding embryos in cichlid fishes and in *Latimeria chalumnae*. Env. Biol. Fish. 32: 249–265. (this volume)

Balon, E.K., M.N. Bruton & H. Fricke. 1988. A fiftieth anniversary reflection on the living coelacanth, *Latimeria chalumnae:* some new interpretations of its natural history and conservation status. Env. Biol. Fish. 23: 241–280.

Bemis, W.E. & R.G. Northcutt. 1991. Innervation of the basicranial muscle of *Latimeria chalumnae*. Env. Biol. Fish. 32: 147–158. (this volume)

Bruton, M.N. 1988. Coelacanth Conservation Council/Conseil pour la Conservation du Coelacanthe, Newsletter no. 1. Env. Biol. Fish. 23: 315–319.

Bruton, M.N. 1989a. Is there a Madagascar coelacanth? Ichthos 23: 25.

Bruton, M.N. 1989b. Does the coelacanth occur in the Eastern Cape? The Naturalist (Port Elizabeth) 33: 3–11.

Bruton, M.N. & M.J. Armstrong. 1991. The demography of the coelacanth *Latimeria chalumnae*. Env. Biol. Fish. 32: 301–311. (this volume)

Bruton, M.N. & R.E. Stobbs. 1991. The ecology and conservation of the coelacanth *Latimeria chalumnae*. Env. Biol. Fish. 32: 313–339. (this volume)

Cloutier, R., H.-P. Schultze, E.O. Wiley, J.A. Musick, J.C. Daimler, M.A. Brown, S.J. Dwyer, L.T. Cook & R.L. Laws. 1988. Recent radiologic imaging techniques for morphological studies of *Latimeria chalumnae*. Env. Biol. Fish. 23: 281–282.

Dugan, J. 1955. The fish. Collier's, 16 September 1955: 64–68.

Fricke, H. 1988. Coelacanths – the fish that time forgot. Nat. Geogr. Mag. 173: 824–838.

Grady, J.E. 1970. Tooth development in *Latimeria chalumnae* (Smith). J. Morph. 132: 377–387.

Hale, R.C., J. Greaves, J.L. Gundersen & R.F. Mothershead II. 1991. Occurrence of organochlorine contaminants in tissues of the coelacanth *Latimeria chalumnae*. Env. Biol. Fish. 32: 361–367. (this volume)

Kamegai, S. 1971. On some parasites of a coelacanth *Latimeria chalumnae:* a new monogenea *Dactylodiscus latimeris* new genus, new species (Dactylodiscide new family) and two larval helminths. Res. Bull. Meguro parasitol. Mus. 5: 1–5.

Khalaf, N.-A.B. 1987. The coelacanth, *Latimeria chalumnae*, in the Science and Natural History Museum, State of Kuwait. Gazelle 15: 1–7.

Locket, N.A. 1980. Some advances in coelacanth biology. Proc. R. Soc. Lond. B208: 265–307.

Martin, R. 1970. D'utiles details sur le coelacanthe. Promo. Al'Camar 16: 25–35.

McAllister, D.E. & C.L. Smith. 1978. Mensurations morphologiques, denombrements meristiques et taxonomie du coelacanthe, *Latimeria chalumnae*. Naturaliste canadien 105: 63–76.

McCosker, J.E. 1979. Inferred natural history of the living coelacanth. Occ. Pap. Calif. Acad. Sci. 134: 17–24.

McCosker, J.E. & M.D. Lagios. (ed.). 1979. The biology and physiology of the living coelacanth. Occ. Pap. Calif. Acad. Sci. 134: 1–175.

Millot, J., J. Anthony & D. Robineau. 1972. État commente des captures de *Latimeria chalumnae* Smith (Poisson, Crossopterygien, Coelacanthide) effectuées jusqu'au mois d'Octobre 1971. Bull. mus. Nat. Hist. nat., Paris, 3e serie, no. 53, Zoologie 39: 533–548.

Ogiso, T. 1986. Island of the coelacanth. Expert 71: 6–10. (In Japanese).

Schultze, H.-P. & R. Cloutier. 1991. Computed Tomography and Magnetic Resonance Imaging studies of *Latimeria chalumnae*. Env. Biol. Fish. 32: 159–182. (this volume)

Smith, C.L., C.S. Rand, B. Schaeffer & J.W. Atz. 1975. *Latimeria*, the living coelacanth, is ovoviviparous. Science 190: 1105–1106.

Smith, J.L.B. 1939. A living coelacanthid fish from South Africa. Trans. R. Soc. S. Afr. 28: 1–106.

Smith, I.L.B. 1953a. A coelacanth in Kenyan waters? S. Afr. Angler 8: 8, 19.

Smith, J.L.B. 1953b. The second coelacanth. Nature 171: 99–107.

Smith, J.L.B. 1956. Old fourlegs. The story of the coelacanth. Longmans & Green, London. 260 pp.

Stobbs, R.E. 1987. The 'galawas' of the Federal Islamic Republic of the Comores, with notes on artisanal fishing and recom-mendations for a maritime museum. Invest. Rep. J.L.B. Smith Inst. Ichthyol. 26: 1–21.

Stobbs, R.E. 1989. The coelacanth enigma. The Phoenix 2(2): 8–15.

Stobbs, R.E. 1990a. Comorian fish names with a preliminary list of Malagasy common names. Invest. Rep. J.L.B. Smith Inst. Ichthyol. 35: 1–34.

Stobbs, R.E. 1990b. N'Galawa: a dugout canoe of the Western Indian Ocean. Skiboat 6(3): 77–81.

Stobbs, R.E. & M.N. Bruton. 1991. The fishery of the Comoros, with comments on its possible impact on coelacanth survival. Env. Biol. Fish. 32: 341–359. (this volume)

Suzuki, N. & K. Tanauma. 1984. Coelacanth fishing by native fishermen. pp. 8–9. *In:* Proceedings of First Symposium on Coelacanth Studies, Tokyo. (In Japanese).

Suzuki, N., Y. Suyehiro & T. Tamada. 1985. Initial report of expeditions for coelacanth – Part 1 – field studies in 1981 and 1983. Sci. Pap. Coll. Arts Sci., Univ. Tokyo 35: 37–79.

Thys van den Audenaerde, D.F.E. 1984. Le coelacanthe des Comores, *Latimeria chalumnae* curiosité zoologique, fossile vivant ou animal aberrant? Afrika-Tervuren 30: 90–103.

Uyeno, T. 1989. Investigating the ecology of the coelacanth based on feeding habits. Newton Graphic Science Magazine 9(13): 20–21. (In Japanese).

Uyeno, T. 1991. Observations on locomotion and feeding of released coelacanths, *Latimeria chalumnae*. Env. Biol. Fish. 32: 267–273. (this volume)

Environmental Biology of Fishes **32**: 391–401, 1991.

Literature relating to fossil coelacanths

Peter L. Forey[1] & Richard Cloutier[1,2]
[1] *Department of Palaeontology, British Museum (Natural History), Cromwell Road, London SW7 5BD, England (present address of* [2])
[2] *Museum of Natural History, Department of Systematics and Ecology, The University of Kansas, Lawrence, KS 66045–2454, U.S.A.*

Received 15.4.1990   Accepted 26.7.1990

Key words: Actinistians, Palaeontology, Bibliography, Publication trends

This paper lists those references dealing, in whole or in part, with fossil coelacanths. As such it stands complementary to Bruton et al. (1991) relating to literature on *L. chalumnae* and provides a bibliography for the article dealing with the listings and primary references of coelacanth species (Cloutier & Forey 1991). It also includes many secondary sources of information about fossil coelacanths.

By way of introducing this list we recognise that any remarks will be eclectic, but to help the reader gain some ideas of availability of literature concerned with coelacanth anatomy, diversity, and evolutionary history we offer the following brief survey of some of the more significant papers of coelacanth studies.

Prior to 1839, a few fossil coelacanth fragments had been assigned to different groups of fishes such as actinopterygians. Agassiz (1844) distinguished an assemblage of fishes which he called 'Ordre Coelacanthes', recognised on the basis of his Permian species *Coelacanthus granulatus* (Agassiz 1839). With *Coelacanthus*, he grouped fishes that we now recognise as placoderms and porolepiforms and it was Thiollière (1858) and Huxley (1861) who more clearly defined the group containing fishes that we recognise today as coelacanths. Huxley's paper, in particular, clearly defined and characterised the group and placed them with lobed-finned fishes. Huxley only knew of a few coelacanths, representing species from the Per-

mian (*Coelacanthus*), the Jurassic (*Undina*) and the Cretaceous (*Macropoma*). Once defined then other nineteenth century workers were able to compare their fossil fishes. In the United States, Newberry (1856) described several species from the Coal Measures. In Europe, Reis (1888) provided descriptions of several species from the famous Upper Jurassic Lithographic Limestone of Bavaria, and he also summarised earlier work on European Carboniferous forms. In 1892 he presented a précis and his own interpretation of the comparative anatomy of coelacanths. The foundations of coelacanth study laid by Agassiz, Huxley and Reis were built upon by Woodward (1891) in the second volume of his *Catalogue of Fossil Fishes in the British Museum (Natural History)*, where a comprehensive listing of coelacanth species and literature is given. Most importantly, this work includes generic and species diagnoses and ideas on synonymy, many of which remain valid today.

Stensiö (1921) provided the first significant research on coelacanths this century when he described species from the Lower Triassic of Spitzbergen, and he followed this (1932) with descriptions of superbly preserved Lower Triassic coelacanths from East Greenland. Subsequently Stensiö (1937) completed this concentration of coelacanth work when he re-investigated the Devonian coelacanths found in Germany in the late nineteenth century. Together these three works outlined the

392

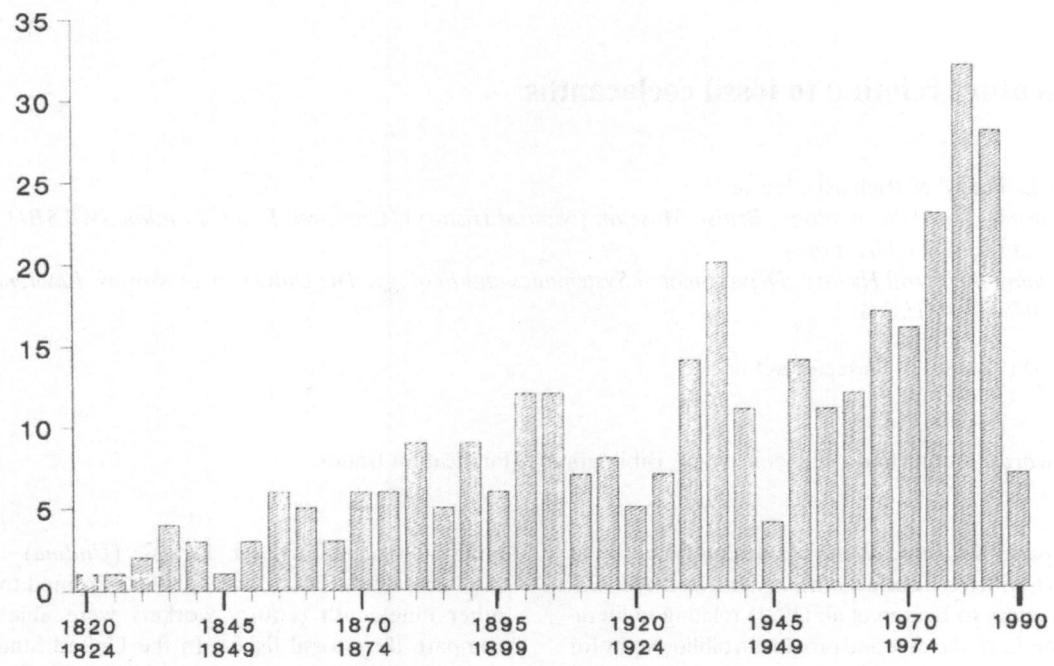

Fig. 1. Histogram showing the number of publications dealing with fossil actinistians since 1820.

history of coelacanth research to date, provided a wealth of new anatomical information – particularly about Devonian coelacanths (the skull of *Diplocercides [= Nesides]* had been sectioned by then) – and had considerably increased the number of species recognised and made detailed comparisons between coelacanths and other fish groups.

In Britain, Moy-Thomas (1937) collated information about the European Carboniferous coelacanths and also redescribed (1935), in more detail, some well-preserved forms from the Lower Triassic of Madagascar.

The discovery of *Latimeria chalumnae* in 1938 (Smith 1939a, b, c) had a knock-on effect for the study of fossil coelacanths with studies of the relationship between the fossils and recent form, and evolutionary rates (see Cloutier & Forey 1991). Studies by Schaeffer (1941, 1948, 1952a, b) included the first attempts to work out an internal phylogeny of coelacanths and to draw the study of coelacanths into the limelight by pointing out some interesting features of evolutionary rates: these stud-

ies have been renewed recently by one of us (Cloutier 1991).

Throughout the 1960's and 1970's work on fossil coelacanths was largely concentrated on describing new forms; and this was continued with the description of several new and unusual species from the Lower Carboniferous of Montana, U.S.A. (Lund & Lund 1985). The emphasis on coelacanth work as a whole, however, had shifted to work on *L. chalumnae*.

But the 1980's have seen a revival of work on fossil coelacanths (Fig. 1), in particular, work which is directed to providing new descriptions of coelacanths in the light of the recent model and to erecting cladistic classifications of coelacanths and comparisons of fossil coelacanths in the study of sarcopterygian interrelationships (Forey 1981, Cloutier & Forey 1991, Cloutier 1990) and these are summarised in Forey (1991) and Cloutier (1991).

Bibliography

Abel, O. 1931. Schwimmfährten von Fischen und Schildkröten im lithographischen Schiefer Bayerns. Natur und Museum 61: 97–106.

Agassiz, L. 1835. Recherches sur les Poissons Fossiles. Vol. 2. 4th livraison. Feuilleton, Neuchâtel: 35–64.

Agassiz, L. 1836. Recherches sur les Poissons Fossiles. 6th livraison. Feuilleton, Neuchâtel: 75–86.

Agassiz, L. 1839. Recherches sur les Poissons Fossiles. Vol. 1. Neuchâtel. 40 pp.

Agassiz, L. 1844. Recherches sur les Poissons Fossiles. Vol. 2. Neuchâtel. 336 pp.

Aldinger, H. 1930. Über das Kopfskellett von *Undina acutidens* Reis und den kinetischen Schädel der Coelacanthiden. Zentralbl. Miner. Geol. Paläontol. B 1930: 22–48.

Aldinger, H. 1931. Ueber karbonische Fische aus Westfalen. Paläontol. Z. 13: 186–201.

Alessandri, G. de. 1910. Studii sui pesci Triasici della Lombardia. Mem. Soc. Ital. Sci. Nat. Mus. Civico Stor. Nat. Milano 7: 1–145.

Allis, E.P. 1922. The myodome and the trigemino-facialis chamber in the Coelacanthidae, Rhizodontidae and Palaeoniscidae. J. Anat. 56: 149–154.

Almeida Campos, D. de & S. Wenz. 1982. Première découverte de coelacanthes dans le Crétacé inférieur de la Chapada do Araripe (Brésil). C.R. Acad. Sci. Paris, Sér. 2, 294: 1151–1154.

Andersson, E. 1916. Über einige Trias-Fische aus der Cava Trefontane, Tessin. Bull. Geol. Instit. Univ. Upsala. 15: 13–34.

Andrews, S.M. 1973. Interrelationships of crossopterygians. pp. 138–177. *In:* P.H. Greenwood, R.S. Miles & C. Patterson (ed.) Interrelationships of Fishes, Zool. J. Linn. Soc. Lond., 53, Suppl. 1.

Andrews, S.M. 1977. The axial skeleton of the coelacanth, *Latimeria.* pp. 227–288. *In:* S.M. Andrews, R.S. Miles & A.D. Walker (ed.) Problems in Vertebrate Evolution, Linn. Soc. Symp. Ser. 4.

Andrews, S.M., B.G. Gardiner, R.S. Miles & C. Patterson. 1967. Pisces. pp. 637–683. *In:* W.B. Harland, C.H. Holland, M.R. House, N.F. Hughes, *et al.* (ed.) The Fossil Record: A Symposium with Documentation, Geol. Soc. London.

Ash, S.R. 1978. Fish scales. Brigham Young Univ. Stud. Geol. 25: 67–68.

Baird, D. 1978. Studies on Carboniferous freshwater fishes. Amer. Mus. Novit. 2641. 1–22.

Balon, E.K. 1991. Probable evolution of the coelacanth's reproductive style: lecithotrophy and orally feeding embryos in cichlid fishes and in *Latimeria chalumnae.* Env. Biol. Fish. 32: 249–265. (this volume)

Balon, E.K., M.N. Bruton & H. Fricke. 1988. A fiftieth anniversary reflection on the living coelacanth, *Latimeria chalumnae:* some new interpretation of its natural history and conservation status. Env. Biol. Fish. 23: 241–280.

Bang, B.S. 1976. Avanceret kemisk praeparationsmetodik eg-

net for palaeontologiske og kulturhistoriske objekter en material v og metodelaere. Geologisk Museum, København. 59 pp.

Barcellos, M.T. 1975. Estudo de escamas e dentes de peixes da Fácies Budó, Sub-Grupo Itararé – R.G.S. Bol. Paran. Geociên. 32: 3–67.

Bassani, F. 1896. La ittiofauna della Dolomia Principale di Pisa, Giffoni (Provincia di Salerno). Palaeontogr. Ital. 1: 169–210.

Belloti, C. 1857. Descrizione di alcune nuove specie di pesci fossili di Perledo e di altre localitá Lombarde. pp. 419–438. *In:* Studii Geologici e Paleontologici sulla Lombardia (Stoppani, A.), Presso Carlo Turati Tipografo – Editore, Milano.

Beltan, L. 1968. La faune ichthyologique de l'Eotrias du N.W. de Madagascar: le neurocrane. C.N.R.S. Paris. 135 pp.

Beltan, L. 1972. La faune ichthyologique du Muschelkalk de la Catalogne. Mem. Real Acad. Cienc. Artes Barcelona 41: 280–325.

Beltan, L. 1979. Eotrias du Nord-Ouest de Madagascar: Etude de quelques poissons dont un est en parturition. Ann. Soc. Géol. Nord, Lille 99: 453–464.

Beltan, L. 1984. Quelques poissons du Muscheskalk superieur d'Espagne. Acta Geol. Hispàn. 19: 117–127.

Bendix-Almgreen, S.E. 1976. Palaeovertebrate faunas of Greenland. pp. 536–573. *In:* A. Escher & W.S. Watt (ed.) Geology of Greenland, Geol. Surv. Greenland, Copenhagen.

Berg, L.S. 1940. Classification of fishes, both recent and fossil. Tr. Zool. Instit. Leningrad 5: 1–517. [Separate English edition by Dover Publ. Co., Ann Arbor, Michigan (1947)].

Berg, L.S. 1955. Classification of fishes and fish-like vertebrates (Second Edition). Tr. Zool. Instit. Leningrad 20: 1–286.

Berger, H.A.C. 1832. Die Versteinerungen der Fische und Pflanzen im Sandsteine der Coburger Gegend. Coburg. 29 pp.

Bjerring, H.C. 1967. Does a homology exist between the basicranial muscle and the polar cartilage. Coll. Internatl. C.N.R.S. 163: 223–268.

Bjerring, H.C. 1971. The nerve supply to the second metamere basicranial muscle in osteolepiform vertebrates, with some remarks on the basic composition of the endocranium. Acta Zool. 52: 189–225.

Bjerring, H.C. 1972. The nervus rarus in coelacanthiform phylogeny. Zool. Scr. 1: 57–68.

Bjerring, H.C. 1973. Relationships of coelacanthiforms. pp. 179–205. *In:* P.H. Greenwood, R.S. Miles & C. Patterson (ed.) Interrelationships of Fishes, Zool. J. Linn. Soc., 53, Suppl. 1.

Bjerring, H.C. 1978. The 'intracranial joint' *versus* the 'ventral otic fissure'. Acta Zool. (Stockholm) 59: 203–214.

Bjerring, H.C. 1985. Facts and thoughts on piscine phylogeny. pp. 31–57. *In:* Evolutionary Biology of Primitive Fishes, NATO ASI series, Ser. A, Life Sci. 103, Plenum Press, New York.

Bock, W. 1959. New eastern American Triassic fishes and Triassic correlations. Nat. Acad. Sci., Geol. Cent. Res. Ser., Philadelphia 1: 1–184.

394

Broili, F. 1933. Der Obere Jura von Montsech (Provinz Lérida) im Vergleich mit den Ob. Jura – Vorkommen von Cerin (Dept. Ain) und von Franken. Congr. Géol. Internatl., 14th Ses., Barcelona, Géologie de la Méditerranée Occidentale 2 (16): 1–11.

Broom, R. 1905. On a species of *Coelacanthus* from the Upper Beaufort Beds of Aliwal North. Rec. Albany Mus. Grahamstown, S. Africa 1: 338–339.

Broom, R. 1909. The fossil fishes of the Upper Karroo Beds of South Africa. Ann. S. Afr. Mus. 7: 251–269.

Brough, J. 1931. On fossil fishes from the Karroo system, and some general considerations on the bony fishes of the Triassic period. Proc. Zool. Soc. London 1931: 235–296.

Bryant, W.L. 1934. New fishes from the Triassic of Pennsylvania. Proc. Amer. Phil. Soc. 73: 319–326.

Carroll, R.L. 1988. Vertebrate paleontology and evolution. W.H. Freeman & Company, New York. 698 pp.

Case, E.C. 1900. Contributions from the Walker Museum. 1. The vertebrates from the Permian bone-bed of Vermillian County, Illinois. J. Geol. 8: 698–729.

Casier, E. 1961. Matériaux pour la faune ichthyologique Eocrétacique du Congo. Annal. Mus. Roy. Afr. Centr., Sci. Géol. 39: 1–96.

Casier, E. 1965. Poissons fossiles de la série du Kwango (Congo). Annal. Mus. Roy. Afr. Centr., Sci. Géol. 50: 1–64.

Chabakov, A.V. 1927. Sur les restes des crossoptérygiens du Carbonifère russe. Bull. Com. Géol. Léningrad 46 (4): 299–309. (In Russian).

Chang, M.-M. 1982. The braincase of *Youngolepis,* a Lower Devonian crossopterygian from Yunnan, south-western China. GOTAB, Stockholm: 1–113.

Chang, M.-M. & X. Yu. 1984. Structure and phylogenetic significance of *Diabolichthys speratus* gen. et sp. nov., a new dipnoan-like form from the Lower Devonian of Eastern Yunnan, China. Proc. Linn. Soc. New South Wales, Sydney 107: 171–184.

Cloutier, R. 1990. Interrelationships of Palaeozoic actinistians: patterns and trends. *In:* M.M. Chang, G. Zhang & Y. Liu (ed.) Proceedings of the Fifth Symposium on Early Vertebrate Studies and Related Problems in Evolutionary Biology, Beijing, People's Republic of China, Science Press, Beijing. (in press).

Cloutier, R. 1991. Patterns, trends, and rates of evolution within the Actinistia. Env. Biol. Fish. 32: 23–58. (this volume)

Cloutier, R. & P.L. Forey. 1991. Diversity of extinct and living actinistian fishes (Sarcopterygii). Env. Biol. Fish. 32: 59–74. (this volume)

Cope, E.D. 1873. On some new Batrachia and fishes from the coal measures of Linton, Ohio. Proc. Acad. Nat. Sci. Philadelphia, 1873: 340–343.

Cope, E.D. 1878. On the Vertebrata of the bone bed in Eastern Illinois. Proc. Amer. Phil. Soc. 17: 52–63.

Cope, E.D. 1882. Third contribution to the history of the Vertebrata of the Permian formation of Texas. Proc. Amer. Phil. Soc. 20: 447–461.

Cope, E.D. 1894. New and little known Palaeozoic and Mesozoic fishes. J. Acad. Nat. Sci. Philadelphia 9: 427–448.

Cope, E.D. & W.D. Matthew. 1915. Hitherto unpublished plates of Tertiary Mammalia and Permian Vertebrata. U.S. Geol. Surv. & Amer. Mus. Nat. Hist., Monogr. ser. 2, 154 pl.

Costa, O.G. 1862. Studii sopra i terreni ad ittiolitti del Regno di Napoli diretti a stabilire l'età geologica dei medesimi. Appendice Agli Alti della Real Accademia delle Scienze 12: 1–44.

Davis, J.W. 1876. On a bone-bed in the Lower Coal-measures, with an enumeration of the fish-remains of which it is principally composed. Quart. J. Geol. Soc. London 32: 332–340.

Davis, J.W. 1883. On the fossil fishes of the Carboniferous Limestone series of Great Britain. Scient. Trans. Roy. Dubl. Soc. 1: 327–600.

Davis, J.W. 1884. On a new species of *Coelacanthus (C. tingleyensis)* from the Yorkshire Cannel Coal. Trans. Linn. Soc. London 2: 427–433.

Davis, J.W. 1890. On *Coelacanthus phillipsii* Agassiz. Geol. Mag. 7: 159–161.

Dean, B. 1895. Fishes, living and fossil. An outline of their forms and probable relationships. Columbia Univ. Biol. Ser. III. MacMillan and Co., New York. 300 pp.

DeCarvalho, M.S.S. 1982. O Gênero *Mawsonia* na ictiofáunula do Cretáceo do Estado da Bahia. Anais Acad. Brasil. Cienc. 54: 519–539.

Deecke, W. 1889. Ueber Fische aus verschiedenen Horizonten der Trias. Palaeontographica 35: 97–138.

Dehm, R. 1956a. Ein Coelacanthide aus dem Mittleren Keuper Frankens. Neues Jahrb. Geol. Paläontol. Mh. 1956: 148–153.

Dehm, R. 1956b. Über einen ersten, verschollenen Coelacanthiden-Fund aus dem Mittleren Keuper Frankens. Neues Jahrb. Geol. Paläontol. Mh. 1956: 525–529.

Demanet, F. 1938. La faune des couches de passage du Dinantien au Namurien dans le synclinorium de Dinant. Mém. Mus. Roy. Hist. Nat. Belgique 84: 1–201.

Demanet, F. 1939. Filtering appendices on the branchial arches of *Coelacanthus lepturus* Agassiz. Geol. Mag. 76: 215–219.

Demanet, F. 1941. Faune et stratigraphie de l'étage Namurien de la Belgique. Mém. Mus. Roy. Hist. Nat. Belg. 97: 1–327.

Dixon, F. 1850. The geology and fossils of the Tertiary and Cretaceous formations of Sussex. Richard & John Edward Taylor, London. 422 pp.

Dolgopol de Saez, M. 1940. Noticias sobre peces fósiles Argentinos. Notas de Museo de la Plata 5: 295–298.

Dziewa, T.J. 1980. Early Triassic osteichthyans from the Knocklofty Formation of Tasmania. Pap. Proc. Roy. Soc. Tasmania 114: 145–160.

Eastman, C.R. 1902. The Carboniferous fish fauna of Mazon Creek, Illinois. J. Geol. 10: 535–541.

Eastman, C.R. 1908a. Notice of a new coelacanth fish from the Iowa Kinderhook. J. Geol. 16: 357–362.

Eastman, C.R. 1908b. Devonian fishes of Iowa. Iowa Geol. Surv. 18: 29–386.

Eastman, C.R. 1911. Triassic fishes of Connecticut. Connecticut Geol. Nat. Hist. Surv., Bull. 18 (4): 1–78.

Eastman, C.R. 1914. Catalogue of the fossil fishes in the Carnegie Museum. Part III. Catalogue of the fossil fishes from the Lithographic Stone of Cerin, France. Mem. Carnegie Mus. 6: 349–388.

Eastman, C.R. 1917. Fossil fishes in the collection of the United States National Museum. Proc. U.S. Nat. Mus. 52: 235–304.

Echols, J. 1963. A new genus of Pennsylvanian fish (Crossopterygii, Coelacanthiformes) from Kansas. Univ. Kans. Mus. Nat. Hist. Publ. 12: 475–501.

Edinger, T. 1927. Ein Coelacanthiden – Flossenträger aus dem Keuper. Senckenbergiana 9: 184–188.

Egerton, P. de M.G. 1850. Family Coelacanthidae, Agassiz. pp. 235–236. In: W. King (ed.) A monograph of the Permian Fossils of England, Palaeontographical Society, London.

Egerton, P. de M.G. 1861a. Holophagus Gulo. p. 19. In: T.H. Huxley (ed.) Preliminary Essay Upon the Systematic Arrangement of the Fishes of the Devonian Epoch, Figures and Descriptions Illustrative of British Organic Remains, Mem. Geol. Surv. U.K., Dec. 10.

Egerton, P. de M.G. 1861b. Tristichopterus alatus. Figures and descriptions illustrative of British organic remains. Mem. Geol. Surv. U.K., Dec. 10: 51–55.

Elliott, D.K. 1987. A new specimen of Chinlea sorenseni from the Chinle Formation, Dolores River, Colorado. J. Arizona-Nevada Acad. Sci. 22: 47–52.

Emmons, E. 1857. American geology, containing a statement of the principles of the science, with full illustrations of the characteristic American fossils. Vol. I, Part. 6, Albany. 158 pp.

Escher, K. 1925. Das Verhalten der Seitenorgane der Wirbeltiere und ihrer Nerven beim Übergang zum Landleben. Acta Zool. 6: 307–414.

Fabre, J., F. de Broin, L. Ginsburg & S. Wenz. 1982. Les vertébrés du Berriasien de Canjuers (Var, France), et leur environment. Géobios 15: 891–923.

Fletcher, H. 1884. Report on the geology of northern Cape Breton. Rep. Prog. Geol. Surv. Canada, Dawson Brothers, Montréal. 10 pp.

Forey, P.L. 1980. Latimeria: a paradoxical fish. Proc. Roy. Soc. Lond. (B) 208: 369–384.

Forey, P.L. 1981. The coelacanth Rhabdoderma in the Carboniferous of the British Isles. Palaeontology 24: 203–229.

Forey, P.L. 1984. The coelacanth as a living fossil. pp. 166–169. In: N. Eldredge & S.M. Stanley (ed.) Living Fossils, Springer, New York.

Forey, P.L. 1987. Relationships of lungfishes. pp. 75–91. In: W.E. Bemis, W.W. Burggren & N.E. Kemp (ed.) The Biology and Evolution of Lungfishes, Cent. Suppl. 1, J. Morphol., Alan R. Liss, New York.

Forey, P.L. 1988. Golden jubilee for the coelacanth Latimeria chalumnae. Nature 336: 727–732.

Forey, P.L. 1989. Le coelacanthe. La Recherche 215: 1318–1326.

Forey, P.L. 1990. The coelacanth fish: progress and prospects. Sci. Progress 74: 53–68.

Forey, P.L. 1991. Latimeria chalumnae and its pedigree. Env. Biol. Fish. 32: 75–97. (this volume)

Forey, P.L., B.G. Gardiner & C. Patterson. 1990. The lungfish, the coelacanth, and the cow revisited. In: H.-P. Schultze & L. Trueb (ed.) Origins of the Major Groups of Tetrapods: Controversies and Consensus, Cornell Univ. Press, Ithaca. (in press).

Forey, P.L., O. Monod & C. Patterson. 1985. Fishes from the Akkuyu Formation (Tithonian), Western Taurus, Turkey. Géobios 18: 195–201.

Forey, P.L. & V.T. Young. 1985. Acanthodian and coelacanth fish from the Dinantian of Foulden, Berwickshire, Scotland. Trans. Roy. Soc. Edinburgh 76: 53–59.

Fritsch, A. 1878. Die Reptilien und Fische der Böhmischen Kreideformation. Fritsch, Prag. 44 pp.

Gall, J.C., L. Grauvogel & J.P. Lehman. 1974. Faune du Buntsandstein. Les poissons fossiles de la collection Grauvogel-Gall. Ann. Paléont. Vert. 60: 129–145.

Gardiner, B.G. 1960. A revision of certain actinopterygian and coelacanth fishes, chiefly from the Lower Lias. Bull. Br. Mus. (Nat. Hist.), Geol. 4: 239–384.

Gardiner, B.G. 1966. Catalogue of Canadian fossil fishes. Contr. 68, Life Sci., Roy. Ontario Mus., Univ. Toronto: 1–154.

Gardiner, B.G. 1973. New Palaeozoic fish remains from southern Africa. Palaeontol. Afr. 15: 33–35.

Gardiner, B.G. 1984. The relationship of the palaeoniscid fishes, a review based on new specimens of Mimia and Moythomasia from the Upper Devonian of Western Australia. Bull. Br. Mus. (Nat. Hist.), Geol. 37: 173–428.

Gaudant, M. 1975. Sur la découverte de deux nouveaux coelacanthes fossiles au Liban et la disparition apparente des actinistiens au Crétacé. C.R. Acad. Sci. Paris, Sér. D, 280: 959–962.

Gorizdro-Kulczycka, Z. 1950. Dwudyszne ryby dewońskie Gór Swiętokrzyskich [Les dipneustes dévoniens du Massif de S-te Croix]. Acta Geol. Polonica 1: 53–105. (In Polish and French).

Gross, W. 1935. Histologische Studien am Aussenskelett fossiler Agnathen und Fische. Palaeontographica (A) 83: 1–60.

Hall, T.M. 1876. Fossil fish in North Devon. Geol. Mag., Dec. II, 3: 410–411.

Hancock, A. & T. Atthey. 1872. Descriptive notes on a nearly entire specimen of Pleurodus Rankinii, on two new species of Platysomus and a new Amphicentrum, with remarks on a few other fish-remains found in the Coal-measures at Newsham. Ann. Mag. Nat. Hist., 4th Ser. 9: 249–262.

Hauff, B. 1952. Das Holzmadenbuch. Öhringen. 54 pp.

Hay, O.P. 1900. Descriptions of some vertebrates of the Carboniferous age. Proc. Amer. Phil. Soc. 39: 96–123.

Hay, O.P. 1902. Bibliography and catalogue of the fossil Vertebrata of North America. Bull. U.S. Geol. Surv. 1902, 179: 1–868.

Hay, O.P. 1929. Second bibliography and catalogue of the fossil Vertebrata of North America. Carnegie Instit. Publ. No. 390, Vol. 1: 1–916.

Heide, S. van der. 1943. La faune ichthyologique du Carbonifère supérieur des Pays-Bas. Meded. geol. Sticht., Ser. C, 4: 1–65.

Heide, S. van der. 1955. La faune du Carbonifère inférieure de l'Égypte. Meded. geol. Sticht. 8: 73–75.

Heimberg, G. 1939. Neue Fischfunde aus dem Weissen Jura von Württemberg. Palaeontographica (A) 97: 75–98.

Heineke, E. 1906. Die Ganoiden und Teleostier des lithographischen Schiefers von Nusplingen. Geol. Paläontol. Abh. 8: 159–213.

Hennig, E. 1951. *Trachymetopon liassicum* Ald., ein Riesen Crossopterygier aus Schwäbischem Ober-Lias. Neues Jahrb. Geol. Paläontol. Abh. 94: 67–79.

Hensel, K. 1986. Morphologie et interprétation des canaux et canalicules sensoriels céphaliques de *Latimeria chalumnae* Smith, 1939 (Osteichthyes, Crossopterygii, Coelacanthiformes). Bull. Mus. Natl. Hist. Nat., Paris, Sér. 4, 8: 379–407.

Hibbard, C.W. 1933. Two species of *Coelacanthus* from the Middle Pennsylvanian of Anderson county, Kansas. Univ. Kans. Sci. Bull. 21: 279–283.

Hussakof, L. 1908. Catalogue of types and figured specimens of fossil vertebrates in the American Museum of Natural History. Part I.-Fishes. Bull. Amer. Mus. Nat. Hist. 25: 1–103.

Hussakof, L. 1911. The Permian fishes of North America. Public. Carnegie Instit. Washington, No. 146: 153–178.

Huxley, T.H. 1861. Preliminary essay upon the systematic arrangement of the fishes of the Devonian epoch. Figures and descriptions illustrative of British organic remains. Mem. Geol. Surv. U.K., Dec. 10: 1–40.

Huxley, T.H. 1866. Illustrations of the structure of the crossopterygian ganoids. Figures and descriptions illustrative of British organic remains. Mem. Geol. Surv. U.K., Dec. 12: 1–844.

Huxley, T.H. 1872. Illustrations of the structure of the crossopterygian ganoids. Figures and descriptions illustrative of British organic remains. Mem. Geol. Surv. U.K., Dec. 13: 1.

Jaekel, O. 1927. Der Kopf der Wirbeltiere. Ergeb. Anat. Entwicklungsgesch. 27: 815–974.

Jain, S.L. 1974. *Indocoelacanthus robustus* n. gen., n. sp. (Coelacanthidae, Lower Jurassic), the first fossil coelacanth from India. J. Paleontol. 48: 49–62.

Jain, S.L. 1980. The continental Lower Jurassic fauna from the Kota Formation, India. pp. 99–124. *In:* L.L. Jacobs (ed.) Aspects of Vertebrate History, Museum of Northern Arizona Press.

Janvier, P. 1974. Preliminary report on Late Devonian fishes from central and eastern Iran. Geol. Surv. Iran, Rep. 31: 5–48.

Janvier, P. 1977. Les poissons dévoniens de l'Iran central et de l'Afghanistan. Mém. H. Sér. Soc. Géol. France 8: 277–289.

Janvier, P. & M. Martin. 1979. Les vertébrés dévoniens de l'Iran central. II – Coelacanthiformes, Struniiformes, Osteolepiformes. Géobios 12: 497–511.

Janvier, P., F. Lethiers, O. Monod & O. Balkas. 1984. Discovery of a vertebrate fauna at the Devonian-Carboniferous boundary in SE Turkey (Hakkari Province). J. Petrol. Geol. 7: 147–168.

Jarvik, E. 1942. On the structure of the snout of crossopterygians and lower gnathostomes in general. Zool. Bidr. Uppsala 21: 235–675.

Jarvik. E. 1954. On the visceral skeleton in *Eusthenopteron* with a discussion of the parasphenoid and palatoquadrate in fishes. K. svenska VetenskAkad. Handl. 5: 1–104.

Jarvik. E. 1955. The oldest tetrapods and their forerunners. Scientific Monthly 80: 141–154.

Jarvik. E. 1964. Specializations in early vertebrates. Ann. Soc. Roy. Zool. Belg. 94: 11–95.

Jarvik. E. 1972. Middle and Upper Devonian Porolepiformes from East Greenland with special reference to *Glyptolepis groenlandica* n.sp. Medd. Grønl. 187: 1–307.

Jarvik. E. 1980a. Basic structure and evolution of vertebrates, Vol. 1. Academic Press, New York. 575 pp.

Jarvik. E. 1980b. Basic structure and evolution of vertebrates, Vol. 2. Academic Press, New York. 337 pp.

Jepsen, G.L. 1949. A natural library. New Jersey State Mus. 3: 1–7.

Jessen, H.L. 1966. Die Crossopterygier des Oberen Plattenkalkes (Devon) der Bergisch-Gladbach – Paffrather Mulde (Rheinisches Schiefergebirge) unter Berücksichtigung von amerikanischem und europäischem *Onychodus*-Material. Ark. Zool. 18: 305–389.

Jessen, H.L. 1973. Weitere Fischreste aus dem Oberen Plattenkalk der Bergisch-Gladbach – Paffrather Mulde (Oberdevon, Rheinisches Schiefergebirge). Palaeontographica (A) 143: 159–187.

Jessen, H.L. 1980. Lower Devonian Porolepiformes from the Canadian Arctic with special reference to *Powichthys thorsteinssoni* Jessen. Palaeontographica (A) 167: 180–214.

Jubb, R.A. & B.G. Gardiner. 1975. A preliminary catalogue of identifiable fossil fish material from southern Africa. Ann. S. Afr. Mus. 67: 381–440.

Keller, G. 1938. Die Fischfauna des Ruhroberkarbons. pp. 135–140. *In:* P. Kukuk (ed.) Geologie des niederrheinisch-westfälischen Steinkohlengebietes in Auftrage der westfälischen Berggewerkschaftskasse zu Bochum, Textband, Verlag von Julius Springer, Berlin. 706 pp.

Kner, R. 1866. Die Fische der bituminösen Schiefer von Raibl in Kärnthen. Sitzungsber. Akad. Wiss. Math.-Naturwiss. Klasse 5: 152–197.

Koenen, A. von. 1895. Über einige Fischreste des norddeutschen und böhmischen Devons. Abh. Königl. Gesellsch. Wiss. Göttingen 40: 1–37.

Lacasa Ruiz, A. 1981. Estudio del yacimiento infracretacico del Montsec de Rubies, «la Pedrera de Meià». Instit. Estud. Ilerdenses Excma. Diput. Prov. Lérida 42: 61–159.

Lambe, L.M. 1916. Ganoid fishes from near Banff, Alberta. Trans. Roy. Soc. Canada, Ottawa, Sect. 4, 10: 35–44.

Lankester, E.R. 1909. A treatise on zoology. Part IX Vertebrata. Craniata: cyclostomes and fishes. Adam & Charles Black, London. 518 pp.

Lehman, J.-P. 1952. Etude complémentaire des poissons de l'Eotrias de Madagascar. K. svenska VetenskAkad. Handl. 2(6): 1–201.

Lehman, J.-P. 1966. Actinopterygii, Dipnoi, Crossopterygii, Brachiopterygii. pp. 1–420. *In:* J. Piveteau (ed.) Traité de Paléontologie, Vol. 4, Part 3, Masson, Paris.

Lelièvre, H. & P. Janvier. 1988. Un actinistien (Sarcopterygii, Vertebrata) dans le Dévonien supérieur du Maroc. C.R. Acad. Sci. Paris. Sér. 2, 307: 1425–1430.

Liu, H.-T. 1964. A new coelacanth from the marine Lower Triassic of N.W. Kwangsi, China. Vert. PalAsiat. 8: 211–214. (In Chinese).

Lund, R. & W.L. Lund. 1984. New genera and species of coelacanths from the Bear Gulch Limestone (Lower Carboniferous) of Montana (U.S.A.). Géobios 17(2): 237–244.

Lund, R. & W.L. Lund. 1985. Coelacanths from the Bear Gulch Limestone (Namurian) of Montana and the evolution of the Coelacanthiformes. Bull. Carnegie Mus. Nat. Hist. 25: 1–74.

Lund, W.L., R. Lund & G.A. Klein. 1985. Coelacanth feeding mechanisms and ecology of the Bear Gulch coelacanths. C.R. Neuv. Congr. Internatl. Stratigr. Géol. Carbonifère 5: 492–500.

Maisey, J.G. 1986a. Heads and tails: a chordate phylogeny. Cladistics 2(3): 201–256.

Maisey, J.G. 1986b. Coelacanths from the Lower Cretaceous of Brazil. Amer. Mus. Novit. 2866: 1–30.

Maisey, J.G. (ed.) 1991. Santana fossils: an illustrated atlas. T.F.H. Publications, Neptune City. 459 pp.

Mantell, G. 1822. The fossils of the South Downs; or illustrations of the Geology of Sussex. Lupton Relfe, London. 327 pp.

Mantell, G. 1836. Descriptive catalogue of the objects of geology, natural history, and antiquity, in the museum attached to the Sussex Scientific and Literary Institution. 6th ed. Relfe & Fletcher, London. 44 pp.

Martin, M. 1981. Les dipneustes et actinistiens du Trias supérieur continental marocain. Stuttg. Beitr. Naturkd., Ser. B, 69: 1–29.

Martin, M. & S. Wenz. 1984. Découverte d'un nouveau Coelacanthidé, *Garnbergia ommata* n.g., n.sp., dans le Muschelkalk supérieur du Baden-Württemberg. Stuttg. Beitr. Naturkd., Ser. B 105: 1–17.

Mawson, J. & A.S. Woodward. 1907. Cretaceous formation of Bahia and its vertebrate fossils. Quart. J. Geol. Soc. 73: 128–139.

Meinke, D.K. 1982. A light and scanning electron microscope study on the dermal skeleton of *Spermatodus* (Pisces: Coelacanthini) and the evolution of the dermal skeleton in coelacanths. J. Paleontol. 56: 620–630.

Meinke, D.K. 1984. A review of cosmine: its structure, development, and relationship to other forms of the dermal skeleton in osteichthyans. J. Vert. Paleontol. 4: 457–470.

Meinke, D.K. & K.S. Thomson. 1983. The distribution and significance of enamel and enameloid in the dermal skeleton of osteolepiform and rhipidistian fishes. Paleobiology 9: 138–149.

Melton, W.G. 1969. A new dorypterid fish from central Montana. Northwest Science, Cheney 43: 196–206.

Miles, R.S. 1977. Dipnoan (lungfish) skulls and the relationships of the group: a study based on new species from the Devonian of Australia. Zool. J. Linn. Soc. London 61: 1–328.

Millot, J. & J. Anthony. 1958. Anatomie de *Latimeria chalumnae*. Tome I. Squelette, muscles et formations de soutien. C.N.R.S., Paris. 118 pp.

Millot, J. & J. Anthony. 1965. Anatomie de *Latimeria chalumnae*. Tome II. Système nerveux et organes des sens. C.N.R.S., Paris. 130 pp.

Millot, J., J. Anthony & D. Robineau. 1978. Anatomie de *Latimeria chalumnae*. Tome III. Appareil digestif – appareil respiratoire – appareil urogénital – glandes endocrines – appareil circulatoire – téguments – écailles – conclusions générales. C.N.R.S., Paris. 198 pp.

Moy-Thomas, J.A. 1935a. A synopsis of the coelacanth fishes of the Yorkshire Coal Measures. Ann. Mag. Nat. Hist., Ser. 10, 15: 37–46.

Moy-Thomas, J.A. 1935b. The coelacanth fishes from Madagascar. Geol. Mag. 72: 213–227.

Moy-Thomas, J.A. 1937. The Carboniferous coelacanth fishes of Great Britain and Ireland. Proc. Zool. Soc. Lond. (B) 3: 383–415.

Moy-Thomas, J.A. 1939. Palaeozoic fishes. Methuens monogr. biol. subjects, Methuen, London.

Moy-Thomas, J.A. & R.S. Miles. 1971. Palaeozoic fishes. 2nd ed. Chapman & Hall, London. 259 pp.

Moy-Thomas, J.A. & T.S. Westoll. 1935. On the Permian coelacanth, *Coelacanthus granulatus* Ag. Geol. Mag. 72: 446–457.

Müller, A.H. 1985. Lehrbuch der Paläozoologie, Band III Vertebraten, Teil 1 Fische im weiteren Sinne und Amphibien. VEB Gustav Fischer Verlag, Jena. 655 pp.

Münster, G. von. 1834. Mitteilungen an Professor Bronn gerichtet. Neues Jahrb. Miner. Geogn. Geol. Petrefakt. 1834: 538–542.

Münster, G. von. 1842a. Beitrag zur Kenntniss einiger neuen seltenen Versteinerungen aus den lithographischen Schiefern in Baiern. Neues Jahrb. Miner. Geogn. Geol. Petrefakt. 1842: 35–46.

Münster, G. von. 1842b. Beiträge zur Petrefakten-Kunde. Buchner'schen Buchhandlung, Bayreuth. 131 pp.

Nelson, G.J. 1970. Subcephalic muscles and intracranial joints of sarcopterygian and other fishes. Copeia 1970: 468–471.

Newberry, J.S. 1856. Descriptions of several genera and species of fossil fish from Carboniferous strata of Ohio. Proc. Acad. Nat. Sci. Philadelphia 8: 96–100.

Newberry, J.S. 1873a. Descriptions of fossil fishes. Rep. Geol. Surv. Ohio 7: 245–355.

Newberry, J.S. 1873b. Notes on the genus *Conchiopsis* Cope. Proc. Acad. Nat. Sci. Philadelphia 1873: 425–426.

Newberry, J.S. 1878. Descriptions of new fossil fishes from the Trias. Ann. N.Y. Acad. Sci. 1: 127–128.

Newberry, J.S. 1889. The Palaeozoic fishes of North America. Monogr. U.S. Geol. Surv. 16: 1–340.

Nielsen, E. 1936. Some few preliminary remarks on Triassic fishes from East Greenland. Medd. Grønl. 112: 1–55.

Northcutt, R.G. 1989. The phylogenetic distribution and innervation of craniate mechanoreceptive lateral lines. pp. 17–

398

78. *In:* S. Coombs, P. Gorner & H. Munz (ed.) The Mechanosensory Lateral Line: Neurobiology and Evolution, Springer Verlag, New York.

Obrucheva, O.P. 1955. Upper Devonian fishes of Central Kazakhstan. Sov. Geol. 45: 84–99. (In Russian).

Olsen, P.E., A.R. McCune & K.S. Thomson. 1982. Correlation of the early Mesozoic Newark Supergroup by vertebrates, principally fishes. Amer. J. Sci. 282: 1–44.

Ørvig, T. 1957. Remarks on the vertebrate fauna of the Lower Upper Devonian of Escuminac Bay, P.Q., Canada, with special reference to the porolepiform crossopterygians. Ark. Zool., Ser. 2, 10: 367–426.

Ørvig, T. 1986. A vertebrate bone from the Swedish Paleocene. Geol. Fören. Stockholm Förhandl. 108: 139–141.

Panchen, A.L. & T.R. Smithson. 1987. Character diagnosis, fossils and the origin of tetrapods. Biol. Rev. 62: 341–438.

Pastiels, A. 1951. Étude du gisement houiller de la Campine. Assoc. Étude Paléontol. Stratigr. Houillères, Bruxelles, No. 7.

Patton, W.W. & I.L. Tailleur. 1964. Geology of the Killik-Itkillik region, Alaska. Geol. Surv. Prof. Pap. 303-G, U.S. Dept. Int.: 409–500.

Priem, F. 1924. Paléontologie de Madagascar. XII. Les poissons fossiles. Ann. Paléontol. 1924: 105–132.

Pruvost, P. 1914. La faune continentale du terrain Houiller du Nord de la France: son utilisation stratigraphique. C.R. 12th Sess. Congr. Géol. Internatl.: 925–938.

Pruvost, P. 1919. La faune continentale du Terrain Houiller du Nord de la France. Mém. Serv. Carte géol. dét. Fr. 1919. 584 pp.

Quenstedt, F.A. 1858. Der Jura. Tubingen. 823 pp.

Reis, O.M. 1888. Die Coelacanthinen, mit besonderer Berücksichtigung der im weissen Jura Bayerns vorkommenden Gattungen. Palaeontographica 35: 1–96.

Reis, O.M. 1892. Zur Osteologie der Coelacanthinen. I. Theil. (Rumpfskelet, Knochen des Schädels und der Wangen, Kiemenbogenskelet, Schultergürtel, Becken, Integument und innere Organe.) Ph.D. Dissertation, K. Ludwigs-Maximilians-Universität, München. 39 pp.

Reis. O.M. 1900. *Coelacanthus lunzensis* Teller. Jahrb. Kaiserl. Königl. geol. Reichsanstalt 50: 187–192.

Reuss, A.E. 1857. Neue Fischreste aus dem Böhmischen Pläner. Denkschr. Akad. Wiss. Wien. Math.-Naturwiss. Klasse 13: 1–10.

Richter, M. 1985. Situaçao da pesquisa paleoictiologica no paleozoico brasileiro. Seçao Paleontol. Estratigrafia, Sér. Geol. 2: 105–110.

Rieppel, O. 1980. A new coelacanth from the Middle Triassic of Monte San Giorgio, Switzerland. Eclogae Geol. Helv. 73: 921–939.

Robineau, D. 1987. Sur la signification phylogénétique de quelques caractères anatomiques remarquables du coelacanthe *Latimeria chalumnae* Smith, 1939. Ann. Sci. nat., Zool., Sér. 13, 8: 43–60.

Romer, A.S. 1933. Vertebrate paleontology. 1st ed. University of Chicago Press, Chicago. 491 pp.

Romer, A.S. 1966. Vertebrate paleontology. 3rd ed. Univ. Chicago Press, Chicago. 468 pp.

Rosen, D.E., P.L. Forey, B.G. Gardiner & C. Patterson. 1981. Lungfishes, tetrapods, paleontology and plesiomorphy. Bull. Amer. Mus. Nat. Hist. 167: 159–276.

Saint-Seine, P. de. 1949. Les poissons des calcaires lithographiques de Cerin (Ain). Nouv. Arch. Mus. Hist. Nat. Lyon 2: 1–357.

Saint-Seine, P. de. 1950. Contribution à l'étude de vertébrés fossiles du Congo Belge. Ann. Mus. Congo Belge, Sci. Géol. 5: 1–32.

Saint-Seine, P. de. 1955. Poissons fossiles de l'étage de Stanleyville (Congo Belge). I. La faune des argilites et schistes bitumineux. Ann. Mus. Roy. Congo Belge, Sci. Géol. 14: 1–126.

Sanz, Y.L., S. Wenz, A. Yebenes, R. Estes, X. Martinez-Delclos, E. Jimenez-Fuentes, C. Diguez, A.D. Buscalioni, L.J. Barbadillo & L. Via. 1988. An early Cretaceous faunal and floral continental assemblage: Los Hoyas fossil site (Cuenca, Spain). Géobios 21: 611–635.

Sauvage, H.E. 1903. Noticia sobre los peces de la Caliza Litografica de la Provincia de Lérida (Cataluña). Mem. Real Acad. Cienc. artes Barcelona 4: 1–32.

Säve-Söderbergh, G. 1933. The dermal bones of the head and the lateral line system in *Osteolepis macrolepidotus* Ag. Nova Acta Regiae Soc. Sci. Upsal., Ser. 4, 9: 1–130.

Säve-Söderbergh, G. 1941. Notes on the dermal bones of the head in *Osteolepis macrolepidotus* Ag. and the interpretation of the lateral line system in certain primitive vertebrates. Zool. Bidrag Uppsala 20: 523–541.

Schaeffer, B. 1941. A revision of *Coelacanthus newarki* and notes on the evolution of the girdles and basal plates of the median fins in the Coelacanthini. Amer. Mus. Novit. 1110: 1–17.

Schaeffer, B. 1948. A study of *Diplurus longicaudatus* with notes on the body form and locomotion of the Coelacanthini. Amer. Mus. Novit. 1378: 1–32.

Schaeffer, B. 1952a. The Triassic coelacanth fish *Diplurus*, with observations on the evolution of the Coelacanthini. Bull. Amer. Mus. Nat. Hist. 99: 25–78.

Schaeffer, B. 1952b. Rates of evolution in the coelacanth and dipnoan fishes. Evolution 6: 101–111.

Schaeffer, B. 1953. *Latimeria* and the history of coelacanth fishes. Trans. N.Y. Acad. Sci., Ser. 2, 15: 170–178.

Schaeffer, B. 1954. *Pariostegus*, a Triassic coelacanth. Notulae Naturae, Acad. Nat. Sci. Philadelphia 261: 1–6.

Schaeffer, B. 1962. A coelacanth fish from the Upper Devonian of Ohio. Sci. Publ. Cleveland Mus. Nat. Hist. 135: 285–342.

Schaeffer, B. 1967. Late Triassic fishes from the western United States. Bull. Amer. Mus. Nat. Hist. 135: 285–342.

Schaeffer, B. 1968. The origin and basic radiation of the Osteichthyes. pp. 207–222. *In:* T. Ørvig (ed.) Current Problems of Lower Vertebrate Phylogeny, Nobel Symposium 4, Almqvist & Wiksell, Stockholm.

Schaeffer, B. 1976. An early Triassic fish assemblage from British Columbia. Bull. Amer. Mus. Nat. Hist. 156: 515–564.

Schaeffer, B. & J.T. Gregory. 1961. Coelacanth fishes from the continental Triassic of the western United States. Amer. Mus. Novit. 2036: 1–18.

Schaeffer, B. & M. Mangus. 1976. An early Triassic fish assemblage from British Columbia. Bull. Amer. Mus. Nat. Hist. 156: 517–563.

Schaeffer, B. & C. Patterson. 1984. Jurassic fishes from the western United States, with comments on Jurassic fish distribution. Amer. Mus. Novit. 2796: 1–86.

Schaumberg, G. 1977. Der Richelsdorfer Kupferschiefer und seine Fossilien, III. Die tierischen Fossilien des Kupferschiefers. 2. Vertebraten. Der Aufschluss 28: 297–352.

Schaumberg, G. 1978. Neubeschreibung von *Coelacanthus granulatus* Agassiz (Actinistia, Pisces) aus dem Kupferschiefer von Richelsdorf (Perm, W.-Deutschland). Paläontol. Z. 52: 169–197.

Schultze, H.-P. 1972. Early growth stages in coelacanth fishes. Nature New Biol. 236: 90–91.

Schultze, H.-P. 1973. Crossopterygier mit heterozerker Schwanzflosse aus dem Oberdevon Kanadas, nebst einer Beschreibung von Onychodontida-Resten aus dem Mitteldevon Spaniens und aus dem Karbon der USA. Palaeontographica (A) 143: 188–208.

Schultze, H.-P. 1980. Eier legende und lebend gebärende Quastenflosser. Natur und Museum 110: 101–108.

Schultze, H.-P. 1985. Reproduction and spawning sites of *Rhabdoderma* (Pisces, Osteichthyes, Actinistia) in Pennsylvanian deposits of Illinois, USA. C.R. Neuv. Congr. Internatl. Stratigr. Géol. Carbonifère 5: 326–330.

Schultze, H.-P. 1987. Dipnoans as sarcopterygians. pp. 39–74. *In:* W.E. Bemis, W.W. Burggren & N.E. Kemp (ed.) The Biology and Evolution of Lungfishes, Cent. Suppl. 1, J. Morphol., Alan R. Liss, New York.

Schultze, H.-P. 1990. Comparison of controversial hypotheses on the origins of tetrapods. *In:* H.-P. Schultze & L. Trueb (ed.) Origins of the Major Groups of Tetrapods: Controversies and Consensus, Cornell University Press, Ithaca. (in press).

Schultze, H.-P. & K.S.W. Campbell. 1987. Characterisation of the Dipnoi, a monophyletic group. pp. 39–74. *In:* W.E. Bemis, W.W. Burggren & N.E. Kemp (ed.) The Biology and Evolution of Lungfishes, Cent. Suppl. 1, J. Morphol., Alan R. Liss, New York.

Schultze, H.-P. & J. Chorn. 1989. The Upper Pennsylvanian vertebrate fauna of Hamilton, Kansas. pp. 147–154. *In:* G. Mapes & R.H. Mapes (ed.) Regional Geology and Paleontology of Upper Paleozoic Hamilton Quarry Area in Southeastern Kansas, Kansas Geol. Surv., Guidebk. Ser. 6.

Schultze, H.-P. & H. Möller. 1973. Wirbeltierreste aus dem Mittleren Muschelkalk (Trias) von Göttingen, West-Deutschland. Paläontol. Z. 60: 109–129.

Schweizer, R. 1966. Ein Coelacanthide aus dem Oberen Muschelkalk Göttingens. Neues Jahrb. Geol. Paläontol. Abh. 125: 216–226.

Sedgwick, A. 1829. On the geological relations and internal structure of the Magnesian Limestone, and the lower portions of the New Red Sandstone Series in their range through Nottinghamshire, Derbyshire, Yorkshire, and Durham, to the southern extremity of Northumberland. Trans. Geol. Soc. London, 2nd Ser., 3: 37–124.

Shainin, V.E. 1943. New coelacanth fishes from the Triassic of New Jersey. J. Paleontol. 17: 271–275.

Smith, J.L.B. 1939a. A living fish of Mesozoic type. Nature 143: 455–456.

Smith, J.L.B. 1939b. A living coelacanthid fish from South Africa. Trans. Roy. Soc. S. Afr. 28: 1–106.

Smith, J.L.B. 1939c. The living coelacanthid fish from South Africa. Nature 143: 748–750.

Stensiö, E.A. 1918. Notes on a crossopterygian fish from the Upper Devonian of Spitzbergen. Bull. Geol. Instit. Univ. Upsala 16: 115–124.

Stensiö, E.A. 1921. Triassic fishes from Spitzbergen. Pt. 1. Holzhausen, Vienna. 307 pp.

Stensiö, E.A. 1922a. Über zwei Coelacanthiden aus dem Oberdevon von Wildungen. Paläontol. Z. 4: 167–210.

Stensiö, E.A. 1922b. Notes on certain crossopterygians. Proc. Zool. Soc. London 1922: 1241–1271.

Stensiö, E.A. 1932. Triassic fishes from East Greenland. Medd. Grønl. 83(3): 1–305.

Stensiö, E.A. 1937. On the Devonian coelacanthids of Germany with special reference to the dermal skeleton. K. svenska VetenskAkad. Handl. 3(16): 1–56.

Stensiö, E.A. 1947. The sensory lines and dermal bones of the cheek in fishes and amphibians. K. svenska VetenskAkad. Handl. 3(24): 1–195.

Stensiö, E.A. 1963. The brain and the cranial nerves in fossil, lower craniate vertebrates. Skr. norske Vidensk-Akad. Mat.-nat. Kl. 1963: 1–120.

Stromer, E. 1936. Ergebnisse der Forschungsreisen Prof. E. Stromers in den Wüsten Ägyptens. VII. Baharîje-Kessel und -Stufe mit deren Fauna und Flora eine ergänzende Zusammenfassung. Abh. bayer. Akad. Wiss. Math.-Naturwiss. Abt., N.F., 33: 1–102.

Tabaste, N. 1963. Étude de restes de poissons du Crétacé Saharien. Mém. Instit. Fr. Afr. Noire 68: 437–485.

Teller, F. 1891. Über den Schädel eines fossilen Dipnoërs *Ceratodus sturii* nov. spec. aus den Schichten der oberen Trias der Nordalpen. Abh. Kaiserl. Königl. geol. Reichsanstalt 15: 1–39.

Thiollière, V. 1854. Description des Poissons fossiles provenant des gisements coralliens du Jura dans le Bugey. J.-B. Baillière, Paris. 27 pp.

Thiollière, V. 1858. Note sur les poissons fossiles du Bugey, et sur l'application de la méthode de Cuvier à leur classement. Bull. Soc. Géol. Fr. 15: 782–793.

Thomson, K.S. 1967. Mechanisms of intracranial kinesis in fossil rhipidistian fishes (Crossopterygii) and their relatives. Zool. J. Linn. Soc. London 46: 223–253.

Thomson, K.S. 1969. The biology of the lobe-finned fishes. Biol. Rev. 44: 91–154.

Thys van den Audenaerde, D.F.E. 1984. Le coelacanthe des Comores *Latimeria chalumnae* curiosité zoologique, fossile vivant ou animal aberrant? Africa-Tervuren 30: 90–101.

400

Tima, V. 1986. Revision of *Macropoma speciosum* Reuss, 1857 (Crossopterygii, Coelacanthiformes). Věstník Ustředniho Ústavu Geologického 61: 209–216.

Traquair, R.H. 1881. Notice of new fish remains from the Blackband Ironstone of Borough Lee, near Edinburgh. Geol. Mag. 1881: 491–494.

Traquair, R.H. 1883. Report on fossil fishes collected by the Geological Survey of Scotland in Eskdale and Liddesdale, Part 1 – Ganoidei. Trans. Roy. Soc. Edinburgh 30: 15–71.

Traquair, R.H. 1890. List of the fossil Dipnoi and Ganoidei of Fife and the Lothians. Proc. Roy. Soc. Edinburgh 17: 385–400.

Traquair, R.H. 1901. Notes on the Lower Carboniferous fishes of eastern Fifeshire. Geol. Mag. 4: 110–114.

Traquair, R.H. 1903. On the distribution of fossil fish remains in the Carboniferous rocks of the Edinburgh district. Trans. Roy. Soc. Edinburgh 40: 687–707.

Traquair, R.H. 1905. Notes on the Lower Carboniferous fishes of Eastern Fifeshire. Proc. Roy. Phys. Soc. Edinburgh 16: 80–86.

Trueb, L. & R. Cloutier. 1990. A phylogenetic investigation of the inter- and intrarelationships of the Lissamphibia (Amphibia: Temnospondyli). *In:* H.-P. Schultze & L. Trueb (ed.) Origins of the Major Groups of Tetrapods: Controversies and Consensus, Cornell University Press, Ithaca. (in press).

Vangerow, E.F. 1957. Mikropaläontologische Untersuchungen in den Kohlscheider im Wurmrevier bei Aachen. Geol. Jahrb. 73: 457–506.

Véran, M. 1988. Les éléments accessoires de l'arc hyoïdien des poissons téléostomes (acanthodiens et osteichthyens) fossiles et actuels. Mém. Mus. Nat. Hist. Nat., Sci. Terre, 54: 1–98.

Vetter, B. 1881. Die Fische aus dem lithographischen Schiefer in Dresdener Museum. Mitt. Königl. miner. geol. Mus. prä-hist. Samml. Dresden 4: 1–118.

Vidal, L.M. 1915. Nota geologicia y paleontologica sobre el Jurasico superior de la Provincia de Lerida. Bol. Instit. Geol. España, 2nd Ser., 16: 17–55.

von Wahlert, G. 1968. *Latimeria* und die Geschichte der Wirbeltiere. Eine evolutionsbiologische Untersuchung. G. Fischer, Stuttgart. 125 pp.

Vorobyeva, E.I. & D.V. Obruchev. 1967. Subclass Sarcopterygii. pp. 420–509. *In:* Y.A. Orlov (ed.) Fundamentals of Paleontology, Vol. XI Agnatha, Pisces, Israel Program for Scientific Translations, Jerusalem. (Russian original 1964).

Wagner, A. 1863. Monographie der fossilen Fische aus den lithographischen Schiefern Bayerns. Abh. bayer. Akad. Wiss. Math.-Naturwiss. Abt. 9: 611–748.

Wang, N. & H. Liu. 1981. Coelacanth fishes from the marine Permian of Zhejiang, South China. Vertebr. PalAsia. 19: 305–312. (In Chinese).

Ward, J. 1875. On the organic remains of the Coal-measures of North Staffordshire, their range and distribution, with a catalogue of the fossils and their mode of occurrence. N. Staff. Nat. Field Club. Annual addresses, papers, etc. 1875: 184–251.

Ward, J. 1890. The geological features of the North Stafford-shire Coal-fields, their organic remains, their range and distribution; with a catalogue of the fossils of the Carboniferous system of North Staffordshire. Trans. N. Staff. Instit. Min. Mechan. Engin. 10: 1–189.

Waterlot, G. 1934. Études des Gîtes Minéraux de la France. Bassin Houiller de la Sarre et de la Lorraine. II. Faune Fossile. Étude de la Faune Continentale du Terrain Houiller Sarro-Lorrain. Imprimerie L. Danel, Lille. 317 pp.

Watson, D.M.S. 1921. On the coelacanth fish. Ann. Mag. Nat. Hist., Ser. 9, 8: 320–337.

Watson, D.M.S. 1926. The evolution and origin of the Amphibia. Phil. Trans. Roy. Soc. London (B) 214: 189–257.

Watson, D.M.S. 1927. The reproduction of the coelacanth fish *Undina*. Proc. Zool. Soc. London 1927: 453–457.

Wehrli, H. 1931. Die Fauna der Westfälischen Strefen A und B der Bochumer Mulde zwischen Dortmund und Kamen (Westfalen). Palaeontographica 74: 93–134.

Weiler, W. 1935. Ergebnisse des Forschungsreisen Prof. E. Stromers in den Wüsten Ägyptens. II. Wirbeltierreste der Baharîje-Stufe (unterstes Cenoman). 16. Neue Untersuchungen an den Fischresten. Abh. bayer. Akad. Wiss. Math.-Naturwiss. Abt., N.F. 32: 1–57.

Wellburn, E.D. 1901. On the fish fauna of the Yorkshire Coal Measures. Proc. Yorkshire Geol. Polytech. Soc. 14: 159–174.

Wellburn, E.D. 1902. On the fish fauna of the Pendleside Limestones. Proc. Yorkshire Geol. Polytech. Soc. 14: 465–473.

Wellburn, E.D. 1903. On some new species of fossil fish from the Millstone Grit rocks, with an amended list of genera. Proc. Yorkshire Geol. Polytech. Soc. 15: 70–78.

Wenz, S. 1975. Un nouveau coelacanthidé du Crétacé inférieur du Niger, remarques sur la fusion des os dermiques. Problèmes actuels de paléontologie – Evolution des Vertébrés. Coll. Internatl. C.N.R.S. 218: 175–190.

Wenz, S. 1979. Découverte d'un coelacanthe dans le Jurassique supérieur de Normandie. Bull. Trimest. Soc. Géol. Normandie amis Mus. Havre 66: 91–92.

Wenz, S. 1980. A propos du genre *Mawsonia*, coelacanthe géant du Crétacé inférieur d'Afrique et du Brésil. Mém. Soc. Géol. France 139: 187–190.

Wenz, S. 1981. Un coelacanthe géant, *Mawsonia lavocati* Tabaste, de l'Albien-base du Cénomanien du sud marocain. Ann. Paléont. Vert. 67: 1–20.

Westoll, T.S. 1937. On the cheek-bones in teleostome fishes. J. Anat. 71: 362–382.

Westoll, T.S. 1939. On *Spermatodus pustulosus* Cope, a coelacanth from the 'Permian' of Texas. Amer. Mus. Novit. 1017: 1–23.

Westoll, T.S. 1943a. The origin of tetrapods. Biol. Rev. 18: 78–98.

Westoll, T.S. 1943b. The origin of the primitive tetrapod limb. Proc. Roy. Soc. Lond. (B) 131: 373–393.

Westoll, T.S. 1960. Recent advances in the palaeontology of fishes. Liverpool Manch. Geol. J. 2: 568–596.

Westoll, T.S. 1979. Devonian fish biostratigraphy. pp. 341–353. In: The Devonian System, Spec. Papers in Palaeontology No. 23.

White, E.I. 1939. A living fossil. Listener, London 21: 663–664.

White, E.I. 1953. The coelacanth fishes. Discovery 14: 113–117.

White, E.I. 1954a. More about the coelacanths. Discovery 15: 332–335.

White, E.I. 1954b. The coelacanth fishes. Ann. Rep. Smithsonian Inst. 1953: 351–360.

White, E.I. & J.A. Moy-Thomas. 1937. The coelacanth genus *Graphiurus* Kner. Geol. Mag. 74: 286.

Wiley, E.O. 1979a. Ventral gill arch muscles and the phylogenetic relationships of *Latimeria*. pp. 56–67. *In:* J.E. McCosker & M.D. Lagios (ed.) The Biology and Physiology of the Living Coelacanth, Occ. Pap. Calif. Acad. Sci. 134.

Wiley, E.O. 1979b. Ventral gill arch muscles and the interrelationships of gnathostomes, with a new classification of the Vertebrata. Zool. J. Linn. Soc. 67: 149–179.

Willemoes-Suhm, R. 1869. Ueber *Coelacanthus* und einige verwandte Gattungen. Palaeontographica 17: 73–89.

Williamson, W.C. 1849. On the microscopic structure of scales and dermal teeth of some ganoid and placoid fish. Phil. Trans. Roy. Soc. London 139: 435–475.

Winkler, T.C. 1871. Mémoire sur le *Coelacanthus harlemensis*. Arch. Mus. Teyler 3: 101–116.

Winkler, T.C. 1880. Description de quelques restes de poissons fossiles des terrains triasiques des environs de Wurzbourg. Arch. Mus. Teyler 5: 109–149.

Woodward, A.S. 1888. Synopsis of the vertebrate fossils of the English Chalk. Proc. Geol. Assoc. 10: 273–338.

Woodward, A.S. 1890. Notes on some ganoid fishes from the English Lower Lias. Ann. Mag. Nat. Hist., Ser. 6, 5: 430–436.

Woodward, A.S. 1891. Catalogue of fossil fishes in the British Museum (Natural History). British Museum (Natural History), London. 567 pp.

Woodward, A.S. 1895. The fossil fishes of the Talbragar Beds (Jurassic?). Mem. Geol. Surv. New South Wales, Palaeontology 9: 1–31.

Woodward, A.S. 1898a. Note on a Devonian coelacanth fish. Geol. Mag. 5: 529–531.

Woodward, A.S. 1898b. Outlines of vertebrate palaeontology for students of zoology. Cambridge University Press, Cambridge. 470 pp.

Woodward, A.S. 1907. On the Cretaceous formation of Bahia (Brazil), and on vertebrate fossils collected therein. II. The vertebrate fossils. Quart. J. Geol. Soc. London 63: 131–139.

Woodward, A.S. 1908. On some fossil fishes discovered by Prof. Ennes de Souza in the Cretaceous formation at Ilhéos (State of Bahia), Brazil. Quart. J. Geol. Soc. London 64: 358–362.

Woodward, A.S. 1909. The fossil fishes of the English Chalk. Part 5. Monogr. Palaeontogr. Soc., London: 153–184.

Woodward, A.S. 1910. On some Permo-Carboniferous fishes from Madagascar. Ann. Mag. Nat. Hist., Ser. 8, 5: 1–6.

Woodward, A.S. 1912. Notes on some fish-remains from the Lower Trias of Spitzbergen. Bull. Geol. Instit. Univ. Upsala 11: 291–297.

Woodward, A.S. 1916. The fossil fishes of the English Wealden and Purbeck formations. Part 1. Monogr. Palaeontogr. Soc., London: 1–48.

Woodward, A.S. 1942. Some new and little-known Upper Cretaceous fishes from Mount Lebanon. Ann. Mag. Nat. Hist., Ser. 11, 9: 537–568.

Woodward, A.S. & C.D. Sherborn. 1890. A catalogue of British fossil Vertebrata. London. 396 pp.

Zidek, J. 1975. Some fishes of the Wild Cow Formation (Pennsylvanian), Manzanita Mountains, New Mexico. New Mexico Bur. Mines Miner. Resour. 125: 1–22.

Zittel, K.A. 1887. Handbuch der Palaeontologie, III, Vertebrata. Druck und Verlag van Oldenbarg, Munich. 900 pp.

Zittel, K.A. von. 1902. Text-book of Palaeontology. (Translated and edited by C.R. Eastman). MacMillan and Co., London. 464 pp.

Zittel, K.A. von. 1932. Text-book of Palaeontology. Vol. II. (Translated and edited by C.R. Eastman; 2nd English ed. revised, with additions, by A.S. Woodward). MacMillan and Co., London. 464 pp.

Environmental Biology of Fishes **32**: 403–433, 1991.

Bibliography of the living coelacanth *Latimeria chalumnae*, with comments on publication trends

Michael N. Bruton, Sheila E. Coutouvidis & Jean Pote
J.L.B. Smith Institute of Ichthyology, Private Bag 1015, Grahamstown, 6140 South Africa

Key words: Ichthyology, Literature, Publication trends, Comoros

Synopsis

A list of published references on the coelacanth *Latimeria chalumnae* is provided. All known publications in the scientific literature are included as well as popular articles and press reports that are considered to provide new information or interpretations. Marked trends are noticeable in the literature as different disciplines have been applied to research on the coelacanth over the past five decades. The bibliography lists a total of 823 publications including 490 papers in journals, 37 books, 3 theses, 45 chapters in books, 166 popular articles, 22 reports and 60 newspaper articles. Studies on taxonomy and morphology initially dominated the literature followed by reports on research in the fields of physiology, behaviour, breeding biology, ecology and conservation as frozen and eventually live specimens became available for study. The literature on the living coelacanth is predominantly in English, French, Japanese and German but references in 12 other languages were also traced. The dominant authors in the first decades of coelacanth research were the French scientists J. Millot and J. Anthony and the South African describer of the first and second coelacanths, J.L.B. Smith. In subsequent years French, British, American, South African, Japanese, Canadian and German authors, among others, have made significant contributions.

Introduction

Since the discovery of the first living coelacanth *Latimeria chalumnae* (Pisces, Coelacanthidae) by scientists in 1938, a rich and diverse literature has been produced on this remarkable animal which possibly rivals that on any other single vertebrate species. Conant (1987a, b) lists 2209 scientific references on the six species of lungfishes, and there are undoubtedly several thousand publications on such well known fish species as the brown trout, rainbow trout, Pacific salmon, largemouth bass, Mozambique tilapia and certain marine angling species, but it is remarkable that over 822 publications have been produced on the single species of living coelacanth, despite the fact that only about

172 specimens have been available for study (Bruton & Coutouvidis 1991).

A bibliography on the coelacanth was previously published by Locket (1980) while extensive lists of cited literature are given by Millot & Anthony (1978), Griffith & Pang (1979), Forey (1980, 1988), Griffith (1980), Anthony (1980) and Balon et al. (1988). A bibliography relating to fossil coelacanths has been published separately by Forey & Cloutier (1991) in this volume. The present bibliography arises out of a decision by the founding members of the Coelacanth Conservation Council (CCC), which was founded in Moroni in April 1987, to produce an updated list of publications on the coelacanth as a service to members of the CCC and to the broader scientific community.

How the bibliography was compiled

The primary source of this bibliography has been the library and archives of the J.L.B. Smith Institute of Ichthyology in Grahamstown, which was started by Margaret and J.L.B. Smith and has major holdings of coelacanth literature. This library continues to receive publications on the coelacanth from authors throughout the world. The Institute of Ichthyology also acts as the Secretariat for the CCC and thus receives regular correspondence from coelacanth researchers. In addition, we consulted the libraries of the Palaeontology and Fish Sections of the British Museum (Natural History), the East London Museum in South Africa, the Centre National du Documentation et de la Recherche Scientifique (C.N.D.R.S.) in Moroni, Comoros, and the Convention on International Trade in Endangered Species (CITES) in Lausanne, Switzerland. We also corresponded with researchers and librarians in museums, universities, research institutes, aquaria, and libraries throughout the world (for list see the 'Acknowledgements'). Use was also made of the following computerised databases: FISHLIT and SABINET (South Africa), DIALOG (U.S.A.) and Aquatic Sciences and Fisheries Abstracts (U.S.A.).

All known publications in scientific journals and other serials and in scientific books have been included in the bibliography as well as selected articles in popular magazines and in daily or weekly newspapers. While we have made every effort to include publications in all languages, it is likely that the primary scientific languages are best represented in the bibliography. Papers, chapters in books or books on extinct coelacanths which make comparative references to *L. chalumnae* have been included. The criterion for the choice of popular and newspaper articles was whether they provided new information or interpretations, or contributed to debates. The J.L.B. Smith Institute of Ichthyology and the East London Museum have large holdings of popular and newspaper articles on the coelacanth story, but most of these articles are repetitive and the ones that have been excluded from the bibliography do not, in our opinion, contribute to a further understanding of the coelacanth. Certain debates in the popular media have, however, been included, such as the debates on coelacanth conservation between J.L.B. Smith and J. Millot in 1956, the further commentary by Smith in 1963 and the more recent debate between the founder members of the Coelacanth Conservation Council and the New York Explorers Club in 1988 and 1989. References to the coelacanth in zoology and biology textbooks have also been excluded unless they contribute new information or interpretations. Their inclusion would have meant that virtually all biology textbooks would have had to be included as the insights provided by the coelacanth for the understanding of evolutionary phenomena are widely cited.

In the analysis of different disciplines covered by the literature, the primary field/s of study of a particular publication is scored except in the case of review articles and books for which several disciplines may be scored.

Analysis of the bibliography

General trends

The 823 publications in the bibliography include 490 papers (including book reviews) in refereed scientific journals, 166 popular articles in magazines and newsletters, 60 articles in daily or weekly newspapers, 45 chapters in scientific books, 37 books, 22 reports and 3 theses. There are 22 books exclusively on the coelacanth, including 11 editions of Smith's book 'Old Fourlegs'. Four major conference proceedings on the coelacanth have been published: McCosker & Lagios (1979), the Proceedings of the Royal Society of London (208: 265–384), the Proceedings of the First Symposium on Coelacanth Studies in Japan (1984), and this volume (Musick et al. 1991).

The 823 publications on the coelacanth were contributed to by 894 authors and editors and the scientific papers were published in 90 different scientific journals; 590 of the publications had single authors, 155 double authors and 78 multiple authors.

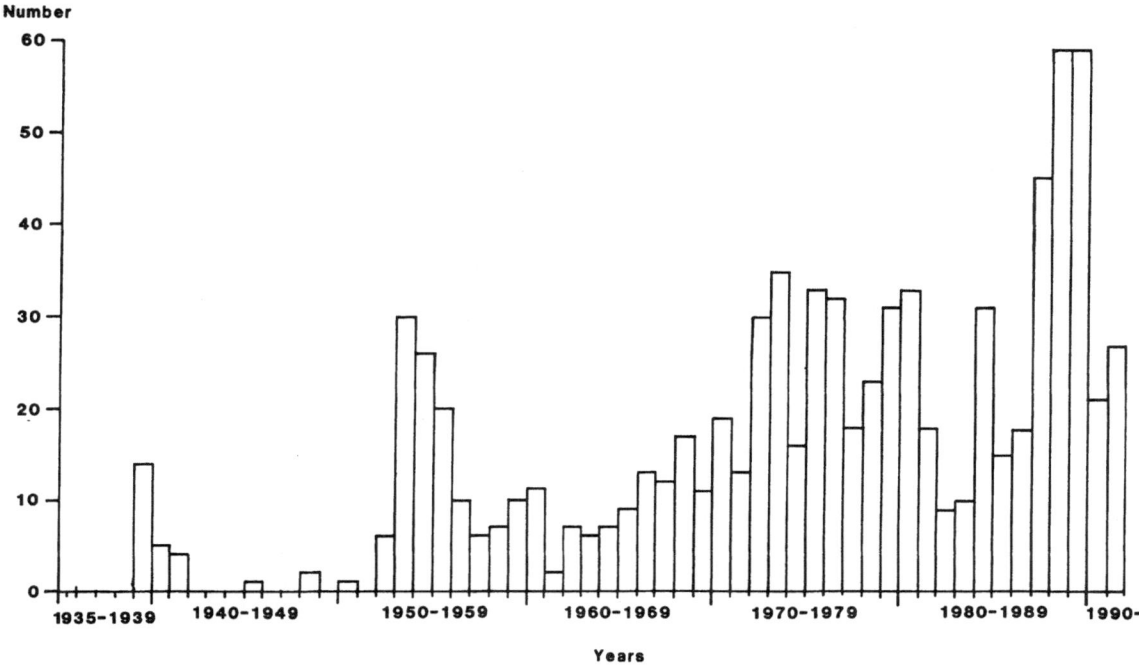

Fig. 1. The number of publications produced on the coelacanth per year from 1939 to 1991 based on information in this bibliography.

The frequency of publication

The number of publications produced on the coelacanth per year is shown in Figure 1. The initial scientific discovery of a living coelacanth stimulated a diverse literature in the early 1940s and the

Table 1. The languages in which literature on the living coelacanth have been published.

Language	No. of publications	% of total
English	607	73.9
French	118	14.2
Japanese	33	4.0
German	31	3.7
Italian	7	0.9
Russian	6	0.7
Slovak	4	0.5
Dutch, Swedish, Polish	3	0.4
Spanish	2	0.2
Afrikaans, Czech, Danish, Estonian, Serbian	1	0.1

capture of additional specimens in the Comoros maintained that interest into the 1950s. Interest appeared to wane in the early 1960s (Fig. 1) but built up again with the British, French and American expeditions of the late 1960s and 1970s. The coelacanth tissues and organs distributed to researchers around the world in the early 1970s following these expeditions further stimulated coelacanth research, but the Soilih regime in the Comoros in the late 1970s inhibited field research and may have dampened research interest at that time. The volume of literature increased again in the late 1980s as a result of new initiatives by, inter alia, Japanese, German, South African, American and Canadian scientists.

The total number of publications produced each decade increased from 13 in 1939 to 10 in the 1940s, 111 in the 1950s, 97 in the 1960s, 247 in the 1970s, 294 in the 1980s and 50 in the first half of 1990 and part of 1991. The years in which most publications on the coelacanth were produced were 1988 and 1989 (59 each) and 1987 (45).

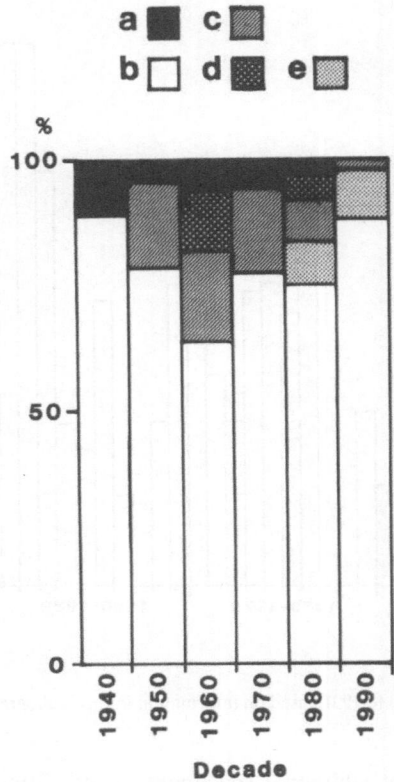

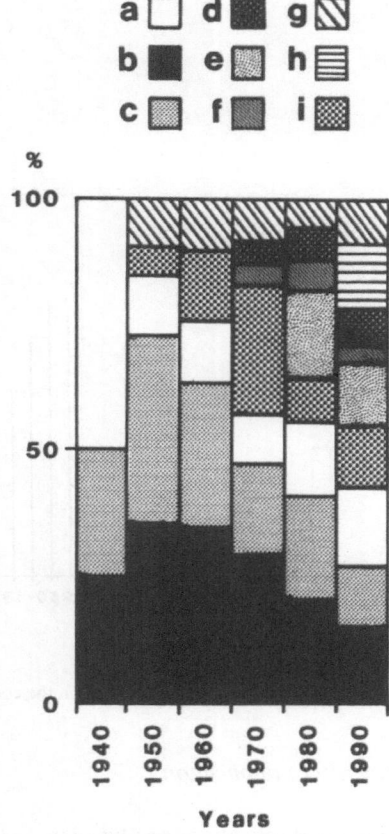

Fig. 2. The proportional contribution of different languages to the coelacanth literature in different decades. The decade '1940' includes 1939, and the decade '1990' includes 1990 and part of 1991. a = other, b = English, c = French, d = German, e = Japanese.

Fig. 3. Analysis of the successive areas of research interest on the coelacanth in different decades as reflected by the literature recorded in this bibliography. The decade '1940' includes 1939, and the decade '1990' includes 1990 and part of 1991. a = taxonomy and evolution, b = morphology and cytology, c = general and popular articles, d = behaviour, e = conservation and fishery, f = breeding biology and early development, g = other, h = ecology, i = physiology.

The languages used

The literature on the living coelacanth is predominantly in English (607 publications, 73.9%) with significant contributions in French (118, 14.2%), Japanese (33, 4.0%) and German (31, 3.7%), but publications in 12 other languages were also traced (Table 1). One of the two publications in Spanish is by the well known science fiction writer Isaac Asimov. The sole publication on the coelacanth known to us in Estonian is the translation of Smith's book 'Old Fourlegs'.

The proportional contribution of different languages in different decades varied widely (Fig. 2).

While English predominated in all decades, its proportional representation was lowest in the 1960s when French and, to a lesser extent, German publications became more prevalent. Japanese contributions began to appear in the mid 1980s.

Frequency of authorship

The author who has published most frequently on

the living coelacanth is J. Millot of France (55 publications), followed by his successor as the leading French student of the coelacanth, J. Anthony (46). The significant contribution made by these pioneers of coelacanth research is further emphasised by the large size and detailed nature of most of their publications. The other major contributors to the coelacanth literature were the describer of *L. chalumnae*, J.L.B. Smith (40 publications), and M.N. Bruton (35, mainly popular articles), K.S. Thomson (20), D. Robineau and E.K. Balon (16 each), R.W. Griffith and H.-P. Schultze (14 each), H. Fricke (12), M.D. Lagios and T. Uyeno (11 each) and P.L. Forey, N. Suzuki and Moya M. Smith (10 each).

The journals used

The journals most frequently used for the publication of scientific papers on the coelacanth have been Environmental Biology of Fishes (Canada, 35 papers), Nature (England, 30), Comte Rendu de l'Academie des Sciences Paris (France, 27), American Zoologist (U.S.A., 16), Occasional Papers of the California Academy of Sciences (U.S.A., 16), Proceedings of the First Symposium on Coelacanth Studies (Japan, 15), Copeia (U.S.A., 12), Science (U.S.A., 10), General and Comparative Endocri-

nology (England, 10) and the Proceedings of the Royal Society of London (England, 8).

Trends in the literature: the changing disciplines

An analysis of the successive areas of research interest on the coelacanth (Fig. 3) reveals interesting trends. Studies on taxonomy dominated the literature in 1939 and the 1940s whereas morphology was the dominant discipline in the 1950s and 1960s. The fascinating insights provided by the study of the physiology of the coelacanth resulted in increased emphasis on this field in subsequent decades. The first observations on living (or rather dying) coelacanths were made in the 1950s and occasional investigations of coelacanth behaviour were possible on captured fishes in the 1960s and 1970s. Coelacanth eggs were first found in 1955 but their viviparous life style was only discovered in 1975 when a female containing yolksac juveniles was dissected. During the late 1970s and 1980s a series of expeditions was launched to the Comoros to study coelacanths, which resulted in a wave of publications on coelacanth behaviour, ecology and conservation.

Observations on free-living coelacanths in their natural environment were, however, only possible with the initiation of research by H. Fricke with his research submersibles in the mid-1980s. Most recently, sophisticated techniques have facilitated studies on tissue microstructure and genetics. Concern for the future survival of the coelacanth began to be expressed in the 1960s and has become an important component of the literature in recent years. A feature of the literature of the 1980s and 1990s is that it now covers a far wider range of disciplines than in the earlier decades.

An analysis of all coelacanth publications over the period from 1939 to 1990 reveals that papers on morphology dominated the literature followed by general and popular publications and then papers on taxonomy, physiology, conservation, behaviour, breeding biology, ecology and genetics (Table 2).

Table 2. The main fields of research on the coelacanth from 1939 to 1990. The primary field/s of study of a particular publication is scored except in the case of review articles and books for which several disciplines may be scored.

Field of study	No. of publications or parts of publications	% of total
Morphology and histology	247	27.3
General and popular publications	212	22.6
Taxonomy and evolution	130	13.8
Physiology	126	13.4
Conservation and fishery	87	9.2
Behaviour	47	5.0
Breeding biology and early development	40	4.3
Ecology and distribution	36	3.8
Genetics	15	1.6

A Living Fish of Mesozoic Type

By Dr. J. L. B. Smith, Rhodes
University College, Grahamstown

EX Africa semper aliquid novi. It is my privilege
to announce the discovery of a Crossopterygian
fish of a type believed to have become extinct by
the close of the Mesozoic period. This fish was
taken by trawl-net at a depth of about 40 fathoms
some miles west of East London on December 22,
1938. It was alive when caught, and shortly after
it died it was handed over to Miss Courtenay-
Latimer, curator of the East London Museum.
Miss Latimer wrote to me, enclosing a sketch and

Fig. 4. The title and opening paragraph of the first scientific
publication by J.L.B. Smith on the coelacanth published in
Nature on 18 March 1939.

The major contributors to the coelacanth literature

The contributions of J.L.B. Smith

The capture of a living specimen of coelacanthid
fish was first announced by J.L.B. Smith in a note
to Nature dated 18 March 1939 (Fig. 4). Before this
note was published Smith sent a preliminary an-
nouncement to the Royal Society of South Africa
which was read at a meeting of the Society in Cape
Town on 15 March 1939. Smith published a detailed description of *Latimeria chalumnae* in the
'Transactions of the Royal Society of South Africa'
in 1940 as well as further descriptions in 1939 and
1940. He announced the discovery of the second
coelacanth in 1953 and continued to produce a
series of scientific and popular articles on the coelacanth in the 1950s, although he concluded his original research with the examination of the second
specimen.

In 1956 Smith published the first edition of his
popular book on the scientific discovery of the first
two coelacanths, entitled 'Old Fourlegs' (Fig. 5).
This book has become a classic of science non-
fiction and has been translated into 8 languages
(Afrikaans, Dutch, Estonian, French, German, Japanese, Russian and Slovak) in addition to the
three editions in English.

In 1963, on the 25th anniversary of the discovery

Fig. 5. The front cover design of the Pan edition of J.L.B.
Smith's book 'Old Fourlegs' published in 1956.

of the coelacanth, Smith made a speech at the East
London Museum on the need to form an international scientific society to conserve the coelacanth
which was published in the local newspaper under
the title 'The atomic bomb and the coelacanth',
surely the most provocative title produced so far?

The contributions of J. Millot and J. Anthony

The monumental contributions of the French scientists J. Millot and J. Anthony dominated the
coelacanth literature in the 1950s and early 1960s
(Fig. 6). Their work has been effectively continued
by D. Robineau, also of the Muséum National

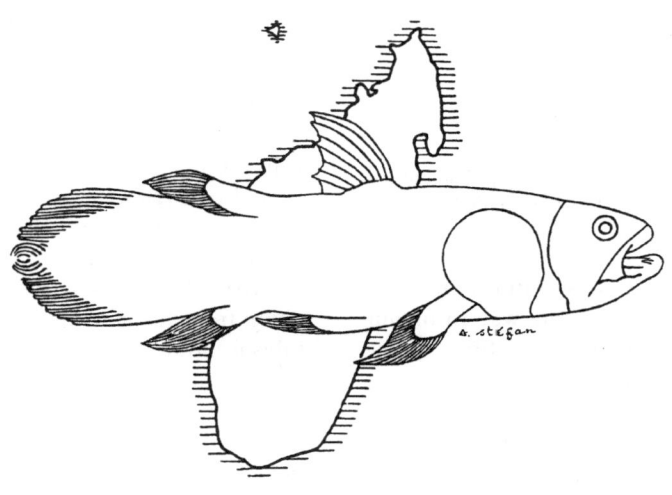

J. MILLOT

LE TROISIÈME CŒLACANTHE

HISTORIQUE · ÉLÉMENTS D'ÉCOLOGIE · MORPHOLOGIE EXTERNE
DOCUMENTS DIVERS

LE NATURALISTE MALGACHE
PREMIER SUPPLÉMENT
1954

Fig. 6. The title page of J. Millot's comprehensive publication on the third coelacanth in 1954.

d'Histoire Naturelle in Paris. Millot and Anthony's research concentrated primarily on taxonomy and morphology, but they also made significant contributions in ethology, ecology and conservation. Their three-part monograph on the anatomy of the coelacanth, published in 1958, 1965 and 1978 (the latter volume with Robineau), laid the foundation for subsequent research on coelacanth morphology, histology and physiology. The French scientists also contributed significantly to coelacanth research by obtaining and properly curating the many early coelacanth specimens that went to museums in Madagascar and France, and they were subsequently also responsible for distributing coelacanth specimens and tissues to scientists in other countries for study.

The international expeditions

The international phase of coelacanth research started in the 1970s. Although a great deal had been achieved through the study of formalin-preserved material, research on cytology and biochemistry was seriously hampered by the lack of suitably preserved study material. As the Comoros

Fig. 7. The logo of the Japanese Scientific Expedition of the Coelacanth (JASEC) during their expeditions in 1981, 1983 and 1986.

is relatively remote and lacks a scientific infrastructure, it was realised that the only way to obtain material would be to organise a scientific expedition to the islands with the intention of collecting tissue samples from a live or at least a deep frozen specimen. An approach from the United States led to the mounting of the 1969 Expedition of the Royal Society of London to Aldabra and Cosmoledo, and the 1972 Royal Society French/British/American and 1975 California Academy of Sciences expeditions to Grand Comoro. A detailed dissection protocol was drawn up by the Royal Society in case specimens became available for study. The members of the 1972 expedition were fortunate as two fresh specimens became available to them (CCC no. 79 and 80, Bruton & Coutouvidis 1991) and tissue samples were sent to scientists worldwide with the assistance of the Society for the Protection of Old Fishes (see reviews of research based on this material by Anthony 1980, Griffith 1980, Hughes 1980, Locket 1980).

In July-August 1981 and September-December 1983 Belgian zoological expeditions from the Royal Museum for Central Africa in Tervuren visited the Comoros to document the fauna and flora and to assist with the establishment of the Comoran National Museum in Moroni. Thys van den Aude-

naerde (1984) summarised their findings on the coelacanth.

More recently, expeditions to the Comoros have been organised by Japanese scientists in 1981, 1983 and 1986 'to observe Coelacanth in its natural habitat, with the purpose also to capture specimens for transport . . . to Japan for morphological, biochemical and biomechanical studies' (Suzuki et al. 1985, Fig. 7). The Japanese researchers made valuable observations on the behaviour and ecology of caught but dying coelacanths, as well as on the Comoran fishery (Suzuki & Tanauma 1984), but they were unable to obtain a live specimen to take to Japan. Several frozen specimens were, however, sent back to Japan for study (see Bruton & Coutouvidis 1991) and a series of papers by, inter alia, Suyehiro, Uyeno, Suzuki, Tamai and Tanaka have emanated from this research, especially following the Japanese conference on the coelacanth held in Tokyo on 9th December 1984. A commercial expedition from Japan in late 1989 funded by the Mitsubishi corporation also attempted to capture a live coelacanth but without success.

In 1986, 1987 and 1988 personnel from the Explorers Club in New York visited the Comoros with the intention of capturing a live coelacanth for an aquarium in New York. Although they were unable to achieve their main objective, two of their four coelacanth specimens in good condition (CCC no. 140 and 141) were obtained for research by a team led by J.A. Musick of the Virginia Institute of Marine Science, U.S.A. The results of this research are reported in this volume (e.g. Hale et al. 1991, Schultze 1991, Schultze & Cloutier 1991) and elsewhere (e.g. Cloutier et al. 1988).

An important new initiative in coelacanth research took place in 1986 when H. Fricke from the Max-Planck-Institut für Verhaltensphysiologie in Germany began his studies on the coelacanth in its natural environment using research submersibles. During subsequent expeditions in 1987, 1988 and 1989 he documented for the first time the behaviour, habitat preferences and relative abundance of free-swimming coelacanths at water depths of about 200 m off Grand Comoro, and also produced several video films that have been shown worldwide.

Fig. 8. The front cover design of the book by Y. Suyehiro on the coelacanth entitled 'Gombessa' which was published in 1988.

A new phase of South African research on the coelacanth was initiated in November 1986 with an expedition from the J.L.B. Smith Institute of Ichthyology led by P.C. Heemstra (Heemstra & Smale 1987). Subsequent expeditions led by M.N. Bruton in April and November 1987, March 1988 and in May 1990 resulted in a re-appraisal of the conservation status of the coelacanth and the factors threatening its survival. Participants in these expeditions included E.K. Balon and C. Flegler-Balon from the University of Guelph in Canada, R. Cloutier from the British Museum (Natural History) and R.E. Stobbs from the Smith Institute.

The popular literature

The mystery of the coelacanth and the drama associated with its discovery by western scientists has fascinated the lay public. Although only 166 popular articles have been cited in the bibliography, many hundreds more were published, and the coelacanth has featured prominently in popular books for adults and children and even in poetry, songs and plays. At least 53 newspaper articles on the coelacanth were written in 1939 alone (archives of the J.L.B. Smith Institute of Ichthyology), only some of which have been included in the bibliography. The discovery of the coelacanth was hailed, rather exaggeratedly, as 'the most important scientific find of the century' (Nancy Spain, Daily Express, 1956) and even as the best fish story in 50 000 000 years (Anon 1939), and has caught the public's imagination far more than, for instance, the discovery of a living lungfish *Neoceratodus forsteri* in 1870 (Bemis & Burggren 1987). The various editions of 'Old Fourlegs', along with other popular books on the coelacanth by Clymer (1963, 'Search for a Living Fossil'), Bell (1969, 'Old Man Coelacanth'), Anthony (1976, 'Operation Coelacanthe'), Brackenbury (1979, 'The Coelacanth') and Suyehiro (1988, 'Gombessa', Fig. 8) served to popularise the dramatic events associated with the discovery of the coelacanth.

The fiftieth anniversary

The fiftieth anniversary of the discovery of the first living coelacanth in December 1988 was commemorated with the publication of several review papers (Balon et al. 1988, Forey 1988, 1989, Greenwood 1988, 1989, Bruton 1988, 1989, van Bruggen 1989). In addition, a special edition of the Ichthos newsletter of the J.L.B. Smith Institute of Ichthyology, which included 34 popular articles on the coelacanth, was published in December 1988. South Africa issued four postage stamps on the coelacanth in February 1989 to commemorate the fiftieth anniversary of the identification of *L. chalumnae* (Bruton 1988).

Controversies in the coelacanth literature

The main controversy that has arisen in the coelacanth literature is concerned with the phylogenetic placement of the Crossopterygii. The two schools of thought – that the coelacanth is a sarcopterygian bony fish or a chondrichthyan – continue to be debated (e.g. Compagno 1979, Lagios 1979, 1982, Griffith 1991, Hillis et al. 1991, Waehneldt et al. 1991). Controversies on the conservation of the coelacanth erupted between J. Millot and J.L.B. Smith in 1956 (Millot 1956, Smith 1956), and between the members of the Coelacanth Conservation Council and the New York Explorers Club in 1988 and 1989 (Hamlin 1988, Anon 1989, Bruton & Stobbs 1991). A more recent debate concerns the evidence of oophagy and placental viviparity in *L. chalumnae* (Balon et al. 1988, Balon 1991, Heemstra & Compagno 1989, Wourms et al. 1980, 1988, 1991).

Conclusions

As this bibliography is incomplete and the coelacanth literature is expanding rapidly, it is hoped that coelacanth researchers will continue to keep the authors informed of additional references. These references can then be incorporated into future editions of the bibliography which will be published in the newsletter of the Coelacanth Conservation Council through the journal 'Environmental Biology of Fishes' as a service to coelacanth researchers.

Acknowledgements

We are grateful to the following who assisted with the compilation of this bibliography: Damir ben Ali and Abdu Shakur Aboud (Centre National de Documentation et de la Recherche Scientifique, Comoros), J.W. Atz (American Museum of Natural History, U.S.A.), E.K. Balon (University of Guelph, Canada), R.N. Bruton (South Africa), R. Cloutier (British Museum [Natural History], England), E. Cambray and M. Crampton (J.L.B. Smith Institute of Ichthyology, South Africa), P.L. Forey (British Museum [Natural History], England), H. Fricke (Max-Planck-Institut für Verhaltensphysiologie, Germany), P.H. Greenwood (British Museum [Natural History], England), P.C. Heemstra (J.L.B. Smith Institute of Ichthyology, South Africa), M. Kenyon (Rhodes University Library, South Africa), Alexandra de Sas Kropiwnicki (East London Museum, South Africa), D.E. McAllister (National Museum of Natural Sciences, Canada), J.E. McCosker (California Academy of Sciences, U.S.A.), J.A. Musick (Virginia Institute of Marine Science, U.S.A.), F. Nadot and D. Robineau (Museum National d'Histoire Naturelle, France), R.E. Stobbs (J.L.B. Smith Institute of Ichthyology, South Africa), N. Teitler (Japan), K.S. Thomson (The Academy of Natural Sciences, Philadelphia, U.S.A.), and T. Uyeno (National Science Museum, Japan). We are particularly grateful to C. Flegler-Balon, R. Cloutier, E.K. Balon and especially J.W. Atz for their editorial assistance.

Literature cited (other than references in the bibliography)

Bemis, W.E. & W.W. Burggren. 1987. Introduction. pp. 3–4. *In:* W.E. Bemis, W.W. Burggren & N.E. Kemp (ed.) The Biology and Evolution of Lungfishes, Alan R. Liss, New York.

Conant, E.B. 1987a. An historical overview of the literature of Dipnoi: introduction to the bibliography of lungfishes. pp. 5–13. *In:* W.E. Bemis, W.W. Burggren & N.E. Kemp (ed.) The Biology and Evolution of Lungfishes, Alan R. Liss, New York.

Conant, E.B. 1987b. Bibliography of lungfishes, 1811–1985. pp. 305–373. *In·* W.E. Bemis, W.W. Burggren & N.E. Kemp (ed.) The Biology and Evolution of Lungfishes, Alan R. Liss, New York.

Bibliography of *Latimeria chalumnae*

Acher, R., J. Chauvet & M.T. Chauvet. 1970. A tetrapod neurohyphophysial hormone in African lungfishes. Nature 227: 186–187.

Adamicka, P. & H. Ahnelt. 1976. Beiträge zur funktionellen Analyse und zur Morphologie des Kopfes von *Latimeria chalumnae* Smith. Annal. naturhist. Mus. Wien. 80: 251–271.

Ahlberg, P.E. 1989. Paired fin skeletons and relationships of the fossil group Porolepiformes (Osteichthyes: Sarcopterygii). Zool. J. Linn. Soc. 96: 119–166.

Alber, T. & G. Kawasaki. 1982. Nucleotide sequence of the triose phosphate isomerase gene of *Saccharomyces cerevisiae* J. mol. appl. Genet. 1: 419–434.

Alexander, R. McN. 1973. Jaw mechanisms of the coelacanth *Latimeria chalumnae.* Copeia 1973: 156–158.

Ali, M.A. & M. Anctil. 1976. Retinas of fishes. Springer-Verlag, Berlin. 284 pp.

Amos, B., I.G. Anderson, G.A.D. Haslewood & L. Tökes. 1977. Bile salts of the lungfishes *Lepidosiren, Neoceratodus* and *Protopterus* and those of the coelacanth *Latimeria chalumnae* Smith. Biochem. J. 161: 201–204.

Anderson, I.G. & G.A.D. Haslewood. 1964. Comparative studies of 'bile salts'. 20. Bile salts of the coelacanth, *Latimeria chalumnae*, Smith. Biochem. J. 93: 34–39.

Andrews, S.M. 1973. Interrelationships of crossopterygians. pp. 137–177. *In:* P.H. Greenwood, R.S. Miles & C. Patterson (ed.) Interrelationships of Fishes, Academic Press, London.

Andrews, S.M. 1977. The axial skeleton of the coelacanth, *Latimeria.* pp. 271–288. *In:* S.M. Andrews, R.S. Miles & A.D. Walker (ed.) Problems in Vertebrate Evolution, Linn. Soc. Lond. Symp. Ser. 4, Academic Press, London.

Andrews, S.M., B.G. Gardiner, P.S. Miles & C. Patterson. 1967. Pisces. pp. 637–683. *In:* W.B. Harland, C.H. Holland, M.R. House & N.F. Hughes (ed.) The Fossil Record: A Symposium with Documentation, Geological Society of London, London.

Anon. 1939. Century's most amazing scientific find is this live fish of 50 000 000 years ago. Life, 3 April 1939: 26–27.

Anon. 1954. Une pêche peu banale celle: du coelacanthe. La Pêche illustrée 46: 214.

Anon. 1963. Society to preserve coelacanth proposed. Eastern Province Herald, 22 December 1963.

414

Anon. 1964. Coelacanth gets a blue label. Peabody Museum News (New Haven) 1: 3–6.

Anon. 1966. A living fossil. The Sciences (N.Y. Acad. Sci.) (6) 6: 17–21.

Anon. 1969. Enter the coelacanth frozen – solid. British Commonwealth Review, October 1969: 22.

Anon. 1969. Plan to capture rare fish. Natal Daily News, 21 November 1964.

Anon. 1970. No holds barred in quest for elusive coelacanths. African Aquarist 3 (7): 13.

Anon. 1972. Living coelacanth caught off the Comoros. Nature 237: 131.

Anon. 1973. Third coelacanth for South Africa. S. Afr. J. Sci. 69: 357.

Anon. 1975. The living coelacanth. Zoology Leaflet No. 10. British Museum (Nat. Hist.), London. 3 pp.

Anon. 1975. Egg of the extant coelacanth fish *Latimeria*. Report on the British Museum (Nat. Hist.), London, 1972–1974: 17.

Anon. 1975. Scripps' scientists study rare fish. The San Diego Union, 30 April 1975: B-3.

Anon. 1976. So the coelacanth does bear live young. Nature 259: 81–82.

Anon. 1977. Our ovovivipare had found a friend. Grapevine, Amer. Mus. Nat. Hist. 34 (4): 1–2.

Anon. 1980. Fish story. Sci. Amer. 243 (4): 88, 90.

Anon. 1981. The coelacanth a living fossil. Royal Scottish Museum, Leaflet 2. 33 pp.

Anon. 1987. Träge Räuber. Der Spiegel 41: 277–278.

Anon. 1987. Der Quastenflosser erstmals im Lebersraum beobachtet. Studentische Zeitung 23: 36.

Anon. 1988. The perfume isles and old four legs. Excellence 4: 44–45.

Anon. 1989. Coelacanth gets new status. Daily Dispatch, 25 November 1989.

Anon. 1989. 'Electronic haggis' bait for coelacanth. Eastern Province Herald, 15 November 1989.

Anon. 1989. Hunt for fossil fish stirs controversy. Earthwatch, December 1989: 6.

Anon. 1990. Taking coelacanths out of the marketplace. Nat. Geogr. 177: 12.

Anthony, J. 1976. Opération coelacanthe. Arthaud, Paris. 201 pp.

Anthony, J. 1976. Quelques secrets du coelacanthe. Connaiss. Homme. 64: 23–34.

Anthony, J. 1979. Jacques Millot 1897–1979. Annals Sci. nat. Zool. Paris, Sér 13, 1: 221–222.

Anthony, J. 1980. Évocation des travaux français sur *Latimeria* notamment depuis 1972. Proc. R. Soc. Lond. B 208: 349–367.

Anthony, J. 1984. Le témoignage du coelacanthe *Latimeria*. Rev. Fr. Aquariol. Herpetol. 11: 33–38.

Anthony, J. & J. Millot. 1972. Première capture d'une femelle de coelacanthe en état de maturité sexuelle. C.R. Acad. Sci. Paris Sér. D 274: 1925–1926.

Anthony, J. & J. Millot. 1972. Variations sexuelles de l'appareil excréteur du coelacanthe, *Latimeria chalumnae*. Connexions

avec l'appareil génital mâle. C.R. Acad. Sci. Paris, Sér. D 276: 2447–2448.

Anthony, J. & J. Millot. 1974. Crossopterygii. Encyclopaedia Britannica (15th ed.): 296–297.

Anthony, J., J. Millot & D. Robineau. 1965. Le cœer et l'aorte ventrale de *Latimeria chalumnae*. C.R. Acad. Sci. Paris, Sér. D 261: 223–226.

Anthony, J. & D. Robineau. 1967. Le cercle céphalique de *Latimeria* (Poisson, Coelacanthidé). C.R. Acad. Sci. Paris, Sér. D 265: 343–346.

Anthony, J. & D. Robineau. 1968. Branchies et artères branchiales de *Latimeria chalumnae*. C.R. Acad. Sci. Paris, Sér. D 266: 375–378.

Anthony, J. & D. Robineau. 1976. Sur quelques caractères juvéniles de *Latimeria chalumnae* Smith (Pisces, Crossopterygii Coelacanthidae). C.R. Acad. Sci. Paris, Sér. D 283: 1739–1742.

Anthony, J. & D. Robineau. 1976. Corrélations anatomo-fonctionelles au niveau de la tête du coelacanthe, *Latimeria chalumnae*. Bull. Soc. Zool. France 101: 981–982.

Anthony, J. & D. Robineau. 1980. Le coelacanthe et son témoignage. pp. 96–105. *In:* F. Nathan (ed.) Mémoires Muséum National d'Histoire Naturelle, Paris.

Arbocco, G. 1962. *Latimeria* in a museum: presentation of a model. Natura 53: 1–4. (In Italian).

Armstrong, W.G., L.B. Halstead, F.B. Reed & L. Wood. 1983. Fossil proteins in vertebrate calcified tissues. Philos. Trans. R. Soc. Lond. B 301: 301–343.

Arratia, G. & H.-P. Schultze. 1990. The urohyal: development and homology within osteichthyans. J. Morphol. 203: 247–282.

Artavanis-Tsakonas, S. & J.I. Harris. 1980. Primary structure of triosephosphate isomerase from *Bacillus stearothermophilus*. Eur. J. Biochem. 108: 599–612.

Ash, S.R. 1978. Fish scales. Bringham Young University Stud. Geol. 25: 67–68.

Asimov, I. 1987. My cousin the coelacanth. El Mercurio, 17 March, 1987, Santiago, Chile. (In Spanish).

Attenborough, D. 1978. Life on Earth. Collins, London. 363 pp.

Atz, J.W. 1973. Comparative endocrinology and systematics. Amer. Zool. 13: 933–936.

Atz, J.W. 1975. How the baby coelacanths were discovered. Spoof Newsletter 15: 1–2.

Atz, J.W. 1976. *Latimeria* babies are born, not hatched. Underwater Naturalist 9 (4): 4–7.

Atz, J.W. 1985. The use of phylogenetic trees in comparative endocrinology. pp. 1143–1145. *In:* B. Lofts & W.N. Holmes (ed.) Current Trends in Comparative Endocrinology, Hong Kong University Press, Hong Kong.

Bailey, R.M. 1960. Coelacanthiformes. pp. 20–21. *In:* R.M. Bailey. Forty-five Articles on Recent Fishes, McGraw-Hill Encyclopedia of Science and Technology, New York.

Baix, Y. & E. Baix. 1983. Comores. Le monstre et le volcan. Océan 124: 35–37.

Baker, V. 1988. Of coral seas and coelacanths. Marung 7: 42, 45, 47, 49.

Ball, I. 1988. Rumpus over world's oldest fish. Daily Dispatch, 30 March 1988.

Balon, E.K. 1975. Reproductive guilds of fishes: a proposal and definition. J. Fish. Res. Board Can. 32: 821–864.

Balon, E.K. 1976. Ecomorphological principles of early ontogeny of *Labeotropheus* Ahl, 1927 (Mbuna, Lake Malawi): advanced protective strategies in fish reproduction. Rev. Trav. Inst. Pêch. Marit. 40: 484–485.

Balon, E.K. 1977. Early ontogeny of *Labeotropheus* Ahl, 1927 (Mbuna, Cichlidae, Lake Malawi), with a discussion on advanced protective styles in fish reproduction and development. Env. Biol. Fish. 2: 147–176.

Balon, E.K. 1981. Additions and amendments to the classification of reproductive styles in fishes. Env. Biol. Fish. 6: 377–389.

Balon, E.K. 1981. Saltatory processes and altricial to precocial forms in the ontogeny of fishes. Amer. Zool. 21: 573–596.

Balon, E.K. 1982. About the courtship rituals in fishes, but also about a false sense of security given by classification schemes, comprehensive reviews and committee decisions. Env. Biol. Fish. 7: 193–197.

Balon, E.K. 1984. Patterns in the evolution of reproductive styles in fishes. pp. 35–53. *In:* G.W. Potts & R.J. Wootton (ed.) Fish Reproduction: Strategies and Tactics, Academic Press, London.

Balon, E.K. 1985. Early life histories of fishes: new developmental, ecological and evolutionary perspectives. Developments in Env. Biol. Fish. 5, Dr W. Junk Publishers, Dordrecht. 280 pp.

Balon, E.K. 1989. The epigenetic mechanisms of bifurcation and alternative life-history styles. pp. 467–501. *In:* M.N. Bruton (ed.) Alternative Life-History Styles of Animals, Perspectives in Vertebrate Science 6, Kluwer Academic Publishers, Dordrecht.

Balon, E.K. 1990. The living coelacanth endangered: a personalized tale. Tropical Fish Hobbyist 38 (February): 117–129.

Balon, E.K. 1990. Tracking the coelacanth: a follow-up tale. Tropical Fish Hobbyist 38 (March): 122–131.

Balon, E.K. 1990. Epigenesis of an epigeneticist: the development of some alternative concepts on the early ontogeny and evolution of fishes. Guelph Ichthyol. Rev. 1: 1–42.

Balon, E.K. 1991. Prelude: the mystery of a persistent life form. Env. Biol. Fish. 32: 9–13. (this volume)

Balon, E.K. 1991. Probable evolution of the coelacanth's reproductive style: lecithotrophy and orally feeding embryos in cichlid fishes and in *Latimeria chalumnae*. Env. Biol. Fish. 32: 249–265. (this volume)

Balon, E.K., M.N. Bruton & H. Fricke. 1988. A fiftieth anniversary reflection on the living coelacanth, *Latimeria chalumnae:* some new interpretations of its natural history and conservation status. Env. Biol. Fish. 23: 241–280.

Balon, E.K. & A. Goto. 1989. Styles in reproduction and early ontogeny. pp. 1–47. *In:* A. Goto & K. Maekawa (ed.) Repro-

ductive Behavior in Fishes: Styles and Strategies, Tokai University Press, Tokyo. (In Japanese).

Bell, S. 1967. Coelacanths a century ago? Field & Tide 9: 8–11.

Bell, S. 1968. A coelacanth for Christmas. Personality, 5 December 1968: 10–17.

Bell, S. 1969. Old man coelacanth. Voortrekkerpers, Johannesburg. 141 pp.

Bell, S. 1983. The coelacanth: a South African scientific adventure. Flying Springbok, July 1983: 45–49.

Bell, S. 1984. South African sea serpents. Flying Springbok, June 1984: 81–83.

Beltan, L. 1975. À propos de l'ichthyofaune triassique de la catalogne espagnole. Colloques Int. C.N.R.S. 218: 273–280.

Beltan, L. 1980. Eotrias du nord-ouest de Madagascar: étude de quelques poissons dont un est un parturition. Annal. Soc. Geol. 99: 453–464.

Bemis, W.E. 1987. Feeding systems of living Dipnoi: anatomy and function. pp. 249–275. *In:* W.E. Bemis, W.W. Burggren & N.E. Kemp (ed.) The Biology and Evolution of Lungfishes, Alan R. Liss, New York.

Bemis, W.E. & T.E. Hetherington. 1982. The rostral [sic] organ of *Latimeria chalumnae:* morphological evidence of an electroreceptive function. Copeia 1982: 467–471.

Bemis, W.E. & R.G. Northcutt. 1991. Innervation of the basicranial muscle of *Latimeria chalumnae*. Env. Biol. Fish. 32: 147–158. (this volume)

Ben Ali, D. 1987. Comores. pp. 9–32. *In:* Ocean Indien. C. Pavard (ed.) L'ile aux Images, Mauritius.

Bently, K. 1989. Back to fossildom for wonder coelacanth. Eastern Province Herald, 12 January 1989.

Berg, L.S. 1939. Appendix. Crossopterygians. Priroda 7: 82. (In Russian).

Berg, L.S. 1940. Classification of fishes, both recent and fossil. Trav. zool. Institute of Leningrad 5: 87–517. (English edition by J.W. Edwards, Ann Arbor, Michigan (1947)).

Bernhauser, A. 1961. Zur Knochen- und Zahnhistologie von *Latimeria chalumnae* (Smith) und einiger Fossilformen. Sitzungber. öst. Akad. Wiss., Wien 170: 119–137.

Birch, D. 1979. Strange creature from the deep. South African Digest, 16 February 1979: 7–8.

Bjerring, H.C. 1972. The nervus rarus in coelacanthiform phylogeny. Zool. Scripta 1: 57–68.

Bjerring, H.C. 1973. Relationships of coelacanthiforms. pp. 179–205. *In:* P.H. Greenwood, R.S. Miles & C. Patterson (ed.) Interrelationships of Fishes, Academic Press, London.

Bjerring, H.C. 1978. The 'intracranial joint' versus the 'ventral otic fissure'. Acta zool. (Stockh.) 59: 203–214.

Bjerring, H.C. 1986. Elecrosensitive organ of *Latimeria chalumnae*. Fauna Flora (Stockholm) 81: 215–222. (In Swedish).

Bock, W. 1953. Secrets of the coelacanths. Frontiers 17: 99–103.

Bolivar y Pieltain, C. 1959. The study of present day coelacanths as a valuable contribution to knowledge. Geo 2: 32–34. (In Spanish).

Bollore, G.A. 1974. Évolution et pêche au coelacanthe. La Palatine, Paris. 91 pp.

Bonaventura, J., R.G. Gillen & A. Riggs. 1974. The hemoglo-

416

bin of the crossopterygian fish *Latimeria chalumnae* Smith – subunit structure and oxygen equilibrium. Arch. Biochem. Biophys. 163: 728–734.

Boulinier-Giraud, G. 1973. Étude morphologique de la pirogue a balancier aux Comores et dans l'ouest de l'océan indien. M.Sc. Thesis, Paris X. 137 pp.

Bowie, B. 1988. Marjorie's blue fish celebrates its jubilee. Personality, June 1974: 18–20.

Brackenbury, R. 1979. The coelacanth. Harvester Press, England. 165 pp.

Braker, W.P. 1960. In which a strange species of fish, believed to have been extinct for some 90,000,000 years, suddenly reappears and gives the experts something to puzzle about! The Aquarium (Phila.) 29: 109–111.

Bräutigam, A. 1989. Analyses of proposals to amend the CITES appendices. Doc. 7.43: No. 48. Submitted to the Seventh Meeting of the Conference of the Parties, Lausanne, Switzerland, 9–20 October 1989. IUCN, Washington. 5 pp.

Brentjes, B. 1972. Eine Vor-Entdeckung des Quastenflossers in Indien? Naturw. Rundsch. 1972: 312–313.

Brien, P. 1967. The African Protoptera: living fossils. African Wildlife 21: 213–233.

Brittan, M.R. 1980. Steinhart Aquarium, a San Francisco jewel. Tropical Fish Hobbyist 28 (12): 4–16.

Brothers, E.B. 1984. Otolith studies. pp. 50–57. *In:* H.G. Moser, W.J. Richards, D.M. Cohen, M.P. Fahay, A.W. Kendall, Jr. & S.L. Richardson (ed.) Ontogeny and Systematics of Fishes, Special Publ. No. 1, Amer. Soc. Ichthyol. Herpetol., Lawrence.

Brown, G.W., Jr. & S.G. Brown. 1966. Intermediary nitrogen metabolism of the coelacanth, *Latimeria chalumnae*. Amer. Zool. 6: 577.

Brown, G.W., Jr. & S.G. Brown. 1967. Urea and its formation in coelacanth liver. Science 155: 570–573.

Brown, G.W., Jr. & S.G. Brown. 1985. On urea formation in primitive fishes. pp. 321–337. *In:* R.E. Foreman, A. Gorbman, J.M. Dodd & R. Olsson (ed.) Evolutionary Biology of Primitive Fishes, Plenum Press, New York.

Brown, G.W., Jr., S.G. Brown & F.P. Solomon. 1975. Coelacanth studies. Research in fisheries. Annual Report of the College of Fisheries, University of Washington, Seattle 444: 52.

Brown, S.G. & G.W. Brown, Jr. 1979. Recapitulation of the discussion at the close of the 15 June 1977 symposium on the relationships of primitive fishes, with particular reference to the coelacanth. Occ. Pap. Calif. Acad. Sci. 134: 170–174.

Browne, M.W. 1987. Photos of rare fish may be due to evolution. The New York Times, 24 September 1987: 1, 19.

Browne, M.W. 1988. Scientists examine two dead fish for ancient and living secrets. New York Times, 8 January 1988: B6.

Browne, M.W. 1988. Post-mortem on life. Sun-Sentinel, 18 January 1988: 7A.

Browne, M.W. 1988. Do scientists pose a threat to rare 'fossil fish'? New York Times, 22 March 1988: C1, C8.

Bruton, C.J. 1988. Famous living fossils. Ichthos, Special edition 2: 13–14.

Bruton, M.N. 1985. The silver coelacanth. Ichthos 8: 17.

Bruton, M.N. 1987. Observations on the living coelacanth. Ichthos 15: 6.

Bruton, M.N. 1987. The life and work of Margaret M. Smith. J.L.B. Smith Institute of Ichthyology, Grahamstown. 12 pp.

Bruton, M.N. 1987. A second East London coelacanth. Ichthos 14: 6.

Bruton, M.N. 1987. Is the coelacanth endangered? Ichthos 13: 1.

Bruton, M.N. 1987. Tribute to Professor Margaret Smith. Grocotts Mail, 11 September 1987: 7.

Bruton, M.N. 1988. Comores – home of the coelacanth. Ichthos, Special edition 2: 15.

Bruton, M.N. 1988. The coelacanth jubilee. Ichthos 18: 1–2.

Bruton, M.N. 1988. A captive coelacanth? Ichthos, Special edition 2: 17.

Bruton, M.N. 1988. Coelacanth Conservation Council. Quagga 24: 12–13.

Bruton, M.N. 1988. The coelacanth stamps. Ichthos, Special edition 2: 19.

Bruton, M.N. 1988. Coelacanth Conservation Council. Env. Biol. Fish. 23: 315–319.

Bruton, M.N. 1988. 'Fossil' fish trade causes concern. Traffic (U.S.A.) 8: 8.

Bruton, M.N. 1988. Coelacanth Conservation Council. Ichthos, Special edition 2: 3–4.

Bruton, M.N. 1988. What is the significance of the coelacanth? Ichthos, Special edition 2: 1–2.

Bruton, M.N. (ed.) 1988. The coelacanth jubilee. Ichthos, Special edition 2: 1–20.

Bruton, M.N. 1989. Fifty years of coelacanths. S. Afr. J. Sci. 85: 205.

Bruton, M.N. 1989. Sensational discovery – 50 years ago. Grocotts Mail, 10 January 1989: 8.

Bruton, M.N. 1989. Coelacanth stamps issued today. The Star, 9 February 1989.

Bruton, M.N. 1989. The living coelacanth fifty years later. Trans. R. Soc. S. Afr. 47: 19–28.

Bruton, M.N. 1989. The coelacanth stamps. The Star, 26 January 1989.

Bruton, M.N. 1989. The coelacanth – can we save it from extinction? WWF Reports, October/November 1989: 10–12.

Bruton, M.N. 1989. Is there a Madagascar coelacanth? Ichthos 23: 25.

Bruton, M.N. 1989. Does the coelacanth occur in the Eastern Cape? The Naturalist (Port Elizabeth) 33 (3): 5–13.

Bruton, M.N. 1989. Japanese scientists pose a threat to the coelacanth. Natal Daily News, 7 September 1989.

Bruton, M.N. 1989. Coelacanth and CITES. Ichthos 24: 4.

Bruton, M.N. & M.J. Armstrong. 1991. The demography of the coelacanth *Latimeria chalumnae*. Env. Biol. Fish. 32: 301–311. (this volume)

Bruton, M.N., C.D. Buxton, R.E. Stobbs & G.R. Hughes. 1989. Recommendations on marine conservation in the Federal Islamic Republic of the Comoros. Invest. Rep. J.L.B. Smith Inst. Ichthyol. 34: 1–103. (Also available in French).

Bruton, M.N. & S.E. Coutouvidis. 1991. An inventory of all known specimens of the coelacanth *Latimeria chalumnae*, with comments on trends in the catches. Env. Biol. Fish. 32: 371–390. (this volume)

Bruton, M.N., S.E. Coutouvidis & J. Pote. 1991. Bibliography of the living coelacanth *Latimeria chalumnae,* with comments on publication trends. Env. Biol. Fish. 32: 403–433. (this volume)

Bruton, M.N. & R.E. Stobbs. 1991. The ecology and conservation of the coelacanth *Latimeria chalumnae.* Env. Biol. Fish. 32: 313–339. (this volume)

Burggren, W.W. & K. Johansen. 1987. Circulation and respiration in lungfishes (Dipnoi). pp. 217–236. *In:* W.E. Bemis, W.W. Burggren & N.E. Kemp (ed.) The Biology and Evolution of Lungfishes, Alan R. Liss, New York.

Burridge-Smith, M. 1987. The coelacanth a real fish tale. Rotunda (Royal Ontario Museum) 20 (2): 50–54.

Burton, M. 1953. The lesson of the coelacanth. Illustrated London News, 7 February 1953: 208.

Burton, M. 1954. Living fossils. Thames & Hudson, London. 282 pp.

Caldwell, D.K. & M.C. Caldwell. 1965. The fossil who stayed alive. Quarterly Los Angeles County Museum 4: 14–15.

Campbell, K.S.W. & R.E. Barwick. 1987. Paleozoic lungfishes – a review. pp. 93–131. *In:* W.E. Bemis, W.W. Burggren & N.E. Kemp (ed.) The Biology and Evolution of Lungfishes, Alan R. Liss, New York.

Carlström, D. 1963. A crystallographic study of vertebrate otoliths. Biol. Bull. 125: 441–463.

Carrington, R. 1957. The fossil that came to life. pp. 155–163. *In:* Mermaids and Mastodons, Chatto & Windus, London.

Castanet, J., F. Meunier, C. Bergot & Y. François. 1975. Données préliminaires sur les structures histologiques du squelette de *Latimeria chalumnae* 1. Dents, écailles, rayons des nageoires. Colloq. Int. C.N.R.S. 218: 161–167.

Černy, K. 1980. *Latimeria chalumnae* in Czechoslovakia for the first time. Živa 28: 185–190. (In Czech).

Chauvet, J.-P. & R. Acher. 1972. Isolation of coelacanth (*Latimeria chalumnae*) myoglobin. FEBS Lett. 28: 16–18.

Chavin, W. 1970. Brain pituitary neurosecretory complex of the Australian lungfish *Neoceratodus forsteri.* Amer. Zool. 10: 496. (abstract).

Chavin, W. 1972. The sarcopterygian thyroid. Amer. Zool. 12: 667. (abstract).

Chavin, W. 1972. Thyroid of the coelacanth, *Latimeria chalumnae* Smith. Nature 239: 340–341.

Chavin, W. 1976. The thyroid of the sarcopterygian fishes (Dipnoi and Crossopterygii) and the origin of the tetrapod thyroid. Gen. comp. Endocrinol. 30: 142–155.

Chui, G. 1989. 'Living fossil' fish, once thought extinct, now really in danger. Montreal Gazette, 8 July 1989.

Chui, G. 1989. Threats put 'living fossil' fish on brink. San José Mercury News, 22 June 1989: 2A.

Cimino, M.C. & G.F. Bahr. 1973. Nuclear DNA content and chromatin ultrastructure of the coelacanth *Latimeria chalumnae.* J. Cell Biol. 59: 55a. (abstract).

Cimino, M.C. & G.F. Bahr. 1974. The nuclear DNA content and chromatin ultrastructure of the coelacanth *Latimeria chalumnae.* Exp. Cell Res. 88: 263–272.

CITES, 1973. Convention on International Trade in Endangered Species of Wild Fauna and Flora, signed at Washington, D.C., on 3 March, 1973. Special Supplement to IUCN Bull. 4: 1–6.

CITES, 1987. Convention on International Trade in Endangered Species of Wild Fauna and Flora. Appendices I & II as adopted by the Conference of the Parties, valid as of 22 October 1987: 1–45.

CITES, 1987. Convention on International Trade in Endangered Species of Wild Fauna and Flora. Amendments to Appendices I & II of the Convention. Proposal for deletion of *Latimeria chalumnae* from Appendix II: 1–2.

CITES, 1987. Convention on International Trade in Endangered Species of Wild Fauna and Flora. Recommendations from the Secretariat. Doc. 6.47. Annex 3: 1.

Clarke, A.C. 1966. Monsters of the night. pp. 125–141. *In:* A.C. Clarke (ed.) The Challenge of the Sea, Dell Publishing Co., New York.

Cloutier, R. 1989. Lungfishes, coelacanths, and tetrapods – beyond a 3-taxa statement. Amer. Zool. 29: 11. (abstract).

Cloutier, R. 1989. Phylogenetic interrelationships of the Actinistians (Osteichthyes: Sarcopterygii): patterns, trends, and rates of evolution. Ph.D. Thesis, University of Kansas, Lawrence. 403 pp.

Cloutier, R. 1990. Interrelationships of Paleozoic actinistians: patterns and trends. pp. 12–26. *In:* Sixth Symposium on Early Vertebrate Studies and Related Problems in Evolutionary Biology, Beijing, October 1990.

Cloutier, R. 1991. Patterns, trends and rates of evolution within the Actinistia. Env. Biol. Fish. 32: 23–58. (this volume)

Cloutier, R. & P.L. Forey. 1991. Diversity of extinct and living actinistian fishes (Sarcopterygii). Env. Biol. Fish. 32: 59–74. (this volume)

Cloutier, R., H.-P. Schultze, E.O. Wiley, J.A. Musick, J.C. Daimler, M.A. Brown, S.J. Dwyer, L.T. Cook & R.L. Laws. 1988. Recent radiologic imaging techniques for morphological studies of *Latimeria chalumnae.* Env. Biol. Fish. 23: 281–282.

Clymer, E. 1963. Search for a living fossil. The story of the coelacanth. Scholastic Book Services, New York. 121 pp.

Colbert, E.H. 1939. A fossil comes to life. Natural History 43: 280–283.

Cole, D.F. 1968. Anterior chamber of the coelacanth eye. Brit. J. Ophthal. 52: 415–418.

Cole, D.F. 1973. Intraocular fluid composition in the coelacanth *Latimeria chalumnae.* Exp. Eye Res. 16: 389–395.

Compagno, L.J.V. 1979. Coelacanths: shark relatives or bony fishes? Occ. Pap. Calif. Acad. Sci. 134: 45–52.

Compagno, L.J.V. 1990. Alternative life-history styles of cartilaginous fishes in time and space. Env. Biol. Fish. 28: 33–75.

Conyngham, J. 1988. To save a coelacanth. Sunday Tribune, 20 November 1988: 21.

Corran, P.H. & S.G. Waley. 1975. The amino acid sequence of

418

rabbit muscle triose phosphate isomerase EC-5.3.1.1. Biochem. J. 145: 335–344.

Courtenay-Latimer, M. 1979. My story of the first coelacanth. Occ. Pap. Calif. Acad. Sci. 134: 6–10.

Courtenay-Latimer, M. 1988. Odd amusing bits of the coelacanth discovery. A retrospective in 1988. J. Border historical Society 26: 19–24.

Courtenay-Latimer, M. 1989. Reminiscences of the discovery of the coelacanth, *Latimeria chalumnae* Smith. Cryptozoology 8: 1–11.

Cowgill, V.M., G.E. Hutchinson & H.C.W. Skinner. 1968. The elementary composition of *Latimeria chalumnae* Smith. Proc. Natl. Acad. Sci. U.S.A. 60: 456–463.

Cracraft, J. 1968. Functional morphology and adaptive significance of cranial kinesis in *Latimeria chalumnae* (Coelacanthini). Amer. Zool. 8: 354.

Croeser, P.M.C. 1981. Another chapter in the coelacanth story. The Naturalist (Port Elizabeth) 25 (1): 40–41.

Crozier, G.F. & D.L. Fox. 1965. Absence or singular specificity of carotenoids in some lower fishes. Science 150: 771–773.

Daniel, G. 1954. Living fossil and missing links. Sunday Times, London, 30 May 1954.

Dartnall, H.J.A. 1972. Visual pigment of the coelacanth *Latimeria chalumnae* Smith. Nature 239: 341–342.

Davies, E.R. & D.M.S. Watson. 1954. Coelacanths. Proc. R. Inst. 35: 1–5.

de Beer, G. 1956. A reply to letter of J.L.B. Smith on conservation of coelacanth. The Times, London, 5 June 1956.

de Lyrot, A. 1954. Coelacanth on view at museum in Paris. New York Herald-Tribune, 20 March 1954.

Denis, C. 1987. Les derniers coelacanthes. Sciences, Mars 1987: 32–33.

de Sylva, D.P. 1966. Mystery of the silver coelacanth. Sea Frontiers 12: 172–175.

Devys, M., A. Thierry, M. Barbier & M.-M. Janot. 1972. Premières observations sur les lipides de l'ovocyte du coelacanthe (*Latimeria chalumnae*). C.R. Acad. Sci. Paris, Sér. D 275: 2085–2087.

Diamond, J.M. 1985. In quest of the wild and weird. Discover (Los Angeles) 6 (3): 34–36, 38–42.

Diamond, J.M. 1985. The biology of coelacanths. Nature 315: 18.

DiJulio, D.H. & G.W. Brown, Jr. 1975. Urea levels in tissues of the coelacanth *Latimeria chalumnae*. Amer. Zool. 15: 801.

Dingerkus, G. 1979. Chordate cytogenetic studies: an analysis of their phylogenetic implications with particular reference to fishes and the living coelacanth. Occ. Pap. Calif. Acad. Sci. 134: 111–127.

Dingerkus, G., H.K. Mok & M.D. Lagios. 1978. The living coelacanth *Latimeria chalumnae* does not have a cloaca. Nature 276: 261–262.

Dollfus, R.P. & G. Campana-Rouget. 1956. Helminthes trouvés dans le tube digestif de coelacanthes. Mém. Inst. Sci. Madagascar sér. A 11: 33–41.

Dugan, J. 1955. The fish. Collier's, 16 September 1955: 64–68.

Dugan, J. 1955. The fish named L.C. Smith. Reader's Digest, December 1955: 147–151.

Dupont, G. 1987. Pour la première fois on a vu nager un coelacanthe. Science et Vie 843: 68–70.

Du Toit, C.A. 1953. Some problems of the coelacanth restated. S. Afr. J. Sci. 49: 332–333.

Dwyer, S.J. 1988. CT and MR images of a living fossil. Radiology August 1988: 318.

Eaton, T.H., Jr. 1945. The symplectic in coelacanthids and Actinopteri. Copeia 1945: 22–25.

Elliott, K. 1989. Water brats in the jaws of extinction. The Independant, London, 21 May 1989: 6.

Elter, O. 1980. The coelacanth. Guide to temporary exhibition, Torina, Maggio-Giugno 1980. Museo regionale di Scienze naturali, Torina. 15 pp. (In Italian).

Endoh, H. & N. Okada. 1986. Total DNA transcription in vitro: a procedure to detect highly repetitive and transcribable sequences with tRNA-like structures. Proc. Natl. Acad. Sci. U.S.A. 83: 251–255.

Epple, A. & J.E. Brinn. 1975. Islet histophysiology evolutionary correlations. Gen. comp. Endocrinol. 27: 320–349.

Ferris, S.D. & G.W. Whitt. 1978. Genetic and molecular analysis of nonrandom dimer assembly of the creatine kinase isozymes of fishes. Biochem. Genet. 16: 811–830.

Fisher, S.E. & G.W. Whitt. 1979. Evolution of the creatine kinase isozyme system in the primitive vertebrates. Occ. Pap. Calif. Acad. Sci. 134: 142–159.

Flegler-Balon, C. 1989. Direct and indirect development in fishes – examples of alternative life-history styles. pp. 71–100. *In:* M.N. Bruton (ed.) Alternative Life-History Styles of Animals, Perspectives in Vertebrate Science 6, Kluwer Academic Publishers, Dordrecht.

Forey, P.L. 1980. *Latimeria:* a paradoxical fish. Proc. R. Soc. B 208: 369–384.

Forey, P.L. 1984. The coelacanth as a living fossil. pp. 166–169. *In:* N. Eldredge & S.M. Stanley (ed.) Living Fossils, Springer-Verlag, New York.

Forey, P.L. 1984. L'origine des tetrapodes. La Recherche 154: 476–487.

Forey, P.L. 1987. Relationships of lungfishes. pp. 75–91. *In:* W.E. Bemis, W.W. Burggren & N.E. Kemp (ed.) The Biology and Evolution of Lungfishes, Alan R. Liss, New York.

Forey, P.L. 1988. Golden jubilee for the coelacanth *Latimeria chalumnae*. Nature 336: 727–732.

Forey, P.L. 1989. Le coelacanthe. La Recherche 20: 1318–1326.

Forey, P.L. 1990. An extraordinary blue fish. Endeavour, N. S. 14(1): 8–13.

Forey, P.L. 1990. The coelacanth fish: progress and prospects. Sci. Progress (Oxford) 74: 53–67.

Forey, P.L. 1991. *Latimeria chalumnae* and its pedigree. Env. Biol. Fish. 32: 75–97. (this volume)

Forey, P.L. & R. Cloutier. 1991. Literature relating to fossil coelacanths. Env. Biol. Fish. 32: 391–401. (this volume)

Forster, G.R. 1970. Diary and report of G.R. Forster on his visit to the Comoro Islands 10 to 16 October 1970. Unpublished report to the Royal Society of London. 11 pp.

Forster, G.R. 1974. The ecology of *Latimeria chalumnae* Smith: results of field studies from Grande Comore. Proc. R. Soc. Lond. B 186: 291–296.

Forster, G.R. 1974. Fish that came in from the past. Manning River Times, August 1974.

Forster, G.R., J.R. Badcock, M.R. Longbottom, N.R. Merrett & K.S. Thomson. 1970. Results of the Royal Society Indian Ocean deep slope fishing expedition, 1969. Proc. R. Soc. Lond. B 175: 367–404.

Fox, D.L. & G.F. Crozier. 1965. Absence or singular specificity of carotenoids in some lower fishes. Science 150: 771–773.

François, Y. 1959. La nageoire dorsale: anatomie et évolution. Année biol. 35: 81–113.

Fraschini, A. 1967. Presence of Paneth cells in *Latimeria chalumnae* and their histochemical characteristics. Boll. Soc. ital. Biol. sper. 43: 1449–1450. (In Italian).

Frey, M. 1989. Zeitgenosse der Vergangenheit. Frankfurter Rundschau, 23 October 1989, 246: 17.

Fricke, H. 1988. Coelacanths: the fish that time forgot. Natl. Geogr. Mag. 173: 824–838.

Fricke, H. 1988. J'ai recontré le fossile vivant. Geo 116: 56–75.

Fricke, H. 1989. Quastie im Baskenland? Tauchen 10: 64–67.

Fricke, H. & K. Hissmann. 1990. Natural habitat of coelacanths. Nature 346: 323–324.

Fricke, H., K. Hissmann, J. Schauer, O. Reinicke, L. Kasang & R. Plante. 1991. Habitat and population size of the coelacanth *Latimeria chalumnae* at Grand Comoro. Env. Biol. Fish. 32: 287–300. (this volume)

Fricke, H. & R. Plante. 1988. Habitat requirements of the living coelacanth *Latimeria chalumnae* at Grande Comore, Indian Ocean. Naturwissenschaften 75: 149–151.

Fricke, H., O. Reinicke, H. Hofer & W. Nachtigall. 1987. Locomotion of the coelacanth *Latimeria chalumnae* in its natural environment. Nature 329: 331–333.

Fricke, H. & J. Schauer. 1987. Im Reich der lebenden Fossilien. Geo 10: 14–34.

Fricke, H., J. Schauer, K. Hissmann, L. Kasang & R. Plante. 1990. Population size of the living coelacanth at Grande Comore. Nature (in press).

Fricke, H., J. Schauer, K. Hissmann, L. Kasang & R. Plante. 1991. Coelacanth *Latimeria chalumnae* aggregates in caves: first observations on their resting habitat and social behavior. Env. Biol. Fish. 30: 281–285.

Fritzsch, B. 1987. Die papilla basilaris von *Latimeria chalumnae* und die Evolution von amphibientypischen auditiven Sinnesepithelien. Verh. Deutsch. Zool. Ges. 80: 145–146. (abstract).

Fritzsch, B. 1987. Inner ear of the coelacanth fish *Latimeria* has tetrapod affinities. Nature 327: 153–154.

Fritzsch, B. 1988. Die Anfange des Hörens liegen im Wasser. Forschung-Mitteilungen der DFG 4: 20–22.

Fukuda, Y., N. Kawamoto, I. Obata & Y. Kanie. 1978. Notes of observation by scanning electron microscope on a scale of *Latimeria chalumnae*. Sci. Rep. Yokosuka City Mus. 24: 45–50. (In Japanese).

Gaffney, E.S. 1979. Tetrapod morphology: a phylogenetic analysis. Bull. Carnegie Mus. Nat. Hist. 13: 92–105.

Gardiner, B.G. 1984. The relationships of the palaeoniscid fishes, a review based on new specimens of *Mimia* and *Moythomasia* from the Upper Devonian of Western Australia. Bull. Br. Mus. (Nat. Hist.) Geol. 37: 173–428.

Gardiner, B.G. & B. Schaeffer. 1989. Interrelationships of lower actinopterygian fishes. Zool. J. Linn. Soc. 97: 135–187.

Gaski, A.L. 1988. 'Fossil' fish trade causes concern. Traffic 8: 9.

Geiger, R.E. 1953. Air's full of strange animals. New York Sunday News, 18 January 1953: 49.

Geraudie, J. 1985. Comparative fine structure of the actinotrichia (elastoidin) in developing teleost and dipnoi fish fins. Life Sci. 93: 451–455.

Geraudie, J. & F.-J. Meunier. 1980. Elastoidin actinotrichia in coelacanth fins: a comparison with teleosts. Tissue & Cell 12: 637–645.

Gill, L. 1939. A fish out of pre-history: why science is excited over East London find. Cape Times, 2 March 1939: 3a.

Giraud, M.-M., J. Castanet, F.-J. Meunier & Y. Bouligand. 1978. Organisation spatiale de l'isopedine des écailles du coelacanthe *Latimeria chalumnae*. Smith. C.R. Acad. Sci. Paris, Sér. D 287: 487–489.

Giraud, M.-M., J. Castanet, F.-J. Meunier & Y. Bouligand. 1978. The fibrous structure of coelacanth scales: a twisted plywood. Tissue & Cell 10: 671–686.

Glaubrecht, M. 1988. Das Ohr des Quastenflossers. Naturw. Rundsch. 41: 275.

Globig, M. 1988. Neue Fahndung nach dem Urfisch. MPG Spiegel 2/88: 1–4.

Gogola, M. 1987/88. Live fossils on film. Proteus 50: 215–218. (In Serbian).

Goldstein, L., S. Harley-DeWitt & R.P. Forster. 1972. Activities of ornithine-urea cycle enzymes and of trimethylamine oxidase in the coelacanth, *Latimeria chalumnae*. Comp. Biochem. Physiol. B 44: 357–362.

Goodman, M., M.M. Miyamoto & J. Czelusniak. 1987. Pattern and process in vertebrate phylogeny revealed by coevolution of molecules and morphologies. pp. 141–176. *In:* C. Patterson (ed.) Molecules and Morphology in Evolution: Conflict or Compromise?, Cambridge University Press, Cambridge.

Goosen, H.P. 1988. Captain Goosen's story: the trip I caught the coelacanth. Ichthos, Special edition 2: 16–17.

Gorden, B.L. 1979. The catch of the century. Nat. Fisherman (Yearbook Issue) 57: 81–84.

Grady, J.E. 1970. Tooth development in *Latimeria chalumnae* Smith. J. Morph. 132: 377–387.

Grant, D. 1990. 'Dinosaur' of the deep. The Globe and Mail, 20 October 1990: D4.

Grant, F.B. & G.E. Pickford. 1966. Ionic composition, urea and serum osmolarity in the blood of the coelacanth *Latimeria chalumnae*. Amer. Zool. 6: 545–546 (abstract).

Greenwell, J.R. 1989. Remembering the coelacanth: a 50th anniversary. Interviews with Marjorie Courtenay-Latimer and Hendrik Goosen. International Society of Cryptozoology, Tucson, Arizona. 18 pp.

420

Greenwell, J.R. 1990. The lady and the coelacanth: remembering the zoological discovery of the century. The Explorers Journal 68: 117–123.

Greenwood, P.H. 1988. A living fossil fish. The coelacanth *Latimeria chalumnae*. Zoology Booklet, British Museum (Natural History), London. 17 pp.

Greenwood, P.H. 1988. The 'sixpenny coelacanth', and some notes on *Latimeria* in a social context. Ichthos, Special edition 2: 5–6.

Greenwood, P.H. 1989. Fifty years a 'living fossil' – the coelacanth fish *Latimeria chalumnae*. Biologist 36: 15–19.

Greenwood, P.H. 1989. Fishy ecological tales. The Guardian, 13 September 1989.

Gregory, W.K. 1941. Grandfather fish and his descendants. Natural History 48: 156–165.

Griffith, R.W. 1973. A live coelacanth in the Comoro Islands. Discovery (New Haven) 9 (1): 27–33.

Griffith, R.W. 1980. Chemistry of the body fluids of the coelacanth, *Latimeria chalumnae*. Proc. R. Soc. Lond. B 208: 329–347.

Griffith, R.W. 1991. Guppies, toadfish, lungfish, coelacanths and frogs: a scenario for the evolution of urea retention in fishes. Env. Biol. Fish. 32: 199–218. (this volume)

Griffith, R.W. & C.J. Burdick. 1976. Sodium-potassium activated adenosine triphosphatase in coelacanth tissues: high activity in rectal gland. Comp. Biochem. Physiol. B 54: 557–559.

Griffith, R.W. & N.A. Locket. 1972. Observations on a living coelacanth. Nature 237: 175.

Griffith, R.W., M.B. Mathews, B.L. Umminger, B.F. Grant, P.K.T. Pang, K.S. Thomson & G.E. Pickford. 1975. Composition of fluid from the notochordal canal of the coelacanth, *Latimeria chalumnae*. J. Exp. Zool. 192: 165–171.

Griffith, R.W. & P.K.T. Pang. 1979. Mechanisms of osmoregulation in the coelacanth, *Latimeria chalumnae*: evolutionary implications. Occ. Pap. Calif. Acad. Sci. 134: 79–93.

Griffith, R.W. & K.S. Thomson. 1972. Observations on a dying coelacanth. Amer. Zool. 12: 730.

Griffith, R.W. & K.S. Thomson. 1973. *Latimeria chalumnae*: reproduction and conservation. Nature 242: 617–618.

Griffith, R.W., B.L. Umminger, B.F. Grant, P.K.T. Pang & G.E. Pickford. 1974. Serum composition of the coelacanth, *Latimeria chalumnae* Smith. J. Exp. Biol. 187: 87–102.

Griffith, R.W., B.L. Umminger, B.F. Grant, P.K.T. Pang, L. Goldstein & G.E. Pickford. 1976. Composition of bladder urine of the coelacanth, *Latimeria chalumnae*. J. Exp. Zool. 196: 371–380.

Grodziński, Z. 1972. The yolk of *Latimeria chalumnae* Smith. Folia histochem. cytochem. 10: 11–18.

Gross, W. 1939. Ein rezenter Crossopterygier entdeckt! Die Naturwissenschaften 14: 226–227.

Grossner, D. 1968. Das Inselorgan des Crossopterygiers *Latimeria chalumnae* J.L.B. Smith. Z. Zellforsch. mikrosk. Anat. 84: 417–428.

Hafeez, M.A. & M.E. Merhige. 1977. Light and electron microscopic study on the pineal complex of the coelacanth, *Latimeria chalumnae* Smith. Cell Tissue Res. 178: 249–265.

Hale, R.C., J. Greaves, J.L. Gundersen & R.F. Mothershead II. 1991. Occurrence of organochlorine contaminants in tissues of the coelacanth *Latimeria chalumnae*. Env. Biol. Fish. 32: 361–367. (this volume)

Hall, M. 1989. The survivor: fifty years in the modern life of an ancient fish. Harvard Magazine 91 (3): 36–42.

Hamlin, J. 1988. Capture of fossil fish also promotes its ultimate survival. The New York Times, 9 April 1988: A30.

Hamoir, G., A. Piront, Ch. Gerday & P.R. Dande. 1973. Muscle proteins of the coelacanth *Latimeria chalumnae* Smith. J. Mar. Biol. Ass. U.K. 53: 763–784.

Harden, M. 1988. Couple isn't just spoofing: Edmonds duo study important evolutionary link in coelacanth. The Herald, 24 February 1988: 1A, 16A.

Haslewood, G.A.D. 1957. Bile salts of a coelacanth, *Latimeria chalumnae*, Smith. Proc. biochem. Soc. 66: 22.

Haslewood, G.A.D. 1964. Bile salts of a coelacanth, *Latimeria chalumnae*, Smith. Biochem. J. 93: 22. (abstract).

Haslewood, G.A.D. 1969. New information about bile alcohols. pp. 151–159. In: L. Schiff, J.B. Carey & J.M. Dietschy (ed.) Bile Salt Metabolism, C. Thomas, Springfield.

Hay, A.W.M. 1974. Comparative aspects of vitamin D transport. Medica International Congress Series 346: 405–407.

Hay, A.W.M. & G. Watson. 1976. The plasma transport proteins of 25-hydroxycholecalciferol in fish, amphibians, reptiles and birds. Comp. Biochem. Physiol. B 53: 167–172.

Hayashida, T. 1977. Immunochemical and biological studies with growth hormone in a pituitary extract of the coelacanth, *Latimeria chalumnae* Smith. Gen. comp. Endocrinol. 32: 221–229.

Hayashida, T. 1979. Immunochemical and biological studies with growth hormone in extracts of pituitaries from existing primitive fishes. Occ. Pap. Calif. Acad. Sci. 134: 128–141.

Hecht, T. 1982. What's so special about the coelacanth? Ichthos 1: 8.

Heemstra, P.C. 1983. More about the coelacanth. Ichthos 4: 12.

Heemstra, P.C. & L.J.V. Compagno. 1989. Uterine cannibalism and placental viviparity in the coelacanth? A skeptical view. S. Afr. J. Sci. 85: 485–486.

Heemstra, P.C. & M.J. Smale. 1987. Fisheries resources of the Islamic Republic of the Comores, with recommendations for the wise use and conservation of the marine fauna of these islands. Invest. Rep. J.L.B. Smith Inst. Ichthyol. 23: 1–34.

Hemenway, D. 1984. Coelacanth: a 'living fossil'. Earth Science 37 (2): 14–15.

Hensel, K. 1986. Morphologie et interprétation des canaux et canalicules sensoriels céphaliques de *Latimeria chalumnae* Smith, 1939 (Osteichthyes, Crossopterygii, Coelacanthiformes). Bull. Mus. Natl. Hist. Nat. Paris, Sér. 4, 8(2): 379–407.

Hetherington, T.E. & W.E. Bemis. 1979. Morphological evidence of an electroreceptive function of the rostral organ of *Latimeria chalumnae*. Amer. Zool. 19: 986.

Heuvelmans, B. 1969. In the wake of the sea-serpents. Hill & Wang, New York. 645 pp.

Highfield, R. 1988. The coelacanth faces second extinction. Daily Telegraph, 11 November 1988.

Hilldén, N.-O. 1983. The coelacanth *Latimeria chalumnae*. Fauna Flora (Stockholm) 78: 243–246. (In Swedish).

Hillis, D.M. & M.T. Dixon. 1989. Vertebrate phylogeny: evidence from 28S ribosomal DNA sequences. pp. 355–367. *In:* B. Fernholm, K. Bremer & H. Jornvall (ed.) The Hierarchy of Life, Molecules and Morphology in Phylogenetic Analysis, Excerpta Medica, Amsterdam.

Hillis, D.M., M.T. Dixon & L.K. Ammerman. 1991. The relationships of the coelacanth *Latimeria chalumnae*: evidence from sequences of vertebrate 28S ribosomal RNA genes. Env. Biol. Fish. 32: 119–130. (this volume)

Hirano, T. 1981. Body fluids of the coelacanth. Seitai no Kagaku 32: 378–381. (In Japanese).

Hobdell, M.H. & W.A. Miller. 1969. Radiographic anatomy of the teeth and tooth supporting tissues of *Latimeria chalumnae*. Arch. Oral Biol. 14: 855–858.

Holčík, J. 1973. Two jubilees of one event. Polovnícto a rybárstvo 25: 36. (In Slovak).

Holčík, J. 1973. The living fossil fish. Vesmír 52: 247. (In Slovak).

Holčík, J. 1990. New data on *Latimeria chalumnae*. Živa 38: 29–32. (In Slovak).

Holmes, E.B. 1985. Are lungfishes the sister group of tetrapods. Biol. J. Linn. Soc. 25: 379–397.

Honma, Y., E. Tamura & A. Chiba. 1984. Some aspects of organs and tissues of the coelacanth with special regard to the surface structure of the third ventricular choroid plexus. pp. 24–25. *In:* Proceedings of the First Symposium on Coelacanth Studies, Tokyo, 1984. (In Japanese).

Hopson, J.L. 1976. Fins to feet to fanclubs: an (old) fish story. Science News 109 (2): 28–30.

Hughes, G.M. 1972. Aspects of the respiration of *Latimeria chalumnae*. Proc. int. Union physiol. Sci., Sidney 1972: 136. (abstract).

Hughes, G.M. 1972. Gills of a living coelacanth, *Latimeria chalumnae*. Experientia 28: 1301–1302.

Hughes, G.M. 1976. On the respiration of *Latimeria chalumnae*. Zool. J. Linn. Soc. 59: 195–208.

Hughes, G.M. 1980. Ultrastructure and morphometry of the gills of *Latimeria chalumnae*, and a comparison with the gills of associated fishes. Proc. R. Soc. Lond. B 208: 309–328.

Hughes, G.M. & Y. Itazawa. 1972. The effect of temperature on the respiratory function of coelacanth blood. Experientia 28: 1247.

Hureau, J.-C. & C. Ozouf. 1977. Détermination de l'âge et croissance du coelacanthe *Latimeria chalumnae* Smith, 1939 (poisson, crossoptérygien, coelacanthide). Cybium Sér. 3, 2: 129–137.

Idyll, C.P. 1965. Living fossils of the deep sea. Sea Frontiers 11: 178–187.

Ihle, J.E.W. & M.E. Ihle-Landenberg. 1948. New research on *Latimeria chalumnae*. Vakbl. Biol. (Helder) 28: 65–71. (In Dutch).

Imai, F., H. Kameda & N. Suzuki. 1986. High deposition of fat in the liver of *Latimeria* – comparisons with human fatty liver. Nippon Iji Sinpou (Japan Medical News) 3252: 43–47. (In Japanese).

Imai, F., I. Takagi, K. Yamazaki & H. Kameda. 1984. Characteristics of the fat deposit in the liver of coelacanth. pp. 14–15. *In:* Proceedings of the First Symposium on Coelacanth Studies, Tokyo. (In Japanese).

Imaki, H. & W. Chavin. 1973. Ultrastructure of integumental melanophores in the coelacanth. Amer. Zool. 13: 1348. (abstract).

Imaki, H. & W. Chavin. 1984. Ultrastructure of mucous cells in the sarcopterygian integument. Scanning Electron Microsc. 1984: 409–422.

Isokawa, S., Y. Toda & K. Kubota. 1968. A histological observation of a coelacanth *Latimeria chalumnae*. J. Nihon Univ. Sch. Dent. 10: 102–114.

Itazawa, Y. 1984. Respiration physiology of the coelacanth. p. 18. *In:* Proceedings of the First Symposium on Coelacanth Studies, Tokyo. (In Japanese).

Jain, S.L. 1974. *Indocoelacanthus robustus* n. gen., n. sp. (Coelacanthidae, Lower Jurassic), the first fossil coelacanth from India. J. Paleont. 48(1): 49–62.

Jarvik, E. 1972. Middle and Upper Devonian Porolepiformes from East Greenland with special reference to *Glyptolepis groenlandica* n. sp. and a discussion of the structure of the head in the Porolepiformes. Meddelelser om Grønland 187: 1–295.

Jarvik, E. 1980. Basic structure and evolution of vertebrates. Volume 1, Academic Press, London. 575 pp.

Jauregui-Adell, J. & J.-F. Pechère. 1978. Parvalbumins from coelacanth muscle. I. General survey. Biochim. biophys. Acta 536: 263–268.

Jauregui-Adell, J. & J.-F. Pechère. 1978. Parvalbumins from coelacanth muscle. III. Amino-acid sequence of the major component. Biochim. biophys. Acta 536: 275–282.

Jessen, H. 1966. Die Crossopterygier des Oberen Plattenkalkes (Devon) der Bergisch-Gladbach-Paffrather Mulde (Rheinisches Schiefergebirge) unter Berücksichtigung von amerikanischem und europäischem *Onychodus* – Material. Arkiv Zoologi, sér. 2, 18(14): 305–389.

Jollie, M. 1986. A primer of bone names for the understanding of the actinopterygian head and pectoral girdle skeletons. Can. J. Zool. 64: 365–379.

Joubert, J. 1987. Couple's work is a monument to study of fish. Weekend Post, 12 September 1987: 10.

Kabasawa, H. 1981. *Latimeria chalumnae*, a frozen specimen on display. Bien. Rep. Keikyu Aburat. mar. Park Aquar. 11: 37–38.

Kähsbauer, P. 1960. Ein lebendiger Fossil Fisch. Naturhistorisches Museum in Wien, Flugblatt 13: 1–2.

Kamegai, S. 1971. On some parasites of a coelacanth *Latimeria chalumnae*: a new monogenea *Dactylodiscus latimeris* new genus, new species (Dactylodiscidae new family) and two larval helminths. Res. Bull. Meguro parasitol. Mus. 5: 1–5.

Kedrova, O.S., N.S. Vladychenskaya & A.S. Antonov. 1983.

422

Molecular hybridization of alligator gar DNA with some fish DNA. Mol. Biol. (Mosc.) 17: 383–391.

Khalaf, N.-A.B. 1987. The coelacanth, *Latimeria chalumnae*, in the Science and Natural History Museum, State of Kuwait. Gazelle 15: 1–7.

Kihira, K., Y. Akashi, S. Kuroki, J. Yanagisawa, F. Nakayama & T. Hoshita. 1984. Bile salts of the coelacanth, *Latimeria chalumnae*. J. Lipid Res. 25: 1330–1336.

Kihira, K., F. Nakayama & T. Hoshita. 1984. Bile salts of the coelacanth, *Latimeria chalumnae*. p. 16. *In:* Proceedings of the First Symposium on Coelacanth Studies, Tokyo. (In Japanese).

Klotchkoff, J.-C. 1989. Les Comores aujourd'hui. Jaguar, Paris. 239 pp.

Kolb, E. & J.I. Harris. 1972. Purification and properties of glycolytic-cycle enzymes from coelacanth (*Latimeria chalumnae*) muscle. Biochem. J. 130: 26. (abstract).

Kolb, E., J.I. Harris & J. Bridgen. 1974. Triose phosphate isomerase EC-5.3.1.1 from the coelacanth. An approach to the rapid determination of an amino-acid sequence with small amounts of material. Biochem. J. 137: 185–197.

Kremers, J.-W.P.M. 1975. The structure of the brain stem of *Latimeria chalumnae*. Acta morphol. 13: 307. (abstract).

Kremers, J.-W.P.M. 1981. The brain of the crossopterygian fish *Latimeria chalumnae* Smith. Ph.D. Thesis, University of Nijmegen, Nijmegen. 125 pp.

Kremers, J.-W.P.M. & R. Nieuwenhuys. 1979. Topological analysis of the brain stem of the crossopterygian fish *Latimeria chalumnae*. J. comp. Neurol. 187: 613–637.

Krishnaswami, S. 1953. The coelacanth. The Hindu, 4 January 1953.

Kumar, P. 1960. The hundred pound fish. The Illustrated Weekly of India, 3 April 1960: 51.

Lagios, M.D. 1972. A partial report and appraisal of participation in Expédition Coelacanthé for the period 23 February through to 13 March 1972. Internal report to the Armed Forces Institute of Pathology, Washington D.C. 14 pp.

Lagios, M.D. 1972. Evidence for a hypothalamo-hypophysial portal vascular system in the coelacanth *Latimeria chalumnae* Smith. Gen. comp. Endocrinol. 18: 73–82.

Lagios, M.D. 1974. Granular epithelioid (juxtaglomerular) cell and renovascular morphology of the coelacanth *Latimeria chalumnae* Smith (Crossopterygii) compared with that of other fishes. Gen. comp. Endocrinol. 22: 296–307.

Lagios, M.D. 1975. The pituitary gland of the coelacanth *Latimeria chalumnae* Smith. Gen. comp. Endocrinol. 25: 126–146.

Lagios, M.D. 1979. The coelacanth and the Chondrichthyes as sister groups: a review of shared apomorph characters and a cladistic analysis and reinterpretation. Occ. Pap. Calif. Acad. Sci. 134: 25–44.

Lagios, M.D. 1979. Reply to the rebuttal of Leonard Compagno. 'Coelacanths: shark relatives or bony fishes'. Occ. Pap. Calif. Acad. Sci. 134: 53–55.

Lagios, M.D. 1982. *Latimeria* and the Chondrichthyes as sister taxa: a rebuttal to recent attempts at refutation. Copeia 1982: 942–948.

Lagios, M.D. & S. Stasko-Concannon. 1979. Presumptive interrenal tissue (adrenocortical homolog) of the coelacanth *Latimeria chalumnae*. Gen. comp. Endocrinol. 37: 404–406.

Laing, M. 1989. Where are our wonderful fossils? Spectrum 27: 25–35.

Lamaud, P. 1985. Le coelacanthe, un fossile vivant. Minéraux Fossiles 116: 14–17.

Lamer, H.I. & W. Chavin. 1975. Ultrastructure of the integumental melanophores of the coelacanth *Latimeria chalumnae*. Cell Tissue Res. 163: 383–394.

Lamont-Brown, R. 1988. The missing link. Practical Fishkeeping, December 1988: 86–87.

Lange, R.H. 1983. Les cristaux de lipovitelline-phosvitine dans l'ovocyte de *Latimeria chalumnae* Smith, 1939 (Coelacanthidae, Pisces). Étude comparative. C.R. Acad. Sci., Paris, Sér. III 297: 393–396.

Langen, R.B. 1987. Visiting ancestor's cousins in a submarine. Crossopterygians filmed for the first time, deep-sea expedition by night. German Research Service 26: 7–8. (In German).

Lauder, G.V., Jr. 1979. (Review of) Anatomie de *Latimeria chalumnae*, Tome III, by J. Millot, J. Anthony & D. Robineau. Copeia 1979: 560–562.

Lauder, G.V., Jr. 1980. (Review of) The biology and physiology of the living coelacanth. Edited by J.E. McCosker & M.D. Lagios. Copeia 1980: 942–944.

Lauder, G.V., Jr. 1980. The role of the hyoid apparatus in the feeding mechanism of the coelacanth *Latimeria chalumnae*. Copeia 1980: 1–9.

Lawsky, D. 1975. Society for protection of old fish. Iberian Sun, 17 September 1975: 5.

Légendre, R. 1980. Jacques Millot (1897–1980). Bull. Soc. zool. France 105: 473–480.

Lehmann, J.-P. 1952. Étude complementaire des poissons de l'eotrias de Madagascar. Handl. K. Svenska Vetensk. Akad. 2: 1–201.

Lehmann, U. 1953. Der zweite rezente Crossopterygier (Quastenflosser). Naturw. Rundsch. 5: 192–193.

Lehmann, U. 1976. Neues über *Latimeria*. Naturw. Rundsch. 1976: 275–276.

Lemire, M. 1970. Étude cytoarchitectonique du rhombencéphale de *Latimeria chalumnae*, Poisson Crossoptérygien, Coelacanthidé. C.R. Acad. Sci. Paris, Sér. D 271: 1994–1997.

Lemire, M. 1971. Étude architectonique du rhombencéphale de *Latimeria chalumnae* Smith, (Poisson, Crossoptérygien, Coelacanthidé). Bull. Mus. Natl. Hist. Nat., Paris, Sér. 3, Zool. 2: 41–95.

Lemire, M. 1971. Étude myéloarchitectonique du rhombencéphale de *Latimeria chalumnae* Smith, Poisson, Crossoptérygien, Coelacanthidé. C.R. Acad. Sci. Paris, Sér. D 272: 425–428.

Lemire, M. 1976. Caractéristiques ultrastructurales des cellules acineuses de la glande post-anale du coelacanthe (*Latimeria chalumnae* Smith). C.R. Acad. Sci. Paris, Sér. D 282: 641–644.

Lemire, M. 1976. Histophysiologie comparée des voies d'excrétion chez les lézards deserticoles. Bull. Soc. zool. France 101: 982.

Lemire, M. 1977. Étude ultrastructurale du parenchyme sécréteur de la glande post-anale du coelacanthe *Latimeria chalumnae* Smith. Ann. Sci. nat. Zool. Sér. 12, 19: 227–245.

Lemire, M. & M. Lagios. 1979. Ultrastructure du parenchyme sécréteur de la glande post-anale de coelacanthe *Latimeria chalumnae* (Smith). Acta Anat. 104: 1–15.

Lenoble, J. & Y. Le Grand. 1954. Le tapis de l'oeil du coelacanthe (*Latimeria anjouanae* (Smith)). Bull. Mus. Natl. Hist. Nat. (Paris) Sér. 2, 26: 460–463.

Liem, K.F. 1968. Coelacanth. Bull. Field Mus. Nat. Hist. 39 (2): 3.

Lizmore, G. 1988. Coelacanth: East London left out in the cold. Coastal Profile, 4 March 1988: 2.

Locket, N.A. 1972. Learning from an anachronistic fish. New Scientist 54: 427–428.

Locket, N.A. 1973. Possible discontinuous retinal rod outer segment formation in *Latimeria chalumnae*. Nature 244: 308–309.

Locket, N.A. 1973. Retinal structure in *Latimeria chalumnae*. Phil. Trans. R. Soc. Lond. B 266: 493–521.

Locket, N.A. 1974. The choroidal tapetum lucidum of *Latimeria chalumnae*. Proc. R. Soc. Lond. B 186: 281–290.

Locket, N.A. 1976. A future for the coelacanth? New Scientist 70: 456–458.

Locket, N.A. 1980. Some advances in coelacanth biology. Proc. R. Soc. Lond. B 208: 265–307.

Locket, N.A. & R.W. Griffith. 1972. Observations on a living coelacanth. Nature 237: 175.

Lombardini, J.B., P.K.T. Pang & R.W. Griffith. 1979. Amino acids and taurine in intracellular osmoregulation in marine animals. Occ. Pap. Calif. Acad. Sci. 134: 160–169.

Long, A. 1975. Living fossils. The Sea 2: 436–438.

Long, J. 1988. The extraordinary fishes of Gogo. New Scientist, 19 November 1988: 40–44.

Long, J.A. 1989. A new rhizodontiform fish from the early Carboniferous of Victoria, Australia, with remarks on the phylogenetic position of the group. J. Vertebr. Paleont. 9: 1–17.

Løvtrup, S. 1977. The phylogeny of vertebrata. John Wiley & Sons, London. 330 pp.

Løvtrup, S. 1987. Darwinism: the refutation of a myth. Croom Helm, London. 469 pp.

Løvtrup, S. 1989. On divergent and progressive evolution. pp. 55–69. *In:* M.N. Bruton (ed.) Alternative Life-History Styles of Animals, Perspectives in Vertebrate Science 6, Kluwer Academic Publishers, Dordrecht.

Lozano Cabo, F. 1970. Ichthyological problems of the continental plateau of Northwest Africa. Rapp. P.-v. Réun. Cons. perm. int. Expl. Mer 159: 149–151.

Lund, R. & W.L. Lund. 1985. Coelacanths from the Bear Gulch Limestone (Namurian) of Montana and the evolution of the Coelacanthiformes. Bull. Carnegie Museum of Natural History 25: 1–74.

Lundberg, J. & E. Marsh. 1976. Evolution and functional anatomy of the pectoral fin rays in cyprinoid fishes, with emphasis on the suckers (family Catostomidae). Amer. Midl. Nat. Monogr. 96: 332–349.

Lutz, P.L. & J.D. Robertson. 1971. Osmotic constituents of the coelacanth *Latimeria chalumnae* Smith. Biol. Bull. (Woods Hole) 141: 553–560.

Lythgoe, G. 1979. Old four legs. Wildlife, Lond. 21 (11): 20–21.

Lythgoe, G. 1980. Coelacanth encounter. Skin Diver 29: 8–11.

MacKenzie, D. 1987. 'Living fossils' come alive on film. New Scientist, 12 February 1987: 20.

Maisey, J.G. 1986. Coelacanths from the Lower Cretaceous of Brazil. Amer. Mus. Novitates 2866: 1–30.

Maisey, J.G. 1987. Notes on the structure and phylogeny of vertebrate otoliths. Copeia 1987: 495–499.

Maizy, P. 1963. The coelacanth. Bulletin de Madagascar 209: 897–899.

Malherbe, E.G. 1981. Never a dull moment. Howard Timmins, Cape Town. 419 pp.

Mangum, C.P. 1991. Urea and chloride sensitivities of coelacanth hemoglobin. Env. Biol. Fish. 32: 219–222. (this volume)

Martin, M. & S. Wenz. 1984. Découverte d'une nouveau coelacanthide, *Garnbergia ommata* n.g. n.sp., dans le Muschelkalk supérieur du Baden-Württemberg. Stuttgarter Beitr. Naturk. B 105: 1–17.

Martin, R. 1970. D'utiles details sur le coelacanthe. Promo. Al'Camar 16: 25–35.

Mattei, X., Y. Siau & B. Seret. 1988. Étude ultrastructurale du spermatozoide du coelacanthe: *Latimeria chalumnae*. J. Ultrastruct. Mol. Struct. Res. 101: 243–251.

Matthieu, J.-M., N. Eschmann, P. Burgisser, J. Malotka & T.V. Waehneldt. 1986. Expression of myelin proteins characteristic of fish and tetrapods by *Polypterus* revitalises long discredited phylogenetic links. Brain Research 379: 137–142.

McAllister, D.E. 1968. The evolution of branchiostegals and classification of teleostome fishes living and fossil. National Museum of Canada, Bulletin 221, Biol. Series 77: 1–239.

McAllister, D.E. 1971. Old fourlegs: a 'living fossil'. Natl. Mus. nat. Sci., Odyssey Sér. 1: 1–25.

McAllister, D.E. 1971. Le vieux quadrupède. Mus. natl. Sci. nat. Coll. Odyssée Ser. 1: 1–25.

McAllister, D E. 1976. (Review of) Opération coelacanthe, by J. Anthony. 1976. Can. Field-Naturalist 90: 511–512.

McAllister, D.E. 1976 (Revue de) Opération coelacanthe, by J. Anthony. 1976. Naturaliste canadien 103: 491–492.

McAllister, D E. 1988. Old fourlegs, the coelacanth. Sea Wind 2 (4): 6–9.

McAllister, D E. & C. Delisle. 1968. Le coelacanthe, un fossile vivant. Le Jeune Scientifique 7 (2): 30–31.

McAllister, D E. & C.L. Smith. 1978. Mensurations morphologiques, dénombrements méristiques et taxonomie du coelacanthe, *Latimeria chalumnae*. Naturaliste canadien 105: 63–76.

424

McAllister, J.A. 1987. Phylogenetic distribution and morphological reassessment of the intestines of fossil and modern fishes. Zool. Jahrb. Anat. 115: 281–294.

McCosker, J.E. 1975. Report of the expedition to the Comores. Spoof Newsletter 14: 1–2.

McCosker, J.E. 1979. Inferred natural history of the living coelacanth, *Latimeria chalumnae*. Occ. Pap. Calif. Acad. Sci. 134: 17–24.

McCosker, J.E. & M.D. Lagios. (ed.). 1979. The biology and physiology of the living coelacanth. Occ. Pap. Calif. Acad. Sci. 134: 1–175.

McCosker, S. & J.E. McCosker. 1976. To the islands of the moon. Pacific Discovery 29 (1): 19–28.

Mednikov, B.M. & A.S. Antonov. 1974. Status of lungfishes (Dipnoi) and their systematic position. Dokl. Acad. Nauk USSR 218: 474–476. (In Russian).

Mee-Mann, C. 1982. The braincase of *Youngolepis,* a Lower Devonian crossopterygian from Yunnan, south-western China. Dept. Geology, University of Stockholm, Stockholm. 113 pp.

Meinke, D.K. 1982. A light and scanning electron microscope study on the dermal skeleton of *Spermatodus pustulosus* (Pisces: Coelacanthini) and the evolution of the dermal skeleton in coelacanths. J. Paleont. 56: 620–630.

Meinke, D.K. 1984. A review of cosmine: its structure, development, and relationship to other forms of the dermal skeleton in osteichthyans. J. Vert. Paleont. 4: 457–470.

Meinke, D.K. 1987. Morphology and evolution of the dermal skeleton in lungfishes. pp: 133–149. *In:* W.E. Bemis, W.W. Burggren & N.E. Kemp (ed.) The Biology and Evolution of Lungfishes, Alan R. Liss, New York.

Meinke, D.K. & K.S. Thomson. 1983. The distribution and significance of enamel and enameloid in the dermal skeleton of osteolepiform rhipidistian fishes. Paleobiology 9: 138–149.

Melmed, R.N. & S.J. Holt. 1975. Ultrastructural observations on exocrine and endocrine pancreatic cells of the coelacanth *Latimeria chalumnae* Smith. Israel J. med. Sci. 11: 405–406.

Menache, M. 1954. Étude hydrogéologique sommaire de la région d'Anjouan en rapport avec la pêche de trois coelacanthes. Mém. Inst. scient. Madagascar A 9: 151–185.

Merle, R. 1953. Le coelacanthide d'Anjouan. Nature, Paris 1953: 18–20.

Meunier, F.-J. 1980. Les relations isopédine – tissu osseux dans le post-temporal et les écailles de la ligne latérale de *Latimeria chalumnae* Smith. Zoologica Scripta 9: 307–317.

Meunier, F.-J. 1984. Spatial organization and mineralization of the basal plate of elasmoid scales in osteichthyans. Amer. Zool. 24: 953–964.

Meunier, F.-J., H. Francillon & J. Castanet. 1975. Histology of the teeth, scales and lepidotriches of the coelacanth. Bull. Soc. Zool. France 100: 257–258.

Meunier, F.-J. & Y. François. 1980. L'Organisation spatiale des fibres collagènes et al minéralisation des écailles des dipneustes actuels. Bull. Soc. Zool. France 105: 215–226.

Meunier, F.-J. & J. Geraudie. 1980. Les structures en contre-

plaque du derme et des écailles des vertébrés inferieurs. Année biol. 19: 1–18.

Milano, E.G. & F. Accordi. 1986. Evolution trends in adrenal gland of anurans and urodeles. J. Morphol. 189: 249–259.

Miller, W.A. 1969. Tooth enamel of *Latimeria chalumnae* Smith. Nature 221: 1244.

Miller, W.A. 1979. Observations on the structure of mineralized tissues of the coelacanth, including the scales and their associated odontodes. Occ. Pap. Calif. Acad. Sci. 134: 68–78.

Miller, W.A. & M.H. Hobdell. 1968. Preliminary report on the histology of the dental and paradental tissues of *Latimeria chalumnae* (Smith) with a note on tooth replacement. Arch. Oral Biol. 13: 1289–1291.

Millot, J. 1953. 'Notre' coelacanthe. Revue Madagascar, Tananarive 17: 18–20.

Millot, J. 1954. Studying the coelacanth. Incomparable witness of distant era of living world. Calcutta Statesman, 8 June 1954.

Millot, J. 1954. Coelacanth under the microscope. Times Weekly Review, London, 10 June 1954.

Millot, J. 1954. Ancestor of man. Scientific study of the coelacanth. Rhodesia Herald, 11 June 1954.

Millot, J. 1954. Further study of the coelacanth. Remains of pulmonary organ. The Times, 3 June 1954: 7.

Millot, J. 1954. Les nouveaux coelacanthes. Nature, Paris 82 (3228): 121–123.

Millot, J. 1954. New facts about coelacanths. Nature 174: 426–427.

Millot, J. 1954. The second coelacanth. Naturaliste malgache 32: 49–52.

Millot, J. 1954. Le coelacanthe, poisson fossile vivant. Science et Nature 2: 13–15.

Millot, J. 1954. Le troisième coelacanthe, historique, éléments d'écologie morphologie externe, documents divers. Naturaliste malgache, suppl. 1: 1–26.

Millot, J. 1955. A propos des coelacanthes. Résponse du professeur J. Millot. Nature, Paris 83 (3241): 202–203.

Millot, J. 1955. First observations on a living coelacanth. Nature 175: 362–363.

Millot, J. 1955. La première femele adulte du coelacanthe. Nature, Paris 83 (3241): 203.

Millot, J. 1955. The coelacanth. Sci. Amer. 193 (6): 34–39.

Millot, J. 1955. Unité spécifique des coelacanthes actuels. Nature, Paris 83 (3238): 58–59.

Millot, J. 1956. A reply to letter by J.L.B. Smith on 'conservation of coelacanths.' The Times, London, 14 June 1956.

Millot, J. 1964. Le diencéphale de *Latimeria chalumnae* Smith (Poisson coelacanthidae). C.R. Acad. Sci. Paris, Sér. D 258: 5051–5055.

Millot, J. & J. Anthony. 1954. Tubes rostraux et tubes nasaux de *Latimeria* (Coelacanthidé). C.R. Acad. Sci. Paris Sér. D 239: 1241–1243.

Millot, J. & J. Anthony. 1955. Considérations physiomorphologiques sur la tête de *Latimeria* (crossoptérygien coelacanthidé). C.R. Acad. Sci. Paris, Sér. D 241: 114–116.

Millot, J. & J. Anthony. 1955. L'articulation intracrânienne de

Latimeria (Coelacanthide). C.R. VIe Congr. Fed. Int. Anat. (Paris): 161–162.

Millot, J. & J. Anthony. 1955. Les canaux sensoriels de la tête chez Latimeria (Coelacanthidae). C.R. VIe Congr. Fed. Int. Anat. (Paris): 161–162.

Millot, J. & J. Anthony. 1956. Considérations préliminaires sur le squelette axial et le système nerveux central de Latimeria chalumnae Smith. Mém. Inst. scient. Madagascar, Sér A, 11: 167–188.

Millot, J. & J. Anthony. 1956. L'organe rostral de Latimeria (Crossoptérygien, Coelacanthidé). Annal. Sci. nat. Zool. Sér. 11, 28: 381–389.

Millot, J. & J. Anthony. 1956. Note préliminaire sur le thymus et la glande thyroide de Latimeria chalumnae (Crossoptérygien coelacanthidé). C.R. Acad. Sci. Paris, Sér. D 242: 560–562.

Millot, J. & J. Anthony. 1958. De l'existence chez Latimeria chalumnae Smith (Coelacanthidae), d'un organe regulateur du courant sanguin supra-branchial. C.R. Acad. Sci. Paris, Sér. D 246: 1600–1602.

Millot, J. & J. Anthony. 1958. Anatomie de Latimeria chalumnae. Tome 1. Squelette, muscles et formations de soutien. C.N.R.S., Paris. 122 pp.

Millot, J. & J. Anthony. 1958. Crossoptérygiens actuels Latimeria chalumnae, dernier des crossoptérygiens. pp. 2553–2597. In: P. Grassé (ed.) Traité de Zoologie 13, Masson, Paris.

Millot, J. & J. Anthony. 1958. Résultats actuels de l'étude du coelacanthe Latimeria chalumnae Smith, dernier des Crossoptérygiens. XV Int. Congr. Zool., London 15: 36–44.

Millot, J. & J. Anthony. 1959. Les neuromastes du système latéral de Latimeria chalumnae. Annal. Sci. nat. Zool. Sér. 12 (1): 317–328.

Millot, J. & J. Anthony. 1960. Appareil génital et reproduction des coelacanthes. C.R. Acad. Sci. Paris, Sér. D 251: 442–443.

Millot, J. & J. Anthony. 1960. Laboratoire d'anatomie comparée du Muséum d'Histoire Naturelle, Paris, France. Science (Paris) 6: 7–20.

Millot, J. & J. Anthony. 1960. Le cloaque chez les coelacanthes. Bull. Mus. Natl. Hist. Nat., Paris, Sér. 2, 32: 287–289.

Millot, J. & J. Anthony. 1960. Le plus vieux poisson du monde. Revue française des sciences et des techniques 6: 7–20.

Millot, J. & J. Anthony. 1960. Un nouvel aspect du coelacanthe: le montage complet de son squelette. Science et Nature 37: 1–3.

Millot, J. & J. Anthony. 1962. Premières précisions sur l'organisation du télencéphale chez Latimeria chalumnae (poisson crossoptérygien, coelacanthidé). C.R. Acad. Sci. Paris, Sér. D 254: 2067–2068.

Millot, J. & J. Anthony. 1965. Anatomie de Latimeria chalumnae, Tome II, système nerveux et organes des sens. C.N.R.S., Paris. 131 pp.

Millot, J. & J. Anthony. 1967. L'organisation générale du prosencéphale de Latimeria chalumnae Smith (poisson crossoptérygien coelacanthidé). pp. 50–60. In: R. Hassler & H. Stephan (ed.) Evolution of the Forebrain, Plenum, New York.

Millot, J. & J. Anthony. 1972. La glande post-anale de Latimeria. Annal. Sci. nat. Zool., Sér. 12, 14: 305–317.

Millot, J. & J. Anthony. 1972. Le pancréas des crossoptérygiens coelacanthidés. Z. Zellforsch. mikrosk. Anat. 123: 215–223.

Millot, J. & J. Anthony. 1973. L'appareil excréteur de Latimeria chalumnae Smith (poisson coelacanthidé). Annal. Sci. nat. Zool., Sér. 12, 15: 293–328.

Millot, J. & J. Anthony. 1973. La position ventrale du rein de Latimeria chalumnae (poisson coelacanthidé). C.R. Acad. Sci. Paris. Sér. D 276: 2171–2173.

Millot, J. & J. Anthony. 1973. Variations sexuelles de l'appareil excréteur du coelacanthe, Latimeria chalumnae. Connexions avec l'appareil génital mâle. C.R. Acad. Sci., Paris, Sér. D 276: 2447–2448.

Millot, J. & J. Anthony. 1974. Les œufs du coelacanthe. Science et Nature 121: 3–4.

Millot, J., J. Anthony & D. Robineau. 1972. État commente des captures de Latimeria chalumnae Smith (poisson crossoptérygien, coelacanthidé) effectuées jusqu'au mois d'octobre 1971. Bull. Mus. Natl. Hist. Nat., Paris, Sér. 3, no. 53, Zoologie 39: 533–548.

Millot, J., J. Anthony & D. Robineau. 1978. Anatomie de Latimeria chalumnae, Tome 3, Appareil digestif, appareil respiratoire, appareil urogénital, glandes endocrines, appareil circulatoire, téguments, écailles, conclusions générales. C.N.R.S., Paris. 198 pp.

Millot, J. & N. Carasso. 1955. Note préliminaire sur l'œil de Latimeria chalumnae (crossoptérygien coelacanthidé). C.R. Acad. Sci. Paris, Sér. D 241: 576–577.

Millot, J., R. Nieuwenhuys & J. Anthony. 1964. Le diencéphale de Latimeria chalumnae Smith (poisson coelacanthidé). C.R. Acad. Sci. Paris, Sér. D 258: 5051–5055.

Millot, J. & A. Policard. 1955. Sur la structure inframicroscopique du tissue conjonctif du coelacanthe. Bull. Microsc. appl. 5: 94–95.

Millot, J. & O. Tuzet. 1959. La spermatogenèse de Latimeria chalumnae Smith (crossoptérygien coelacanthidé). Annal. Sci. nat. Zool., Sér. 12, 1: 61–69.

Mills, G.L. & C.E. Taylaur. 1973. The distribution and composition of serum lipoproteins in the coelacanth (Latimeria). Comp. Biochem. Physiol. 44B: 1235–1241.

Mok, H.K. 1981. The posterior cardinal veins and kidneys of fishes, with notes on their phylogenetic significance. Jap. J. Ichthyol. 27: 281–290.

Mommsen, T.P. & P.J. Walsh. 1989. Evolution of urea synthesis in vertebrates: the piscine connection. Science 243: 72–75.

Monastersky, R. 1987. 'Living fossils' display unusual behaviour. Science News 132(14): 213.

Monod, T. 1954. Sur une larve de gnathiide (Praniza milloti nov. sp.), parasite du Latimeria chalumnae (coelacanthe). Mem. Inst. scient. Madagascar, Sér. A 9: 91–94.

Morris, E. & A. Morris. 1973. In pursuit of the coelacanth. Pacific Discovery 26 (2): 29–32.

Mouton, P. 1990. A la decouverte des derniers coelacanthes. Le Monde de la Mer 48: 32–39 (with photographs by H. Fricke and R. Plante).

426

Munnion, C. 1987. Fighting to save 'old four legs'. Daily Telegraph, 7 September 1987: 11.

Munnion, C. 1988. Remembering old four legs. Optima 36: 42–51.

Munro, I.W.T. 1953. Out in the open, coelacanth queries. Christchurch Star, New Zealand, 17 March 1953.

Murdoch, W. 1989. Uncrowned king of the Comores. Sunday Times Magazine, 23 July 1989.

Musel, R. 1953. The coelacanth expert finds little spectacular in much publicised rare fish. Montreal Daily Star, 2 February 1953.

Musick, J.A., M.N. Bruton & E.K. Balon. 1991. Introduction: the recent chronology and contributions. Env. Biol. Fish. 32: 15–19. (this volume)

Musick, J.A., M.N. Bruton & E.K. Balon (ed.) 1991. The biology of Latimeria chalumnae and evolution of coelacanths. Dev. Env. Biol. Fish. 12, Kluwer Academic Publishers, Dordrecht. 446 pp.

Myers, G.S. 1953. The living-fossil coelacanth fishes. Aquarium J. 24: 66–68.

Myking, L.M. 1977. Old four legs: the living fossil. Sea Frontiers 23: 334–341.

Nelson, G.J. 1969. Gill arches and the phylogeny of fishes, with notes on the classification of vertebrates. Bull. Amer. Mus. Nat. Hist. 141: 475–552.

Nelson, G.J. 1970. Subcephalic muscles and intracranial joints of sarcopterygian and other fishes. Copeia 1970: 468–471.

Nelson, G.J. 1973. Relationships of clupeomorphs, with remarks on the structure of the lower jaw in fishes. pp. 333–349. In: P.H. Greenwood, R.S. Miles & C. Patterson (ed.) Interrelationships of Fishes, Academic Press, London.

Nelson, J.S. 1984. Fishes of the world. 2nd ed. John Wiley & Sons, New York. 523 pp.

Nevenzel, J.C., W. Rodegker, J.F. Mead & M.S. Gordon. 1966. Lipids of the living coelacanth, Latimeria chalumnae. Science 152: 1753–1755.

Newman, C. 1988. The Comores – a fishing frontier. pp. 87–92. In: C. Norman (ed.) African Angler, Southern Book Publishers, Johannesburg.

Newman, M. 1971. Search for a coelacanth. Vancouver Public Aquarium Newsletter 15 (6): 5 pp. (unnumbered).

Nichols, H.A. 1952. Latimeria the 'living fossil'. Fish Culturist 32: 1–3.

Nichols, H.A. 1953. The second coelacanth. Fish Culturist 32: 49–56.

Nichols, H.A. 1954. The third coelacanth. Fish Culturist 33: 49–51.

Nichols, H.A. 1954. Naming the coelacanth. Wat. Life Aquar. World, April 1954.

Nichols, H.A. 1955. First female coelacanth caught. Fish Culturist 34: 58.

Nielsen, E. 1954. Living fossil fish waits for mother and wife. Vor viden. 122: 545–553. (In Danish).

Nieuwenhuys, R. 1962. The forebrain of some groups of fishes. Anat. Rec. 142: 262. (abstract).

Nieuwenhuys, R. 1965. The forebrain of the crossopterygian fish Latimeria chalumnae Smith. J. Morphol. 117: 1–23.

Nieuwenhuys, R. 1975. Topological analysis of the brain stem. A general introduction. J. comp. Neurol. 156: 255–276.

Nieuwenhuys, R., J.-W.P.M. Kremers & C. van Huijzen. 1975. The brain of the crossopterygian fish Latimeria chalumnae. Acta Morph. neerl.-scand. 13: 306. (abstract).

Nieuwenhuys, R., J.-P.M. Kremers & C. van Huijzen. 1977. The brain of the crossopterygian fish Latimeria chalumnae: a survey of its gross structure. Anat. Embryol. 151: 157–169.

Nishimura, H. & M. Ogawa. 1973. The renin-angiotensin system in fishes. Amer. Zool. 13: 823–838.

Nishimura, H., M. Ogawa & W.H. Sawyer. 1973. Renin-angiotensin system in primitive bony fishes and a holocephalian. Amer. J. Physiol. 224: 950–956.

Norman, J.R. 1939. An account, illustrated with lantern-slides, of 'A living coelacanth from South Africa; a fish believed to have been long extinct'. Proc. Linn. Soc. Lond., 151st Session, 16 March 1939: 142–145. (abstract).

Norman, J.R. 1975. A history of fishes, 3rd edition, revised by P.H. Greenwood. Ernest Benn, London. 467 pp.

Northcutt, R.G. 1977. Retino fugal projections in the lepidosirenid lungfishes. J. comp. Neurol. 174: 553–574.

Northcutt, R.G. 1980. Central auditory pathways in anamniotic vertebrates. pp. 79–118. In: A.N. Popper & R.R. Fay (ed.) Comparative Studies of Hearing in Vertebrates, Springer-Verlag, New York.

Northcutt, R.G. 1981. Anatomical evidence of electroreception in the coelacanth, Latimeria chalumnae. Anat. Histol. Embryol. 9: 289–295.

Northcutt, R.G. 1987. Lungfish neural characters and their bearing on sarcopterygian phylogeny. pp. 277–297. In: W.E. Bemis, W.W. Burgren & N.E. Kemp (ed.) The Biology and Evolution of Lungfishes, Alan R. Liss, New York.

Northcutt, R.G. & T.J. Neary. 1975. Observations on the optic tectum of the coelacanth Latimeria chalumnae. Amer. Zool. 15: 806.

Northcutt, R.G., T.J. Neary & D.G. Senn. 1978. Observations on the brain of the coelacanth Latimeria chalumnae: external anatomy and quantitative analysis. J. Morphol. 155: 181–192.

Ogiso, T. 1986. Island of the coelacanth. Expert 71: 6–10. (In Japanese).

O'Reilly, J. 1952. Air race to save dead fish stirs scientists here. New York Herald Tribune, 30 December 1952.

Otterstal, K. 1989. Re-evaluation of tasselfin (coelacanth)! Forskning & Framsteg 3/89: 409. (In Swedish).

Panchen, A.L. & T.R. Smithson. 1987. Character diagnosis, fossils and the origin of tetrapods. Biol. Rcv. 62: 341–438.

Panchen, A.L. & T.R. Smithson. 1988. The relationships of the earliest tetrapods. pp. 1–32. In: M.J. Berton (ed.) The Phylogeny and Classification of the Tetrapods, Vol. 1, Amphibians, Reptiles, Birds, Clarendon Press, Oxford.

Panchen, A.L. & T.R. Smithson. 1990. The pelvic girdle and hind limb of Crassigyrinus scoticus (Lydekker) from the Scottish Carboniferous and the origin of the tetrapod pelvic skeleton. Trans. R. Soc. Edinburgh: Earth Sciences 81: 31–44.

Pang, P.K.T. & A. Epple. (ed.). 1980. Evolution of vertebrate endocrine systems. Graduate Stud., Texas Tech. Univ. 21: 1–404.

Pang, P.K.T., R.W. Griffith & J.W. Atz. 1977. Osmoregulation in elasmobranchs. Amer. Zool. 17: 365–377.

Papadopol, A. 1982. Des formes nouvelles, perfectionées et modernisées de présentation de quelques pièces anciennes de grand valeur scientifique-muséistique au Muséum d'Histoire naturelle 'Grigore Antipa'. Trav. Mus. Hist. nat. 24: 285–288.

Parker, E. 1964. East London nearly spurned an amazing event. Evening Post, Weekend Mag., 4 January 1964: 1, 7.

Parmalee, P.W. 1957. Coelacanth – living fossil. Science 19: 245–246.

Pechère, J.-F., H. Rochat & C. Ferraz. 1978. Parvalbumins from coelacanth muscle. II. Amino acid sequence of the 2 less acidic components. Biochim. biophys. Acta 536: 269–274.

Pedersen, R.A. 1971. DNA content, ribosomal gene multiplicity and cell size in fish. J. Exp. Zool. 177: 65–78.

Pegueta, V.P. 1968. Enchondral ossification in Latimeria chalumnae Smith. Dokl. Akad. Nauk Ukr. RSR 30: 653–656. (In Russian).

Perlman, D. 1975. A rare fish fillet: scientists split a 'living fossil'. San Francisco Chronicle, 28 May 1975.

Pfeiffer, W. 1968. Über die Epidermis von Latimeria chalumnae J.L.B. Smith 1939 (Crossopterygii, Pisces). Z. Morph. Ökol. Tiere 63: 419–427.

Pfeiffer, W. 1970. Sinnesorgane 'niederer Knochenfische'. Umschau. Wiss. Tech. 70: 414.

Pickford, G.E. & F.B. Grant. 1967. Serum osmolarity in the coelacanth Latimeria chalumnae: urea retention and ion regulation. Science 155: 568–570.

Plante, R. 1987. Rapport sur sa participation à la mission du N.O. Metoka (Décembre 1986 – May 1987). Centre d'Océanologie de Marseille, Marseille. 6 pp.

Plumb, R.K. 1953. Strange fish from the ocean deep should cast new light on the process of evolution. New York Times, 4 January 1953.

Poll, M. 1969. Les Agnathes et les Poissons. pp. 34–105. In: P.-P. Grassé (ed.) La Vie des Animaux: la Montée vers l'Homme, Librairie Larousse, Paris.

Poplin, C. 1981. Les homologies du pont prootique chez les osteichthyens. Cybium 5: 3–17.

Pote, J. 1988. 'Old fourlegs' – the story of the coelacanth (the book that knew no barriers). Ichthos, Special edition 2: 18.

Prailaune, S. de. 1955. Sur l'hémoglobine du coelacanthe. C.R. Soc. Biol. Paris, 149: 655–658.

Pritchard, S. 1981. The coelacanth Latimeria chalumnae. Aquarist Pondkpr 46 (9): 26–27.

Prosperi, F. 1957. Vanished continent. An Italian expedition to the Comoro Islands (transl. D. Moore). Hutchinson, London. 233 pp.

Protzman, F. 1989. Effort to capture 'fossil fish' draws fire. The New York Times, Tuesday September 1989: C4.

Rand, C.S. 1977. Blood cells and hemopoietic tissues of Latimeria chalumnae (adult embryo). Amer. Assoc. Adv. Sci., 58th annual meeting of the Pacific Division, Abstracts: 4.

Randall, J.E. 1980. Conserving marine fishes. Oryx, Journal of the Fauna Preservation Society 15: 287–291.

Rasmussen, L.E. 1979. Some biochemical parameters in the coelacanth: ventricular and notochordal fluids. Occ. Pap. Calif. Acad. Sci. 134: 94–110.

Rasmussen, L.E. & R.A. Rasmussen. 1978. Effect of transitory hyperuremia on lactate dehydrogenase levels in selected marine elasmobranchs and chimeras: an apparent urea requirement. Comp. Biochem. Physiol. 61A: 309–319.

Renwick, M. 1974. Coelacanth revisited. Vancouver Public Aquarium Newsletter 18: 1–3.

Richter, M. 1989. Acregoliathidae (Osteichthyes, Teleostei), a new family of fishes from the Cenozoic of Acre State, Brazil. Zool. Scripta 18: 311–319.

Ricqles, A. de. 1970. Fil d'Ariane des tetrapodes. La Recherche 1: 686–687.

Rieb, J.-P. 1971. Aufstellung des Urweltfisches im zoologischen Museum von Strassburg. Praeparator 17: 129–132.

Roberto, A. 1962. The Latimeria at the museum: presentation of a model. Natura 53: 3–4. (In Italian).

Robineau, D. 1973. Signification fonctionnelle de l'articulation intracrânienne chez Latimeria chalumnae (poisson, crossoptérygien, coelacanthidé). C.R. Acad. Sci. Paris, Sér. D 277: 1341–1343.

Robineau, D. 1975. Le système de la veine jugulaire et ses homologies chez Latimeria chalumnae (Pisces, Crossopterygii, Coelacanthidae). C.R. Acad. Sci. Paris, Sér. D 281: 45–48.

Robineau, D. 1976. Les organes regulateurs de la pression artérielle céphalique chez Latimeria chalumnae (Crossopterygii, Coelacanthidae). Rev. Trav. Inst. Pêch. Marit. 40: 730–732.

Robineau, D. 1986. Les affinités phylogénétiques du coelacanthe Latimeria chalumnae: évocation des principaux travaux récents s'y rapportant. Oceanis 12: 151–160.

Robineau, D. 1987. Sur la signification phylogénétique de quelques caractères anatomiques remarquables du coelacanthe Latimeria chalumnae Smith, 1939. Ann. Sci. Nat. Zool. Sér. 13, 8: 43–60.

Robineau, D. 1989. Les diverticules artériels ou ORPACS du coelacanthe. pp. 2–7. In: Encyclopédie de l'Artère 1, Sandoz ed., Paris.

Robineau, D. & J. Anthony. 1971. Le problème de la veine cave postérieure chez Latimeria chalumnae (poisson crossoptérygien). C.R. Acad. Sci. Paris, Sér. D 273: 689–692.

Robineau, D. & J. Anthony. 1973. Biomécanique du crâne de Latimeria chalumnae (poisson, crossoptérygien, coelacanthidé). C.R. Acad. Sci. Paris, Sér. D 276: 1305–1308.

Robinson, T. 1970. Coelacanth No. 70 awaits a buyer. Eastern Province Herald, 30 December 1970.

Romer, A.S. 1955. Herpetichthyes, Amphibioidei, Choanichthyes or Sarcopterygii? Nature 176: 126.

Romer, A.S. 1959. (Review of) Anatomie de Latimeria chalumnae. Tome I. Squelette, muscles et formations de soutien, by J. Millot & J. Anthony. Copeia 1959: 181–184.

Romer, A.S. 1969. (Review of) Latimeria und die Geschichte

428

der Wirbeltiere: Eine evolutions biologische Untersuchung, by G. von Wahlert. Quart. Rev. Biol. 44: 416–417.

Rosen, D.E., P.L. Forey, B.G. Gardiner & C. Patterson. 1981. Lungfishes, tetrapods, paleontology, and plesiomorphy. Bull. Amer. Mus. Nat. Hist. 167: 159–276.

Rosenthal, E. 1961. Wonders of South Africa – 5. A fish from 2000 000 000 years ago. Die Huisgenoot, September 1961: 22–23. (In Afrikaans).

Rosiński, F.M. 1976. New data on the fish Latimeria chalumnae. Wszechświat 11: 280–281. (In Polish).

Rossi, P. 1970. La découverte des coelacanthes. Revue med. vet. 121: 1033–1044.

Roux, G.H. 1941. The microscopic anatomy of the Latimeria scale. S. Afr. J. med. Sci. 7: 1–18.

Sasagawa, I., M. Ishiyama & H. Kodera. 1984. Fine structure of the pharyngeal teeth in Latimeria chalumnae. pp. 26–27. In: Proceedings of the First Symposium on Coelacanth Studies, Tokyo. (In Japanese).

Sasagawa, I., M. Ishiyama & H. Kodera. 1984. Fine structure of the teeth in Latimeria chalumnae (Coelacanthidae). Acta Anat. Nippon 59: 496. (abstract).

Schaeffer, B. 1940. (Review of) A living coelacanthid fish from South Africa, by J.L.B. Smith. Copeia 1940: 144–145.

Schaeffer, B. 1941. A revision of Coelacanthus newarki and notes on the evolution of the girdles and basal plates of the median fins in the Coelacanthini. Amer. Mus. Novitates 1110: 1–17.

Schaeffer, B. 1948. A study of Diplurus longicaudatus with notes on the body form and locomotion of the Coelacanthini. Amer. Mus. Novitates 1378: 1–32.

Schaeffer, B. 1952. The Triassic coelacanth fish Diplurus, with observations on the evolution of the Coelacanthini. Bull. Amer. Mus. Nat. Hist. 99: 25–78.

Schaeffer, B. 1952. Rates of evolution in the coelacanth and dipnoan fishes. Evolution 6: 101–111.

Schaeffer, B. 1953. Latimeria and the history of coelacanth fishes. Trans. N.Y. Acad. Sci., Ser. 2, 15: 170–178.

Schaeffer, B. 1959. A study of Latimeria chalumnae. Natural History 68: 426–429, 484–486.

Schaeffer, B. 1959. (Review of) Anatomie de Latimeria chalumnae, Vol. 1, by J. Millot & J. Anthony. Science 129: 1124–1125.

Schaeffer, B. 1962. A coelacanth fish from the Upper Devonian of Ohio. Scientific Publications of the Cleveland Museum of Natural History, New Series 1: 1–13.

Schaumberg, G. 1978. Neubeschreibung von Coelacanthus granulatus Agassiz (Actinistia, Pisces) aus dem Kupferschiefer von Richelsdorf (Perm, W.-Deutschland). Paläont. Z. 52: 169–197.

Schrafft, J.W. 1988. Quest for the coelacanth. Explorers Journal 66: 126–129.

Schreibman, M.P. 1980. Adenohypophysis: structure and function. Graduate Studies, Texas Techn. University 21: 107–131.

Schultze, H.-P. 1972. Early growth stages in coelacanth fishes. Nature new. Biol. 236: 90–91.

Schultze, H.-P. 1980. Review of 'The Biology and Physiology of the Living Coelacanth'. Syst. Zool. 29: 226–228.

Schultze, H.-P. 1980. Eier legende und lebend gebärende Quastenflosser. Natur und Museum 110: 101–108.

Schultze, H.-P. 1984. Juvenile specimens of Eusthenopteron foordi Whiteaves, 1881 (osteolepiform rhipidistian, pisces) from the Late Devonian of Miguasha, Quebec, Canada. J. Vert. Paleont. 4: 1–16.

Schultze, H.-P. 1985. Reproduction and spawning sites of Rhabdoderma (Pisces, Osteichthyes, Actinistia) in Pennsylvanian deposits of Illinois, USA Neuvième Congr. Int. Stratigr. Géol. Carbonifère, Washington 1979, Compte Rendu 5: 326–330.

Schultze, H.-P. 1987. Dipnoans as sarcopterygians. pp. 39–73. In: W.E. Bemis, W.W. Burggren & N.E. Kemp (ed.) The Biology and Evolution of Lungfishes, Alan R. Liss, New York.

Schultze, H.-P. 1988. Notes on the structure and phylogeny of vertebrate otoliths. Copeia 1988: 257–259.

Schultze, H.-P. 1990. A new acanthodian from the Pennsylvanian of Utah, U.S.A., and the distribution of otoliths in gnathostomes. J. Vert. Paleont. 10: 49–58.

Schultze, H.-P. 1991. CT scan reconstruction of the palate region of Latimeria chalumnae. Env. Biol. Fish. 32: 183–192. (this volume)

Schultze, H.-P. & G. Arratia. 1989. The composition of the caudal skeleton of teleosts (Actinopterygii: Osteichthyes). Zool. J. Linn. Soc. 97: 189–231.

Schultze, H.-P. & K.S.W. Campbell. 1987. Characterization of the Dipnoi, a monophyletic group. pp. 25–37. In: W.E. Bemis, W.W. Burggren & N.E. Kemp (ed.) The Biology and Evolution of Lungfishes, Alan R. Liss, New York.

Schultze, H.-P. & R. Cloutier. 1991. Computed Tomography and Magnetic Resonance Imaging studies of Latimeria chalumnae. Env. Biol. Fish. 32: 159–182. (this volume)

Schwartz, I. 1988. Tiefseejagd auf den Urfisch aus fernen Zeiten. Wochenende 12/13. 3. 1988.

Schwarzbach, M. 1981. Das Museum von East London mit dem ersten Fund von Latimeria (Südafrika). pp. 304–307. In: Berühmte Statten geologischer Forschung, Wissenschaftliche Verlagsgesellschaft mbH, Stuttgart.

Scoones, P. 1980. Coelacanth encounter. Skin Diver 29 (1): 8–9, 11.

Setter, A.L. & G.W. Brown, Jr. 1981. Enzymes of the coelacanth Latimeria chalumnae evidenced by starch gel electrophoresis. Env. Biol. Fish. 32: 193–198. (this volume)

Shellis, R.P. & D.F.G. Poole. 1978. The structure of the dental hard tissues of the coelacanthid fish Latimeria chalumnae. Arch. Oral Biol. 23: 1105–1113.

Shipp, H. 1981. The coelacanth a living fossil. Royal Scottish Museum, Leaflet 2. 4 pp.

Shvanvich, B.N. 1939. A live 'fossil' fish. Priroda 1939 (7): 81–82. (In Russian).

Smith, C.L., C.S. Rand, B. Schaeffer & J.W. Atz. 1975. Latimeria, the living coelacanth, is ovoviviparous. Science 190: 1105–1106.

Smith, J.L.B. 1939. A living fish of mesozoic type. Nature 143: 455–456.

Smith, J.L.B. 1939. A living fossil. Cape Naturalist 1 (6): 187–194.

Smith, J.L.B. 1939. A surviving fish of the order Actinistia. Trans. R. Soc. S. Afr. 27: 47–50. (Reprinted in Trans. R. Soc. S. Afr. 47: 1–7 (1989) to commemorate the fiftieth anniversary).

Smith, J.L.B. 1939. A 'living fossil' from South Africa. Discovery, London: 240–242.

Smith, J.L.B. 1939. The living coelacanthid fish from South Africa. Nature 143: 748–750.

Smith, J.L.B. 1940. A living coelacanthid fish from South Africa. Trans. R. Soc. S. Afr. 28: 1–106.

Smith, J.L.B. 1941. A living fossil (*Latimeria chalumnae* J.L.B. Smith). Smithsonian Report, 1940: 321–327.

Smith, J.L.B. 1949. Sub-class Choanichthyes, Order Crossopterygii, Sub-order Coelacanthini, coelacanth fishes. pp. 79–80. *In:* J.L.B. Smith, The Sea Fishes of Southern Africa, 1st ed., Central News Agency, Cape Town.

Smith, J.L.B. 1953. Why two coelacanths stirred the world. Daily Dispatch, 13 January 1953.

Smith, J.L.B. 1953. The second coelacanth. Nature 171: 99–101.

Smith, J.L.B. 1953. Problems of the coelacanth. S. Afr. J. Sci. 49: 279–281.

Smith, J.L.B. 1953. The two coelacanths. Bull. S. Afr. Mus. Ass. 5: 206–210.

Smith, J.L.B. 1953. Professor Smith tells his own story of the coelacanth. S. Afr. Angler 7: 6, 23.

Smith, J.L.B. 1953. My plans for the next coelacanth expedition. S. Afr. Angler 8: 9.

Smith, J.L.B. 1953. A coelacanth in Kenyan waters? S. Afr. Angler 8: 8, 19.

Smith, J.L.B. 1953. *Malania*, a living fossil. Elseviers Weekblad, 26 December 1953: 15. (In Dutch).

Smith, J.L.B. 1953. The primitive fishes of southern Africa. pp. 7–9. *In:* R.A. Dart (ed.) Africa's Place in the Human Story, Bernard Price Inst. Palaeontological Research, Johannesburg.

Smith, J.L.B. 1953. The incredible coelacanth. Times Weekly Review, London, 8 January 1953.

Smith, J.L.B. 1953. Investigation of the coelacanth. Letter to the editor. Nature 174: 389.

Smith, J.L.B. 1953. Coelacanth may reveal evolutionary secrets. Natal Daily News, 7 January 1953.

Smith, J.L.B. 1953. Coelacanth. London Calling, 26 February 1953: 11.

Smith, J.L.B. 1955. Live coelacanths. Nature 176: 473.

Smith, J.L.B. 1956. Old fourlegs. The story of the coelacanth. Longmans, Green, London. 260 pp.

Smith, J.L.B. 1956. The search beneath the sea. The story of the coelacanth. Henry Holt & Co., New York. 260 pp.

Smith, J.L.B. 1956. Conservation of the coelacanth. The Times, London, 4 June 1956.

Smith, J.L.B. 1957. Vergangenheit steigt aus dem Meer. Günther Verlag, Stuttgart. 253 pp.

Smith, J.L.B. 1958. Old fourlegs. Second edition. Pan Books Ltd, London. 284 pp.

Smith, J.L.B. 1960. À la poursuite du coelacanth. Librairie Plon, Paris. 325 pp.

Smith, J.L.B. 1962. Old man coelacanth. State Publisher of Geographical Literature, Moscow. 216 pp. (In Russian).

Smith, J.L.B. 1963. The atomic bomb and the coelacanth. The Daily Dispatch, 10 December 1963.

Smith, J.L.B. 1964. Old fourlegs. Eesti Riiklil Kirjastus, Tallin. 286 pp. (In Estonian).

Smith, J.L.B. 1965. Old fourlegs. Tafelberg-Uitgewers, Cape Town. 249 pp. (In Afrikaans).

Smith, J.L.B. 1967. Old fourlegs. Mladé Leta, Bratislava. 225 pp. (In Slovak).

Smith, J.L.B. 1968. The coelacanth. pp. 49–57. *In:* J.L.B. Smith (ed.) Our Fishes, Voortrekkerspers, Johannesburg.

Smith, J.L.B. 1968. Coelacanth story. pp. 19–22. *In:* J.L.B. Smith (ed.) High Tide, Books of Africa, Cape Town.

Smith, J.L.B. 1968. The lost home of the coelacanths. Stywe Lyne – Tight Lines 9: 19–23.

Smith, J.L.B. 1970. Coelacanth. Encyclopaedia Britannica 6: 17.

Smith, J.L.B. 1971. Marine fishes. Animal life in southern Africa (abstracts from the Standard Encyclopaedia of Southern Africa): 231–261.

Smith, J.L.B. 1973. Old fourlegs. Pimpi-Publisher, The Hague. 350 pp. (In Dutch).

Smith, J.L.B. 1981. Old fourlegs: a story of the coelacanth. Tuttle-Mori Agency, Japan. 240 pp. (In Japanese).

Smith, Margaret M. 1954. The treatment and preservation of a coelacanth for specialised study. Turtox News 32: 174–178.

Smith, Margaret M. 1970. The search for the world's oldest fish. Oceans 3 (6): 26–36.

Smith, Margaret M. 1976. Surprise – a baby! Eastern Cape Naturalist 59: 4–6.

Smith, Margaret M. 1976. The coelacanth rises from the past. Rhodesian Aquarist 7: 2–7.

Smith, Margaret M. 1977. Surprise, surprise – a baby! Afr. Wildl. 31: 30–31.

Smith, Margaret M. 1979. The influence of the coelacanth on African ichthyology. Occ. Pap. Calif. Acad. Sci. 134: 11–16.

Smith, Margaret M. 1986. Latimeriidae. pp. 152–153. *In:* M.M. Smith & P.C. Heemstra (ed.) Smiths' Sea Fishes, MacMillan, Johannesburg. (second printing 1988, Southern Book Publishers, Johannesburg).

Smith, Moya M. 1975. The microstructure of the scales of *Latimeria chalumnae*. Biol J. Linn. Soc. 7: 298–299.

Smith, Moya M. 1977. The microstructure of the dentition and dermal ornament of three dipnoans from the Devonian of Western Australia: a contribution towards dipnoan interrelations, and morphogenesis, growth and adaptation of the skeletal tissues. Philos. Trans. R. Soc. Lond. B 281: 29–72.

Smith, Moya M. 1978. Enamel in the oral teeth of *Latimeria chalumnae* (Pisces, Actinistia): a scanning electron microscope study. J. Zool., Lond. 185: 355–369.

Smith, Moya M. 1979. SEM of the enamel layer in oral teeth of fossil and extant crossopterygian and dipnoan fishes. Scanning Electron Microscopy 1979 (2): 483–489.

Smith, Moya M. 1979. Scanning electron microscopy of odontodes in the scales of a coelacanth embryo, *Latimeria chalumnae* Smith. Arch. Oral Biol. 24: 179–183.

Smith, Moya M. 1989. Distribution and variation in enamel structure in the oral teeth of sarcopterygians: its significance for the evolution of a protoprismatic enamel. Historical Biology 3: 97–126.

Smith, Moya M. & M.H. Hobdell. 1973. Comparisons between the microstructure of scales of *Latimeria chalumnae* and extant dipnoan and teleostean scales. J. Dent. Res. 52: 957. (abstract).

Smith, Moya M., M.H. Hobdell & W.A. Miller. 1971. The structure of the scales of *Latimeria chalumnae* Smith. J. Dent. Res. 50: 1169–1170.

Smith, Moya M., M.H. Hobdell & W.A. Miller. 1972. The structure of the scales of *Latimeria chalumnae*. J. Zool., Lond. 167: 501–509.

Smith, W.M. 1988. William Smith remembers. Ichthos, Special Edition 2: 3.

Snyman, L. & V. Miros. 1989. Comores: the perfume isles. Getaway, October 1989: 8–13, 15.

Solomon, F.P., F.M. Utter & G.W. Brown. Jr. 1977. Enzymes of the coelacanth *Latimeria chalumnae*. 1976 Research in fisheries. Ann. Rep. Coll. Fish. Univ. Wash. Contr. 460: 60. (abstract).

Soulé, G. 1966. The ocean adventure. Science explores the depths of the sea. Appleton-Century, New York. 278 pp.

Spicer, A. 1989. Pseudo-scientists threaten living fossil. Observer, 10 September 1989.

Stevens, J. 1971. Le coelacanthe: un fossile vivant. Science et Avenir 289: 230–235.

Stobbs, R.E. 1987. 'Galawas' of the Federal Islamic Republic of the Comores, with notes on artisinal fishing and recommendations for a maritime museum. Invest. Rep. J.L.B. Smith Inst. Ichthyol. 26: 1–21.

Stobbs, R.E. 1988. Coelacanth mythology. Ichthos, Special edition 2: 18–19.

Stobbs, R.E. 1989. Laxative lipids and the survival of the living coelacanth. S. Afr. J. Sci. 85: 557–558.

Stobbs, R.E. 1989. The coelacanth enigma. Phoenix 2: 8–15.

Stobbs, R.E. 1990. N'Galawa: a dugout canoe of the Western Indian Ocean. Skiboat 6: 77–81.

Stobbs, R.E. & M.N. Bruton. 1991. The fishery of the Comoros, with comments on its possible impact on coelacanth survival. Env. Biol. Fish. 32: 341–359. (this volume)

Stock, D.W., K.D. Moberg, L.R. Maxson & G.S. Whitt. 1991. A phylogenetic analysis of the 18S ribosomal RNA sequence of the coelacanth *Latimeria chalumnae*. Env. Biol. Fish. 32: 99–117. (this volume)

Suyehiro, Y. 1983. Some views on the dissected specimens of coelacanth. Bienn. Rep. Keikyu Aburat. mar. Park Aquar. 12: 12.

Suyehiro, Y. 1988. Gombessa. Be-Pal Books, Tokyo. 254 pp. (In Japanese).

Suyehiro, Y., T. Tsutsumi & T. Uyeno. 1982. On dissection of the coelacanth *Latimeria chalumnae*. pp. 12–13. *In:* Proceedings of the First Symposium on Coelacanth Studies, Tokyo. (In Japanese).

Suyehiro, Y., T. Uyeno & N. Suzuki. 1984. Coelacanth: dissecting a living fossil. Newton Graphic Science Magazine 2 (8): 82–93. (In Japanese).

Suzuki, N., A. Azuma & T. Hamada. 1984. Body shape and swimming locomotion. pp. 32–33. *In:* Proceedings of the First Symposium on Coelacanth Studies, Tokyo. (In Japanese).

Suzuki, N. & T. Hamada. 1989. Examination of the modern coelacanth with an x-ray photography and an x-ray computed tomography. Sci. Pap. Coll. Arts Sci., Univ. Tokyo 39: 73–96.

Suzuki, N. & T. Hamada. 1989. Coelacanth: study of functional aspects using new technology. Newton Graphic Science Magazine 9(13): 10–13. (In Japanese).

Suzuki, N. & T. Hamada. 1990. Three dimensional analysis of coelacanth body structure by computer graphics and x-ray CT images. Sci. Pap. Coll. Arts Sci. Univ. Tokyo 40: 49–60.

Suzuki, N. & T. Hamada. 1990. Analyzing the living fossil, the coelacanth. Iden (Genetics) 44(9): 30–34. (In Japanese).

Suzuki, N., T. Hamada & S. Yoshimura. 1984. Experimental function of latimerian heart. pp. 20–21. *In:* Proceedings of the First Symposium on Coelacanth Studies, Tokyo. (In Japanese).

Suzuki, N., Y. Suyehiro & T. Hamada. 1985. Initial report of expeditions for coelacanth – Part 1 – Field studies in 1981 and 1983. Sci. Pap. Coll. Arts Sci., Univ. Tokyo 35: 37–79.

Suzuki, N., Y. Suyehiro & K. Tanauma. 1984. Scientific expedition of coelacanth in the Comoro Islands. pp. 2–3. *In:* Proceedings of the First Symposium on Coelacanth Studies, Tokyo. (In Japanese).

Suzuki, N., S. Tada, H. Iwai, H. Takahashi, T. Tsutsumi & T. Hamada. 1984. X-ray CT of three specimens of *Latimeria*. pp. 10–11. *In:* Proceedings of the First Symposium on Coelacanth Studies, Tokyo. (In Japanese).

Suzuki, N. & K. Tanauma. 1984. Coelacanth fishing by native fishermen. pp. 8–9. *In:* Proceedings of the First Symposium on Coelacanth Studies, Tokyo. (In Japanese).

Svetovidov, A.N. 1956. First observations on a 'living fossil'. Priroda 1956 (5): 104–105. (In Russian).

Sylvestre, J.-P. 1990. Le coelacanthe fossile vivant. La Pêche en Mer 58: 84–87.

Szarski, H. 1977. Sarcopterygii and the origin of tetrapods. pp. 517–540. *In:* M.K. Hecht, P.C. Goody & B.M. Hecht (ed.) Major Patterns in Vertebrate Evolution, Plenum Press, New York.

Szarski, H. 1980. Actual knowledge of the anatomy of *Latimeria*. Przegląd Zool. 24: 5–14. (In Polish).

Takahashi, K. 1990. A DNA-level approach to a living fossil. Newton Graphic Science Magazine 9(13): 24–25. (In Japanese).

Talbot, F.H. 1966. The coelacanth, living relic of 50 million years ago. Australian Nat. Hist. 15: 137–140.

Tamai, Y. 1985. Evolution of glycolipids in the vertebrate nervous system – phylogenetic distribution of myelin glycolipids and its physiological significances. Biochemistry 57: 1249–1270. (In Japanese).

Tamai, Y. & K. Abe. 1984. Lipid compositions of various tissues of the coelacanth. p. 17. *In:* Proceedings of the First Symposium on Coelacanth Studies, Tokyo. (In Japanese).

Tamai, Y., H. Kojima & K. Abe. 1984. Chemical characterization of the brain stem of a coelacanth. pp. 22–23. *In:* Proceedings of the First Symposium on Coelacanth Studies, Tokyo. (In Japanese).

Tamai, Y., H. Kojima & K. Abe. 1986. Chemical characterization of the brain of a coelacanth *Latimeria chalumnae*. Comp. Biochem. Physiol. 83B: 295–299.

Tamai, Y., H. Kojima, K. Abe & A. Suzuki. 1983. Evolutionary change of glycolipids and myelin-proteins in the brain of the primitive bony fishes. 26th Annual Meeting of Japanese Neurochem. Res. 9: 1120. (In Japanese).

Tanaka, Y. 1985. An anatomical study on the spleen of archaic fishes 1. Coelacanthiformes and Dipneusti. Acta haematol., Japan 48: 710–723. (In Japanese).

Taylor, J.W. 1953. Coelacanth is not the missing link says Professor Smith. Mombasa Times, 28 January 1953.

Thenius, E. 1973. Fossils and the life of the past. Springer-Verlag, Berlin. 194 pp.

Thiruvathukal, K.V. 1975. Microscopic analysis of selected tissues of *Latimeria*. Amer. Zool. 15: 807.

Thomson, K.S. 1966. Intercranial mobility in the coelacanth. Science 153: 999–1000.

Thomson, K.S. 1966. Mobility of the skull and fins in the coelacanth (*Latimeria chalumnae* Smith). Amer. Zool. 6: 565–566. (abstract).

Thomson, K.S. 1966. The history of the coelacanth. Discovery (New Haven) 2 (1): 27–32.

Thomson, K.S. 1967. Mechanisms of intracranial kinetics in fossil rhipidistian fishes (Crossopterygii) and their relatives. J. Linn. Soc. Zool. 46: 223–253.

Thomson, K.S. 1967. Notes on the relationships of the rhipidistian fishes and the ancestry of the tetrapods. J. Paleont. 41: 660–674.

Thomson, K.S. 1969. The biology of the lobe-finned fishes. Biol. Rev. 44: 91–154.

Thomson, K.S. 1970. Research voyage in the Western Indian Ocean. Discovery (New Haven) 5: 88–96.

Thomson, K.S. 1970. Intracranial movement in the coelacanth *Latimeria chalumnae* Smith (Osteichthyes, Crossopterygii). Postilla 149: 1–12.

Thomson, K.S. 1973. Secrets of the coelacanth. Nat. Hist. 82 (2): 58–65.

Thomson, K.S. 1973. New observations on the coelacanth fish *Latimeria chalumnae*. Copeia 1973: 813–814.

Thomson, K.S. 1981. The capture and study of two coelacanths off the Comoro Islands, 1972. Natl. Geogr. Soc. Res. Rep. (1972) 13: 615–622.

Thomson, K.S. 1986. A fishy story. Amer. Sci. 74: 169–171.

Thomson, K.S. 1989. The second coelacanth. Amer. Sci. 77: 536–538.

Thomson, K.S., J.G. Gall & L.W. Coggins. 1973. Nuclear DNA contents of coelacanth erythrocytes. Nature 241: 126.

Thoney, D.A. & W.J. Hargis, Jr. 1991. Juvenile anisakine parasites from the coelacanth *Latimeria chalumnae*. Env. Biol. Fish. 32: 281–283. (this volume)

Thornton, D.A. 1989. Letter from South Africa, David Thornton tells a fishy story. Gibbons Stamp Monthly, June 1989: 33.

Thys van den Audenaerde, D.F.E. 1984. Le coelacanthe des Comores, *Latimeria chalumnae*, curiosité zoologique, fossile vivant ou animal aberrant? Afrika-Tervuren 30: 90–103.

Tietz, R.M. 1988. Coelacanth on the brink of extinction 50 years after discovery. Custos 17: 40–41.

Tietz, R.M. 1988. Saving the first coelacanth. Ichthos, Special edition 2: 12–13.

Tippet, P. & P. Teesdale. 1973. Limited blood group tests on samples from two coelacanths *Latimeria chalumnae*. Vox Sang. 24: 175–178.

Todd, T.N. 1975. The lungfishes – living fossils. Tropical Fish Hobbyist 24 (1): 19, 23–24, 26, 28, 32.

Torchio, M. 1960. Review of 'A la poursuite du coelacanthe' by J.L.B. Smith. Revista di Scienze naturali, Natura: 118–119. (In Italian).

Trewavas, E. 1958. The coelacanth yields its secrets. Discovery 19: 196–205.

Trewavas, E. 1959. Anatomy of a coelacanth. Nature 183: 566.

Trewavas, E., E.I. White, N.B. Marshall & D.W. Tucker. 1955. (Response to A.S. Romer's note). Herpetichthyes, Amphibioidei, Choanichthyes or Sarcopterygii? Nature 176: 126–127.

Truscott, B. 1980. Corticosteroids of the coelacanth *Latimeria chalumnae* Smith: a provisional study on their identity. Gen. comp. Endocrinol. 41: 287–295.

Tsuyuki, H. & S. Itoh. 1970. Composition of fatty acids in coelacanth oil. Bull. Jap. Soc. scient. Fish. 36: 788–790. (In Japanese).

Tuzet, O. & J. Millot. 1959. La spermatogenèse de *Latimeria chalumnae* Smith (crossoptérygien coelacanthidé). Annal. Sci. nat. Zool., Sér. 12, 1: 61–69.

Uyeno, T. 1984. Age estimation of coelacanth by scale and otolith. pp. 28–29. *In:* Proceedings of the First Symposium on Coelacanth Studies, Tokyo. (In Japanese).

Uyeno, T. 1988. Coelacanths. Newton Graphic Science Magazine 8(12): 40–47. (In Japanese).

Uyeno, T. 1989. Dissecting a coelacanth. National Science Museum News 244: 6–9. (In Japanese).

Uyeno, T. 1989. Investigating the ecology of the coelacanth based on feeding habits. Newton Graphic Science Magazine 9(13): 20–21. (In Japanese).

Uyeno, T. 1991. Observations on locomotion and feeding of released coelacanths, *Latimeria chalumnae*. Env. Biol. Fish. 32: 267–273. (this volume)

Uyeno, T., T. Hamada, S. Murakami, K. Takahashi, C. Olhagaray & N. Suzuki. 1989. Coelacanth; Japanese research to date. Newton Graphic Science Magazine 9(13): 10–31. (In Japanese).

Uyeno, T. & T. Tsutsumi. 1991. Stomach contents of *Latimeria*

432

chalumnae and further notes on its feeding habits. Env. Biol. Fish. 32: 275–279. (this volume)

Uyeno, T., T. Tsutsumi, N. Suzuki & T. Hamada. 1984. Distribution and ecology. p. 34. *In:* Proceedings of the First Symposium on Coelacanth Studies, Tokyo. (In Japanese).

Van Bruggen, A.C. 1989. Fifty year coelacanth. Biovisie Mag. Vakblad voor Biologen 69: 5–6. (In Dutch).

Van der Merwe, C. 1988. Can we save the coelacanth? Personality, 22 August 1988: 46–47.

Van der Merwe, C. 1988. The sea of Zanj. Flying Springbok, December 1988: 19–33.

Van Kemenade, J.A.M. 1976. Anatomy and cytology of the pituitary gland of the coelacanth fish *Latimeria chalumnae* Smith. Gen. comp. Endocrinol. 29: 264. (abstract).

Van Kemenade, J.A.M. & J.-W.P.M. Kremers. 1975. The pituitary gland of the coelacanth fish *Latimeria chalumnae* Smith: general structure and adenohypophysial cell types. Cell Tissue Res. 163: 291–311.

Van Oordt, P.G.W.J. 1980. A typology of the gnathostome adenohypophysis with some emphasis on its gonadotropic function. Gen. comp. Endocrinol. 40: 308–309.

Vialli, M. 1957. The quantity of deoxyribonucleic acid in the nucleus of the erythrocytes of *Latimeria*. Rc. Ist. lomb. Sci. Lett. A91: 680–686. (In Italian).

Vialli, M. 1963. Presence of heterochromatine cells in *Latimeria chalumnae*. Monit. zool. ital. 70–71: 313–319. (In Italian).

Vincent, M. 1988. The filming of the story of the coelacanth. Ichthos, Special edition 2: 8–9.

Vogel, S. 1988. Face to face with a living fossil. Discovery (Los Angeles) 9 (3): 56–57.

Vorobyeva, E.I. 1977. Morphology and nature of evolution of crossopterygian fishes. Trudy palaeont. Inst. Akad. Nauk SSSR 163: 1–239. (In Russian).

Vorobyeva, E.I. 1981. Cartilage evolution in the skeleton of ancient anamnia. Arkh. Anat. Gistol. Embriol. 80: 83–87.

Vorobyeva, E.I. & D.V. Obruchev. 1967. Subclass Sarcopterygii. pp. 420–509. *In:* Y.A. Orlov (ed.) Fundamentals of Paleontology, Vol. XI, Agnatha, Pisces, Israel Program for Scientific Translations, Jerusalem.

Waehneldt, T.V. 1986. Appearance of myelin proteins during vertebrate evolution. Neurochem. Int. 9: 463–474.

Waehneldt, T.V. & J. Malotka. 1989. Presence of proteolipid protein in coelacanth brain myelin demonstrates tetrapod affinities and questions a chondrichthyan association. J. Neurochem. 52: 1941–1943.

Waehneldt, T.V., J. Malotka, G. Jeserich & J.-M. Matthieu. 1991. Central nervous system myelin proteins of the coelacanth *Latimeria chalumnae*: phylogenetic implications. Env. Biol. Fish. 32: 131–143. (this volume)

Waehneldt, T.V., J.-M. Matthieu & G. Jeserich. 1986. Appearance of myelin proteins during vertebrate evolution. Neurochem. Int. 9: 463–474.

Wahlert, G., von. 1967. Demonstration von *Latimeria chalumnae* J.L.B. Smith. Verhandlungen der deut. Zool. Ges. 34: 527–529.

Wahlert, G., von. 1968. *Latimeria* und die Geschichte der Wir-

beltiere. Eine evolutionsbiologische Untersuchung. G. Fischer Verlag, Stuttgart. 125 pp.

Wahlert, G., von & H. von Wahlert. 1962. Funktion und biologische Bedeutung der Quastenflosser. Natur und Museum 92: 7–12.

Wahlert, G., von & H. von Wahlert. 1967. Bau und Funktion der paddelformigen Unpaarflossen von *Latimeria chalumnae* J.L.B. Smith (Actinistia, Osteichthyes). Stuttg. Beitr. Naturk. 172: 1–3.

Wake, M.H. 1987. Urogenital morphology of dipnoans, with comparisons to other fishes and to amphibians. pp. 199–216. *In:* W.E. Bemis, W.W. Burggren & N.E. Kemp (ed.) The Biology and Evolution of Lungfishes, Alan R. Liss, New York.

Watson, D.M.S. 1954. Coelacanths. New Biology 16: 86–92.

Watson, D.M.S. 1955. Coelacanths. Proc. R. Inst. Great Britain 35: 782–786.

Weber, R.E., J.F. Bol, K. Johansen & S.C. Wood. 1973. Physicochemical properties of the hemoglobin of the coelacanth *Latimeria chalumnae*. Arch. Biochem. Biophys. 154: 96–105.

White, E. 1953. The coelacanth fishes. Discovery 14: 113–117.

White, E. 1954. More about the coelacanths. Discovery 15: 332–335.

White, E. 1954. The coelacanth fishes. Annual Report Smithson. Inst. Washington 1953: 351–360.

White, E.I. 1939. One of the most amazing events in the realm of natural history in the twentieth century: the discovery of a living fish of the coelacanth group, thought to have been extinct 50 million years, off South Africa. The Illustrated London News Suppl., March 1939.

Wiley, E.O. 1978. The phylogenetic relationships of *Latimeria chalumnae*. Trans. Kansas Acad. Sci. 81: 93–94.

Wiley, E.O. 1979. Ventral gill arch muscles and the phylogenetic relationships of *Latimeria*. Occ. Pap. Calif. Acad. Sci. 134: 56–67.

Wiley, E.O. 1979. Ventral gill arch muscles and interrelationships of gnathostomes, with a new classification of the Vertebrata. Zool. J. Linn. Soc. 67: 149–179.

Williams, H. 1975. Very ancient fish deals science yet another surprise. The Seattle Times. 7 December 1975.

Willis, J.B. 1953. The second coelacanth. Victorian Naturalist 70: 43.

Willox, R. 1989. Madagascar and Comores. Lonely Planet Publications, Berkeley. 183 pp.

Witkowski, A. & W. Szymczyk. 1976. Have Crossopterygii been ovoviviparous? Przegl. Zool. 20: 174–178. (In Polish).

Wood, L. 1976. Comoro Islands coelacanth stamp. The Christian Science Monitor, 20 January 1976.

Wood, S.C., K. Johansen & R.E. Weber. 1972. Haemoglobin of the coelacanth. Nature 239: 283–285.

Woods, L.P. 1953. Coelacanth, 'front page fish', is in museum. Chicago Nat. Hist. Mus. Bull. 24 (2): 8.

Woodward, A.S. 1940. The surviving crossopterygian fish, *Latimeria*. Nature 146: 53–54.

Woodward, A.S. 1940. The existing coelacanth fish, *Latimeria*. Nature 146: 590.

Wourms, J.P., J.W. Atz & M.D. Stribling. 1991. Viviparity and the maternal-embryonic relationship in the coelacanth *Latimeria chalumnae*. Env. Biol. Fish. 32: 225–248. (this volume)

Wourms, J.P., B.D. Grove & J. Lombardi. 1988. The maternal-embryonic relationship in viviparous fishes. pp. 1–134. *In:* W.S. Hoar & D.J. Randall (ed.) Fish Physiology, Vol. 11b, Academic Press, San Diego.

Wourms, J.P., M.D. Stribling & J.W. Atz. 1980. Maternal fetal nutrient relationships in the coelacanth, *Latimeria*. Amer. Zool. 20: 962. (abstract).

Xylander, W. & L. Nevermann. 1987. *Latimeria*. Sporttaucher 12/87: 32–33.

Yancey, P.H. & G.N. Somero. 1979. Counteraction of urea destabilisation of protein structure by methylamine osmoregulatory compounds of elasmobranch fishes. Biochem. J. 183: 317–324.

Yancey, P.H. & G.N. Somero. 1980. Methylamine osmo-regulatory solutes of elasmobranch fishes counteract urea inhibition of enzymes. J. exp. Zool. 212: 205–214.

Yuan, P.M. & R.W. Gracy. 1984. Primary structure of human triose phosphate isomerase. J. biol. Chem. 259: 11.

Zupanc, G.K.H. 1985. Fish and their behaviour. 1st ed. Tetra-Press, Melle. 188 pp.

Cryptozoology, 8, 1989, 1–11
© 1989 International Society of Cryptozoology

Special Guest Article in Celebration of the 50th Anniversary of the Discovery of the Coelacanth

REMINISCENCES OF THE DISCOVERY OF THE COELACANTH, *LATIMERIA CHALUMNAE* SMITH

Based on Notes from a Diary Kept at the Time

Marjorie Courtenay-Latimer
6 Lake Street, Vincent, 5247 East London,
Republic of South Africa

The Editor is pleased to present this special guest article on the discovery of the coelacanth by the discoverer herself, Marjorie Courtenay-Latimer, retired director of the East London Museum. Miss Courtenay-Latimer, Hon. Doc. Phil., an Honorary Member of the International Society of Cryptozoology since its founding in 1982, graciously consented to prepare this article, based on notes from the diary which she kept at the time. We are privileged to be able to publish this account—written in her own words—50 years after what many have called the zoological discovery of the century.—Editor.

This Article Is Dedicated to the Memory of Captain Hendrik Goosen, Who Died on January 5, 1990.

Could this day, December 22, 1988, really be 50 years since the greatest trauma in my life's career had occurred? Oh, yes! Reading through my old diary, I am again woken to the facts revealed therein.

I therefore pen these facts from my diary as a tribute to my friend and colleague Captain Hendrik Goosen, to whom the world owes a debt of gratitude; for without his interest and energy, there never would have been a coelacanth to astound the world of natural science.

'Tell me, why did our cousins bother to evolve into humans?'

Species and subject index

Environmental Biology of Fishes **32**: 437–446, 1991.

440

446

Vertical migration 293, 298, 330
  movement 318, 319
Visceral skeleton 179
Vitelline duct 233, 243
Vitellogenesis 249–265
Vitellogenin 251
Viviparity 12, 225–265, 299, 321, 329,
    333, 335, 378, 407, 412
  placental 246, 412
Volcanic eruption 291
Volcano 291, 314, 345
Vomer 183–192

WAGNER program 26
Water balance 207
  depth, coelacanths 287, 288, 291–296,
    316–318, 330
  loss 199–218
  pressure 330, 335

temperature 288, 292–295, 298, 318,
    330, 342
Weight frequency, coelacanths 303–305
Westoll's method 25
Whale 313
Whiteia 67, 80, 85, 86, 92, 258
  africanus 67, 77, 96
  groenlandica 77, 96
  tuberculata 26, 27, 53, 67, 77, 91, 96
  woodwardi 26, 27, 53, 67, 77, 81, 96
Wimania 26, 35, 67, 69, 70, 85, 258
  multistriata 69, 77, 97
  sinuosa 27, 53, 67, 77, 97
World Heritage Site 332
  Convention 327

Xenobiotics 361–367
Xenocongridae 279
Xenopus 105, 108–110, 119, 122–123,
    126, 141

laevis 101, 103, 120, 139, 208, 209
X-ray 160, 161, 167, 378, 380, 386

Yolk 12, 225–265, 304, 335, 336
  crystal 225, 237
  density 250, 259
  platelets 225, 247
  syncytium 237
  synthesis 249
  volume 249–251, 258, 259
Yolksac 249–265, 317, 380
  juvenile 257, 304, 321, 335, 378, 380,
    407
  placenta 243–245
Youngichthys 68
  xinhuanisis 68, 96
Youngolepiformes 24, 25, 191
Youngolepis 84, 186
  praecursor 189